BIOGEOGRAPHY AND ECOLOGY IN ANTARCTICA

MONOGRAPHIAE BIOLOGICAE

EDITORS

W. W. WEISBACH
Den Haag

P. VAN OYE
Gent

VOLUMEN XV

SPRINGER-SCIENCE+BUSINESS MEDIA, B.V. - 1965

BIOGEOGRAPHY AND ECOLOGY IN ANTARCTICA

EDITED BY

J. VAN MIEGHEM   and   P. VAN OYE
Brussels   Gent

ASSOCIATE EDITOR
J. SCHELL
(Gent)

SPRINGER-SCIENCE+BUSINESS MEDIA, B.V. - 1965

ISBN 978-94-015-7206-4   ISBN 978-94-015-7204-0 (eBook)
DOI 10.1007/978-94-015-7204-0

CONTENTS

PREFACE

This book is the third in a series of publications devoted to the biogeographical and ecological research in the Southern Hemisphere, published in the "Monographiae Biologicae". After dealing with Australia (vol. VIII) and Southern Africa (Vol. XIV) it was thought essential to include Antarctica in this series.

Ever since the expedition of the "Belgica" made the first successful wintering within the antarctic circle in 1898 and brought back a very rich harvest of scientific data, Belgium kept a vivid interest in Antarctica and took an active part in the modern and international exploration of this vast continent. As part of their programs for the International Geophysical Year (I.G.Y.) twelve nations established permanent or semi-permanent bases on the Antarctic Continent or on subantarctic islands. Thus a new era of vast and free international scientific collaboration in the Antarctic was opened and it culminated in the formulation and the signing of the Antarctic Treaty (Washington 1959). It was recognized and accepted that "Antarctica" shall be used for peaceful purposes only and "Freedom of scientific investigation in Antarctica and cooperation toward that end, as applied during the I.G.Y., shall continue.."

In order to organize this collaboration e.g. by full exchange of programs and results a "Special Committee on Antarctic Research" (S.C.A.R.) was founded in 1957.

As president of the Belgian Centre National de Recherches Polaires (C.N.R.P.B.) and member of the special group on Biology of the C.N.R.P.B., the Editors of this book were actively concerned with the international scientific activities in the Antarctic. They felt the time had come for a general survey of what had been done thus far in the way of biogeographical and ecological research. Indeed very few publications of this kind had appeared and the "Monographiae Biologicae" were considered to be an ideal medium to give this review a wide and truly international attention. It is hoped that this book will serve the dual purpose of giving a thorough summing-up of the present status of knowledge which is expanding very rapidly, and providing a sound basis for the planning of future research by indicating the many and wide gaps still existing in our knowledge and the points of special interest deserving further attention.

It is realised that this survey has a very preliminary character since it is by no means complete. Many such important groups as the Marine Algae, Protozoa, Porifera, Echinodermata, Vermes (e.g.

VIII

Polychaeta), Rotifera, Nematoda, Tunicata, Pinnipedia, Cetacea a.o. had to be omitted because otherwise the publication would have been delayed for too long a time and also because biological research in the Antarctic has really only just started.
Other recent publications on these topics are:
Antarctic Research Series Vol. I
  "Biology of the Antarctic Seas" Ed. MILTON O. LEE.
  American Geophysical Union, Washington, D.C. 1964
"Information Bulletin of the Soviet Antarctic Expeditions"
  Elsevier (Translation) Vol. I to III.
  Covering Bulletins 1-30. The American Geophysical Union will continue this series on a serial basis.
"Discussion on the Biology of the Southern Cold Temperate Zone"
  Proc. Roy. Soc. Lond.,B, **152**, 1960
"The Life Sciences in Antarctica"
  Report by the Committee on Polar Research, part. I.
  Nat. Acad. Sciences, Washington, D.C. (prib. no. 839) 1961
"Symposium on Antarctic Biology"
  Paris, September 1962. (Abstracts of many papers can be found in *Polar Record*, II, **72**, 1962).
CARRICK, R., HOLDGATE M.W. & PREVOST, J. (Eds.) 1964.
  "Biologie Antarctique" Proc. 1st Symposium on Antarctic Biology. (Paris, Sept. 1962) Paris – Hermann.
GRESSITT, J. L. (Editor) 1964.
  "Pacific Basin Biogeography: A Symposium". Bishop Museum Press, Honolulu
SLIJPER, E. J., Whales.

Whatever great and important achievements may have been made in the past, it seems quite beyond discussion that good scientific research, especially in the Antarctic, can only be pursued on a basis of active and well organized international collaboration, free of political interferences.

It is hoped that this book will clearly illustrate and contribute to this goal, both by the true international character of the list of contributors and by its content from which, it is hoped, it will be obvious that the results and observations described get their true significance only when placed in a larger context.

ACKNOWLEDGMENTS

The editors are indebted to the contributors for the patience with which they have accepted to bring their manuscripts up to date after too long a period had elapsed between reception and actual publication of the manuscripts.

To Miss Paulette DOYEN they are greatly indebted for her continuous help with the redaction of this book.

INTRODUCTORY REMARKS

BY

J. SCHELL

When discussing biogeographical problems of the Antarctic it is necessary to draw the borders of the region under discussion. As accepted by the S.C.A.R. the northern limit of Antarctica is set by the so called Antarctic Convergence or Oceanic Polar Front (see p.119). This is a remarkably sharp zone where cold surface waters flowing north converge with and dive under warmer waters flowing generally south, resulting in a sudden and abrupt change of some physical characteristics of the surface waters such as temperature, salinity, and speed of flow. The Antarctic Convergence turned out to be not only a marked physical boundary of the Southern ocean but also, and in an even more pronounced way, a biological boundary.

Not only on theoretical grounds is the Antarctic Convergence a good limit for the Antarctic region, but it also is a very useful boundary to work with since it is reasonably stable and stationary from season to season and over the years, and its position can be determined with great ease either by determining the temperature or salinity of the surface waters or by investigating the biological content of those waters. Moreover, since the Antarctic Convergence is situated between 50° and 60° south latitude the Antarctic region thus defined includes most of the important Antarctic islands and runs in between the Kerguelen and Heard islands. A second region will be considered in this book: the Subantarctic region with the Antarctic Convergence as southern limit and extending northward to about 40°, where another transition in flow, temperature and salinity exists, designated as the Subtropical Convergence, which also marks the approximate limit of the northward drift of ice from the Antarctic. This region, the northern limit of which is less stable and sharp and therefore not so well defined as the Antarctic Convergence, is important biogeographically mainly because it includes several important islands as well as the southern extremity of the American continent and part of New Zealand.

Two main parts can be recognized in this book: first a detailed description of the different aspects of the environment, including the geology and morphology, the climatology and the oceanography of the regions under discussion. This part is to provide a base for the understanding of the specific and extreme environmental conditions prevailing in the Antarctic in order to be able to realize

X

the role these extreme conditions played in the distribution, past and present, of living organisms in the Antarctic.

The second part is devoted to the discussion of some of the main plant, microorganism and animal groups of these regions.

Since the present status of knowledge is very incomplete and varies widely from group to group it was not possible to discuss ecological and biogeographical problems, such as seasonal movements, distribution, cold adaptation etc. in general, but rather in connection with particular groups in the hope that the comparison between several groups would point to some general conclusions.

Short History of Antarctic Research Expeditions

A succinct description of the history of Antarctic expeditions and research has its place in this book because many references are made to different expeditions and because several geographical places (mountains, coasts, seas, islands etc.) have derived their names from well known expeditions and explorers, or from the countries and people that organised the expeditions.

Brief descriptions of the history of research on special aspects of the Antarctic are given in different chapters in this book, e.g. by O. WILSON: Human adaptation, p. 690; M. HIRANO: Freshwater algae, p. 127; J. McN. SIEBURTH: Microbiology, p. 267; P. M. DAVID Chaetognata, p. 297; M. D. ROGICK: Bryozoa, p. 401 and A. P. ANDRIASHEV: Fishes, p. 491.

a. The heroic era

The long accepted image of a vast and rich "Terra Australis Incognita" was only definitively abandonned after JAMES COOK had shown, in the years 1772 to 1775, by a complete circumnavigation at high southern latitudes, that a wide southern continent could only exist around the Pole and south of 60° southern latitude. The first to chart part of the Antarctic mainland was Edward BRANSFIELD, who discovered the Antarctic Peninsula in January 1820. Later in the same year the American Nathaniel PALMER and the important Russian expedition led by Thaddeus Gottlieb VON BELLINGSHAUSEN also discovered the Antarctic Peninsula. Traces of the violent polemics that followed these discoveries are still to be found on some charts where the Antarctic Peninsula is either called: "Palmerland" by the Americans or "Grahamland" by the English. In the following period whalers and sealers have made no small a contribution to the discovery of the Antarctic continent. In 1823 James WEDDELL, with the "Jane", discovered a vast sea reaching deep south into the continent and which was named after him. Subsequently the brothers ENDERBY, directors of an important Britisch Whaling Company and especially Charles ENDERBY, who was one of the founders of the "Royal Geographical Society", asked their captains to complete their hunting seasons by geographical exploration. Thus John BISCOE, on board "Tula" (1831) discovered "Enderbyland", Peter KEMP (1833) "Kempland" and John BALLENY with his ships "Sabrina" and "Eliza Scott" discovered the "Balleny Islands" and the "Sabrina Coast" in 1839. Meanwhile several countries had organised special scientific expeditions. The French expedition of DUMONT D'URVILLE with its two ships, the "Astrolabe" and the "Zelée" explored in 1838 the Weddell Sea and the Antarctic Peninsula and discovered in 1840 a new coast which was named after D'URVILLE's wife. It is this "Adelie Land" that still forms the actual French sector in the Antarctic. At about the same time an American expedition led by Charles WILKES explored the same region.

One year later, in 1841, an English expedition under the command of James Clark Ross, with two ships especially built for navigation in the ice, the "Erebus" and the "Terror", succeeded in crossing the pack-ice and discovered a vast and open sea, free of ice, reaching deep inwards the continent. This was the Ross Sea, counterpart of the Weddell Sea at the other side of the Pole and the deepest bay into the continent. On his way in this sea, Ross also discovered the only active volcano on the Antarctic continent, the top of which reaches 4,000 m above sea-level, and gave it the name of his ship, the "Erebus". Finally the expedition was stopped by an enormous ice wall of 60 m height. This was the limit of the great ice-shelf now called the "Ross ice-shelf".

During half a century attention was focused to the exploitation of the Subantarctic islands, but with the turn of the century purely scientific interest resumed and several expeditions with increasingly high scientific standards left for the Antarctic continent. In 1895 the first men to set foot on the continent itself, at Cape Adare, N-E extremity of Victoria Land, were two Norwegians: C. BORCHGREVINK and the captain of the "Antarctic" L. KRISTENSEN. The first expedition to perform a full year-cycle of scientific observation was the Belgian expedition led by A. DE GERLACHE 1897-1899, on board the ship "Belgica". For the first time an international staff of officers, scientists and sailors wintered at 72 °S in the Bellingshausen Sea after their ship was accidentally trapped by pack-ice. It was here that R. AMUNDSEN, who was first lieutenant on board the "Belgica", had his first experience with dog teams performing raids during the Spring. With a remarkable and heroic continuity and notwithstanding the very bad conditions the staff went on with the scientific program and the expedition brought back a very rich harvest of observations and biological collections.

In 1899 BORCHGREVINK and a party of nine men returned to Cape Adare and performed the first planned wintering on the continent. Besides the many observations and the raids during the summer the main importance of this expedition was to show that it was possible to spend the winter on the Antarctic continent in a wooden hut, while the ship, the "Southern Cross", came back the next Summer to releave the wintering party.

Several important expeditions were now about to start: in 1901 a Swedish expedition under command of O. NORDENSKJÖLD on board the "Antarctic", in the same year a German expedition led by E. VON DRYGALSKY on board the "Gauss" and in 1902 the Scottish National Antarctic Expedition with the "Scotia" under command of Dr. BRUCE. The interest of these expeditions was manyfold: apart from important geographical and cartographical work: discovery of "Kaiser Wilhelm II" coast in Kempland by the "Gauss" expedition and of "Coats" land in the Weddell sea by the "Scotia" expedition, these expeditions brought back very important scientific documents as a result of prolonged observations both during the summer cruises and during the winter periods.

In 1903 a first French expedition under Dr. J. CHARCOT on board the "Français" left from Brest to try to rescue. O. NORDENSKJÖLD whose ship the "Antarctic", had disappeared in the pack-ice. After CHARCOT heard that NORDENSKJÖLD had been rescued by the Argentin warship "Uruguay" he went on to complete the work of the "Belgica" in the De Gerlache Detroit. Immediately after his return in 1904, CHARCOT organised a new expedition and a second French expedition under his command started in 1908 on board the ship "Pourquoi pas?". Both expeditions fulfilled very important cartographical work on the west coast of the Antarctic Peninsula und sustained a valuable scientific observation program during summer cruises and continental raids as well as during the winter period.

With their different continental raids NORDENSKJÖLD, V. DRYGALSKY, DE GERLACH and CHARCOT opened an epoch of fierce competition for the conquest of the South Pole since they proved that it was possible both to winter on the continent and to travel on it with ski's and sledges.

This adventurous conquest is dominated by three names: R. AMUNDSEN, R. F. SCOTT and E. SHACKLETON. Apart from the well known tragic competition between AMUNDSEN and SCOTT, who both reached the South Pole during the Summer 1911-1912, these expeditions, especially those by SCOTT and SHACKLETON, played an important scientific role. SCOTT organised a first expedition in 1901 with a good ship, the "Discovery", large governmental support and distinguished collaborators. (e.g. SHACKLETON, WILSON and WILD). The principal aim was to carry out regular geological, glaciological, magnetic and meteorological observations as far south into the continent as possible. This program was largely fulfilled during two succesive winterings in the Ross Sea at the McMurdo Sound. In the Summer of 1902 SCOTT, SHACKLETON and WILSON reached 82°17' South during a big inland raid. SHACKLETON's expedition of 1908 on board the "Nimrod" in the Ross Sea carried out an important scientific program as well as an attempt to reach the South Pole during which SHACKLETON reached 88°23' South. Meanwhile a party led by Dr. P. MAWSON, the famous geologist, discovered the Magnetic Pole at 72°25'S and 155°16' E.

In 1910 SCOTT started a new expedition with the "Terra Nova". The aim was to reach the South Pole as well as to accomplish a heavy scientific program. After wintering in the McMurdo Sound, SCOTT and his party reached the South Pole in January 1912 only to find that AMUNDSEN had got there first a few weeks earlier. The Australian expedition of the "Aurora" led by D. MAWSON (1911-1914) did very important cartographical and geological work along the coasts of Adelie and Land the Shackleton Ice-Shelf and during continental raids in this area.

With the two unsuccessful but epic expeditions by the German Dr. FILCHNER (1911-1912) on board the "Deutschland" and by E. SHACKLETON (1914-1917) with the "Aurora" and the "Endurance", who both tried to make a junction between the Weddell and the Ross Seas, one can say that, what has been called the "heroic era" of exploration in Antarctica, came to an end.

b. The modern era

The modern era in Antarctic exploration is characterized by the increasing use of aviation for cartographical and geographical exploration and also as a means of mechanical transport. Apart from extensive exploration and mapping of the continent much of the effort was concentrated on investigating the scientific problems of this vast and still largely unknown area.

After the first flights by WILKINS this "Modern era" was really opened by R. BYRD's first (1928-30) and second (1934) expeditions. During November 1929, starting from his base "Little America" on the Ross Ice-Shelf, BYRD performed the first flight over the South Pole. On the other hand the several Norwegian expeditions organised by L. CHRISTENSEN with the "Norvegia" from 1927 onwards, the British Australian New Zealand Antarctic Research Expedition (B.A.N.Z.A.) (1929) led by D. MAWSON and the new "Discovery" expedition (1930) played an important role in this period, exploring and mapping such important regions as the Enderby, Kemp, Wilkes and King George V lands as well as the present Prince Olav Coast. Very important were also the regular expeditions organised by the DISCOVERY COMMITTEE founded in 1923 to organise and control whaling in the Falklands sector. Using three ships: the old "Discovery", the "Discovery II" and the "William Scoresby" as well as several whalers, very important oceanographical work was done in the English sector ever since 1925. Meanwhile L. CHRISTENSEN with several of his ships including the "Norvegia", the "Thorshavn" and the "Thorshammer" and with the use of airplanes piloted by captain Riiser LARSEN and Nils LARSEN, explored the Australian and Norwegian sectors between 1931 and 1937.

In total 4,000 km of coasts were explored and 2,000 km photographed. About 2,000 aerial photo's were taken covering 80,000 km² between 81°50'E and 26°E.

An other name to remember is that of the American L. ELLSWORTH who was the first to make an epic air junction between the Weddell and Ross Seas in 1936.

With the expedition of the "Schwabenland" (1938-39) Germany reentered the Antarctic and performed an extensive aerial cartographical survey covering a region between 69°S and 75°S and 10°W and 20°E. Important meteorological, oceanographical, biological and geophysical observations were also obtained.

The most extensive aerial exploration was due to the American operation "High Jump" in 1946-1947 under command of admiral R. BYRD, using 12 warships with 4,000 officers and sailors. In total 8,000 km of coasts and 900,000 km² of the continent were photographed and mapped.

c. Era of expansive exploration

With the Norwegian-British-Swedish expeditions of 1949-52 the new era of expansive and really international exploration, now underway, was opened. An important base "Maudheim" was established and the Queen Maud Land was extensively surveyed by aerial exploration and by major traverses into the interior, giving valuable information about the nature of the rock surfaces lying under the ice cover. An area of 1,000 km coast line and 500,000 km² of land was covered. Furthermore an important meteorological, glaceological, topographical and geological program was carried out. During the Summer cruises the "Norsel" also carried out oceanographical research.

At the same time started the French expeditions organised by P. E. VICTOR. They established a base in Adelie Land. After "Port Martin" burned down in 1952, a new base at "Pointe géologie" was established. Very important results on the biology of Adelie and Emperor penguins were obtained.

The Norwegian-British-Swedish expedition laid the foundation for the Antarctic phase of the International Geophysical Year in 1957-58. It is to the I.G.Y. that one can trace the present rapid advance of science into the Antarctic. In the context of the I.G.Y., twelve nations dispatched expeditions to the Antarctic, manning some 40 stations on the mainland as well as on antarctic and sub-antarctic islands.

Preparations for the I.G.Y. started already during the years 1954-1955-1956 by America, (operations "Task Force" and "Deep Freeze"), Russia (cruises of the "Ob"), Norway (establishment of a base in 1956 at 70°30'S, 2°32'W) and England (preparation of the Commonwealth Transantarctic expedition by M. Fuchs and Sir E. Hillary). During the I.G.Y. Argentina, Australia, Belgium, Chili, the United States, France, Japan, New Zealand, Norway, the South African Republic, U.R.S.S. and the United Kingdom, all had permanent bases in the Antarctic. Attention was focused primarily to meteorology, upper-atmosphere physics, geomagnetism, glaciology and seismology (very important ground traverses were organised; e.g. the Transantarctic expedition by Fuchs and Hillary). Fortunately it was soon recognized that this highly succesful work had to continue which led to the foundation in 1957 of the Special Committee on Antarctic Research (S.C.A.R.), the purpose of which was to co-ordinate the scientific programs of the twelve nations involved, during at least eleven years (a solar cycle 1958-1969). Scientists of the participating nations agreed to co-ordinate their national programs, to standardize their instruments, to exchange personnel and results and to give each other mutual assistance on logistical problems. Although some of the nations involved had to close their stations temporarely, it can be said that it is thanks to this vast international effort that scientific investigation is now going on uninterrupted ever since 1957. The present programs are a continuation of the I.G.Y. program, but such important branches as geology, biology, medicine, geodesy, catography and oceano-graphy have received increasing attention. This unprecedented international scientific collaboration set the path for an international diplomatic conference on the Antarctic, held in Washington in 1959, which resulted in the signing of a Treaty ratified by all the nations involved in Antarctic research. This treaty calls for the demilitarization of all national bases, it bans nuclear explosions and the dumping of radioactive wastes and, above all, it reserved the Antarctic as a vast continent open to all nations interested in peaceful scientific research. It was thus brilliantly shown that international scientific collaboration can open the way to major international political agreements and to a better understanding between nations.

With the reopening of the "Roi Baudouin" base by a mixed Belgian-Dutch expedition and of the "Syowa" base by the Japanese, twelve nations will maintain permanent stations in the Antarctic during the period 1965-1966.

Short Review of the Different Chapters

PART I. Environmental Factors

Both on morphological and geological grounds Antarctica can be divided in two parts: *East Antarctica* or the Gondwana province and *West Antarctica* or the Andean province. H. J. HARRINGTON in *Chapter I* gives a detailed description of the geological aspects of the different parts of these provinces and of the Antarctic and Subantarctic islands with an emphasis on the stratigraphical (p.8-30) and the paleontological (p.30-57) records. Using the available evidence the author discusses paleographical problems and theories, giving a speculative modification of both continental drift and land-bridge theories. (See p.XXIV)

Chapter II by M. J. RUBIN is devoted to the climatology of the Antarctic. The duration of daylight with the long summerdays and a five-month period of darkness, the high mean elevation of the mainland ($\pm2,000$ m), its continental character, the thick ice-sheet and the freezing of a large area of the surrounding ocean, are the main factors influencing the Antarctic climate. The air contains very little dust and other pollutants and is mainly characterized by its very low water vapour concentration.

There is a very important difference in the radiation balance of rock and soil surfaces.

Large scale air circulation tends to be zonal and is strongest from fall to late winter. A discussion of the weather and climatic phenomena shows that a favourable environment for plant and animal life on the continent only exists in some areas at the air-surface boundary layer.

The combination of intensive storms over the ocean, the strong cooling of the surface layers of air and the steep slopes of the continent result in strong winds especially in the coastal zones. Most important here are the drainage winds carrying great quantities of drifting snow (Blizzards).

Generally precipitation is very low and so is the relative humidity giving the Antarctic Continent the character of a "cold-desert".

In *Chapter III* F. OSTAPOFF discusses the Antarctic oceanography.

After describing the horizontal and vertical dimensions together with the bottom topography of the Antarctic ocean the author discusses the main chemical and physical variables.

Both the horizontal and vertical temperature, salinity, density and oxygen distributions are given and their relations to the main water circulations discussed. The most important of these currents is the Polar Front or Antarctic Convergence. Several possible criteria for the definition of the Polar Front are critically discussed.

The following general conclusions can be drawn from the first three chapters with regard to the environmental conditions influencing life in the Antarctic.

The Land

Most of the Antarctic mainland is a "ice-desert" due to the vast ice-sheet that covers 11.5 million square kilometers of its surface. The mean thickness of this ice-sheet must be between 2,000 and 2,500 m reaching a maximum of over 4,000 m.

The region of maximum precipitation (the Antarctic Peninsula) only receives 20 inches a year and moreover most of it under the form of snow and therefore only available to plant life if melted. A very low content of organic material, a very dry atmosphere, very low mean temperatures, strong cold and dry winds and a five month-period of constant darkness, are the main factors influencing life on this "cold-desert". Due to the enormous ice cover, formed by hard-packed snow and glacier ice, many parts of the underlying rock are depressed and lie under the sea level. The only rock or soil surfaces emerging are formed by outcroppings in low coastal areas, clipps forming the inner boundary of ice-shelves (e.g. Ross Ice Shelf) and high mountain peaks (so called nunataks).

Such soils as Antarctica affords for the growth of plants and for the animal life depending on it are mainly alkaline and composed of finely crumbled and unweathered rock and sand. There is little or no organic carbon except near bird rookeries and below patches of growing mosses and lichens.

Life only has a chance in shallow ponds, formed by melted ice or snow, and on soil and rock surfaces free of ice-cover. Very important here is the formation of relatively warm microclimates at the air-surface boundary layer due to adsorbtion or radiation heat especially in places sheltered from the strong winds.

The main adaptations necessary to survive under these conditions appear to be protection against extreme dessication and cold, rigorous economy of food chains and survival during long periods when active life is impossible. Due to these very severe conditions the Antarctic life community presents an ideal model for the study of the interdependence of living forms at the limits of adaptation to dessication and cold especially by comparison with the Antarctic and Subantarctic islands where conditions gradually become more favourable.

The Ocean

From the biological point of view the difference between the Antarctic ocean and continent is very striking. Whereas the continent sustains the poorest life community of any continent, the Antarctic ocean quantitatively has one of the richest living communities. The possible adverse effect of the low mean temperatures (near or below zero) is largely compensated by a very high turnover of waters set in motion by various currents. Due to the upwelling of bottom waters, surface waters are rich in nitrates, nitrites and phosphate salts. Another important factor is the high oxygen content.

Due to its enormous dimensions the Antarctic ocean is an efficient distribution barrier to many terrestrial plants and animals. Due to this isolation and to the action of strong and regular circumpolar currents the whole region tends towards considerable uniformity in the distribution of its living forms.

PART II. Discussion of some of the main plant, micro-organism and animal groups

Freshwater Algae are the most abundant plants in the Antarctic and they grow on most of the available habitats: on open ground, in ice and snow as well as in water. In *Chapter IV* M. HIRANO gives lists of species found and describes the finding places of freshwater algae collected in Antarctic and Subantarctic regions. Most species found in the Antarctic mainland are widely distributed. From one third to half of all Antarctic forms are also found on other continents. The distribution of the most important algae is discussed in relation to the different habitats in which they occur.

Serial changes in distribution appear to be a promising field of research.

A comparison between the algal flora of the Antarctic and Arctic as well as European regions reveals that, although these floras are similar, which merely reflects the distribution of cosmopolitan genera in the algal kingdom, they are not closely related. Differences and similarities are discussed in function of the different environmental conditions prevailing.

Lichens are the plants that best withstand the cold dessication of the Antarctic climate. C. W. DODGE *(Chapter V)* discusses the morphological and physiological adaptations that enable them to do so. The lichens show a relatively high degree of endemism and have probably migrated into the Antarctic from Fuegia long ago.

The importance of N. WACE's chapter on Vascular Plants *(Chapter VI)* resides in the first place in his discussion of the plant fossil record and distribution in both the Antarctic and Subantarctic regions (cfr. with the paleontological records of chap.I). Paleozoic, Mesozoic and Tertiary floras are listed and geographical relationships discussed. The morphology and the leaf sizes point to a moist and temperate subtropical climate for the continental Tertiary flora. The problem of the past and present distribution of *Nothofagus* and *Podocarpaceae* is discussed and it seems probable that the Antarctic continent has played an important part in the spread of these plants. The author attempted to compile and discuss a list of plants for which Antarctica might have been a distribution centrum.

To describe the present zonal distribution of Antarctic and Circum-Antarctic vegetation, WACE recognizes four zones using criteria independent of climatic, geographic or of the supposed floristic origins of the plants involved and calls them temperate, Subantarctic, Low-Antarctic and High-Antarctic zones. As criteria he uses the main structure features of the plant communities.

Finally the author describes the present day vegetation in the different zones considered and indicates the probable factors responsable for the extreme paucity of the present phanerogamic vegetation.

Chapter VII by J. McN. SIEBURTH is devoted to the microbiology of Antarctica.

All the different habitats show a very low bacterial content. Although most samples from air, snow and ice were found to be sterile, the possibility that air-borne organisms account for most of the bacterial activities in the Antarctic, by forming autochthonous snow and ice floras under favourable conditions, is still attractive, especially since most of the identified organisms are very common forms.

Most soil samples were found to contain relatively few organisms.

An important continental habitat is formed by the Cryptogamic flora. Here too no specific forms were found, the bacterial flora of mosses was quite similar to the epiphytes found in temperate climates.

The marine habitat offers an apparent paradox since the very productive marine basins have a very poor bacterial content. Since bacteria play an essential role in completing food cycles, one must probably accept that a functional bacterial microflora has not yet been discovered except if it turns out that strong ocean currents transport the organic materials to more temperate zones for decay. Phosphates, nitrites and nitrates could then be replenished by upwelling from bottom waters from these regions.

Another very interesting problem is the production of antibiotics by several representatives of the Antarctic phytoplancton.

Finally the Avian, Mammalian, fish and invertebrate habitats are discussed, together with some pathological aspects.

Promising prospects are the study of biochemical activities in relation to well defined physical, chemical an biological environments. In this context the control of temperature on nutrient requirements is unexpected and very interesting.

The Chaetognata of the southern ocean are discussed by P. M. DAVID in *Chapter VIII*. This important group of the Zooplancton can be used as an indicator of hydrographical conditions. Both the horizontal and vertical distributions of most of the known species are given.

What appears to be a general rule for the Antarctic fauna and flora holds true for this group composed of small number of species with high numbers of individuals. The basic distribution pattern is circumpolar. Seasonal vertical movements during the life history of most of the species could play a role as a protection against adverse conditions. The southern Chaetognata do not seem to have been separated from subtropical forms since long which indicates a colonization from the North.

Decapod Crustacea *(Chapter IX* by J. C. YALDWYN) are almost completely absent from the Antarctic benthic fauna. This paucity of the Antarctic benthic decapods in comparison e.g. with the Arctic, constitutes a very interesting biogeographical problem for which no good explanation is as yet available. Such species as have been found are mostly restricted to the Antarctic. The Subantarctic fauna is very different from the Antarctic one since it is relatively rich and varied and contains many temperate zone forms extending into the Subantarctic region.

Several important biogeographical problems are illustrated by the study of the Mollusca *(Chapter X* by A. W. B. POWELL). Contrary of what is generally the case for invertebrate faunas, the Antarctic Convergence does not form a sharp boundary between Antarctic and Subantarctic faunas since many species range over both zones. No terrestrial forms are known. The West Wind Drift is the main distributing agent especially for species associated with algae. The present molluscan fauna appears to be in an active stage of colonization mostly by way of S. America and the Scotia Arc. An earlier fauna, largely exterminated by relatively recent ice-ages but of which some examples may still constitute an element in the recent fauna, is thus replaced by new colonists. Biogeographical provinces and quadrants are described but do not seem to have a true biological significance.

In his notes on the free-living, marine Copepoda *(Chapter XI)* W. VERVOORT discusses the problem of the depletion of pelagic animal life by northward currents. Copepods are amongst the most important forms in the Antarctic and Subantarctic Zooplancton and can serve as a model to study the influence of hydrological conditions on this Zooplancton. Vertical migrations coupled with reproductory phenomena is one explanation to the problem of depletion but does not seem to apply to all forms. The distribution and vertical migrations of standing crops of Zooplancton are discussed. Many species have a circumpolar distribution under the influence of the West Wind Drift.

More than half of the 321 known Bryozoa *(Chapter XII* by the late M. D. ROGICK) appear to be endemics for Antarctica.

In *Chapter XIII* P. DALENIUS deals with the Antarctic Acaridae. This group gives an opportunity to discuss the general aspects of the terrestrial life on the Antarctic continent.

Terrestrial biotopes are characterised by very short foods chains. Of the three main factors influencing the soil faunas: nourishment, temperature and humidity, the latter one appears to be the most limiting one. It is pointed out that there might be an interaction between resistance to cold and to dessication: in both cases it is necessary to bind water.

Arthropods *(Chapter XIV)* appear to have at present the most southern distribution and are the most important resident land-animals of Antarctica. This chapter by J. L. GRESSITT turns out to contain the most complete account of life on Antarctica from the ecological and biogeographical viewpoint.

Three ecologically different regions are considered: the fringe of the continent, outer- and middle Antarctica. It would appear that some species are remnants of a once rich fauna although they appear to have a restricted distribution, whereas others are recent immigrants but have a wide spread distribution.

Air dispersal, by air currents and by birds, appears to have been the most important distributing factor.

Adaptations to the environment are examplified by short food chains and resistance to cold and dessication. Some interesting physiological experiments with Collembola are reported. Wing reduction appears to be an adaptation to resist strong winds.

The general picture is one of an erratic distribution as a result of a dispersal by the prevailing air- and sea-currents as well as by animal transport and depending on the availability of appropriate ecological niches. Endemism is moderate.

Chapter XV by A. P. ANDRIASHEV is devoted to a review of the Antarctic Fishes. In many respects this fauna is peculiar. It includes families and genera of which some are similar but others very different with regard to their origin and degree of endemism.

An important factor in the vertical distribution of fishes in the coastal area is the "Sunken" character of the Antarctic continental shelf.

Special physiological aspects, such as cold adaptation, reproduction, feeding, growth rates and especially the occurrence of white-bloodedness among Chaenichthyid fishes, are dealt with in some detail. Many Antarctic fishes are true stenothermic forms since their metabolism has a low temperature optimum. Whiteblooded fishes seem to be adapted to cold waters with a high oxygen content.

The recent finding of well preserved remains of benthic invertebrates and a number of large fish-remains on the Ross Ice Shelf points to the possibility of marine life even under the permanent ice-shelves.

Zoogeographically seven types of distribution are recognized and described by the author. Four more or less isolated faunal units are proposed as subdivisions of the Antarctic region.

As a general conclusion it is said that all Antarctic fish genera and species are peculiar and nearly all belong to the group of the Nototheniiformes characteristic for and almost completely restricted to the Antarctic and Subantarctic zones. This points to a long isolation of the Antarctic fish fauna. The basis of the fauna consists of ancient autochthonic elements completed by some more recently immigrated notal genera related to the Patagonians and Falkland fauna.

The best known, and in many ways the most typical Antarctic birds are the Penguins (J. PRÉVOST and J. SAPIN-JALOUSTRE, *Chapter XVI*). Attention is focused primarily on the two truely Antarctic forms: the Adelie and the Emperor penguin. A thorough description and discussion of specific environmental conditions is given. Special attention is given to chill values and the formation of natural and artificial microclimates.

The authors describe in some detail the species characteristics, the activities, the ethology, the life cycle and the main mortality factors of both species. It appears that the Emperor penguin can be regarded as the bird best and most completely adapted to the Antarctic continent.

Physiological and behavioural adaptations towards winter breeding and possible difficulties and advantages related to it, are indicated. A short comparison between the Emperor and King penguins is included.

Concluding it appears that the two fundamental problems of alimentation and cold resistance have been solved in quite different ways by the two Antarctic species that live under different ecological conditions and have very different ethological characteristics.

Antarctic Birds *(Chapter XVII* by K. H. Voous) spend the greater part of their lives at sea on which they completely depend for their food. Apart from the penguins, nests of no more than eight species have been found on the mainland, belonging mainly to Petrels and Skuas.

Some species were originally colonists from the Northern Hemisphere and seem to have subsequently recolonized the North Atlantic. The distribution is mainly circumpolar as far as identical conditions are available. Infiltration took place mainly by way of the Antarctic Peninsula but also by way of the Kerguelen and the New-Zealand and Macquarie islands. The distribution is zonal but not sectoral. Both summer and winter breeding seasons are known. Differences in breeding seasons appear to have played a role in species formation.

The majority of Antarctic birds get their food from the pelagic animal life of the surface of the sea although several are predators and scavengers feeding on eggs, young and even on some small adults of other bird species.

Only one species is known to feed on inland freshwater crustaceans and insects on the Kerguelen.

To escape the extremely severe conditions of the Antarctic winter. most birds migrate northwards. Very spectacular in this context is the Wilson Storm Petrel. This bird migrates every year to and back from deep into the Northern Hemisphere. The Antarctic bird fauna is augmented by summer visitors and passage migrants.

The book ends with a discussion of human adaption problems to life in the Antarctic *(Chapter XIV* by O. Wilson)

Climatic factors such as wind, chill, radiation heat exchanges and absence of light as well as food, housing, sanitation and clothing problems are discussed. Most of these problems have been satisfactorily solved in modern expeditions. As far as medical aspects are concerned it is interesting to note that very few infective diseases occur in Antarctica. The majority of cases are caused by injuries and by teeth trouble. Many possible physiological adaptations to cold are reviewed. The conclusion is that no convincing evidence of a definite process of adaptation has been observed. The artificial formation of a suitable microclimate is the main factor in cold resistance.

Finally a brief review of the psychological problems caused by isolation and promiscuity, such as mental stresses, psychological reactions and adjustement is given.

Review of Some General Biogeographical and Ecological Problems
in Relation to the Antarctic and Subantarctic Regions

1°) Continental Drift and Land-Bridges

The Antarctic holds a key situation in any theory on historical biogeography. This problem has been discussed in several chapters in this book.

Geological and paleontological data are reviewed by H. J. HARRINGTON (Chap. I). It appears that East Antarctica has a paleontological record from the Cambrian to the Jurassic, indicating that life appeared on this continent as early as in other continents. Antarctica seems to have had a normal biota of cold and cold temperate type during the Phanerozoic period. The lower Devonian fossils of Antarctica differ from those of the Northern hemisphere and resemble more closely the South African and South American faunas of the same period. Similar associations of tillite and Permian coal deposits have been found in East Antarctica and India, Madagascar, South Africa, South America, New Zealand and Australia, providing evidence that an East Antarctic glaciation occurred contemporaneously with the late Paleozoic glaciations affecting peninsular India and all the southern Hemisphere continents. No evidence for other cold climates is to be found until the beginning of the present glacial conditions in the late Cenozoic. Throughout the Tertiary the climate remained sufficiently warm to support the growth of sead-bearing plants as *Araucaria* and *Nothofagus* (See part on Tertiary vascular plants in Chapter VI). But although Antarctica thus remained in the mainstream of floral evolution it seems to have somehow missed part of the animal evolution since no mid-Tertiary mammal fossils were found.

Geological and paleontological evidence is in good agreement with the existence of a land-bridge or island chain explaining the dispersal of living forms from the Andes via the Scotia Arc to West Antarctica during the Mesozoic and Cenozoic periods, but no other land connections are satisfactorily documented. To explain all the problems raised by the evidence of warm climates, Cretaceous and lower Tertiary vascular flores, Permian Gymnosperm Woods, Devonian brachiopods and Cambrian Archaeocyathinae in East Antarctica, HARRINGTON (p.59) proposes an hypothesis based on both the Continental drift and land-bridges theories, whereby E. Antarctica would be part of a former Gondwana land that drifted together with S. America until both continents hit at some time between Cretaceous and mid-Tertiary, an "island Arc" that remained relatively fixed in position throughout history. West Antarctica would then be part of this "Island Arc".

In a recent article J. DARLINGTON* reviews the geographical, geological, glacial, paleomagnetic and biogeographic evidence in relation to the theory of Continental drift and concludes that Africa and S. America were probably united and drifted apart not later than the Triassic and that the gaps between the other Southern Continents and peninsular India might have been narrower before these continents drifted to their present positions. Paleomagnetic together with geological evidence seems the most convincing and indicates that all southern continents, except E. Antarctica, probably drifted Northwards.

The Schemes presented by HARRINGTON (Chap. I, p.60) and by DARLINGTON* are in good agreement.

In Chapter VI N. W. WACE concludes from a description of Antarctic and Subantarctic Tertiary floras that a land (or island) chain link between the Antarctic Peninsula and the Southern Andean Chain is in agreement with the distribution of Tertiary vascular plants. Other land connections are very problematical. The author also reports experimental studies on the present dispersal capacities of New-Zealand podocarps and *Nothofagus*, indicating that dispersal on or over sea is extremely improbable.

J. L. GRESSITT (Chapter XIV, p.464-465) concludes that too few groups are involved at present on Antarctica so that no direct evidence on the problem can be drawn from the present continental Arthropod fauna. Moreover air-dispersal appears to explain adequately the present distribution.

From a review of Antarctic fishes A. P. ANDRIASHEV (Chap. XV) concludes that neither marine nor freshwater fishes support the theory that the Antarctic continent connected S. America with Australia during the Tertiary period.

2°) Bipolarity

Species that have a distribution pattern common to both the Arctic and the Antarctic but are absent in the intermediate latitudes are said to have a "Bipolar" distribution.

Several authors discuss bipolar distributions. After a short review of environmental conditions in the Arctic and Antarctic in relation to the Algae flora, M. HIRANO concludes (Chapter IV, p.159-165) that only very few Algae have a true bipolar distribution.

Several molluscan genera (Chapter X, p.338) show a bipolar distribution but for the most part the resemblance appears to be more ecological than morphological. Examples supporting several different hypotheses, (e.g. relics of former cosmopolitan fauna, deep-

---

* J. DARLINGTON. *Proc. Nat. Acad.Sci.Wash.*, **52,** *1084,* 1964.

XXVI

water migrations and independant parallel development) can be
found among molluscan genera.

Bryozoa (Chapter XII, 402) offer no support for bipolarity, although some very cosmopolitan bryozoans range from high northern latitudes down to the Antarctic.

Several tens of genera of fishes (Chapter XV, p.544-545) have a
bipolar distribution, deep-water migration seems to be the most
likely explanation.

3°) Cold Adaptation

The degree of adaptation to cold varies from group to group.

Cyanophyceae (Chap. IV, p.147-149) seem to be able to adapt to
several different types of extreme conditions. Their simple life-
cycles and autotrophic character (both for nitrogen and carbon)
and the several favourable microclimates formed during the summer
seem to explain adequately their presence on the Antarctic Continent. Highly organised algae are completely absent from the Cryovegetation, among simple algae truely cryophilous (p.157) forms
were found.

Except some small species of lichens (Chapter V) that live in
sheltered conditions in crystalline rock cracks, most species are
protected from wind erosion and cold dessication by the formation
of thick-walled cortical hyphae or by the cementing of the hyphae
in a layer impervious to water vapour but easily permeated by
water. In the Umbilicanieceae e.g. (p.195) a gelified amorphous
layer is formed by dead fastigiate cortex cells. In contrast to the
algae, sexual reproduction seems to be the rule. Most forms have
relatively large spermogonia, producing more spermatia than temperate forms.

Microorganisms (Chap. VII). No truely psychrophilic bacteria
have been found until now. Among the yeasts an obligate psychrophile was often isolated: *Candida scotii*. Several other psychrophilic yeast species were isolated and also most of the fungal isolates
turned out to grow at 5°C but not at 30°C (p.274). It would thus
appear that yeasts and fungi are better adapted to the Antarctic than
are bacteria. Experimental studies (p.289-290) indicate that nutrient requirements are to some extent controlled by temperature.
Multiple temperature optima, probably due to different temperature
optima for different enzyme systems, have been observed and could
explain the influence of temperature on nutrient requirement.

Laboratory studies, along with field studies, on the influence of
temperature on biochemical processes look very promising. Seasonal vertical movements, coupled with reproductory phenomena,
may be the answer to the cold problem for Chaetognata (Chap. VIII)
and for some Copepoda (Chap. XI) and may be more in general for

part of the zooplancton. The cold winter months are spend in deeper and warmer layers, whereas active growth takes place in Spring and Summer in the top layers.

Influence of temperature on terrestrial life in the Antarctic is best studied for Arthropoda (Chapters XIII and XIV). P. DALE-NIUS (p.426-427) argues that cold resistance must be connected with resistance to dessication since in both cases it is necessary to bind water. The formation of relatively warm microclimates (up to 27.8 °C) is a very important factor in the survival of this group. The very important question of: how do most terrestrial animal and plant forms survive the winter period, remains an open field of investigation.

Laboratory and field tests on temperature tolerances of Spring-tails are described on p.448. These studies seem to indicate that the best temperatures for these organisms range between +6.5 °C and +17.5 °C with an optimum around 11 ° to 12 °C.

Many Antarctic fishes (Chap. XV p.522) appear to be true ste-notherm forms since the optimum temperatures for oxidative meta-bolism range between −1.9 °C and 0 °C. White blooded fishes (p.527-530) seem to be adapted to cold waters with a high oxygen content.

J. PRÉVOST and J. SAPIN-JALOUSTRE have made a thorough study of the cold problem (Chap. XVI p.561) in relation to the penguins. It appears that apart from the protective skin layers penguins make use from the more favourable microclimates existing in the air-layers just above the ice-surface and from artificial micro-climates produced by special group formations (especially during the winter breeding of the Emperor penguin).

No convincing evidence of physiological cold adaptation in humans has been found except locally in the hands and maybe in the face and the feet. The formation of a suitable artificial micro-climate is the most important factor in the cold resistance of humans.

GEOLOGY AND MORPHOLOGY OF ANTARCTICA

BY

H. J. HARRINGTON

Department of Geology, University of New England, Armidale, N.S.W., Australia

(with 7 figs.)

MORPHOLOGY

Two morphologies must be considered for Antarctica, one the bedrock morphology and the other the glacial morphology which is geologically ephemeral (figs. 2 and 3). The continent has been criss-crossed by oversnow traverse parties and aircraft during and since the International Geophysical Year of 1957—58, so that there is now a reconnaissance knowledge of both morphologies, which have been shown on a map of Antarctica, at a scale of 1 to 5 million with contours at intervals of 500 metres, published by the American Geographical Society in 1962. There is now international agreement that the metric scale will be used in Antarctica and that maps will be oriented so that the Prime Meridian (of Greenwich) is at the top of the map.

Bedrock Morphology

Where it is hidden beneath the ice sheets the bedrock morphology has been determined in broad reconnaissance fashion by gravity and seismic-sounding traverses (BENTLEY, 1962). The contours (fig. 3) are for present elevations and are not adjusted for the iso-static uplift and sea-level changes that would occur if the ice were to melt. When studied in conjunction with the geological map (fig. 4) the contours reflect a primary division of the continent into two sub-continents East and West Antarctica, also known as Greater and Lesser Antarctica, or the Gondwana and Andean Provinces (ADIE, 1962). There appears to be a secondary subdivision into a few broad segments trending roughly parallel to the meridians 20 °W and 160 °E.

A large part of East Antarctica is, like southern Africa, a high-standing irregular continental platform, and if the covering ice were to disappear would rise even higher, by isostatic adjustment. A broad zone of bedrock highs, probably rising to over 4000 m in Queen Maud Land extends through the Pole of Inaccessibility to the western part of Wilkes Land (BUGAYEV & TOLSTIKOV, 1960; KAPITZA, 1960). On the northern flank of this platform there is a considerable amount of block-faulting in the vicinity of the Amery

2

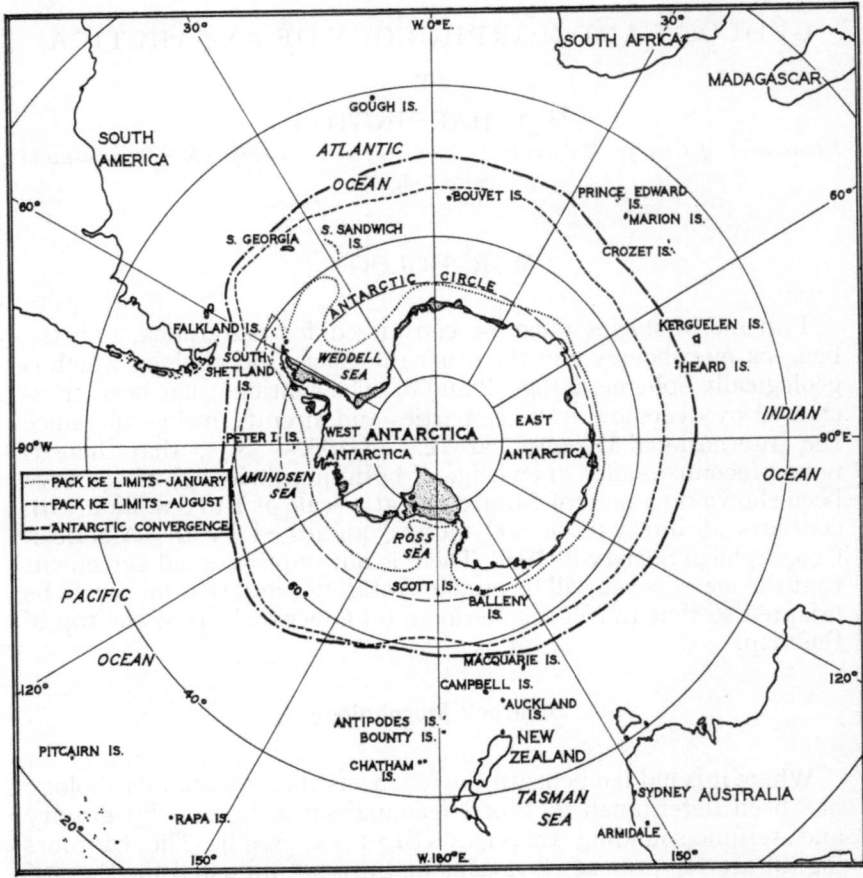

Fig. 1. Antarctica in relation to surrounding continents and islands.

Ice Shelf, the great Lambert Glacier (possibly the largest glacier in the world) and the Prince Charles Mountains (CROHN, 1959). By analogy with features in East Africa, this faulted region can be regarded as part of an ice-filled "rift valley". The valley possibly extends inland to the bedrock saddle or trough found to the west of the Pole of Inaccessibility by Soviet traverse parties. There is a suggestion of another deep depression extending westwards from the Lambert Glacier towards the Princess Ragnhild Coast, which might have the effect of separating Enderby Land as an island. Many similar but smaller depressions and mountain ranges occur towards the coast in other regions, notably inland from Mirny. Towards the coast also the ice sheet becomes thinner, and the edges of the bedrock platform are exposed in scattered mountain ranges, nunataks, coastal cliffs`and near-shore islands.

The Polar and Wilkes Subglacial Basins descend to about 500 m below sea-level and are probably sedimentary basins filled by soft rocks that could have been partially scoured out by glacial erosion. The northward limits of the Polar Basin in the direction of Coats Land are not known, but the line of soundings by the Trans-Antarctic Expedition indicates that it does not trend towards the Filchner Ice Shelf. It might continue to the region east of the Theron Mountains.

This zone of basins is separated from the Ross Sea and Weddell Sea by a high-standing segment of the continent which projects through the ice to form one of the main mountain chains on the surface of the earth, the Trans-Antarctic Mountains. These mountains are often over 3000 m in height, and have a maximum height of a little over 5000 m in the Queen Maud Range at the head of the Ross Ice Shelf. They extend continuously for over 2000 km from the Admiralty Mountains near Cape Adare to the Horlick Mountains, and appear through the ice discontinuously for another 1250 km to the Theron Mountains in Coats Land. For most of its length the chain rises steeply from the shores of the Ross Sea and the Weddell Sea, forming a feature that may be called the Trans-Antarctic Escarpment (fig. 4), which is a tectonic lineament on the line of the boundary between East and West Antarctica except possibly near the Ellsworth Mountains and Berkner Island, which could be an unusual part of East Antarctica. It is a zone of faulting and monoclinal warping. The Ross Sea section of the escarpment is approximately collinear with the eastern edge of the Antarctandean Ridge but the significance of this is not known. It should be mentioned that DAVID & PRIESTLEY (1914) used the term Antarctic Horst for the Trans-Antarctic Mountains in Victoria Land. At that time it was more customary than at present to interpret mountain features in terms of blockfaulting, usually on the basis of geomorphological evidence with or without supporting structural evidence. No significant faulting is known on the western side of Victoria Land where the Beacon System dips gently towards the Polar Basin and Wilkes Basin. Consequently there has been a strong tendency to abandon the term Antarctic Horst.

Since the first geological work by ARCTOWSKI (1895), CHARCOT, and NORDENSKJÖLD (1913) the nature of the bedrock morphology of the West Antarctic region to the north of the Trans-Antarctic Escarpment has been the subject of a large speculative literature which has been summarized by FAIRBRIDGE (1952). It has been accepted that West Antarctica is geologically a Paleozoic, Mesozoic and Cenozoic orogenic zone or mobile zone (Table I) quite different from the epeirogenic East Antarctica region. The speculation has centred mainly on two related problems, whether the Ross Sea and the Weddell Sea are linked by a sub-glacial channel, and whether the mountains of the Antarctic Peninsula cross West Antarctica directly to Byrd Land or whether they have some other pattern. The essential facts are now being gathered by a superb series of oversnow traverses and sea and air expeditions by United States parties (ANDERSON, 1960; BENTLEY et al., 1960; BENTLEY & OSTENSO, 1961; CRADDOCK & HUBBARD, 1961; CRARY, 1961; DOUMANI, 1960; DRAKE, 1962; LONG, 1962; SCHMIDT, 1962; THIEL, 1961) which have continued and extended work commenced by the early Byrd expeditions (GOULD, 1935; WADE, 1945; WARNER, 1945; PASSEL, 1945).

Nunataks and mountain groups have been found in the Ellsworth Highland region athwart the position of the supposed broad channel between the Ross Sea and the Weddell Sea. There can be only a very narrow channel possible between the Whitmore Mountains and the Horlick Mountains (THIEL, 1961) but it could be geologically highly significant, even if it is morphologically unimpressive. The Ross Sea depression is continued eastwards by the Byrd Subglacial Basin and the Bentley Subglacial Trench descending to more than 2500 m below sea-level. On one side of the basin the Ellsworth Mountains rise spectacularly in the Ninson massif to 5140 m probably the highest point on the continent. There is a relief of 7600 m between the summit of the massif and the floor of the trench. The Antarctic Peninsula and adjoining islands may be grouped together as the Antarctandean Ridge, by adaptation of a name first used by ARCTOWSKI (1895). To the north of the basin there is another bedrock high here termed for convenience the Usarp Ridge since the region has been explored by expeditions of the United States Antarctic Research Program. The ridge has a maximum height of 3677 m at one of the volcanic peaks of the Crary Mountains, and all its highest points are in fact large central volcanoes rising from a basement of metasediments and granitic rocks that is mainly subglacial and in part below sea-level. According to the map by BEHRENDT & PARKS (1962) the southern end of the Antarctandean Ridge curves morphologically towards the Usarp Ridge, being separated from it at the eastern end of the Eights Coast by a sub-glacial channel only about 500 m or less below sea-level, whereas the Antarctic Peninsula is separated from the Ellsworth Mountains by a deeper trough that is an extension of the Byrd Subglacial Basin. This trough possibly links with one under the Filchner Ice Shelf. Topographically therefore the Ellsworth Mountains seem to be more closely linked to East Antarctica than to the Antarctic Peninsula, but their geological affinities are quite uncertain pending the publication of the results of current surveys. One possibility is that the Antarctandean Ridge continues through the Usarp Ridge both forming one geanticlinal

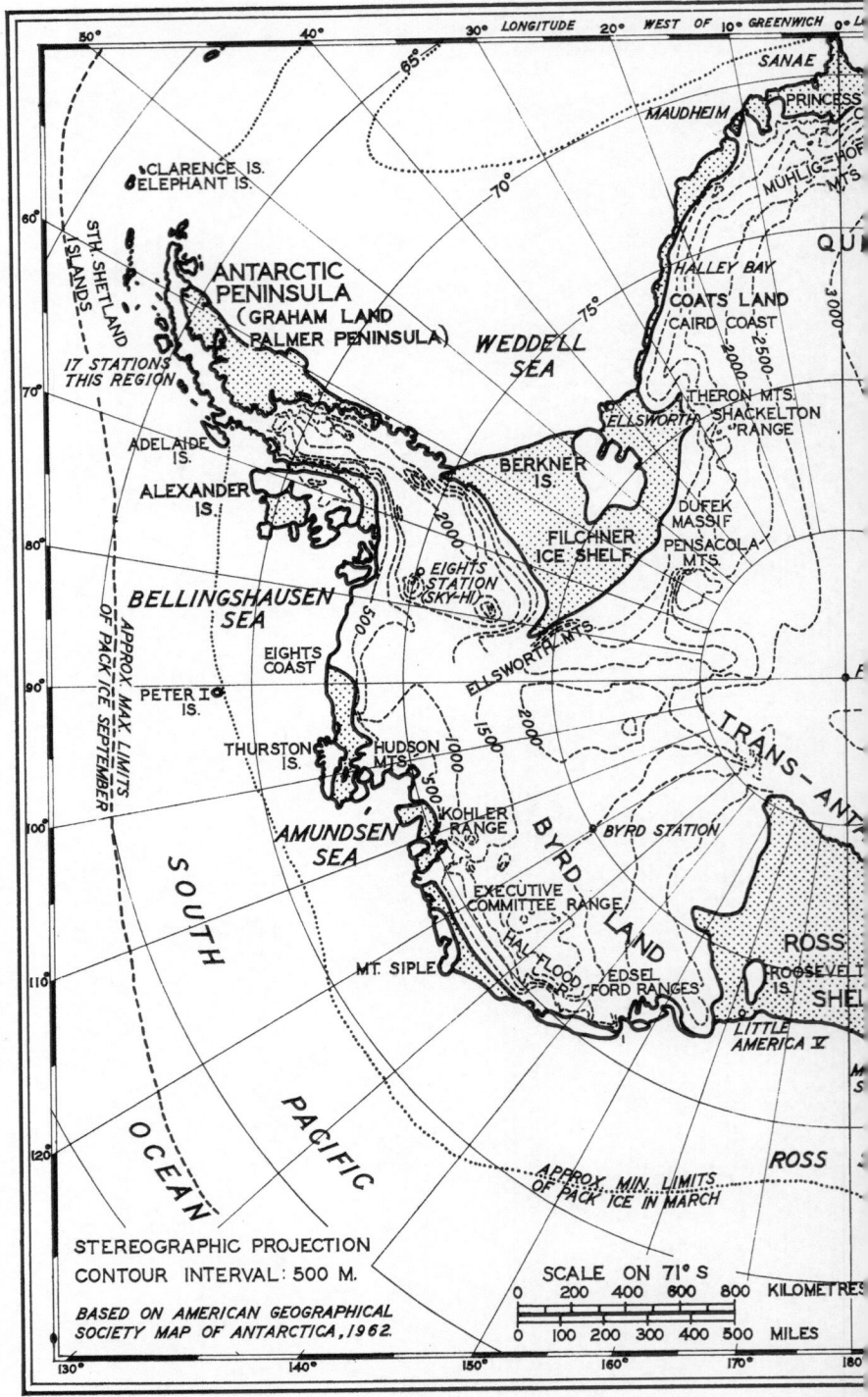

Fig. 2. Morpholo

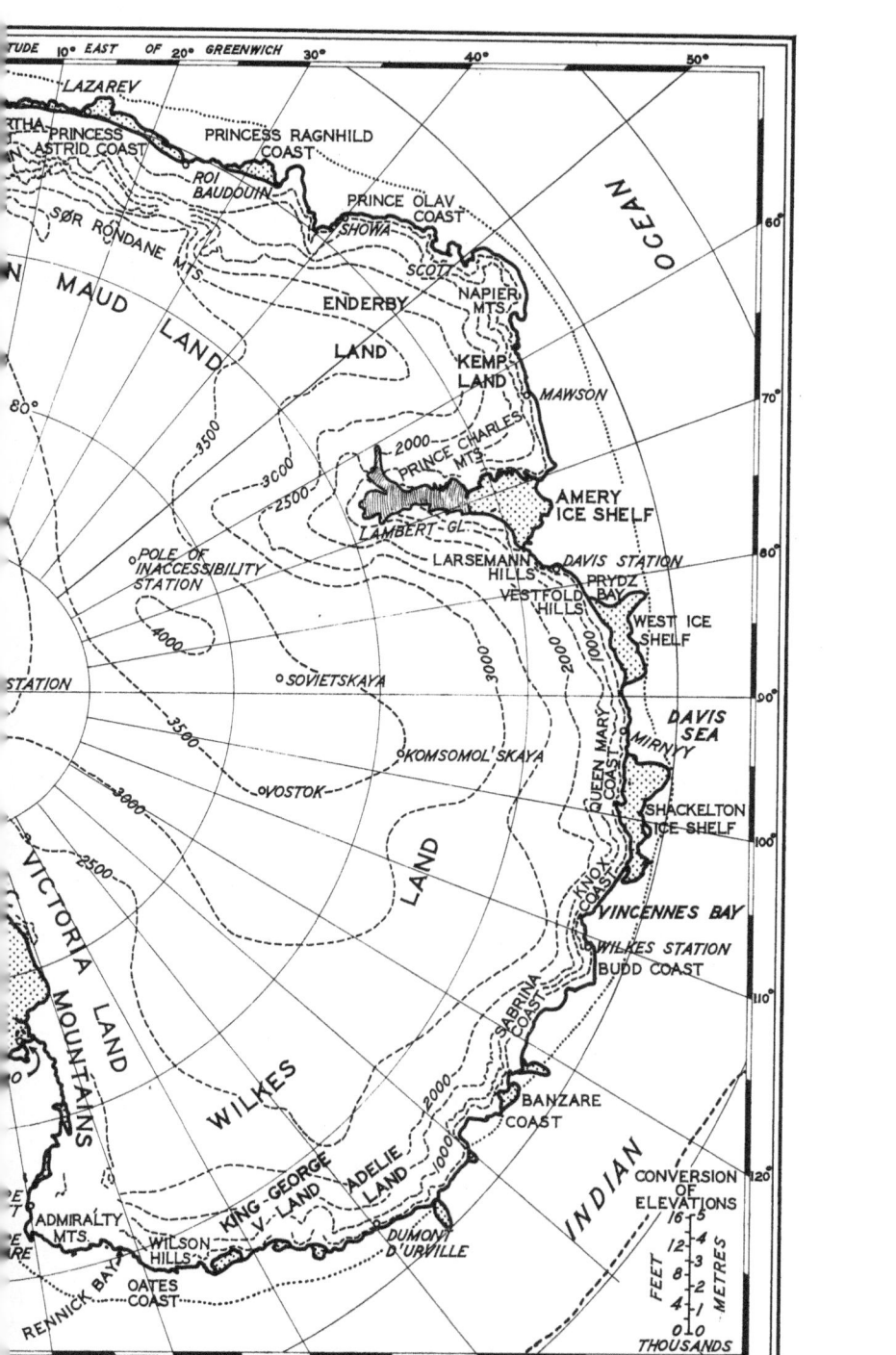

Antarctica.

feature, and another possibility is that the ridges are two geanti-
clines lying en echelon and separated by a rather narrow trough of
Late Mesozoic and Cenozoic sediments. If the second possibility is
correct, there is also an additional possibility that the Antarctandean
Ridge geologically includes the Ellsworth Mountains, but has been
separated from them by development of a transverse Late Cenozoic
structural depression. It seems more likely, however, that the
Ellsworth Mountains are a transitional folded zone that is really
a part of East Antarctica, and comparable with the region of Cape
folds in southernmost Africa. If that is so the Byrd Subglacial Basin
is the boundary between East and West Antarctica. Irrespective of
geological affinities it is clear that if the ice were to disappear the
Antarctandean and Usarp ridges and the Ellsworth Mountains would
form chains of mountainous islands with fringing archipelagos,
falling away to the deeper areas of the Byrd Basin and the South
Pacific Ocean. The relationship between these "highs", and between
them and East Antarctica, is one of the most puzzling and intriguing
morphological and structural problems on the surface of the earth,
and an explanation is attempted later in this review.

Glacial Morphology

The glacial morphology is controlled in its broad features by the
subglacial morphology. In East Antarctica the surface of the ice
sheet forms a dome rising to over 4000 m, and with its broad crest
above the bedrock high of the Queen Maud Land-Wilkes Land
region. It falls away, slowly at first, towards the coast and the Polar
Basin-Wilkes Basin depression. Its surface is very often diversified
by small snow dunes, and by similar features, called sastrugi, up to
2 m high, carved from semi-compacted snow by wind-erosion and
oriented parallel to the prevailing gravity winds (karabatic winds).
Snow dunes and sastrugi are small, but have serious effects on
surface travel. Other surface irregularities are controlled by irregularities in the underlying bedrock. In places there are undulations
in the surface with wavelengths measurable in kilometres and ampli-
tudes measurable in tens of metres. Crevassing occurs wherever
there is sufficient irregularity in ice flow or basement and is therefore
most common towards the margin of the ice sheet or generally
where bedrock is relatively close to the surface. Parts of the sheet
feed ice streams flowing more rapidly than surrounding regions,
usually because of funneling controlled by bedrock morphology
(SWITHINBANK, 1959). There can be considerable diversification in
coastal regions. Where funneling is pronounced the ice may push
into the sea as long ice tongues, and these may be welded together
by contact and by local snow falls to form coastal ice shelves. In
other places the ice may reach the sea as a broad sheet with a

6

cliffed edge about 30 m high. Such cliffs are characteristic of many long stretches of the East Antarctic coastline. In a few places the ice sheet terminates a short distance from the coast, leaving bare rock areas and "oases" that provide suitable sites for permanent scientific stations, and in those regions there may be an ablation zone rising to altitudes of 600 m or more with running streams and rivers during the warmer parts of summer days. Part of the ice escapes to the Ross and Weddell seas through the Trans-Antarctic Mountains

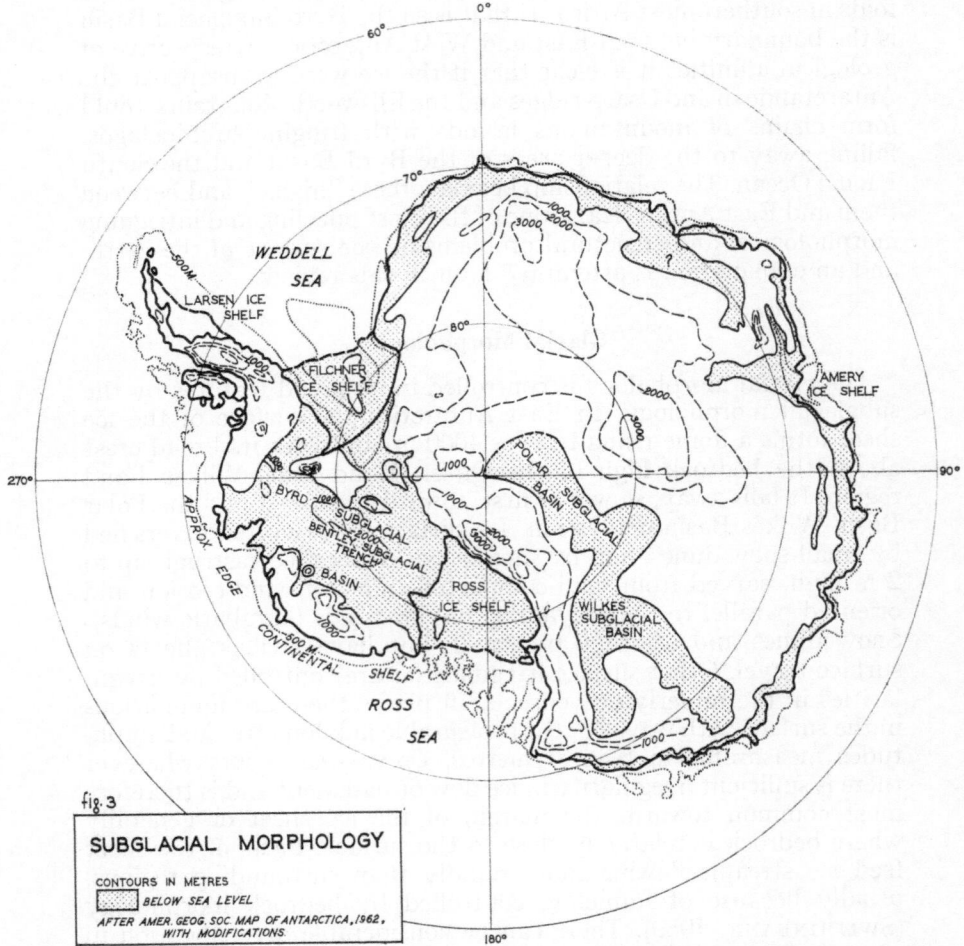

Fig. 3. Subglacial or bedrock morphology, after an inset in the American Geographical Society's map of Antarctica 1962. Areas below sea-level are shown by a dotted ornament. The 500 m bathymetric contour indicates the approximate position of the outer edge of the continental shelf.

via some of the steep outlet glaciers, up to 25 km wide and 150 km long.

In West Antarctic there are two main ice domes, again centred over bedrock highs, one in the Ellsworth Highland, and the other in the Usarp Ridge north of Byrd Station. There is a less regular third ice sheet over the Antarctandean Ridge north of the Ellsworth Mountains. The Ross and Filchner ice shelves (and the Amery Ice Shelf in East Antarctica) are nourished mainly by the surrounding ice sheets and outlet glaciers and in part by local snow fall.

In terms of bedrock morphology the outer edge of Antarctica is the outer edge of the continental shelf (fig. 3). It was once thought that Antarctica might lack a continental shelf, but it is now known to be simply deeper than usual (LAW, 1961), the outer edge being often at depths of 300 to 500 m (fig. 2). In terms of glacial morphology the outer limits of the continent vary with the seasonal variation in sea-ice cover. The January and August limits of the sea-ice or pack-ice are shown on fig. 1, and are discussed by OSTAPOFF in Chapter III of this volume.

GEOLOGY

Any small-scale geological map of a continent must be schematic and interpretative, and this is particularly so in the case of Antarctica where 95% of the bedrock is hidden beneath ice. Many parts of fig. 4 are therefore inferential and it is stressed that they are to be regarded as predictions rather than established facts, and a mixture of the specific and the general. Thus there are gaps in exposures and knowledge of the terrain between the Enderby Land mass, and the scattered mountains of Queen Maud Land, but the nature of the bedrock is inferred or predicted. Again large regions in West Antarctica are shown as "9, Jurassic, Cretaceous and Cenozoic (Depressions)", but rocks of that classification are known *in situ* only in the Antarctic Peninsula region, and have been found as erratics at Minna Bluff in McMurdo Sound. They are not exposed in the enormous intervening sub-continental region. This part of the map pattern is therefore only an interpretation of sub-glacial morphology, based on a knowledge of other Circum-Pacific regions.

The geology of East Antarctica and West Antarctica are treated separately below, and with emphasis on the stratigraphical and paleontological record as seems appropriate in a volume of this type. The descriptions are in terms of the rock units shown on fig. 4, using the same numbering system.

EAST ANTARCTICA

1, 2, Basement Complex, Ross System, Granite Harbour Intrusive Complex

The exposed margins of East Antarctica consist essentially of pre-Devonian highly folded meta-sediments, metamorphics and intrusives, and it has become apparent that these rocks can be separated into two main associations or "systems" or "complexes".

An older more highly metamorphosed part, here termed the Basement Complex, extends eastwards from the Princess Martha Coast at about the longitude of Greenwich through Enderby Land to Adelie Land, and possibly as far as the Oates Coast at longitude 153°E. A younger part, named the Ross System or Supergroup crops out in the Trans-Antarctic Mountains between the Oates Coast and Coats Land, its limits being within the known limits of occurrence of the Beacon System. Possible fragments of the Basement Complex occur also in the regions occupied by the Ross System.

1. Basement Complex

Recent accounts of the distribution, structure and lithology of the older polymetamorphic Precambrian Basement Complex have been provided by VORONOV (1958), who appears to have been the first to recognize its distinctness, and by RAVICH & VORONOV (1958), McLEOD (1959), and other geologists of the Australian, Belgian, British-Norwegian-Swedish, French, Japanese, Soviet and United States expeditions. Earlier petrographical descriptions have been listed by STEWART (1956). The sector between Maudheim and the Sör Rondane Mountains is not yet very well known since there are few coastal exposures, and access to the interior nunataks and mountains is difficult. Inland from Maudheim there are banded gneisses, amphibolites, schists and pegmatites (ROOTS, 1953). In the Sör Rondane Mountains catazonal and deep mesozonal biotite and amphibole gneisses, migmatites and rare marbles are intruded by granite and dioritic rocks (PICCIOTTO, MICHOT & MICHOT, 1960). From Enderby Land eastwards coastal exposures are more abundant and there is correspondingly more information. In the vicinity of Showa Station there is the first record of charnockitic gneisses, so abundant eastwards, together with granite and pegmatite (KIZAKI, 1962; NICOLAYSEN et al., 1961). On the coast in the vicinity of Mawson Station and in the Framnes Mountains to the south of it there is a porphyritic charnockitic "granite" moderately foliated and with numerous inclusions. This "granite" appears to be a result of the mobilisation of granulitic charnockite which is widespread in Enderby Land and Kemp Land (McLEOD, 1959; CROHN, 1959) and

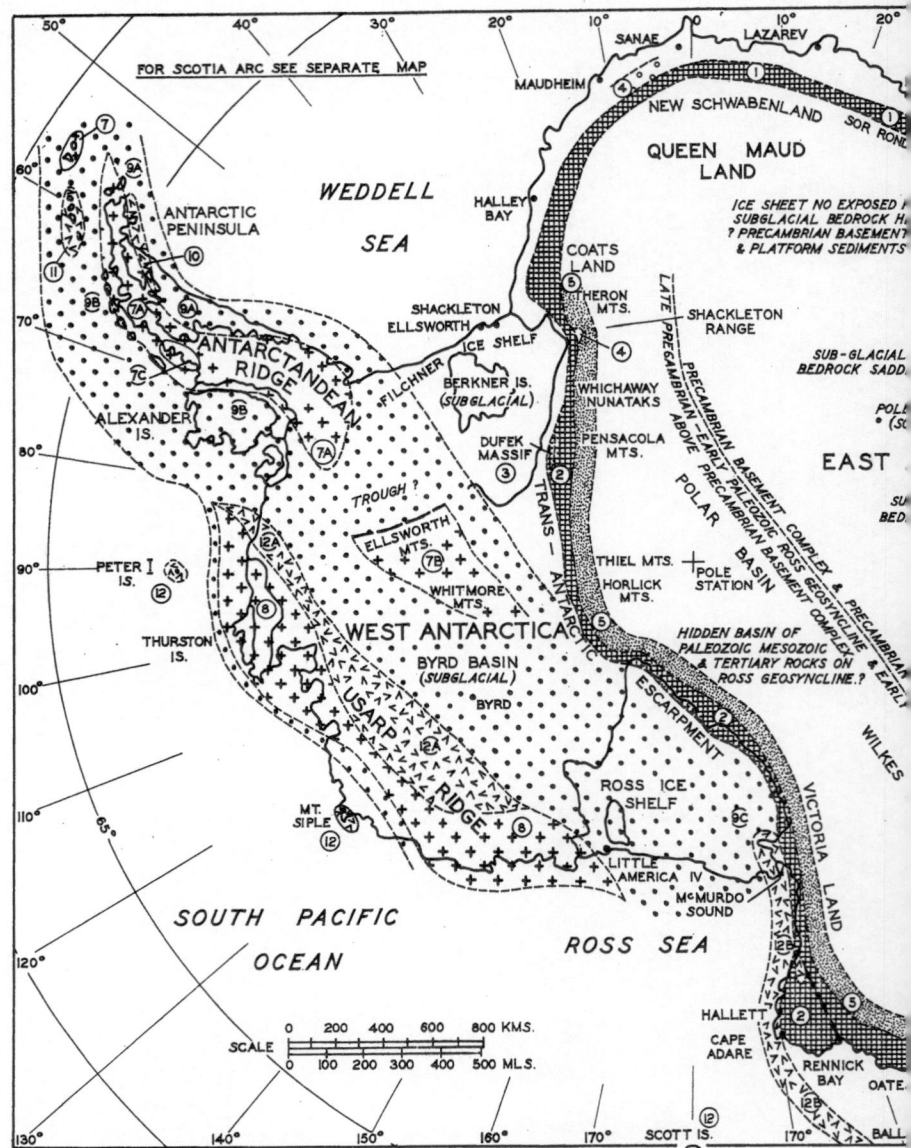

Fig. 4.

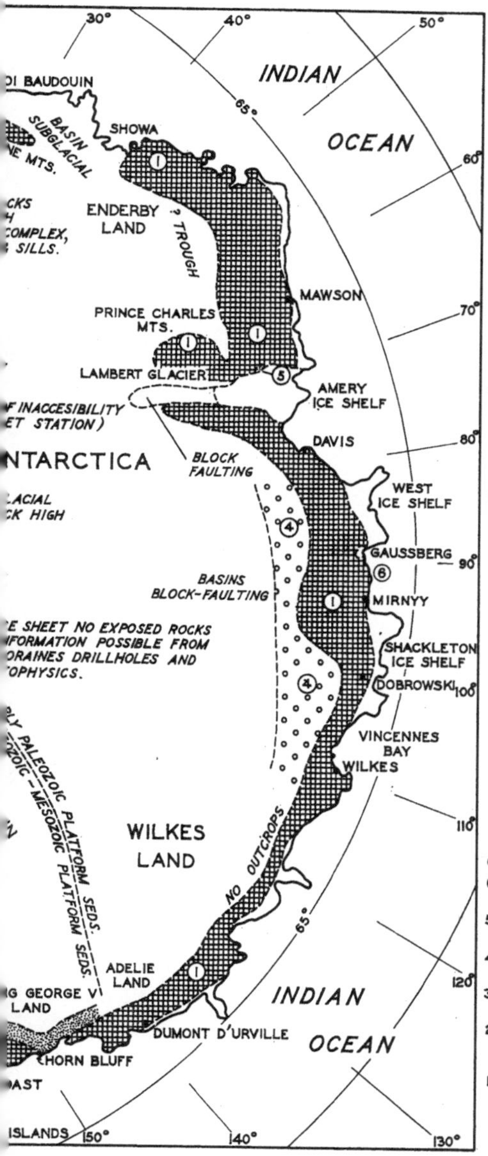

FIG. 4.

GEOLOGY
SCHEMATIC
WEST ANTARCTICA

QUATERNARY	SEDIMENTS, MORAINES, NOT SHOWN.
10, 11 CENOZOIC 12. VOLCANICS.	12B. MᶜMURDO VOLCANICS. UPPER CENOZOIC ALKALINE OLIVINE-BASALT ASSOC.
	12A. BYRD VOLCANICS. UPPER CENOZOIC ALKALINE OLIVINE-BASALT ASSOC.
	11 SOUTH SHETLAND ISLANDS. UPPER MIOCENE BASALTS, AND QUATERNARY BASALT—ANDESITE—DACITE ASSOCIATION (AND JURASSIC VOLCANICS.)
	10 JAMES ROSS VOLCANICS. UPPER MIOCENE BASALTS.
9. JURASSIC, CRETACEOUS, AND CENOZOIC (DEPRESSIONS.)	9A. VOLCANICS AND SHELF SEDIMENTS OF EAST. SIDE OF ANTARCTIC PENINSULA.
	9B. VOLCANICS AND SEDIMENTS OF WEST SIDE OF ANTARCTIC PENINSULA.
	9C. ROSS SEA LOWER TERTIARY AND UPPER CRETACEOUS SEDIMENTS.
7, 8 PRECAMBRIAN TO LOWER TERTIARY (ANTICLINAL RIDGES).	8 COMPLEX OF PALEOZOIC (?) METASEDIMENTS AND GRANITOID INTRUSIVES OF UNCERTAIN AGE IN THE USARP RIDGE.
	7A COMPLEX OF PALEOZOIC AND (?) TRIASSIC. METASEDIMENTS, AND LOWER TERTIARY ANDEAN GABBRO—GRANITE INTRUSIVES IN ANTARCTIC PENINSULA.
	7B PALEOZOIC METASEDIMENTS OF ELLSWORTH MTS., SCHISTS ETC. OF WHITMORE MTS.
	7C PRECAMBRIAN METAMORPHICS AND EARLY PALEOZOIC VOLCANICS AND INTRUSIVES OF MARGUERITE BAY, ANTARCTIC PENINSULA.

N.B. THIS IS ONLY ONE OF SEVERAL POSSIBLE SCHEMATA FOR WEST ANTARCTICA. IT IS EQUALLY POSSIBLE THAT THE ELLSWORTH MTS. ARE AN UNUSUAL PART OF EAST ANT- ARCTICA, SEE TEXT. THE USARP RIDGE MIGHT BE A DIRECT EXTENSION OF THE ANTARCTIC PENINSULA, AND COULD BE CROSSED BY NORTH-TRENDING FAULTS.

EAST ANTARCTICA

QUATERNARY	SEDIMENTS, MORAINES, NOT SHOWN.
6 CENOZOIC	GAUSSBERG BASALTS. UPPER MIOCENE LEUCITE – BASALTS.
5 DEVONIAN TO JURASSIC	BEACON SYSTEM INCL. FERRAR VOLCANICS, FERRAR DOLERITE SILLS, & MAWSON TILLITE.
4 PROTEROZOIC & EARLY PALEOZOIC	SANDOW BEDS, SEDIMENTS AND METASEDIMENTS.
3 EARLY PALEOZOIC	DUFEK MASSIF, ULTRABASIC AND NORITIC INTRUSIVE.
2 PRECAMBRIAN TO EARLY PALEOZOIC	ROSS SYSTEM, & GRANITE HARBOUR INTRUSIVES METASEDIMENTS AND METAMORPHICS AND GRANITIC INTRUSIVES.
1. PRECAMBRIAN	"BASEMENT COMPLEX" OF METAMORPHICS & GRANITIC ROCKS (IN PART CHARNOCKITES.)

...logy, schematic.

there is evidence of at least one high-grade metamorphism imposed on earlier metamorphic rocks. Some of the earlier paragneisses appear to have formed from a sequence of arkosic and pelitic sediments with igneous intercalations, and there are some metamorphosed basic dykes and sills. The Prince Charles Mountains consist of migmatised high-grade metamorphics in their northern parts, but the southern parts contain extensive low-grade rocks of the green-schist facies, mainly mica-schists, slates and quartzites with carbonate rocks (TRAIL, 1963).

The Lambert Glacier and Amery Ice Shelf occupy a zone of major block-faulting (CROHN, 1959) and to the east of this zone charnockites are less abundant, the most common rocks being sillimanite, garnet and biotite gneisses, with intrusive granite, and oligoclase-amphibole and pyroxene gneisses and granulites. Charnockitic rocks reappear in force in the regions of Mirny and Wilkes Stations and the Sabrina Coast (NOCKOLDS, STARIK et al., RAVICH, CAMERON et al.) but have not been reported further east. In Adelie Land (HEURTEBIZE, 1952) and King George V Land (MAWSON, STILLWELL, KLEEMAN, GLASTONBURY, see STEWART, 1956) there are, however, high-grade gneisses and amphibolites as well as phyllites, with granitic intrusives. The Basement Complex may be taken tentatively to end at about longitude 153° where it appears to be faulted against the Berg 'Series' of phyllitic schists (KLIMOV & SOLOVYEV, 1958, 1960; SOLOVYEV, 1959) which extends to longitude 157°. The Soviet geologists and GUNN (1963) seem to consider that the Berg 'Series' is an approximate equivalent of the Robertson Bay Group to the east of Rennick Bay, but there is no direct correlation because in the Wilson Hills between Rennick Bay and longitude 157° there is an intervening region of biotite gneisses and granite that could be part of either the Basement Complex or the Granite Harbour Intrusive Complex or both (see later). There is a distinct possibility that the rocks of the Wilson Hills will be found to trend to Terra Nova Bay and McMurdo Sound via the upper reaches of the Rennick Glacier.

About 180 isotope dates have been measured for rocks from the Basement Complex (CAMERON et al., 1959; STARIK et al., 1961; PICCIOTTO, 1961; NICOLAYSEN et al., 1961; SAITO et al., 1961). Nearly all the determinations have been by the potassium-argon method, a large number being by the Soviet workers who have used whole rocks for speed, and state that "it is possible that our values are somewhat too low, but the maximum reduction probably does not exceed 20 to 25 per cent". Other dates have been determined using potash-feldspar which can suffer serious argon leakage. The frequency of the available dates is shown on fig. 5, kindly supplied by E. E. ANGINO, and it is interesting that none are greater than about 1800 million years (m.y.). This does not mean that no really old

10

rocks occur in East Antarctica but only that there have been repeated metamorphisms. STARIK et al. (1961) suggest that the dates fall into groups with mean ages at 1440, 1090, 735, 585 and 460 m.y., which indicates at least five tectono-magmatic cycles. Similar cycles are recognised by ANGINO (1963). Dates of 400 to 600 m.y. indicate Lower Paleozoic metamorphism and intrusion, continuing to the latest Silurian or earliest Devonian. Rocks of this age are so widespread as to suggest the possibility that the whole of the peri-

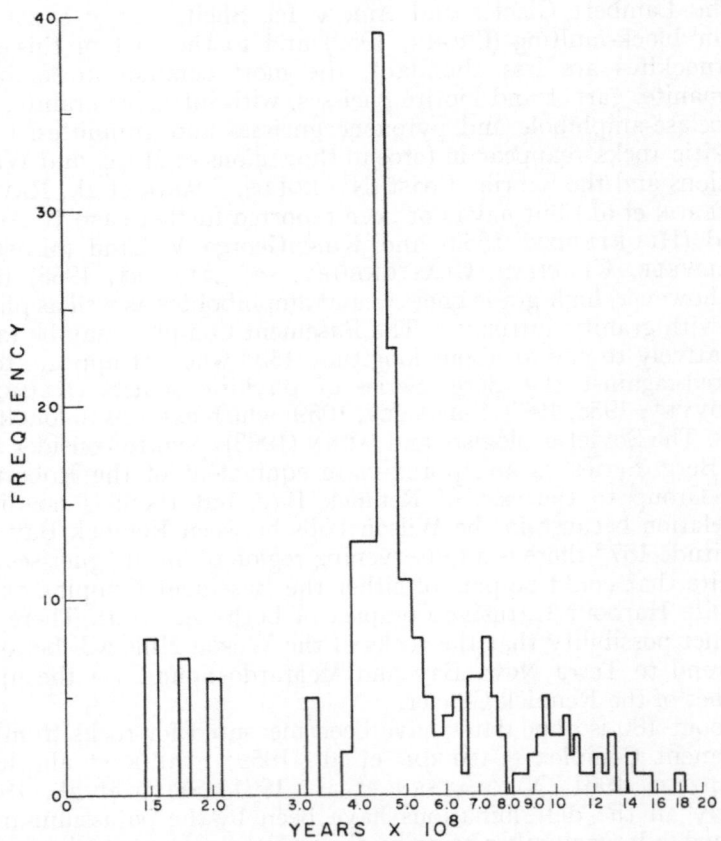

Fig. 5. Frequency of isotope dates, Antarctica. The youngest dates are omitted.
Supplied by E. E. ANGINO. Semi-log-scale.

phery of East Antarctica was affected by the Lower Paleozoic "Caledonian" orogeny that caused the deformation of the Ross System (PICCIOTTO & COPPEZ, 1963). There is no evidence as to whether the interior of the sub-continent was also affected, but proof of a purely marginal orogeny would be significant in assessing continental drift hypotheses, since it would suggest that East Antarctica had its present general shape and entity as early as the Lower Paleozoic. The paper by STARIK et al. (1961) is in a volume containing the latest lists of isotope dates for Precambrian and Lower Paleozoic rocks of other continents. With a view to testing drift hypotheses an attempt was made to match the dates for Antarctic rocks with those for rocks in surrounding continents. There is a broad similarity, but the sound data are still too few for reliable conclusions (see however, HAMILTON 1962).

2. Ross System and Granite Harbour Intrusive Complex

Ross System or Supergroup

When in 1958 an informal committee of Antarctic geologists considered the nomenclature of rock units in the Ross Sea region it was apparent that an extensive group of pre-Beacon (pre-Devonian) sediments in the Trans-Antarctic Mountains could be provisionally grouped together under the name Ross System (HARRINGTON, 1958a).

The nomenclatural committee realised very clearly that the proposal of terms such as Ross System and Beacon System, by adaptation of the term System as used in South Africa, was not in conformity with Stratigraphic Codes in use in Australia and the United States, but no other terms were available. The words "Sequence" and "Supergroup" have been coined since then to solve comparable problems elsewhere (SLOSS, 1963). Similarly groups and formations were clearly recognisable but could not be defined formally by the committee in conformity with the two national codes mentioned above. In any case British and some other workers would have used the generic term 'series' for them. At that time it was better to have *any* single nomenclature than the confusion that would result if for example, different workers simultaneously published the specific name "Ross' for rocks as distinct as those of the McMurdo Volcanics and the Ross System. The report of the committee served its purpose, but it is now time to have a stratigraphic code for Antarctica, perhaps by adoption of the International Stratigraphic Code at present under consideration (HEDBERG, 1961).

Extensive reconnaissance work has tended to confirm the earlier impression that the Ross "System", "Sequence" or "Supergroup" extends for at least 2250 km from the Admiralty Mountains in northernmost Victoria Land to the Leverett Glacier (85° 30' S, 145° W). There is also a working hypothesis among Australian and New Zealand geologists that it extends onwards another 2500 km to the mountains of Coats Land, but this hypothesis really rests only on strongly suggestive facts, such as that Archaeocyathine limestones occur in the Ross System in the Nimrod Glacier region at 82° 20′ S, 160° 20′ E (LAIRD & WATERHOUSE, 1962; LAIRD, 1963), and are abundant in moraines in the Shackleton Range (STEPHENSON, 1959). The hypothesis involves the idea that the sediments were deposited in the Late Precambrian and Lower Paleozoic "Ross Geosyncline" (GUNN & WARREN, 1962; GUNN, 1963b) which is pictured as extending along the edge of an East Antarctic Precambrian platform or craton consisting of the Basement Complex. Soviet geologists have independently developed a very similar idea (VORONOV, 1958; VIALOV, 1959; KLIMOV & SOLOVYEV, 1959) and in addition appear to consider that the Sandow Beds (fig. 4) are craton or platform deposits to be correlated with the geosynclinal Ross System. A comparable hypothesis has been advanced independently also by HAMILTON (1960).

The nomenclatural committee mentioned earlier was very uncertain as to whether the metasediments of the Edsel Ford Ranges and Rockefeller Mountains in the Usarp Ridge, which are in folds with axial trends of 305°, should be included in the Ross System. They appear to resemble the Robertson Bay Group in which the fold axis strikes at 290° in the Admiralty and Victory Mountains. This matter is still uncertain, and affects ideas concerning the development, nature, content and spatial limits of the Ross Geosyncline (GUNN, 1963) and the geological distinctions between East and West Antarctica. GUNN (1963) recognises the Berg, Robertson Bay, Skelton, Beardmore, Byrd, Edsel Ford and Sentinel groups, but it is doubtful whether the Berg, Edsel Ford and Sentinel groups should be included in the Ross System.

At Robertson Bay near Cape Adare PRIESTLEY in 1911 found a formation of unfossiliferous "slates and greywackes" which are now known to be widespread in the Admiralty Mountains where they have a thickness of at least several thousands of metres and are disposed in angular concertina folds around sub-horizontal axes trending at 110° (HARRINGTON et al., 1963). Small-scale cross-bedding and graded bedding are common in the sediments, which can be called miogeosynclinal metagreywackes and argillites, though they have been slightly metamorphosed mainly to the albite-epidote hornfels facies (B. M. GUNN, pers. comm.). Gently deformed sheets of the post-orogenic Tucker Granodiorite, one of which is over 2000 m thick, cut across the axial planes of the folds and are responsible for most of the low-grade regional thermal metamorphism. In the sequence LE COUTEUR & LEITCH (1963) have found a group of orthoquartzites and quartz-conglomerates of similar structural trend and style of folding and containing organic tracks and borings. The Robertson Bay Group, judging from an inspection of U.S. Navy aerial photographs, continues southwards with a similar strike to the vicinity of Wood Bay and Terra Nova Bay. The strike of fold axes,

110°, is anomalous because everywhere south of Terra Nova Bay the dominant fold axes in the Ross System appear to trend within a few degrees of north (GUNN & WALCOTT, 1962). To elucidate the meaning of the change in strike is one of the pressing problems for future field work.

The metamorphic and plutonic rocks between Terra Nova Bay and the McMurdo Sound region were examined in several areas by geologists of the Scott and Shackleton expeditions early in this Century (FERRAR, 1907; DAVID & PRIESTLEY, 1914) and their collections have been described petrographically by a number of specialists (see FAIRBRIDGE, 1952; STEWART, 1956). Most of the region has been mapped in a major reconnaissance by GUNN & WARREN (1962), and other complementary studies of the basement rocks in the deglaciated valleys west of McMurdo Sound have been made by WEBB & MCKELVEY (1959), MCKELVEY & WEBB (1962), ANGINO et al. (1962), HAMILTON & HAYES (1960, 1961), HAMILTON (1961) and BLANK et al. (1963). It had been shown that the rocks are in scattered masses of several formations of metagreywacke, limestones, marbles and hornfelses, and schists and gneisses, having been "overwhelmed" by pre-, syn-, and post-tectonic intrusives. A potassium-argon date of 520 m.y. for one of the pre-tectonic granite gneisses was reported by GOLDICH et al. (1960), and later revised to 500 m.y. by CAMERON et al. (1960), this date, like others obtained since then, indicating not the age of the rock but of a late Cambrian or earliest Ordovician metamorphic or intrusive phase. GUNN & WARREN (1962) consider that metamorphism increases southwards in Victoria Land from Robertson Bay to Granite Harbour and then decreases southwards to the Skelton Glacier, but GUNN (1963) now doubts that conclusion. There is a distinct possibility that some of the more metamorphosed rocks are not part of the Ross System but of an older basement complex. This suggestion is supported by a single date of 1000 m.y. for a porphyry dyke reported by DEUTSCH & WEBB (1963), and by relationships in southern Victoria Land.

South of McMurdo Sound and extending through southern Victoria Land to the Beardmore Glacier district, the sediments and metamorphics are classified in the Skelton, Beardmore and Byrd groups. The Beardmore Group was proposed by GUNN & WALCOTT (1962) to include metagreywackes of the Goldie Formation, and higher-grade metamorphics of the Miller Formation, but there is a suggestion that the former lie uncomformably on the latter, so that the greater part of the Miller Formation has been removed from the group and named Nimrod Group (GRINDLEY, 1963; GRINDLEY et al., 1963). The Goldie Formation consists of greywackes and argillites with low-grade phyllites and schists, isoclinally folded on northerly-trending axes, and in part, showing close lithological resemblance to the Robertson Bay Group. It extends for several hundreds of kilometres through the ranges of southern Victoria Land and is often thermally altered by the Granite Harbour Intrusives to spotted slates and hornfelses of the albite-epidote and hornblende hornfels facies. The Nimrod Group consists of folded amphibolite schists and tremolite marbles, and high-grade gneisses probably much older than the Beardmore Group (GRINDLEY et al. 1963). It is conceivably part of the Basement Complex. The Skelton Group at the southern end of the McMurdo Sound district (GUNN & WARREN, 1962) is in part metamorphosed to the amphibolite facies (BLANK et al., 1963), and broken up by intrusives into areas of lithologies like those of all the other groups. Within it, the Koettlitz Marble near McMurdo Sound contains vague impressions suggestive of Archaeocyathinae (LAIRD, 1963). The discovery of Byrd Group (LAIRD, 1963; GRINDLEY, 1963) marks one of the major recent advances in Antarctic geology, because the stratigraphic position of the often discussed and mysterious Archaeocyathid—bearing limestones has at last been established.

The group lies above Goldie Formation apparently as infolded and infaulted synclinal strips, and consists mainly of several thousands of feet of sediments of a shelf facies in which two formations can be recognised (LAIRD, 1963). The Shackleton Limestone consists of at least 11,000 feet (3,500 m) of grey, black and cream coloured limestones with local lenses of conglomerate, sandstone and shale, and at several localities contains Early to Middle Cambrian Archaeocyathinae and calcareous algae, so far undescribed. The Starshot Formation consists of a similar thickness of calcareous conglomerate, grit, shale and sandstone with ripple-marks, current-bedding and load-casting, and minor rhyolitic or trachytic volcanics. LAIRD suggests that in the Byrd Group the Shackleton Limestone could represent a reef facies and the Starshot Formation a fore-reef or back-reef facies. Sediments younger than these two formations probably occur but have not yet been differentiated.

Archaeocyathids were first discovered, more or less accidentally by GRIFFITH TAYLOR in specimens collected from the Buckley nunatak at the head of the Beardmore Glacier by F. WILD of Shackleton's expedition (GRIFFITH TAYLOR in DAVID & PRIESTLEY, 1914) and were re-collected from moraines at the foot of the glacier by the Pole party of Scott's second expedition who dragged them for several hundreds of kilometres until the party died. Their discovery was of particular interest to TAYLOR and DAVID who were familiar with the archaeocyathids of the Adelaide Geosyncline in South Australia, and has since excited great interest because of the paleoclimatic and paleogeographic significance of the fauna so close to the South Pole. The later dredging of similar material from the floor of the Weddell Sea, 2,800 km from the Beardmore Glacier

showed that the parent rocks probably had a wide distribution
(GORDON, 1920). The fauna so far described includes *Archaeocyathus*,
Dikidocyathus, sponge spicules, and a calcareous alga *Epiphyton*
(GORDON, SKEATS, CHAPMAN). New collections, including material
collected from moraines in Coats Land by STEPHENSON (1960) and
by Dr. R. L. OLIVER from the Beardmore moraines, are being studied
by Dr. D. HILL.

South-east of the Beardmore Glacier there are gneisses, schists
and granites below the Beacon System in the Queen Maud Mountains
(GOULD, 1935). In the Horlick Mountains granitoid rocks are overlain nonconformably by the Lower Devonian strata at the base of
the Beacon System (TREVES, 1963).

The Ellsworth Mountains trending north-north-west are puzzling
(see section on bedrock morphology) and a brief account of their
geology by CRADDOCK et al. (1963) indicates that they consist of
6,100 m of quartzite, conglomerate which might be tillitic, and
pelite with a very thick carbonate unit near the base. Folds are
overturned and recumbent, form an asymmetric anticlinorium, and
trend roughly parallel to the mountains. Fossils from four units
include trilobites, gastropods, a probable cephalopod and other
forms. "The youngest bedrock formation is no older than middle
Palaeozoic". It seems therefore that most of the assemblage might
be of the same age as the upper part of the Ross System particularly
the Byrd Group, but it is not impossible that higher parts could be a
geosynclinal folded equivalent of the Taylor Group of the Beacon
System, and therefore comparable with the Cape System of the
Cape folded belt of southern Africa.

The Pensacola Mountains include the large ultrabasic intrusion
of the Dufek Massif (q.v.) but appear to consist mainly of metasediments like those of southern Victoria Land, and a carbonate
unit, overlain by the Beacon System and Ferrar Dolerites (SCHMIDT
et al., 1963). Archaeocyathids have not been reported from the
carbonate unit.

The mountains of Coats Land have been reconnoitred by STE-
PHENSON (1960 and pers. comm.) At the south-western end of the
Shackleton Range he found a complex of strongly cleaved slates and
quartzites, the Eclipse Metamorphics, which are closely folded on
axes trending at 100°. At the western end of the range there is a
complex, the Shackleton Metamorphics, of marbles, mica, garnet
and amphibole schists and gneisses. Both complexes may be compared with the Beardmore and Nimrod groups of southern Victoria
Land. Unconformably above the Shackleton Metamorphics, but
with unknown relationships to the Eclipse Metamorphics, there is a
formation about 3000 m thick of tilted sandstones, grits and conglomerates named the Novices Beds, and possibly related to the Sandow
Beds (q.v.). The Theron Mountains and Whichaway Nunataks on

either side of the Shackleton Range consist of Permian Beacon sediments with Jurassic Ferrar Dolerites. Prolific in moraines at the Whichaway Nunataks are cream to white coloured limestones, some of which contain Archaeocyathinae and trilobite fragments. These have apparently been derived from the south-east, and are so abundant that they indicate a "most extensive region of Cambrian limestones" (STEPHENSON), but no outcrops are known, and the parent rocks are probably beneath the ice.

Beyond Coats Land the metamorphics and granites in nunataks inland from Maudheim (ROOTS, 1953; SWITHINBANK, 1959) are included arbitrarily in the Basement Complex.

Granite Harbour Intrusive Complex

Within the pre-Beacon rocks of the Trans-Antarctic Mountains, there are abundant pre-, syn-, and post-tectonic batholiths, stocks, bosses, sheets and dykes, originally grouped together as the Admiralty Intrusives (HARRINGTON, 1958a), but now known as the Granite Harbour Intrusive Complex (GUNN & WARREN, 1962). The pre- and syn- tectonic intrusives are most abundant in the McMurdo Sound region of central Victoria Land from the Darwin Glacier to Terra Nova Bay, but the post-tectonic masses are found in all districts beneath the Kukri Peneplain from northern Victoria Land to the Thiel Mountains. Their distribution and petrography is discussed in the reports on areal geology listed in the description of the Ross System, so that only partial documentation need be given here.

In Victoria Land their distribution and petrography has been a particular interest of Dr. B. M. GUNN (GUNN & WARREN, 1962; GUNN & WALCOTT, 1962; GUNN, 1963b; HARRINGTON et al., in Press), and the petrography of specimens collected by the early Scott and Shackleton expeditions has been a particular interest of Dr. CAMPBELL SMITH (1924, and other reports). In the McMurdo Sound region some gneissic granites and granodiorites and metadiorites including the Olympic Granite-gneiss of McKELVEY & WEBB (1962) are classed as pre-tectonic mainly because of their gneissic structure. The pre-tectonic gneisses include some rocks that could be much older than the Ross System (see earlier). A major weakly gneissic syn-tectonic batholith extends for 400 km through the coastal hills and islands northwards from McMurdo Sound nearly to the Nordenskjöld Ice Tongue north of Granite Harbour. It has been called the Theseus Granodiorite by McKELVEY & WEBB (1962) and the Larsen Granodiorite by GUNN & WARREN (1962). It has been described petrographically by GUNN and by earlier writers, and is mainly a biotite-granodiorite with many inclusions recognisably from older rocks ranging in lithology from granite to limestone. The post-tectonic plutons of the McMurdo Sound region of central Victoria Land are grouped in three formations. At the northern and southern ends respectively of the region there are small plutons of the Gauss Granodiorite and Skelton Granodiorite (GUNN & WARREN, 1962). The Vida Granite of McKELVEY & WEBB (1962), called the Irizar Granite by GUNN & WARREN (1962), is the "pink granite" of the pioneer geologists of the Scott and Shackleton expeditions (SMITH, 1924) and forms named steep-sided isolated plutons and bosses, of roughly oval and circular outline, characterized by coarse potash feldspars and hornblende. At several localities other distinct bosses occur, are classified as Delta Diorite, and consist of hornblende-andesine diorite. Suites of dykes of microgranite, microdiorite, and lamprophyres also occur, as they do elsewhere in the Ross System, and at some places in the "Dry Valleys" they form dyke swarms and anastomosing networks showing very prominently in aerial views. So far there are about 20 chemical analyses available of the Granite Harbour Intrusives, mainly rocks from central and northern Victoria Land (HAMILTON, 1961; HARRINGTON et al., in Press).

In northern Victoria Land the Tucker Granodiorite forms an extensive gently-deformed sheet 1500 m thick cutting cleanly across the axial surfaces of folds in the Robertson Bay Group (HARRINGTON et al., 1963 and in Press). It is a hornblende-granodiorite, in places dioritic, resembles the Skelton Granodiorite petrographically (B. M. GUNN, pers. comm.) and has associated with it acidic and basic dykes classified as Edisto Granites.

In southern Victoria Land from the Beardmore Glacier district northwards, there are several intrusions ranging in composition from hypersthene-granodiorite to biotite-muscovite-microcline granite, all at present classified under the name Hope Granite. Some of them resemble the Skelton Granodiorite, except for the absence of hornblende (GUNN & WALCOTT, 1962; GUNN, 1963b; LAIRD, 1963; OLIVER, 1963). About 280 km south of McMurdo Sound at the Darwin Glacier several pre-, syn-, and post-tectonic intrusions ranging from granite to granodiorite in composition have been mapped along with meladiorite, lamprophyre and pegmatite dykes (HASKELL et al., 1963). This district is very similar to the McMurdo Sound region.

The region from the Beardmore Glacier to Coats Land is the most inaccessible sector of the Trans-Antarctic Mountains so that mapping and petrographic work started later than in Victoria Land. TREVES (1963) has stated that in the Ohio Range of the Horlick Mountains the basement consists primarily of a pink, slightly gneissic, porphyritic quartz-monzonite intrusive into a closely-related grey-white gneissic granodiorite. In the Thiel Mountains a very extensive cordierite-bearing hypersthene-quartz-monzonite-porphyry, possibly forming either a folded shallow sheet-like intrusive body or a volcanic pile, is intruded by granodiorite and quartz-monzonite, with lead-alpha ages of 470 ± 50 m.y. and 560 ± 60 m.y., that is, late Cambrian to early Ordovician (FORD, 1963). Lead-alpha measurements indicate that the porphyry is late Precambrian in age. Locally it is recrystallized and has a granulitic appearance. The description is most unusual, and it is possible to wonder, conjecturally, whether it might be an acidic metavolcanic of the ignimbrite family.

Acidic plutonic rocks have not been reported from the Ellsworth Mountains (CRADDOCK et al., 1963) or from the Pensacola Mountains (SCHMIDT et al., 1963), or from Coats Land (STEPHENSON, 1960), though the Pensacola Mountains do contain the large ultrabasic Dufek Massif, and a variety of mafic dykes, distinct from the Ferrar Dolerites, and of uncertain ages.

3. Dufek Massif

The Dufek Massif, on the edge of the Filchner Ice Shelf, is one of the northern ranges in the Pensacola Mountains. AUGHENBAUGH (1961) and WALKER (1961) state that it consists of a basic igneous complex, in which layering is so pronounced as to give the impression of a pile of flat-lying sediments in a distant view. The rocks are of the gabbro-norite-anorthosite-pyroxenite association characteristic of certain large basic and ultrabasic layered intrusions such as the Bushveld and Stillwater complexes in South Africa and Montana.

There is no direct evidence concerning the age and correlation of this intrusion, but one possibility is that it is related to the Jurassic Ferrar Dolerites, and another that it is a part of the pre-Beacon basement. There is a distinct possibility of correlation with the Precambrian Bushveld Intrusion and the Great Dyke of southern Africa (DU TOIT, 1954), and comparative isotope dates would be valuable.

4. Sandow Beds

At Mt. Sandow (67° 22′ S, 100° 22′ E) inland from the Bunger Hills there is a sequence of red and green coloured phyllites, slates, quartzites and arenites which are also found sparsely in moraines at the Vestfold Hills, Mirny and Vincennes Bay. Samples were obtained initially by VIALOV from moraines near Mirny, and TIMO-FEYEV extracted 10 forms of microfossils of kinds resistant to hydrofluoric acid and assessed their age as Proterozoic (VIALOV & TIMO-FEYEV, 1959). Later the rocks were found *in situ* at Mt. Sandow, and from samples taken there and from moraines at Farr Bay near Mirny and the Vestfold Hills (fig. 2), a marine microfossil assemblage of 20 sphaeroligotriletids, 4 leiosphaerids, a hystrichosphaerid and a diacrodid has been determined, and assigned to the Lower Cambrian (KOROTKEVICH & TIMOFEYEV, 1959).

The recognition of the Sandow Beds is one of the major new discoveries in Antarctic geology. VIALOV, and other Soviet geologists, appear to regard them as Proterozoic and Lower Paleozoic sediments deposited on the East Antarctic craton or platform at the same time as the Ross System was being deposited in the Ross Geosyncline.

It is possible that the Sandow Beds or correlatives of them occur elsewhere in East Antarctica. Specimens of sandstones have been retrieved from ice-bergs and dredged from the sea-floor at several places near the coast of East Antarctica, and have been found in moraines near Gaussberg (ZIRKEL & REINISCH, 1905; REINISCH, 1912). It can no longer be considered automatically that erratics of that type were derived from the Beacon System (FAIRBRIDGE, 1952). Inland from Maudheim ROOTS (1953) found an extensive sequence of almost flat-lying greywacke, siltstone, shale, mudstone and conglomerate, overlain by, and intercalated with, altered andesitic lava. These sediments are intruded by basic sills and dykes, so that other writers have included them in the Beacon System, but they do not contain the *Glossopteris* flora or other fossils and they are slightly metamorphosed, so that they could be approximate correlatives of the Sandow Beds. In the same region a continental sequence containing supposed glacigene rocks and intruded by mafic sills on a grand scale has been examined by NEETHLING (1963) who regards it as probably of Beacon age, but states that he did not find diagnostic fossils.

In the Shackleton Range in Coats Land STEPHENSON (1960, and pers. comm.) found a sequence about 3000 m thick of tilted sandstones, grits and conglomerates which he has called the Novices Beds. The only fossil found in them is a *Lingula* which is Cambrian or younger, and too long-ranging to give additional age information. It is possible that this thick sequence is related to the Sandow Beds, but there is as yet no information about its relationship to the Cambrian archaeocyathid-bearing limestones that are so abundant in moraines in the same region, being presumably correlatives of the Byrd Group in southern Victoria Land.

Conjecturally it may be wondered whether the Sandow Beds are correlatives of the Nama System of South Africa (DU TOIT, 1954), and whether correlatives of other South African "Proterozoic" platform or cratonic sequences occur in East Antarctica. It would be of very great economic interest if correlatives of the Witwatersrand System with its gold-bearing "bankets" were found in East Antarctica. The Sandow Beds, the Beacon System, and other platform sequences if they occur, could pass laterally to thicker more intensely deformed and metamorphosed geosynclinal rocks, as happens in the shield regions of Australia.

5. Beacon System, Ferrar Dolerites, Ferrar Volcanics, and Mawson Tillite

These rocks are all closely-associated in the Trans-Antarctic Mountains and cannot be separated on a small map. The Beacon sediments form a sequence about 2500 m thick which can be divided in central Victoria Land into two groups, the lower of Devonian and perhaps Silurian age, and the upper of Permian to Jurassic age. In a wider context it is convenient to recognise a Lower Beacon System and an Upper Beacon System. The whole system is intruded by Jurassic sills of the Ferrar Dolerites, and at the top includes the thick and extensive Ferrar Volcanics which were probably derived from the same magma as the sills. A highly irregular erosion surface cut in the Beacon rocks is overlain by the Mawson Tillite which contains basaltic volcanics and is apparently of Jurassic age, though it could be Quaternary.

Beacon System

Introduction

In 1907 FERRAR gave the name Beacon Sandstone to a sequence of lightly-indurated and gently-dipping quartzose and arkosic sandstones and shales with plant fossils and coaly streaks occurring in Ferrar Valley, dipping gently westwards, and intruded by dolerite sills. SCOTT later found that the apparently continuous glacier occupying the Ferrar Valley is in fact two glaciers, the lower one being the true Ferrar Glacier, and the upper one the Upper Taylor Glacier, which leaves the valley a little downstream from Beacon Heights. Later work by the pioneer expeditions of SCOTT, SHACKLETON, MAWSON, BYRD and others showed that *Glossopteris*-bearing sediments like those of the Beacon Sandstone are widespread in East Antarctica. Shortly after the commencement of the new cycle of geological work that began with the IGY of 1957—58, a British Commonwealth nomenclatural committee suggested that all these sediments could be named Beacon System, with those in Victoria Land being called Beacon Sandstone (Group), and those elsewhere being given local group names (HARRINGTON, 1958a). It was seen clearly that this suggestion did not conform with the United States and Australian stratigraphic codes, as mentioned earlier, but did conform basically with usage in South Africa and elsewhere. (The United States and Australia were then, and are now, the only countries with formal published codes, and their views are disregarded or opposed in some countries with different geological environments and philosophies.) Since then the term Beacon Sandstone (Group) has been shortened to Beacon Group by several workers, and it is shown below that further changes are necessary following the recognition of at least two "groups" or "supergroups" or "lithosystems" in central Victoria Land. The continued use of the term "System" can be defended, partly on the grounds that the geologies of East Antarctica and southern Africa are remarkably similar, and partly because it can be prefixed by the term "Litho" (TRUTER in HEDBERG, 1961).

Content, Distribution and Lithology

1. Central Victoria Land

Following the pioneer work by FERRAR (1907), DAVID & PRIESTLEY (1914), DEBENHAM (1921) and STEWART (1934) the system has been studied in central Victoria Land by a number of workers (McKELVEY & WEBB, 1959, 1961; ZELLER, ANGINO & TURNER, 1961; ANGINO & OWEN, 1962; GUNN & WARREN, 1962; HAMILTON & HAYES, 1960, 1963; HARRINGTON & SPEDEN, 1962; SHAW, 1962; VIALOV, 1962; ALLEN & GIBSON, 1962; ALLEN, 1962; WEBB, 1963). A monograph on the plant fossils has been published by PLUMSTEAD (1962).

The Taylor-Ferrar, Wright and Victoria valleys west of McMurdo Sound are easily accessible from the United States and New Zealand bases on Ross Island, so that nearly all the field parties have worked in them and have studied the lower part of the Beacon sequence. GUNN & WARREN (1962) made a major reconnaissance of the surrounding district in 1957—58, and because they were using dog-teams did not work in the deglaciated Lower Taylor, Wright and Victoria valleys. They are the only workers who have described the higher parts of the Beacon sequence in this region. In some recent seasons several field parties have been at work simultaneously in different but overlapping districts and have published their results independently. Consequently, the literature is in a conflicting and confused state, and is almost indecipherable without some personal knowledge of the region. While attempting to make the literature intelligible for readers of this review, it was realised that a disconformity in the middle of the sequence is of previously unsuspected importance, and enables the Beacon sediments in central Victoria Land to be divided into two major parts here called the Taylor Group and the Victoria Group.

A modified and consolidated consequential nomenclature for other units is compared in Table I with five earlier classifications. It will be a long time before the subdivisions and nomenclature of the Beacon sediments are finally stabilised and it is anticipated that further work will show that the Knobhead Formation and Boreas Subgreywacke, and possibly the Aztec Siltstone, should be removed from the Taylor Group, and that the Victoria Group will be subdivided also. This possibility becomes apparent as soon as the sections in central Victoria Land are compared with those elsewhere in the Trans-Antarctic Mountains (Table II), and it is suggested therefore that in this wider context the terms Lower Beacon System and Upper Beacon System (or Lithosystem) are useful (Table III), the latter including the Ferrar Volcanics.

The magnificent section extending for 25 km along the south wall of the Taylor-Ferrar valley is, by tautology, the type section of the Beacon sediments, because it includes the twin peaks called Beacon Heights. In this section HAMILTON & HAYES (1963) recognized several mappable units which they named informally. At the top and base of their "Sandstone of Pyramid Mountain", here termed Pyramid Sandstone, they recognised disconformities or erosion surfaces. The higher surface was considered to be beneath the Aztec Siltstone of WEBB (1963), but this unit is missing where the section was examined by HAMILTON & HAYES. The unit is important because its correlatives contain Devonian freshwater fish (GUNN & WARREN, 1962), but it is thin, and is missing at a number of localities where it could be expected. If the upper erosion surface of HAMILTON & HAYES is considered to cut across the top of the Aztec Siltstone, thus accounting for the patchy distribution of that unit, then many other features of rock distribution, thickness, lithologies and fossil content fall into a pattern.

The formations below the erosion surface, here termed the Maya Erosion Surface, are Devonian and have a unity of lithological aspect, which is partly a result of relative scarcity of carbonaceous material, and perhaps partly a reflection of climatic environment, and of an interdigitation of marine and non-marine sediments. This assemblage is called the Taylor Group of the Beacon System.

The strata above the Maya Erosion Surface are dirtier sediments with a higher content of matrix in many of them. They are characterized by carbonaceous material, and their feldspars are often kaolinised. They were clearly deposited in a continental environment and were derived from a source area covered by an abundant vegetation. All the *Glossopteris* floras have been obtained from beds above the Maya Erosion Surface. This assemblage is called the Victoria Group of the Beacon System. The superficial similarity of the two groups is indicated by the fact that it has taken so long to recognise the division.

One of the problems of the Beacon sediments in central Victoria Land has been since 1959 to pin-point the stratigraphic position equivalent to the Buckeye, Pagoda and Darwin tillites that are present in the sequence in southern Victoria Land and the Horlick Mountains. It can now be claimed that the Maya Erosion Surface is at that position being probably a product of glacial and periglacial erosion of the underlying Taylor Group. This erosion could ac-

ZELLER, ANGINO & TURNER, 1961. FERRAR VALLEY	GUNN & WARREN, 1962 CENTRAL VICTORIA LAND	ALLEN, 1962 VICTORIA VALLEY	HAMILTON & HAYES, 1963 TAYLOR-FERRAR VALLEY	WEBB, 1963 TAYLOR & WRIGHT VALLEYS	THIS REVIEW	AGE
	JURASSIC SANDSTONES				JURASSIC SANDSTONES	JURASSIC
	TRIASSIC PLANT BEDS				? TRIASSIC PLANT BEDS ?	TRIASSIC
	GLOSSOPTERIS SANDSTONES, UPPER ARENITE	MOUNT BASTION COAL MEASURES	SANDSTONE OF FINGER MOUNTAIN	WELLER SANDSTONE	BASTION COAL MEASURES	(UPPER PERMIAN & "PERMO-CARB."
	DEVONIAN SILTSTONES	(? ERODED)	(ERODED)	AZTEC SILTSTONE	AZTEC SILTSTONE	U. OR M. DEVONIAN
MEMBER "A1"	LOWER ARENITE	FORTRESS SANDSTONE	SANDSTONE OF PYRAMID MOUNTAIN	BEACON HEIGHTS ORTHOQUARTZITE	? ? PYRAMID SANDSTONE	LOWER DEVONIAN
		WEBB SANDSTONE	SANDSTONE OF NEW MOUNTAIN		? ? ODIN ARKOSE	LOWER DEVONIAN
TERRA-COTTA MOUNTAIN MEMBER WINDY GULLY MEMBER	BASAL BEDS	(? NOT PRESENT)	(TERRA-COTTA MOUNTAIN MEMBER) (WINDY GULLY MEMBER)	ODIN ARKOSE	KNOBHEAD FORMATION (TERRA COTTA MT. & WINDY GULLY MEMBERS)	? EARLY DEVONIAN
				BOREAS SUBGREYWACKE MEMBER	? ? BOREAS SUBGREYWACKE (OR ? BOREAS TILLITE)	? SILURIAN
BASEMENT	BASEMENT	BASEMENT	BASEMENT	BASEMENT	BASEMENT	CAMBRIAN AND OLDER

(VICTORIA GROUP and TAYLOR GROUP)

Table I.

Subdivisions of the Beacon System in Central Victoria Land.

FOSSILS	AGE	CENTRAL VICTORIA LAND	SOUTHERN VICTORIA LAND	HORLICK MOUNTAINS	COATS LAND
CONIFERS, OTOZAMITES, CONCHOSTRACANS	JURASSIC	JURASSIC SANDSTONES (VICTORIA GROUP)			
DICROIDIUM, CYCADOPHYTES, AND ? RHEXOXYLON	TRIASSIC	TRIASSIC PLANT BEDS	FALLA FORMATION		
GLOSSOPTERIS BUT NO GANGAMOPTERIS	UPPER PERMIAN	BASTION COAL MEASURES	BUCKLEY COAL MEASURES	MT. GLOSSOPTERIS FORMATION	THERON MOUNTAINS
GLOSSOPTERIS AND GANGAMOPTERIS	"PERMO—CARB."				WHICHAWAY NUNATAKS
? SPORES		(? NOT DEPOSITED)	MACKELLAR FORM.	DISCOVERY RIDGE FORMATION	
SPORES	? UPPER CARB.		PAGODA TILLITE	BUCKEYE TILLITE	
FRESHWATER FISH	U. OR M. DEVONIAN	AZTEC SILTSTONE			
HAPLOSTIGMA AND STEMS; MARINE PROBLEMATICA	LOWER DEVONIAN	PYRAMID SANDSTONE (TAYLOR GROUP)	ALEXANDRA FORMATION	(? ERODED)	
MARINE PROBLEMATICA AND ? PLANTS		ODIN ARKOSE			
BRACHIOPODS, PSILOPHYTES	LOWER DEVONIAN	KNOBHEAD FORM.		HORLICK FORM.	
MARINE PROBLEMATICA	? EARLY DEVONIAN	BOREAS SUBGREYWACKE			
	? SILURIAN				

Table II.

Correlation of the Beacon System in the Trans-Antarctic Mountains.

AGE	EAST ANTARTICA THIS REVIEW	SOUTH AFRICA DU TOIT 1954	FALKLAND ISLANDS ADIE 1963 (MODIFIED)
JURASSIC	FERRAR DOLERITES	KARROO DOLERITES	DOLERITE DYKES
JURASSIC	FERRAR VOLCANICS. JURASSIC SANDSTONES.	STORMBERG SERIES — DRAKENSBERG VOLCANICS	
		CAVE SANDSTONE	
TRIASSIC		RED BEDS	WEST LAFONIAN BEDS
		MOLTENO BEDS	
	TRIASSIC PLANT BEDS	BEAUFORT SERIES — UPPER BEAUFORT	
UPPER PERMIAN		MIDDLE BEAUFORT	BAY OF HARBOURS BEDS
PERMIAN	BASTION COAL MEASURES	LOWER BEAUFORT	
		ECCA SERIES — U. ECCA SHALES	CHOISEUL SOUND BEDS
		COAL MEASURES	LAFONIAN SANDSTONE
		L. ECCA SHALES	
			BLACK ROCK SLATES
LOWER PERMIAN	DISCOVERY RIDGE FORMATION	DWYKA SERIES — WHITE BAND	
	~~ ? ~~	U. DWYKA SHALES	
? EARLY PERMIAN & U. CARB.	BUCKEYE TILLITE	DWYKA TILLITE	LAFONIAN TILLITE
M. OR U. DEVONIAN	AZTEC SILTSTONE (TAYLOR GROUP)	WITTEBERG SERIES — U.WITTEBERG SHALES	BLUFF COVE BEDS
LOWER DEVONIAN	PYRAMID SANDSTONE	WITTEBERG SERIES	PORT STANLEY BEDS
	ODIN ARKOSE	BOKKEVELD SERIES — U. BOKKEVELD	PORT PHILOMEL BEDS
LOWER DEVONIAN	HORLICK FORM. KNOBHEAD FORM.	L. BOKKEVELD	FOX BAY BEDS
		TABLE MOUNTAIN SERIES — U. SANDSTONE	PORT STEPHENS BEDS
		U. SHALES	
SILURIAN	BOREAS SUBGREYWACKE (BOREAS TILLOID)	GLACIAL ZONE	
		L. SANDSTONE	
		L. SHALES	

(East Antarctica: SYSTEM — VICTORIA GROUP / TAYLOR GROUP; BEACON — UPPER / LOWER. South Africa: KARROO SYSTEM; CAPE SYSTEM.)

Table III.

Correlation of Gondwana sequences in Southern Lands.

count for the lack of a transition from Devonian beds to Permian strata, that is, the absence of the Lower Carboniferous could be a result of glacial erosion, and the absence of all or most of the Upper Carboniferous and the tillite and shales equivalent to the Discovery Ridge Formation could be a result of non-deposition during the glaciation or glaciations, and perhaps for a time afterwards.

It may be noted also, that *after* the two groups had been separated it was realised that they correspond to the Karroo System and upper part of the Cape System in southern Africa (Du Toit, 1954). The Maya Erosion Surface corresponds with the uncomformity that is present below the Dwyka Tillite in many parts of southern Africa. Moreover the Pyramid Sandstone and fish-bearing Aztec Siltstone, described below, seem to be lithologically very similar to the sandstones and fish-bearing upper shales of the Witteberg Series. The Odin Arkose could be an equivalent of either the lower part of the Witteberg or the upper part of the Bokkeveld Series. No equivalent of the lower Bokkeveld is known in central Victoria Land, except possibly Knobhead Formation, nor of the thick Table Mountain Series (but see later mention of the Boreas Subgreywacke).

The sequence of strata in the Beacon System in central Victoria Land (Table I) can now be described in abbreviated tabular form, from the top downwards:—

Ferrar Volcanics (q.v.)

Victoria Group

Jurassic Sandstones: informal name (Gunn & Warren, 1962) for 30 m of brown micaceous silty sandstones and siltstones interbedded with thin dark plant-bearing cherts in the isolated Carapace Nunatak at the head of the Mackay Glacier; overlain by Mawson Tillite; contact with Triassic beds not known, and it is difficult to predict a place to search for it in central Victoria Land.

Triassic Sandstones: informal name (Gunn & Warren, 1962) for 300 m of distinctive thin-bedded brown micaceous sandstones and siltstones with plant fossils occurring in nunataks and the westernmost spurs of the mountains between the heads of the Upper Taylor and Victoria valleys. At Shapeless Mountain the Triassic beds are probably in sequence with the Bastion Coal Measures but the section has not been described.

Bastion Coal Measures: these are the Mount Bastion Coal Measures of Allen (1963) and are at least 900 m thick in the type section in the Victoria Valley, though possibly thinner in other sections. There are numerous thin dirty coal seams in the carbonaceous lower part of the section, but higher beds are less carbonaceous and are characterized by ash-grey quartzose and arkosic sediments with kaolinised feldspars. This formation has yielded all the *Glossopteris* floras of central Victoria Land. The Glossopteris Sandstones of Gunn & Warren (1962) are only about 100 m thick, contain several seams of clean coal 1 m to 2.5 m thick, and are probably equivalent to a portion of the lower coal-bearing sequence in Victoria Valley. The coal seams were probably originally sub-bituminous but many have been baked by Ferrar Dolerite sills. Quality and analyses of some of the seams of the Glossopteris Sandstones has been discussed by Schopf (1962) and Mulligan et al. (1963). The Weller Sandstone of Webb (1963) at the head of the Upper Taylor Glacier must be equivalent to a lower part of the type section of the Bastion Coal Measures but not necessarily the base, because of irregularities in the Maya Erosion

Surface. In the type section the erosion surface is underlain by the Fortress Sandstone of ALLEN (1963) which is probably the Pyramid Sandstone.

Maya Erosion Surface: see comments above.

Taylor Group

A continuous section is exposed along the south wall of the Taylor-Ferrar Valley, through all the gently westward-dipping type beds of the Taylor Group except the Boreas Subgreywacke, which is known only on the north side of the Wright Valley.

Aztec Siltstone: local areas of red and green siltstone facies, 0 to 100 m, with arkose and subarkose beds; desiccation polygons, and concentrations of worm borings. Called Devonian Siltstones by GUNN & WARREN (1962) who found fresh-water fish remains at three localities and two horizons, and also noted a thin local limestone. Missing at localities where Maya Erosion Surface cuts into underlying beds.

Pyramid Sandstone: quartz sandstone 340 m thick at Pyramid Mountain (HAMILTON & HAYES, 1963, but note an error in the meters scale in their fig. 5) and rather thicker at Beacon Heights West where it includes unit III, the upper 1000 feet approximately (300 m) of unit I, and probably unit V of HARRINGTON & SPEDEN (1962). It is also equivalent to the upper 500 m approximately of the Beacon Heights Orthoquartzite of WEBB (1963). It is a cliff-forming formation of repeated units of massive and cross-bedded quartz sandstones grading up to limonitic quartz sandstone capped by dark siltstones, as shown in a photograph by HARRINGTON & SPEDEN (1962, fig. 6). These workers and their associates SHAW (1962) and VIALOV (1962), unlike others who have studied the Beacon sequence, maintained that the section at Beacon Heights is predominantly marine. The present writer considers in addition that the abundance of worm borings, animal tracks and ripple-marks indicates that the sandstones are certainly subaqueous and probably of shallow-water marine origin. The dark shales capping each sandstone could be coastal marsh deposits, a suggestion supported by the occurrence of sparse plant fossils in the beds. It is considered that there were repeated slight rises in sea-level, each rise being followed by a flush of well-sorted and reworked sand, and that as deposition slowed down the sand supported a population of soft-bodied burrowing organisms, and was finally covered by shales deposited in semi-saline or brackish waters. A progressive upward change towards a non-marine environment in the Pyramid Sandstone is suggested by the occurrence of freshwater fish and abundant desiccation polygons in the overlying Aztec Siltstone (McKELVEY & WEBB 1962; WEBB, 1963; GUNN & WARREN, 1962). Descriptions suggest that the lower part of the Odin Arkose and Knobhead Formation are also predominantly marine.

Arena Erosion Surface: This is the lower disconformity of HAMILTON & HAYES (1963). Its nature and significance are not clear but it might represent a significant time interval because it separates predominantly feldspathic and quartzose sediments of the Odin Arkose from overlying predominantly quartzose sediments. The Odin Arkose could represent a flood of sediment stripped from a crystalline basement source, and an unknown thickness of it could have been eroded at the disconformity. It could have been reworked repeatedly in some other district and on the sea-floor, with concentration of its quartzose fraction, before being transported to contribute to formation of the Pyramid Sandstone.

Odin Arkose: This unit of WEBB (1963) is extended up to the Arena Erosion Surface and now includes the lower 100 m approximately of his Beacon Heights Orthoquartzite, and the lower 300 m of unit I of HARRINGTON & SPEDEN (1962). It is the same as the "sandstone of New Mountain" of HAMILTON & HAYES (1963) who have subdivided it into their units 2 to 9, some of which might be recognised as members by further work. It also includes Member A1 of ZELLER et al. (1961). It consists of rhythmically-repeated units of feldspathic and quartzose sandstones with ripples and worm-casts passing up to brightly-coloured shales. On the north side of the Taylor Valley it lies directly on the basement.

Windy Gully Erosion Surface: A distinct disconformity was reported by ZELLER et al. (1961) above their Terra Cotta Mountain Member. Its regional significance is not clear, but underlying beds of the Taylor Group are patchy in distribution, and the surface might represent a considerable time interval. It is possible that correlatives of the Horlick Formation (see later) have been eroded, and later work might show that the Taylor Group should not be extended below this surface.

Knobhead Formation: The beds between the Kukri Peneplain and the Windy Gully Erosion Surface were divided by ZELLER et al. (1961) into two members, which are here combined as Knobhead Formation. The lower or Windy Gully Member rests on an irregular erosion surface, consists of 2 to 30 m of sandstones with boulder and cobble beds and fucoidal markings, and is overlain by the Terra Cotta Mountain Member consisting of about 30 m of alternating sandstones and shales of varied colours. "Basal Beds" occurring elsewhere in the Beacon sequence have been described by GUNN & WARREN (1962), and vary in lithology from place to place.

? Erosion Surface

Boreas Subgreywacke (or? Boreas Tillite): A basal and local subgreywacke (lithic arenite) facies 0 to 100 m thick on the north side of the Wright Valley has been described by McKELVEY & WEBB (1961) and WEBB (1963), and consists of dark breccias and conglomerates, sandstones and siltstones with graded and cross-bedding and small-scale slump structures. Messrs McKELVEY & WEBB (pers. comm.) suggest that the unit should be re-examined because it is more indurated than overlying beds and might be a tillite. The Table Mountain Series of the Cape System of South Africa contains a Glacial Zone that is not younger than Lower Devonian and probably Silurian (Table III). The word Tilloid in Table III should be changed to Tillite.

2. Horlick Mountains and Queen Maud Range

Quite different sections have been found far to the south in the Horlick Mountains by LONG (1962, 1963) who has described the following sequence through 1200 m of beds in the Buckeye Range at 84° 45′S, 114°W:

Permian Mt. Glossopteris Formation
> 600 m coal measures, arkose, sandstone, shale, coal, with *Glossopteris* and other plants.

Permian? Discovery Ridge Formation
> 168 m; lower platy shale with invertebrate trails 46 m; upper carbonaceous shale 122 m.

Permian? Buckeye Tillite
> 270 m ±, tillite, with striated pavements and striated pebbles, ice movement west to east; spores.

Lower Devonian Horlick Formation
> 0 to 46 m, coarse sandstone and dark shale, fossiliferous.

Erosion Surface on Monzonite

The Mt. Glossopteris Formation can be compared and correlated with the Bastion Coal Measures of the Victoria Group in the Mc Murdo Sound district, but the lower strata are quite different, and their discovery was a major advance in Antarctic geology. The Buckeye Tillite is undoubtedly glacial, with superb striated pavements, and soled and striated pebbles, and can be correlated with Upper Paleozoic glacigene rocks in other Gondwana continents.

No equivalent of the Horlick Formation, with its marine brachiopods and other fossils is known, but it is possible that if correlatives were deposited in central Victoria Land they were later eroded at the Windy Gully Erosion Surface. Correlatives of the Odin Arkose and Pyramid Sandstone are also missing and the erosion surface below the tillite could be an extension of the Maya Erosion Surface. Above this surface the Discovery Ridge Formation is known to have an equivalent in southern Victoria Land, but has not been reported in central Victoria Land.

In the Queen Maud Range Beacon sediments about 2200 m thick were found by GOULD (1935), and LONG (1963) states that beds resembling the Discovery Ridge Formation and Mount Glossopteris Formation are present, though the Horlick Formation and Buckeye Tillite have not been found.

3. Southern Victoria Land

North of the Beardmore Glacier GRINDLEY (1962) and others have found the following sequence of 2150 m of Beacon sediments, overlain by the Jurassic or late Triassic Kirkpatrick Basalts:

Falla Formation with *Dicroidium*
Buckley Coal Measures with *Glossopteris*
Mackellar Formation of dark carbonaceous shales and sandstones
Pagoda Tillite
Alexandra Formation of quartz sandstones

The beds below the tillite must be correlated with some part of the Taylor Group, and the Pagoda Tillite with the Buckeye Tillite. The Mackellar Formation is almost identical with the Discovery Ridge Formation, and shows a strong resemblance to sediments at this stratigraphic position in other Gondwana lands, such as the Dwyka Shales above the Dwyka Tillite in South Africa, and the Black Rock Formation of the Falkland Islands (Table III). The thin pyritic carbonaceous bed called the White Band at the top of the Upper Dwyka Shales is identical with the Iraty Shales of Brazil. In both continents it contains species of a curious free-swimming reptile *Mesosaurus*, half a metre long. Neither it, nor abundant Triassic reptiles have been found in Antarctica, so far. The beds above Mackellar Formation are a sequence closely equivalent to the Victoria Group.

In the Darwin Glacier district only 280 km south of McMurdo Sound, 600 m of white quartz sandstone at the base of the Beacon sequence is overlain by 28 m of Darwin Tillite, 9 m of varvoid rocks, and 120 m of quartzose coal measures with *Gangamopteris* (HASKELL et al., 1963).

4. Other Regions

North of the McMurdo Sound region the Beacon System extends to a line between Terra Nova Bay and the Rennick Glacier, but not further north (HARRINGTON, 1963; HARRINGTON et al., 1963). Inland from Terra Nova Bay it has been found as thin erosional remnants overlain by Ferrar Basalts (GAIR, 1963; RICKER, 1963), and thence continues to the west side of the Rennick Glacier (WEIHAUPT, 1960, 1961). These last occurrences probably link with those at Horn Bluff (68° 23' S, 149° 48' E), King George V Land, and Adelie Land (MAWSON, 1940; SOLOVYEV, 1959; PAVLOV, 1958). Sandstones found by MAWSON at 143°E might include fragments of the Sandow Beds. Beacon sediments with *Glossopteris* have been found in the Pensacola Mountains by SCHMIDT et al. (1963). In Coats Land the Beacon sediments occur at the Theron Mountains and Whichaway Nunataks (STEPHENSON, 1960) and their *Glossopteris* floras have been described by PLUMSTEAD (1962). Possible occurrences in Queen Maud Land have been mentioned in the discussion of the Sandow Beds.

At the "Amery Locality", beside the Amery Ice Shelf, CROHN (1959) reports a small down-faulted block, with an area of 750 sq. km and mainly hidden under moraine and ice, containing carbonaceous mudstone and low-rank coal, which have yielded Upper Permian plant microfossils. Correlatives of the Beacon System could be extensive in the block-faulted depression or rift valley occupied by the Amery Ice Shelf and Lambert Glacier for TRAIL (1963) has found erratics with *Glossopteris* in moraines near the southern side of the Prince Charles Mountains. In this connection it is interesting to note that in continental-drift reconstructions this depression *can* be made collinear with the Darling Fault and the Perth Basin in Western Australia. The Perth Basin contains a very thick sedimentary sequence which includes glacigene rocks.

Paleontology and Age

1. Devonian

The oldest known fossil assemblage in the Beacon System is probably a marine shelly fauna in the Horlick Formation. It contains a terebratuloid brachiopod discussed in detail by BOUCOT et al. (1963) and assigned to the Lower Devonian. The formation also includes remains of primitive land plants (psilophytes) which indicate a littoral swampy environment for the beds (LONG, 1963).

Freshwater fish remains found by GRIFFITH TAYLOR's party of 1911—12 in shale erratics on the Mackay Glacier north of McMurdo Sound were assigned to the Upper Devonian by WOODWARD (1921). New specimens have been found *in situ* at three localities and two horizons at the stratigraphic position of the Aztec Siltstone by

GUNN & WARREN (1962), who report that they have been assigned to the Upper or Middle Devonian, or possibly both, by Dr. ERROL WHITE of the British Museum, after a preliminary examination. It can be expected that the fish will prove to be similar to those in shales at the top of the Witteberg Series in South Africa. One stem impression definitely from the Pyramid Sandstone, and four others possibly from the Odin Arkose or the Pyramid Sandstone, were collected by HARRINGTON & SPEDEN (1962), who report preliminary identification of them by Dr. E. PLUMSTEAD as lycopod stems of late Carboniferous age. They have now been described in detail by PLUMSTEAD (1962) and identified as *Haplostigma irregulare*, a primitive lycopod found in South Africa, and to a form cf. *Protolepidodendron lineare* found in New South Wales. Both are restricted to Lower and Middle Devonian horizons. The abundance of organic problematica such as worm borings and animal trails in the Taylor Group has been mentioned earlier. The writer believes that this group is predominantly marine and that a detailed search for identifiable marine fossils and better plant fossils in the Taylor Group would be justified, because even the section in the Taylor-Ferrar Valley has been examined only in reconnaissance fashion.

2. Carboniferous and Permian

The absence of fossils from the Buckeye, Pagoda and Darwin tillites, except for unidentified spores mentioned by LONG, is unfortunate because the date of glaciation cannot at present be closely fixed. KING (1958) has developed the idea that the Upper Paleozoic tillites of southern lands are diachronous, along with the sediments of other Paleozoic climatic belts. The possibility that the tillities do cross time planes in Antarctica cannot be overlooked, along with the possibility that there are tillites at several horizons, as in Australia. CAMPBELL (1962) concluded after a study of marine fossils, that in New South Wales the glaciations must have commenced in the late Namurian or early Westphalian ("Mid" Carboniferous) and continued to the early Sakmarian (Early Permian), and that there is now no evidence of Dinantian (Lower Carboniferous) glaciation anywhere in the southern hemisphere.

Fossils have not been reported from the Discovery Ridge Formation or Mackellar Formation, but the shales are carbonaceous and should contain plant microfossils.

The Bastion Coal Measures have yielded many plant fossils since *Glossopteris* was first collected from them by FERRAR, and by other members of early expeditions. Those collected by GUNN & WARREN, along with those collected by STEPHENSON in Coats Land, have been described by PLUMSTEAD (1962) in a major monograph. She notes that the lowest *Glossopteris* horizons, often associated with tillites in other countries, have not been found, but "Permo-Carboniferous" and Permian horizons have yielded 20 "genera" and 40 "species" of fossil plants with a preponderance of Glossopteridae, mainly *Glossopteris* and *Gangamopteris*, a number of roots and stems including *Vertebraria*, fructifications, scale leaves, and the silicified woods *Dadoxylon* and *Taeniopitys*, with the Equisetales, *Phyllotheca australis* and *Annularia*. There are also insect wings, and problematica such as fungal spots and algal markings or worm tracks. Lower floras containing *Gangamopteris* are 'Permo-Carboniferous' (?Mid Permian), but higher floras lacking this form are Permian. She states that the assemblages form a pure *Glossopteris* flora with an absence of lycopods or *Sphenophyllum*, and a paucity of evidence of Pteridospermae and Cordaitales and other northern 'Arcto-Carbonic' elements. Distributions in other Gondwana countries are listed and zonal correlations are discussed, but PLUMSTEAD concludes that at present the floras can only be correlated broadly with those of the Lower Gondwana beds of southern lands such as the Coal Measures or Middle Ecca of the Karroo System, the Damuda System of India from the Karharbari Stage to the Raniganj Series, the Lower Bowen Series of Queensland, the Greta Coal Measures and lower Newcastle Coal Measures of New South Wales, and the Bonito Group of the Tubarao Series of Brazil (for outlines of Gondwana stratigraphy see GIGNOUX 1955; KING, 1958; and GRINDLEY, 1963; but note that all three contain errors in inter-continental correlations).

The *Glossopteris* floras of the Mount Glossopteris Formation in the Ohio Range of the Horlick Mountains have been described by SCHOPF (1961) and CRIDLAND (1963), and assigned to the Permian, after examination of both microfossils and macrofossils. This determination is supported by the identification by DOUMANI & TASCH (1963) of two species of conchostracans (bivalve crustaceans) of Middle to Upper Permian age, specifically Lower Beaufort. SCHOPF recognises 20 types of spores, three referred to *Accinctisporites* and one to *Striatites*, the latter being compared with and distinguished from SEWARD'S *Pityosporites*. His residues also contain abundant fusinized fragments of secondary xylem. Macrofossils include leaves of *Glossopteris* species, seeds *(Samaropsis)*, an (?) arthrophyte stem and invertebrate trails. Wood specimens are tentatively referred to the Permian araucarian *Dadoxylon* group of gymnosperms though they could be classed with the Triassic *Rhexoxylon*. They have well-developed growth rings indicating rapid growth in a climate of continental temperate type, and this is significant paleoclimatically because the fossils are now only 5° from the South Pole. CRIDLAND (loc. cit.) also examined the Mount Glossopteris floras, and recognises several *Glossopteris* species and a possible *Gangamopteris*, and gives attention to scale leaves and *Samaropsis* seeds.

Microfossils from the Beacon sediments of King George V Land have been described by PAVLOV (1958), and from the Amery locality, a Permian assemblage of 14 spore types including *Pityosporites* and *Florinites* has been listed by CROHN (1959), the dominance of *Leuckisporites* suggesting an Upper Permian age. *Glossopteris* has been found on a tributary of the upper Lambert Glacier (TRAIL, 1963).

3. Triassic

Of Triassic floras only those collected by GUNN & WARREN (see earlier) in central Victoria Land have been described (PLUMSTEAD, 1962). She states that the small Triassic floras from four sites near the head of the Upper Taylor Glacier are dominated by the poorly-preserved cycadophytes *Zamites* (4 spp.), *Nilssonia* and *?Williamsonia*, with several ribbed equisitaceous stems *(Schizoneura* and *Neocalamites)*, and a few fern or pteridosperm stems and fronds, along with pieces of wood and fructifications. Fronds of *Dicroidium odontopterioides* (formerly *Thinnfeldia)* are important, because this plant is virtually a southern hemisphere Triassic zone fossil, and by reference to the Molteno Beds of South Africa, is not later than Middle Triassic. It has been found also in the Beardmore Glacier region (GRINDLEY et al., 1963).

The *Dicroidium* flora (TOWNROW, 1957) has been recorded with certainty only from the same regions as the southern *Glossopteris* flora and from New Zealand, and is mainly older than the Rhaetic northern floras with *Thinnfeldia* though overlapping them to some extent in age. PLUMSTEAD notes that a Triassic age can be given also to the piece of peculiar wood found long ago on a moraine of the lower Priestley glacier at Terra Nova Bay, and named *Antarcticoxylon priestleyi* by SEWARD (1914), and later considered by WALTON to be generically identical with *Rhexoxylon* wood occurring in the Molteno beds. It has been suggested that *Rhexoxylon* is the wood of *Dicroidium.* PLUMSTEAD states that the same species as the Antarctic specimen is found with other *Rhexoxylon* species in South Africa and provides an important Triassic link between the two lands. On the other hand SCHOPF (1962) considers that it could be called *Dadoxylon*, and others have also discussed the original Antarctic specimen including KRÄUSEL (in PLUMSTEAD, 1962), and CRIDLAND (1963) who does not agree that it can be classed with *Rhexoxylon* and states that he cannot differentiate it from *Dadoxylon.* The Antarctic specimen has been very troublesome in another way also, because SEWARD (1914, 1930, 1933) extracted from it a winged pollen grain *Pityosporites antarcticus* which was later classed as a shrivelled pith cell, but is now once again a pollen grain after much uncertain debate (CRANWELL, 1959; SCHOPF, 1962, following MANUM; and others). The important point is that SEWARD believed the grain was shed by a member of the Abietineae or Podocarpineae, conifers with no megafossil record in the Paleozoic. This belief in turn was one of the factors affecting opinions about the vexed question of the upper age limit of *Glossopteris* floras, because it was believed that the original *Pityosporites* must have been derived from *Glossopteris*-bearing beds. It is now known that there are probably Triassic beds in the basin of the Priestley Glacier. The discussion has become less important because *Pityosporites* occurs in Paleozoic beds (CRIDLAND, 1963), including the Beacon System at the Amery locality.

4. Jurassic

Jurassic plants were found by GUNN & WARREN (1962) in chert at only one locality on the upper Mackay Glacier, and all are gymnosperms with a predominance of conifers according to PLUMSTEAD (1962) who has identified four genera *Otozamites*, *Brachyphyllum*, *Pagiophyllum* and *Elatocladus*. Conchostracans occur with the plants. The horizon is definitely Jurassic and probably Liassic, and the plants again show affinity with other southern hemisphere floras, and with the flora at Hope Bay in West Antarctica (q.v.), as discussed in detail by PLUMSTEAD. The distribution of these similar fossil floras from Antarctica to India, across all the present climatic zones, is very difficult to explain.

Ferrar Dolerites

The Beacon System is intruded by dolerite (diabase) sills and feeder dykes and pipes wherever it is known, exept at the Amery locality. In parts of Victoria Land the intrusions are so abundant and voluminous that the Beacon sediments are literally broken into great blocks "floating in a medium of dolerite". As a result of their hardness the intrusions are ridge, summit and cliff formers in the topography. Glacierized mountain landscapes cut from the complex of sills and Beacon sediments often rival in appearance and grandeur the Grand Canyon of the Colorado, with black sills set against a background of white snow and ice and rusty red or multi-coloured Beacon sediments. The sills are not confined to the Beacon sediments, but extend a short distance down into the underlying basement, a fact that raises difficult problems concerning the mechanics of intrusion. Isotope dates of 150 to 190 m.y. (fig. 5) show that they were intruded in the Lower and Middle Jurassic, and are correlatives of the Tasmanian dolerites and Karroo Dolerites (STARIK et al. 1961; EVERNDEN & RICHARDS, 1962; McDOUGALL, 1963), and probably of the basic lavas and intrusives of Brazil (DU TOIT & HAUGHTON, 1954; JENKS, 1956; GIGNOUX, 1955).

The type region of the Ferrar Dolerites, in Victoria Land west of McMurdo Sound, has been mapped by McKELVEY & WEBB (1961, and other papers), who have shown that the sills have a total thickness of about 1000 m in a section with a thickness of about 2100 m.

The lowest or "Basement Sill" is an undulating sheet, 245 m or more thick, usually intruded into the basement rocks below the Kukri Peneplain. A younger "Peneplain Sill" of comparable thickness has mainly followed the Kukri Peneplain and the base of the Beacon sediments but in places rises higher or moves lower into the basement. Few of the higher sills in the Beacon sediments exceed 125 m in thickness, and their pattern of emplacement was extremely irregular. They frequently rise abruptly to higher levels or split into thinner units and are associated with massive feeder dykes. Some sills are composite, consisting of more than one intrusion, but this is not always obvious without microscopic examination.

The Basement and Peneplain sills might extend intermittently or even continuously for 1000 km or more along the mountains of Victoria Land, for sills at this position have been seen on the ground and from the air at a number of localities (GUNN & WALCOTT, 1962). Together, they represent the intrusion of an enormous volume of basic tholeiitic magma, perhaps of the order of 36 million cubic km. With the addition of the higher sills the total volume of magma represented by the Ferrar Dolerites and Ferrar Basalts between the Theron Mountains and Horn Bluff could be two or three times greater, perhaps of the order of 100 million cubic km, or even much more if other intrusions, such as the dyke-swarm of the Vestfold Hills, and the sills of Queen Maud Land (NEETHLING, 1963) are Jurassic (see Sandow Beds). Mafic sills have been reported from the Ellsworth Mountains (SCHMIDT, 1962) and might be of Jurassic age.

Early petrographic descriptions of the intrusions in Victoria Land by BENSON, and BROWNE, have been followed by those of McKELVEY & WEBB (1961), and HAMILTON & HAYES (1963), and by detailed mineralogical, modal, and chemical studies by GUNN (1962, 1963, and papers in preparation). He has shown that marked differentiation controlled by early crystallisation and gravitational settling of orthopyroxene has occurred in several intrusions, such as the Basement Sill, leading to the formation of some cumulate rocks, and in the residuum to depletion in magnesia from 22% to under 1%, slight increase in total iron, increase of silica from 47% to 67%, and the production of alkali and silica-rich pegmatoids and granophyres. In the Peneplain Sill at Solitary Rocks the only pyroxenes are augite and pigeonite, and the differentation is minimal.

Ferrar Volcanics

Puzzling tuff and basalt erratics of unknown age were found by PRIESTLEY in 1911 in moraines of the Priestley Glacier at Terra Nova Bay. It has been discovered very recently that both late Cenozoic volcanics and probable Jurassic volcanics are present in that region.

In the Mawson Tillite at the head of the Mackay Glacier west of

Granite Harbour GUNN & WARREN (1962) found a series of interbedded basaltic flows, pillow lavas and tuffs. In the same season
lavas were found by topographical surveyors 320 km further south
in the western Darwin Mountains (GUNN, 1962). More recently the
Kirkpatrick Basalts have been found at the top of the Beacon
Sequence in the Beardmore region (GRINDLEY, 1963) and have a
thickness of approximately 1000 m with intercalated sediments, but
the assemblage is poorly known. In the district at the head of the
Priestley Glacier 250 m of basaltic flows have been found (RICKER,
1963), and a little farther north they are at least 1375 m thick and
interbedded with carbonaceous and arkosic Beacon sediments containing tree trunks (GAIR, 1963).

These volcanics are closely associated with the Ferrar Dolerites
and have been classified with them in the Ferrar Group (GRINDLEY,
1963), but in this review are classified as the top formation of the
Upper Beacon System to agree with South African classifications.
They are younger than the Jurassic Sandstones of the Beacon Group
in central Victoria Land, but GRINDLEY (1963) considers that in the
Beardmore district local volcanism might have started in the late
Triassic, with the main outpourings in the Jurassic. Sill intrusions
and basaltic volcanism commonly terminated Gondwana sedimen-
tation in other southern countries, and probable correlatives are the
Drakensberg and Batoka basalts of southern Africa, the Serra Geral
Basalts of Brazil and similar basalts in Argentina, and the Rajmahal
Basalts of India. Several of these assemblages overlie Rhaetic beds.

At all the Antarctic localities the lavas are amygdaloidal, and
petrographically include a wide range of basaltic rock types including hypersthene-basalts, and a pigeonitic lava, having affinities
with the Ferrar Dolerites (GUNN & WARREN, 1962; GUNN 1962).

In western Queen Maud Land basic sills have been intruded on
a grand scale (NEETHLING, 1963) and are associated with basic
volcanics (BRUNN, 1963). These might be correlatives of the Ferrar
Volcanics (but see Sandow Beds).

Mawson Tillite

About 450 m or more of horizontal indurated tillite has been found
by GUNN & WARREN (1962) north of the head of the Mackay
Glacier west of Granite Harbour in Victoria Land. It is known in
three nunataks and has been followed for 40 km, and possibly
extends further into unexplored country. The tillite overlies both
the Jurassic Sandstones and the coal measures (Glossopteris Sandstones) of the Beacon System, since a highly irregular erosion surface
was cut in them before its deposition. It is an aggregate of unsorted
rock fragments, possibly a product of eastward-moving ice, and
at its southerly occurrences contains tuffs and irregular basaltic
pillow lavas. There is no fossil evidence of its age except that it is

younger than the plant-bearing Jurassic Sandstones, but the lavas are correlated with the Ferrar Volcanics, which in turn are considered indirectly to be Jurassic in age. There is a chance of circular argument in this situation, and isotope dates are needed. The induration of the formation is consistent with a Jurassic or Cretaceous age, but is not conclusive evidence against a Cenozoic age. Moreover Quaternary tillites of this type could be expected in Antarctica. No similar formation is known elsewhere in Antarctica and there are no suggested correlatives in other southern lands. The deposit might be a product of a fairly limited montane glaciation connected with high land formed by the eruption of the Ferrar Volcanics, but equally it could be a product of continental glaciation.

6. Cenozoic Basalts

Leucite basalts and tuffs exposed over a small area at Gaussberg (66° 48'S, 89° 19'E) have been described by REINISCH (1906) and are the only truly East Antarctic Cenozoic volcanics that are known, except parts of the McMurdo Volcanics which lap over the Trans-Antarctic Escarpment and are described with West Antarctic rocks. Gaussberg is linked to the volcanic Heard and Kerguelen islands by a submarine ridge. A sample of its basalt has an isotopic date of 20 m.y. equivalent to Lower Miocene (STARIK et al., 1961). Leucite basalt plugs are known also in the Canning Basin in West Australia.

Quaternary Sediments

See under West Antarctica.

WEST ANTARCTICA

7a, 7b, 8. Precambrian to Lower Tertiary (Anticlinal Ridges)
7a, 7b. Antarctandean Ridge

The rocks included in this grouping on fig. 4 are:
i) Precambrian gneisses of Marguerite Bay,
ii) Early Paleozoic volcanics and intrusives of Marguerite Bay,
iii) the Trinity Peninsula Series of Upper Paleozoic metasediments,
iv) the Early Tertiary Andean intrusives of the Antarctic Peninsula.

These four rock groups form the core of the Antarctandean Ridge, and are flanked by Jurassic and younger rocks. The rocks of the Ellsworth Moutains have been mentioned under Ross System but their true position is uncertain. The same applies to schist and granitic rocks in nunataks between the Ellsworth and Horlick Mountains (THIEL, 1961).

38

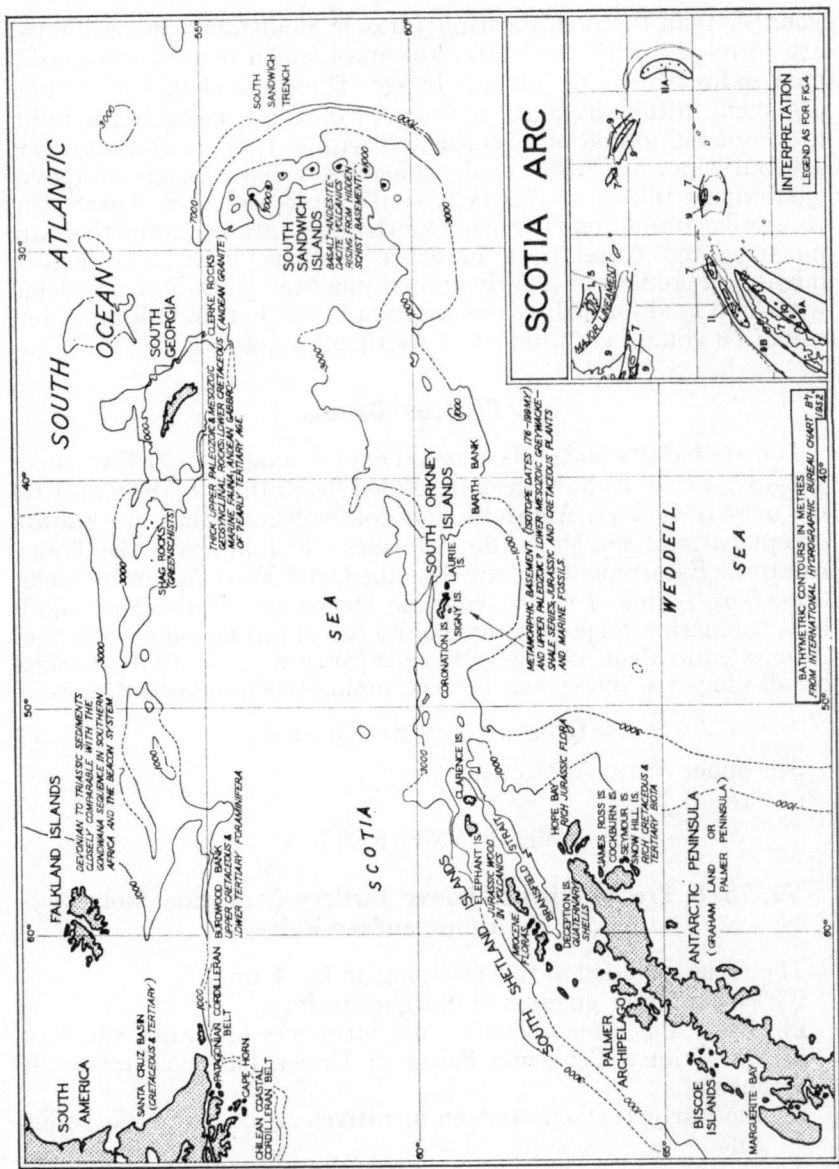

Fig. 6. Scotia Arc. In the South Sandwich Trench there are depths of over 8000 m.
The inset shows an interpretation of the geology to accompany fig. 4, the significance of the numerals being: 1, Precambrian metamorphics; 5, Devonian to Triassic
geosynclinal sediments and Early Tertiary Andean Intrusives; 9, Jurassic and
younger sediments; 10, James Ross Volcanics; 11, Volcanics of South Shetland
Islands; 11A, Volcanics of South Sandwich Islands.

The geology of the Antarctic Peninsula and adjoining islands is dominated by a backbone of igneous rocks of the Andean Granite-Gabbro Intrusive Suite, which is estimated to make up 80% or more of the total area. ANDERSSON (1906, p. 59) expressed this aptly in his statement that "the Jurassic beds of Hope Bay form one of the small patches of sedimentary rocks which are scattered on the mainland and adjacent islands in a vast sea of eruptives". The intrusives are predominantly dioritic and have been described by ADIE (1955), who states that they are all crystallization-differentiation products of a common parental magma and form a normal calc-alkaline series. Their age is uncertain, but they are generally regarded as late Cretaceous or early Tertiary by analogy with very similar rocks forming the core of the Andes, and especially the western Patagonian cordillera (fig. 6), and this opinion is reinforced by isotope dates, obtained by HALPERN (1963), which indicate an age on the Cretaceous-Tertiary boundary.

Around the margins of the peninsula there is a great variety of other rocks, broken into fragmentary areas by sea, ice, and the Andean Intrusives, and there is no systematic regional description of them. The boundary between units 7 and 9 on fig. 4 is schematic, being in fact much more irregular, especially on the western flank of the peninsula.

In the Marguerite Bay region in the southern part of the peninsula there is a succession of 1000 m of Basement Complex gneisses referred to the "Archaean", and another complex of plutonic and volcanic rocks that are considered to be Early Paleozoic probably Cambrian (ADIE, 1954, 1962). A hidden occurrence of these rocks in the northern part of the peninsula is suggested by erratics at James Ross Island.

Elsewhere in the Antarctic Peninsula and particularly in its northern part, there are thick highly-deformed geosynclinal metasediments called the Trinity Peninsula Series, which contain a few fragmentary and obscure plant remains (ANDERSSON, 1906) and are considered on this slender evidence to be Carboniferous. They have been extensively metamorphosed by the Andean intrusives to biotite, cordierite, chiastolite and sillimanite hornfelses (ADIE, 1957c).

8. Metasediments and Intrusives of the Usarp Ridge

The Usarp Ridge is a bedrock high consisting of Cenozoic volcanics lying on a basement, mainly subglacial, of metasediments, metamorphics and acidic plutonics. The basement rocks have been mapped towards the Ross Sea end of the ridge in the Edsel Ford and Rockefeller mountains, and in the King Edward VII Peninsula, where a sequence of folded and slightly metamorphosed metasediments at least 5000 m thick has been extensively intruded by granite, granodiorite and monzonite (WADE, 1945; WARNER, 1945; PASSEL, 1945). Fold axes trend at about 305°, and the smaller folds appear to be part of a major synclinorium plunging in the same direction. The sediments are unfossiliferous, so that their age is uncertain, but possibly somewhere in the interval Late Precambrian to Jurassic. There is also no evidence of the age of the plutonic intrusives, and isotope dating is needed. They might be Paleozoic, or could be of much the same age as the Late Mesozoic-Early Tertiary Andean intrusives.

Farther to the east, most of the exposed rock masses of the Usarp Ridge are Cenozoic volcanics, but DOUMANI (1960) reports diorite below the volcanics in an unnamed mountain at 76°S, 125°W, and metasediments and pink granite in the Mount Petras district (76°S, 126°W) between the Hal Flood and Executive Committee ranges. Schist, diorite and gneissic diorite have been found at Thurston Island and the Eights Coast (CRADDOCK & HUBBARD, 1961; DRAKE, 1962).

9. Jurassic, Cretaceous and Cenozoic (Depressions)

9A. Sediments and Volcanics of East Side of Antarctic Peninsula

On the east flank of the northern end of the Antarctic Peninsula there is the following succession of rocks:

Late Pliocene or Early Quaternary "Pecten Conglomerate" of Cockburn Island, 20 m
    Unconformity
Upper Miocene and (?) Pliocene James Ross Volcanics, 500 m +
Upper Oligocene (?) – Lower Miocene Seymour Island Beds, c. 250 m
    Unconformity
Late Cretaceous Snow Hill Island Beds, c. 500 m
    Unconformity
(Upper?) Jurassic volcanics, c. 2400 m
Mid-Jurassic Mount Flora Beds, 200 m
    Unconformity
Carboniferous (?) Trinity Peninsula Series

Many formations of the Jurassic and younger rocks are fossiliferous, so that this district has attracted more attention from pioneer geologists than any other part of Antarctica except McMurdo Sound.

The main reports on the geology are in the publications of the Swedish Antarctic Expedition of 1901–3 and more recently of the Falkland Islands Dependencies Survey (now British Antarctic Survey). Valuable summaries of the literature which is mainly paleontological, have been provided by FAIRBRIDGE (1952) and ADIE (1952b, 1957a, 1957b, 1958, 1962), and the best account of the field geology is still in many ways that by ANDERSSON (1906).

The British Antarctic Survey has in hand the preparation of new geological maps and modern stratigraphical and lithological descriptions (R.J. ADIE, pers. comm.), and a sketch map of Hope Bay has been published by ADIE (1957c), who has also (1962) provided a table of correlations between the Antarctic Peninsula, and the Andean and Magellan geosynclines of the Patagonian Cordillera. Detailed surveys have been undertaken also by the expeditions of the Argentine and Chile.

Jurassic Mount Flora Beds

At a small locality on the sides of Mt. Flora in Hope Bay (fig. 6) the Trinity Peninsula Series is overlain unconformably by the following succession which is about 500 m thick and disposed in an isolated shallow syncline (ANDERSSON, 1906; ADIE, 1952b; ADIE 1957c):

Dark-coloured vertically-jointed rhyolites;
Acid tuffs, white, light-grey and dark;
Fissile shales with abundant plant remains, frequently pyritic;
Basal conglomerate, coarse and unstratified, some plants.

The vertical jointing in the rhyolites suggests that they might be ignimbrites. HALLÉ (1913) identified 61 forms including 19 cycads and 13 conifers from the lacustrine plant beds and referred them to the Middle Jurassic, the genera reported being *Araucarites*, *Brachyphyllum*, *Carpolithus*, *Cladophlebis* (7 spp.), *Coniopteris* (3 spp.), *Cycadolepis*, *Dictyophyllum*, *Elatocladus* (4 spp.), *Equisetites*, *Nilssonia*, *Otozamites* (6 spp.), *Pachypteris*, *Pagiophyllum* (4 spp.), *Pseudoctenis* (3 spp.), *Ptilophyllum*, *Sagenopteris*, *Schizolepidella*, *Scleropteris*, *Sphenolepidium*, *Sphenopteris* (8 spp.), *Stachyopitys*, *Thinnfeldia* (now *Dicroidium*), *Todites*, *Williamsonia* and *Zamites* (4 spp.).

This assemblage has been rediscussed in an important comparative paper by FLORIN (1940), and SEWARD (1959, p. 368—371, a reprint of a 1933 edition) has indicated the affinities with other Jurassic floras in many parts of the world, and has given comments on the systematics. (For the change in nomenclature from *Thinnfeldia* to *Dicroidium* see TOWNROW, 1957, and the section of this review on the paleontology and age of the Beacon System.) The flora grew vigorously under optimum conditions. Its discovery was a paleontological sensation, and has been a factor in subsequent paleoclimatic and paleobotanical writing and speculation.

The plant beds have yielded also two species of aquatic freshwater beetles, *Grahamelytron crofti* and *Ademosynoides antarctica* which suggest a connection with Australia (ZEUNER, 1959), and freshwater gastropods and fish (ADIE, 1962).

Jurassic Volcanics

The acidic volcanics above the Mount Flora Plant Beds are mainly tuffaceous, and extend southwards along the coast to Cape Longing and Cape Sobral (64° 33'S, 58° 50'S, 59° 40'W) and attain a great thickness of about 2400 m. In places they contain beds of carbonised woods derived from former forests (ADIE, 1957a), and since they overlie the Mount Flora Plant Beds they are presumably of Middle or Upper Jurassic age. They were continuous originally with the Upper Jurassic volcanics that are widespread on the western flank of the Antarctic Peninsula, including ash-beds with belemnites at Alexander Island (ADIE, 1962). They might have been contemporaneous also with the basic Ferrar Volcanics of Victoria Land.

Late Cretaceous Snow Hill Island Beds

Some 80 km south of Hope Bay in the James Ross Archipelago (fig. 6) there is the best-known sequence of Cretaceous and Tertiary strata. As shown above in tabular form, the Upper Jurassic is represented by volcanics, and Lower Cretaceous and Early Tertiary are missing at unconformities. The gap is possibly connected with the injection of the Andean Intrusives (ADIE, 1963), but nevertheless the Upper Jurassic and Lower Cretaceous are represented by sediments on the west side of the peninsula. Lower Cretaceous occurs also at South Georgia.

At Snow Hill, James Ross, Seymour and Cockburn islands the Cretaceous strata, called the Snow Hill Island Beds, consist of fine-grained loose soft fossiliferous sandstones more or less glauconitic and in places concretionary. Lithologically they resemble fossiliferous sediments of about the same age deposited in the diachronous outer littoral or inner neritic facies zones in New Zealand (FLEMING, 1959) and in the Magellan Geosyncline of southern South America. Their thickness is not recorded but cannot be much more than about 500 m because they are almost flat-lying, dipping only gently to the east or east-south-east. Fossil collections have not been well-located stratigraphically in the sequence and this causes difficulties in interpretation. Ammonites are the most abundant fossils and ANDERSSON (1906) states that "next to them in importance come the bivalves and gastropoda; fish-remains and corals are also tolerably common, while echinids and decapoda are but rarely found". Some pieces of wood were collected also by the Swedish expedition.

Revisions of the ammonite fauna have been made by SPATH (1953) and HOWARTH (1957). The former assigns them to the Upper Campanian but HOWARTH states that they are all of Lower and Middle Campanian age and his decision has been adopted. (See Table IV for standard sequence of Cretaceous epochs.) He places most of his 70 ammonites in the Kossmaticeratidae, notably

Maorites tuberculatus, and the Pachydiscid species *Eupachydiscus grossouvrei*, and shows that they can be correlated with genera that have been collected bed by bed in a sequence in western Madagascar.

WILCKENS (1910), who studied the pelecypods and gastropods, found the genera *Lahillia, Lima, Pinna, Ostrea, Trigonia, Turritella,* and *Fusus.* A new study of the brachiopods, lamellibranchs and gastropods is in hand (R. J. ADIE, pers. comm). The corals include the genera *Cycloseris, Parasmilia* and *Oculina* and the echinoids include *Cyathocidaris.* The Foraminifera have been described by HOLLAND, the fish-remains by SMITH-WOODWARD, the lobsters and worms by BALL (1960) and a remarkable cirripede by WITHERS (1951). A vertebra of a Plesiosaur-like reptile found at the Naze, James Ross Island, is mentioned by ADIE (1957, p. 457, 1958, p. 9—10). He also records that Mesozoic sediments and volcanics have been mapped on Joinville Island, and at Seal Nunataks (65° 03′S, 60° 18′W), though he does not list their fossil content.

As mentioned below, CRANWELL (1959) has obtained some information on plant microfossils including *Nothofagus* apparently derived from Upper Cretaceous beds.

Upper Oligocene – Lower Miocene Seymour Island Beds

The Tertiary beds rest unconformably on the Campanian beds at the northern end of Seymour Island and at Cockburn Island, but there can only be a slight angular difference for the break was formerly called a disconformity. The beds are tuffaceous, and in part glauconitic, littoral conglomerates and sandstones, and are of slightly different facies on each island. The fossils have been assigned to the Upper Oligocene and Lower Miocene, with a slight tendency to favour the latter.

From the lower part of the Seymour Island succession the Swedish Antarctic Expedition of 1901/03 obtained a macroflora from which DUSEN (1908) identified *Araucaria imponens*, some fragments tentatively placed in *Sequoia* and more numerous angiosperms, including *Drimys, Fagus* (2 spp., see later comment), *Knightia, Laurelia, Lomatia, Myrica, Nothofagus* (2 spp.), and some others (often leaf or stem fragments only) assigned more doubtfully to the Aquifoliaceae, Cyperaceae, Lauraceae, Leguminosae, and the warmth-loving Melastomaceae *(Miconiiphyllum).* A seed was also tentatively identified as podocarpaceous, and four living genera of ferns were recognized. GOTHAN (1908) recovered wood of *Nothofagus,* of *Araucaria (Dadoxylon),* and of *Phyllocladus* or *Podocarpus* type.

In his revision of coniferous types in these Tertiary sediments FLORIN (1940, p. 32) confirms the identification of *Araucaria* and of podocarps but points out that the "Sequoia" is partly made up of *Acmopyle* (a podocarp now living only in New Caledonia and Fiji). GOTHAN's *Phyllocladoxylon* he thinks most likely to be wood of

44

Table

The Fossil Record

Era	Period	Epoch	Absolute Age (Beginning of epoch or period in years × 10^6 after KULP)
Cenozoic	Quaternary	Holocene	0.02
		Pleistocene	0.8
	Tertiary	Pliocene	13
		Miocene	25
		Oligocene	36
		Eocene	58
		Paleocene	63
		Danian	
Mesozoic	Cretaceous	Maestrichtian	72
		Campanian	
		Santonian	84
		Coniacian	
		Turonian	90
		Cenomanian	110
		Albian	120
		Aptian	
		Neocomian	
	Jurassic (Lias) (Rhaetic)		166 / 181
	Triassic		230
Paleozoic	Permian		280
	Carboniferous		345
	Devonian		405
	Silurian		500
	Ordovician		
	Cambrian		(600)
Precambrian (Cryptozoic)	"Proterozoic"		(1,500)
	"Archaean"		(4,000)

IV.

in Antarctica

Sub-fossil penguins, many rookeries; sub-fossil shells Falkland Islands, McMurdo Sound etc.
Shell beds, Deception Island (late Pleistocene?); shell beds McMurdo Sound, and Cockburn Island (early Pleistocene or late Pliocene).

Miocene plants in James Ross Volcanics of the James Ross Archipelago, and at the South Shetlands;
Lower Miocene or Upper Oligocene plants and rich marine faunas at Seymour and Cockburn islands, east side Antarctic Peninsula.

Spores and pollen *(Nothofagus* etc.) and marine microfossils in erratics, McMurdo Sound.
Lower Tertiary and Upper Cretaceous foraminifera, Burdwood Bank, Scotia Arc; derived pollen in Upper Oligocene Seymour Island Beds.

Rich biota of Lower and Middle Campanian marine invertebrates, and plants, at James Ross Archipelago, Antarctic Peninsula, east side.

Aptian plants and marine fauna, Alexander Island, Aptian ammonites, lamellibranchs etc., South Georgia; Cretaceous plants, brachiopods etc, South Orkney Islands.

Plants and marine fauna, Alexander Island; wood, South Shetlands.
Rich plant beds, Hope Bay, Antarctic Peninsula; plants in Victoria Land with bivalve crustacea (conchostracans).

Plants, Beacon System *(Dicroidium)*

Wood, leaves, microfossils, *Glossopteris* flora, and invertebrate trails many localities, Beacon System (Glaciation in Upper Carboniferous)
Mid or Upper Devonian freshwater fish remains, and Lower Devonian plants, with marine problematica, Victoria Land. Lower Devonian spore assemblage and rich marine fauna, Horlick Mountains.
Supposed Ordovician graptolite, South Orkneys, now considered obscure plant fragment.
Cambrian Archaeocyathid sponges, trilobites, algae, etc. in limestones of Ross System; Cambrian microfossils in Sandow Beds.

Acmopyle. He is inclined to admit a species of *Podocarpus* (section *Stachycarpus*).

A further revision of the macroflora as a whole has been undertaken by SELLING (see ADIE, 1958, p. 7), and CRANWELL (1959) has succeeded in extracting a microflora from a small sample of pale-coloured calcareous vitric tuff or tuffaceous limestone collected by the Swedish expedition. Most important has been the discovery that all the fagaceous pollen grains belong to the southern beeches *(Nothofagus)*. No traces of the tricolporate grains of northern beech *(Fagus)* were found. Further, she states that there is a contrast in the state of preservation of the grains which suggests a history of abrasive water-carriage for some and a gentle air-carriage for others. The damaged grains could be derived from older (Cretaceous) sediments, possibly the Campanian Snow Hill Beds, and though the preparations thus may include remains of both Cretaceous and Mid-Tertiary floras they are very valuable. The pollen extract is dominated by conifers and species of *Nothofagus*, as in other widespread fossil and living southern hemisphere assemblages. The conifers include *Araucaria*, perhaps *Agathis*, and bisaccate grains of the *Podocarpus* – *Acmopyle* – *Dacrydium* – *Phyllocladus* complex. Small podocarpoid trisaccate pollen grains occur but are rare in the specimen available to CRANWELL (pers. comm.). All the *Nothofagus* belongs to the distinctive *fusca* and *cranwellae (brassi)* groups, not to the more fragile *menziesii* group. Other angiospermous families represented include Proteaceae, Myrtaceae, Loranthaceae, Cunoniaceae (or Elaeocarpaceae) and possibly Winteraceae. Pteridophytes are weakly represented by spores of Cyatheaceae and possibly Schizaeaceae. Many units are rare, or broken. No doubt this record would be greatly improved if more stratigraphically controlled samples were available for analysis from both the Tertiary beds and the thin Campanian coal seams and plant beds. CRANWELL emphasises the importance of the absence of *Fagus* and the presence of *Nothofagus*, now proved through finds of pollen, leaves and wood. The absence of pollen of the Centrolepidaceae, Compositae, Cyperaceae (though DUSEN reported a *Scirpus*-like fragment), Gramineae, Pandanaceae and Restionaceae is also important, but cannot be used in attempted datings until richer matrices have been checked.

Above the plant beds lies a littoral facies in which marine fossils include a vertebra probably belonging to an Archaeocete (fossil whale) and gigantic penguins. The penguin bones, of which some 75 have now been collected, have been discussed by WIMAN (1905), SIMPSON (1946) and MARPLES (1953). The two latter authors recognise four genera, but differ in their assigment of them to subfamilies. The resemblance between the fossil penguins of Seymour Island and New Zealand is emphasised by MARPLES. BUCKMAN (1908) described the brachiopods from these beds, and WILCKENS (1912) the Mollusca, which are of widespread genera such as *Cucullaea, Lahillia, Modiola, Nassa, Nucula, Polynices* and *Venus.*

9B. Jurassic and Cretaceous Sediments and Volcanics of West Side of Antarctic Peninsula

The Upper Jurassic is very widespread on the western flank of the Antarctic Peninsula. The rocks are mainly basaltic, andesitic and rhyolitic volcanics occurring between the Danco Coast, and the Palmer Archipelago, and the hinterland of Port Lockroy in the north, and Adelaide Island and the adjoining Fallières Coast behind Marguerite Bay in the south (HOOPER, 1962). The volcanism reached its maximum development in the Marguerite Bay region, and there exhibits a classical evolutionary trend from basic to acid (ADIE, 1957, p. 457). The fact that the Andean Intrusives show a similar trend, and intrude the volcanics, suggests that the latter might have been derived from the intrusive magmas. While volcanism proceeded to the north there was active sedimentation on the eastern coast of Alexander Island of tuffaceous and fossiliferous Upper Jurassic and Lower Cretaceous sediments as discovered by W. L. FLEMING, and investigated later in more detail by V. E. FUCHS and R. J. ADIE (ADIE, 1957a, 1958) who collected belemnites, ammonites, gastropods, lamellibranchs and annelids.

The Jurassic and Cretaceous sequences at Alexander Island are far from clear because of a series of thrusts, but they are separated by 240 m of conglomerate which marks an unconformity or disconformity. HOWARTH (1958) states that in the collection of 30 ammonites, nine are comparable with the species *Perisphinctes (Orthosphinctes) transatlanticus* of the lower part of the Kimeridgian stage of the Upper Jurassic, and the remainder are the species *Sanmartinoceras patagonicum, Ancyloceras patagonicum* and *Silisetes trajani* of the Aptian stage of the Lower Cretaceous (Table I). Among the Cretaceous macrofossils Cox (1953) found a gastropod *Aporrhais (Tessarolax) antarctica,* and an abundance of the lamellibranchs *Aucellina radiato-striata* and two species of "*Pecten*", along with *Cyprina* and *Thracia,* and two species of the annelid *Rotularia.* An indeterminable species of *Inoceramus* is present .This genus could be of very great value for correlation with the Jurassic and Cretaceous in New Zealand and elsewhere if better specimens are found. Access to Alexander Island is difficult but it is clear that further geological field work is very desirable. Mapping of the southern end of the Antarctic Peninsula is proceeding from a new United States station called Sky-Hi, and the discovery of fossiliferous marine sandstone has been reported (BEHRENDT & PARKS, 1962).

9C. Lower Tertiary and late Cretaceous sediments of the Ross Sea

On the assumption that the geology of West Antarctica is broadly as shown on fig. 4, I. G. SPEDEN and the writer in 1958/59 searched for Cretaceous and Tertiary sediments in moraines in the McMurdo Sound district. The moraines consist mainly of hard metasediments and igneous rocks, since softer sediments are destroyed readily during glacial erosion and transport. Nevertheless a considerable number of erratics of sediments comparable lithologically with certain facies of the New Zealand Cretaceous and Lower Tertiary, and of the Papuan Geosyncline in New Caledonia and Papua, were found at White Island and Minna Bluff, along with rhyolitic and ignimbritic volcanics, which could have been derived either from an assemblage like the Jurassic volcanics of the Antarctandean Ridge, or from the poorly-known acidic volcanics of the Byrd Group of the Ross System. The erratics include a large disintegrated boulder of very glauconitic sandstone, and fairly numerous small pieces of light-coloured calcareous sandy mudstone and calcilutite some of them containing small shell fragments. One contains part of a worm-tube, *Rotularia* sp., and from this rock G. H. SCOTT extracted four genera of poorly-preserved Foraminifera, including an *Angulogerina* suggesting an Eocene or Lower Oligocene age, and a *Globoquadrina* (?), a genus which first appears in the Lower Oligocene of New Zealand and Australia. CRANWELL succeeded in extracting a pollen component and a range of hystrichosphaerids, dinoflagellates, and other microplankton of kinds resistant to hydrofluoric acid (CRANWELL, HARRINGTON & SPEDEN, 1960). The microfossils appear to include forms derived from older sediments, probably Cretaceous. The assemblage is probably Lower Tertiary, and if so is the only assemblage of that age known from Antarctica. It is however similar to but not identical with that found by CRANWELL (1959) in a specimen from the Upper Oligocene-Lower Miocene Seymour Island Beds. Pollen grains of species of *Nothofagus* predominate, with those of conifers less abundant. Palmoid and proteaceous types occur also. The absence of Compositae, as at Seymour Island, might be taken to indicate a pre-Miocene age. The paleobotanical significance of the *Nothofagus* species in indicating links on the one hand with South America, and on the other with the Australasian region is emphasized by CRANWELL (in preparation). The plant fossils are discussed also by WACE in Chapter VI of this volume.

The discovery of these sediments in moraines, and the fact that the Sandow Beds were also first found as fragments in moraines, suggest that much more attention should be given in Antarctic geology to the careful examination of terrestrial and submarine erratics, and to subsequent palynological analysis of suitable sediments. Apocryphal erratics could occur locally, for there might be truth in the recent story that a visitor to McMurdo Sound was found throwing exotic fossiliferous rocks on moraines, and when asked why, replied "I *hate* geologists". In view of this story it should be mentioned that the Minna Bluff and White Island erratics are too abundant and scattered to be forgeries, and moreover some were found a foot or more inside moraine heaps tens of square miles in area at localities not visited previously by *anyone*.

10, 11, 11a, 11b. Cenozoic Volcanics
10. James Ross Volcanics, Miocene

The Middle and Upper Miocene is represented in the Antarctandean Ridge by the basaltic James Ross Volcanics which occur not only on James Ross Island but also over a wide region on the northeastern side of the peninsula and in the South Shetland Islands (BAECKSTROM, 1915). There were at least two lava and three explosive phases, and in the South Shetlands there are perfectly preserved plant remains in water-deposited tuff on Nelson Island, and at Ezcurra Inlet and Dufayel Island (ADIE, 1953, p. 16) which have been studied by SELLING, though the results are not yet published (ADIE, 1953, p. 6). Other forms have been mentioned by BARTON (see below).

11. South Shetland Islands

This group seems to form a distinct geanticlinal ridge e n e c h e l o n with the Antarctandean Ridge, and separated from it by a trough under Bransfield Strait. Elephant and Clarence islands consist of a Basement Complex of schists and phyllites of supposed Precambrian age (ADIE, 1962). At King George Island (HAWKES, 1961a) and Livingston Island (HOBBS, 1959) the schists also occur, and are followed by Carboniferous (?) greywacke, by Jurassic volcanics containing wood and intruded by the Early Tertiary Andean Intrusives, and by an uncomformable cover of Oligocene and Miocene volcanics. Deception Island (HAWKES, 1961b) has a breached crater caldera that provides a much-used harbour with hot springs and steam jets on its shore, and its volcanics can be divided into a Pliocene Pre-Caldera Series, and a Quaternary Post-Caldera Series. They are petrographically a basalt-andesite-dacite association, as described by TYRRELL (1945), and are comparable with those of the South Sandwich Islands described under the heading Scotia Arc. Deception Island includes unusual alkaline rocks. Bridgeman Island has several times been observed "smoking". The younger Cenozoic rocks of the group include a Pecten Conglomerate (see later), and raised beaches.

BARTON (1963) has reported that in the Oligocene and Miocene volcanics of King George Island there are two entirely separate floral assemblages, one with very common *Nothofagus* having re-placed a slightly older and warmer flora in which that genus is absent. He suggests that the *Nothofagus* flora invaded the locality from South America, but the present ecology of that genus in New Zealand and other South-West Pacific regions suggests that it might have existed elsewhere in the same region at a higher altitude or colder or more misty site.

12a. "Byrd" Volcanics

Small and local occurrences of basic volcanics were found in the Edsel Ford Ranges of the Usarp Ridge by the early Byrd expeditions, and in the last few years members of United States oversnow traverses have found that major volcanic piles occur between the Ross Sea and the Eights Coast (DOUMANI, 1960; ANDERSON, 1961; CRADDOCK et al., 1963). The southern masses appear to form a "Byrd Volcanic Arc" on the south side of the Usarp ridge, but this is uncertain because field work has been done under difficult conditions and only brief notes have been published. The rocks were originally excluded from the McMurdo Volcanics by an informal nomenclatural committee (HARRINGTON, 1958a) and they are here termed for convenience the "Byrd Volcanics", pending detailed description and naming by the United States geologists. In places as shown by photographs, the volcanics build great central volca-noes, such as Mt. Takahe, similar to Mt. Erebus, and rising to about 4000 m. In age and petrography the rocks seem to be similar to or identical with the McMurdo Volcanics, though rhyolites have been reported, probably incorrectly (DOUMANI & EHLERS, 1962). On the oceanic side of the Usarp Ridge, there are two more large central volcanoes Mt. Siple and Peter I Island, which are very similar in appearance to the larger central volcanoes of the McMurdo and Byrd arcs. These masses are also poorly described and might actually lie along northerly-trending faults or lines of eruptive centres connected across the Usarp Ridge with the southern arc.

12b. McMurdo Volcanics

An active volcano in the Balleny Islands was discovered as early as 1839, and another in the Ross Sea region was discovered in January 1841 during the first southward voyage of H.M.S. *Erebus* and H.M.S. *Terror* under the command of Sir JAMES CLARK ROSS. The work that has been done since then on the volcanoes and basaltic rocks be-tween the Balleny Islands and McMurdo Sound has been summarised by HARRINGTON et al. (in Press).

In the McMurdo Sound region the volcanic forms are atypical, being distributed along radial lines from two main centres, Mt. Erebus and Mt. Discovery. There are also small and scattered occurrences of lava flows and cinder piles in the Koettlitz and Taylor valleys. The rocks have been fairly closely studied where they are accessible from the shores of McMurdo Sound (FERRAR, 1907; DAVID & PRIESTLEY, 1914; SMITH, 1954) but in other districts are very poorly described or undescribed. Mt. Erebus, rising to nearly 4000 m, has had a complex history and is still active, its steam plume being familiar to all who have worked at McMurdo Sound. All the lavas of the McMurdo Volcanics tend to build physiographic forms with convex slopes rather than concave, so that there are fewer cones than domes, but this characteristic is obscured in the McMurdo Sound region where the upper parts of Mt. Erebus, Mt. Discovery and Mt. Terror rise commandingly from their lower dome-like pedestals. In this region also numerous small centres of eruption are dotted over the flanks of the larger masses wherever they are deglacierised, and glaciated trachytic tholoids or plugs form prominent lesser hills such as the well-known Observation Hill at the United States' McMurdo Station.

Farther north the volcanics are found in four distinct districts, which may be called the Franklin Island, Mt. Melbourne, Hallett Coast and Balleny Islands lines. Franklin Island has not been described but the writer has observed in distant views that it is a shield volcano consisting of a meridional line of 3 or 4 high overlapping basalt domes. Mt. Melbourne, on the coast north of Terra Nova Bay is a dominating cone of the Fujiyama type, and it has been found recently (GAIR, 1963) that a previously unsuspected line of volcanoes extend northwards from it inland from the Lady Newnes Ice Shelf. On the Hallett coast of northern Victoria Land, a line of 10 major overlapping domes grouped in three shield volcanoes rising to 1500 m trends southwards from Cape Adare to Cape Jones, and is there offset slightly to the east to the subsidiary line of the two domes forming Coulman Island (HARRINGTON et al., 1953, and in press). The line of domes and cones in the Balleny Islands has not been explored geologically but a few rock specimens have been described by MAWSON (1940). Views from the sea show that Scott Island at the entrance to the Ross Sea is basaltic, but it has not been described. It is so small that a considerable part of it could be swept by the sea in a storm. A dissected volcano, apparently Cenozoic (but possibly part of the Ferrar Volcanics), has been reported in the Horlick Mountains by LONG, and some of the basic dykes in the Pensacola Mountains could be Cenozoic or Jurassic.

The volcanoes are considered to be of Pliocene to Holocene age mainly on physiographic evidence. Magnificent cliff sections through the lavas and tephra deposits are exposed at many places on the shores of the Ross Sea, but no plants or other fossils, and no soil-horizons, have been noticed in them. Isotope dates are needed.

The rocks are an alkaline olivine-basalt association of olivine-basalt, basalt, basanite, limburgite, mugearite, trachytes, and phonolite, including the intriguing rhomb-porphyry incorrectly known as kenyte. Several petrographic and chemical studies have been made, the latest and most important being those by SMITH (1954) and GUNN (in HARRINGTON et al., in press), both presenting chemical analyses and variation diagrams.

The two centres at McMurdo Sound and the four lines further north can be regarded in a broad way as segments of the major McMurdo Volcanic Arc. No real progress has been made in understanding the causes of the volcanism or its tectonic control, though there has been unpublished speculation. At present it can only be noted that the Balleny Islands rise from a branch of the mid-ocean ridge system, and that further south the McMurdo Volcanics lie close to the zone of contact between East Antarctica and the Ross Sea. The southern parts of the "Byrd Volcanics" are in a similar position if the boundary between East and West Antarctica is in fact along the Byrd Subglacial Basin and the Bentley Subglacial Trench. The northern parts of the Byrd Volcanics could be along the zone of contact between West Antarctica and the Pacific Ocean, and they could be regarded also as being linked to Scott Island and a spur of the mid-ocean ridge system.

Quaternary sediments (of East & West Antarctica)

Some leading features of the Quaternary history of both West and East Antarctica are summarized under this heading.

There is as yet no sedimentary record of the history of East Antarctica from the later Jurassic to the Pliocene, but by analogy with Australia it is possible that it suffered long-continued peneplanation accompanied by the infilling of broad interior basins with Cretaceous and Tertiary sediments, both marine and non-marine. This stable phase was ended, possibly in the later Cenozoic, by epeirogenic uplift and block-faulting in the Trans-Antarctic Mountains, the Amery Ice Shelf – Lambert Glacier – Enderby Land – Prince Charles Mountains region, and elsewhere. The stratigraphy and paleontology of West Antarctica indicate that the present glacierization probably commenced with the world-wide onset of cooling in the Pliocene, but the continental ice sheets might not have formed until the Early Quaternary and could have built up to their maximum size in 50,000 years, which is less than one-sixth of Quaternary time (HOLLIN, 1962). Glacial erratics and moraines have been found at many places in Antarctica at heights up to 750 m or more above present ice levels, and indicate the former maximum extent of Quaternary glaciation. HOLLIN suggests that fluctuations in the size of the ice sheets were small when compared with the great fluctuations of northern hemisphere Quaternary ice sheets but

nevertheless they were very significant. Some valleys in Victoria
Land west of McMurdo Sound are at present deglaciated and in
them there is a record of four Quaternary glaciations of decreasing
intensity (PÉWÉ, 1960; BULL, MCKELVEY & WEBB, 1962). During
the two earlier glaciations nearly all the present bare-rock areas of
the Antarctic Continent were covered by ice, and the Ross and
Weddell Seas were largely filled by ice. This must have caused almost
complete expulsion of the land fauna and flora, including probably
the smaller penguins, and a considerable modification of the marine
environment, partly through world-wide falls in sea-level, and world-
wide lowering of sea temperatures. There is evidence, however, of
pre-glacial or early interglacial marine colonization, in the occurrence of shelly deposits at Cockburn Island at the north end of the
Antarctic Peninsula, at the South Shetland Islands, and in Mc
Murdo Sound.

At Cockburn Island the "Pecten Conglomerate" is a tiny remnant of a beach deposit lying unconformably on the James Ross
Volcanics at a height of about 300 m. It contains numerous shells of
Pecten (Myochlamys), *Hemithyris* and *Magasella*, as well as *Balanus*,
a foraminiferal fauna, and a dozen polyzoans (bryozoans), which
have been described by HENNIG (1911) and SOOT-RYEN (1952). The
Pecten Conglomerate occurs also in the South Shetlands (q.v.).
A tiny remnant of a similar elevated deposit occurs on White Island
in McMurdo Sound (SPEDEN, 1962). It has been almost destroyed by
Quaternary glacial erosion, and widely distributed as erratics in
the McMurdo Sound district. Shells, possibly derived from that
formation or from the floor of the Ross Sea, have been found by
C. G. JOHNSON and the writer in moraines at heights of over 650 m
on the western flank of Mt. Bird. The present elevation of the shelly
beaches occurring *in situ* at White Island and Cockburn Island could
be in part a result of isostatic uplift of the continent following its
progressive deglaciation in the Late Quaternary (SPEDEN, 1960),
or could be in part a result of tectonism, including at McMurdo
Sound, block-faulting on the line of the Trans-Antarctic Escarpment. More speculatively the uplift might be related to continental
drift, or even, following recent Soviet reports, to changes in the
shape of the earth. Antarctica is notably non-seismic at present.

The study of cores of submarine sediments has provided some
information about Quaternary history, and eventually might provide the most complete evidence (HOUGH, 1950), but most samples
obtained so far have been dredgings (STEWART, 1963). Some banks
on the continental shelf have been interpreted as Quaternary end-
moraines (FAIRBRIDGE, 1952, following MAWSON; and see also
LISITZIN & ZHIVAGO, 1960).

About 30 C[14] dates which have a bearing on the problem of the
invasion, expulsion and re-invasion of the biota in the McMurdo

Sound region during the Late Quaternary have been reported in the literature, but all those based on marine organisms are now either suspect or are known to be incorrect because it has been found that present-day Antarctic sea-water has a considerable C^{14} age. Thus a date of 1210 ± 70 years before 1950 for the Adélie Penguin colonization of Cape Hallett (HARRINGTON & McKELLAR, 1958) and a date of 1150 ± 45 years for the colonization of Beaufort Island should be revised to "modern" (T. A. RAFTER, pers. comm.). Geologically it could have been expected that the present invasion or re-invasion by Adélie Penguins occurred no later than the "Thermal Maximum" of 5000 B.C. to 2000 B.C., but the available evidence suggests that penguins did not settle in the Ross Sea region until a little before the time of Captain Cook! There must be improvements in the C^{14} method of dating if progress is to be made on this problem.

SCOTIA ARC

The Scotia Arc (fig. 6) is essentially a well-defined submarine ridge linking South America and the Antarctandean Ridge. The islands rising from it provide a fragmentary record of sedimentation and tectonic activity since at least the Upper Paleozoic, and possibly since the Precambrian. The loop is associated with the South Sandwich Trench, with seismic activity, and with active volcanism in the South Sandwich Islands and the South Shetland Islands.

The geology of the arc has been reviewed recently by MATTHEWS (1959) and bibliographies and tabular comparisons of the islands have been provided by ADIE (1957b, 1958, 1962). The Antarctic Peninsula and the South Shetland Islands have been described under West Antarctica.

In South Georgia there is a Precambrian Basement Complex of schists, unconformably overlain by two main lithofacies, that are probably in part true stratigraphical units of Upper Paleozoic and Mesozoic age. The beds of the Sandebugten 'Series' are quartzose metasediments that have been intensely deformed on axes parallel to the length of the island. They are possibly Carboniferous, and resemble the Greywacke-Shale Series in the South Orkneys and the Trinity Peninsula Series of the Antarctic Peninsula. The metasediments of the Cumberland Bay Series are characterised by volcanic material, and contain Upper Aptian (Lower Cretaceous) ammonites, lamellibranchs, fish scales and a cirripede (WILCKENS, 1937, 1947). There is apparently an unconformity between the two series. At the south-eastern end of the island and at the Clerke Rocks there are Andean Intrusives and Tertiary basic dykes, as well as erratics of greywacke.

The South Orkney Islands have a long rock succession of some complexity, starting with a metamorphic Basement Complex of garnet schists, marble and amphibolites grading westwards to less

varied rocks of lower grade, the Moe Island Series, which closely
resemble the Cape Valentine Series of Elephant Island. The ages of
nine mica concentrates from schists of the Basement Complex have
been determined by the potassium-argon method and are grouped in
the range 176 to 199 m.y., that is, Late Triassic to Mid Jurassic
(MILLER, 1960). They are interpreted as indicating the date of folding
of the succeeding Greywacke-Shale Series. The metamorphic complex is followed by the Greywacke-Shale Series containing obscure
plant fragments that were originally identified as an Ordovician
graptolite. The rocks resemble the Upper Paleozoic Trinity Peninsula Series, and the Sandebugten beds of South Georgia. Fold axes
in the Basement Complex trend north, and in the Greywacke-Shale
Series trend N.N.W. Above the greywackes and metamorphics there
is an almost flat-lying conglomerate more than 500 m thick which
contains unidentified plants. Locally at the base of the conglomerate
there is a black shale containing minute lamellibranchs and gastropods that are probably Cretaceous. Furthermore the conglomerate
contains rare fragments of calcareous grits in which there are brachiopods, lamellibranchs, belemnites and plants, all Mesozoic and
probably Jurassic in age. These fragments, along with pieces of
rhyolite and felsite (?Jurassic) came from a near source, called the
Derived Series, which has not been found *in situ* and is probably
beneath the sea. Upper Jurassic dolerite dykes and Eocene conglomerates and shales have been reported (ADIE, 1962).

The South Sandwich Islands consist of an andesite-dacite association (BAECKSTROM, 1915) and at least five volcanoes are still
active. The volcanics are apparently underlain by Lower Paleozoic
dolomitic mudstones and Precambrian schist (ADIE, 1962, Table 4;
HOLDGATE, 1962; GASS et al., 1963).

The Falkland Islands (fig. 6) are not part of the Scotia Arc, but
they are important because they consist mainly of a sequence of
Devonian to Triassic rocks (HALLÉ, 1912) showing a most remarkable similarity to the succession in South Africa on the opposite side
of the South Atlantic, as shown in Table III (ADIE, 1952a, 1952c,
1962). The resemblance was one of the important factors in the
enthusiastic adoption of the continental drift hypothesis by DU
TOIT (1937).

PAN-ANTARCTIC ISLANDS

Several of the islands rising from the mid-ocean ridges lying
between Antarctica and other southern lands (fig. 1) are basaltic
volcanoes, but the Kerguelen group, Heard Island and Macquarie
Island include non-volcanic rocks, and provide a valuable but
limited amount of paleontological information.

The Kerguelen Islands, linked to Antarctica by the Kerguelen-

Gaussberg Ridge consist of Tertiary basalt volcanoes (DE LA RUE, 1932) but locally interbedded with the volcanics there are lignites and large fossil tree-trunks. J. D. HOOKER was fascinated by the occurrence of fossil trees in a glacierized and extremely isolated island group. His interest infected Charles DARWIN, and was the origin of subsequent speculation about the role of Antarctica as a centre of dispersal for the floras and faunas of southern lands. It seems that the trees grew in place, and did not float to the island from more northern latitudes. EDWARDS, and SEWARD, have iden- tified the coniferous woods as belonging to *Araucarioxylon* and *Cupressinoxylon* which probably belongs to the Podocarpaceae (FLORIN, 1940). From the lignite COOKSON (1947) extracted a typical southern hemisphere pollen and spore assemblage. FLETCHER (1938) described Tertiary marine fossils. (See also WACE, chapter VI).

Heard Island (LAMBETH, 1952; LAW & BURSTALL, 1953) also rises from the Gaussberg-Kerguelen Ridge, and is of particular interest because of the occurrence of the Laurens Peninsula Limesto- nes at its northern end. About 100 m of these beds are exposed in a syncline with an easterly-trending axis. The sediments are thinly- bedded, fine-grained, even-textured pelagic limestones of white, grey, blue and brown colours, interbedded with thin soft tuffaceous shales, and contain abundant foraminifera indicating a Paleogene (Lower Tertiary) age. More work is needed on the microfossils of this formation which has been re-examined recently by P. J. STEPHEN- SON. It is overlain unconformably by the thick and more extensive Drygalski Agglomerate containing the scallop *Chlamys (Zygochla- mys) heardensis* (FLEMING, 1957) which indicates a Neogene (Upper Cenozoic) age, probably post-Miocene, perhaps Pliocene. *Zygochla- mys* is a sub-genus of wide distribution on post-Miocene Antarctic and Subantarctic shelves. The greater part of Heard Island consists of Quaternary volcanics of an alkaline olivine-basalt association building up the high central volcano, Big Ben.

Macquarie Island is only a little closer to Tasmania and New Zealand than it is to Antarctica, and is 650 km from Auckland and Campbell Islands. It justifies special attention because of the possi- bility that it is the only visible part of a disrupted submarine ridge or arc that might have provided a Cenozoic link between Antarctica and New Zealand (see later). The island has been described geologi- cally by MAWSON (1943), IVANAC (1948) and LAW & BURSTALL (1956), and there are three significant groups of rocks. The oldest consists of a thick sequence of basalt flows and sills, and tuffaceous greywacke, intruded by a layered gabbroic and ultrabasic complex. This association could be of any pre-Tertiary age, though on insuffi- cient evidence MAWSON compared the intrusives with the Upper Paleozoic Dun Mountain ultramafic belt in New Zealand (GRINDLEY et al., 1959). The second group of rocks consists of submarine basic pillow lavas, dykes, breccias and agglomerates, with globigerina ooze filling interstices between the pillows. These were compared with lavas at Oamaru, New Zealand, now known to be of Eocene age, but the lithological comparison is of no value, and paleontological evidence is needed. Moraines and erratics form the third significant group of rocks. MAWSON, using the observations of BLAKE, stated that a line of harzburgite erratics derived from the layered ultra-basic intrusion on the west coast extends right across the island to the east coast. This was taken as evidence that the island had been over-ridden by a Quaternary ice-sheet moving from the west, that is, from the sea. This postulate is so unusual that it has been doubted, but it is accepted by IVANAC (1948), who was the only experienced geologist who had seen the evidence until a recent re-examination by an ANARE party (report in preparation). If the suggestion is correct it seems to require the existence of a landmass to the west, or at least a shallow marine bench, a possibility that is not entirely im-probable, for Macquarie Island appears to be a large fault-block, and is in a seismically active area. GUTENBERG & RICHTER (1941, p. 39) plotted seismic epicentres in this region, and claimed that they outline a "Macquarie Island Loop". As shown on their map the loop is convex to the west, that is, towards the Indian Ocean (this is peculiar) and is comparable in size with the Scotia Arc (fig. 6). Bathymetric surveys have been started by Australian and New Zealand oceanographers.

Campbell Island, Auckland Island, the Antipodes, Bounty and Chatham Islands can be regarded geologically as an extension of New Zealand, perhaps via the "Chathams Arc" (see later) since they lie on the Campbell Plateau and contain a varied assemblage of rocks, including Paleozoic schist and granite, and fossiliferous marine and terrestrial Cenozoic sediments and volcanics like those of New Zealand (OLIVER et al., 1950; FLEMING, 1959; ADAMS, 1962).

PALEOGEOGRAPHY

Some readers of this volume will be interested in the geology of the continent insofar as it can provide information about the paleobio-geography. They will want to know the extent to which Antarctica could have functioned as a migration route, as a dispersal centre, and as a home for an endemic biota. The available paleontological data have been given earlier and are summarized in Table IV. Part of the evidence and some of the inferences from it have been consid-ered in the description of Quaternary sediments.

Certain things are plain. East Antarctica has a paleontological record from the Cambrian to the Jurassic, and West Antarctica from the Paleozoic to the Quaternary. Considering the small percentage of exposed rock, and the incomplete geological work, the paleonto-

58

logical record is surprisingly full, and surprisingly "normal". Antarctica seems to have had a normal biota of cold-temperate and temperate type for nearly the whole of Phanerozoic (Cambrian and later) time. The Buckeye, Pagoda and Darwin tillites of the Beacon System provide evidence of an East Antarctic glaciation that was approximately contemporaneous with the late Paleozoic continental glaciations that affected Peninsular India and all the southern hemisphere continents. The Mawson Tillite is possibly Jurassic, but could be a product of a later Cenozoic montane glaciation. There is no other indication of cold climates in the Phanerozoic until the commencement of the present glacial conditions in the Late Cenozoic, perhaps as early as the Miocene but more probably in the Pliocene. Since the days of Hooker and Darwin biogeographers have wanted an Antarctica with 'normal' vegetation and a 'normal' history and the geological evidence is that Antarctica has been such a continent except for the comparatively short time since the Late Cenozoic or Early Quaternary (Table II). Dispersal both ways between West Antarctica and the Andes via the Scotia Arc seems to pose no serious difficulties for Mesozoic and Cenozoic time, but when asked to indicate other dispersal routes the geologist is in difficulties. There is no obvious permanent or intermittent connection with other lands, except for a conjectural link with Campbell Island and New Zealand, via the hypothetical Macquarie Island Loop (see under Pan-Antarctic Islands). Volcanic islands like those rising from the present mid-ocean ridges could have provided some stepping-stones to and from other southern lands, but they are very widely-spaced.

There are also other problems that cannot be answered by the orthodox geological theory of land-bridges and consequently one has to consider some of the unorthodox theories. For example, the occurrence of "normal" climates and "normal" fossil biotas in a continent at the South Pole is basically abnormal. Similarly how does one explain the occurrence of Cretaceous and Lower Tertiary floras dominated by southern conifers and other plants that are mainly evergreen, such as *Nothofagus*, in regions of such high latitude that there are several months of darkness and semi-darkness each year? How does one explain the occurrence of Permian gymnospermous wood with well-developed annual growth rings, or of Devonian brachiopods and Cambrian Archaeocyathinae only 5° from the South Pole? Answers are provided by the unorthodox theories of continental drift and polar wandering, but these are speculative theories that have as many geological opponents as they have supporters (Du Toit, 1937; Mayr, 1952; Carey, 1958). It is only too easy to select evidence in favour of these theories, and to overlook unfavourable evidence. The serious student who wants to make an independent assessment of the southern hemisphere geological evidence can, however, be referred to the regional geologies provided by CHILDS & BEEBE (1963), DAVID & BROWNE (1950), KRISHNAN (1956), VOISEY (1959), DU TOIT & HAUGHTON (1954), GRINDLEY, HARRINGTON & WOOD (1959), FLEMING (1962) and JENKS (1956), supplemented by GIGNOUX (1955), KING (1958), TEICHERT (1952) and MAYR (1952). That an assessment is not easy can be realised quickly by comparing the papers by CASTER and DUNBAR in MAYR (1952).

In the writer's opinion, however, the southern hemisphere geological evidence for some form of combined continental drift and polar wandering is continually becoming stronger, and perhaps inescapable. One of the strongest arguments seems to be provided by the paleomagnetic evidence, and in Antarctica, by paleomagnetic studies on the Ferrar Dolerites (BULL & IRVING, 1960; BRIDEN & OLIVER, 1963). Supporting evidence is provided by the astonishing lithological and paleontological similarity between the Beacon System and the Devonian to Mesozoic sequences in the Falkland Islands and southern Africa (Table III) and the eastern parts of South America. These similarities have become stronger with each season's field work in recent years, and have had a profound effect on some Australasian geologists who have been traditionally very sceptical of continental-drift theories, but have had to contemplate them after seeing the Antarctic sequences. The similarities between some southern shield areas are literally astonishing and without parallel in either the northern hemisphere or in the southern mobile geosynclinal belts such as New Zealand and the Andes. The similarities have been mentioned in this review and have been emphasized by PLUMSTEAD (1962) as a result of her study of the floras, and by GRINDLEY (1962) and ADIE (1962). It must be noted, however, that the Devonian to Permian sequences in the Beacon System in Antarctica and the Tasman Geosyncline of eastern Australia are both known to the writer, and the similarities in this case are only broad and general. There is a closer resemblance between the Ross Geosyncline and the Adelaide Geosyncline of South Australia, and between the *Upper* Beacon System and the post-Devonian rocks of Tasmania. The strong similarities seem to be between East Antarctica, southern Africa, the Falkland Islands and eastern South America on the one hand, and between Australia and India on the other hand.

Since orthodox geology, and even the usual schemes of continental drift and polar wandering do not provide simple and straight-forward answers to major paleogeographic problems, perhaps the writer can end this review by proposing a frankly speculative modification of both drift theories and land-bridge theories, which have been regarded as mutually exclusive.

A glance at fig. 4 is sufficient to suggest that East Antarctica

Fig. 7. Paleogeography. The Circum-Pacific zone of mobile belts and the southern lands with Gondwana affinities. For oceanic Islands see fig. 1.

could have drifted into West Antarctica, the line of "collision" being at the foot of the Trans-Antarctic Escarpment or the Byrd Subglacial Basin (fig. 2). East Antarctica could be regarded as a portion of a former Gondwanaland, but West Antarctica could be regarded as a segment of a Circum-Pacific system of mobile belts (fig. 7) or "island arcs" that have remained relatively fixed in position throughout their history, which ranges back to the Precambrian. Possibly they had a comparatively smooth great-circle outline until they were "hit" by East Antarctica and the South American portions of Gondwanaland, at some time between Cretaceous and mid-Tertiary. After the hypothetical collisions, the mobile belts could have yielded plastically, being moulded to the shapes of the leading edges of the moving segments of Gondwanaland, and being bent to form the Scotia Arc. Between the Ross Sea and the Campbell Plateau the zone of mobile belts was possibly disrupted, perhaps by stretching or by boudinage, so that a former direct connection was broken. This region in any case is now crossed by the mid-ocean ridge, from which rise the basalt volcanoes of Scott Island and the Balleny Islands. The existence of a seismic arc in the vicinity of Macquarie Island has been mentioned earlier under Pan-Antarctic Islands. There is no space here to discuss the extensive geological literature on the other islands to the south and east of New Zealand but it has been summarised by FLEMING (1959) under the names of the individual islands (and see also OLIVER et al., 1950; FLEMING, 1962; and ADAMS, 1962). It can be mentioned that the possibility of the existence of a Chathams Arc (fig. 7) linking the islands did not originate with the writer. It has been considered for many years by a few New Zealand geologists, though the idea in its various forms has not been published. Indicating an arc on fig. 7 serves to show that schists and other metamorphics, granitic rocks, and thick fossiliferous clastic sediments do occur in the islands rising from the Campbell Plateau and are similar to those in New Zealand, but it is not intended to suggest that the arc is a proved geological feature.

The Alpine Fault in New Zealand has a strike-separation of 480 km which the writer considers is a proved strike-slip displacement. This large movement must be considered when attempting to decide whether continental drift is possible. If the New Zealand Recurved Arc and the Chathams Arc are regarded as folds on the megascale then the Alpine Fault and its complex of splay-faults could be regarded as a zone of shearing in one septum (axial-surface) of the New Zealand Recurved Arc. It is then possible to suggest that shear-folding on a megascale provides a genetic link between the arcs, the faulting, and the Hikurangi, Kermadec and Tonga submarine trenches.

It is still quite speculative also to suggest that the circum-Pacific mobile belt was once continuous between Antarctica and Macquarie Island, and that the other southern lands of Gondwana affinities were once much closer together or even joined in one supercontinent which was not in contact with the mobile belt. Nevertheless that speculative hypothesis, involving both land-bridges and continental drifting, is the best that the writer can offer to biogeographers. It allows movement of the biota in both directions along islands of a mobile-belt system extending from the Andean cordillera through West Antarctica to Campbell Island and New Zealand during the Paleozoic, Mesozoic and Lower Tertiary, and perhaps the Miocene. It also allows southern Africa, the eastern part of South America, the Falkland Islands and Madagascar, and also Peninsular

India and Australia, to have been linked until the late Jurassic or Cretaceous in a supercontinent possibly having no direct land-bridge connection with the northern hemisphere landmass, and probably none with the circum-Pacific mobile belts. A paleogeography of that kind allows solutions to many otherwise inexplicable features in the distribution of the fossil and living biota of the southern hemisphere. It also means that the western cordillera of South America (and possibly North America) should be separated by a major discontinuity from the cratonic continental region to the east, but, so far as the writer is aware, a discontinuity of that kind has not been proposed and has not been reported in South America or North America.

ACKNOWLEDGEMENTS

The writer is indebted to many geologists for assistance in the form of discussion, or correspondence, or exchange of publications, including R. J. ADIE, E. E. ANGINO, C. CRADDOCK, B. M. GUNN, W. HAMILTON, W. E. LONG, I. R. McLEOD, R. J. OLIVER, T. PÉWÉ, P. J. STEPHENSON, C. SWINTHINBANK, O. S. VIALOV, and geological companions on two Antarctic expeditions, A. C. BECK, C. G. JOHNSON, G. J. LENSEN, I. C. McKELLAR, I. G. SPEDEN, and B. L. WOOD. Several read a draft of the manuscript in 1962 and special thanks are tendered to them. I am also indebted to my colleague B. C. McKELVEY for frequent discussions, and to L. M. CRANWELL for correspondence and paleobotanical information. A review of Antarctic geology by FAIRBRIDGE (1952) has also been of great assistance over a period of years.

BIBLIOGRAPHY

ADAMS, R. D., 1962. Thickness of the Earth's Crust Beneath the Campbell Plateau. *N.Z. J. Geol. Geophys. 5* (1), *74—85*.

ADIE, R. J., 1952a. Representatives of the Gondwana System in the Falkland Islands. *XIX Int. geol. Congr. Algiers, Symposium sur les Séries de Gondwana, 385—392*.

ADIE, R. J., 1952b. Representatives of the Gondwana System in Antarctica. *Ibid., 393—399*.

ADIE, R. J., 1952c. The Position of the Falkland Islands in a Reconstruction of Gondwanaland. *Geol. Mag. 89* (6), *401—410*.

ADIE, R. J., 1954. The Petrology of Graham Land, I, The Basement Complex and Early Paleozoic Plutonic and Volcanic Rocks. *Falk. Is. Dep. Surv. sci. Rep. 11*.

ADIE, R. J., 1955. The Petrology of Graham Land, II, The Andean Granite-Gabbro Intrusive Suite. *Ibid. 12*.

ADIE, R. J., 1957a. Geological Research in Graham Land. *Advanc. Sci. 53, 454—460*.

ADIE, R. J., 1957b. Geological Investigations in the Falkland Islands Dependencies Before 1940. *Polar Rec. 8* (57) *502—513*.

ADIE, R. J., 1957c. The Petrology of Graham Land, III, Metamorphic Rocks of the Trinity Peninsula Series. *Falk. Is. Dep. Surv. sci. Rep. 20*.

ADIE, R. J., 1958. Geological Investigation in the Falkland Islands Dependencies Since 1940. *Polar Rec. 9* (58), *3—17*.

ADIE, R. J., 1962. The Geology of Antarctica. *Amer. geophys. Un., geophys. Mon. 7, 26—39*.

ALLEN, A. D., 1962. Formations of the Beacon Group in the Victoria Valley Region. *N.Z. J. Geol. Geophys. 5, 278—294*.

ALLEN, A. D. & GIBSON, G. W., 1962. Outline of the Geology of the Victoria Valley Region. *N.Z. J. Geol. Geophys.* **5**, *234—242.*

ANDERSON, V., 1960. Geographic Features Observed on the Marie Byrd Land Traverse 1957—1958. *Amer. geog. Soc. IGY glac. Rep.* **4**, *143—150.*

ANDERSSON, J. G., 1906. On the Geology of Graham Land. *Geol. Inst. Upsala Bull.* **7**, *19—71.*

ANGINO, E. E., 1963. Antarctic Orogenic Belts as Delineated by Absolute Age Dates. *Polar Rec.* **11** (75), *780* (Abstract).

ANGINO, E. E. & OWEN, D. E., 1962. Sedimentologic Study of Two Members of the Beacon Formation, Windy Gully, Victoria Land, Antarctica. *Kansas Acad. Sci. Trans.* **65**, *61—69.*

ANGINO, E. E., TURNER, M. D. & ZELLER, E. J., 1962. Reconnaissance Geology of Lower Taylor Valley, Victoria Land, Antarctica. *Geol. Soc. Amer. Bull.* **73**, *1553—1562.*

ARCTOWSKI, H., 1895. Observations sur l'intérêt que présente l'exploration géologique des Terres Australis. *Soc. géol. France Bull.* **23**, Ser. 3, *589—591.*

AUGHENBAUGH, N. B., 1961. Preliminary Report on the Geology of the Dufek Massif. *Amer. geog. Soc. IGY Glac. Rep.* **4**, *155—193.*

BAECKSTROM, O., 1915. Petrographische Beschreibung einiger Basalte von Patagonien, Westantarktika und den Süd-Sandwich-Inseln. *Bull. geol. Inst. Univ. Upsala* **13** (1), *115—182.*

BALL, H. W., 1960. Upper Cretaceous Decapoda and Serpulidae from James Ross Island, Graham Land. *Falk. Is. Dep. Surv. sci. Rep.* **24**.

BARTON, C. M., 1963. The Significance of Two Separate Tertiary Plant Assemblages from King George Island, South Shetland Islands. *Polar Rec.* **11** (75), *784—785* (Abstract).

BEHRENDT, J. C. & PARKS, P. E., 1962. Antarctic Peninsula Traverse. *Science* **137** (3530), *601—603.*

BENTLEY, C. R., 1962. Glacial and Subglacial Geography of Antarctica. *Amer. Geophys. Un., geophys. Mon.* **7**, *11—25.*

BENTLEY, C. R., CRARY, A. P., OSTENSO, N. A. & THIEL, E. C., 1960. Structure of West Antarctica. *Science* **131**, *131—136.*

BENTLEY, C. R. & OSTENSO, N. A., 1961. Glacial and Sub-Glacial Topography of West Antarctica. *J. Glaciol.* **3**, *882—911.*

BLANK, H. R., COOPER, R. A., WHEELER, R. H. & WILLIS, I. A. G., 1963. Geology of the Koettlitz-Blue Glacier Region. Southern Victoria Land, Antarctica. *Trans. roy. Soc. N.Z. Geol.* **2** (5), *79—100.*

BOUCOT, A. J., CASTER, K. E., IVES, D. & TALENT, J. A., 1963. Relationships of a New Lower Devonian Terebratuloid (Brachiopoda) from Antarctica. *Bull. Amer. Pal.* **46** (207), *81—151.*

BRIDEN, J. C. & OLIVER, R. L., 1963. Paleomagnetic Results from the Beardmore Glacier Region, Antarctica. *N.Z. J. Geol. Geophys.* **6** (3), *388—394.*

BRUNN, V. VON, 1963. Note on Some Basic Rocks in Western Dronning Maud Land. *Polar Rec.* **11** (75), *770* (Abstract).

BUCKMAN, S. S., Antarctic Fossil Brachiopoda Collected by the Swedish South Polar Expedition, 1901—03. *Wiss. Ergeb. Schwed. Südpolar Exped. 1901—03,* **3**, *1—43.*

BUGAYEV, V. A. & TOLSTIKOV, E. I., 1960. Principal Features of the Relief of East Antarctica. *Sov. Antarc. Exped. Inform. Bull.,* *11—15.*

BULL, C. & IRVING, E., 1960. The palaeomagnetism of Some Hypabyssal Intrusive Rocks from South Victoria Land, Antarctica. *Roy. Astron. Soc. Geophys. J.* **3** (2), *211—224.*

BULL, C., McKELVEY, B. C. & WEBB, P. N., 1962. Quaternary Glaciations in Southern Victoria Land, Antarctica. *J. Glaciol.* **4**, *63—78.*

CAMERON, R. L., GOLDICH, S. S. & HOFFMAN, J. H., 1960. Radioactivity of Rocks from the Windmill Islands, Budd Coast, Antarctica. *Acta Univ. Stockholmiensis, Contr. in Geology* **6**, *1—6*.

CAMPBELL, K. S. W., 1962. Marine Fossils from the Carboniferous Glacial Rocks of New South Wales. *J. Paleontol.* **36**, *38—52*.

CAREY, S. W., Editor, 1958. Continental Drift, A Symposium. *Univ. Tasmania Geol. Dept. Publ.*

CHILDS, O. E. & BEEBE, B. W., Editors, 1963. Backbone of the Americas-Tectonic History from Pole to Pole. *Amer. Assoc. Petrol. Geol. Mem. 2.*

COOKSON, I. C., 1947. Plant Microfossils from the Lignites of the Kerguelen Archipelago. *Brit.-Aust.-N.Z. Antarc. Res. Exped. Rep. Ser. A.* **2.**

COX, L. R., 1953. Lower Cretaceous Gastropoda, Lamellibranchia, and Annelida from Alexander I Land. *Falk. Is. Dep. Surv. sci. Rep.* **4.**

CRADDOCK, C., ANDERSON, J. J. & WEBERS, G. F., 1963. Geological Outline of the Ellsworth Mountains. *Polar Rec.* **11** (75), *755—756* (Abstract).

CRADDOCK, C., BASTIEN, T. W. & RUTFORD, R. H., 1963. Geology of the Jones Mountains Area. *Polar Rec.* **11** (75), *756* (Abstract).

CRADDOCK, C. & HUBBARD, H. A., 1961. Preliminary Geologic Report on the 1960 U.S. Expedition to the Bellingshausen Sea, Antarctica. *Science* **133**, *886—887*.

CRANWELL, L. M., 1959. Fossil Pollen from Seymour Island, Antarctica. *Nature, Lond.* **184**, *700—702*.

CRANWELL, L. M., 1962. Antarctica: Cradle or Grave for its Nothofagus?. *Pollen et Spores* **4** (1), *190—192*. (see also *Proc. 10th Pac. Sci. Congr.*, Honolulu, Gressitt Symposium, in Press.)

CRANWELL, L. M., HARRINGTON, H. J. & SPEDEN, I. G., 1960. Lower Tertiary Microfossils from McMurdo Sound, Antarctica. *Nature, Lond.* **186**, *700—702*.

CRARY, A. P., 1961. Marine-sediment Thickness in the Eastern Ross Sea Area, Antarctica. *Geol. Soc. Amer. Bull.* **72**, *787—90*.

CRIDLAND, A. A., 1963. A Glossopteris Flora from the Ohio Range, Antarctica. *Amer. J. Bot.* **50** (2), *186—195*.

CROHN, P. W., 1959. A Contribution to the Geology and Glaciology of the Western Part of Australian Antarctic Territory. *Austral. Nat. Antarc. Res. Exped. Rep. Ser. A.* **3.**

DAVID, T. W. E. (& BROWNE, W. R. B.), 1950. "Geology of the Commonwealth of Australia". Arnold, London.

DAVID, T. W. E. & PRIESTLEY, R. E., 1914. Geology, Physiography, Stratigraphy and Tectonic Geology of South Victoria Land. *Brit. Antarc. Exped. 1907—9, Repts. sci. Inves., Geology,* **1**, *1—319*.

DAVIES, W. E., 1956. Antarctic Stratigraphy and Structure. *Amer. geophys. Un. Publ.* **462** *(geophys. Mon.* 1), *44—51*.

DEBENHAM, F., 1921. The Sandstone Etc. of the McMurdo Sound, Terra Nova Bay, and Beardmore Glacier Regions. *Brit. Antarc. ("Terra Nova") Exped. 1910, Geology,* **1** (4a), *103—119*.

DE LA RUE, E. A., 1932. Étude géologique et géographique de l'Archipel de Kerguelen. *Rev. Géogr. Phys. Géol. dynamique,* **5** (1 and 2), *1—224*.

DEUTSCH, S. & WEBB, P. N., 1963. Sr/Rb Dating on Basement Rocks from Victoria Land; Evidence for a 1000 Million Year Old Event. *Polar Rec.* **11** (75), *781* (Abstract).

DOUMANI, G. A., 1960. Geological Observations in West Antarctica During Recent Oversnow Traverses. *Amer. geophys. Un. Trans.* **41**, *IGY Bull.* **41**, *6—10*.

DOUMANI, G. A. & EHLERS, E. G., 1962. Petrography of Rocks from Mountains in Marie Byrd Land, West Antarctica. *Geol. Soc. Amer. Bull.* **73**, *877—882*.

65

DOUMANI, G. A. & LONG, W. E., 1962. The Ancient Life of the Antarctic. *Sci. Amer.* **207** (3), *169—184.*

DOUMANI, G. A. & TASCH, P. 1963. Leaiid Conchostracan Zone in Antarctica and its Gondwana Equivalents. *Science* **142** (3592), *591—592.*

DRAKE, A. A. JR., 1962. Preliminary Geologic Report on the 1961 U.S. Expedition to Bellingshausen Sea, Antarctica. *Science* **135**, *671—672.*

DUSEN, P., 1908. Über die Tertiäre Flora der Seymour-Insel. *Wiss. Erg. Schwed. Südpolarexped., 1901—3,* **3** (3).

DU TOIT, A. L., 1937. "Our Wandering Continents". Oliver & Boyd, Edin. & London.

DU TOIT, A. L. (& HAUGHTON, S. H.), 1954. "The Geology of South Africa". Oliver & Boyd, Edin. & London.

EVERNDEN, J. F. & RICHARDS, J. R., 1962. Potassium-Argon Ages in Eastern Australia. *Geol. Soc. Austral. J.* **9**, *1—49.*

FAIRBRIDGE, R. W., 1952. The Geology of the Antarctic, in "The Antarctic Today", N.Z. Antarc. Soc., & A. H. & A. W. Reed, Wellington, *56—101.*

FERRAR, H. T., 1907. Report on the Field Geology of the Region Explored during the 'Discovery' Antarctic Expedition 1901—4. *Nat. Antarc. Exped., nat. Hist.* **1**, *Geol., 1—100.*

FLEMING, C. A., 1957. A New Species of Fossil *Chlamys* from the Drygalski Agglomerate of Heard Island, Indian Ocean. *Geol. Soc. Austral. J.* **4**, *13—19.*

FLEMING, C. A., 1959. New Zealand. *Lexique strat. Int.* **6** (4).

FLEMING, C. A., 1962. New Zealand Biogeography: A Paleontologist's Approach. *Tuatara* **10**, *53—108.*

FLETCHER, H. O., 1938. Marine Tertiary Fossils and a Description of a Recent *Mytilus* from Kerguelen Island. *Brit.-Aust.-N.Z. Antarc. Res. Exped. Rep. Ser. A,* **2** (6), *101—116.*

FLORIN, R., 1940. The Tertiary Fossil Conifers of South Chile and Their Phytogeographical Significance, with a Review of Fossil Conifers in Southern Lands. *Kungl. Svensk. Vet. Handl., Ser. 3,* **19** (2), *1—107.*

FORD, A. B., 1963. Cordierite-bearing Hypersthene-Quartz-Monzonite-Porphyry in the Thiel Mountains. *Polar Rev.* **11** (75), *771* (Abstract).

FOURCADE, N. H., 1960. Estudio geológico-petrográfico de Caleta Potter, Isla 25 de Mayo, Islas Shetland del Sur. *Inst. Antart. Argentino Publ.* **8**.

GAIR, H. S., 1963. Summary of the Geology of the Polar Plateau Between the Rennick, Aviator and Campbell Glaciers, Northern Victoria Land. *Polar Rec.* **11** (75), *757* (Abstract).

GASS, I. G., HARRIS, P. G. & HOLDGATE, M. W., 1963. Pumice Eruption in the Area of the South Sandwich Islands. *Geol. Mag.* **100** (4): *321—330.*

GIGNOUX, M., 1955. "Stratigraphic Geology." W. H. Freeman Co., San Francisco.

GORDON, W. T., 1920. Cambrian Organic Remains from a Dredging in the Ross Sea. *Roy. Soc. Edinburgh Trans.* **52**, *681—714.*

GOTHAN, W., 1908. Die fossilen Hölzer von der Seymour- und Snow Hill-Insel. *Wiss. Erg. Schwed. Südpolar-Exped., 1901—03,* **3** (8).

GOULD, L. M., 1935. Structure of the Queen Maud Mountains, Antarctica. *Geol. Soc. Amer. Bull.* **46**, *973—984.*

GRINDLEY, G. W., 1963. The Geology of the Queen Alexandra Range, Beardmore Glacier, Ross Dependency, Antarctica; With Notes on the Correlation of Gondwana Sequences. *N.Z. J. Geol. Geophys.* **6** (3), *307—347.*

GRINDLEY, G. W., HARRINGTON, H. J. & WOOD, B. L., 1959. The Geological Map of New Zealand. *N.Z. geol. Surv. Bull. n.s.* **66**.

GRINDLEY, G. W., McGREGOR, V. R. & WALCOTT, R. I., 1963. Outline of the Geology of the Nimrod-Beardmore-Axel Heiberg Glaciers Region, Ross Dependency. *Polar Rec.* **11** (75), *757—758* (Abstract).

66

GUNN, B. M., 1962. Differentiation in Ferrar Dolerites, Antarctica. *N.Z. J. Geol. Geophys.* **5** (5), *820—863*.

GUNN, B. M., 1963a. Layered Intrusions in the Ferrar Dolerites, Antarctica. *Min. Soc. Amer. Spec. Pap.* **1**, *124—133*.

GUNN, B. M., 1963b. Geological Structure and Stratigraphic Correlation in Antarctica. *N.Z. J. Geol. Geophys.* **6** (3), *423—443*.

GUNN, B. M. & WALCOTT, R. I., 1962. The Geology of the Mt. Markham Region, Ross Dependency, Antarctica. *N.Z. J. Geol. Geophys.* **5**, *407—426*.

GUNN, B. M. & WARREN, G., 1962. Geology of Victoria Land Between the Mawson and Mulock Glaciers, Ross Dependency, Antarctica. *N.Z. geol. Surv. Bull. n.s.* **71**; also published as *Trans-Antarc. Exped. Sci. Rep.* **11**.

GUTENBERG, B. & RICHTER, C. F., 1941. Seismicity of the Earth. *Geol. Soc. Amer. Spec. Pap.* **34**.

HALLÉ, T. G., 1912. On the Geological Structure and History of the Falkland Islands. *Bull. geol. Inst. Univ. Upsala* **11**, *115—229*.

HALLÉ, T. G., 1913. The Mesozoic Flora of Graham Land. *Wiss. Erg. der Schwed. Südpolar-Exped. 1901—03*, **3** (14).

HALPERN, M., 1962. Potassium-Argon Dating of Plutonic Bodies in Palmer Peninsula and Southern Chile. *Science* **138** (3546), *1261—1262*.

HALPERN, M., 1963. Cretaceous Sedimentation in "General Bernado O' Higgins" Area of North-West Antarctic Peninsula. *Polar Rec.* **11** (75), *763—764* (Abstract).

HAMILTON, W., 1960. New Interpretation of Antarctic Tectonics. *U.S. geol. Surv. Prof. Pap.* **400—B**, *379—380*.

HAMILTON, W., 1962. Antarctic Tectonics and Continental Drift. *Soc. econ. Geologists Paleontologists Spec. Publ.* "Polar Wandering and Continental Drift", *74—93*.

HAMILTON, W., 1963. Tectonics of Antarctica. *Amer. Assoc. Petrol. Geol. Mem.* **2**, *4—15*.

HAMILTON, W. & HAYES, P. T., 1960. Geology of Taylor Glacier — Taylor Dry Valley Region, South Victoria Land, Antarctica. *U.S. geol. Surv. Prof. Pap.* **400—B**, *376—378*.

HAMILTON, W. & HAYES, P. T., 1963. Type Section of the Beacon Sandstone of Antarctica. *U.S. geol. Surv. Prof. Pap.* **456—A**.

HARRINGTON, H. J., 1958a. Nomenclature of Rock Units in the Ross Sea Region, Antarctica. *Nature, Lond.* **182**, *290*.

[HARRINGTON, H. J.], 1958b. Geological and Survey Work (in Ross Dependency). *Ann. Rep. N.Z. Dept. sci. ind. Res. for the Year Ended 31 March, 1958*, *14—16*.

HARRINGTON, H. J., 1963. Recent Exploration of Victoria Land North of Terra Nova Bay. *Geog. J.* **129** (1), *36—52*.

HARRINGTON, H. J. & MCKELLAR, I. C., 1958. A Radiocarbon Date for Penguin Colonization of Cape Hallett, Antarctica. *N.Z. J. Geol. Geophys.* **1**, *571—576*.

HARRINGTON, H. J. & SPEDEN, I. G., 1962. Section Through the Beacon Sandstone at Beacon Height West, Antarctica. *N.Z. J. Geol. Geophys.* **5** (5): *707—717*.

HARRINGTON, H. J., WOOD, B. L., MCKELLAR, I. C. & LENSEN, G. J., 1963. Tucker Glacier; Geological Map of Ross Dependency, Antarctica, 1 : 250,000 (with text on back). *N.Z. Dept. sci. ind. Res. Publ.*

HARRINGTON, H. J., WOOD, B. L., MCKELLAR, I. C. & LENSEN, G. J., in Press. Topography and Geology of the Cape Hallett District, Victoria Land, Antarctica. *N.Z. geol. Surv. Bull.*

HASKELL, T. R., KENNETT, J. P. & PREBBLE, W. M., 1963. Basement and Sedimentary Geology of the Darwin Glacier Area. *Polar Rec.* **11** (75), *765—766* (Abstract).

HAWKES, D. D., 1961a. The Geology of the South Shetland Islands I, The Petrology of King George Island. *Falk. Is. Dep. Surv. sci. Rep.* **26.**

HAWKES, D. D., 1961b. Idem, II, The Geology and Petrology of Deception Island. *Ibid.* **27.**

HEDBERG, H. D., 1961. Statement of Principles of Stratigraphic Classification and Terminology (by International Subcommission). *Rep. 21st. int. geol. Congr., Norden. Pt.* **25.**

HENNIG, A., 1911. Le Conglomérat Pleistocène à Pecten de l'Isle Cockburn. *Wiss. Erg. Schwed. Südpolar-Exped. 1901—03.* **3** (10), *1—72.*

HEURTEBIZE, G., 1952. Sur les formations géologiques de la Terre Adélie. *C.R. Acad. Sci. France* **234,** *1380—1382.*

HOLDGATE, M. W., 1962. Observations in the South Sandwich Islands. *Polar Rec.* **11** (73), *394—405.*

HOLLIN, J. T., 1962. On the Glacial History of Antarctica. *J. Glaciol.* **4** (32), *173—195.*

HOOPER, P., 1962. The Petrology of Anvers Island and Adjacent Islands. *Falk. Is. Dep. Surv. sci. Rep.* **34.**

HOUGH, J. L., 1950. Pleistocene Lithology of Antarctic Ocean-Bottom Sediments. *J. Geol.* **58,** *254—260.*

HOWARTH, M. K., 1958. Upper Jurassic and Cretaceous Ammonite Faunas of Alexander Land and Graham Land. *Falk. Is. Dep. Surv. sci. Rep.* **21.**

IVANAC, J. F., 1948. Geological Observations on Macquarie Island. *Austral. Bur. Min. Res. Geol. Geophys. Rep. 1948/39, Geol. Series* **15** (mimeographed.)

JENKS, W. F., 1956 (Editor). Handbook of South American Geology. *Geol. Soc. Amer. Mem.* **65.**

KAPITZA, A. P., 1960. New Data on Ice Cover Thickness of the Central Region of Antarctica. *Sov. Antarc. Exped. Inform. Bull.* **19,** *10—14.*

KING, L. C., 1958. Basic Palaeogeography of Gondwanaland during the Late Palaeozoic and Mesozoic Eras. *Quart. J. geol. Soc. London* **114** (I), *47—70.*

KIZAKI, K., 1962. Structural Geology and Petrology of East Ongul Island, East Antarctica, 1, Structural Geology. *Antarc. Rec. Tokyo,* **14,** *1147—1155.*

KLIMOV, L. B. & SOLOVYEV, D. S., 1958. Some Characteristics of the Geological Structure of Wilkes Land. *Doklady Akad. nauk SSSR* **123** (In Russian).

KLIMOV, L. B. & SOLOVYEV, D. S., 1960. Correlation of Geological Formations of the Shore of the Ross Sea and Oates Coast. *Sov. Antarc. Exped. Inform. Bull.* **16.** (In Russian)

KOROTKEVICH, E. S. & TIMOFEYEV, B. V., 1959. About the Age of East Antarctic Rocks. *Sov. Antarc. Exped. Inform. Bull.* **12,** *41—46.* (In Russian)

KRISHNAN, M. S., 1956. "Geology of India and Burma". Higginbothams, Madras, 555 pp.

LAIRD, M. G., 1963. Geomorphology and Stratigraphy of the Nimrod Glacier-Beaumont Bay Region, Southern Victoria Land, Antarctica. *N.Z. J. Geol. Geophys.* **6** (3), *465—484.*

LAIRD, M. & WATERHOUSE, J. B., 1962. Archaeocyathine Limestones of Antarctica. *Nature, Lond.* **194,** *861.*

LAMBETH, A. J., 1952. A Geological Account of Heard Island. *Roy. Soc. N.S.W. J. Proc.* **85,** *14.*

LAW, P., 1961. The Edge of the Antarctic Continental Shelf. *Polar Rec.* **10** (67), *415.*

LAW, P. G. & BURSTALL, T., 1953. Heard Island. *Austral. nat. Antarc. Res. Exped. Interim Rep.* **7.**

LAW, P. G. & BURSTALL, T., 1956. Macquarie Island. *Ibid* **14.**

LE COUTEUR, P. C. & LEITCH, E. C., 1963. Preliminary Report on the Geology of an Area South-west of the Upper Tucker Glacier, Northern Victoria Land. *Polar Rec.* 11 (75), *758* (Abstract).

LISITZIN, A. P., 1962. Bottom Sediments of the Antarctic. *Amer. geophys. Un., geophys. Mon.* 7, *81—88.*

LISITZIN, A. P. & ZHIVAGO, A. V., 1960. Marine Geological Work of the Soviet Antarctic Expedition, 1955—1957. *Deep-Sea Res.* 6 (2), *77—87.*

LONG, W. E., 1962. Sedimentary Rocks of the Buckeye Range, Horlick Mountains, Antarctica. *Science* 136, *319—321.*

LONG, W. E., 1963. The Stratigraphy of the Horlick Mountains. *Polar Rec.* 11 (75), *766* (Abstract).

MARPLES, B. J., 1953. Fossil Penguins from the Mid-Tertiary of Seymour Island. *Falk. Is. Dep. Surv. sci. Rep.* 5.

MATTHEWS, D. H., 1959. Aspects of the Geology of the Scotia Arc. *Geol. Mag.* 96, *425—441.*

MAWSON, D., 1940. Sedimentary Rocks. *Australas. Antarc. Exped. 1911—1914, sci. Rep. Ser. A,* 4, *Geol.* 2, *347—367.*

MAWSON, D., 1943. Macquarie Island, its Geography and Geology. *Ibid Ser. A,* 5.

MAWSON, D., 1950. Basaltic Lavas of the Balleny Islands. *Roy. Soc. South Austral. Trans.* 73, *223—231.*

MAYR, E., 1952 (Editor). The Problem of Land Connections Across the South Atlantic, with Special Reference to the Mesozoic. *Amer. Mus. nat. Hist. Bull.* 99, *art. 3, 79—258.*

McDOUGALL, I., 1963. Potassium-Argon Age Measurements on Dolerites from Antarctica and South Africa. *J. geophys. Res.* 68 (5), *1535—1545.*

McKELVEY, B. C. & WEBB, P. N., 1959. Geological Investigations in South Victoria Land, Antarctica, 2, Geology of the Upper Taylor Glacier Region. *N.Z. J. Geol. Geophys.* 2, *718—728.*

McKELVEY, B. C. & WEBB, P. N., 1961. Geological Reconnaissance in Victoria Land, Antarctica. *Nature, Lond.* 189, *545—547.*

McKELVEY, B. C. & WEBB, P. N., 1962. Geological Investigations in Southern Victoria Land, Antarctica, 3, Geology of Wright Valley. *N.Z. J. Geol. Geophys.* 5, *143—162.*

McLEOD, I. R., 1959. An Outline of the Geology of the Western Portion of Australian Antarctic Territory. Antarctic Symposium, Buenos Aires, November, 1959, Paper Presented, and *Austral. Bur. Min. Res. Records 1959/130;* (see also *Int. Un. Geodesy Geophys. Mon.* 5, 1960).

McLEOD, I. R., 1963. Geological Investigations along the Antarctic Coast between Longitudes 108°E and 160°E in 1960 and 1961. *Austral. Bur. Min. Res. Geol. Geophys. Rec. 1963/4.*

MILLER, J. A., 1960. Potassium-Argon Ages of Some Rocks from the South Atlantic. *Nature, Lond,* 187, *1019—1020.*

MULLIGAN, J. J., et al., 1963. Willett Range Coal Deposits, Mackay Glacier Area, Victoria Land, Antarctica. *U.S. Bur. Mines Rep. Invest.* 6331.

NEETHLING, D. C., 1963. The Geology of the "Zukkertoppen Nunataks", Ahlmannryggen, Western Dronning Maud Land. *Polar Rec.* 11 (75), *767—768* (Abstract).

NICOLAYSON, L. O., BURGER, A. J., TATSUMI, T. & AHRENS, L. H., 1961. Age Measurement on Pegmatites and a Basic Charnockite Lens Occurring near Lutzow-Holm Bay, Antarctica. *Geochim. Cosmochim. Acta* 22, *94—98.*

NORDENSKJÖLD, O., 1913. Antarktis. *Handbuch d. reg. Geol.* 8 (6).

OLIVER, R. L., 1963. Some Basement Rock Relations in Antarctica. *Polar Rec.* 11 (75), *759* (Abstract).

OLIVER, R. L., FINLAY, H. J. & FLEMING, C. A., 1950. The Geology of Campbell Island. *N.Z. Dep. sci. ind. Res., Cape Exped. Series Bull.* 3.

PAVLOV, V. V., 1958. Results of Palynological Analysis of Specimens from the Beacon System, (Horn Bluff). *Inst. Geol. Arctic & Antarc. Leningrad, Collected Papers (Sbornik) on Pal. & Biostrat.*, 12, 77—79 (in Russian).

PEARN, W. C., ANGINO, E. E. & STEWART, D., 1963. New Isotopic Age Measurements from the McMurdo Sound Area, Antarctica. *Nature, Lond.* 199 (4894), *685.*

PÉWÉ, T. L., 1960. Multiple Glaciation in the McMurdo Sound Region, Antarctica. *J. Geol.* 68, *498—514.*

PHILIPPI, E., 1906. Geologische Beschreibung des Gaussberges. *Deutsche Süd-Polar Exped. 1901—3, II, Kartographie und Geologie,* 2.

PICCIOTTO, E., 1961. Quelques résultats scientifiques de l'Expédition Antarctique Belge 1957—1958. *Univ. Libre de Bruxelles, Ciel et Terre* 4—5—6, *1—43.*

PICCIOTTO, E. & COPPEZ, A., 1963. Bibliographie des Mesures d'âges Absolus en Antarctique. *Ann. Soc. géol. Belg.* 85, *Bull.* 8, *B263—308.*

PICCIOTTO, E. MICHOT, J. & MICHOT, P., 1960. Reconnaissances géologiques et petrographiques de Monts Sor-Rondane (Terre de la Reine Maud). *Soc. Belge Géol. Paléontol. Hydrol. Bull.* 69 (2).

PLUMSTEAD, E. P., 1962. Fossil Floras of Antarctica. *Trans-Antarc. Exped. sci. Rep.* 9.

RAVICH, M. G. & VORONOV, P. S., 1958. Geological Structure of the Coast of East Antarctica Between 55° and 110° E. *Sovetskaya Geologiya* 2. (In Russian).

REINISCH, R., 1906. Petrographische Beschreibung der Gaussberg-Gesteine. *Deutsche Südpolar Exped. 1901—03, 2 Geog. und Geol.,* (1), *73—87.*

REINISCH, R., 1912. Erratische Gesteine (besonders aus Eisbergen). *Ibid.* 2 (7).

RICKER, J. F., 1963. Outline of the Geology Between the Mawson and Priestley Glaciers, Victoria Land. *Polar Rec.* 11 (75), *759* (Abstract).

ROOTS, E. F., 1953. Preliminary Note on the Geology of Western Dronning Maud Land. *Norsk. geol. Tids.* 32, *18—33.*

SAITO, N., TATSUMI, T. & SATO, K., 1961. Absolute Age of Euxenite from Antarctica. *Antarc. Rec. Tokyo,* 12, *1057—1062.*

SCHMIDT, P. G., 1962. Geological Reconnaissance of the Ellsworth Mountains. *U.S. Antarc. Proj. Officer Bull.* 3 (8), *17—20.*

SCHMIDT, D. L., FORD, A. B., DOVER, J. H. & BROWN, R. D., 1963. Preliminary Geology and Structure of the Patuxent Mountains. *Polar Rec.* 11 (75), *760* (Abstract).

SCHOPF, J. M., 1962. A Preliminary Report on Plant Remains and Coal of the Sedimentary Section in the Central Range of the Horlick Mountains, Antarctica. *Ohio State Univ. Res. Foundation, Inst. polar Studies Rep.* 2.

SEWARD, A. C., 1914. Antarctic Fossil Plants. *Brit. Antarc.* ("Terra Nova") *Exped. 1901, Brit. Mus. (Nat. Hist.), Geol.* 1 (1), *1—49.*

SEWARD, A. C., 1959. "Plant Life Through the Ages." Hafner Publishing Co., Inc. N.Y. (Facsimile copy of second edit. 1933, Camb. Univ. Press).

SHAW, S. E., 1962. Petrography of Beacon Sandstone Samples from Beacon Height West, Upper Taylor Glacier, Antarctica. *N.Z. J. Geol. Geophys.* 5 (5), *733—739.*

SIMPSON, G. G., 1946. Fossil Penguins. *Amer. Mus. nat. Hist. Bull.* 87, *5—99.*

SKINNER, D. N. B., 1963. A Summary of the Geology of the Region Between the Byrd and Starshot Glaciers, South Victoria Land. *Polar Rec.* 11 (75), *761* (Abstract).

SLOSS, L. L., 1963. Sequences in the Cratonic Interior of North America. *Geol. Soc. Amer. Bull.* 74. *93—114.*

SMITH, W. C., 1924. The Plutonic and Hypabyssal Rocks of South Victoria Land. *Brit. Antarc.* ("Terra Nova") *Exped. 1910, nat. Hist. Rep., Geol.* 1 (6), *167—227.*

SMITH, W. C., 1954. The Volcanic Rocks of the Ross Archipelago. *Brit. Antarc. ("Terra Nova") Exped. 1910, nat. Hist. Rep. Geol.* 2 (1), 1—107.

SOLOVYEV, D. S., 1959. Lower Paleozoic Metamorphic Schists of the Oates Coast. *IGY sci. Invest. Inst. Arctic. Geol. Min. Geol. Conserv. Resources USSR, 113, Sov. Antarc. Exped.* 147.

SOOT-RYEN, T., 1952. *Laternula elliptica* (King and Broderip 1831) from the Pecten-Conglomerate, Cockburn Island. *Arkiv. f. Zool., Ser.* 2, 4 (9), 163—4.

SPATH, L. F., 1953. The Upper Cretaceous Cephalopod Fauna of Graham Land. *Falk. Is. Dep. Surv. sci. Rep.* 3.

SPEDEN, I. G., 1960. Post-Glacial Terraces near Cape Chocolate, McMurdo Sound, Antarctica. *N.Z. J. Geol. Geophys.* 3 (2), 203—217.

SPEDEN, I. G., 1962. Fossiliferous Quaternary Marine Deposits in the Mc-Murdo Sound Region, Antarctica. *N.Z. J. Geol. Geophys.* 5 (5), 746—777.

STARIK, I. YE., KRYLOV, A. YA., RAVICH, M. G. & SILIN, YU. I., 1961. The Absolute Ages of East Antarctic Rocks. *Ann. New York Acad. Sci.* 91, art. 2, Geochronology of Rock Systems, 576—582.

STEPHENSON, P. J., 1959. Geology in the Weddell Sea Sector, Antarctica. Antarc. Symposium, Buenos Aires, November, 1959, Paper Presented.

STEPHENSON, P. J., 1960. Geology in the Weddell Sea Sector, Antarctica. *Int. Un. Geodesy Geophys. Mon.* 5, Antarc. Sympos. Buenos Aires, 80.

STEWART, D., 1934. The Petrography of the Beacon Sandstone of South Victoria Land. *Amer. Min.* 19, 351—359.

STEWART, D., 1956. On the Petrology of Antarctica. *Amer. Geophys. Un. Publ.* 462, *(geophys. Mon. 1), Antarctica in the International Geophysical Year*, 52—74.

STEWART, D., 1963. Petrography of Some Dredgings Collected by Operation Deep Freeze IV. *Proc. Amer. philos. Soc.* 107 (5), 431—442.

SWITHINBANK, C., 1959. The Morphology of the Inland Ice Sheet and Nunatak Areas of Western Dronning Maud Land. *Norweg.-Brit.-Swed. Antarc. Exped., 1949—52, sci. Results,* III D (Norsk Polarinst., Oslo).

TAYLOR, G., 1940. Antarctica. *Reg. Geol. der Erde,* 1, *Die Alten Kerne,* 8, Leipzig.

TEICHERT, C., 1952, (Editor). Symposium sur les Séries de Gondwana. *Rep. XIX internat. geol. Congr. Algiers.*

THIEL, E., 1961. Antarctica: One Continent or Two? *Polar Rec.* 10 (67), 335—348.

THIEL, E. C., 1962. The Amount of Ice on Planet Earth. *Amer. geophys. Un. Mon.* 7, 172—175.

TOWNROW, J. A., 1957. On *Dicroidium,* Probably a Pteridospermous Leaf, and other Leaves Now Removed from this Genus. *Trans. geol. Soc. S. Afr.* 60, 21—60.

TRAIL, D. S., 1963a. Low-Grade Metamorphic Rocks from the Prince Charles Mountains, East Antarctica. *Nature, Lond.* 197 (4867), 548—550.

TRAIL, D. S., 1963b. Schist and Granite in the Southern Prince Charles Mountains. *Polar Rec.* 11 (75), 775—776 (Abstract).

TREVES, S. B., 1963. Igneous and Metamorphic Geology of the Ohio Range, Horlick Mountains. *Polar Rec.* 11 (75), 776—777 (Abstract).

TYRRELL, G. W., 1945. Report on Rocks from West Antarctica and the Scotia Arc. *Discovery Rep., Cambridge,* 23, 37—102.

VIALOV, O. S., 1959. Tectonics and History of Development of Antarctica. *Rep. (Doklady) Acad. Sci. Ukr. Repub.* 8, 878—880. (In Ukrainian, Russian and English.)

VIALOV, O. S., 1962. Problematica of the Beacon Sandstone at Beacon Height West, Antarctica. *N.Z. J. Geol. Geophys.* 5 (5), 718—732.

VIALOV, O. S. & TIMOFEYEV, B. V., 1959. First Find of Ancient Spores in Antarctica. *Acad. Sci. Ukr. Republic Rep.* **10**, *1133—1135.* (In Ukrainian, Russian and English.)

VOISEY, A. H., 1959. Australian Geosynclines. *Aust. J. Sci.* **22**, *188—198.*

VORONOV, P. S., 1958. Structural Diagram of Antarctica. *Sov. Antarc. Exped. Inform. Bull.* **1**, *21—25.*

WADE, A., 1941. The Geology of the Antarctic Continent and its Relationship to Neighbouring Land Areas. *Roy. Soc. Queensland Proc.* **52**, *24—35.*

WADE, F. A., 1945. The Geology of the Rockefeller Mountains, King Edward VII Land, Antarctica. *Amer. philos. Soc. Proc.* **89** (1), *67—77.*

WALKER, P. T., 1961. Study of Some Rocks and Minerals from the Dufek Massif, Antarctica. *Amer. geog. Soc. IGY glac. Rep.* **4**, *195—213.*

WARNER, L. A., 1945. Structure and Petrography of the Southern Edsel Ford Ranges, Antarctica. *Amer. philos. Soc. Proc.* **89** (1), *78—122.*

WEBB, P. N., 1962. Isotope Dating of Antarctic Rocks, A Summary — I. *N.Z. J. Geol. Geophys.* **5** (5), *790—796.*

WEBB, P. N., 1963. Geological Investigations in Southern Victoria Land, Antarctica, Part 4, Beacon Group of the Wright Valley and Taylor Glacier Region. *N.Z. J. Geol. Geophys.* **6** (3), *361—387.*

WEBB, P. N. & MCKELVEY, B. C., 1959. Geological Investigations in South Victoria Land, Antarctica, Part I, Geology of Victoria Dry Valley. *N.Z. J. Geol. Geophys.* **2**, *120—136.*

WEIHAUPT, J. G., 1960. Reconnaissance of a Newly Discovered Area of Mountains in Antarctica. *J. Geol.* **68**, *669—673.*

WEIHAUPT, J. G., 1961. Geophysical Studies in Victoria Land, Antarctica. *Univ. Wisconsin geophys. and polar Res. Center Rep. Ser.* **1**.

WEXLER, H., 1961. Ice Budgets for Antarctica and Changes in Level. *J. Glaciol.* **3**, *867—872.*

WILCKENS, O., 1910. Die Anneliden, Bivalven, und Gastropoden der Antarktischen Kreide-Formation. *Wiss. Erg. der Schwed. Südpolar-Exped. 1901—03,* **3**, (12), *1—132.*

WILCKENS, O., 1912. Die Mollusken der Antarktischen Tertiärformation. *Ibid.* **3** (13), *1—42.*

WILCKENS, O., 1924. Die tertiäre Fauna der Cockburn-Insel (Westantarktika). *Further zool. Res. Schwed. Antarc. Exped. 1901—3,* **1** (5).

WIMAN, C., 1905. Über die Alttertiären Vertebraten der Seymourinsel. *Wiss. Erg. Schwed. Südpolar. Exped. 1901—03,* **1**, *1—37.*

WITHERS, T. H., 1951. Cretaceous and Eocene Peduncles of the Cirripede *Euscalpellum. Brit. Mus. (Nat. Hist.) Bull. Geol.* **1**, *149—162.*

WOODWARD, A. S., 1921. Fish Remains from the Upper Old Red Sandstone of Granite Harbour, Antarctica. *Brit. Antarc. ("Terra Nova") Exped. 1910, Nat. Hist. Rep. Geol.* **1** (2), *51—62.*

WOOLLARD, G. P., 1961. Some Significant Results and Comments on the IGY Inter-Disciplinary Program in Earth Sciences in Antarctica. *U.S. Nat. Acad. Sci., Nat. Res. Council Publ.* 878. *Science in Antarctica, II The Physical Sciences,* *109—113.*

ZELLER, E. J., ANGINO, E. E. & TURNER, M. D., 1961. Basal Sedimentary Section at Windy Gully, Taylor Glacier, Victoria Land, Antarctica. *Geol. Soc. Amer. Bull.* **72**, *781—786.*

ZEUNER, F. E., 1959. Jurassic Beetles from Grahamland, Antarctica. *Paleontology* **1**, *407—409.*

ZIRKEL, F. & REINISCH, R., 1905. Petrographie, I, Untersuchung des von Enderby Land gedreschten Gestein-Materials. *Wiss. Erg. Deutsche Tiefsee-Expedition auf dem Dampfer Valdivia 1898—1899,* **10**, *1—10.*

ANTARCTIC CLIMATOLOGY

BY

MORTON J. RUBIN
(with 7 figs.)

Astronomical, Geodetic and Topographic Factors

In regard to the climatology of the polar regions, the primary
control is that which is fixed by astronomical considerations, namely
the changing seasonal effects as the earth revolves around the sun.
However, a basic difference exists between the Antarctic and Arctic
regions in that the Southern Hemisphere — and Antarctica —
during the current phase of the 21,000-year period of cyclic variation
of the line of apsides is at perihelion during its summer season
(January) rather than at aphelion (July) as is the Northern
Hemisphere, and is about 5×10^6 km closer to the source of solar
heat at the time when the sun's rays are most direct. Thus, the top
of the atmosphere over the polar region of the Southern Hemisphere
receives approximately 7% more radiation at the summer solstice
than comparable latitudes in the Northern Hemisphere. Of course,
other fixed as well as variable factors such as topography, oceanic
distribution, surface character, etc. are influential in determining
the climate, once the solar radiation factor is fixed.

As a result of the earth's rotation around its axis and its inclina-
tion with respect to the plane of the ecliptic, the period of daylight
varies in the polar regions. The factor of the long summer days and
equally long polar nights is important in the climate and ecology
of the region. Figure 1 illustrates graphically the varying duration
of daylight, in hours, at latitudes from $50°$ to $90°$S throughout the
year. The twilight effect tends to increase the period of daylight
somewhat so that at the South Pole there is only a five-month
period of darkness, although the sun is below the horizon for the
full six months. Even on a mid-winter's day there is at least one
hour of twilight per day as far south as about $72°$S; on the Antarctic
Circle there is a period of about two hours of daylight plus an
additional two hours of twilight. Although the amount of direct
radiation available to warm the atmosphere and the earth surface
is negligible during twilight hours, the fact that there is some light
is of significance for outdoor activities.

A third factor in the Antarctic climate is its physiography. Antarctica is the highest continent on earth, with a mean elevation of
about 2000 meters. The mere fact of its elevation would cause it to be
about $12°$C colder, on the average, than the Arctic regions which

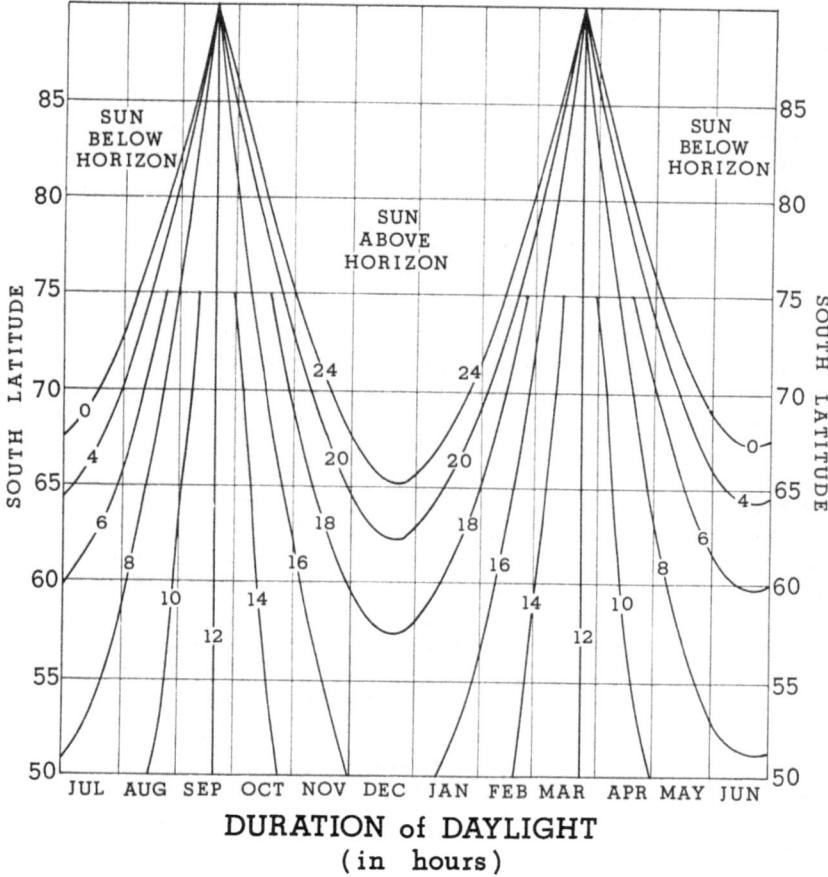

Fig. 1. Duration of daylight, in hours, for 50°—90°S latitude at sea level.

are at sea level. The fact that Antarctica, in effect, is a cold central core of continental mass making up about 10% of the non-oceanic area of the globe and more than 90% of the glaciated area, surrounded by a relatively warm ocean, gives its climate a distinctive character not shared by the Arctic. In addition, during the course of an average winter season the freezing of about 18×10^6 km^2 of the ocean more than doubles the area of the "effective continent", as compared with its minimum area of about 14×10^6/km^2 (OSTAPOFF, 1965). The resulting great decrease in the flux of heat from the ocean to the atmosphere and the removal, even, of appreciable sources of heat are of significance in the thermal regime and climate.

74

Atmospheric Composition and Constituents

The composition of the air over and around Antarctica in the lowest 100 km is not different from that of other regions of the world, except that there is very much less dust and other pollutants, and only about one-tenth the concentration of water vapor that exists in middle latitudes. Water vapor is the most important variable of the atmospheric constituents because of its strong absorption of outgoing long-wave terrestrial radiation. The amount of water vapor that can be held in the atmosphere is primarily a function of its temperature. As in other regions of the world, the highest concentration of water vapor is in the lowest layers of the atmosphere.

Another important constituent is carbon dioxide, the amount of which is thought to have experienced a world-wide increase during recent decades due to the combustion of fossil fuels. However, the role of the ocean as a carbon dioxide sink or reservoir, and the fixing of carbon dioxide as carbonates is not fully determined. Despite the lack of combustion, land vegetation and exposed rock in the Antarctic region, and, indeed, nearly everywhere poleward of about 40°S latitude in the Southern Hemisphere, the carbon dioxide concentration in Antarctica is about 310 to 315 parts per million — about the same as in other parts of the world (KEELING, 1960). As one would expect, there does not appear to be a seasonal variation in carbon dioxide content in Antarctica as there is in other parts of the world. The fact of the relatively uniform distribution of carbon dioxide on the earth, both in latitude and within the troposphere, attests to the efficiency of the atmosphere in dispersing and diffusing its constituents. This is of importance in the radiation climate of the earth, and Antarctica in particular, as carbon dioxide is relatively opaque to outgoing terrestrial radiation and thereby serves to retard the radiative cooling of the polar region, just as does water vapor.

Ozone, another variable constituent of the atmosphere, occurs mainly in the stratosphere where its maximum concentration over Antarctica occurs at about 15 to 20 km. In mid-latitudes of the Northern Hemisphere and in the Arctic regions the maximum concentration is at about 30 km. Ozone effectively absorbs and attenuates the ultraviolet radiation from the sun and is instrumental in the rise in temperature at the stratospheric levels from —85°C in late winter to —35°C in early spring, although advection of warm air and vertical motions play an important role in this phenomenon. Its effects at the earth's surface are not so great nor so dramatic, but they are of interest in their biological effects and as a tracer for atmospheric circulation. Surface ozone values are of the order of 45 micrograms per cubic meter of air, about 25% higher than in middle latitudes of the Northern Hemisphere (WEXLER et al.,

1960). Surface ozone values as well as total ozone values tend to be highest in the late winter season, when essentially no ozone is formed in the polar regions. This attests again to the efficiency of the atmospheric circulation in carrying ozone from far distant source regions. Recent balloon-borne ascents of ozone measuring instruments have confirmed that the atmospheric ozone values are high in the winter season. It seems most likely, then, that ozone is transported from lower latitudes, probably in the troposphere through the agency of the great cyclonic storms which form on the polar front and where there is strong mass exchange between the stratosphere and the troposphere. There is also the possibility that discharges of static electricity caused by the frequent and wide-spread storms of drifting and blowing snow contribute to the surface ozone amounts.

The incidence of several naturally formed isotopes is of interest in possibly tracing climatic conditions in the recent past. The ratio of oxygen-18 to oxygen-16 is a function of the temperature at which the water vapor was condensed; lower condensation temperatures result in a lower O^{18}/O^{16} ratio, which is helpful in establishing the annual cycle of snowfall amounts in the layers deposited on the continent. Through analysis of snow samples and density determinations the annual accumulation can be determined. There is still some question of technique and interpretation, which may be subjective, particularly if wind erosion has removed or depleted some of the annual layers.

Measurements of radioactivity in the air and in the snow have been made at several locations in Antarctica. The level of natural radioactivity is low, and even until the end of 1958 the fission products in the Antarctic atmosphere and precipitation averaged only about 0.12 disintegrations per minute per cubic meter, about 2% of the Northern Hemisphere concentration (LOCKHART, 1960; PICCIOTTO et al., 1962). The appearance of isotopes from known thermonuclear explosions in the air can be used to trace inter-hemispheric exchange of air masses and to date the formation of particular layers of snow at some later time. The fission products which enter the stratosphere seem to have their maximum fallout in the spring season. During 1958 there were several thermonuclear tests, the first as early as March. The monthly mean level of concentration of fission products in the air measured at the Roi Baudouin station (70°26'S, 24°19'E) had decreased from just under 2 disintegrations per minute per cubic meter in February 1958 to about 0.05 per minute in July 1958, and then rose to 0.2 per minute by January 1959. Recent observations in the Southern Hemisphere seem to indicate that the present level of radioactivity in the Southern Hemisphere is approximately equal to that of the Northern Hemisphere.

76

Solar and Terrestrial Thermal Radiation

Except for insignificant amounts of geothermal and thermal nuclear energy, the energy source for this planet is the sun. Measurements of the amount of heat transmitted upward at the bottom boundary of the sea are about 10^{-5}gcal/cm^2/min[1]. The annual total amount of solar energy that reaches the top of the earth's atmosphere is an inverse function of latitude (Fig. 2). This is not true as regards the three months of the summer season (Fig. 2), which is the only season in which solar radiation is of appreciable proportions; because of the increase of daylight hours with latitude, poleward of the Antarctic Circle, the summertime total solar energy

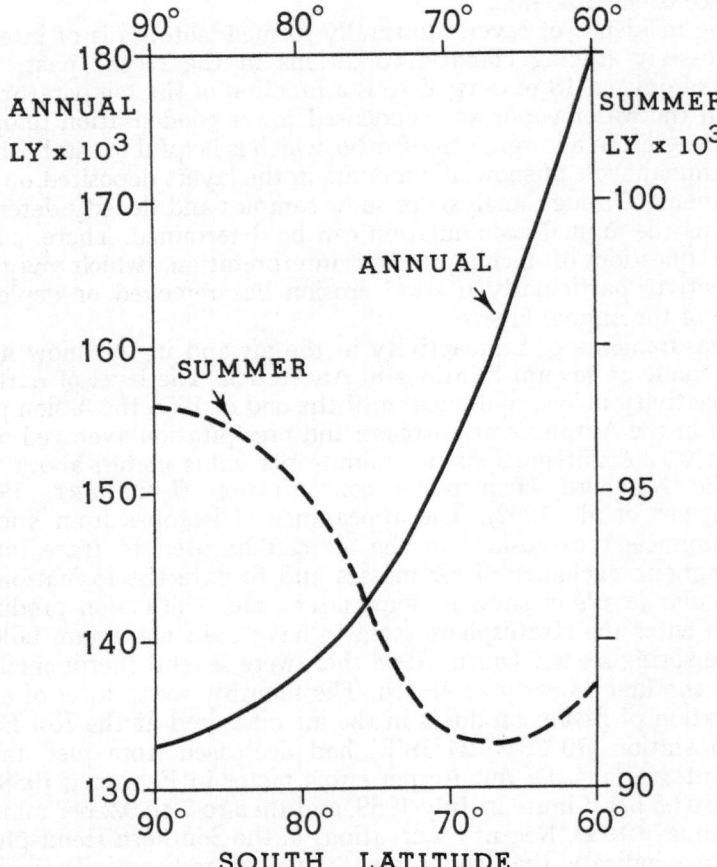

Fig. 2. Latitudinal variation of the total annual and summer season extraterrestrial radiation from 60°—90°S, in langleys (cal/cm²) × 10³.

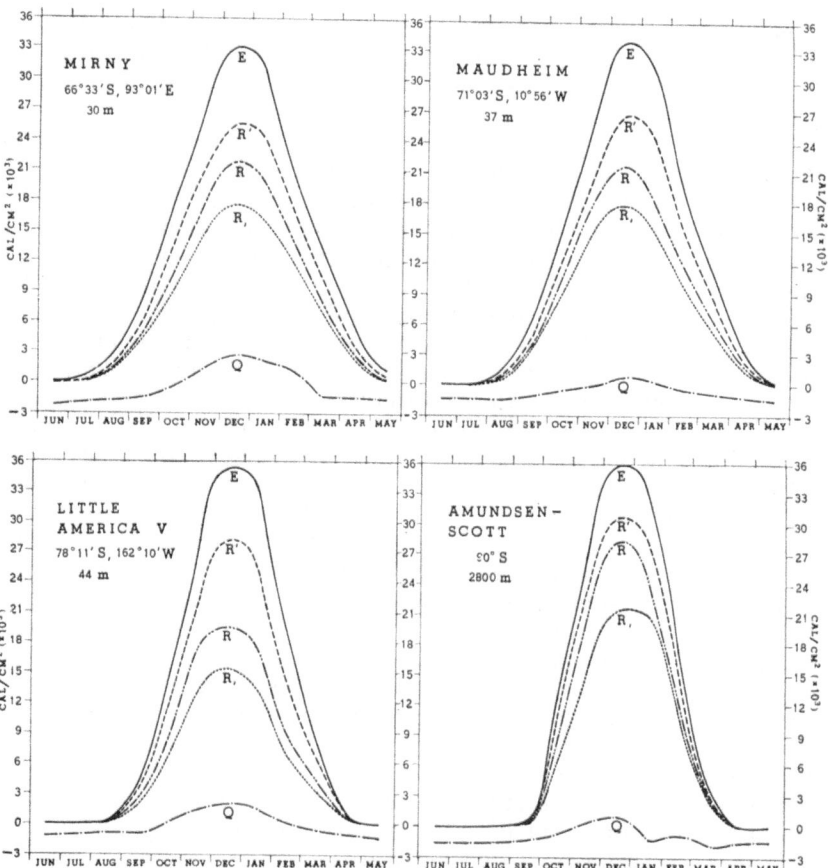

Fig. 3. Annual variation of radiation components at four Antarctic stations, in cal/cm² × 10³. E — total extraterrestrial radiation on a horizontal surface; R¹ — radiation on a horizontal plane at the earth's surface with clear sky; R — radiation on a horizontal plane at the earth's surface with average cloudiness; R_1 — radiation reflected from earth's surface (albedo); Q — net all-wave radiation at earth's surface.

reaching the top of the atmosphere also increases with latitude. At the South Pole about 73% of the annual total of extraterrestrial solar radiation is received during the summer season, while only 60% is received at 70°S. The solar constant is assumed to be approximately 2 gcal/cm²/min, on a surface normal to the sun's rays at the top of the atmosphere. Variations in the visible radiation, which constitutes about one-half of the energy reaching the earth, are currently considered to be minor. In the ultraviolet, however, variations are thought to be considerable. Figure 3 shows the radiation curves for four representative Antarctic stations. The incoming solar energy (E) is depleted through absorption, selective scattering by the atmosphere, reflection from clouds, absorption by atmospheric gases, etc., so that a considerably decreased amount of solar radiation reaches the earth's surface during both clear-sky conditions (R^1) and average cloudiness conditions (R), where it is most effective in raising the temperature of the earth and its atmosphere. The amount of solar radiation that reaches the earth's surface, of course, is primarily dependent upon the amount of cloudiness and the turbidity of the atmosphere. At Little America, for example, where the average cloudiness for December is 70%, the solar radiation that reaches the surface is 56% of the extraterrestrial; at the South Pole where the average cloudiness for December is 46%, the solar radiation is 79% of the extraterrestrial. On the basis of annual totals, the percentages are 56 and 73, respectively. However, much of the incident radiation is reflected by the surface (R_1) and is not effective in heating the surface, and thereby the atmosphere, through long-wave terrestrial radiation. The ratio of the reflected to the incident radiation is the albedo of the surface. The albedo is also a function of the wavelength of the incident radiation; the ultraviolet and shorter wavelengths in the visible portion of the spectrum are reflected more than the infrared and longer wavelength light. For instance, the albedo of fresh clean snow is 0.83 for radiation at 4000 Å units, but only 0.63 at 8400 Å units. At Little America the December value of the albedo is 79% while the annual value is 77. The South Pole albedo value is 76% for December and 81% for the year. The variation in albedo tends to be seasonal since the smoothness and crystal structure of the surface layer changes due to wind scouring, melting, recrystalization, etc. The albedo of a soil and rock surface averages only about 15 to 20%.

The snow which acts as an excellent reflector for short-wave solar radiation is likewise an excellent absorber and radiator of long-wave radiation. The transfer of heat between the atmosphere and snow is accomplished largely through the medium of long-wave radiation. Sublimation, melting and evaporation are of relatively minor importance in the heat budget of the snow surface. The radiation budget of the snow surface (Q) is generally negative throughout the year at all latitudes (Fig. 3), and only in summer is it positive for a few months; i.e. there is a net gain of heat by the snow surface through radiation. Since the annual mean temperatures of the top layers of snow do not vary much over several years and, thus, have a balanced heat budget, the radiation deficit has to be made up by other means. This is accomplished mainly through the turbulent exchange of sensible and latent heat. HANSON & RUBIN (1962) calculate that at the South Pole this exchange amounts to 8400 cal/cm² for the dark half-year and 4400 cal/cm² for the light half-year. It is the loss of heat through the long-wave radiation from the snow surface and the lowest layers of the atmosphere which results in the characteristic temperature inversion over the polar snow fields. Instead of the elsewhere normal decrease of temperature with height, the inversion is characterized by an increase of temperature with height, as much as 30 °C in the lowest 1000-meter thickness at the South Pole in mid-winter. At plateau stations the temperature inversion may average from 10° to 20 °C in 700 meters in winter, but only about 5 °C during the warm season.

Because of their low albedo, the radiation balance of rock and soil surfaces is significantly different from that of snow surfaces. For instance, for two locations at sea level near the Antarctic Circle (RUSIN, 1961), the station on ice has a negative radiation budget for eight months of the year, while the station on rock and soil has a radiation deficit for only five months. For the year, the station over snow may have a radiation budget deficit of 5000 cal/cm^2, while the station on rock or soil may have a surplus of more than 37,000 cal/cm^2, with an individual summer month contributing as much as 10,000 to 12,000 cal/cm^2. This has a great effect on the temperature of the surface and lowest few meters of the air, as will be shown later.

Sensible and Latent Heat

The Antarctic snow cover has a negative annual radiation budget and the Antarctic atmosphere loses heat by radiation from the top of the atmosphere to space. The fact that Antarctica over relatively long periods seems not to have become colder because of the radiative losses means that enough heat is brought into the region to approximately balance the annual heat budget of the snow and the atmosphere. At about the 10-meter level the temperature of the snow layer changes insignificantly during the course of the year. Since the upward flow of heat from the interior of the earth is infinitesimal, the temperature evidence thus points to a balanced budget with the exchange of heat occurring almost entirely between the atmosphere and the upper 10 meters of snow; this refers only to continental ice and not to ice which floats over the ocean. Also, there appears to be a long-term balance in the heat budget of the atmosphere as shown by the relatively minor year-to-year variations in the temperature at all levels in the troposphere and the stratosphere. A balance is achieved through the transport of heat energy from lower and warmer latitudes and partly through the release of potential and latent heat energy. The net sensible heat transport is simply the net result of the movement of warmer air into Antarctica as colder air moves out. In the middle troposphere (about 4500—5000 m) the monthly mean temperature during the coldest month at a coastal station in East Antarctica is only about

8 °C colder than the warmest month. Even at the South Pole the temperature difference amounts to less than 10 °C. This is not true at the surface nor in the stratosphere; at the surface the range of monthly mean temperature may be less than 20 °C at coastal stations and more than 30 °C at interior stations, while the range in the lower stratosphere may be 35 °C at coastal stations and 45 °C at interior stations (WEXLER & RUBIN, 1961). Were the radiative losses not compensated for by advection of warmer air, the temperature would be very much colder (RUBIN, 1953). The net transport of water vapor into Antarctica and the precipitation process also releases heat which helps to balance the budget. From calculations of the net annual accumulation (RUBIN, 1961) the amount of latent heat released during the course of one year is about 1.4×10^{21} cal. This is about 12% of the net annual sensible heat (11.5×10^{21} cal) transport into Antarctica. The release of latent heat by direct sublimation is thought to be a minor factor, as is the expenditure of heat energy in evaporation. The annual heat budgets of the snow cover and the atmosphere are in balance, therefore the energy lost through radiation at the top of the atmosphere equals the net transport into the region. This is about 12.9×10^{21} cal (RUBIN, 1961).

The exchange of energy and mass between each component of the ocean-atmosphere-ice system is complex. Each component can serve as a source of energy, a medium for transport of energy, as well as a sink over greater or lesser periods of time. This is related to year-to-year meteorological variations as well as long-term changes of climate. The atmosphere cannot store appreciable amounts of water. In general, it merely transports water. The world's oceans contain about 14×10^{17} tons, while the ice in Antarctica contains about 22×10^{15} tons (SHUMSKII, 1961; THIEL, 1961), which is approximately 1/60th of the ocean mass. The earth's atmosphere, however, only holds about 3×10^{13} tons and even a large increase in the mean temperature of the atmosphere would not significantly increase the amount of water vapor. Thus, any transfer process of water mass from ocean to glacier is a slow one. If there were no ice loss at all from Antarctica, and if the average annual net precipitation were 15 cm, it would take over 10,000 years to deposit the present mass of ice on the Antarctic continent. This, in turn, would lower the present worldwide ocean level by about 60 meters. Conversely, the melting of all of Antarctica's ice would raise the ocean level by the same amount.

Circulation and Storm Tracks

The circulation of the atmosphere results essentially from differential heating of the surface of the earth, although topographic influences alter the large-scale motions. The climate of a particular region, then, is the result of the astronomical, topographical and atmospheric factors that operate to bring about a composite characteristic of the particular location. The circulation of the atmosphere, which is a statistical depiction of the mean motions, is of particular interest around Antarctica.

Antarctica is a heat sink; only the reaction of the atmosphere to the radiationally-caused temperature gradient keeps the temperatures from dropping even lower than they are in Antarctica. The atmosphere tries to make up the heat budget deficit through the transport of sensible heat, through the transport and release of latent heat in precipitation and by the utilization of potential heat through subsidence. The first two processes are effective mainly in the troposphere, while the last is effective in the stratosphere as well. The large-scale circulation tends to be zonal, and becomes increasingly strong from fall to late winter. For the yearly average some recent computations (GRUZA, 1960) show that the kinetic energy of the Southern Hemisphere circulation is about twice that of the Northern Hemisphere, while the meridional component of the kinetic energy is about the same in the two hemispheres. In winter the low-level easterly winds are extensive around the coast and inland to about 75 °S in East Antarctica, while they extend to latitudes further south in the region of the Ross and Weddell Seas.

During the year the great cyclonic storms tend to follow long arc-like clockwise paths north of the continent and, except for low-lying regions in West Antarctica, seldom penetrate it. These storms which circle the continent, and which can affect deep layers of the troposphere, account for much of the precipitation, heat transport and strong winds along the coast and on the slopes.

The atmosphere is inconstant in Antarctica, as elsewhere, with respect to short as well as long periods. Short-period variations of the order of hours or a day are due to minor perturbations in the atmosphere, sometimes due to local influences and day-to-night differences. Others, of longer duration, perhaps a few days to a week, are due to the translatory movement and life cycle of long-wave patterns of troughs and ridges in the troposphere, often associated with low-level cyclonic disturbances. These often upset the seasonal pattern of climate one would expect from purely radiational controls. However, over the months they tend to average out and generally do not change the seasonal climate drastically from one year to another. However, if a particular circulation pattern becomes statistically fixed for a month or more, rather interesting weather statistics occur. In general, however, the pattern varies from month to month and from year to year. Thus, the same month in different years may have rather different weather conditions (RUBIN, 1960). This is seen in Figure 4 in which are presented

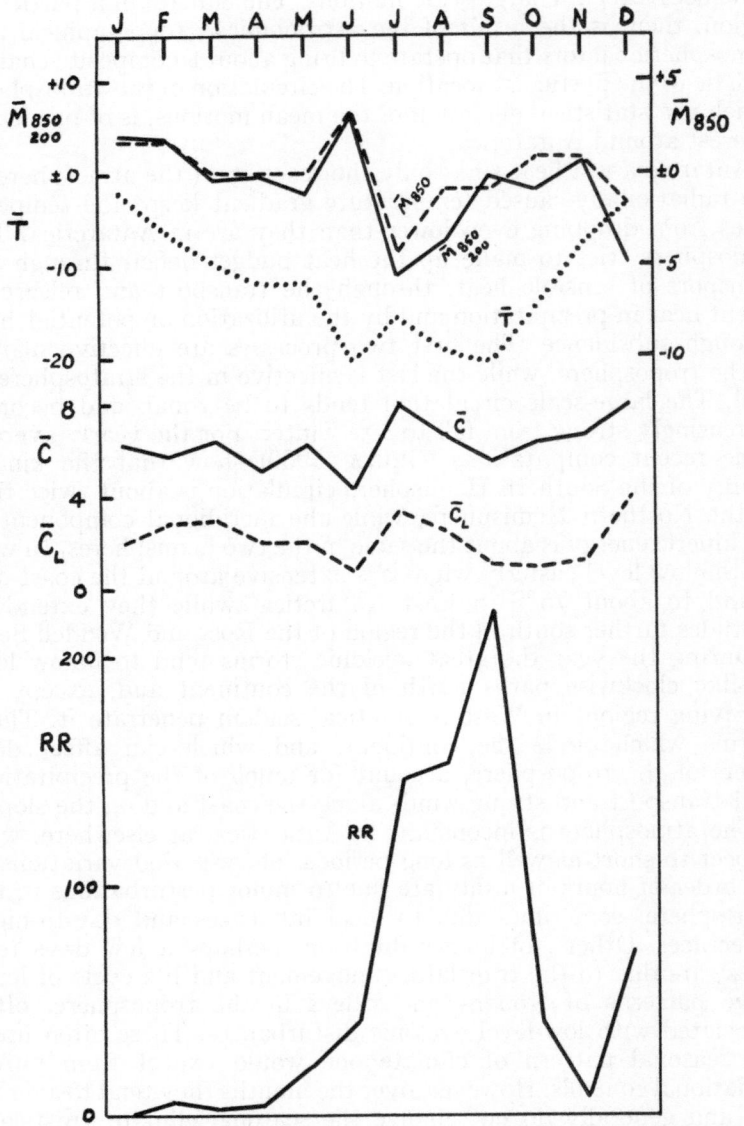

Fig. 4. Curves showing the monthly mean meridional mass-transport at 850 mb
($\overline{M}$ 850) and in the layer 850 to 200 mb ($\overline{M}\,_{200}^{850}$) and monthly mean values of temperature (T°C), total cloudiness ($\overline{c}$), low cloudiness ($\overline{c}_L$) (octas) and precipitation
(RR) (mm) at Mirny during 1958.

several curves to show the interrelationships of temperature, cloudiness, precipitation and meridional exchange of air and sensible heat. *A priori* one would expect to find certain relationships existing between the several parameters, particularly during the winter period when the influence of solar radiation is slight. The trend of monthly mean meridional mass-transport during the year has large variations, especially marked from May to June, and again from June to July. The temperature trend is in complete accord with the advection; when the net meridional transport component was from the north, or when the south component was decreased, the monthly mean temperature always rose. The same held true for the transport at the 850-mb level. For the period from February to October the trend of total cloudiness correlated well with the mass-transport curve. Cloudiness, particularly low cloudiness, is a significant factor in the surface temperature regime. Because of the frequent and prolonged blizzards in Antarctica which result in much drifting and blowing snow, there is a large element of uncertainty in measuring the amount of precipitation, or even in determining whether snow is falling at all. However, in the figure there seems to be a tendency for the precipitation to have been in excess of 10 mm during these months when the net transport was from the north.

One manifestation of the "constant inconstancy" of the atmosphere in polar regions is the so-called "kernlose" temperature phenomenon which is characterized by one or more reversals of the expected decrease of monthly mean temperature during the winter (Fig. 4). The first reversal of the downward trend of temperature seems to be a reaction of the atmosphere to the strong temperature gradient. As radiational cooling takes over again, the temperature decreases, aided by an extension of the ice around the continent until another minimum is reached, at which point the atmosphere again reacts, successfully this time since it is aided by the return of the sun and warming of the surface of the earth at lower latitudes.

Weather and Climatic Phenomena

The weather and climate of Antarctica are the result of several independent, yet interrelated geographical and physical phenomena. The basic control is that of latitude which results in a period of winter darkness with little or no solar radiation, and a summer period with long hours of daylight and solar radiation. The continuous daylight overcompensates for the low angle of the sun so that during the summer season the total extraterrestrial radiation within the polar circle is a maximum at the South Pole and a minimum at the Antarctic Circle. This is the reverse of the distribution if the annual value of radiation is considered (Fig. 2). The geographical

84

control of the cold continental core surrounded by a relatively warm ocean is a second factor of significance. The third factor is the great reflective power of the snow cover and its strong radiative characteristics. The fourth important factor is the great height of the continent and its steep slopes, which may be considered as both a cause and a result of the distinctive climate since much of Antarctica's elevation is due to the accumulated snow and ice, over 2000 meters thick in many places.

Temperature

The annual mean temperature of Antarctica is the lowest of any continent in the world. Along the coasts, where the tempering influence of the ocean and the advection of warm air occur even in

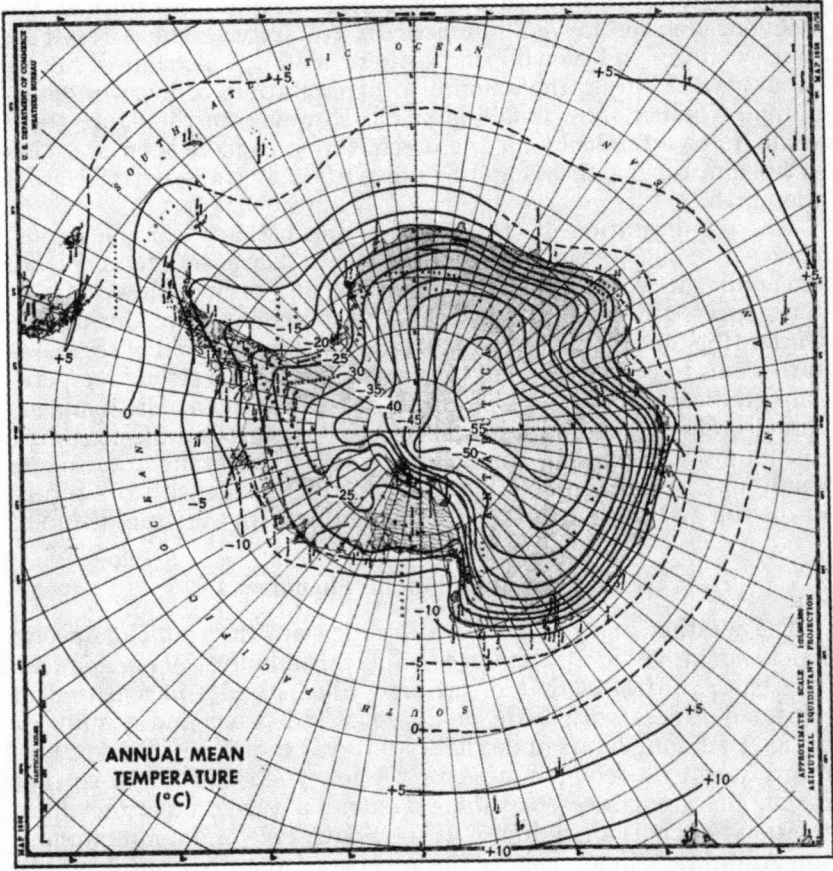

Fig. 5. Distribution of annual mean temperature in Antarctica. Isotherms in °C.

mid winter, the mean temperature at sea level is considerably higher than the temperatures in the interior of the continent (Fig. 5). The range of temperature at a particular location is greatest (more than 30°C) at interior stations at the higher elevations, and least (less than 20°C) at coastal stations. Table I gives the approximate range

Table I.

Approximate range of annual and seasonal mean temperature (°C) at sea level, at Vostok (78°27′S, 106°52′E, 3420 M) and at South Pole (90°S, 2800 M)

Latitude (°S)	Winter	Spring	Summer	Fall	Annual
60—65*	— 8/—20°	— 3/—11°	+1/— 2°	— 2/—14°	— 3/—12°
65—70	—11/—25°	— 8/—18°	0/— 5°	— 6/—16°	— 5/—13°
70—75	—25/—30°	—15/—25°	—2/—10°	—13/—20°	—10/—18°
75—80	—25/—35°	—18/—28°	—5/—12°	—20/—30°	—18/—28°
Vostok	—69°	—57°	—37°	—62°	—56°
South Pole	—59°	—49°	—32°	—57°	—49°

* Antarctic Peninsula

of mean temperatures at sea level around the coast of Antarctica and at two interior stations. The higher temperatures occur mostly in the Antarctic Peninsula region where the marine influence is greatest and the continental influence is at a minimum. The temperature on islands or on drifting ice (latitude 60—65°) is higher also because of marine influences. Individual months will be colder in winter and warmer in summer than indicated for the seasonal average. At interior stations above the 3000-m elevation, the coldest month may have an average temperature of about —68°C, while the warmest month may reach only —25°C. On the coast, excepting the Antarctic Peninsula, the coldest month may reach only as low as —18°C, while the warmest month can average about the freezing point. In the Antarctic Peninsula summer temperatures may be somewhat higher. For shorter periods, of the order of days, lower temperatures in winter and higher temperatures in summer will be experienced. At the high interior stations the absolute minimum is quite low (e.g. Vostok —88.3°C on 24 August 1960). On the coast the minimum temperature rarely goes below —40°C, and summer maxima may be as much as +9°C.

The effect of altitude and latitude on the slope of the continent results in a decrease of slightly more than 1°C per hundred meters rise in elevation. Beyond the region of the slope, on the plateau, there is still a decrease in temperature, mainly due to the latitude effect, but also influenced somewhat by the continued decrease in elevation. Figure 6 illustrates the relationship between temperature,

86

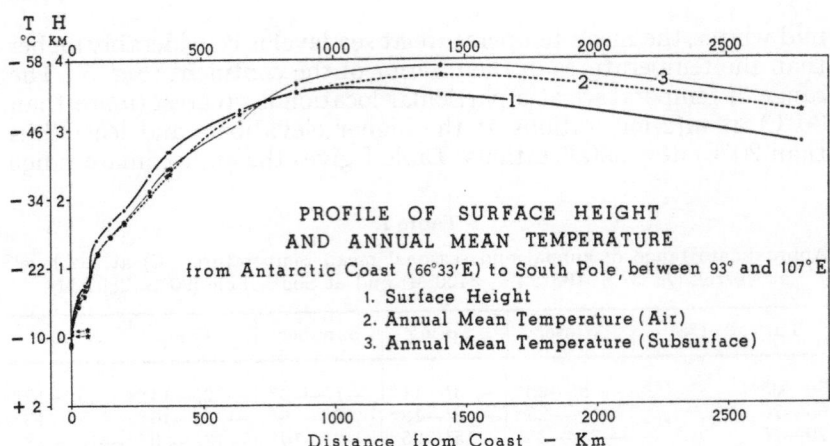

Fig. 6. Profile of surface height and annual mean temperature from Antarctic Coast (66°33′E) to South Pole, between 93° and 107°E. 1. surface height; 2. annual mean temperature (air); 3. annual mean temperature (subsurface).

elevation and distance from the sea. On the slope, where the major influence is the rapid rise in elevation, and where the effect of changing latitude is minor, the rate of decrease in temperature is about the same as that of air rising freely in the atmosphere.

A characteristic feature of the Antarctic temperature regime is the strong temperature inversion (increase of temperature with height). This is the result of strong radiational cooling at the surface because of the physical properties of the snow and is most pronounced during periods of relatively light winds. The inversion is strongest in winter, but is present also in the summer season, or at least as an isothermal layer (no change in temperature with height). This is not true over the ocean in the cases where cold air from the continent moves over the relatively warm surface (RUBIN, 1958). The inversion is broken down during periods of strong wind and blizzards, and the temperature may even rise to within a few degrees of the freezing point in mid-winter at coastal stations. The wind-chill factor, however, increases greatly and the hurricane-force winds combined with drifting and blowing snow make conditions precarious for any creatures exposed to the elements.

It is of interest to note the very significant vertical stratification of temperature in Antarctica as well as the effect of the underlying surface on this temperature stratification. Usually the temperature measured at about 2 meters above the surface is given as the "air temperature" in meteorological reports. This temperature may be considerably higher or lower than the surface; the temperature difference is a function, primarily, of the season of the year, the time of day, and the character of the underlying surface. Table II

Table II.

Monthly mean values of the temperature (°C) of the air and the underlying surface at Mirny (66°33'S, 93°01'E, 35 M), Vostok-I (72°08'S, 96°35'E, 3140 M), and Bunger Oasis (66°16'S, 100°45'E, 29 M) during 1957.

	J	F	M	A	M	J	J	A	S	O	N	D	Year
Mirny													
Air Temp.	—0.6	—4.2	—9.3	—8.6	—14.2	—15.5	—16.8	—14.4	—14.6	—10.8	—7.3	—1.9	—9.9
Soil Temp.	—	—5.0	—9.9	—9.4	—15.4	—17.1	—17.7	—15.3	—15.5	—11.1	—6.9	—1.0	—11.3
Air-Soil	—	0.8	0.6	0.8	1.2	1.6	0.9	0.9	0.9	0.3	0.4	0.9	1.4
0100 Air $\bar{T}$	—2.0	—6.3	—10.3	—8.8	—14.5	—15.2	—16.7	—14.6	—15.2	—12.1	—10.1	—3.7	—10.8
0100 Soil $\bar{T}$	—3.3	—7.0	—11.3	—9.9	—15.7	—16.6	—17.7	—15.5	—16.4	—13.4	—12.0	—5.4	—12.0
Air-Soil	1.3	0.7	1.0	1.1	1.2	1.4	1.0	0.9	1.2	1.3	1.9	1.7	1.2
1300 Air $\bar{T}$	—0.5	—2.8	—7.4	—8.0	—13.9	—15.5	—16.8	—13.8	—13.4	—8.8	—4.3	—0.4	—8.7
1300 Soil $\bar{T}$	—	—2.4	—6.9	—8.3	—15.1	—17.2	—17.7	—14.2	—13.8	—7.4	—1.1	—2.6	—9.2
Air-Soil	—	0.4	0.5	0.3	1.2	1.7	0.9	0.4	0.4	1.4	3.2	3.0	0.5
Vostok-I													
Air Temp.				—48.0	—52.6	—51.7	—58.0	—53.0	—48.7	—43.3	—36.6		
Snow Temp.				—48.7	—54.3	—53.6	—59.7	—54.4	—49.7	—44.4	—		
Air-Snow				0.7	1.7	1.9	1.7	1.4	1.0	1.1	—		
0100 Air $\bar{T}$				—48.0	—52.2	—50.8	—58.1	—53.1	—50.1	—47.1	—43.5		
0100 Snow $\bar{T}$				—48.4	—53.9	—52.8	—60.1	—54.5	—50.9	—48.2	—		
Air-Snow				0.4	1.7	2.0	2.0	1.4	0.8	1.1	—		
1300 Air $\bar{T}$				—47.0	—52.8	—52.3	—58.1	—52.6	—46.3	—38.6	—31.0		
1300 Snow $\bar{T}$				—48.3	—54.6	—54.1	—59.9	—54.0	—47.2	—40.0	—		
Air-Snow				1.3	1.8	1.8	1.8	1.4	0.9	1.4	—		
Bunger Oasis													
Air Temp.	2.2	—1.6	—6.0	—6.3	—13.6	—18.2	—17.4	—12.7	—13.1	—8.6	—3.7	0.7	—8.2
Soil Temp.	8.2	—3.2	—4.7	—7.3	—16.1	—23.0	—20.6	—14.9	—14.0	—7.0	—2.2	—7.3	—7.2
Air-Soil	—6.0	—4.8	—1.3	—1.0	—2.5	—4.8	—3.2	—2.2	—0.9	—1.6	—5.9	—6.6	—1.0
0100 Air $\bar{T}$	0.7	—2.1	—6.9	—6.8	—13.8	—17.7	—17.4	—13.3	—14.1	—10.0	—6.0	—1.0	—9.0
0100 Soil $\bar{T}$	1.5	—1.7	—7.7	—8.4	—16.4	—23.2	—20.6	—15.6	—15.6	—11.0	—6.5	—2.3	—10.6
Air-Soil	—0.8	—0.4	0.8	1.6	2.6	5.5	3.2	2.3	1.5	1.0	0.5	1.3	1.6
1300 Air $\bar{T}$	3.6	—0.4	—4.5	—5.3	—13.0	—18.3	—17.4	—11.6	—11.1	—6.8	—1.6	2.3	—7.0
1300 Soil $\bar{T}$	17.3	9.7	1.4	4.9	—15.4	—22.4	—20.0	—13.0	—10.6	1.2	12.0	17.3	2.5
Air-Soil	—13.7	—10.1	—5.9	0.4	2.4	4.1	2.6	1.4	0.5	5.6	13.6	—15.0	4.5

contains data pertinent to this aspect of the temperature discussion. The three stations listed — Mirny, Vostok-I and Bunger Oasis — are situated, respectively, on permanent continental ice with rock outcroppings near sea level, on permanent continental ice on the high plateau, and on a soil and rock surface near sea level, only occasionally snow-covered.

The Mirny mean air temperature is lower than the soil-surface temperature during only about three months of summer and averages 1.4 °C colder for the year. In general the temperature at this latitude evidences a diurnal variation; the minimum temperature occurs near 0100 hours and the maximum near 1300 hours local time. If we consider the air-soil surface temperature difference for these hours, the Mirny monthly mean air temperature is higher than the soil temperature at 0100 hours during every month and averages 1.2 °C higher; at 1300 hours the air is warmer than the soil during only six months, and averages only 0.5 °C higher. Thus, we see that the boundary-layer temperature inversion is most pronounced in winter and during the nighttime hours. During no month, except possibly January, is the monthly mean soil-surface temperature as high as 0 °C. However, during mid-afternoon in summer the soil-surface temperature is as high as 0 °C, on the average, during about two months; the absolute maximum soil temperature even reaches 12.5 °C in December and may be above 0 °C on occasion during five months of the year. At plateau stations, such as Vostok-I, the snow-surface temperature never approaches the melting point.

At coastal stations, such as Bunger Oasis, situated on rock and soil near the Antarctic Circle, there are striking deviations from the normal Antarctic radiation and temperature regime. These differences are due primarily to the nature of the surface. As was pointed out earlier, the station has a surface radiation budget surplus of more than 37,000 cal/cm². During six months of the year the soil surface is warmer than the air, and on the yearly average it is 1 °C higher. The diurnal variation is normal in that the lowest temperatures occur at night and the highest during mid-afternoon. Only in midsummer, during the period of almost continuous sunshine, is the soil surface warmer than the air at 0100 hours local time. At 1300 hours the soil averages warmer than the air during eight months of the year. During six months of the year the maximum surface temperature for the month is above 0 °C, the absolute maximum for the year having reached the amazingly high value of 32.9 °C. Conversely, the minimum soil temperature for every month is below the freezing point, although in January the minimum value was only —3.0 °C. In winter the minimum soil temperature may be 5 °C lower than the minimum for the air, while in summer it may be 5 °C higher than the minimum for the air.

From observations at Mirny and Bunger Oasis (GRIGORYEV, 1959) it is apparent that there exists a seasonal thawing layer in the Antarctic soil, at least in the coastal regions at the lower latitudes. At Bunger Oasis, for example, the thawing of the subsurface begins during the first days of November and continues until the second half of January. About 10—15 cm depth of soil is thawed during the first month, although an influx of cold air at the surface may, on occasion, reverse the thawing trend and even result in refreezing the entire layer. In December, however, the thaw penetrates at the rate of 2—3 cm/day. About 60% of the depth of the seasonal thaw layer is formed in the second month, and 30% during the third month. The zero degree isotherm was observed to reach a depth of 1.5 m in 1957 and 1.4 m in 1958. The refreezing of the layer occurs during the third decade of February. At Mirny, where there are several outcroppings of rock and soil, the soil thaw starts at the beginning of November, and refreezing occurs by the end of February. The maximum depth of thawing is about 2 m, and is reached at the end of January.

From the above it may be concluded that a favorable environment for specialized plant and animal life exists in the air-surface boundary layer at some Antarctic locations.

Another kind of typical surface in Antarctica is a beach consisting of consolidated sand, rock and pebbles, guano and penguin carcasses, e.g. Hallett Station (72°18'S, 170°18'E). Some subsurface temperature data are available (BENES, 1960) which show that even at high latitude locations relatively warm soil conditions exist in summer. In Table III we see that during the two summers the mean monthly air temperature was above the freezing point during only one month, while the monthly mean temperature at —10 cm, and presumably at the surface, was above freezing during four months. Daily temperature data show that the —10-cm temperature was above the freezing point (0°C) as early as November 24 in 1958, was continuously above 0°C for 26 days in December 1958, and again was above freezing in January 1959. During the austral summer of 1959—1960 the —10-cm temperature was above freezing from 13 December 1959 through January 1960. At no time, however, did the thaw penetrate to the —50-cm level.

Table III.

Monthly mean air and subsurface (—10 cm) soil temperatures (°C) at Hallett Station (72°18'S, 170°18'E) 1958—1959 and 1959—1960.

	Dec. 1958	Jan. 1959	Dec. 1959	Jan. 1960
Air	—3.5	+0.8	—1.8	—2.1
—10 cm	2.2	1.4	0.9	0.4

90

At Scott Base (77°51'S, 166°48'E), HATHERTON (1961) reports that the temperature of the ground surface reached 0°C for about three weeks during January 1958. At no time was the temperature above 0°C at the 6-inch (15-cm) level.

The air temperature and surface temperature are subject to control by cloudiness and wind (Fig. 4). In winter the lower clouds often may be warmer than the surface; the movement of such clouds over a region tends to raise the temperature of the surface by reducing radiative loss from the surface and by downward long-wave radiation from the cloud. In summer the clouds tend to lower the temperature of the surface by reducing the incoming short-wave radiation; although the clouds still reduce outgoing radiation, the decrease in solar radiation has the greater effect. The wind factor is most effective in winter when the strong low-level temperature inversion forms. Strong winds tend to stir up the boundary layer of air and bring down warmer air to mix with the radiatively cooled surface air, thus increasing the temperature. Sharp and appreciable temperature rises are noted on such occasions, due only to increases in wind.

Wind

The Antarctic continent, and particularly the coastal region and slopes of East Antarctica, is one of the windiest regions on earth. The combination of intense storms over the ocean, the strong cooling of the surface layers of air, and the steep slope of the continent are factors in the strong and persistent winds. The drainage (katabatic) winds are only a few hundred meters in depth and generally flow rather uniformly. At times, however, when critical velocities are reached, they become very gusty and carry great quantities of drifting and blowing snow. According to DRALKIN (1960) the strong katabatic wind component decreases linearly with distance from the coast, i.e. from about 18 m/sec to zero at about 25 km; the ordinary katabatic wind decreases from about 14 m/sec to zero within about 20 km. The gustiness of the wind also decreases linearly and reduces to insignificance within 15 km. The snow transported by the wind seems to be carried well away from the coast and is deposited on the sea ice mainly in the zone between about 10 and 30 km, the maximum deposition occurring in the zone between about 15 and 25 km. The presence of individual large icebergs or concentrations of icebergs near the coast may alter the wind and precipitation regime markedly in the vicinity of the icebergs.

Table IV presents the approximate range of mean wind speed, by latitude-zone and season, for the coastal region of Antarctica. Except for a minimum in summer, all seasons of the year seem to have an equivalent range of wind speed; however, individual stations generally show strongest monthly mean winds in the winter season.

Table IV.

Approximate range of seasonal and annual mean wind speed (m/sec) at coastal stations in several latitude zones

Latitude (°S)	Winter	Spring	Summer	Fall	Annual
60—65°*	4— 9	4— 8	3— 6	4— 9	4— 8
65—70°	4—12[1]	4—11[2]	3—10[3]	5—12[4]	4—10[5]
70—75°	1— 7	2— 7	3— 8	3—13	2— 8
75—80°	5— 9	5— 9	5— 6	6— 9	5— 7

Note: In all zones some locations may experience stronger winds because of terrain factors.

1. up to 22 m/sec   4. up to 23 m/sec
2. up to 20 m/sec   5. up to 21 m/sec
3. up to 17 m/sec

* Antarctic Peninsula

Persistent downslope winds have a warming effect as they are heated adiabatically by 1 °C per 100-m vertical descent. They also have a drying effect because of the concomitant reduction in relative humidity. In particular locations such as the dry valleys in Victoria Land leading to McMurdo Sound, the dwindling remnants of glaciers are thought to be due to these factors of warm and dry foehn type winds.

At some coastal stations in East Antarctica the wind may reach hurricane force due to intense storms on more than seven occasions during the course of a winter season; on about 220 days per year the wind will attain a speed of more than 15 m/sec, and even in the summertime the wind attains a strength of 25 m/sec at least once during every month. The wind regime at coastal stations located at the edge of an extensive ice shelf, e.g. Little America (78°11'S, 162°10'W) or at an interior plateau station, e.g. Vostok (78°27'S, 106°52'E, 3420 m), is rather different from the regime at stations near the steep slope of the continental ice cover. Table V presents a comparison of the wind regimes at Mirny, Little America and Vostok.

Cloudiness and Sunshine

In general the winter is the season of minimum cloudiness at Antarctic stations; continental stations have less cloudiness than coastal stations. Table VI gives the approximate range of cloud cover, in per cent, at coastal stations in several latitude zones and at the South Pole. Topographic features, prevailing winds and oceanic influences are factors in the cloud regime in a particular region, as well as the large-scale influences of general circulation and storminess.

Table V.

Mean wind speed, maximum wind speed, and average number of days with wind speed greater than 15 m/sec at Mirny and Little America for two-year period (1957 and 1958), and for Vostok (1958 only).

	J	F	M	A	M	J	J	A	S	O	N	D	Year
Mirny													
Mean	7.8	9.5	11.8	14.1	14.5	13.8	15.0	15.8	13.4	11.2	9.3	8.2	12.0
Maximum	20	25	20	32	40	39	51	51	46	40	25	30	51
No. of Days	8	13	20	22	27	22	26	26	18	20	11	6	219
Little America													
Mean	4.3	4.3	5.5	4.9	6.4	6.3	5.4	5.9	5.4	6.1	4.3	5.5	5.4
Maximum	17	13*	17	19	27	24	17	34	25	23	17	22	34
No. of Days	i	0*	4	3	5	6	4	6	4	4	1	2	40
Vostok*													
Mean	4.6	3.7	5.1	4.0	5.2	4.9	5.0	4.7	5.4	4.0	4.2	3.8	4.6
Maximum	14	9	10	10	15	25	13	13	13	12	13	11	25
No. of Days	0	0	0	0	1	1	0	0	0	0	0	0	2

* 1958 only

The amount of sunshine at a particular location is, of course, the converse of the cloudiness. Interior stations, far removed from coastal storms, have a high percentage of sunshine. Some oceanic islands and the Antarctic Peninsula receive as little as 3% of the possible sunshine in winter. Table VII gives the range of observed percentage of possible sunshine for coastal stations in five-degree latitude zones from 60° to 80°S.

Table VI.

Approximate range of cloudiness (per cent) at coastal stations in several latitude zones and at the South Pole

Latitude (°S)	Winter	Spring	Summer	Fall	Annual
60—65°*	48—78	61—85	64—87	59—81	71—81
65—70°	43—74	44—77	59—88[1]	51—73	49—74
70—75°	51—63	55—75	70—88	52—73	61—74
75—80°	48—56	64—73	66—74	54—64	61—66
South Pole	32	51	50	34	42

* Antarctic Peninsula
1. over ocean.

Table VII.

Approximate range of observed percentage of possible sunshine at coastal stations in several latitude zones

Latitude (°S)	Winter	Spring	Summer	Fall	Annual
60—65°*	3 —20	14—30	15—27	6 —21	10—24
65—70°	4*—22	15—50	20—47	11—32	15—35
70—75°	6 —44	37—60	28—32	22—27	27—41
75—80°	14—42	29—39	32—54	14—29	30—38

* Antarctic Peninsula

Precipitation

Because of the high frequency of blizzards and blowing snow in Antarctica, the measurement of actual precipitation is difficult to accomplish. The best estimates seem to come from measurements of net accumulation around arrays of thin stakes set out in relatively undisturbed areas, and from stratigraphic studies made in special glaciological pits. Some studies of accumulation have also been made using ratios of O^{18}/O^{16} and other isotopes. Generally precipitation in Antarctica is very low and it can be rightly called a "white desert". The low temperatures which prevail over and around the

continent result in low absolute humidity and reduce the possibility
of considerable amounts of precipitation.

The maximum amounts of precipitation occur near the coast, on
the slopes of the continental ice sheet and around topographic
features which disturb the wind flow. Because of local topography
and wind drift there may be significant small-scale variations in the
measured accumulation. Figure 7 is an estimate of the large-scale

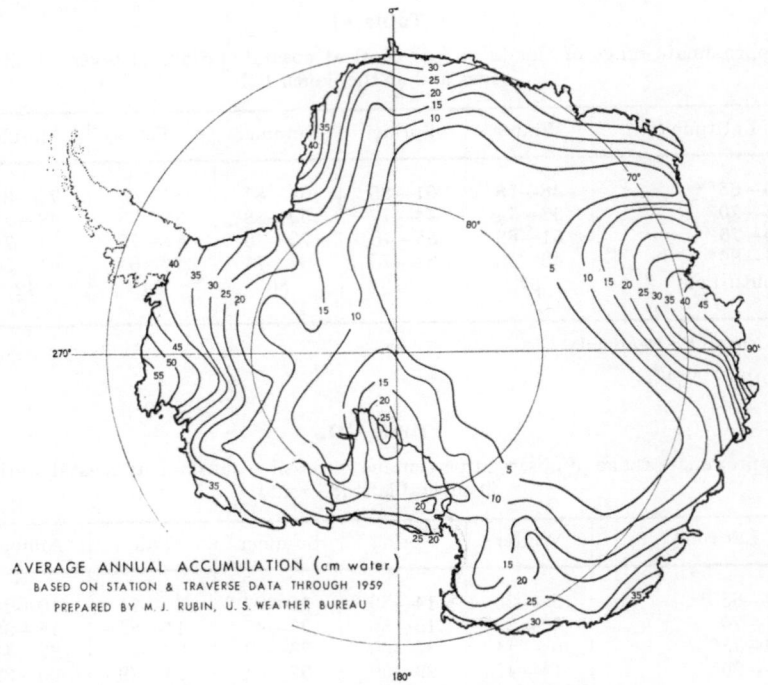

AVERAGE ANNUAL ACCUMULATION (cm water)
BASED ON STATION & TRAVERSE DATA THROUGH 1959
PREPARED BY M. J. RUBIN, U. S. WEATHER BUREAU

Fig. 7. Annual snow accumulation map of Antarctica.

distribution of snow accumulation (in cm of water-equivalent) over
Antarctica (RUBIN, 1961). The values are based upon snow-stake
and pit stratigraphic studies. The average annual net *accumulation*
over Antarctica is 14.5 cm of water equivalent; allowing for loss
of snow from the surface through drifting snow, surface melting
and evaporation — all of which have not yet been exactly determined — the estimate of the average annual *precipitation* is between 14.6 and 19.2 cm of water equivalent.

Remarks

This discussion of Antarctic climatology has been primarily concerned with the mainland. For a discussion of the climate of the Antarctic ocean and the sub-Antarctic islands see VAN ROOY, M. P. (editor), *Meteorology of the Antarctic*, Weather Bureau, Pretoria, South Africa, 1957.

ACKNOWLEDGEMENT

The author wishes to express his appreciation to his colleague Mr. EDWIN C. FLOWERS for the computations used in preparing Figure 3.

REFERENCES

BENES, N. S., 1960. Soil temperatures at Cape Hallett, Antarctica, 1958. *Monthly Weather Rev.* **88**, 6.

DRALKIN, A. G., 1960. Preliminary results of the work of the Fourth Continental Antarctic Expedition. (In Russian). *Problems of the Arctic and Antarctic*, **5**.

GRIGORYEV, N. PH., 1959. Some results of permafrost research in East Antarctica. *Bull. Soviet Antarctic Exped.* **7**.

GRUZA, G., 1960. On some zonal characteristics of the general circulation of the atmosphere. *Reports of NAUK, USSR, Geophys. Series,* **1**. (In Russian).

HANSON, K. J. & RUBIN, M. J., 1962. Heat exchange at the snow-air interface at the South Pole. *J. geophys. Res.,* **67**, 8.

HATHERTON, T., 1961. New Zealand IGY Antarctic Expeditions, Scott Base and Hallett Station. *New Zealand Dept. Sci. Industr. Res., Bull.* **140**, Wellington.

KEELING, C. D., 1960. The concentration and isotopic abundances of CO_2 in the atmosphere. *Tellus,* **12**, 2.

LOCKHART, L. B., Jr., 1960. Atmospheric radioactivity in South America and Antarctica. *U. S. Naval Res. Lab. Rep. No. 5526*, Washington, D. C.

OSTAPOFF, F., 1965. Antarctic oceanography. *Biogeography and Ecology in Antarctica. Monogr. Biol.,* **15**.

PICCIOTTO, E., WILGAIN, S., KIPFER, P. & BOULENGER, R., 1962. Radioactivité de l'air dans l'Antarctique en 1958 et profil radioactif entre 60°N et 70°S. *Radioisotopes in the Physical Sciences and Industry*, International Atomic Energy Agency, Vienna.

RUBIN, M. J., 1953. Seasonal variations of the Antarctic tropopause. *J. Meteorol.* **10**, 2.

RUBIN, M. J., 1958. An occurrence of steam fog in Antarctic waters. *Weather,* **13**, 7.

RUBIN, M. J., 1960. Advection across the Antarctic boundary. *Proc. of Symposium on Antarctic Meteorology held in Melbourne, February 1959.* Published by Pergamon Press Ltd.

RUBIN, M. J., 1961. Atmospheric advection and the Antarctic heat and water budget. *Proc. of Matthew Fontaine Maury Memorial Symposium on Antarctica.* 10th Pacific Science Congress, Honolulu, August 1961. To be published by American Geophysical Union.

RUSIN, N. P., 1961. *The Meteorology and Radiation Regime of Antarctica.* (In Russian) Hydrometeorological Press, Leningrad, 447 pages.

SHUMSKII, P. A., 1961. Glaciology of Antarctica. *Proc. of Matthew Fontaine Maury Memorial Symposium on Antarctica.* 10th Pacific Science Congress, Honolulu, August 1961. To be published by American Geophysical Union.

96

THIEL, E., 1961. The amount of ice on Planet Earth. *Proc. of Matthew Fontaine Maury Memorial Symposium on Antarctica*. 10th Pacific Science Congress, Honolulu, August 1961. To be published by American Geophysical Union.

WEXLER, H., MORELAND, W. B. & WEYANT, W. S., 1960. A preliminary report on ozone observations at Little America, Antarctica. *Monthly Weather Rev.*, 88, 2.

WEXLER, H. & RUBIN, M. J., 1961. *Antarctic Meteorology. Science in Antarctica, Part II: The Physical Sciences in Antarctica*. Publication 878, National Academy of Sciences — National Research Council, Washington, D. C.

ANTARCTIC OCEANOGRAPHY*

BY

FEODOR OSTAPOFF

U. S. Weather Bureau, Washington D.C.

(with 15 figs.)

Horizontal and vertical dimensions of the Antarctic Ocean

The name "Antarctic Ocean" or "Southern Ocean" is used frequently for convenience in oceanographic literature to designate the circumpolar water ring around the Antarctic Continent, from which the three major oceans spread northward. It was internationally (INT. HYDR. BU., 1937) agreed to extend the three major oceans to the Antarctic Continent and to choose the meridians running through the south tips of the respective land masses as boundaries between these oceans: the meridian of Cape Agulhas (20 °E), the meridian of South East Cape in Tasmania (147 °E) and the shortest connection between Cape Horn — South Shetland Island — Deception Island — Antarctic Peninsula or Palmer Peninsula (in British usage also: Graham Land). Thus, we may distinguish between the Atlantic, Indian and Pacific sectors of the Antarctic Ocean, respectively.

In general, the northern limit of the Antarctic region has been placed at the Oceanic Polar Front (WÜST, 1928; DEFANT, 1928), also known as the Antarctic Convergence. For the purpose of this survey we will include in our discussion areas north of the Polar Front in order to describe conditions in the Antarctic Ocean in proper relation to those in the adjacent oceanic areas.

First, it may be useful to recall the general distribution of land and water in the Southern Hemisphere. Although the Southern Hemisphere as a whole is by about 20% more water-covered than the Northern Hemisphere, the area from 90 °S to approximately 75 °S is entirely occupied by land (and solid ice), whereas 99.7% of the area between 65 ° and 60 °S is oceanic (Fig. 1). This contrast is increased enormously during late winter season, when the ice coverage extends to these latitudes forming quasi-continental conditions much further north. The seasonal variation of the mean northern limit of pack-ice is considerable (MACKINTOSH & HERDMAN, 1940) and, in a climatic sense, has the effect of a "pulsating" continent. The area covered by pack-ice in winter (mean northern limit of the pack-ice for September) has been included in Figure 1.

* The material on which this review is based was derived, in part, from work supported by grants from the National Science Foundation.

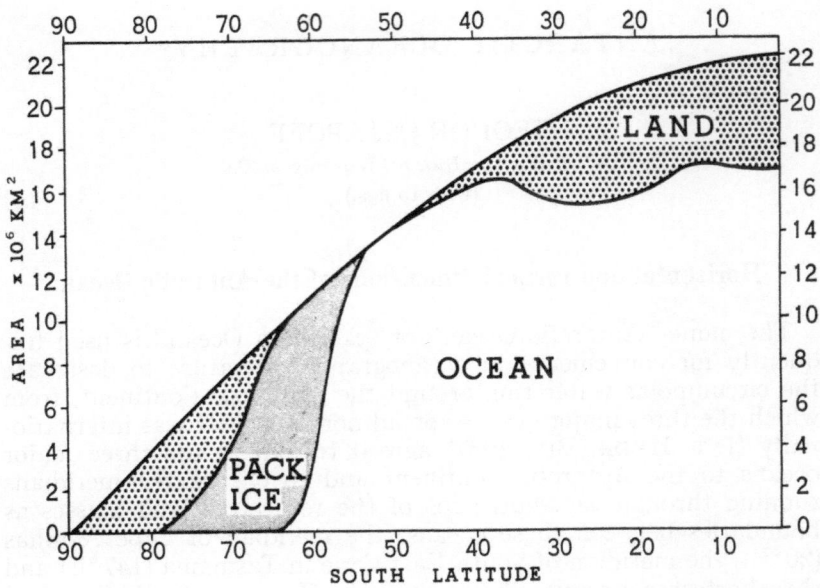

Fig. 1. Distribution of land and water areas in the Southern Hemisphere for 5° zones (after Kossina, as quoted by Wüst et al., 1954) and pack-ice cover in September (according U. S. Hydr. Office; 1957).

The areal ice-cover of the southern polar cap in winter is double that in summer.

The bottom topography of the Antarctic Ocean is closely related to the major morphological features of the adjacent oceans, as seen from Figure 2. The extension of the Mid-Atlantic Ridge (I) forms the Atlantic-Indian Rise (II) in approximately 50—55°S, south of Africa and protruding into the Indian sector. Proceeding eastward the next prominent elevations of the sea floor are the rises on which Prince Edward Islands and Crozet Island (III) are located. One of the most pronounced ridges of the Antarctic Ocean is the Kerguelen-Gaussberg Ridge (IV), almost 20 latitude degrees in extension and separated from the Antarctic shelf in about 90°E by a relatively narrow and deep channel. The Indian-Antarctic Ridge (V), with its major axis oriented more or less in zonal direction leads into the Pacific sector. South of Tasmania and New Zealand the bottom topography is very irregular. The Pacific-Antarctic Ridge (VI) represents the major feature of the Pacific sector. The Pacific Ocean is separated from the Atlantic Ocean by the narrowest constriction of the entire circumpolar water ring, the Drake Passage (about 600n.m.) (1112 km). The topography east of the spasage where the Scotia Arc (VII) connects geologically the South American Continent and the Antarctic Peninsula is very complex and exerts profound influence on the hydrographic conditions in and around Antarctica. Just east of the Scotia Arc the deepest depressions of the Atlantic Ocean are found ($\approx$ 8200 m).

This system of ridges and the distribution of continental land masses determine the deep-sea basins and major indentations of the Antarctic Coast. Three important seas shall be mentioned: Bellings-hausen Sea, Ross Sea and Weddell Sea, the latter one oceano-graphically the most important one as the main area of formation of Antarctic bottom water. These morphological features modify profoundly the oceanographic, and with it, climatic and biological conditions impressing considerable departures from strict zonality. Thus, at Campbell Is. ($52^1/_2$°S, 169°E) the average water and air temperatures are considerably higher than those at Marion Is.

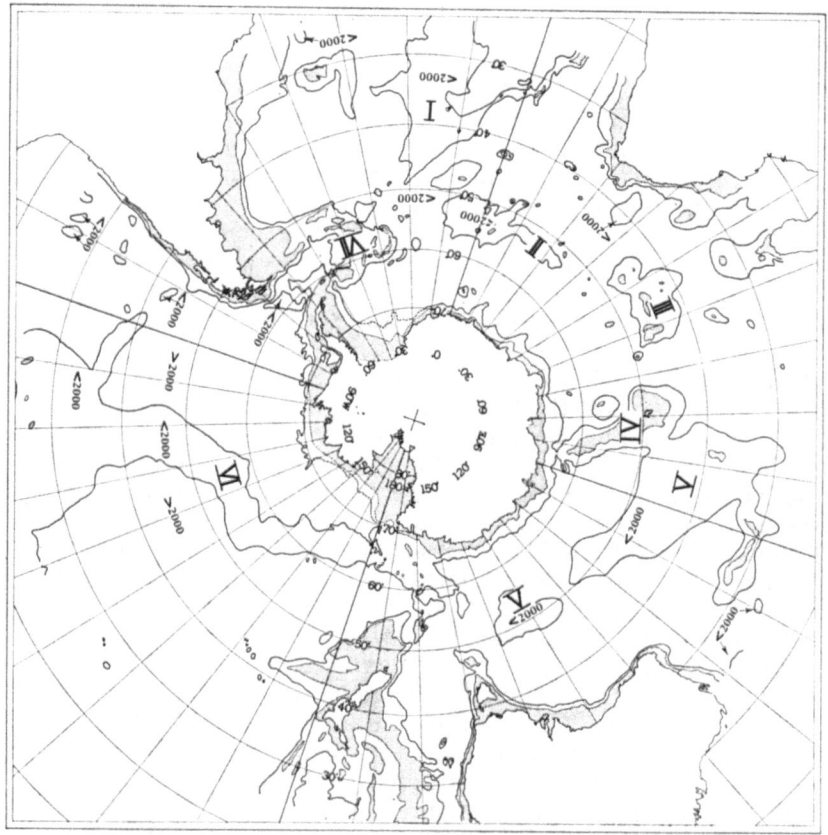

Fig. 2. The major topographical features as revealed by the 2000-fathoms line (solid) and the 1000-fathoms line (shaded area depth less than 1000 fathoms). (After U.S. Hydrographic Office charts H. O. Misc. 15,254-9/10/11/12.)

(47 °S, 38 °E) although the latter one lies more than 5° of latitude further north (DREYER, 1961).

Distribution of physical and chemical variables

The fundamental variables, temperature, salinity and pressure, determine the physical properties of sea water, such as compressibility, thermal expansion, freezing point, temperature of maximum density, etc. These relations are discussed in detail, for example, by SVERDRUP et al. (1942). In this survey the fields of temperature and salinity around Antarctica will be described and their characteristic features outlined. The density field, important for dynamic considerations, is derived from the temperature and salinity distribution. As far as other physical or chemical components are concerned we will restrict ourselves mainly to the discussion of the oxygen distribution.

Temperature at the sea surface

At a particular point in the ocean, the temperature at the sea surface is determined by the local heat exchange (see for example DIETRICH & KALLE, 1957) and depends on the incoming net radiation, on the sensible heat transfers both from the atmosphere to the water and vice versa as well as that associated with precipitation, on the heat loss due to evaporation and heat gain due to condensation of water vapor, on the advection of heat by ocean currents and vertical motion, or melting (or freezing) of ice and cooling effects by the presence of icebergs, on the heat of oxidation by chemical-biological processes and the heat of assimilation. Although not very important, the heat flux through the ocean floor (geothermal heat), heat due to radioactive decay and heat of friction due to dissipation of kinetic energy must also be included in the complete budget. Only the radiation term, the sensible and latent heat terms and the advection term are generally considered important (SVERDRUP et al., 1942). A discussion of all these terms can be found in standard oceanographic textbooks (for example: SVERDRUP et al., 1942; DIETRICH & KALLE, 1957; DEFANT, 1961), also at some detail in a paper by LAEVASTU (1961).

All these different processes, some more pronounced in particular latitudes than others, produce the global temperature distribution at the sea surface. This temperature distribution is indicated in Figure 3 (curve 1) which presents the zonally averaged temperature distribution as function of latitude. As this hemisphere is covered by water to a large extent (Figure 1), the temperature distribution presented in Figure 3 gives also some indication about the temperature distribution which would exist on a water-covered globe and is used as such by DIETRICH & KALLE (1957). The general distri-

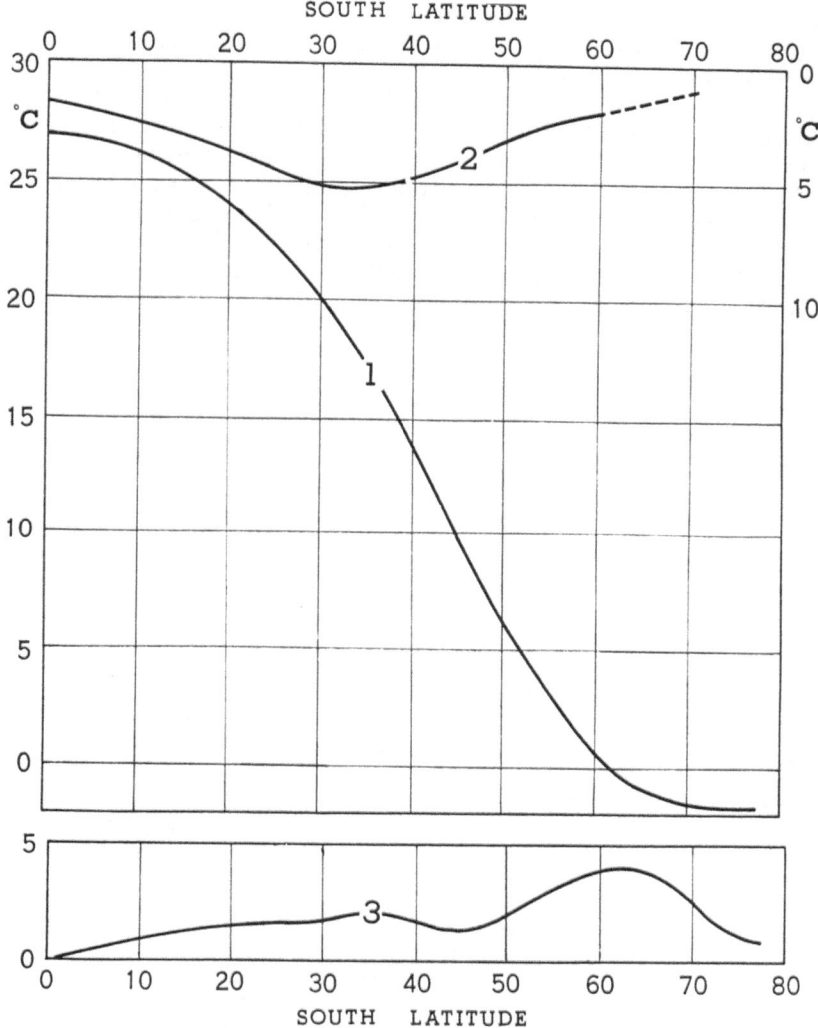

Fig. 3. Average zonal temperature distribution in the Southern Hemisphere, (curve 1, after Wüst et al., 1954), the annual temperature range (curve 2, after Sverdrup et al., 1942) and the temperature difference Northern Hemisphere minus Southern Hemisphere (curve 3, from Wüst et al., 1954).

bution with latitude follows closely the net radiation or radiation balance with latitude (Mosby, 1936; Budyko, 1955). Curve 2 in Figure 3 shows the annual temperature variation, also as function of latitude. In general, this variation is small because of the remarkable ability of the ocean to disperse heat. The maximum value

between 30° and 40°S amounts to about 5°C and in Antarctic
waters it drops to about 2°C, a little less in the Atlantic and Indian
sectors and somewhat more in the Pacific sector. Also, the water
south of the continents exhibits larger variations (about 4°C).
Curve 3 has been added in Figure 3 to show the temperature difference between the Northern Hemisphere and the Southern Hemisphere for the corresponding latitudes. In all corresponding latitudes
the surface water in the Northern Hemisphere is warmer than
that in the Southern Hemisphere. The greatest difference is found
in the latitudinal belt 60°—65°. Similar temperatures in the Northern Hemisphere are encountered some 10° of latitude further

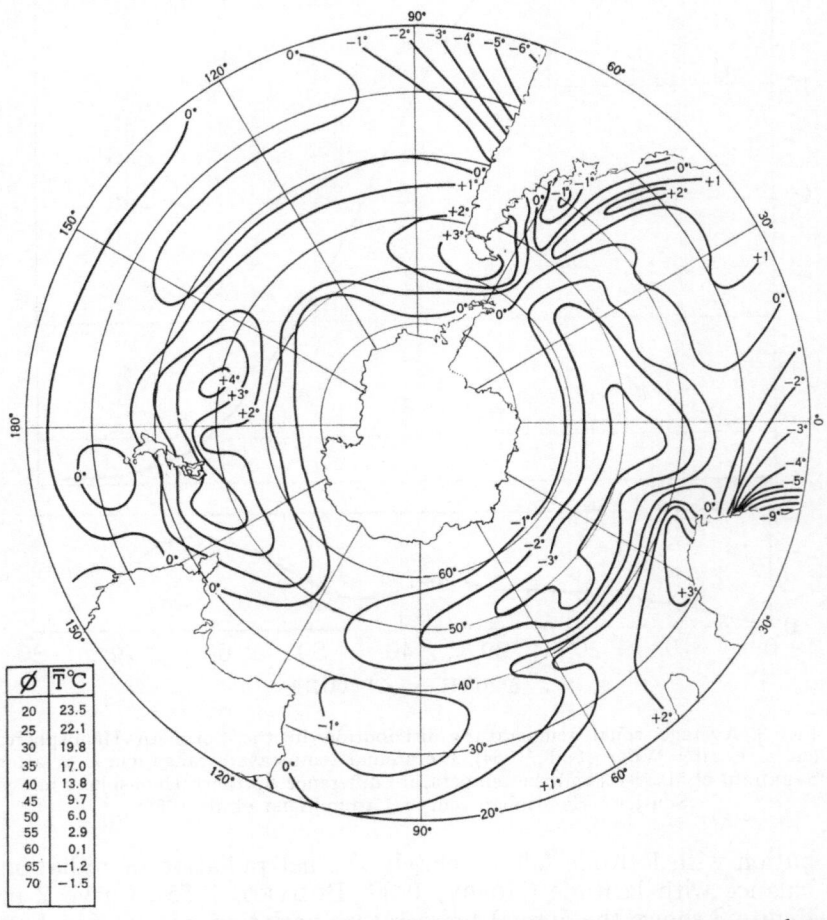

∅	T°C
20	23.5
25	22.1
30	19.8
35	17.0
40	13.8
45	9.7
50	6.0
55	2.9
60	0.1
65	−1.2
70	−1.5

Fig. 4. Deviation of the sea surface temperature from the zonal mean values,
which are shown in the lower left-hand corner (according to DIETRICH & KALLE,
1957).

north. The position of the Antarctic Circumpolar Current in the Atlantic sector would correspond in the Northern Hemisphere to the latitude of France (!) with the Antarctic Polar Front "located" at about the English Channel.

Whereas the zonally averaged temperatures (Figure 3) show a very smooth distribution with latitude as well as isothermal conditions from the ice limit to the Polar Front, the individual oceans and sectors exhibit almost as large deviations from this average as those found anywhere else in the world ocean. This stresses the considerable departures from strict zonal distributions of temperature and also of the other properties of sea water. Figure 4 shows a map of sea surface temperature anomalies from the zonal mean (after DIETRICH & KALLE, 1957) which is tabulated in the insert of Figure 4. Such representation reveals at a glance the thermal regime of an area with respect to "average" conditions. It is seen that, for example, the Atlantic sector and the Indian sector are much cooler than the Pacific sector, a fact also true for the air temperature. The cold pole does not lie at the geographical pole but is displaced toward East Antarctica (reference is made to the section on climatology in this volume) to the vicinity of the Pole of Inaccessibility which is the point farthest away from any coast (about 83°S and 63°E). From Figure 4 the actual mean temperature for a particular region can be read off using the zonal mean temperatures.

So far we have discussed only the large scale temperature distribution. Superimposed on the climatological average are fluctuations of various periods. Neglecting possible very long period changes over decades or centuries, the largest temperature wave is the one with the annual cycle. Its range is given in Figure 3 by curve 2. In this discussion the diurnal changes may be considered as the smallest cycle. In middle latitudes the diurnal range amounts to about 0.5°C on the average (MOSBY, 1933). In higher latitudes the range becomes even smaller owing primarily to the existing radiative conditions (reference is made to the section on climatology in this volume). It is further modified by cloudiness and velocity, i.e. with an increase in either of these parameters the diurnal range decreases. Since the area of our concern is known to have a high frequency of clouds (VOWINCKEL, 1957) as well as high wind velocities, only a very small diurnal temperature variation, if any, should be expected.

Between these two periods all kinds of fluctuations may be expected depending mainly on the scales at the atmospheric motions. More complicated conditions are experienced near islands, where local conditions may play a considerable role. The example in Figure 5 illustrates such fluctuations in the vicinity of South Georgia in East Cumberland Bay. This figure presents graphically the temperature, salinity and density fluctuations at the surface and 75 meters over a period of almost 5 months at a station in East

104

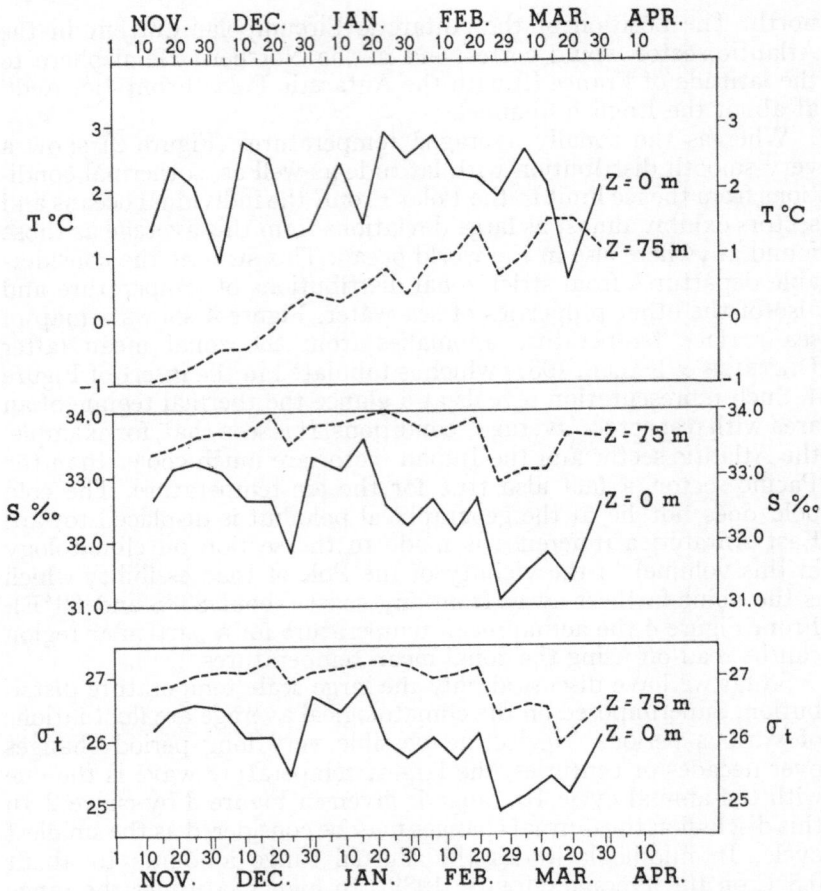

Fig. 5. The temperature, salinity and density σ_t at 0 m and 75 m in East Cumberland Bay, South Georgia between November 12, 1930 and March 29, 1931 (data from *Discovery Reports*, vol. 4, 1932).

Cumberland Bay of South Georgia. The observations were made in 1930—31 at weekly intervals (Discovery Reports, vol. 4, 1932). Whereas the sea surface temperature varies greatly and erratically, the temperature at 75 meters exhibit less variations and a pronounced seasonal trend.

Salinity at the sea surface

Whereas the temperature at the sea surface is determined by the local heat exchange the salinity depends on the local water exchange. The exchange with the atmosphere is determined by the evaporation minus precipitation. Excess of evaporation increases the salinity,

while the excess of precipitation lowers the salinity. The salt flux
into a particular region is given by the net advection of salt and by
mixing. In high latitude regions, as near the Antarctic continent,
freezing and melting become very important. Finally, in coastal
regions and, in particular, adjacent seas the run-off from the continent must be included in the budget. The zonal average salinity
distribution is (Figure 6, curve 1) highly correlated with the atmospheric exchange rate, evaporation minus precipitation, as shown

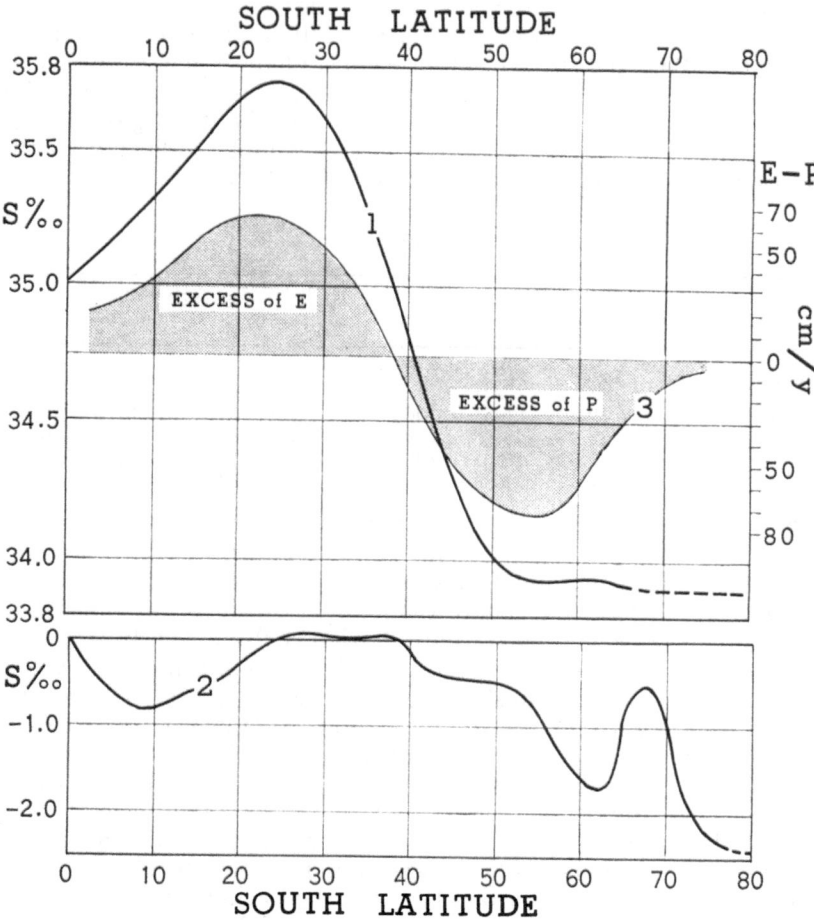

Fig. 6. Average zonal salinity distribution at the sea surface in the Southern
Hemisphere (curve 1, after WÜST et al., 1954), the salinity difference Northern
Hemisphere minus Southern Hemisphere (curve 2, from WÜST et al., 1954) and
zonal distribution of evaporation minus precipitation (curve 3, after WÜST et al.,
1954).

by Wüst (1954) (Figure 6, curve 3). Particular attention should be placed on the belt of maximum excess of precipitation, to which we will refer further in connection with the salinity distribution at the 200-meter level. Curve 2 in Figure 6 gives the average zonal salinity in the Northern Hemisphere minus that in the Southern Hemisphere. South of 40°S the salinity becomes higher than that in corresponding latitudes in the Northern Hemisphere. Maximum differences exist in very high latitudes due to the very low salinity in the Arctic Ocean which is semi-enclosed and effectively separated from the world oceans by land barriers. According to Figure 6 and other presentations (Sverdrup et al., 1942; Dietrich & Kalle, 1957; Defant, 1961) the surface salinity south of the Polar Front seems to have a constant value around $33.9\%_0$. However, a particular observation may show a value of as low as about $31\%_0$ or as high as $34.5\%_0$. These large variations, as for example, indicated by the stations Ob 191 and Ob 192, exist on a small scale.

Table I.

	OB 191					OB 192			
$\varphi = 68°18'S$		$\lambda = 73°08'E$				$\varphi = 67°57'S$	$\lambda = 71°39'E$		
Depth	t°C	S‰	σ_t	$O_2\%$		t°C	S‰	σ_t	$O_2\%$
0	1.21	30.58	24.51	106		1.37	34.15	27.36	103
10	2.52	32.87	26.25	113		1.21	34.20	27.41	105
25	3.16	33.62	26.93	108		0.04	34.29	27.55	101
50	—1.48	34.33	27.65	93		—1.48	34.45	27.74	90

These two stations are only 40 n. miles (74 km) apart, the surface salinity difference being more than $3.5\%_0$. However, at a depth of only about 30 to 35 meters both stations exhibit nearly similar S values. Numerous other examples could be found in the IGY and Discovery data. Obviously, the great lowering of the salinity is produced by ice melt in the vicinity of the stations. The Oceanographic Atlas of the Polar Seas, Part I Antarctica (1957) shows for the summer month of February a belt around Antarctica, where the surface salinity frequently ranges between 32.0 and $34.5\%_0$. This belt is limited to the south by the ice limit and extends north to about 60°S in the Atlantic and Indian sectors to about 100°E, and is relatively narrow around 70°S in the Pacific sector. Its largest northward extent (56°S) is found at about 15°W. The Atlantic and Indian sectors are also the regions with greatest ice release from the Antarctic Continent (Maksimov, 1961). Therefore, maps of average surface salinity are difficult to construct inasmuch as quite a large number of observations must be available to obtain representative averages. Furthermore, such presentation should also contain the range of the variations in order to be meaningful.

In coastal regions day-to-day fluctuations of large amplitude are encountered at times. This is illustrated in Figure 5. Large temperature fluctuations of almost 2 °C at the surface were quite often observed. At the 75-meter level the more smooth seasonal trend prevails. Comparing the salinities and densities reveals that at this low temperature the density distribution follows quite closely the salinity distribution. This example strikingly illustrates how meteorological and glaciological factors influence greatly the oceanographic properties of a particular limited region.

The salinity distribution at 200 meters

Since the range in surface salinity is so large and, as of now, the observations not abundant enough to present meaningful mean maps, the 200-meter level was chosen to describe the salinity distribution around Antarctica. This level is below the direct effects of atmospheric forces and air-water exchange processes; therefore, there is some hope of deducing from the scanty data a representative picture of the average salinity distribution at this level. Figure 7 presents this map. It was constructed from some 2800 salinity observations (pre-IGY and IGY data), more than half of which fall into the Atlantic sector. Surprisingly, quite a large body of observations was obtained at high latitudes near the Antarctic shores or ice boundary, whereas, in particular in the Indian and Pacific sectors, the zone between 50 °S and 30 °S (the northern limit of this map) is barely surveyed. It is hoped that the International Indian Ocean Expedition, commencing at this time, will close the gap in the Indian sector, while the activities of the new research vessel Eltanin, operated by the U.S. National Science Foundation, will extensively cover the Pacific sector south of and near the Polar Front in its first year of operation (1962).

Most data was obtained in the summer season. However, as the seasonal fluctuations at the 200-meter level are rather minute, all data can be treated as representative of the annual conditions, in particular south of the Polar Front. Some large scatter is to be expected near the Polar Front and north of it resulting from seasonal fluctuations and possible meandering of the Circumpolar Current. So far the evidence presented (DEACON, 1937, p. 24) seems to indicate that these fluctuations are small, if the position of the Polar Front is taken as an indicator.

The main features of the salinity distribution at 200 meters manifest themselves in the continuous belt of low salinity near 50 °S in the Atlantic sector and a belt of high salinity near 60 °S in the Pacific sector. Near the Antarctic coast the salinity values

108

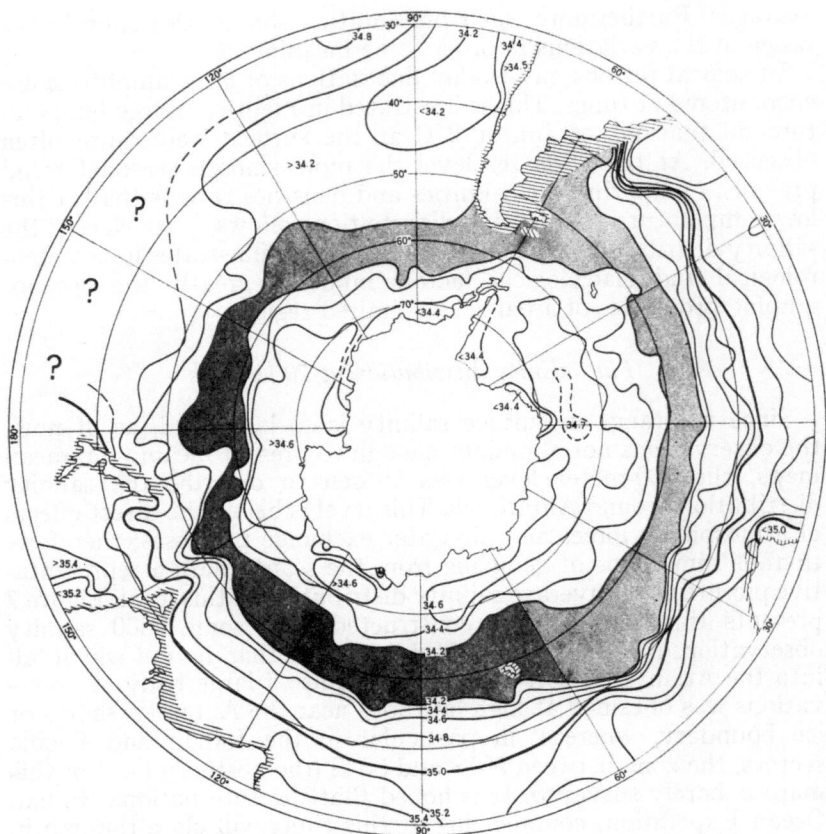

Fig. 7. Salinity distribution at the 200-meter level based on averages for $2\frac{1}{2}$°-fields.

decrease again. The low salinity belt results from a combination of atmospheric influences (excess of precipitation over evaporation in and south of the westerly belt (Figure 6, curve 3)) and vertical motion associated with the currents. North of the Polar Front the water sinks into deeper and deeper layers while spreading northward. This can be seen also from the vertical longitudinal section of salinity distribution (see Figure 8). The belt of maximum salinity near the Antarctic Continent at 65°S coincides very closely with the zone of divergence according to KOOPMANN (1953), who from the wind distribution calculated the position of the convergences and divergences around Antarctica. Finally, the low salinity near the ice edge is undoubtedly the result of increased melting and downward motion which KOOPMANN calls the "Festlandskonvergenz" or Continental Convergence.

The temperature and salinity distribution in the vertical

Studying the observational facts of the various expeditions it is always surprising to find the constant recurrence of certain characteristic features in the vertical distribution of the hydrological variables irrespective of the particular section under consideration. This pattern repeats itself in the Atlantic, Indian and Pacific sectors as well as in the connecting passages. Thus, except for continental influences, we may discuss the general distribution of the variables along a particular cross section south of South Africa as obtained by the Discovery expedition. Figure 8 presents the temperature

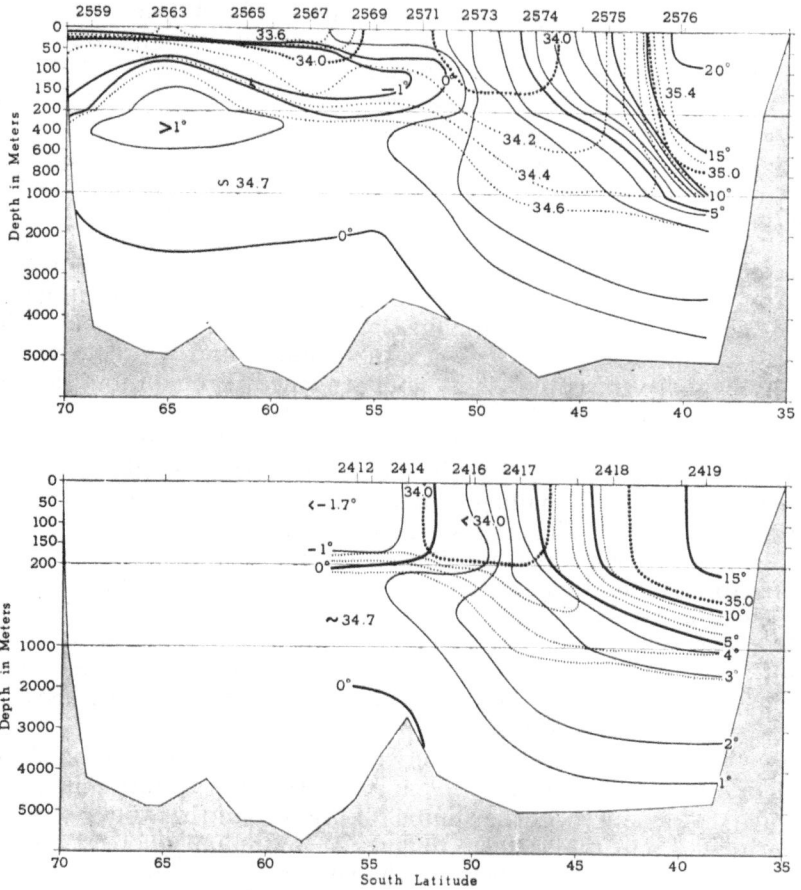

Fig. 8. Meridional section between Antarctica and South Africa along 20 °E, of the temperature (solid lines) and salinity (broken lines) distributions. Top figure summer conditions, bottom figure winter conditions. Numbers at the top refer to the Discovery stations used. Note change in vertical scale at 200 and 1000 meters.

distribution (solid lines) and the salinity distribution (broken lines) in summer (top) and winter (bottom). The temperature distribution is characterized by the very cold sub-surface tongue of less than —1°C in the core. This tongue extends northward approximately to the Polar Front and has been used in part by MACKINTOSH (1946) to identify this phenomenon. Underneath the cold tongue the relatively warm Deep Water "flows" southward, as indicated in Figure 8 by the bending of the 2°-isotherm. In the summer case we find at station 2569 a surface temperature of 1.28°C, —1.19°C at 150 meters and 0.92°C at 400 meters. Thus, a warm layer lies on top of the very cold tongue which is the remnant of the winter water. As seen in the bottom section (winter) the 0° and —1° isotherms extend to the surface forming homogeneous conditions. In summer owing to warming from the top a warm surface layer is formed. Vertical mixing due to wind action does not extend below 100 meters and thus does not affect the cold sub-layer. Sometimes, especially in the late summer season, the cold sub-layer tends to warm up, but never to the degree that the layer of sub-surface minimum temperature is destroyed. Ob sections obtained during the IGY also show that, occasionally, the cold water tongue degenerates into a cold water pouch in the region where the isohalines begin to slope downward. Figure 8 gives evidence that the conditions in the sub-layer between the Antarctic Continent and, in this cross section, 45°S do not change appreciably between the summer and winter seasons.

The salinity distribution is more stratified. From Antarctica to about 50°S most of the changes in salinity occur in the top 200 meters. Approximately at the latitude where the Polar Front is located (in Figures 8 and 10 about 50°S) the salinity gradient decreases and the isohalines drop to deeper levels, as apparent from the 34.5 isohaline. This is also the region where isotherms and isohalines cross each other and form a strong solenoidal field in contrast to the conditions further north. This solenoidal field seems to be characteristic of the Circumpolar Current, especially its poleward side. This kind of thermohaline structure is also found on the left side of the Gulf Stream as is the sub-surface cold water accumulation in a corresponding position with respect to the current (STOMMEL, 1958). At the surface, north of the Polar Front, a belt of low salinity water surrounds Antarctica. In Figure 8 it is seen north of 50°S. According to this example it seems that the salinity minimum is more pronounced in the summer section than the winter section (bottom). From the surface, tongues of low salinity spread downward and northward indicating the sinking of Antarctic Intermediate Water (SVERDRUP et al., 1942), which can be traced, for example in the Atlantic sector, by its characteristics of low salinity and high oxygen content to at least the Equator (WÜST, 1936) (Figure 12).

The density distribution

As the density is a function of temperature, salinity and pressure, a few remarks are appropriate about the influences of these variables on the density. The pressure effect is commonly eliminated by referring the density values to atmospheric pressure. The functional relation between temperature and salinity on the density is a complicated one. In general, the density increases with decreasing temperature and increasing salinity. However, the rate of change in temperature to produce the same density change differs depending on the temperature range we are dealing with. Figure 9 shows graphically how the "density, σ_t" (defined as $\sigma_t = (\rho_{s,t}-1)\cdot 1000$, where $\rho_{s,t}$ is the density of water at the temperature t and salinity s) varies with temperature and salinity. The shaded area in the figure bounded by the 2° and —2° isotherms and the 30‰ and 35‰ isohalines denotes approximately the range of water found around Antarctica. It is seen that at these low temperatures the density is almost entirely determined by the salinity, which is not true for water at higher temperatures. Thus, it is not surprising that the density distribution follows closely the salinity distribution in Antarctic regions. As a matter of fact, as we have seen in the discussion of the temperature distribution, the very low temperatures

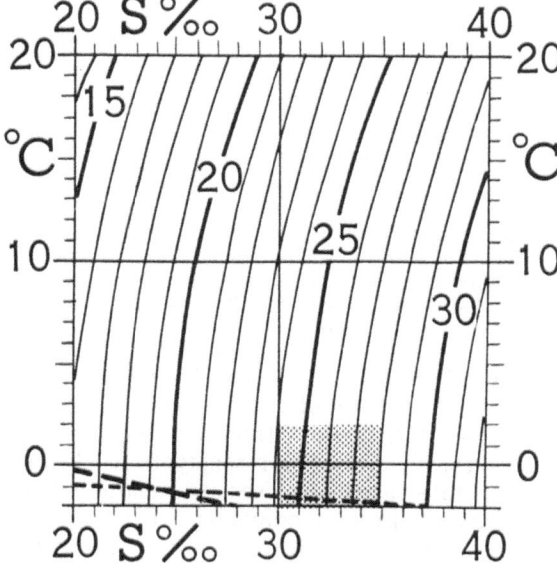

Fig. 9. The density σ_t as function of temperature and salinity. Heavy broken line indicates the density maximum, light broken line indicates the freezing point, both as function of temperature and salinity.

112

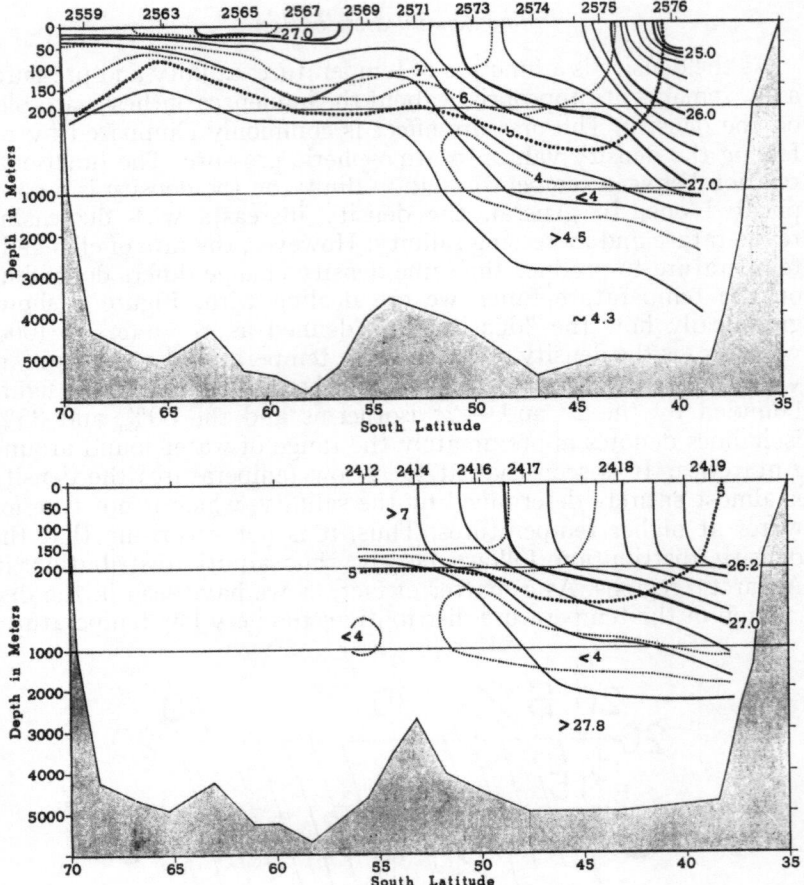

Fig. 10. Meridional section between Antarctica and South Africa along 20°E, of the density σ_t (solid lines) and oxygen content (broken lines) distributions. Top figure summer conditions, bottom figure winter conditions. Numbers at the top refer to the Discovery stations used. Note change in vertical scale at 200 and 1000 meters.

at 150 meters depth (winter water) can be maintained only because of this behavior and, in spite of the 3° colder water lying on top of the warm Deep Water, the stratification is still stable. Furthermore, the term "thermocline" so extensively used in oceanography at low and moderate latitudes loses its meaning here. There the vertical temperature gradient coincides with the vertical density gradient separating the warm top layer from the cold deep layers. This is not the case in Antarctica and here we should speak about the "halocline" or better "pycnocline," because the density distribution is the important quantity in dynamic considerations.

The distribution of density in the Antarctic Ocean will be discussed by means of Figure 10 (solid lines) which shows the same cross sections as Figure 8 along 20°E, south of South Africa. Comparison of the two figures shows that between the Antarctic Continent and station 2574, at about 47°S, the density field is indeed extremely similar to the salinity field, whereas north of 47°S the increased importance of the temperature begins to be felt. South of the Polar Front, and below 200 meters the water is of almost uniform density, σ_t, around 27.8. Above 200 meters we encounter the transition layer of the pycnocline reaching a surface density, σ_t, of from 27.0 to 27.3. Also seen in Figure 10 is the slight increase of σ_t between 51°S (just south of the Polar Front) and 57°S, the decrease between 57° and 62°S and the increase again near 65°S.

Around 50°S (Polar Front) the isopycnals slope downward to the north indicating a strong current, the Circumpolar Current. Between stations 2575 and 2576 (top), and 2418 and 2419 (bottom) another strong slope of the isopycnals is associated with the return branch of the Agulhas Current (DIETRICH & KALLE, 1957) which carries warm water from the Indian Ocean around the south tip of Africa (Figure 4).

This general density distribution is found in more or less modified form all around Antarctica, especially the part of the section which covers the Antarctic regime and the Circumpolar Current.

The oxygen distribution

In addition to the distribution of temperature and salinity, the oxygen distribution is of paramount importance for the physical as well as the biological oceanographer. Not only is oxygen necessary for all respiratory processes, but it is also considered as an indicator of the "age" of a water mass. Therefore, it is not surprising that the oxygen content is used by the physical oceanographer as an aid in determining the relative motions in the ocean.

The distribution of oxygen depends on the various sources and sinks in the hydrosphere. The main supply of oxygen is provided by the atmosphere and to a lesser degree by photosynthetic processes within the photic zone. Thus, the surface and the near-surface layers are regions of sources. On the other hand, oxygen depletion may occur at all depths by respiratory processes and decomposition of organic material. The actual distribution of oxygen is determined by the interplay of supply and depletion, and the redistribution by horizontal and vertical motions in the ocean. Assuming that the rate of O_2 depletion is nearly constant in the vertical from the surface to the bottom, the rate of O_2 supply will depend on the distance between the surface and the place in question as well as on the velocity of the water and the rate of mixing. The rate of depletion is a function of the speed of oxidation of organic substances (i.e. primarily a function of temperature) and of the amount and kind of organic substances available for oxidation.

The capability of water to absorb gaseous oxygen depends mainly on the temperature and only slightly on the salinity. The lower the temperature the higher is the saturation value. Therefore, the highest oxygen values are found in the cold waters surrounding Antarctica. Figure 10 shows superimposed on the density distribution the oxygen distribution (dashed lines) in the vertical along a cross section at 20°E, for a summer section (top) and winter section (bottom). In the surface layer south of 50°S (Polar Front) generally an oxygen content of more than 7 ml/l is found which decreases rapidly below the strong pycnocline between 100 and 200 meters. The next most characteristic feature of the oxygen distribution is the existence of an oxygen minimum layer at mid-depth, at 45°S at about 1500 meters. The minimum slopes upward in the vicinity of the Polar Front at 50°S and remains at about 500 meters south of this front. This general distribution is found all around Antarctica with only slight regional modification.

If the oxygen content is given in units of percent saturation, then near the Antarctic Continent we find mostly near saturation conditions. Many of the Ob observations in the Indian sector indicate extensive supersaturation in the surface layers, generally to about 25 meters (see Table I). However, sometimes relatively very thick layers are supersaturated as at Ob 188 (68°56'S, 77°58'E), where the saturation value is reached at about 100 meter depth with an intense supersaturation maximum of 118% at 11 meters. The source may be attributed to the abundant plankton growth. In addition to this biological cause it is possible to obtain supersaturation by mixing of two water masses of different temperatures, both saturated with oxygen, as pointed out by RAKESTRAW & EMMEL (1938). A third possibility, which could contribute to the oxygen supersaturation, is: if very cold water saturated with oxygen is seasonally warmed in the surface layer and the oxygen content does not change by biological processes, then the temperature effect alone could produce supersaturation. It is possible that all these processes are active in Antarctic water. However, the degree and extent of supersaturation in the surface layers and its relation to other nutrient components needs more investigation.

Three cases of vertical distribution of oxygen content are shown in Figure 11: Ob station 408 is located south of the Polar Front, Ob station 413 north of but near the Polar Front and Ob station 415 well north of the Polar Front. It is clear that the surface layer of constant oxygen content increases in thickness with decreasing latitude. The transition layer between 200 meters and about 1000 meters becomes enriched in oxygen in the same direction, as indi-

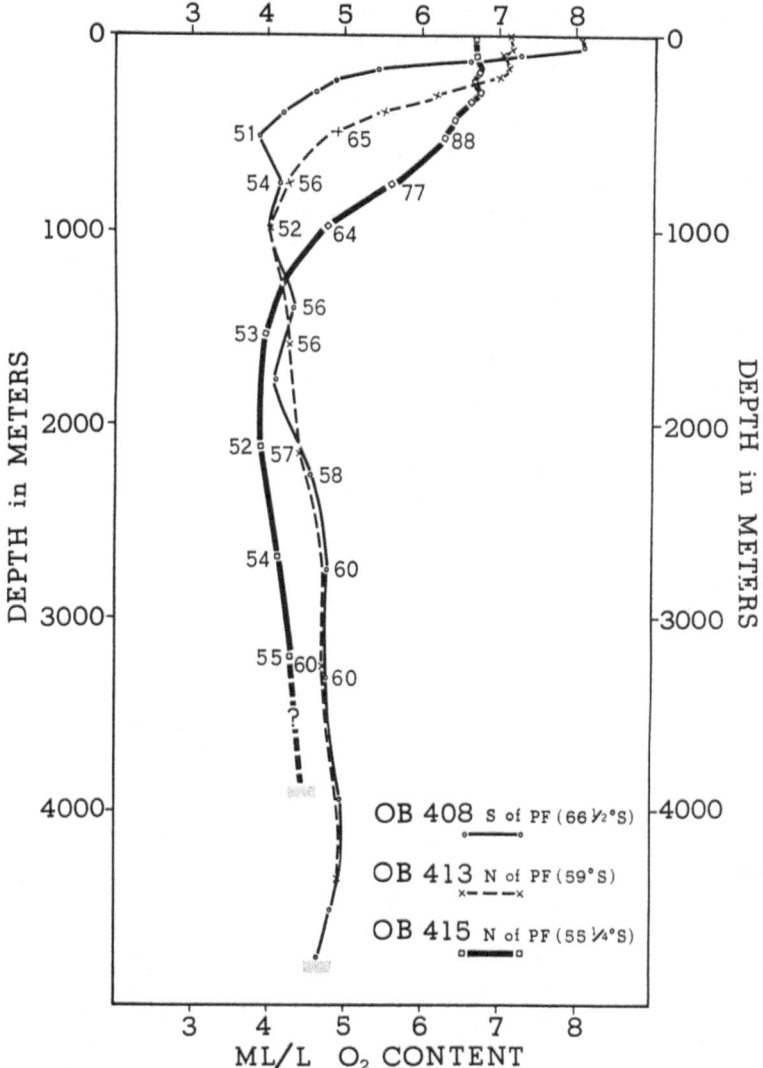

Fig. 11. Oxygen content as function of depth for three selected Ob stations. Numbers along the curves give the percentage of saturation.

cated by the values of relative oxygen content (% of saturation). The minimum is found at station 408 in 500 meters, at station 413 in 1000 meters and at station 415 near 2000 meters.

In general, it can be stated that a good relation between the oxygen and phosphate distribution exists, i.e. low O_2 values are

116

usually associated with high P values. However, it seems that this relation is much better developed in the Antarctic Ocean and the Circumpolar Current than in other regions of the world ocean. WATTENBERG (1938) pointed out that on the basis of the first two Meteor sections, which were the most southern ones, it was suspected that the depth of the O_2 minimum coincides with depth of the P maximum everywhere. However, observations from moderate and tropical regions revealed that there the P maximum is located several hundred meters *below* the O_2 minimum. This fact formed the basis for the conclusion that in most parts of the ocean chemical and biological processes play an important role in the formation of the O_2 minimum. On the basis of Meteor data WATTENBERG (1957, p. 152) found that between 30°S and 50°S the O_2 minimum lies *below* the P maximum. Subsequent IGY observations (Ob data) showed this to be true in many cases also in very high latitudes. This exception in higher southern latitudes led WATTENBERG (1957, p. 152) to conclude that here "kinematic processes are so strong down to mid-depths that they determine the distribution of oxygen and phosphate."* Thus, the vertical distribution of O_2 and P has been used as supplementary evidence in finding the so-called reference level for calculating the geostrophic currents through the Drake Passage (OSTAPOFF, 1960).

The circulation in the Antarctic Ocean

The horizontal and vertical circulation is derived primarily from the distribution of the variables discussed above and, in the case of surface currents, from ships' records. Recently, surface currents also have been determined by means of the Geomagnetic Electrokinetograph (GEK) (LEDENEV, 1961).

With the exception of currents near Antarctica, the surface water moves around Antarctica in general from west to east with maximum velocity near and north of the Polar Front. The actual speed in the core of the Circumpolar Current is difficult to assess. In the Drake Passage calculations of the geostrophic current, assuming a reference level of zero velocity at 3000 meters**, indicate maximum velocity of about 27 cm/sec (OSTAPOFF, 1960). This is an average velocity over a meridional distance of about 100 km. Probably the core velocity amounts to 40 to 60 cm/sec. Figure 12 shows the velocity distribution along the 0° meridian based on hydrographic observations of the Discovery II and the 2000 meter reference level as an average across the entire cross section. The curves have

---

* Translation from German by author.
** As, at present, the absolute pressure field in the ocean cannot be determined from observations alone, calculations of the geostrophic current yield a *relative* velocity distribution. These relative values are reduced to absolute values by assuming a certain velocity (commonly zero velocity) at a particular level.

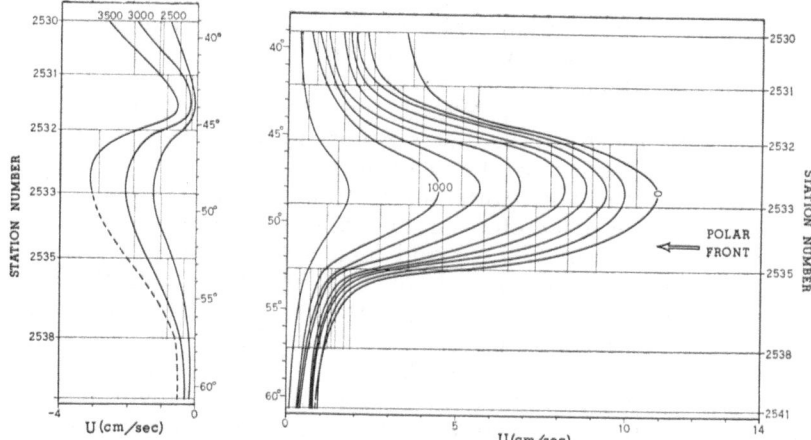

Fig. 12. Geostrophic velocity distribution along the 0° meridian assuming a 2000 d-bar reference surface for the following levels: 0, 100, 200, 300, 400, 600, 800, 1000, 1500, 2500, 3000 and 3500 meters. Station numbers refer to Discovery stations.

been fitted in such a way that their averages between the corresponding stations equal the computed average value (vertical thin lines). In this case the stations are about 400 km apart and the maximum velocity at the surface is approximately 10 cm/sec. The family of curves represents the velocity distribution at selected levels. It should be emphasized that this method of calculating the zonal velocity component assumes that there is a balance between the pressure forces and the Coriolis force and that the inertia and friction terms are considered to be of minor importance. Investigations by WYRTKI (1960) support this notion for the zonal velocity components, whereas the meridional velocity components seem to be governed more by the frictional terms.

South of the Polar Front the eastward zonal velocity component diminishes rapidly and there is some indication that a northly flow is superimposed upon it. Around 65°S, about where the Antarctic Divergence zone is located, the direction of the flow reverses and we find westward flow near the Antarctic Continent in accordance with the general wind distribution. The Weddell Sea, Ross Sea and the Bellingshausen Sea exhibit motions in a clockwise direction. As the western shores of the Ross Sea and Bellingshausen Sea are less protruding than that of the Antarctic Peninsula in the Weddell Sea, these coasts remain relatively ice free in summer.

The meridional circulation is mainly deduced from the temperature and salinity distribution along meridional cross sections. Figure 13 illustrates schematically in the form of a block diagram*

---

* The author wishes to gratefully acknowledge the kindness of Prof. WÜST in supplying this revised diagram.

118

the important circulation features. North of the Polar Front the Antarctic Intermediate Water sinks and spreads northward as is indicated by the tongue of low salinity. Underneath the Intermediate Water, the Deep Water, located at about 2000 to 3000 meter depth north of the Polar Front, moves slowly southward and begins to rise near the Polar Front to about 500 to 1000 meters, thus, forming the huge water mass around Antarctica below the pycnocline. It is characterized by high temperatures (the intermediate t-maximum, Fig. 8) and high salinity. Surface water in winter becomes dense enough to sink and mix with the Deep Water forming a new water type, the Antarctic Bottom Water. Freezing of surface water increases the salinity, forming water with a density, σ_t, of 27.89 (34.62‰ and —1.9°C) which is heavier than the adjacent circumpolar water. While sinking the heavy surface mixes with the deep water, which is more saline and warmer forming the bottom water of 34.66‰ and —0.4°C. This water flows down the continental slope (SVERDRUP et al., 1942). The coldest water of —0.92°C is found at the western edge of the Weddell Sea. Smaller areas of formation of bottom water seem to be present near the Antarctic Continent between S. Orkney Island and S. Shetland Island [DEACON (1937) concluded this from obser-

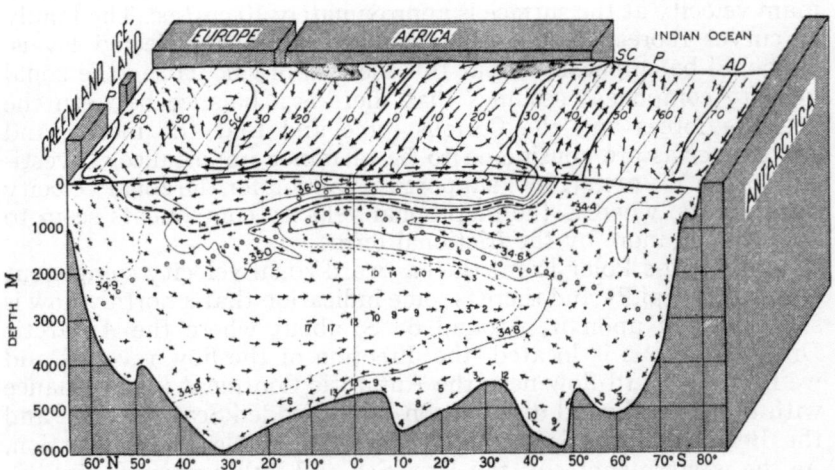

Fig. 13. Schematic block diagram of the deep circulation and salinity distribution along the west side of the Atlantic Ocean and the currents at the surface (after WÜST, 1951 and 1957).

Legend:

SC = Subtropical Convergence; P = Polar Front; AD = Antarctic Divergence; ∼ Physical sea level; - - - - - Boundary of warm water sphere (> 8°); — Isohalines; ⊛ ⊛ ⊛ Reference Level; ⠿⠿ Upwelling. *Surface:* → 5—40 cm/sec; ➡ 40—170 cm/ sec; *Depth:* → Deep currents with calculated velocities (cm/sec); ⇢ Convection and advection; o ⊛ o practically no currents. Vertical exaggeration X 1300.

vations by Discovery II] and along the Antarctic Continent in the Indian sector [based on observations of B.A.N.Z.A.R. Expedition (SVERDRUP, 1940)]. As the main area of formation of bottom water lies in the Weddell Sea from where it moves northward along the eastern side of the Scotia Arc it is most conspicious in the western basins of the Atlantic Ocean (WÜST, 1936) and can be traced readily by the potential temperature distribution. According to DIETRICH & KALLE (1957) there are also indications that bottom water penetrates into the Indian Ocean and the western Pacific Ocean, but in smaller amounts than in the Atlantic Ocean.

The Polar Front

The Polar Front, also called Antarctic Convergence, is often used to separate the Antarctic regime from the sub-Antarctic regime (SVERDRUP et al., 1942). It can be defined in several ways depending on the notions about the nature of this phenomenon. MEINARDUS (1923) first recognized the existence of the Polar Front as a circumpolar phenomenon. Analyzing the observations made on Gauss in 1901 to 1903 he found that in the Indian sector the meridional sea surface temperature gradient increased considerably in about 50 °S. Comparing the observed temperature distribution with the expected climatological distribution he noticed the anomalously cold water south of 50 °S (for example, Figure 4). Using the observations of the Challenger, Valdivia and Gauss, MEINARDUS determined the position of the Polar Front between the meridians 105°W and 80°E embracing the Atlantic sector. DEACON (1937) and MACKINTOSH (1946) expanded this work using all available Discovery data. This work leads to the conclusion that the nature of this front is extraordinarily stable and stationary from season to season and year to year. DEACON (1937, p. 24) reports that in the Falkland area, where enough observations were collected to permit general conclusions, the position of the Polar Front does not vary very much: — "probably not more than 60 miles between its extreme positions, and it shows no regular seasonal change." This is indeed very remarkable.

Naturally, the front was defined in terms of the surface temperature distribution as the surface temperature is the most easily measurable quantity. However, DEACON (1937) already pointed out that certain characteristics associated with the front seem to be as well pronounced in the deep layers as in the surface layer. The stability of the phenomenon as well as the features of the deep layers lead DEACON to conclude that the position of the Polar Front is controlled by the deep and bottom water. At the front the heavier Antarctic surface water moving north-eastward sinks to levels corresponding to its density. South of the front the deep water rises toward the south and prevents the Antarctic surface water from sinking. But when the Antarctic water reaches the latitude where the deep water begins to ascend, the surface "Antarctic water flows over the steep ascent of the warm deep water like a stream over a waterfall". Earlier, DEACON (1934a) and SVERDRUP (1933, 1934) seemed to agree that the wind field is primarily the responsible force for the existence of the front. Lately, both explanations have been invoked to explain some characteristics associated with the front (WYRTKI, 1960).

There are several possible criteria for the Polar Front. The first one relies entirely on the sea surface temperature distribution. The front is placed between 2° and 6°C where the sharpest drop in temperature is observed. WEXLER (1959) noticed that continuous

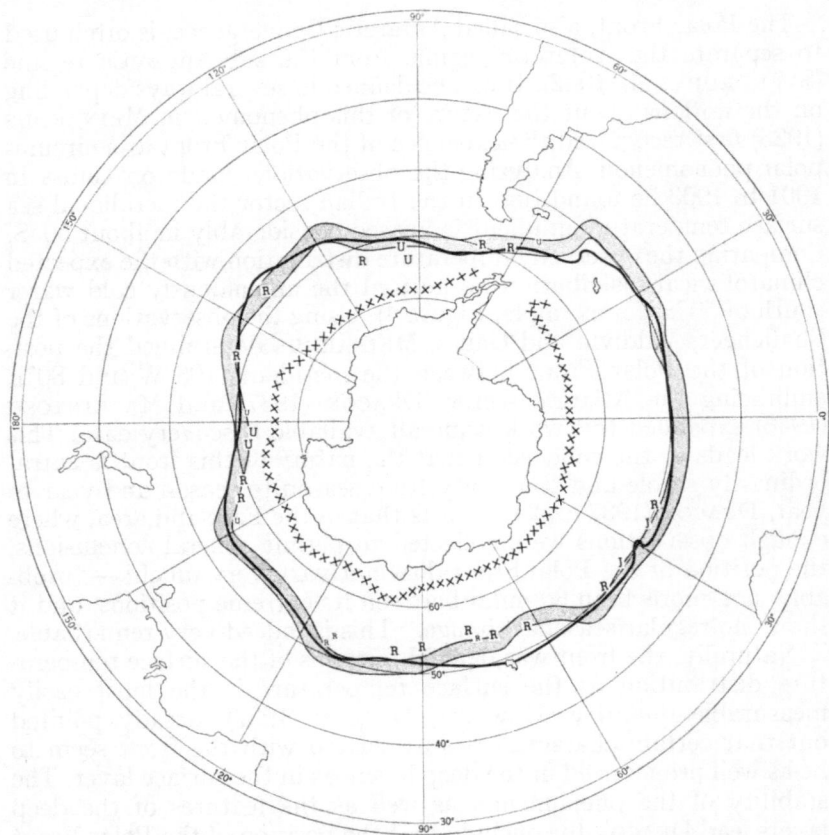

Fig. 14. The band of the salinity minimum (34.00 to 34.2 ‰), indicated by the shaded area, and the band of the salinity maximum (>34.6‰), indicated by the crosses. Continuous line represents the Polar Front according to MACKINTOSH (1946). Letters indicate the position of the Polar Front as found during IGY, R — Russian, J — Japanese, U — United States observations.

traces of surface temperature frequently showed a minimum temperature near the Polar Front and used this feature in finding the position of the Polar Front. These definitions have some drawbacks, because frequently a pronounced gradient at the surface may be masked by the (unknown) meteorological conditions prior to the time of recording. Looking at the problem from the synoptic point of view, we have to start from the fact that climatological mean charts will represent long period motions but daily weather maps soon convince us that the belt of westerlies should lie to the north of the cyclone tracks and winds will alternate their direction with the frequency of cyclone passages. Associated with the atmospheric high and low pressure cells are areas of surface convergence and divergence, respectively, as well as advection to the north or south depending on the direction of the wind. Thus, the sea surface temperature will depend to a great degree on the past meteorological conditions over the area under investigation and a unique definition of the position of the front on this basis becomes at times very difficult.

MACKINTOSH (1946) still retained the surface criterion, but added a further feature of the thermal structure near the front: he chose the surface temperature gradients between two hydrographic stations where one station showed the sub-surface temperature minimum above 200 meters and the other below 200 meters indicating the region where the surface water begins to descend. On this basis he constructed the map of the mean position of the Polar Front or Antarctic Convergence. This line is drawn in Figure 14. Superimposed is a band of shaded area which is obtained from the salinity distribution at 200 meters (Figure 7), representing the axis of the minimum salinity. Since the water of low salinity originates at the surface its presence at the 200-meter level must be the result of downward advection and we are inclined to use this criterion to place the mean position of the Polar Front at the southward side of this band. As not enough observations exist for the Pacific sector, here the position of the Polar Front must be considered as approximate. Although, in general, the two presentations, MACKINTOSH's and ours, agree quite well in most areas, there are some deviations which deserve our attention. The most conspicious deviation is found in the Indian sector. Near Kerguelen Island our convergence line lies about 5° of latitude further south than those of DEACON and MACKINTOSH who let the convergence run through Kerguelen Island, however MACKINTOSH did not have observations for a section of about 60° of longitude. Russian IGY observations support the more southern location of the Polar Front, although MAKSIMOV (1961), on the basis of surface temperature gradients, places the Polar Front still further south. Using MACKINTOSH's definition the position of the Polar Front, as found during IGY

122

cruises, has been indicated in Figure 14 by capital letters (R — Russian, J — Japanese, U — United States). In almost all instances the convergence was found slightly south of the minimum salinity at 200 meters.

The general downward motion north of the Polar Front can be illustrated also by the oxygen distribution at the 200-meter level as function of latitude. Figure 15 shows this based on stations of the Discovery expedition and the Ob expedition. Curve 1 is the average distribution based on 8 sections at 20° E of the Discovery obtained from 1938 to 1939. Curve 2 and 3 show the O_2 distributions along 160°W and 110°W, respectively, based on Ob observations. The respective latitude scales have been shifted in such a way that all three sections are oriented similarly with respect to the Polar Front (arrow). This presentation shows that the oxygen content increases slightly south of the Polar Front, as defined above. Furthermore, the distribution seems to indicate that prior to the rise in oxygen content a minimum of O_2 exists some 2° to 3° of latitude south of the Polar Front. Such a minimum could indicate a divergence near the Polar Front and could explain the temperature minima at the surface found in the same region by WEXLER (1959) in many bathythermograph records.

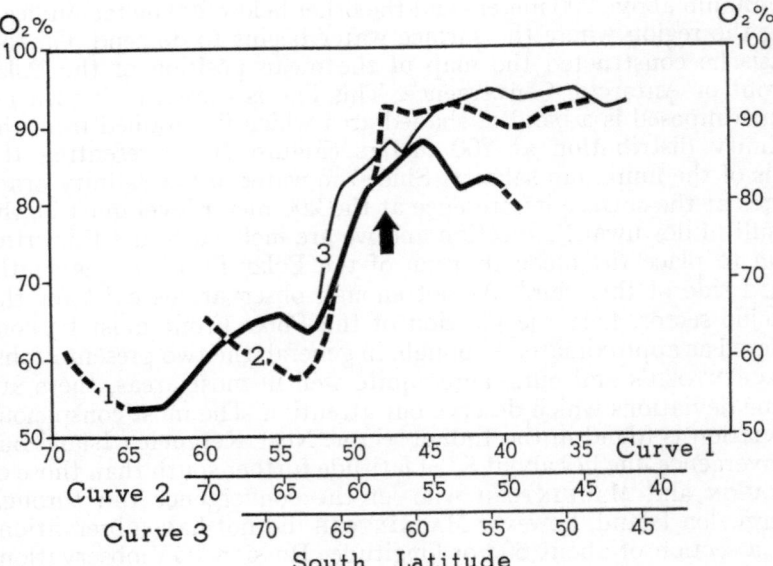

Fig. 15. Oxygen distribution (in percent of saturation) at 200 meters along selected meridians. Curve 1 along 20°E represents the average based on 8 Discovery sections, curve 2 along 160°W based on OB section and curve 3 along 110°W based on OB section. All three curves are oriented with respect to the Polar Front (arrow).

A second major singularity is found near 65°S and termed the Antarctic Divergence. It is recognizable in Figure 14 by the belt of high salinity at 200 meters. This divergence in the surface layer coincides with the divergent belt as computed by KOOPMANN (1953) from the mean atmospheric pressure distribution. His calculations also reveal two secondary divergences, the Bouvet Divergence and the Inserted Divergence south of New Zealand. These divergences seem to be closely related to the wind field and are supported by oceanographic evidence. KOOPMANN found in these regions increased silicate content and uses this as indicator for wind induced divergences. But also the salinity distribution is in agreement with these investigations. The area of salinity maximum at 200 meters (Fig. 7) is broader in extent in the region of the Bouvet Divergence as well as in the region south of New Zealand.

Finally, near the Antarctic Continent at the ice boundary a continental convergence is found by KOOPMANN. Evidence for such a convergence can be seen in Figures 8 and 10, where near the continent all isolines are bent downward. Figure 15, curve 1 shows also an increase of O_2 content at the 200-meter level in agreement with the other oceanographic parameters. Such a descent of cold surface water near the continent has been made responsible for a case of mass mortality of fish as reported by BRONGERSMA-SANDERS (1957) to have occurred on March 22, 1952 west of Adelaide Island in 67°04'S and 71°03'W. In this case the environmental conditions of the deep water, in which the fish (*Pleurogramma antarcticum* BLGR.) lives, have been drastically changed by the sinking of cold near-surface water.

There exist several detailed studies of the different aspects of Antarctic oceanography. In addition to the work already mentioned in this survey, we wish to cite the more important ones. In the field of physical oceanography the pioneering work of the Discovery Commission must be mentioned first. The probably most detailed description of oceanographic conditions in Antarctic and sub-Antarctic waters is due to the work of DEACON (1933, 1937). Recently, an important contribution concerning the hydrology of the water south of New Zealand has been published by BURLING (1961). The distribution of phosphate and silicate has been discussed by CLOWES (1938). The published results and observations by the Discovery Commission are still one of the primary sources of information. As far as the Atlantic sector is concerned, reference is made to the atlas volumes of the Meteor expedition (1936a and 1936b; 1940) showing the distributions of various physical and chemical variables.

A lengthy description of the oceanographic conditions around Antarctica was published by MAKEROV (1956) and some results of the Soviet Antarctic expeditions during the IGY and post-IGY periods

124

can be found in the publication series *Information Bulletin of the Soviet Antarctic Expedition*, published by the "Morskoi Transport" in Leningrad since 1958. The horizontal distributions of temperature and oxygen content at 3000 meters are presented in a paper by MAKSIMOV (1961), those of temperature, salinity and oxygen content at 4000 and 5000 meters in an unpublished atlas by STROUP & STOMMEL (1961). The *Oceanographic Atlas for the Polar Seas*, issued by the U.S. Navy Hydrographic Office in 1957 can be consulted about information on tides at the coasts of Antarctica as well as in the Antarctic Ocean. Various aspects of marine ecology are treated comprehensively in a volume published by the Geological Society of America (1957) [see reference BRONGERSMA-SANDERS (1957)] where Antarctic problems are discussed within a world-wide frame.

Recently, the *Committee on Polar Research* (1961) summarized the results of biological and geophysical research in Antarctica and made recommendations concerning future scientific endeavors in that area.

REFERENCES

BRONGERSMA-SANDERS, M., 1957. Mass mortality in the sea. *Geol. Soc. Am.*, *Mem.* 67, 1.

BUDYKO, M. I., 1955. *Atlas of the heat balance*. Leningrad.

BURLING, R. W., 1961. Hydrology of circumpolar waters south of New Zealand. *N.Z. Dep. Sci. Industr. Res. Bull.* 143, *1—66*.

Committee on Polar Research, 1961. *Science in Antarctica*. Part I. The life sciences in Antarctica. Part II. The physical sciences in Antarctica. Nat. Acad. of Sci.-National Res. Council, Pub. **878**, Washington, D.C.

CLOWES, A. J., 1938. Phosphate and silicate in the Southern Ocean. *Discovery Reports*, 19, *1—120*.

DEACON, G. E. R., 1933. A general account of the hydrology of the South Atlantic Ocean. *Discovery Reports*, 7, *173—238*.

DEACON, G. E. R., 1934. Die Nordgrenzen antarktischen und subantarktischen Wassers im Weltmeer. *Ann. Hydr. u. mar. Met.*, 62, *129—136*.

DEACON, G. E. R., 1937. The hydrology of the Southern Ocean. *Discovery Reports*, 15, *1—124*.

DEFANT, A., 1928. Die systematische Erforschung des Weltmeeres. *Z. ges. Erdkunde*, Sonderband, *459—505*.

DEFANT, A., 1961. *Physical Oceanography*, vol. 1, Pergamon Press, London.

DIETRICH, G. & K. KALLE, 1957. *Allgemeine Meereskunde*. Gebr. Borntraeger, Berlin-Nikolassee.

DREYER, A. J., 1961. Air and sea temperatures for Marion Island. *News Letter*, Weather Bureau of South Africa, No. **151**, *162—164*.

International Hydrographic Bureau, 1937. Limites des Océans et des Mers. *Spec. Publ. no.* 23.

KOOPMANN, G., 1953. Entstehung und Verbreitung von Divergenzen in der oberflächennahen Wasserbewegung der antarktischen Gewässer. *Dtsch. hydr. Zeit.*, Ergänzungsheft 2, *1—38*.

LAEVASTU, T., 1960. Factors affecting the temperature of the surface layer of the sea. *Soc. Sci. Fennica. Comm. Phys.-Math.* **XXV**, *1*.

LEDENEV, V. G., 1961. On the surface currents in the Pacific sector of Antarctica. *Bull. Sov. Antarctic Exp.*, no. **27**, *18—24*.

MACKINTOSH, N. A. & H. F. P. HERDMAN, 1960. Distribution of the pack-ice in the Southern Ocean. *Discovery Reports*, **19**, *285—296*.

MACKINTOSH, N. A., 1946. The Antarctic Convergence and the distribution of surface temperatures in Antarctic waters. *Discovery Reports*, **22**, *177—210*.

MAKEROV, IU. V., 1956. *Antarktica*. Part II. Gidromet. Izd., Leningrad, 119 pages (in Russian).

MAKSIMOV, I. V., 1961. The front of the Antarctic convergence and the long period changes of the northern limit of icebergs in the Southern Ocean. *Problemi Arktiki i Antarktiki*, no. **8**, *47—52*.

MEINARDUS, W., 1923. Ergebnisse der Seefahrt des "Gauss." *Dtsch. Südpolar-Expedition 1901—03*. III. Meteorologie, **1**, 1, *544*.

MOSBY, H., 1933. The sea-surface and the air. *Sci. Results of the Norwegian Antarctic Exp. 1927—1928*, Oslo.

MOSBY, H., 1936. Verdunstung und Strahlung auf dem Meere. *Ann. Hydr. mar. Met.*, **54**, *281—286*.

OSTAPOFF, F., 1960. On the mass transport through the Drake Passage. *J. geoph. Res.*, **65**, 9, *2861—2868*.

RAKESTRAW, N. W. & V. M. EMMEL, 1938. The relation of dissolved oxygen to nitrogen in some Atlantic waters. *J. mar. Res.*, **1**, 3, *207—216*.

STOMMEL, H., 1958. *The Gulf Stream*. Univ. of California Press, 202 p.

SVERDRUP, H. U., 1933. On vertical circulation in the ocean due to the action of the wind with application to conditions within the Antarctic Circumpolar Current. *Discovery Reports*, **7**, *139—170*.

SVERDRUP, H. U., 1934. Wie entsteht die Antarktische Konvergenz? *Ann. Hydr. mar. Met.*, **62**, *315—317*.

SVERDRUP, H. U., 1940. Hydrology. *B.A.N.Z. Antarctic Res. Exp. 1929—31*. Ser. A, v. **3**, 2, Adelaide.

SVERDRUP, H. U., M. W. JOHNSON & R. H. FLEMING, 1942. *The Oceans*. Prentice-Hall, New York.

U. S. Hydrographic Office, 1957. *The Oceanographic Atlas for the Polar Seas*. Part I. Antarctic. H.O. Pub. No. **705**, Washington, D. C.

VOWINCKEL, E. & H. VAN LOON, 1957. Das Klima des Antarktischen Ozeans. III. Die Verteilung der Klimaelemente und ihr Zusammenhang mit der allgemeinen Zirkulation. *Arch. Met. Geoph. Biokl.*, Serie B, **8**, 1, *75—102*.

WATTENBERG, H., 1938. Die Verteilung des Sauerstoffs im Atlantischen Ozean. *Wiss. Erg. dtsch. Atl. Exp. "Meteor" 1925—27*, **9**, 1, 132 pages.

WATTENBERG, H. (ed. by K. KALLE) 1957. Die Verteilung des Phosphats im Atlantischen Ozean. *Wiss. Erg. Atl. Exp. "Meteor" 1925—27*, **9**, 2, 48 pages.

WEXLER, H., 1959. The Antarctic Convergence — or Divergence? *The Rossby Memorial Volume*, New York.

WÜST, G., 1928. Ursprung der Atlantischen Tiefenwässer. *Z. ges. Erdkunde*, Sonderband, *506—534*.

WÜST, G., 1936. Schichtung und Zirkulation des Atlantischen Ozeans. *Wiss. Erg. dtsch. Atl. Exp. "Meteor" 1925—27*, **6**, 1.

WÜST, G., 1951. Über die Fernwirkung antarktischer und nordatlantischer Wassermassen in den Tiefen des Weltmeeres. *Naturwiss. Rundschau*, **3**, *91—108*.

WÜST, G., 1954. Gesetzmässige Wechselbeziehungen zwischen Ozean und Atmosphäre in der zonalen Verteilung von Oberflächensalzgehalt, Verdunstung und Niederschlag. *Arch. Met., Geoph. Biokl.*, A, **7**, *305—328*.

126

Wüst, G., W. Brogmus & E. Noodt, 1954. Die zonale Verteilung von Salzgehalt, Niederschlag, Temperatur und Dichte an der Oberfläche der Ozeane. *Kieler Meeresforschungen*, **10**, 2, *137—161.*

Wüst, G., 1957. Stromgeschwindigkeiten und Strommengen in den Tiefen des Atlantischen Ozeans. *Wiss. Erg. dtsch. Atl. Exp. "Meteor" 1925—27*, v. **6**, 2, *261—420.*

Wyrtki, K., 1960. The Antarctic Circumpolar Current and the Antarctic Polar Front. *Dtsch. hydr. Z.*, **13**, 4, *153—174.*

FRESHWATER ALGAE IN THE ANTARCTIC REGIONS

BY

MINORU HIRANO

Introduction

From the beginning of the nineteenth century onwards many European nations organised research expeditions to the Antarctic. Knowledge of freshwater algae in the Antarctic and the surrounding regions has gradually accumulated in the published reports of these expeditions. However, the information available at present is far from complete and insufficient to permit any general interpretation. Such a general interpretation, if it could be made, would be of considerable interest and value.

History of the studies of Antarctic Freshwater Algae

J. D. HOOKER, a member of the British expedition of 1839–43, led by JAMES CLARK ROSS, made collections at the Kerguelen Islands and Cockburn Island (lat. 64 °S. and 57 °E. long.) off the Antarctic Continent. Three species of freshwater forms were reported from Cockburn Island: *Phormidium autumnale* (AG.) GOM., *Microcoleus vaginatus* (VAUCH.) GOM. and *Prasiola crispa* (LIGHTF.) MENEGH. (HOOKER & HARVEY, 1847).

The collection made at Christmas Harbour in the Kerguelens revealed the following species: *Vaucheria Dillwynei* AG., *Conferva angulata* HOOK. & HARV., *C. quadratula* HOOK. & HARV., *C. podagraria* HOOK. & HARV., *Oscillatoria purpurea* HOOK. & HARV., *Calothrix olivacea* HOOK. & HARV., *Ulva cristata* HOOK. & HARV., *Mastodia tesselata* HOOK. & HARV., *Trypothallus anastomosans* HOOK. & HARV., *Nostoc commune* VAUCH. and *N. microscopicum* CARM. (HOOKER & HARVEY, 1847).

Later the British ship "Venus" visited the islands of Kerguelen and established a base at Observation Bay from October 11,1874 to February 27,1875. Material collected by A. E. EATON, a member of the expedition, was examined by P. E. REINSCH and in a preliminary report (REINSCH 1876) he presented data on 31 new species and varieties. In a more detailed report (REINSCH 1878) he identified 106 species of freshwater algae belonging to 67 genera all indigenous to Rodriguez and Kerguelen. These comprised 21 species of diatoms, 33 species of blue-green algae, 50 species of green algae and one species each of red and brown algae.

REINSCH also published a report on the collections made at S. Georgia during the German expedition (1882–83).

He identified 73 species of freshwater algae consisting of 49 species of green algae, 5 species of blue-green algae and 19 species of diatoms.

In addition, DICKIE reported *Prasiola Sauteri* MENEGH. which he found while studying marine algae in EATON's collection. This *Prasiola* species was obtained from the stones of a stream 400 feet above sea-level, to the west of Swain Bay. Later GAIN (1912) compiled a list of algae found in the Kerguelens based on the reports of J. D. HOOKER, the Venus Expedition, the Challenger Expedition, and the results of the Second French Antarctic Expedition, of which he was a member. GAIN altered the proper names of the algae to those that are in common use today.

H. N. MOSELEY, a member of the Challenger Expedition, collected marine algae during a stay at the Kerguelens from January 7 to 31, 1874, and obtained some freshwater forms. ARCHER (1876) studied this collection and found the following species:

Spirogyra quinnia (?), *S. condensata* (?), *Mesocarpus (Staurospermum)*, *Oedogonium* (three forms), *Zygogonium delicatulum, Conferva bombycina, Vaucheria* sp., *Microthamnion Kützingianum, Pediastrum boryanum, P. ellipticum, Closterium parvulum, Cosmarium Meneghinii* var., *C. crenatum, Nostoc* sp., *Schizothrix* sp., *Tolypothrix distorta.*

M. E. RACOVITZA, a member of the Belgian Expedition, led by A. DE GERLACHE, collected some algae at Graham Land during the visit of the "Belgica" in the years 1897—1899. His findings were studied by DE WILDEMANN, (1900) and were found to be two species of freshwater alga namely *Chlamydomonas* sp. and *Racovitziella antarctica* DE WILDEM.

In the years 1898—1900, the Norwegian Expedition of the "Stella Polare", led by C. E. BORCHGREVINK, explored the area around Victoria Land, and collections were made. WILLE (1902) reported two freshwater species from that material: *Merismopedia punctatum* MEYEN and *Prasiola crispa* (LIGHTF.) MENEGH.

The first French Antarctic Expedition to Graham Land in the years 1903—1905 was led by J. CHARCOT. M. TURQUET, a member of this expedition, collected some algae, and HARIOT (1908) studied the freshwater forms from this collection and identified the following four species: *Lyngbya nigra* AG. f. *antarctica* GOM., *Phormidium Charcotianum* GOM., *Vaucheria* sp. *and Prasiola crispa* (LIGHTF.) MENEGH.

The report on the Belgian expedition "Belgica", VAN HEURCK (1909) gives a detailed account of Antarctic diatoms in two parts. The first is based on material from the Antarctic continent and its vicinity, and the second part on material from Kerguelen. The diatoms are chiefly marine forms, and he provided a table showing the distribution of diatoms in both regions.

In the years 1907—1909 the Nimrod Expedition led by SIR ERNEST SHACKLETON visited Victoria Land. JAMES MURRAY, a member of the expedition, made collections chiefly at Cape Royds (77° 32′ S. lat. and 166° 12′E. long.) and at South Victoria Land. The material of this British Expedition was studied by WEST & WEST (1911). It consists of 15 species of green algae belonging to 6 genera, 30 species of diatoms belonging to 16 genera, and 39 species of blue-green algae belonging to 11 genera. Of these, the following are new species:

Pleurococcus frigidus W. & G. S. WEST, *Pl. antarctica* W. & G. S. WEST, *Tropidoneis laevissima* W. & G. S. WEST, *Navicula glaberrima* W. & G. S. WEST, *N. muticopsiforme* W. & G. S. WEST, *N. peraustralis* W. & G. S. WEST, *N. Murrayi* W. & G. S. WEST, *N. cymatopleura* W. & G. S. WEST, *N. shackletoni* W. & G. S. WEST, *Nostoc antarctica* W. & G. S. WEST, *Lyngbya shackletoni* W. & G. S. WEST, *L. Murrayi* W. & G. S. WEST, *L. erebi* W. & G. S. WEST, *Phormidium glaciale* W. & G. S. WEST, *Ph. antarcticum* W. & G. S. WEST, *Oscillatoria subproboscidea* W. & G. S. WEST, *O. producta* W. & G. S. West, *O. priestleyi* W. & G. S. WEST, *O. deflexa* W. & G. S. WEST, *Microcystis chroococcoidea* W. & G. S. WEST, *Asterocystis antarctica* W. & G. S. WEST.

The collections were made chiefly in freshwater ponds but the *Prasiola* species were obtained near the Penguin rookery. Specimens of snow and ice algae were not brought back. The pond waters of these regions are said to be very saline, and this high salinity seems to arrest the growth of green algae. In fact green algae are rare, especially desmids, but marine forms of diatoms were found in the Green Lake.

FRITSCH (1910) reported the results of the Scotia Expedition (1902—1904) based on materials obtained at South Orkney. He was the first to report the algae forming yellow and red snow, (FRITSCH 1912) and later (FRITSCH 1912b) in the Report of the scientific results of the voy. of S.Y. "Scotia" III. In these papers he reported that yellow and red snow appeared in the snow-field of South Orkney and he identified the new genera and new species of snow algae. Yellow snow consists of 18 species of algae and two species of fungi. Algae forming red snow consist of five species of green algae and five species of diatoms.

FRITSCH has published a report based on the materials obtained by the Discovery Expedition (1901—1904). The collections were made at Cape Adare (lat. 71°S.) and around the McMurdo Sound (almost lat. 78°S.). These collecting places are further south than South Orkney. FRITSCH observes that the Antarctic algal flora is characterised by the paucity of green algae. On the other hand NORDSTEDT studied *arctic* algal flora and showed that it is rich in desmids, though blue-green algae are not so prominent as in Antarctica they are dominant and abundant. *Chlamydomonas caudata* and *C. subcaudata* occur in Antarctica and North Europe, but have not been found as yet in other regions of the world.

Chloromonas alpina and *Eucapsis* species are distributed in alpine regions and are found in Antarctica.

Snow algae were not obtained by the Discovery expedition (1901—1904), but material from ice-fields was brought back for study, and these specimens comprise collections from two places: one from Cape Adare, and the other from the Bay of McMurdo. FRITSCH gives a list of the algae comprising the ice-flora as follows:

Cape Adare	McMurdo Bay
Chloromonas alpina WILLE	*Microcystis parasitica* KÜTZ.
Eucapsis minuta FRITSCH	var. *glacialis* FRITSCH
Navicula muticopsis VAN HEURCK	*M. parasitica* KÜTZ.
N. cymatopleura W. & G. S. WEST	*M. merismopedioides* FRITSCH
N. seminulum GRUN.	*Phormidium frigidum* FRITSCH
Surirella angusta KÜTZ.	**Pleurococcus antarcticus* W. & G. S. WEST
	**Pl. dissectus* NÄG.
	**Nostoc disciforme* FRITSCH
	Melosira sp.
	Fragilaria tenuicollis HEIB.
	var. *antarctica* W. & G. S. WEST
	Navicula seminulum GRUN.
	N. *muticopsis* VAN HEURCK
	N. *cymatopleura* W. & G. S. WEST
	N. *shackletoni* W. & G. S. WEST
	N. *globiceps* GREG.

The asterisk denotes algae found as epiphytes growing on *Phormidium* bodies and they are not true ice algae. The blue-green algae predominantly grow in ponds or are imbedded in the ice forming a huge sheet. These large sheets consists chiefly of *Phormidium* species and sometimes of *Lyngbya* species. This large growing mass is a favourable environment for the growth of other blue-green algae. The Discovery collections are richer in heterocystous blue-green algae, compared with those of the Nimrod Expedition. *Prasiola* species and the many species of blue-green algae make up a characteristic feature of the Antarctic algal flora. The epiphytic algae found on the *Phormidium* sheet are listed by FRITSCH in a table. The periodicity of growth of the Antarctic algae was not shown. The materials obtained in the Gap pond of Winter Harbour during the years 1902—1904 may be divided into four time periods, but these materials present no differences. Rapid growth is impossible under the severe climatic conditions of Antarctica. Multiplication probably takes place gradually and several seasons must elapse before same algae become mature. However, the situation is different for simple unicellular forms of green algae. Cyst-formation in *Chlamydomonas caudata* demonstrates an alternation of dormancy and swimming periods.

The French Second Antarctic Expedition was carried out in Graham Land in the years 1908—1910 and GAIN, a member of the expedition, collected specimens of terrestrial and snow algae. The terrestrial algae were collected on the faces of rocks, among mosses, and in waters caused by thawing. Of these algae seven species out of a total of 27 were new. The snow algae, studied by WILLE, comprised 11 species, four of them new. GAIN listed the terrestrial algae obtained at Jenny Island (67° 43′S. lat. and 68° 35′E. long.); they were collected chiefly among moss vegetation and in streams caused by thawing. Those places lie on a slope about 150 metres above the sea. The collections obtained at Jenny Island contain 8 species of Conjugatae which are not found in MURRAY's collection from Victoria Land and Ross Island. GAIN found red and green snows from December, 1908 to March, 1909. The snow algae appeared especially on the surface of snow hardened by the traffic of birds near the Penguin Rockery, and disappeared in April. According to GAIN's record the first of March, 1909 was a warm day; the average temperature was up to 6° C and the snow algae appeared over several hectares: it was *Chlamydomonas antarcticus* WILLE.

The Swedish Antarctic Expedition, led by OTTO NORDENSKIÖLD, made collections from 1901 to 1903 in the Antarctic regions, South Shetland, Graham Land, South Georgia, and Falkland. CARLSON studied these specimens and identified 25 species of blue-green algae, 80 species of diatoms, 28 species of Conjugatae, 23 species of green algae and two other algae. The green snow algae found at South Shetland consist of one species of blue-green algae, 13 species of diatoms and five species of green algae.

The German Antarctic Expedition led by E. DRYGALSKI was held in the years 1901—1903, and E. VANHÖFFEN, a member of the expedition, collected algae in the territory surrounding Mt. Gauss. N. WILLE studied this collection and reported 10 species and one variety of blue-green algae and 8 species of green algae.

This German expedition stayed at Station Bay of the Kerguelens and collections were made by E. VANHÖFFEN and E. WERTH from the beginning of January, 1902 to November 18, 1902. The position of the station was the same as that of the previous Venus Expedition. WILLE studied these collections and identified the algae from these places. The majority are cosmopolitan species and some are proper to these regions. However, among cosmopolitan species the Antarctic forms are somewhat different from the forms found in other areas of the world. These differences are probably due to adaptation to the severe environment of the Antarctic. The 146 species obtained from the Kerguelen Islands consist of 46 species of blue-green algae, 36 species of Conjugatae (of these 28 species are desmids) and 64 species of green algae and others, and one red algae.

DE WILDEMANN (1900) examined RACOVITZA's collection obtained during the travels "Belgica" of the Belgian Antarctic Expedition and identified twenty species of freshwater forms, consisting of five species of blue-green algae and 15 species of green algae and others. The species reported by him are:

Phormidium Racovitzae DE WILD., *Lyngbya aestuarii* (MERTENS) LIEBM., *Scytonema Racovitzae* DE WILD., *Stigonema hormoides* (KÜTZ.) BORN. & FLAH., *Hapalosiphon fontinalis* (AG.) BORN., *Stichococcus bacillaris* NÄG., *Raphidonema nivale* LAGERH., *Prasiola tessellata* KÜTZ., *Schizogonium murale* KÜTZ., *Conferva bombycina* AG., *Racovitziella antarctica* DE WILD., *Protococcus viridis* AG., *Trochiscia nivalis* LAGERH. *Chlamydomonas sanguinea* LAGERH., *Zygnema ericetorum* (KÜTZ.) HANSG., *Hyalotheca dissiliens* (SM.) BRÉB., *Closterium malinvernianum* DE NOT., *Cosmarium botrytis* (BORY) MENEGH., *C. subspeciosum* NORDST. var. *validius* NORDST., *Staurastrum muticum* BRÉB.

BOURRELLY & MANGUIN (1954) have studied the collection obtained by R. ARÉTAS during the years 1949—1950 from the Kerguelen islands. The collecting places consisted of two lakes, streams, and moors. They identified 226 species in all, consisting of 160 species of diatoms, 14 species of blue-green algae, 44 species of green algae, including 22 species of Conjugatae and 17 species of Chlorococcales and 8 species of other algae. The number of algae excluding diatoms is generally small; moreover, the majority of these algae were obtained in a single place. 99 species out of total 160 species of diatoms are cosmopolitan and this corresponds to 61.8% of all diatoms. There are 55 species of diatom proper to the Kerguelens and this corresponds to 29.3% of the total. The species confined in distribution are 11, and eight of them are alpine, two are from Antarctic South America; only one is Antarctic. The algae except diatoms are mostly cosmopolitan but three are species proper to the Antarctic.

Algal Flora of the Antarctic Continent and its adjacent islands

a) Graham Land. This area extends from the Antarctic Continent towards the north, and is part of the so-called Palmer peninsula. Its northernmost part reaches to lat. 63°S., and the main part of the peninsula occupies lat. 63°S. to lat. 74°S. The First and Second French Expeditions led by J. CHARCOT, and the Swedish Expedition led by O. NORDENSKIÖLD (1901—1903) explored this area and collected much material. The reports published by the Second French Expedition and the Swedish expedition are important in helping us to understand the algal flora of this area. The Second French Expedition passed the Antarctic winter at Petermann Island from February to November 1909, and collected freshwater algae at Jenny Island during that time. Jenny Island (lat. 67°43′S. and 68°35′E. long.) lies to the south of the closed base, and is composed of black coloured lava, forming a rugged slope which is about 150 metres above the sea. It has moss-vegetation and plenty of small streams caused by thawing. GAIN gives the following list of algae from this area.

Chroococcus turgidus (KÜTZ.) NÄG.	*Cylindrocystis crassa* DE BARY
C. macrococcus (KÜTZ.) RABENH.	*Cosmarium antarcticum* GAIN
Gloeocapsa Janthina NÄG.	*C. crenatum* RALFS
Oscillatoria amphibia AG.	*C. undulatum* CORDA
O. tenuis AG.	*Pleurotaeniopsis pseudoconnata*
Lyngbya Erebi W. & G. S. WEST	(NORDST.) LAGERH.
L. antarctica GAIN	*Trochiscia hystrix* REINSCH
Lyngbya sp.	*T. tuberculifera* GAIN
Phormidium autumnale (AG.) GOM.	*Pteromonas Willei* GAIN
Nostoc minutum DESM.	*P. Penardii* GAIN
N. Borneti GAIN	*Prasiola crispa* (LIGHTF.) MENEGH.
N. pachydermaticum GAIN	*Ulothrix flaccida* KÜTZ.
Anabaena oscillarioides BORY	*Conferva glacialis* KÜTZ.
Calothrix sp.	

The Swedish expedition stayed on Snow Hill Island (lat. 64°30′S. and 57°W. long.) from February 12 to 13, 1902, and collected green sheets in the pools caused by thawing, and identified four species of algae: *Oscillatoria subproboscidea* W. & G. S. WEST, *O. amphibia* AG., *Microcoleus cryophilus* CARLSON, and *Penium curtum* BRÉB. The green algae listed by GAIN include all members of the common genera found in Antarctica. It is especially interesting to find some desmids, five species of which grow in this area (lat. 68°S. and lat. 64° 30′S).

b) Victoria Land. This region lies on the opposite side of the Pacific ocean to Graham Land, on the western part of the continent, facing the Ross Sea. Several expeditions entered the Ross Sea and explored interior parts of the continent at the beginning of the twentieth century. JAMES MURRAY, a member of the Nimrode Expedition led by E. SHACKLETON, collected freshwater algae in various parts of Victoria Land and these were studied by W. & G. S. WEST (1911). The collection was made on Ross Island (lat. 77° 32′S. and 166° 12′E. long.) and the base camp was at Cape Royds where another collection was made around the camp. W. & G. S. WEST (1911) report in detail the various places where specimens were collected, so we know the algal flora of the region.

1) Pony Lake. This was surveyed by J. MURRAY, on April 22, 1908, and again on January 4, 1909. January is midsummer in Antarctica, and the water temperature of the lake becomes slightly higher than freezing point. The maximum temperature in the lake water was recorded as 35°F. The algae found in this lake were two

134

species of *Chlamydomonas*: *Chlamydomonas subcaudata* WILLE and *Ch. intermedia* CHODAT forma *antarctica* W. & G. S. WEST, and *Phormidium autumnale* (AG.) GOMONT; but the lake was frozen on the surface and the algae had markedly decreased in quantity when he visited it in April. Beside these algae *Phormidium antarcticum* W. & G. S. WEST and *Pleurococcus frigidus* W. & G. S. WEST were recorded.

2) *Ponds at Cape Royds*. The temperature of pond-waters is 60° F. at the end of December. There is some pond to pond variation in the algae found in this area, but in general *Nostoc* is predominant together with some diatoms; in another pond *Oscillatoria telebriformis* AG. forma *tenuis* W. & G. S. WEST and *O. amphibia* AG. var. *robusta* W. & G. S. WEST were growing, accompanied by *Ulothrix subtile* KÜTZ. var. *variabilis* (KÜTZ). KIRCHN.

3) *Green Lake*. This lake lies near the camp at Cape Royds and is characterised by its high salinity. The surface of Green Lake partly broke up on November 29, 1908, and the temperature of the open water rose to 36° F. on January 3, 1909. The temperature of lake water was 35° F. at the beginning of freezing, on February 2, and then gradually decreased. It was recorded as 21° F. on June 26, and as 8° F. on August 6. In this lake 26 species of algae were found including abundant blue-green algae. The blue-green algae, in general, grow in the form of a sheet, but in this lake the blue-green algae were floating. The diatoms obtained were chiefly marine forms, but freshwater forms were also found. *Navicula muticopsis* V. HEURCK and *N. globiceps* GREG. are the freshwater forms and these two species are known to exist in some places which show a slight degree of salinity. W. & G. S. WEST list the algae obtained in this lake as follows:

Ulothrix aqualis KÜTZ. forma	*Trachyneis aspera* (EHRENB.) CLEVE
U. tenerrima KÜTZ. f. *antarctica*	*Tropidoneis laevissima* W. & G. S.
W. & G. S. WEST	WEST
Pleurococcus angulosus LAGERH.	*Cocconeis litigiosa* V. HEURCK
forma	*Lyngbya limnetica* LEMM.
Pl. antarcticus W. & G. S. WEST	*L. murrayi* W. & G. S. WEST
Pl. dissectus (KÜTZ.) NÄG.	*Phormidium fragile* (MENEGH.) GOM.
Trochiscia aspera (REINSCH) HANSG.	*Oscillatoria deflexa* W. & G. S. WEST
Coscinodiscus lentiginosus JAN.	*O. chlorina* KÜTZ.
Hemiaulus ambiguus JAN. var.	*O. limosa* AG.
Triceratium arcticum BTW.	*Chroococcus minutus* (KÜTZ.) NÄG.
Fragilaria obliquecostata V. HEURCK	*C. cohaerens* (BRÉB.) NÄG.
F. . . forma *maxima* V. HEURCK	*C. minor* (KÜTZ.) NÄG. forma
Navicula muticopsis V. HEURCK	*minima* WEST
N. globiceps GREG.	*Microcystis chroococcoides* W. & G. S. WEST
	Asterocystis antarctica W. & G. S. WEST

4) The ground near the Penguin Rookery at Cape Royds was polluted by the excrement of birds, and specimens of *Prasiola crispa* (LIGHTF.) MENEGH. and *Navicula muticopsis* V. HEURCK were found there.

5) *Blue Lake.* This lake lies 1.5 miles from the camp, and is the largest lake in this area. The material obtained from beneath the ice layer consisted of a small amount of *Phormidium glaciale* W. & G. S. WEST, *Ph. inundatum* KÜTZ., *Lyngbya martensiana* MENEGH. and *Pleurococcus antarcticus* W. & G. S. WEST forma *robusta* W. & G. S. WEST; the cysts of *Chlamydomonas nivalis* (BAUER) WILLE were found mixed with these algae. *Prasiola crispa* and *Navicula muticopsis* were found near the shore. Diatoms were relatively scarce.

6) The coastal lake was surveyed and open water was visible on November 28, 1908; it became quite free of ice in January 1909, and then the surface of the lake was closed by ice again at the beginning of February. The temperature inshore rose to 47° F., on December 4, 1908, and was recorded as 40° F., on January 2, 1909. The material collected in January consisted entirely of blue-green algae, which was floating. The floating masses of these algae were *Oscillatoria* species. W. & G. S. WEST recorded six species of *Oscillatoria* in this lake: *O. deflexa* W. & G. S. WEST, *O. cortiana* MENEGH., *O. formosa* BORY, *O. tenuis* AG., *O. subproboscidea* W. & G. S. WEST, and *O. limosa* AG. and added a number of other blue-green algae such as *Microcystis stagnalis* LEMM., *Chroococcus minutus* (KÜTZ.) NÄG. var. *obliteratus* (RICHTER) HANSG., and *Lyngbya Kützingii* SCHMIDLE.

7) There are a number of ponds on the slope of Mt. Erebus and two ponds were surveyed. The material collected by D. MAWSON, on March 28, 1908, consisted of well-developed colonies of *Lyngbya erebi* W. & G. S. WEST, which forms a cartilaginous patch, 3—5 mm thick. *Lyngbya Kützingii* SCHMIDLE and *Oscillatoria limosa* BORY were found to be mixed with this algae there. There were used colonies of *Pleurococcus dissectus* (KÜTZ.) NÄG., and some species of diatoms, such as *Tabellaria flocculosa* (ROTH) KÜTZ., *Cymbella pusilla* GRUN., *Melosira varians* AG. and *Navicula radiosa* KÜTZ. The material collected on December 31, 1908, was a sheet of filamentous bacteria-like *Leptothrix*, and mixed in with this sheet was *Phormidium autumnale* (AG.) GOMONT, *Oscillatoria producta* W. & G. S. WEST, and *Pleurococcus dissectus* (KÜTZ.) NÄG.

8) The moraines of Mt. Erebus are situated about 300—400 feet above the sea. The material collected in this place, in December 1908, consisted of *Ulothrix implexa* KÜTZ. and a large quantity of diatoms, especially *Navicula muticopsis* V. HEURCK. The material collected in January 1909 was chiefly *Prasiola antarctica* KÜTZ., and the less frequent species were *Gloeocapsa Shuttleworthiana* KÜTZ. and *Aphanocapsa montana* Cramer. *Prasiola antarctica* and *Aphanocapsa montana* were found covering the faces of stones. The same

expedition also collected material from two places in South Victoria Land. The material was collected by R. E. PRIESTLEY in January, 1909, in a lake lying on the western side of McMurdo Sound, lat. 77° 45′S., 25 miles from the camp at Cape Royds. Red Rotifers were swimming in the lake, and the dominant algae found there were *Oscillatoria Priestleyi* W. & G. S. WEST. Subsidiary species were *Phormidium autumnale* (AG.) GOM., *Oscillatoria deflexa* W. & G. S. WEST, *Chroococcus pallidus* NÄG. and *Ch. minor* (KÜTZ). NÄG. Many diatoms were also found in this lake, of species similar to those on Ross Island. The material collected by T. W. E. DAVID, on October 23, 1908, contained various forms of *Prasiola crispa* (LIGHTF.) MENEGH.: mainly *Schizogonium* and *Hormidium* types. The place where they were found lies on the west side of the Ross Sea and is 5.5 miles from Cape Irizar, corresponding to lat. 75° 40′S. The algal flora of Ross Island and South Victoria Land summarised above is based on papers by W. & G. S. WEST (1909, 1911) 84 species of algae, including some ill-defined species, were recorded in these regions.

The material of the British National Antarctic Expedition was collected at Cape Adare (lat. 71° S.) and around McMurdo Sound (lat. 78°S.). FRITSCH (1912) studied this collection and records that 91 species of algae were found in these regions. The principal part of the collection consists of blue-green algae. It has been shown that the bulk of blue-green algae which form a huge sheet can be utilised by other algae as a substratum for growth.

It is notable that the higher blue-green algae of heterocystous form were found among them. This is in contrast to the report on Ross Island by WEST. No species of Chlorococcales were found among these collections, but species of *Ulothrix* and *Prasiola* are known to occur in these places. *Phormidium* flourishes there and together with *Lyngbya* forms a large sheet in the lakes and ponds; and a number of species of epiphytic cyanophycean algae attach themselves to this substratum. These are species of *Microcystis*, *Chroococcus, Oscillatoria, Lyngbya, Nostoc, Anabaena, Calothrix*, and *Pleurococcus antarcticus* W. & G.S. WEST, *Pl. frigidus* W. & G. S. WEST, *Pl. koettlitzii* FRITSCH and *Pl. dissectus* (KÜTZ.) NÄG. Such an epiphytic flora is typical of any lake or pond in this region, and is essentially the same as the diatom flora. Diatom species such as *Navicula muticopsis* V. HEURCK, *N. globiceps* GREG., *N. Schackletoni* W. & G. S. WEST, *N. cymatopleurus* W. & G. S. WEST are commonly found and flourish in the ponds and lakes of this region. It is worth noting that *Chloromonas alpina* from alpine regions of Europe, and *Eucapsis minuta* FRITSCH are present in these regions. Neither species has been found so far in any other part of the world except Europe. The diatoms *Fragilaria tenuicollis* HEIB. var. *antarctica* W. & G. S. WEST, *Hantzschia amphioxys* (EHRENB.) GRUN. and *H. elongata* (HANTZSCH) GRUN. are commonly found in abundance in these regions. FRITSCH lists the epiphytic flora growing in the sheets of blue-green algae as follows:

Pond half-way between "Black" and "Brown" Islands.	Gap pond, Winter Harbour.	Pond, Granite Harbour.
Ulothrix subtilis var. *variabilis*	*Ulothrix subtilis* var. *variabilis*	*Ulothrix subtilis* var. *variabilis*
Pleurococcus antarcticus	*Pleurococcus antarcticus*	*Pleurococcus antarcticus*
Pl. koettlitzi	*Pl. koettlitzi*	
Prasiola (young stage)	*Prasiola* (young stage)	*Prasiola* (young stage)
Dactylococcopsis rhaphidioides	*Chroococcus minutus*	
	Ch. minor	
	Gloeocapsa rupicola	*Gloeocapsa rupicola*
Microcystis parasitica	*Microcystis parasitica*	*Microcystis parasitica*
		Microcoleus vaginatus
Lyngbya Lagerheimi	*Lyngbya aestuarii* var. *antarctica*	*Lyngbya martensiana*
L. attenuata	*L. attenuata*	*L. Scotti*
	L. aerugineo-caerulea	
Oscillatoria tenuis	*Oscillatoria irrigua*	*Oscillatoria tenuis*
	O. koettlitzi	
Calothrix antarctica	*Calothrix intricata*	
	C. gracilis	
Nostoc sphaericum	*Nostoc disciforme*	
	N. longstaffi	
	N. hederulae	
Nodularia quadrata	*N. fuscescens*	
	Anabaena antarctica	
	Nodularia spumigena var. *minor*	
Navicula muticopsis	*Navicula muticopsis*	*Navicula muticopsis*
N. cymatopleurus	*N. Murrayi*	*N. Shackletoni*
N. Shackletoni		*Pinnularia borealis*
N. globiceps	*Hantzschia elongata*	*Hantzschia amphioxys*

This indicates that blue-green algae are an important element of the Antarctic algal flora. In short the algal flora of Cape Adare is similar to that of Ross Island in the number of species, and in the dominance of the blue-green algae. Cape Adare and Ross Island have many species of algae in common. However, there are some differences and the Cape Andare list contains new species which were not found previously in Ross Island. Some freshwater algae were brought back by the British Terra Nova Expedition, led by R. F. SCOTT, from Cape Adare and Cape Sustruzi (lat. 75°S.) in the year 1910. The material collected by R. E. PRIESTLEY, in December 1911, contained *Prasiola crispa* (LIGHTF.) MENEGH. and its *Hormidium* stage, the fragmental part of a *Phormidium* sheet consisting of three species, *Ph. autumnale* GOMONT, *Ph. Priestleyi* FRITSCH, *Ph. fragile* GOMONT forma *tenuis* W. & W. G. WEST, and *Pleurococcus* and four species of diatoms. The *Pleurococcus* species are *Pl. antarcticus* W. & G. S. WEST and *Pl. dissectus* NÄG., and *Protoderma Brownii* FRITSCH and *Navicula muticopsiforme* W. & G. S. WEST, *N. muticopsis* V. HEURCK, *N. Murrayi* W. & G. S. WEST, *N. dicephala* EHRENB. were also found. The material obtained from Cape Sustruzi consisted chiefly of a sheet of *Phormidium* with a number of epiphytic algae on it not essentially different from those reported by WEST (1911) and FRITSCH (1912). *Schizothrix antarctica* FRITSCH and *Nostoc fuscescens* FRITSCH var. *mixta* FRITSCH were obtained on these *Phormidium* sheets. F. E. FRITSCH notes that it is important in the collection of algae from Antarctica to pay particular attention to the *Phormidium* sheets.

c) Kaiser-Wilhelm II Land. During the German Antarctic Expedition algae were collected by E. VANHÖFFEN around Mt. Gauss of Kaiser-Wilhelm II Land, and WILLE studied them (WILLE 1902). The algae obtained from material buried in pond-ice were: *Chroococcus cohaerens* (BRÉB.) NÄG. var. *antarctica* WILLE, *Phormidium glaciale* W. & G. S. WEST forma *longiarticulata* WILLE, *Nostoc sphaericum* VAUCH. and *Chlorococcum infusionum* MENEGH. forma *antarctica* WILLE. Found attached among the polsters of *Bryum filicaule* BROTH were the following: *Chroococcus minutus* (KÜTZ.) NÄG. and var. *amethystacea* WILLE, *Phormidium autumnale* (AG.) SCHMIDLE, *Ph. tenue* (MENEGH.) GOMONT, *Lyngbya Scotti* FRITSCH, *Nostoc sphaericum* VAUCH., *N. pachydermaticum* GAIN, *Calothrix* sp., *Scotiella antarctica* FRITSCH, *Chlorella Koettlitzi* (FRITSCH) WILLE, *Pleurococcus tectorum* TREVIS forma *antarctica* WILLE, *Chlorella conglomerata* (ARTARI) OLTM., and *Stichococcus bacillaris* NÄG. Mt. Gauss at lat. 66° 47'48" S., is a volcano composed of lava and tufa. The soil on the mountain is generally dried up, except during the thaw in the antarctic summer. The algal flora is poor in species and growth of the algae is generally feeble due to insufficient nutrition. N. WILLE concluded that this was due to the absence of nitrogen-fixing microorganisms in the soil, and was in part due to a deficiency of N-containing compounds in the water.

Snow and ice-algae of Antarctica and its neighbouring islands

GAIN collected material from various parts of the snow-fields at Petermann, Wiencke, Booth-Wandel, Jenny Island, and Cape Tuxen. Snow algae seem to have appeared frequently in the lowlands of the snow-fields when the temperature rose, and the surface of the snow melted. This is reported to occur on the surface of snow-fields hardened by the traffic of the birds near the nests of Penguins. The warmest day was January 1, 1909, and the average temperature rose to 6° C. At that time coloured snows appeared over several hectars of snow surface, and was chiefly due to the occurrence of *Chlamydomonas antarcticus* WILLE. The surface of the snow-fields appeared to have been dyed a rose colour when seen from a distance. The coloured snows were chiefly green and red in colour.

Examples of green snow were collected at Petermann Island on March 4, 1909. It appeared in large quantities over the snow fields of the lowlands on the island. The algae found were: *Chlorella ellipsoidea* GERN. forma *antarctica* WILLE, *Stichococcus bacillaris* NÄG., *Mycacanthococcus antarcticus* WILLE and *Ulothrix subtilis* KÜTZ. var. *tenerrima* (KÜTZ.) KIRCHN. forma *antarctica* WILLE. Green snow was observed in the snow-fields which were thawing rapidly at Wiencke Island December 27, 1908. The snow algae were: *Ulothrix subtilis* KÜTZ. var. *tenerrima* (KÜTZ.) KIRCHN. forma *antarctica* WILLE, *Mycacanthococcus cellaris* HANSG. forma *antarctica* WILLE, *M. ovalis* WILLE, *Pseudotetraspora Gainii* WILLE and *Raphidonema nivale* LAGERH. forma *minor* WILLE. Green snow was collected on the ice-cliff fronting Salpetrière Bay on Booth-Wandel Island, February 22, 1909. The algae obtained were: *Stichococcus bacillaris* NÄG., *Chlamydomonas antarcticus* WILLE, *Pseudotetraspora Gainii* WILLE, *Pleurococcus vulgaris* MENEGH. var. *cohaerens* WITTR., *Mycacanthococcus antarcticus* WILLE, and *M. ovalis* WILLE.

Examples of red snow were collected from the snow-fields and from the ice-cliffs near the Penguin Rookery at Wiencke Island on December 27, 1908. The red snow contained the following species: *Raphidonema nivale* LAGERH. forma *minor* WILLE, *Pleurococcus vulgaris* MENEGH. var. *cohaerens* WITTR., *Stichococcus bacillaris* NÄG. forma *minor* WILLE, *Ancylonema Nordenskiöldii* BERGR., *Chlamydomonas antarcticus* WILLE and *Mycacanthococcus cellaris* HANSG. forma *antarctica* WILLE and *Pteromonas Willei* GAIN. Red snow obtained at Petermann Island appeared in the lowland snow-fields on March 1, 1909. The algae were: *Chlamydomonas antarcticus* WILLE, *Stichococcus bacillaris* NÄG. and *Pseudotetraspora Gainii* WILLE. Both green and red snows appeared separately, but sometimes they were found together; Both snow floras contain characteristic species, but have some species in common. LAGERHEIM found that red snow collected by the Swedish expedition at Lonis Phillipe peninsula on Graham Land was composed of *Chlamydomonas nivalis* (BAUER) WILLE.

Specimens of ice algae were obtained by the Discovery Expedition at Cape Adare on January 9, 1909, and at McMurdo Bay on September 13, 1902. FRITSCH studied these ice-algae and identified the species listed in the table (page 137).

140

Freshwater Algae of the Subantarctic Regions

1) *South Shetlands*

These islands are situated nearest to the Palmer peninsula of
Antarctica. The Swedish expedition led by NORDENSKIÖLD made a
collection at Nelson Island, belonging to this island group on January 12, 1902. Nelson Island is at lat. 62° 20'S. and 59°W. long. The
material obtained from rocks was *Hormiscia penicilliformis*, but
samples of green snow contained a number of diatoms as follows:

Oscillatoria fracta CARLSON, *,** *Coscinodiscus minor* EHRENB., *Diploneis subovalis*
CLEVE, *Navicula muticopsis* V. HEURCK, *N. Murrayi* W. & G. S. WEST var. *elegans*
W. & G. S. WEST, *N. austroshetlandica* CARLSON, *Pinnularia Brébissonii* (KÜTZ.)
RABENH., **Cocconeis costata* GREG. var. *pacifica* GRUN., *Denticula antarctica*
(CASTR.) CARLSON, *Diatomella balfouriana* (GREV.) AG., **Grammatophora angulosa*
EHRENB., *Tabellaria flocculosa* (ROTH) KÜTZ., *Cylindrocystis cohaerens* CARLSON,
Pleurococcus vulgaris MENEGH., *Trochiscia reticularis* (REINSCH) HANSG., *Prasiola
crispa* (LIGHTF.) MENEGH.

Those marked with an asterisk are of brackish form, and double
asterisks denote marine specimens. According to CARLSON's report
the diatoms obtained from green snow are known only as frustules
so that it is doubtful whether diatoms had really lived in the snow.

2) *South Georgia*

This island is situated at lat. 54°—55°S. and long. 35° 50'—38°
10'W. and is a long island, lying in a N—S direction. It is well to the
north, compared with the position of the northernmost part of the
Palmer peninsula which is in lat. 63°S., while the outer margin of
the Continent is at about lat. 66°S. The algal flora of the island has
been reported by REINSCH (1890) based on material collected by the
German expedition (1882—1883). The Swedish expedition led by
O. NORDENSKIÖLD made a collection at Cumberland Bay and at
Royal Bay, on the east side of the island, during their stay from
April to May of 1902. Material was obtained from various kinds of
habitat e.g. among the moss-polster, in streams, and in wet soils.
CARLSON has described the algal flora of South Georgia based on the
material of the Swedish expedition and the report by REINSCH.
He lists 13 species of blue-green algae, 55 species of diatoms, 50
species of green algae (including 30 species of desmids), and one
species of Heterokontae. The Swedish expedition made the collection during the poor season for algal growth, from the beginning of
autumn to winter. If the collection had been made during the
optimal season for algae, a greater number of species would have
been obtained. Blue-green algae do not constitute an important
part of the flora, but green algae are represented by many species
especially desmids such as *Closterium* and *Cosmarium*. Cosmopolitan genera belonging to Chlorococcales such as *Eudorina, Pedia-*

strum and *Scenedesmus* occur and CARLSON quotes the algal flora of a stagnant pool in autumn as an example.

3) *Kerguelens*

The Kerguelen Islands consist of the main island and a number of smaller islands in the Indian ocean. Its position is lat. 48°—50°S. and long. 68°—70°E. The islands have been visited by many expeditions in the part so that the algal flora of the islands is well known. E. VANHÖFFEN and E. WERTH, members of the German expedition led by E. DRYGALSKI made collections during their stay on the island from the beginning of January, 1902 to November 18 of the same year. Material was collected from the lakes, ponds, moors, and from the nests of penguin around the base, and was examined by WILLE (1902). The collection consists chiefly of material obtained from stagnant water and not from streams or from among the mosses. There were 46 species of blue-green algae, 63 species of green algae, 36 species of Conjugatae containing 28 species of desmids, one species of red algae and one other species. Diatoms are not treated in WILLE's 1902 report. Blue-green algae are predominant as in S. Georgia. The algal flora of the Kerguelens contains a number of *Closterium* and *Cosmarium*, and is rather different in that the species of *Staurastrum* and *Euastrum* are few. Species of green algae belonging to Chlorococcales are common while the filamentous forms of green algae are scarce. R. ARÉTAS collected material during the summer 1949—1950 from the lakes, moors and streams which were reported on by BOURRELLY & MANGUIN (1954). They report 14 species of blue-green algae, 44 species of green algae (of which 22 species are Conjugatae and 17 species belong to Chloroccocales) and 8 others (7 flagellate forms and one red algae). In addition, 160 species of diatoms are reported. Altogether 47 new species and new forms are recorded.

The material obtained from the rheophilous circumstance contained 84 species of diatoms and four species of Chlorococcales and Conjugatae. The Conjugatae may have been brought here from

Table I.

Name of lakes	Lake Simone	Lake Marie-Nicole
Algal groups		
Diatoms	65	91
Blue-green algae	4	3
Volvocales	2	–
Chlorococcales	7	6
Conjugatae	4	6
Others	–	5

neighbouring moors. There are large quantities of *Fragilaria vires-cens* RALFS var. *elliptica* HUSTEDT and *Synedra rumpens* var. *familiaris* KÜTZ. in this habitat. The plankton collected in the streams comprised 48 species of diatoms and three species of Conjugatae, and several other species; however a true potamoplankton is not found. Material was obtained from Lake Simone and Lake Marie-Nicole and a number of species belonging to Chlorococcales were found. Table I as shown by BOURRELLY & MANGUIN (1954) gives the content of algal flora by the number of species.

It is interesting that there are some species of euplankton such as *Botryococcus braunii* KÜTZ. and *Melosira distans* (EHRENB.) KÜTZ. which are tychoplankton, and also *Fragilaria Vaucheriae* (KÜTZ.) B. PETERS. var. *longissima* BOURR. & MANG. and var. *tenuis* BOURR. & MANG. was found as limnoplankton. The terrestrial algal flora from among wet rocks and among the mosses and *Sphagnum* was:

Aphanothece castagnei (BRÉB.) RABENH., *Nostoc muscorum* KÜTZ., *Petalonema densum* (A.BR.) MIGULA, *Pediastrum integrum* NÄG., *Cylindrocystis Brébissonii* MENEGH., *Netrium digitus* (EHRENB.) ITZIG & ROTHE.

The alpine species found in the Kerguelen islands are *Cylindrocystis Brébissonii* MENEGH. var. *minor* W. & G.S. WEST, *Pediastrum Braunii* WARTM. and *Enallax costatus* (SCHMIDLE) PASCHER. Diatoms that seem to prosper in moor habitats are:

Achnanthes Aretasii BOURR. & MANG., *A. confusa* BOURR. & MANG., *A.Germainii* BOURR. & MANG., *A. Manguinii* HUSTEDT, *Neidium Aubertii* BOURR. & MANG., *Stauroneis obtusa* LAGERST., *Navicula corrugata* BOURR. & MANG., *N. linearis* (O. MÜLL.) FRENG., *N. portomontana* CLEVE, *Pinnularia Backebergii* BOURR. & MANG., *P. borealis* EHRENB. var. *cuneorostrata* BOURR. & MANG., *P. circumducta* BOURR. & MANG., *P. Kolbei* BOURR. & MANG., *P. lata* (BRÉB.) W. SM. var. *minor* GRUN., *Cymbella kerguelensis* BOURR. & MANG., *C. lacustris* (AG.) CLEVE var. *australis* BOURR. & MANG., *Nitzschia ignorata* KRASSKE forma *longissima* BOURR. & MANG.

In short, 99 species out of the 160 are cosmopolitan (61.8%); 55 species are endemic in the Kerguelen islands (29.3%) and 11 species are confined in distribution. These confined species are a group of alpine origin. They are:

Melosira Dickiei THWAIT., *Achnanthes lanceolata* (BRÉB.) GRUN. var. *elliptica* SCHULZ, *Stauroneis obtusa* LAGERST., *Navicula bryophila* B PETERS. var. *lapponica* HUSTEDT, *Pinnularia divergentissima* (GRUN.) CL. and var. *capitata* HUSTEDT, *P. lata* (BRÉB.) W. SM. var. *minor* GRUN. and *Cymbella cesatii* (RABENH.) GRUN. of Antarctic American origin is *Navicula linearis* (O. MÜLL.) FRENG., *N. portomontana* CLEVE.

Of Antarctic origin is *Navicula muticopsis* V. HEURCK forma *capitata* CARLSON. A total of 260 species and forms are known in the Kerguelen islands at present. Of these about 40 seem to be endemic to this region but this estimate is high due to the fact that it includes the indistinct species described by REINSCH. In all, 75 species were found.

4) *South Orkney.*

South Orkney is in the Atlantic Ocean and further south than South Georgia. The islands lie rather nearer to South America than to Africa in latitude 61 °S. RUDMOSE BROWN, a member of the Scottish National Antarctic Expedition, collected red and yellow snows from the snow-fields. In addition other material was obtained from pools, streams, and in the wet soil. This material was studied by FRITSCH (1910 and 1912) who identified 68 species of algae. One genus, and seven species were new. The filamentous forms of green algae are not common, but blue-green algae and green algae belonging to Chloroccocales are the principal elements of the flora. The cosmopolitan genera of green algae such as *Scenedesmus, Pediastrum, Closterium,* and *Cosmarium* are absent; however the filamentous Conjugatae such as *Zygnema* and *Mougeotia, Oedogonium* were present and species of these genera might be dominant in some seasons. The collecting period lasted from March 27, 1903 to February 7, 1904, and most of the material was collected from late spring to the end of summer. Yellow snow was collected on March 27, 1903, and on February 7, 1904. The former collection is poor in algae and corresponds to that found at the end of summer. There are no diatoms in the yellow snow. FRITSCH recorded the following snow algae:

Protoderma brownii FRITSCH	*Raphidonema nivale* LAGERH.
Chlorosphaera antarctica FRITSCH	*Raphidium pyrenogerum* CHODAT
Scotiella antarctica FRITSCH	*Ulothrix subtilis* KÜTZ.
S. polyptera Fritsch	*Oedogonium* sp.
Pteromonas nivalis CHODAT	*Pleurococcus vulgaris* MENEGH.
Chodatella brevispina FRITSCH	*Chlamydomonas caudata* WILLE
Oocystis lacustris CHODAT	*Mesotaenium endlicherianum* NÄG.
forma *nivalis* FRITSCH	*Nostoc minutissimum* KÜTZ.
Sphaerocystis schröteri CHODAT	
forma *nivalis* FRITSCH	
Trochiscia antarctica FRITSCH	

The algae forming the yellow snow may be transported by wind or birds as spores or in the resting stage, and the substratum formed by the growth of the first transportation could supply a favourable growth habitat to other algae which are transported later. FRITSCH states that *Protoderma brownii* FRITSCH is likely to be such a pioneer alga. It is interesting that the algae compromising the yellow snow are planktonic forms. Various amounts of fat have been observed in the cell contents of algae causing yellow snow. The pigments which form the yellow colour appear to be contained in the fat bodies which sometimes occupy the whole cell. The pigments become dissolved in preservative fluid so that it is impossible to ascertain the nature of the pigment. Spectroscopic examination of an alcohol solution of the pigment reveals a marked absorption band

144

in the violet. This correspondents to the absorption spectrum of carotine or xanthophyll, so that this pigment may be similar to the carotenoids. Fat storage cannot be seen in the cells during the reproductive period, because it is probably utilized as an energy-source for cell division. Fat storage is certainly an adaptation to habitat in the cold season.

Red snow. Red snow does not occur as frequently in the snow-fields of South Orkney as yellow snow. It contains relatively few species. There are some similarities and some differences between the two coloured snows. The main constituent of yellow snow is found also in red snow, i.e. *Scotiella antarctica* FRITSCH and *Raphidonema nivale* LAGERH. and contains the yellow-coloured fat. There are two marked differences between red and yellow snow. A number of spherical red cells are found in red snow. These cells are probably the resting stages of various species belonging to the Chlamydomonadaceae. These resting cells are often provided with a gelatinous envelope. This gelatinous sheath is sometimes 2—3 times as broad as the cell. The red colour of the cells is due to some constituent in the fluid of the cell, which diffuses into the sheath and dyes the sheath red. Another difference between red and yellow snow is the presence of diatoms in red snow which might have been brought to the red snow by the wind or by penguins. Some of the diatoms are certainly of marine form. This suggests transportation from the sea coast. FRITSCH determined the constitution of red snow as follows:

Chlamydomonas nivalis (SOMMERF.) WILLE	*Melosira sol* KÜTZ.
Scotiella antarctica FRITSCH	*Coscinodiscus radiatus* EHRENB.
Raphidonema nivale LAGERH.	*Pinnularia borealis* EHRENB.
Oedogonium sp.	*Amphora ovalis* KÜTZ.
Zygnema sp.	*Triceratium arcticum* BRIGHT?

FRENGUELLI (1943) published a report on the diatoms in South Orkney, based on material collected in the years 1914 and 1917 in Uruguay Bay, Laurie Island, an island in the South Orknies. The material, obtained near the sea-shore, contained a number of freshwater diatoms among the marine forms. They were:

Achnanthes coarctata forma *constricta* KRASSKE, *A. lanceolata* BRÉB. var. *dubia* GRUN., *A. trinodis* EHRENB., *Amphora ovalis* KÜTZ., *Anomoeoneis serians* (BRÉB.) CLEVE, *Asterionella gracillima* (HANTZ.) HEIB., *Ceratoneis arcus* (EHRENB.) KÜTZ., var. *amphioxys* (RABENH.) GRUN., *Cymbella affinis* KÜTZ., *C. cuspidata* KÜTZ., *Diatoma hiemale* var. *mesodon* (EHREBNB) GRUN., *Diploneis ovalis* (HILSE) CLEVE, *Encyonema lunatum* (W.SM.) V. HEURCK, *E. ventricosum* (KÜTZ.) GRUN., *Eunotia exigua* (BRÉB.) GRUN., var. *bidens* HUSTEDT, *E. lunaris* (EHRENB.) GRUN., *E. monodon* EHRENB., var. *major* (W.SM.) HUST., *E. tenella* (GRUN.) A. CLEVE, *Gomphonema gracile* EHRENB., *G. parvulum* (KÜTZ.) GRUN., var. *micropus* (KÜTZ.) CLEVE, *Hantzschia amphioxys* (EHRENB.) GRUN., *Melosira distans* var. *alpigena* GRUN., *Meridion circulare* var. *constricta* (RALFS) V. HEURCK, *Navicula contenta* GRUN., *N. minima* GRUN., *N. mutica* KÜTZ., *N. radiosa* KÜTZ.,*Nitzschia fonticola* GRUN., *N. gracillima* HEID.-KOLBE, *Pinnularia borealis* EHRENB., *P. divergentissima* (GRUN.) CLEVE, *P. graciloides* HUSTEDT, *P. Hartleyana* GREV., *Surirella linearis* W.SM., *Vanheuckia rhomboides* (EHRENB.) BRÉB., var. *elliptica* HUSTEDT, *V. vulgaris* (THW.) V. HEURCK.

These diatoms were probably growing in inland waters nearby.

5) *Falkland*

These islands lie at lat. 51 41′ S. and long. 57 51′ W., east of the southernmost part of South America. Collections were made from soils, rocks, and among the mosses in the neighbourhood of Port Stanley and Berkeley Sound from April 8, to August 12, 1902. GAIN (1914) lists the algae found at that time. He edited the results of HOOKER, HARIOT, and SPEGAZZINI. CARLSON published another list of algae based on material collected by the Swedish expedition. The algae found on Falkland recently, through CARLSON's studies are listed in the appendix.

6) *Antarctic South America*

The southernmost part of South America, i.e. the southern part of Patagonia and the Fuego district, extends from lat. 50 °S. to 56 °S. The algal flora of these regions has been studied by many previous workers including O. MÜLLER, J. FRENGUELLI. FRENGUELLI (1923, 1924) has identified 135 species and forms of diatoms found in the Fuego district. Most of the species are cosmopolitan, but 27 are new. FRENGUELLI notes that they did not find *Amphiprora*, *Campylodiscus*, nor *Cyclotella*.

Recently THOMASSON (1955) reported on the phytoplankton in some lakes of the Fuego district based on material obtained by R. SANTESSON during his expedition to South America in the years 1939—1941. The lakes concerned vary from the oligotrophic to the eutrophic type. The flora of the phytoplankton found in these lakes is similar to that of corresponding regions in the temperate parts of the North Hemisphere, and contains representatives of the main groups of green algae. In oligotrophic lakes the desmid plankton is rich in species and dominant, while in eutrophic ones the Chlorococcales are dominant and accompanied by the diatoms. Blue-green algae are not an important feature of the planktonic-flora of this region. It is clear that the algal flora of Fuego and southern Patagonia in Antarctic South America, is rich in various kind of algae, especially desmids and chlorococcaceous algae and diatoms. This is similar to that of temperate zones and also of other cold parts of the world, such as North Europe and south Greenland.

Characteristic features of Antarctic Algal Flora

The Antarctic Continent is surrounded by vast oceans, separated from the Continents of the temperate parts of the world. The Antarctic with its adjacent islands of the subantarctic regions and the southernmost part of South America lies generally south of latitude 50°S, with the exception of islands of the Kerguelen. We cannot exclude the Kerguelen islands and the southern part of South America and the prolongation of the continent itself, even though these districts are situated partly north of the 50° south latitude circle, because the algal flora of Antarctica is probably related to that of the adjacent islands of the subantarctic or the near continent and derive some of its floral components from these adjacent regions. I have attempted to show the variations in the algal components of the flora in the high latitudes of Antarctica in Table II, based upon the results of various past expeditions. Our knowledge of the Antarctic Algal Flora at the present time is not complete, particularly concerning the various algal groups in these regions.

Table II.

Variations in the algal components of the flora
in the high latitude of the Antarctica

Regions	Kerguelens	S.Georgia Falkland Fuego	S.Orkney S.Shetland Graham L.	Kaiser-Wilhelm Land	Victoria Land
Latitude	48–50°S.	50–60°S.	60–65°S.	66–70°S.	70–80°S.
Algal groups					
Cyanophyceae	56	25	39	9	87
Diatoms	80	184	46	–	48
Heterokontae	7	5	4	0	0
Filamentous Conjugatae	9	3	2	0	0
Desmids	41	43	10	0	1
Chlorophyceae	44	41	43	8	20
Rhodophyceae	11	3	0	0	0

The figures represent the number of species.

It is too early to form any conclusions about the distribution of algae in the Antarctic based upon the data given above. However, we may point out some general trends in algal distribution. In the high latitude of the Antarctic, a rapid decrease of the Conjugatae such as filamentous Zygnemales and Desmids presents a remarkable feature since these are the commonest members of the algal flora in the temperate regions. Similar are the cosmopolitan genera of the Chlorococcales, such as *Scenedesmus* and *Pediastrum*, members of the Rhodophyceae and Heterokontae. With blue-green algae, diatoms, and green algae, the decrease in number of species is not so conspicuous as that in other algal groups in the Antarctic, although their habitats, the bare land or melting snow areas in warm season, diminish as one nears the south pole. For this reason, these three groups of algae gradually diminish in number of species according to the decrease of suitable habitats, but their decrease is not so rapid as that shown in the distribution of Desmids. In the Antarctic blue-green algae are exceedingly dominant, as demonstrated by the number of species in Victoria Land (see table II). FRITSCH has pointed out that the habitats where blue-green algae grow are the best places to search for other algae. There may be many reasons why blue-green algae grow so well under the severe conditions in the Antarctic.

1. A complicated life cycle and reproduction do not favour rapid growth in a short Antarctic summer, so that it is an advantage for blue-green algae to be able to reproduce asexually. The higher plants in the Antarctic, such as mosses, are multiplied exclusively asexually and not sexually. Antarctic mosses do not exhibit capsule and spore formation. This may be due to preserved material, or more probably because of adaptation to the severe climatic conditions.

Blue-green algae probably multiply only by individual cell division. There are relatively few cells in a number of species of the higher forms such as the heterocystous form. It may be noted that members of Stigonemataceae and Scytonemaceae are either lacking, or are rarely found in Antarctic Cyanophycean Flora.

2. The blue-green algae, more than other algae are adapted to a greater range of environments both in the water and on land.

3. The blue-green algae are capable of assimilating free nitrogen and can adapt to oligotrophic soil. FRITSCH has pointed out that there are plenty of epiphytic algae in the Antarctic flora and they are especially rich in green algae mixed with blue-green algae. This is due to the epiphytic nature of the green algae and indicates that they depend on nutrients supplied by the blue-green algae.

4. The microclimate for the growth of microorganisms in the Antarctic is not so severe as may at first be supposed. There are records from past expeditions showing that the temperature of stagnant water may rise above 10° C in warm periods (Antarctic summer) when the maximum air temperature of a fine day is about 2—3° C. W. & G. S. WEST (1909) have published records of the water temperature in the high latitude inland waters of Victoria Land:

FUKUSHIMA (1959) observed the temperature of pond waters at Ongul Island at the Syowa Base from February 1 to 4, 1959. His data concerning observations at seven ponds is presented here although there are about twenty or more ponds on Ongul Island.

Table III.

Localities	Time	Water temp.	Note
Pony Lake, Cape Royds	December 4, 1908	35 °F	partly with ice
Pond, Cape Royds	End of Dec., 1908	60 °F	open, free from ice
Coast Lake, Cape Royds	December 4, 1908	47 °F	partly closed with ice
Pond at Cape Barne	December 12, 1908	54 °F	??

Most of them are about 5.6—30 m in diameter and are shallow, generally about 30 cm in depth; during the warm season they are almost permanent and do not dry up. About two-thirds of the shores of some of the ponds are covered with lingering snow. The air temperature of Ongul Island is generally low, but the water temperature of the ponds is relatively high due to the radiant heat on a fine day. Water temperature may rise to 10.9 ° C at 16.30 hours. The difference between water temperature and that of the air is 9.7 ° C maximum, and the difference becomes maximal in the afternoon, and becomes minimal just before and after sunset. The diurnal variations at each pond are not known in detail, so that we do not know how the change of water temperature influenced the aquatic life of microorganisms.

Table IV.

Temperature and pH in ponds on Ongul Island, Syowa Base, observed by H. FUKUSHIMA February, 1–4, 1959.

No. of pond	Air temp.	Water temp.	pH	Time
1	−0.7 °C	7.8 °C	7.0	10.00
2	1.2 °C	5.3 °C	6.9	10.40
3	0.0 °C	8.4 °C	7.0	11.50
4	0.4 °C	8.3 °C	−	16.10
5	0.4 °C	10.1 °C	7.6	16.30
6	−2.1 °C	7.2 °C	7.1	19.00
7	−1.2 °C	4.6 °C	7.2	20.50

5. Surveys of Antarctic Algae have been made chiefly in the coastal regions and not in the interior of the continent. W. & G. S. WEST reported an extremely high salt content in the pools and lakes at Ross Island, these are filled with water by the thawing of snow or ice, and they note that Green Lake is very saline. They described it as follows: "The fluid obtained from under the ice at the time when it was thickest did not freeze until the temperature was redu-

ced to 7° F." This may be due to the considerable amount of salt carried by blizzards from the sea. In Japan, which is frequently struck by typhoons, it is frequently observed that grasses, shrubs and road-side trees facing the direction of the sea breeze are withered and exhibit leaf-edge discoloration. K. SUGAWARA analyzed pool water on East Ongul Island at the Syowa Base, and demonstrated a considerable chloride content:

Table V.

	No. 1	No. 2	No. 3	Average
Cl (mg/1)	134.7	136.3	204.2	158.4
Na (mg/1)	74.0	74.0	85.0	77.7
K (mg/1)	2.46	3.3	5.5	3.75
Mg (mg/1)	9.65	13.4	13.7	12.3
Ca (mg/1)	7.7	14.0	11.0	10.9
Sr (mg/1)	0.13	0.36	0.21	0.23
SO₄ (mg/1)	36.2	16.4	25.2	25.9

SUGAWARA pointed out that this may be caused by the transport of sea-spray.

The inlands waters of the coastal regions of the Antarctic Continent are more or less halophilous and support the growth of blue-green algae better than other algae. Blue-green algae are capable of adapting themselves to a broad range of environments at relatively low or high water temperatures, and also can utilise the nutrients found there. Blue-green algae are the most important element of the flora in hot springs and rely on the high salinity of the springs. There are many reports that marine and brackish species are often found in the inlands waters of the Antarctic. CARLSON reports the following marine algae:

Denticula antarctica (CASTR.) CARLSON, Pseudonitzschia migrans (CLEVE) V. HEURCK, Podosira montagnei KÜTZ., Ulothrix flacca (DILLW.) THUR., Hormiscia penicilliformis (ROTH) FRIES.

FRENGUELLI also reports marine species of diatoms from the Fuego district of South America. They are:

Actynoptychus undulatus (EHRENB.) RALFS, Cocconeis distans A. SCHM., var. minima H. PER., C. scutellum EHRENB., Coscinodiscus excentricus EHRENB., C. minor EHRENB., C. radiatus EHRENB., Melosira sulcata (KÜTZ.) EHRENB., var. crenulata GRUN., Nitzschia constricta RALFS, var. similis GRUN., N. panduriformis GREG., var. parva FRENG., Pleurosigma strigosum W.SM., Rhophoneis amphiceros EHRENB., var. rhombica GRUN., Surirella striatula TURP., Pleurosigma Wansbeckii DONK.

It may be supposed that diatoms would have a wide distribution on land in the Antarctic and its neighbouring islands; in water e.g.

lakes, and pools, wet rocks, or among mosses. Only two papers on diatoms have appeared in the reports of past expeditions. One is by W. & G. S. WEST from Victoria Land, and the other is by CARLSON from the islands surrounding the Antarctica. Recently (1943) FRENGUELLI has published detailed reports on diatoms from Fuego and south Patagonia of Antarctic South America and from South Orkney. FRENGUELLI identified many species of diatom, especially from Fuego and S. Patagonia, however, the main elements in the diatom flora are related closely to those of South America and other Continents, and have little affinity to the Antarctic. Only 27 forms are found on land in the Antarctic. The relationship between Antarctic South America and other regions is shown in Table VI. The districts marked with an asterisk are not fully surveyed for diatoms; those marked with a double asterisk are even less well studied.

Table VI.

Regions	* Antarctica	*S.Orkney	*S.Georgia	*S.Shetland	*Falkland
No. of forms	27	8	35	4	22
	Kerguelens	Africa	S.America	** Australia & New Zealand	N.America
	20	107	107	57	122
	Europe	Asia			
	194	118			

CROSBY & FERGUSON WOOD and others have shown that the forms in common between Australia and Antarctic South America have markedly increased in number. At present the diatom flora of Antarctic South America (Fuego and the South Patagonia district) seems to have a close affinity with that of Europe. This is probably because Europe is adequately surveyed for diatoms and other algae. 56 forms including species and varieties are known to be endemic in Antarctic South America, but these forms have not been found as yet in the Antarctic main land or in its neighbouring islands. The islands of Kerguelen are among the best surveyed for freshwater algae of all the subantarctic islands, and are isolated from the main-land by the ocean. BOURRELLY and MANGUIN (1954) analyzed the alga-flora of these islands and showed that of the total number of diatoms obtained (160 forms), 99 are cosmopolitan, (61.8%). 55 are endemic (29.3%); two species *Navicula linearis* (O. MÜLL.) FRENG. and *N. portomontana* CLEVE occur also in Antarctic South America. *N. muticopsis* V. HEURCK forma *capitata* CARLSON was found in Graham Land. The total number of species previously known from Antarctica is 63. The diatom species (excepting marine forms) endemic to the Antarctica are:

Fragilaria tenuicollis HEIB. var. *antarctica* W. & G. S. WEST, *Navicula glaberrimum* W. & G. S. WEST, *N. muticopsis* V. HEURCK, *N. Shackletoni* W. & G. S. WEST, var. *pellucida* W. & G. S. WEST, *N. stauropteroides* FRITSCH, *N. Murrayi* W. & G. S. WEST, var. *elegans* W. & G. S. WEST, *N. cymatopleura* W. & G. S. WEST, *Tropidoneis laevissima* W. & G. S. WEST, *Nitzschia acicularis* (KÜTZ.) W.SM.

Of these species *Navicula muticopsis* W. & G. S. WEST seems to be characteristic of the Antarctic and its neighbouring islands, however it has recently been reported from Africa, by FRITSCH, and from Pamir, by HUSTEDT. Most species found in the Antarctic main land are widely distributed over the world. Their distribution is shown in Table VII.

Table VII.

Region	Australia & NewZealand	Africa	S.America	N.America	Europe	Asia	Arctic
No. of forms	11	31	23	26	40	31	31

From one-third to one-half of all Antarctic forms are found on other Continents, and this percentage will probably be further increased by future studies in Antarctica. The diatom species obtained from the Syowa Base and its neighbouring regions have been reported by FUKUSHIMA, NEGORO, and the author. The appendix table does not include those Japanese results obtained by expeditions in the present International Geophysical Year. These results are summarized in the following species list:

Melosira varians AG., **M. sol* (EHRENB.) KÜTZ., **Triceratium arcticum* BRIGHTW., **Biddulphia aurita* (LYNGB.) BRÉB. & CODEY, **Coscinodiscus planus* KARSTEN, **C. polyradiatus* CASTR., *Asterionella gracillima* (HANTZ.) HEIB., *Diatoma vulgare* BORY var. *linearis* GRUN., *D. hiemale* (LYNGB.) HEIB., *Synedra ulna* (NITZSCH) EHRENB., var. *splendens* (KÜTZ.) BRUN., var. *oxyrhynchus* (KÜTZ.) V. HEURCK, *S. affinis* KÜTZ., **Fragilariopsis antarctica* (CASTR.) HUST., **F. cylindricus* (GRUN.) HUST., **F. obliquecostata* (V. HEURCK) HEID. & KOLBE, *Ceratoneis arcus* KÜTZ., *Eunotia pectinalis* (KÜTZ.) RABENH. var. *minor* (KÜTZ.) RABENH., **Cocconeis costata* GREG. var. *pacifica* GRUN., **C. pinnata* GREG., *Achnanthes brevipes* AG. var. *intermedia* (KÜTZ.) CLEVE, *Frustulia rhomboides* (EHRENB.) DETONI, var. *saxonica* (RABENH.) DETONI, *Caloneis bacillum* (GRUN.) MERESCH. var. *fontinalis* (GRUN.) MAYER, **Diploneis stigmosa* HEID. & KOLBE, *Stauroneis anceps* EHRENB., **Navicula longa* (GREG.) RALFS, **N. cancellata* DONK var. *Gregorii* (RALFS) GRUN., *N. seminulum* GRUN., *N. subtilissima* CLEVE, *N. cryptocephala* KÜTZ., var. *intermedia* GRUN., *N. mutiscopsis* V. HEURCK, *N. perpusilla* GRUN., *N. austroshetlandica* CARLSON, **Trachyneis aspera* (EHRENB.) CLEVE, *Pinnularia quadratarea* A.SCHM. var. *fluminen-*

152

sis (GRUN.) CLEVE, *P. bicuneata* HEID. & KOLBE, *P. lanceolata* HEID. & KOLBE var.
interrupta (A.CL.) HIRANO, *P. biglobosa* (SCHUM.) A. CLEVE, *P. borealis* EHRENB.,
Amphora ovalis KÜTZ., *A. ovalis* var. *libyca* (EHRENB.) CLEVE, *Cymbella comta*
(EHRENB.) KÜTZ., *C. tumida* (BRÉB.) V. HEURCK, *C. ventricosa* KÜTZ., *C. cymbi-
formis* (KÜTZ.) V. HEURCK, *Gomphonema kerguelensis* BOURR. & MANG., *G. olivaceum*
(LYNGB.) KÜTZ., *G. parvulum* (KÜTZ.) GRUN., *Tropidoneis laevissima* W. & G. S.
WEST, *Denticula antarctica* (CASTR.) CARLSON, *Rhopalodia gibba* (EHRENB.) O. MÜLL.,
R. gibberula (EHRENB.) O. MÜLL. var. *vanHeurckii* O. MÜLL., *Hantzschia amphioxys*
GRUN., *Nitzschia dubia* W.SM., *N. amphibia* GRUN., *N. palea* (KÜTZ.) W.SM., *N.
romana* GRUN.

In early explorations observation was centered on macrophytic algae, especially marine algae, and some large freshwater algae were collected at that time, see HOOKER & HARVEY (1874).

Prasiola species with marine algae were found on the sea-cost. *Prasiola* species and filamentous green algae (such as the *Ulothrix* group) are among the most important members of the Antarctic Algal Flora. Both genera are adapted to freshwater and marine habitats and include freshwater and marine species. *Prasiola* habitats can be classified into the following groups: one, on wet rocks, on the ground near the nesting places of sea birds, in bird droppings or in guano; two, on wet cliffs near the sea, and three, in the clear water of the upper part of streams in mountain regions, and in the polluted water of the plains. The *Prasiola* species found in the first two groups are adapted to terrestrial life, and in the third to aquatic life. On the other hand, *Prasiola* species may be classified into two groups by their nutrition. First the species growing on marine cliffs, in guano, on the ground near birds nests, and in the polluted water of plains, are adapted to eutrophic habitats. Second the species growing in mountain streams are oligotrophic. *Prasiola* grow in various habitats in the Antarctic except at low temperature, and during the long dark winter months. *Prasiola* is the characteristic element of the Antarctic Algal Flora and four species are known to occur on the Antarctic mainland and in the surrounding islands, including subantarctic regions. Of these, *Prasiola crispa* (LIGHTF.) MENEGH. is widely distributed in many regions of Antarctica and also in many other regions of the world. *Prasiola crispa* is adapted to terrestrial life. Three other species of *Prasiola* were thought to be confined in distribution to the Antarctic mainland; but, *Prasiola antarctica* KÜTZ. has been reported from Antarctic South America.

Ulothrix, like *Prasiola*, comprises both freshwater and marine species. Freshwater species are found mostly in cold streams, so it is not surprising to find them in streams and pools on the Antarctic mainland and its adjacent islands. The distribution of *Ulothrix* species is limited to habitats where there is a fixed base permitting growth and as one goes further into the interior *Ulothrix* species become rarer. *Ulothrix* species may be found distributed along the coast and in adjacent islands when more detailed explorations are made. 11 *Ulothrix* species occnr in the Antarctic regions. Other filamentous green algae, excluding *Ulothrix*, are found only in the subantarctic islands, e.g. the Kerguelens, therefore we may conclude that other green algae with a high degree of organization (filamentous or foliose forms) cannot grow in the high latitudes of Antarctica. One characteristic feature of the Antarctic Algal Flora is the presence of various groups of Chlorococcales, especially the coccoid forms such as *Chlamydomonas* and *Pleurococcus*. Examples of high organization among green algae in number of species and genera as one goes south; but there are various Chlorococcales even in places where genera with highly developed organizations scarcely occur. Although some highly organized members of the green algae may grow as far as the southernmost limits of Antarctica such as the species of *Ulothrix* and *Prasiola* as mentioned above. It may be noted however, that the cosmopolitan genera among the Chlorococcales, which are very common and abundant in waters of the temperate zone of both hemispheres, are quite lacking. While such genera as *Pediastrum, Scenedesmus, Sorastrum, Ankistrodesmus, Kirchneriella, Dictyosphaerium, Characium, Crucigenia* and *Coelastrum* are found in the Kerguelen Islands, and there are a number of other species particularly of *Pediastrum, Scenedesmus* and *Coelastrum*. These three genera are particularly rich in the number of species in waters of the temperate zones. The Kerguelen Islands lie north of the latitude 50°S. and are visited by many explorers.

On the other hand, Antarctic South America lies in about the same latitude as the Kerguelen Islands, and while no detailed survey has yet been made of the algal flora, there seem to be remarkably few species of chlorococcaceous algae. However, it is likely that there will be a considerable increase when further studies are carried out. THOMASSON (1955) has reported finding phytoplankton in these regions, and this suggests the occurrence of various kinds of algae, especially Chlorococcales. South Georgia lies in the same latitude as both Kerguelen and the southernmost part of South America; nevertheless, some algae, such as species of *Coelastrum, Pediastrum, Scenedesmus, Closterium* and *Cosmarium* which belong to cosmopolitan genera, have been reported by REINSCH from this region. These Chlorococcales and Desmids are aquatic and generally grow in swamps and moor-bogs. Such habitats occur up to about 50° south latitude and further south. It is interesting to follow the reduction of these cosmopolitan genera in relation to changes in their habitats, and also to follow the changes in aquatic species of Chlorococcales. The ideal place for such research into these serial changes of distribution with the change of latitude, would be along the Palmer Peninsula, which extends from 60° to 65° South Latitude and extends into the open sea. Island groups are the only confined places in which to study fragmental materials and, consequently, South Georgia and South Orkney and other islands of the surrounding regions of the Antarctica would be suitable places for research into serial changes in distribution. Desmids are freshwater green algae, and their distribution is unusual in that their habitat appears to be confined to peat-bogs. Many reports from the Kerguelen Islands, South Georgia, and Antarctic South America, (BOURRELLY & MANGUIN (1954) REINSCH (1890), WILLE (1924), THOMASSON (1955)) indicate that conditions in the region of 50° South Latitude permit desmids to grow; however, desmids decrease rapidly towards the south. This corresponds well with the distribution of chlorococcaceous algae. GAIN (1912) has reported some species of desmids from the Palmer peninsula and neighbouring islands, while WEST & WEST (1911) have reported only one species in the lakes of Ross Island (77° South Latitude). The material examined by GAIN was collected from among algae in lakes and pools and many forms of benthic diatoms were obtained, so it may be supposed that the collected material had contained some desmids if conditions had been suitable for them. The data of WEST & WEST (1911) suggest that desmids decrease rapidly in number of species in the high latitudes of Antarctica. The Palmer Peninsula, and places intermediate between a subantarctic position and the high latitudes of the Antarctic would be suitable places for future research in plant geography.

Ross Island habitats would probably not be suitable for desmids since there is a considerable amount of salt in its lake waters. The habitats which are suitable for blue-green algae at Ross Island and other places are not always suitable for desmids. Microorganisms are difficult for the Antarctic explorer to collect and require the skill of a specialist. The absence of desmids and Chlorococcales from past collections may not mean the actual absence of these algae from the collecting region. It is interesting that GAIN (1912) obtained 10 species of desmids in the intermediate regions of the Palmer peninsula. Another characteristic of the Antarctic Algal Flora is the presence of many coccoid forms of lower green algae. Some of them are subaerial or terrestrial and some are aquatic. *Chlamydomonas* species belong to the latter, and *Pleurococcus* to the former, but the difference between them is not absolute because species of both genera can withstand severe conditions such as dryness. It is difficult to identify the resting stages of *Chlamydomonas*, *Pleurococcus* and other coccoid forms of unicellular green algae and fix their exact taxonomic position. Furthermore, only one stage of the whole life-cycle of these algae is present in the preserved material brought back from the Antarctic. Unicellular coccoid species of green algae become modified by external conditions. Their simple and unicellular form changes when preservative fluids such as alcohol or formalin are added. Globular cells may be the resting cells of *Chlamydomonas* or the vegetative cells of *Pleurococcus* or

Chlorella. They are often found in material taken from wet ground or in collections taken from pools and streams in the Antarctic. But determination of their taxonomic position is difficult. It is preferable to examine living material and to culture it in a pure state in the laboratory.

Snow and ice algae of the Antarctica have been reported by FRITSCH (1912) and GAIN (1912) from the Antarctic mainland and South Orkney. These cryophilous algae present a varied appearance against the monotonous white land-scape of the Antarctic. Their organization is simple and for this reason it is difficult to identify the species exactly as was pointed out above for the coccoid forms of the lower green algae. They appear suddenly in a specific habitats. Red and green snow were found in the hollows, on the edge of snow-fields in contact with rocks, and in various localities of the Antarctic islands by GAIN (1911) during the Second French Antarctic Expedition in the summer of 1908. On Petermann Island, cryovegetation (snow algae) appeared frequently when the temperature rose, and when the surface layer of snow-fields was softened by thawing after several warm days. Snow algae were also found in fields near the Penguin Rookery.

FRITSCH (1910) has reported snow algae in the South Orkney, collected by RUDMOSE BROWN at "The Beach" in Scotia Bay, and states that it often appears in climatically alpine regions of the temperate and tropical zones and in the arctic. Yellow snow outside Antarctica is known only from the snow-fields of the Carpathians (ROSTAFFINSKI). This turns out to be a species of *Chlamydomonas*. The species forming yellow snow in S. Orkney are the same as those in the Carpathians. Sometimes species of red snow are mixed with yellow snow. Other yellow snows contain only species characteristic of yellow snow. They may also contain such species as *Protoderma brownii* FRITSCH, which supplies a suitable substratum for the growth of snow algae. There are some characteristic pigments in the cell contents of snow and ice algae. The pigments in yellow snow algae have been described previously (see p. 143/144), the red pigment may be the haematochrome of *Chlamydomonas nivalis* (BAUER) WILLE, or phycoporphyrin forming the violet or purple pigments of *Ancylonema Nordenskiöldii* BERGGR. These pigments absorb the heat rays of the sun and so serve to protect the chlorophyll against heat damage. The yellow snow obtained from the South Orkney consists of 18 species of algae, and two species of fungi. The main constituent of yellow snow is *Chlorosphaera antarctica* FRITSCH, together with two species of *Scotiella*. *Chlorosphaera antarctica* FRITSCH varies in size from groups of small cells to isolated large cells, which are enveloped in a characteristic hyaline, stratified, mucilaginous often irregular sheath. Isolated large cells appear to contain round fat bodies of a bright, golden colour, together with green pigments. The single large cells probably divide into groups of small cells when reproduction takes place. There is a series of cells intermediate in size between the single large cells and the groups of small cells. The author has observed these same cell types in a collection from the Lang Hovde area near the Syowa Base. It could not be ascertained from the dry specimens whether any reproductive motile forms, e.g. zoospores are produced or whether division is solely vegetative. The specimens from the Lang Hovde area were collected from under the snow. These specimens had probably once grown in a snow-field. The *Scotiella* species causing yellow snow described by FRITSCH vary from elliptic to broad fusiform, with longitudinal ridges extending from end to end. There are many small fat bodies in the cell contents, which are coloured red by haematochrome. No flagellae can be seen. According to FRITSCH, *Pteromonas nivalis* CHODAT may be the same as *Scotiella*. No algae with flagellae occur in yellow snow. All of the species appear to reproduce by simple cell-division. Red snow did not occur as frequently as yellow snow, and was not so rich in species, all of which were in the resting stage.

Recently, KOL (1944) suggested that *Chlamydomonas nivalis* (BAUER) WILLE was a compound name that is representative of the *Chlamydomonas* species which cause the red colour in snow. *Chlamydomonas* species, identified as *Chlamydomonas nivalis* (BAUER) WILLE in Antarctic reports, are probably *Chlamydomonas sanguina* LAGERH., though *Chlamydomonas antarctica* WILLE is the dominant species in Antarctica. The constituents of green snow in Antarctica are species belonging to the so-called *Stichococcus* type, and are different from those of the Northern Hemisphere, which are known as the *Raphidonema* type. The constituents of Antarctic green snow are *Stichococcus bacillaris* NÄG. forma *minor* (NÄG.) RABENH., forma *maior* (NÄG.) RABENH., and *Chlorella ellipsoidea* GOROSCH. forma *antarctica* WILLE (according to KOL (1942) (1944)). These species are dominant in Antarctica, but are not as yet known in the Northern Hemisphere. In addition to these species *Raphidonema nivale* LAGERH., *Mycacanthococcus antarcticus* WILLE, *Pleurococcus vulgaris* MENEGH. var. *cohaerens* WITTR. are also found and these species are well known in the green snow of the Northern Hemisphere. The first is the representative species of the green snow which appears in the Northern Hemisphere.

The algae which occur in snow and ice have a relatively simple organization, and so can withstand the changing environment. The resting stages of *Chlamydomonas* which appear in red snow may exist as motile forms with flagellae under favourable conditions. The surface layers of snow and ice thaw slowly during the fine days of an Antarctic summer and hollows fill with water. Thus the temperature of the water above the upper layers of snow and ice is slightly raised above freezing point, and *Chlamydomonas* may swim under such conditions. However, they must be able to change rapidly from the motile form to a resting stage to withstand the cold conditions, in blizzards and at night. Meteorological observations in Antarctic stations indicate that water-temperatures probably drop below freezing point at night. Data from Mawson and Davis stations are given in Table VIII.

We have no detailed knowledge of the habitats of coloured snow and ice. Nor do we know whether these algae multiply by cell fission or by sexual reproduction. *Chlamydomonas* cells in the resting stage may change motile forms under experimental conditions. They may exploit the extreme limits of their environment for multiplication. Algal adaptation in response to changes in environment may be observed in cultures. Collections obtained from cryovegetation should look as if they were alive, because the characteristic colour and features of snow and ice algae disappear when specimens are preserved in fluid and some of the cell contents disappears. This makes it difficult to determine the exact taxonomic position of specimens preserved in fluid, unless there is some recognizable characteristic feature. These difficulties will be partly removed at the permanent observatory stations.

The constituents of cryovegetation consist of two types of algae, one is truely cryophilous, and the other is facultatively transported by agencies, such as wind and animals. The characteristic features of cryoalgae are determined by the cryophilous members. They respond readily to changes in environment and for this reason retain a simple organization which is capable of adapting to severe conditions, and the responding to the short Antarctic summer. It is not therefore surprising that highly organized algae are completely absent from cryovegetation. The evidence from preserved material, such as fat bodies, the red, violet, and purple pigments of haematochrome and phycoporphirin in the cell contents of snow and ice algae, is not fully explained.

There is an hypothesis that the accumulation of fat is an adaption to severe cold, and it may be a source of energy for growth.

The specific pigments, haematochrome and phycoporphyrin may act as screens against heat rays, or may absorb heat, and facilitate the thawing of snow and ice. Will snow and ice algae that multiply in the summer tolerate the severe cold and darkness during the long Antarctic winter? These cryoalgae may remain in a resting stage in snowdrifts until the next warm season. Conditions under snowdrifts may not be so severe as might be supposed because the algae are protected under a thick blanket of snow. The microorganisms which live in Antarctica are obviously better able to tolerate severe cold than those which live in warmer areas. The development of cryovegetation is due to the establishment of

Table VIII.

The daily variations of temperature in
Mawson and Davis stations in Antarctica.

Mawson Latitude 67° 36′ S. Longitude 62° 53′ E. Height above M.S.L. 27 feet

Hour	1957 January		1957 February		1957 December	
	Dry Bulb	Wet Bulb	Dry Bulb	Wet Bulb	Dry Bulb	Wet Bulb
02	30.5	26.5	21.5		28.3	
05	30.2	25.9	21.0		28.2	
08	32.2	27.4	21.9		29.9	
11	35.5	29.9	25.9		32.9	
14	36.4	30.9	28.7		34.2	
17	36.1	30.8	28.4		33.9	
20	34.9	30.1	26.4		32.7	
23	32.3	28.1	23.0		30.5	

	Mean Max.	Extr. Max.	Mean Min.	Extr. Min.	Mean Terr. Min.	Extr. Terr. Min.
1957 January	37.8	43.1 on 7th	28.5	21.7 on 22th	27.2	20.2 on 22th
1957 February	30.0	40.2 on 2nd	18.8	6.6 on 27th	17.5	5.3 on 27th
1957 December	35.6	41.3 on 16th	26.2	15.9 on 3rd	25.1	11.0 on 3rd

Davis Latitude 68°35′S. Longitude 77°58′E. Height above M.S.L. 40 feet

Hour	Mean Temperature°F		Temperature°F				
	1957 November	1957 December	Mean Max.	Extr. Max.	Mean Min.	Extr. Min.	
	Dry Bulb	Dry Bulb					
02	17.2	30.0	25.0	32.7	15.1	5.3	1957
05	18.2	30.2		on		on	Nov.
08	20.5	32.2		25th		4th	
11	22.5	33.6		30th			
14	23.0	34.4					
17	23.1	33.2	36.5	43.7	28.4	23.4	1957
20	21.6	32.0		on		on	Dec.
23	18.9	31.2		19th		1st	

A.N.A.R.E. Reports series D, Vol. 10, 1960 Meteorology

indigenous species of cryoalgae in certain places followed by other species transported from neighbouring localities by wind and birds. The presence of foreign particles in cryovegetation indicates that aerial transportation is possible. Resting spores of those algae transported by outside agencies may grow on the surface layers of snowfields after thawing, if the water temperature becomes high enough. Condition in cold streams produced by thawing are just those conditions which permit the growth of cryoalgae. FUKUSHIMA (1962) found the cryoalgae *"Raphidonema nivale"* in wet soil produced after thawing.

Comparison of the Algal Flora of Arctic and Antarctic Regions

Comparison of the algal flora of the Arctic with that of the Antarctic helps in understanding the characteristic features of antarctic species. The algal flora of both polar regions are similar; however some differences in environment exist between the two regions. Arctic regions are topographically and oceanographically complex. In Antarctica the land to the north beyond the antarctic circle, consists of part of the Antarctic Continent and its neighbouring islands in the Indian Ocean, and the Palmer peninsula, which projects to the north opposite Antarctic South America. These northern parts of Antarctica are isolated from the other continents by sea. In arctic regions, however, the northern parts of the Continents of Europe and Asia and North America extend far into the arctic circle. These lands within the arctic circle are parts of the Continents themselves, and there is no sea barrier to distribution. If we exclude the northern extensions of the land masses of the Continents of Europe, Asia and North America, the arctic region consists of Greenland, Spitzbergen, Novaja-Semlja, Franz Josefs Land, and the Canadian arctic islands.

In Antarctica, there is no warm ocean current, so that the circle of the northern limit of pack-ice spreads north far beyond the antarctic circle. In arctic regions, on the other hand, a warm current, the North Atlantic Drift, penetrates north beyond the arctic circle and then separates into two branch currents. One of these flows along the western side of Greenland, and the other reaches up to the western sea coast of Spitzbergen. This means that the southern limit of the pack-ice lies north well within the arctic circle in those regions influenced by this warm current. Thus Iceland is situated outside the southern limit of the pack-ice. Most of the Antarctic main land, together with neighbouring regions, does not rise above 32° F. during even the warmest season of the year, with the exception of the northern part of the Palmer peninsula. In the arctic, on the other hand, the area in which the temperature in summer does not

160

Table

Monthly Climatic Data for the World
Vol. 13,

Station	Latitude	Longitude	Elevation meters	Temperature		
				June	July	August
Greenland						
Nord	81 36 N	16 40 W	36	−0.2	4.7	1.8
Thule A.B.	76 31 N	68 50 W	11	2.9	7.1	3.3
Egedesminde	68 42 N	52 52 W	48	3.9	8.7	8.3
Prins Christians Sund	60 03 N	43 12 W	77	4.7	7.0	7.1
Danmarkshavn	76 46 N	18 46 W	18	0.9	4.1	2.5
Angmagssalik	65 37 N	37 39 W	36	0.5	7.3	6.5
Franz Josef Land						
Bukhta Tikhaya	80 19 N	52 48 E	6	−0.6	1.4	−0.1
Spitzbergen						
Isfjord Radio	78 04 N	13 38 E	9	2.0(+0.1)	6.1(+1.5)	5.0(+0.7)
Bear Island						
Bjornoya	74 31 N	19 01 E	14	2.5(+0.7)	6.5(+2.2)	4.4(−0.2)
Jan Mayen	71 01 N	08 28 W	39	2.6(+0.3)	5.3(+0.3)	5.8(+0.2)
Iceland						
Reykjavik	64 08 N	21 56 W	18	10.0(+0.4)	12.2(+0.9)	11.2(+0.6)
Faeroes						
Thorshavn	62 03 N	06 45 W	26	10.6(+1.7)	11.8(+1.3)	11.2(+0.9)
Vardo (Norway)	70 22 N	31 06 E	15	6.7(+0.9)	13.5(+4.6)	10.5(+1.6)
Murmansk (USSR)	68 58 N	33 03 E	46	10.2	19.4	12.6
Arkhangelsk (USSR)	64 35 N	40 30 E	13	12.9	21.2	14.3
Leningrad (USSR)	59 58 N	30 18 E	4	16.8	20.2	16.6
Alaska						
Barter Island	70 07 N	143 40 W	15	1.1(−0.7)	2.9(−2.0)	2.9
Barrow	71 18 N	156 47 W	2	1.0(−0.1)	3.1(−1.2)	1.2(−2.4)
Nome	64 30 N	165 26 W	14	6.2(−1.3)	12.6(+2.9)	10.4(+0.8)
Anchorage	61 10 N	149 59 W	40	12.8(+0.7)	14.2(+0.1)	13.5(+0.5)
Juneau	58 22 N	134 35 W	7	10.6(−1.4)	12.4(−0.2)	11.8(−0.5)
Arctic Canada						
Aklavik	68 14 N	135 00 W	9	10.7(+1.3)	10.5(−3.1)	9.2(−0.9)
Coopermine	67 49 N	115 15 W	9	5.0(+1.4)	10.2(+0.8)	8.6(+0.5)
Frobisher Bay	63 45 N	68 33 W	21	3.6(+0.4)	8.3(+0.7)	6.9(+0.0)
(Baffin Ld.)						
Fort Churchill	58 45 N	94 04 W	35	7.2(+1.4)	11.3(−1.3)	11.5(−0.2)
(Hudson Bay)						
Goose Bay (Labrador)	53 19 N	60 25 W	44	13.1(+2.9)	15.8(+0.0)	15.3(+0.9)
Gander	48 57 N	54 34 W	147	13.2(+2.1)	16.9(+0.7)	16.3(+0.2)
(New Foundland)						
Petropavlovsk Na	52 58 N	158 45 E	70	6.8	12.2	14.6
(Kamchatka)						

IX.

U.S. Department of Commerce Weather Bureau
No. 6–9, 1960

September	Relative humidity				Precipitation Total mm			
	June	July	August	Sept.	June	July	August	September
−7.9	87	79	86	82				
−3.3	70	41	76	77	4	20	3	20
3.1		83	87	84				
5.5	81	79	88					
−4.1	83	84	84	83				
4.7	78	75	82	82				
−2.2					4	20	20	4
3.5(+2.1)	87(+0)	88(—1)	86(—2)	91(+6)	20(+3)	60(+34)	40(—1)	90(+58)
5.1(+1.9)	86(—6)	84(-10)	85(—8)	85(—6)	40(+17)	20(—5)	30(—5)	50(+4)
5.9(+2.3)	90(+2)	90(+1)	85(—3)	87(+2)	20(—4)	40(+2)	30(—26)	100(+24)
9.1(+1.3)	84	78	78	84	80(+30)	40(—11)	10(—42)	60(—31)
10.2(+1.2)	86(+4)	88(+2)	86(+0)	87(+2)	60(+8)	60(—10)	80(—2)	130(+23)
7.8(+1.3)	86(+1)	80(—8)	86(—2)	87(+2)	30(—12)	10(—26)	40(—4)	50(—18)
7.7	60	60	75	83	—	30	40	30
7.2(—1.0)	74	60	75	91(+8)	140	20	60	90
10.2(—0.4)	70	73	76	83(+3)	50	70	70	40
−1.4		88	94	95	10(—1)	20(—10)	20(—10)	20
−4.1	88	89	93	93	6(—1)	5(—16)	30(+10)	10
3.6(—1.9)	77	76	82	75	10(—19)	50(—17)	50(—40)	70(—2)
8.4(—0.4)	60	71	75	77	6(—13)	70(+28)	70(+1)	120(+52)
9.6	80	81	83	89	90(+9)	110(—8)	120(—11)	220
2.4(—1.0)	67(—9)	78(+1)	79(—6)	84(—3)	3(—18)	40(+5)	50(+13)	60(+36)
2.4(+0.0)	91(+2)	74(-11)	83(—3)	87(—3)	20(—3)	20(—19)	90(+47)	30(—2)
2.6(+0.7)	73(—6)	74(—2)	81(+3)	81(+0)	50(+24)	30(—11)	90(+39)	70(+27)
5.4(—0.8)	77(—6)	77(—2)	77(—6)	82(—6)	140(+108)	50(—10)	70(+4)	30(—27)
9.6(+0.0)	64(—3)	65(—2)	64(—5)	68(—2)	90(+29)	130(+48)	80(+9)	110(+53)
12.3(+0.2)	72(—1)	69(—6)	69(—8)	74(—6)	110(+39)	30(—62)	40(—51)	50(—43)
—	78	82	80	—	40	120	60	—

Table

Monthly Climatic Data for the World
Vol. 13,No. 12 –

Station	Latitude	Longitude	Elevation Meters	Temperature; Antarctic summer		
				December	January	February
Antarctic region				December	January	February
Norway Base	70 20 S	02 00 W		−5.9	5.8	−8.7
Amundsen-Scott Station	90 00 S		#2,800	−27.1	−30.6	−41.4
Ellsworth IGY Station	77 43 S	41 07 W	40		−8.6	−13.1
Byrd IGY Station	80 00 S	120 00 W	#1,500	−14.9	−13.1	−20.0
Williams Air Operations Facility, Mc Murdo Sound	77 50 S	166 36 E	45	−4.7	−27.0	−8 3
Cape Hallett	72 18 S	170 18 E	5	−2.7	0.0	−3.5
Wilkes IGY Station	66 15 S	110 35 E	#11–12	−1.4	−0.6	−1.8
Syowa Base	69 00 S	39 35 E	15	−1.9	−2.2	−2.2
Mawson	67 36 S	62 53 E	14	−1.5	−0.6	−0.1
Camp Dumont d'Urville	66 40 S	140 01 E	40	−3.0	−1.6	−5.0
Davis	68 33 S	77 56 E	14	−1.5	−4.3	−2.6
Observatorio Naval I. Orcadas	60 44 S	44 44 W	4	0.3	1.2	0.5(+0.3)
Destacamento Naval Decepcion	62 59 S	60 43 W	7			1.0
Argentine Island	65 15 S	64 16 W	10	−1.1		
Stanley (Falkland)	51 42 S	57 52 W	53	7.2		
South America						
Puerto Santa Cruz	50 01 S	68 34 W	111	12.6	13.6(—0.7)	14.0(+0.7)
Rio Gallegos	51 40 S	69 16 W	22	11.3	11.2(—1.2)	11.8(—0.1)
Ushuaia (Fuego)	54 48 S	68 19 W	7	6.9(—1.7)		8.5(—0.6)
Punta Arenas (Chile)	53 10 S	70 54 W	8	8.6(—1.6)	9.8(—1.3)	9.8(—0.8)
Porto-aux-Français Iles Kerguelen	49 20 S	70 13 E	14	5.8	5.9(—1.2)	6.4(—1.3)
Campbell Island	52 33 S	169 07 E	23	7.9	8.8	8.1
Macquarie Island	54 30 S	158 57 E	6	6.5	6.2	5.2
Marion Island	46 53 S	37 52 E	26	6.1(+0.4)	6.6(—0.4)	6.8(—0.7)

T Less than one millimeter
approximate

IX. (continued)

U.S. Department of Commerce Weather Bureau
Vol. 14, No. 13, 1961

1960-1961	Relative humidity average %				Precipitation Total mm			
March	Dec.	Jan.	Febr.	March	December	January	February	March
–10.8	78	79	74	72	–	–	T	–
-57.1					3	T	T	T
-22.3		72	62	55				
-29.7	85	78	75		T	5	1	10
					5	20	–	
-19.4	77	79	73		100	20	40	–
–9.3	83	66	59	61	4	20	10	20
–4.9		77	80	84	T	2	2	40
	69	73	62		–	–	–	–
	58	59	68	71	–	–	–	–
–8.1	64	66	63	61	–	–	–	–
–1.3	49	48	46	64	0	–	–	–
–0.7	88(+2)	90(+4)	90(+3)	89(+3)	30(+3)	50	100(+60)	–
–2.1			92	87	–	–	20	40
	77				–	–	–	–
					–	–	–	–
10.3(—1.4)	51	51	49	49	10	20(+2)	10(—4)	20(—1)
8.9(—0.6)	49	59	58	57	40	60(+29)	20(—4)	20(—8)
6.5(—1.1)	–	–	–	–	110		90(—37)	100(+37)
7.8(—1.1)	60(—6)	71(+5)	61(—6)	72(+1)	–	60(+27)	30(+2)	20(—16)
6.6(+0.4)	67	74	77(+.6)	72(—1)	30	70(—11)	120(+59)	80(+14)
7.5	84	91	94	85	120	70	140	150
5.2	92	95	85	90	80	70	60	110
6.8(—0.6)	82(+1)	83(+1)	81(—2)	84(+1)	230(+15)	180(—23)	210(+45)	400(+186)

164

exceed 32° F. is confined to the interior part of Greenland and neigh-
bouring parts of the North Pole. Table IX shows the temperatures
at main stations in both polar regions during the years 1960—1961.
The monthly average summer temperatures in Greenland and
Spitzbergen are clearly higher than those in the same latitudes in
Antarctica, where the average temperatures in December and
January (the Antarctic summer) do not rise above freezing point.
It is not valid to compare the environments in antarctic habitats
with those in the arctic merely by their latitude, because of the in-
fluence of ocean currents. Iceland lies south, beyond the south limit
of the pack-ice. Scandinavia and northern Europe, west of Novaja
Semlja, face the Barents Sea, and are affected by the North Atlantic
Drift so these regions are excluded from consideration of the arctic.
The Arctic Lands in this limited sense consists of: Greenland,
Spitzbergen, Novaja Semlja, Franz Josefs Land, Jan Mayen and
Bear Island and their adjacent islands. The arctic regions in which
freshwater algae have been fairly thoroughly surveyed are Greenland and Spitzbergen only; in other regions algal surveys are not
complete. Summarized in Table X is an outline of the distribution of
arctic algae.

Table X.

	Greenland, east side	Greenland, west side	Spitzbergen	*Franz Josefs Land	Novaja-Semlja	*Jan Mayen	*Bear Island
Latitude	65°30'–76°	64°10'–70°17'	76°50'–81	80°–82°	70°50'–77°	70°50'–71°50'	72°90'–74°35'
Cyanophyceae	11	55	31	6	24	–	–
Chlorophyceae includ. fil. Conjugatae	38	61	36	15	40	–	–
Desmids	197	172	136	16	105	–	–
Heterokontae	5	5	3	3	4	–	–
Chrysophyceae	1	15	1	0	0	–	–
Diatoms	165	385	250	108	–	47	75
Rhodophyceae	0	4	1	–	–	–	–

The Asterisk indicates those districts not fully surveyed for algae.

Geographically, the Arctic is at almost the same latitude north
as the southern latitude of the coast of the Antarctic Continent.
The algal flora of arctic regions differs from that of the Antarctic in
the richness of its green algae, and an abundance of desmids. Arctic
blue-green algae are not so numerous in species and genera as they
are in Antarctica. It is well established that there are more than 150

species of desmids in lat. 60—70°N. on either side of Greenland. Less than ten species have been found in the same corresponding latitude of Antarctica. Further about 136 species of desmids have been found in lat. 76° 50′—81°N. on Spitzbergen while only one species was found in Victoria Land in lat. 70°S. in Antarctica. We do not know to what extent species of desmids decrease in number in the northern latitudes of Greenland, particularly further north than lat. 80°N., because there has been no collection of material. It would be instructive to know the composition of the algal flora of Greenland, particularly in latitude 80°N. and northwards. There are 16 species of desmids in regions from 80°—82°N. lat. in Franz Josefs Land. This number of species is interesting, especially since the region has not yet been fully surveyed for desmids. It is probable that this number will increase when fuller explorations are carried out. There are some sterile forms of filamentous Conjugatae in high latitudes of the arctic regions, and the total flora almost includes all the cosmopolitan genera of the filamentous Conjugatae. Whether this is true in the corresponding high latitudes of the Antarctic is not yet known. There are no essential differences in green algae between arctic and antarctic regions of the same latitude. The cosmopolitan genera: *Pediastrum, Scenedesmus, Ankistrodesmus, Closterium, Spirogyra, Zygnema* are common in the waters in temperate zones, and occur in the Fuego district of Antarctic South America, South Georgia, and the Kerguelen Islands. These genera are distributed as far as 60°S. lat. but are very rare south of that latitude. Some species of a few genera are found further south. On the other hand the genera *Pleurococcus, Trochiscia, Mycacanthococcus, Chlamydomonas* have gradually increased in number, and many species of *Ulothrix* are found everywhere. In Spitzbergen, Greenland and Novaja Semlja species of *Pediastrum* and *Scenedesmus*, occur in places 65°N. lat. and further north, and in these regions desmids are not rare. Higher forms of green algae, such as *Oedogonium* and *Bulbochaete*, have also been recorded from places in high latitudes in arctic regions. Diatoms have been well studied in many places in the arctic, especially in Greenland. Recent reports by Foget on the northern and western parts of Greenland have considerably increased our knowledge of the diatoms in these districts. Hustedt (1937) and Krasske (1938) have studied the diatom-flora of Spitzbergen, and about 250 species have been recorded. Hence it is reasonable to suppose that numerous diatoms occur in other regions of the arctic and await detailed surveys in the future. In contrast there are only a few records of freshwater diatoms in Antarctica. These may have been seriously underestimated as discussed earlier. Algae which are distributed in the arctic and antarctic regions and which have not been found in other regions of the world are very few.

Table XI.

Frequency of algal species in the Arctic and Antarctic

Algae	Number of species in Antarctica	Number of Antarctic species also found in Arctic	Percentage Antarctic species found in Arctic
Blue-green algae	155	35	23
Diatoms	304	81	27
Green algae (except Conjugatae)	142	35	25
Desmids	44	25	64

It is clear from the Table XI that at least one quarter of all the species distributed in Antarctica are also found in the arctic. Furthermore, the majority of species which are commonly distributed in the arctic are cosmopolitan, and are widely distributed in the temperate zone. However, some species which are mainly distributed in the bipolar regions, are largely confined to alpine regions in the temperate zone. For instance, *Eucapsis alpina* CLEMENTS & SHANTZ is distributed chiefly in the alpine regions of Europe and North America, and it is also found in the alpine moors of Japan; but in Antarctica, another species of *Eucapsis* (*Eucapsis minuta* FRITSCH) is found. *Chlamydomonas* species are widely distributed all over the world, and six species occur in Antarctica. *Chlamydomonas nivalis* (BAUER) WILLE is one of the most representative members of the cryovegetation that shows a worldwide distribution in cold regions of the Northern Hemisphere (the arctic snow-fields of the European alps and North American mountains). KOL (1944) suggests that the *Chlamydomonas nivalis* recorded previously in Antarctica is not a true species of *Ch. nivalis*, but more likely *Ch. antarctica* WILLE. In future studies, the number of *Chlamydomonas* species from the Antarctic is likely to be reduced by reclassification. If we accept the suggestion of KOL (1944) then there is no evidence that the same species of *Chlamydomonas* are distributed in the Antarctic and in the Arctic. Species of *Trochiscia*, *Scotiella* and *Pleurococcus* occur in the ice and snow-fields of both polar regions, but the species found in the Antarctic are not the same as those in the alpine and other cold regions of the Northern Hemisphere. The species from snow and icefields in alpine and polar regions are not usually found in lowland waters, but may occur in the snow-fields of temperate districts when it thaws, towards the end of the spring. 11 species of *Ulothrix* are known from Antarctica, and seven of these are also reported from the Arctic. *Ulothrix* commonly grows in cold waters, and contains marine and freshwater species. Both the Arctic and Antarctic have many *Ulothrix* species. *Prasiola* has the same type of distribution and grows in brackish water as well as in fresh water. It has aquatic and terrestrial species. 8 species of *Pleurococcus* occur in Antarctica, and *Pleurococcus* is one of the important members of the Antarctic algal flora, however it is seldom found in the Arctic. *Pleurococcus* is terrestrial and occurs in other regions of the world: it is especially noticeable in the barren land of the Antarctic because other green algae are rare. There is no very close relationship between the Antarctic and the Arctic algal floras except that some genera of green algae are common to both, for example species of *Ulothrix, Prasiola* and snow growing members of *Scotiella, Trochiscia*. In both polar regions members of the cosmopolitan genera of green algae, and golden algae like *Hydrurus*, are lacking.

Comparison between the Algal floras of the Antarctic and Europe

The algal flora of Europe has been comprehensively studied. The species reported from Antarctica which are also found in Europe are shown in the appendix. Table XII shows the number of Antarctic species of algae which are also found in Europe. This table shows clearly that more than half of all Antarctic species of algae are also found in Europe. This points to a close relationship between the algal floras of Europe and Antarctica. There is a similar relationship between the algal floras of other well-surveyed regions of the world. But this close relationship is more apparent than real, it merely reflects the distribution of cosmopolitan genera in the algal kingdom. In Central Europe 653 species of blue-green algae are known, and in Britain 667 species of desmids, a large number for such a restricted region. But the European species of algae have not yet been fully recorded. 86 of the cyanophycean species of middle Europe are found in Antarctica, this correspondents to 13%. Only 38 species (6%) of desmids from the British list occur in Antarctica. Considering Europe as a whole, the percentage of Antarctic species present in

Table XII.

Frequency of algal species in common between the Antarctic and Europe

Algae	Number of species in Antarctic	Number of Antarctic species also found in Europe	Column 3 expressed as percentage of column 2
Blue-green algae	155	95	61
Diatoms	304	176	58
Desmids	44	40	90
Green algae	142	80	56

Europe increases. The Antarctic algal flora is not closely related to the European (see the appendix table). The table also points to the same conclusion for the absence of a close relationship to the algal floras of Africa, South America, and even of Australia and New Zealand (which lie close to Antarctica). In fact, the same comparison may be made with the algal flora from any other region of the world. There are some genera in common between the algal floras of Antarctica and Europe. There are many species of *Trochiscia* common to both regions, this reflects the similarity of natural conditions between Antarctica and the alpine regions of Europe which are the habitats of *Trochiscia*. The differences in habitat between the plains of Europe and Antarctica lead to a deficiency of cosmopolitan genera of green algae. Since climatic conditions become milder and conditions for the growth of algae become more like those in Europe as one progresses north, away from the antarctic circle; the cosmopolitan genera gradually increase in number. One good example of this is seen in the Kerguelen Islands and in the Fuego district of subantarctic regions. Habitats in Europe similar to those in Antarctica are the snow and ice-fields ot the alphine regions, running water and cold springs. There are common genera in such places with many species, for example *Chlamydomonas*, *Ancylonema*, *Pleurococcus* and *Trochiscia* in alpine snow and ice-field and *Ulothrix* species in cold springs.

Some Problems concerning the study of Antarctic Algae

Problems concerning the future study of Antarctic Algae can be divided into two classes. First, a systematic investigation and critical appraisal of all the known species based on preserved specimens. Second, surveys of the unexplored regions both geographically and ecologically. Concerning the first problem a list of all the Antarctic fresh- and brackish-water algae is presented in the appendix table. Some of the names in the list are probably incorrect and may be subject to reclassification in the future. Some genera are rich in species, but this may be due to too minute a subdivision into species. Some so-called species may be merely adaptive forms due to habitat or may be merely points of the life cycle of other species. Blue-green algae are difficult to identify because identification is based on the inadequate features of their simple morphology. Collections made by past expeditions are insufficient, and were not always carried out by skilled workers, so it is necessary to compare material collected by modern methods and preserved material. By comparison of old and new material leading to the exclusion of questionable species and perhaps to new additions improvement to the provisional list in the appendix may follow.

The coast-line of the Antarctic Continent is very extensive, and there is much barren land along the coastal plain and in the interior of the continent. Most of this land still remains unexplored biogeographically and ecologically. Islands adjacent to the continent, and islands in subantarctic regions strangely enough have not as yet been well investigated. Future studies should concentrate on the regions of Antarctica not yet surveyed for algae. Suitable places for the growth of algae are known, for example, the innumerable pools, ponds and lakes caused by thawing found by antarctic explorers. However collections from these areas are extremely poor and there remain merely a photographic record. Past expeditions laid emphasis first on geography, and even to-day biological surveys of Antarctica are subsiduary; Too short a time was devoted to this purpose in Japanese expeditions. Very little attention has been paid to the ecology and limnology of antarctic inland waters in the past. It is important to clarify the algal distribution and the changes in composition of the flora with changes of latitude, as mentioned above. There are some barren regions in the high latitudes of Antarctica, even in the interior, where innumerable pools and ponds caused by thawing during the warm season occur. Such places almost certainly contain algae. It is particularly important to study the algal flora of the subantarctic islands around Antarctica because these subantarctic islands lie in the path of algal transportation to the Antarctic, from other continents. We have no knowledge of the freshwater algal flora of Auckland, nor of the Campbell Islands that lie to the north. The Kerguelen Islands are the most surveyed islands in the subantarctic regions, but the districts surveyed are only parts of the islands, and phycological knowledge of the islands is not complete. Detailed surveys of the Palmer peninsula might be useful in explaining changes in floral composition, because the Palmer peninsula lies in the range 63° to 67° lat. S. Direct research into algal distribution in the Palmer peninsula would be preferable to surveys of the islands lying randomly in the oceans of the subantarctic. Previous studies have been based on preserved materials, and older collections.

It is now possible to study fresh material in stations at the site and to obtain material as required. It is simpler now to find snow and ice algae, to study the ecological conditions, the causes of the occurrence and disappearance of cryovegetation, and to search further for the details of the complete life-cycles of algae. Studies on the influence of environmental factors on antarctic algae have hitherto been inadequate, except for the report by WEST & WEST (1911) on the ponds and lakes of Ross Island. The daily and monthly variations in temperature and the chemical constituents of water on the whole antarctic continent are not satisfactorily reported, and they have a direct influence on aquatic plant life. Comparative studies on the habitats of algae, especially pools, lakes, and running water

170

are necessary to help explain the adaptation of algae to the Antarctic environment. In addition, studies on the relationship between alpine conditions and Antarctic conditions must be promoted by the accumulation of regional data. It is likely that aquatic areas around the bases in Antarctica are gradually being polluted by human activity. Vigorous growth of blue-green algae in polluted water has been observed around some bases. Some of these algae might not have been ranked previously as proper members of the Antarctic flora, and others should perhaps be excluded as members of the flora.

Biological Laboratory
of the Yoshida College,
Kyoto University

Appendix

List of Antarctic and Subantarctic Algae known from previous expeditions, and their geographical distribution

	Antarctica	S.Shetland	S.Orkney	S.Georgia	Falkland	Fuego and S. Patagonia	Kerguelen	Arctica	Australia and New Zealand	Africa	S.America	N.America	Europe	Asia
Microcystis chroococcoidea W. & G. S. West	+								+					
M. coerulea Dickie forma Wille							+							
M. ichthyoblabe Kütz. forma Wille							+						+	
M. kerguelensis Wille							+							
M. marginata (Menegh.) Kütz	+			+			+			+		+	+	+
M. merismopedioides Fritsch	+		+				+			+				
M. olivacea Kütz			+										+	
M. parasitica Kütz	+			+			+			+			+	+
M. — var. glacialis Fritsch	+													
M. stagnalis Lemm.	+											+	+	
Aphanocapsa delicatissima W. & G.S.West						+	+					+	+	+
A. montana Cramer	+							+					+	+
Aphanothece castagnei (Bréb.) Rabenh.							+						+	+
A. microscopica Näg.							+	+			+	+	+	+
A. prasina A.Br.	+										+	+		
A. — forma minor Wille							+							
A. saxicola Näg.				+			+	+				+	+	+
A. — var. aquatica Wittr. & Nordst.							+	+						
Chroococcus cohaerens (Bréb.) Näg.	+									+		+	+	+
C. — var. antarctica Wille	+													
C. consociatus Hariot						+								
C. helveticus Näg.	+							+				+	+	+
C. kerguelensis Wille							+							
C. macrococcus (Kütz). Rabenh.	+						+	+					+	+
C. minor (Kütz.) Näg.	+			+			+			+		+	+	+
C. — forma minima West	+											+		
C. minutus (Kütz.) Näg.	+			+	+		+	+		+	+	+	+	+
C. — var. obliteratus (Richt.) Hansg.	+											+		
C. — var. amethystacea Wille	+													
C. pallidus Näg.	+							+		+		+	+	+
C. turgidus (Kütz.) Näg.	+							+		+		+	+	+
C. — forma minor Wille							+							
Gloeocapsa didyma Kütz. forma Wille							+						+	
G. janthina Näg.	+							+		+				
G. livida (Carm.) Kütz. forma Wille							+							
G. quaternaria (Bréb.) Kütz. f. Wille							+						+	
G. rupicola Kütz.	+												+	
G. shuttleworthiana Kütz.	+												+	
Gloeothece tepidariorum (A.Br.) Lagerh.					+								+	+
Gomphosphaeria aponina Kütz.		+				+	+		+	+	+	+	+	+
G. lacustris Chodat var. compacta Lemm.								+				+	+	+
Coelosphaerium Kützingianum Näg.		+					+	+	+	+	+	+	+	+
C. Nägelianum Unger						+							+	
Eucapsis minuta Fritsch	+													

	Antarctica	S.Shetland	S.Orkney	S.Georgia	Falkland	Fuego and S. Patagonia	Kerguelen	Arctica	Australia and New Zealand	Africa	S.America	N.America	Europe	Asia
Merismopedia glauca (Ehrenb.) Näg.			+			+	+	+	+	+	+	+	+	+
M. punctata Meyen	+					+				+	+	+	+	+
M. tenuissima Lemm.	+	+				+	+			+		+	+	+
Synechococcus aeruginosus Näg.			+				+	+		+		+	+	+
S. elongatus Näg. forma minor Wille							+							
S. kerguelensis Wille						+	+							
S. maior Schr.							+						+	
Dactylococcopsis antarctica Fritsch	+													
D. irregularis G.M.Smith						+								
D. rhaphidioides Hansg.	+									+		+	+	+
Entophysalis granulosa Kütz.				+									+	
Chamaesiphon confervicola A.Br. forma Wille							+	+				+	+	
Lyngbya aerugineo-caerulea (Kütz.) Gomont	+						+			+		+	+	+
L. aestuarii (Mert.) Liebm.	+			+						+		+	+	+
L. — var. antarctica Fritsch	+													
L. antarctica Gain	+												+	
L. attenuata Fritsch	+													
L. erebi W. & G. S. West	+													
L. Kützingii Schmidle	+							+		+		+	+	+
L. — var. distincta (Nordst.) Lemm.	+									+		+		
L. Lagerheimii (Möbius) Gomont	+									+	+	+	+	+
L. limnetica Lemm.	+								+	+		+	+	+
L. martensiana Menegh.	+									+		+	+	+
L. Murrayi W. & G. S. West	+													
L. nigra Ag. forma antarctica Gom.	+													
L. perelegans Lemm.							+			+		+	+	+
L. purpurea (Hook. & Harv.) Gom.							+			+	+			
L. Scottii Fritsch	+												+	+
L. — var. minor Fritsch	+													+
L. Shackletonii W. & G. S. West	+													+
Oscillatoria Agardhii Gomont forma Wille							+							
O. americana Kütz.						+								
O. amoena (Kütz.) Gomont							+	+		+		+	+	+
O. amphibia Ag.	+						+	+		+		+	+	+
O. — var. robusta W. & G. S. West	+													
O. brevis Kütz.	+		+						+			+	+	+
O. chlorina Kütz.	+								+	+	+	+	+	+
O. cortiana Menegh.	+									+		+	+	+
O. deflexa W. & G. S. West	+												+	
O. formosa Bory	+						+			+		+	+	+
O. fracta Carlson		+												
O. irrigua Kütz.	+					+			+	+			+	+
O. koettlitzii Fritsch	+													
O. limosa Ag.	+					+		+	+	+	+	+	+	+
O. nigroviridis (Thwaites) Gom.					+					+		+	+	
O. priestleyi W. & G. S. West	+						+			+			+	
O. proboscidea Gomont	+									+	+	+	+	+

	Antarctica	S.Shetland	S.Orkney	S.Georgia	Falkland	Feugo and S. Patagonia	Kerguelen	Arctica	Australia and New Zealand	Africa	S.America	N.America	Europe	Asia
O. producta W. & G. S. West	+													
O. prolifica (Grev.) Gomont						+						+	+	+
O. sancta Kütz.	+		+					+		+		+	+	+
O. simplicissima Gom. var. antarctica Fritsch	+													
O. splendida Grev.			+							+		+	+	+
O. subproboscidea W. & G. S. West	+						+			+		+	+	+
O. subtilissima Kütz.			+							+		+	+	+
O. tenuis Ag.	+	+	+			+		+		+		+	+	+
O. — forma sordida Kütz.		+	+											
O. telebriformis Ag.	+									+		+	+	
O. — forma tenuis W. & G. S. West	+													
Spirulina subtilissima Kütz.			+		+					+		+	+	+
Phormidium angustissima W. & G. S. West	+									+		+	+	+
Ph. antarcticum W. & G. S. West	+									+		+	+	
Ph. autumnale (Ag.) Gomont	+													
Ph. Charcotianum Gomont	+				+		+	+		+		+	+	+
Ph. fragile (Menegh). Gomont	+								+	+	+	+	+	+
Ph. — forma tenuis W. & G. S. West	+													
Ph. frigidum Fritsch	+												+	+
Ph. glaciale W. & G. S. West	+													
Ph. — forma longiarticulata Wille	+													
Ph. inundatum Kütz.	+									+		+	+	+
Ph. laminosum Gomont	+									+		+	+	+
Ph. Priestleyi Fritsch	+													
Ph. retzii (Ag.) Gomont	+							+		+		+	+	+
Ph. tenue (Menegh.) Gomont	+							+		+		+	+	+
Schizothrix antarctica Fritsch	+													
S. kerguelensis Wille							+							
Hydrocoleum Eatoni Reinsch.							+							
Microcoleus cryophilus Carlson	+													
M. Friesii Thur						+								
M. paludosus (Kütz.) Gomont							+					+	+	+
M. vaginatus (Vauch) Gomont	+						+	+		+		+	+	+
Nostoc antarcticum W. & G. S. West	+													
N. bornetii Gain	+		+											+
N. commune Vauch.	+					+	+	+	+	+	+	+	+	+
N. — forma antarctica W. & G. S. West	+													
N. disciforme Fritsch	+													
N. hydrocoleoides Reinsch	+													
N. fuscescens Fritsch	+													
N. kihlmani Lemm.							+	+						
N. longstaffii Fritsch	+													
N. microscopicum Carmer						+				+		+	+	+
N. minutum Desm.	+					+						+	+	+
N. minutissimum Kütz.			+										+	
N. pachydermaticum Gain	+						+							
N. paludosum Kütz.			+	+		+	+		+	+		+	+	+
N. punctiforme (Kütz.) Hariot	+		+	+		+	+		+	+	+	+	+	+

	Antarctica	S.Shetland	S.Orkney	S.Georgia	Falkland	Fuego and S. Patagonia	Kerguelen	Arctica	Australia and New Zealand	Africa	S.America	N.America	Europe	Asia
N. sphaericum Vauch.	+							+		+		+	+	+
Anabaena antarctica Fritsch	+													
A. cylindrica Lemm.							+			+			+	
A. flos-aquae (Lyngb.) Bréb.						+	+			+	+	+	+	+
A. Hassallii (Kütz.) Wittr. var. cyrtospora Wittr.								+						
A. involuta Reinsch.								+						
A. oscillarioides Bory	+				+	+				+		+	+	+
A. — forma kerguelensis Wille							+							
Nodularia spumugena Mert. var. minor Fritsch	+					+						+	+	+
N. quadrata Fritsch	+													
Isocystis infusionum (Kütz.) Bornet.			+	+										
Desmonema Wrangelii (Ag.) Born. & Flash.							+		+			+	+	
Aulosira implexa Born. & Flah.							+			+	+			+
A. minor Wille							+							
Dichothrix austrogeorgica Carlson				+										
D. Baueriana (Grun.) Born. & Flah.							+					+	+	+
D. olivacea (Hooker & Harvey) Born. & Flah							+			+	+	+	+	
Gloeotrichia pisum (Ag.) Thur.							+		+	+	+	+	+	+
Calothrix aeruginea Thur.			+		+	+	+						+	+
C. antarctica Fritsch	+													
C. epiphytica W. & G. S. West	+						+			+	+	+	+	+
C. fusca (Kütz.) Born. & Flah.				+			+			+		+	+	+
C. gracilis Fritsch	+													
C. intricata Fritsch	+													
C. juliana (Menegh.) Born. & Flah.							+			+		+	+	+
Plectonema notatum Schmidle	+											+	+	+
Petalonema densum (A.Br.) Migula							+					+	+	+
Tolypothrix conglutinata Borzi	+												+	+
T. tenuis Kütz.				+		+	+	+	+	+	+	+	+	+
T. — forma australica Möbius							+		+					
Microchaete tenera Thur.				+			+			+				+
Scytonema ocellatum Lyngb.							+			+		+	+	+
Vanhöffeana antarctica Wille							+							
Batrachospermum atrum (huds.) Harv.							+							
B. claviceps Kütz.						+								
B. Dillenii Bory					+	+						+	+	
B. vagum C.Ag.												+	+	+
Asterocystis antarctica W. & G. S. West	+													
Phacus aenigmaticus Drez.							+						+	
Anisonema acinus Duj.							+						+	
Peridinium umbonatum Stein var. inaequale Lemm.							+			+			+	
Uroglena europaea (Pascher) Congr.							+						+	
Dinobryon utriculus Stein							+						+	
Hyalobryon ramosum Lauterb.							+						+	
Chlorobotrys regularis Bohlin							+				+		+	+

175

	Antarctica	S.Shetland	S.Orkney	S.Georgia	Falkland	Fuego and S. Patagonia	Kerguelen	Arctica	Australia and New Zealand	Africa	S.America	N.America	Europe	Asia
Characiopsis minuta (A.Br.) Borzi var. disciliferum Wittr.							+							
C. subulata (A.Br.) Borzi							+							
Ophiocytium parvulum (Perty) A.Br.			+			+			+	+	+	+	+	
Tribonema bombycinum (Ag.) Derb. & Sol.		+	+	+		+	+			+		+		+
T. minus (Wille) Hazen						+						+	+	+
T. utriculosum (Kütz.) Hazen						+						+		
Racovitziella antarctica Wildemann	+													
Botrydium granulatum (L.) Grev.						+				+	+	+	+	+
Vaucheria caespitosa deCandolle					+									
V. sessilis deCandolle var. repens (Hass.) Hansg.							+			+				
Melosira distans (Ehrenb.) Kütz.	+			+		+	+	+		+	+	+	+	+
M. — var. alpigena Grun.						+				+	+		+	
M. granulata (Ehrenb.) Ralfs						+			+	+	+	+	+	+
M. laevis (Ehrenb.) Grun. var. fuegiana Freng.						+								
M. lineata Grun. var. patagonica O.Müller						+								
M. nummuloides (Dillw.) Ag.					+	+	+						+	
M. Roeseana Rabenh. var. epidendron Grun							+			+			+	
M. sol Kütz.	+		+				+			+				
M. sulcata (Kütz.) Ehrenb.						+						+	+	
M. — forma coronata Grun.						+		+					+	
M. — forma radiata Grun.						+		+					+	
M. — var. biseriata Grun.						+		+					+	
M. — var. crenulata Grun.						+		+					+	
M. varians Ag.	+		+			+		+		+	+	+	+	+
Hyalodiscus radiatus (O'Meara) Grun.						+	+	+			+	+		
Coscinodiscus decipiens Grun.	+					+							+	
C. excentricus Ehrenb.	+			+		+		+		+	+	+	+	+
C. griseus Greville var. gallapagensis Grun.											+			
C. lacustris Jan.	+					+		+		+			+	
C. lentiginosus Jan.	+							+						
C. minor Ehrenb.						+						+	+	+
C. radiatus Ehrenb.	+		+			+						+	+	+
C. subtilis Ehrenb.	+			+		+	+					+	+	+
Cyclotella operculata Kütz.	+									+		+	+	
Actynoptychus undulatus (Ehr.) Ralfs						+				+	+		+	
Triceratium arcticum Brightw.	+					+		+					+	
T. scitulum Brightw.						+					+	+	+	
Hemiaulus ambiguus Janisch	+												+	
Asterionella gracillima (Hantzsch) Heiberg						+				+	+		+	+
Tabellaria flocculosa (Roth) Kütz.	+	+				+	+		+	+	+	+	+	+
Diatoma elongatum Ag.	+		+	+		+		+		+	+		+	
D. — var. densestriatum Grun.			+											
D. — var. Ehrenbergii (Kütz.) W.Sm.	+		+	+		+		+					+	
D. — var. hybridum Grun.			+											
D. — var. minus Grun.			+					+					+	

	Antarctica	S.Shetland	S.Orkney	S.Georgia	Falkland	Fuego and S. Patagonia	Kerguelen	Arctica	Australia and New Zealand	Africa	S.America	N.America	Europe	Asia
D. — var. tenuis (Ag.) Van Heurck						+				+		+	+	
D. vulgare Bory var. Ehrenbergii (Kütz.) Grun.				+									+	
Meridion circulare (Grev.) Ag. var.														
M. constricta (Ralfs) Van Heurck						+								+
Fragilaria capucina Desm.				+						+			+	+
F. — var. acuminata Grun.				+									+	
F. — var. genuina Grun.				+		+							+	
F. — var. lanceolata Grun.				+						+			+	+
F. construens (Ehrenb.) Grun.						+	+			+	+	+	+	+
F. — var. binodis (Ehrenb.) Grun.						+				+		+	+	
F. — var. minor Freng.						+								
F. — var. trigona (Grun.) Freng.						+							+	
F. — var. venter (Ehrenb.) Grun.						+	+			+	+		+	+
F. curta Van Heurck	+													
F. linearis Castr.	+													
F. obliquecostata Van Heurck	+													
F. pinnata Ehrenb.				+		+	+			+	+	+	+	+
F. — var. intercedens Grun.						+							+	
F. — var. lancettula (Schum.) Hustedt						+					+	+		+
F. — var. lapponica Grun.						+							+	
F. — var. minor (Grun.) Freng.						+								
F. — var. tetragona Freng.						+								
F. rumpens Grun.				+	+	+							+	
F. tenuicollis Heib.var. antarctica W. & G. S. West	+													
F. Vaucheriae (Kütz.) B. Peters. var. capitellata Grun.							+				+	+		
F. — var. longissima Bourr. & Manguin.							+	+	+					
F. — var. tenuis Bourr. & Mang.							+	+						
F. virescens Ralfs	+					+		+		+	+	+	+	+
F. — var. exigua Grun.						+							+	
F. — var. fuegiana Freng.						+								
F. — var. subsalina Grun.								+					+	
Synedra acus (Kütz.) Grun. var. angustissima Grun.						+						+	+	+
S. — var. delicatissima (W.Sm.) Grun.				+		+				+			+	
S. affinis Kütz. var. acuminata Grun.		+		+			+						+	
S. — var. tabulata (Ag.) Heurck					+								+	
S. fulgens (Kütz.) W.Sm. var. mediterranea Grun.	+					+			+	+	+	+	+	+
S. pulchella (Ralfs) Kütz.						+				+			+	+
S. — var. Grunowii Freng.						+								
S. — var. lanceolata O'Meara						+						+		
S. — var. major Grun.						+							+	
S. ulna (Nitzsch) Ehrenb.	+					+			+	+	+	+	+	+
S. — var. aequalis (Kütz.) Hust.										+			+	
S. — var. danica (Kütz.) Heurck						+		+		+		+	+	+
Ceratoneis arcus (Ehrenb.) Kütz.				+				+				+	+	+

	Antarctica	S.Shetland	S.Orkney	S.Georgia	Falkland	Fuego and S. Patagonia	Kerguelen	Arctica	Australia and New Zealand	Africa	S.America	N.America	Europe	Asia
Eunotia gracilis W.Sm.			+			+		+	+	+	+	+	+	
E. monodon Ehrenb. var. major (W.Sm.) Hust. f. elongata Bourr. & Mang.							+							
E. — — f.robusta Bourr. & Mang.							+							
E. nymanniana Grun.														
E. polydentula Brun. var. mediotumida Bourr. & Manguin						+			+	+			+	
E. — var. perpusilla Grun.							+							
E. praerupta Ehrenb.									+				+	+
E. — var. bidens (Ehrenb.) Grun.				+		+		+	+	+	+	+	+	+
E. — var. curta Grun.						+		+	+		+	+	+	+
E. — var. laticeps Grun.						+		+					+	
E. — var. tridentata (Ehrenb.) Freng.						+		+					+	
Cocconeis costata Greg.		+	+	+		+		+	+			+	+	
C. — var. pacifica Grun.		+		+		+		+			+		+	+
C. distans A. Schmidt var. minima H. Per.						+		+			+		+	+
C. kerguelensis Bourr. & Manguin							+							
C. litigiosa Van Heurck	+													
C. pediculus Ehrenb.					+	+		+	+	+	+	+	+	+
C. placentula Ehrenb.						+		+	+	+	+	+	+	+
C. — var. lineata (Ehrenb.) Heurck						+		+			+	+	+	+
C. scutellum Ehrenb.	+				+	+		+		+		+	+	+
Achnanthes abundans Bourr. & Manguin						+								
A. — var. elliptica Bourr. & Manguin						+								
A. Bourginii Bourr. & Manguin						+								
A. brevipes Ag.	+				+				+	+			+	+
A. — var. intermedia (Kütz.) Cleve	+				+			+		+			+	+
A. coarctata (Bréb.) Grun. forma falklandica Carlson					+	+			+					
A. confusa Bourr. & Manguin								+						
A. — var. atomoides Bourr. & Manguin								+						
A. delicatula (Kütz.) Grun. var. australis Bourr. & Manguin								+						
A. — var. magellanica Freng.						+								
A. germainii Bourr. & Manguin								+						
A. lanceolata (Bréb.) Grun.			+	+	+	+	+		+	+	+	+	+	+
A. — var. dubia Grun.			+	+	+					+			+	+
A. Manguinii Hustedt						+								
A. — var. elliptica Bourr. & Manguin						+								
A. minutissima Kütz. var. cryptocephala (Näg.) Grun.						+		+		+	+	+	+	+
A. modesta Bourr. & Manguin						+								
A. Mülleri Carlson			+	+	+									
A. pseudolanceolata Hustedt										+				
A. staurastroides Bourr. & Manguin						+								
A. trinodis Ehrenb.													+	
Diatomella balfouriana (Grev.) Ag.	+			+		+	+				+	+	+	+
D. Hustedtii Bourr. & Manguin						+								

	Antarctica	S.Shetland	S.Orkney	S.Georgia	Falkland	Fuego and S. Patagonia	Kerguelen	Arctica	Australia and New Zealand	Africa	S.America	N.America	Europe	Asia
Frustulia pulchra Germ. var. lanceolata Bourr. & Manguin							+							
F. rhomboides (Ehrenb.) Cleve	+			+			+	+		+	+	+	+	+
F. — var. crassinervis forma antarctica Van Heurck	+			+			+							
F. vulgaris (Thwaites) Cleve						+	+	+	+	+	+	+	+	+
Caloneis austrogeorgica Carlson														
C. fasciata Lag. var. fontinalis (Grun.) Freng.						+			+	+	+	+		
C. — var. gigantea M. Per.						+								
C. macloviana Carlson				+										
C. Marnieri Bourr. & Manguin							+							
C. pandriformis Carlson				+										
C. patagonica Cl. var. Schmidtii Freng.						+								
C. silicula (Ehrenb.) Cl.						+				+	+	+	+	+
C. — var. parva Freng.						+								
C. — var. patagonica Freng.						+								
C. — var. peisonis Hustedt							+			+	+		+	
C. — var. semicruciata Freng						+								
C. — var. ventricosa (Ehrenb.) Cl.						+		+			+		+	+
Vanheurckia interposita (Lewis) De Toni var. incomperta (Lewis) Perag.						+								
V. vulgaris (Thw.) Heurck						+								
Gyrosigma acuminatum (Kütz.) Grun.						+				+		+	+	+
G. attenuatum (Kütz.) Rabenh.					+				+	+	+	+	+	+
G. — forma subbalticum Carlson					+									
G. wansbeckii Donk.						+							+	
Pleurosigma strigosum W.Sm.						+							+	
Neidium affine (Ehrenb.) Cleve						+		+		+			+	+
N. — var. ampliata Ehrenb.						+								
N. — var. undulata (Grun.) Cleve						+						+	+	
N. Aubertii Bourr. & Manguin							+							
N. bisulcatum Lag.						+		+			+	+	+	
N. iridis Ehrenb.						+		+	+	+	+	+	+	+
N. — var. amphigomphus (Ehrenb.) Heurck						+		+						+
N. — var. dubia (A.Schm.) Freng.						+								
N. — var. firma (Kütz.) Heurck						+					+	+		+
N. magellanica Cleve						+								
N. — var. Candelariae Freng.						+								
N. sauramoi Mölder						+							+	
Diploneis elliptica Kütz.						+			+	+	+	+	+	+
D. Smithii Bréb. var. argentina Freng.						+					+			
D. subovalis Cleve		+		+	+	+	+			+			+	+
Stauroneis acuta W. Sm.				+		+		+		+		+	+	+
S. anceps Kütz.	+			+		+	+	+	+	+	+	+	+	+
S. — var. abnormis Freng.						+								
S. — var. amphicephala Kütz.	+			+		+		+	+			+	+	+

	Antarctica	S.Shetland	S.Orkney	S.Georgia	Falkland	Fuego and S. Patagonia	Kerguelen	Arctica	Australia and New Zealand	Africa	S.America	N.America	Europe	Asia
S. — var. hyalina Br. & Per.						+		+	+	+		+	+	+
S. — var. siberica Grun.			+											+
S. Boudetii M. Per.	+													
S. legumen Ehrenb. var. integra Bourr. & Manguin							+							
S. obtusa Lagerst.							+						+	
S. perminuata Grun.			+			+								
S. phoenicenteron Ehrenb.			+			+								
S. — var. amphilepta (Ehrenb.) Cl.						+		+	+	+		+	+	+
S. — var. gracilis Cleve						+						+	+	+
S. quadrata Hérib.						+								
Navicula anglica Ralfs						+		+			+	+	+	+
N. — var. subsalina Grun.						+							+	
N. arcuata Heid. & Kolbe							+							
N. austroshetlandica Carlson		+												
N. avenacea Bréb.							+			+		+	+	
N. bacilliformis Grun.						+			+	+		+	+	+
N. bicephala Hustedt						+			+	+				
N. bryophila B. Peters.						+	+			+	+		+	+
N. bryophiloides Bourr. & Manguin						+	+							
N. Charcotii M.Per.						+				+				
N. — var. magellanica Freng.						+				+				
N. cincta Ehrenb.						+		+		+		+	+	+
N. — var. cari (Ehrenb.) Cleve						+		+	+				+	
N. — var. Heufleri Grun.						+				+		+	+	+
N. corrugata Bourr. & Manguin							+							
N. cryptocephala Kütz.					+	+	+			+		+	+	+
N. — var. exilis (Kütz.) Grun.						+	+			+			+	+
N. — var. veneta (Kütz.) Rabenh.						+	+			+			+	+
N. cuspidata Kütz.				+		+	+		+	+		+	+	+
N. — var. danaica Grun.						+	+							
N. — var. gracilis M.Per.						+								
N — var. major Freng.						+								
N cymatopleura W. & G. S. West	+													
N. cymbula Donk.						+							+	
N. dicephala Ehrenb.						+		+		+	+	+	+	+
N. — forma australis Bourr. & Manguin							+							
N. — var. subcapitata Grun.						+							+	
N. elegans W.Sm. var. cuspidata Cleve							+							
N. equiornata Bourr. & Manguin							+							
N. excellens Carlson					+									
N. expeditionis Freng.						+								
N. fuegiana Freng.						+	+							
N. — var. rostrata Freng.						+								
N. geniculata Germ.							+							
N. glaberrima W. & G. S. West	+													
N. gracilis Ehrenb.						+				+	+	+	+	+
N. gregaria Donk.						+		+		+	+	+	+	+
N. heterostauron Germ. var. rostrata Germ.								+						

	Antarctica	S.Shetland	S.Orkney	S.Georgia	Falkland	Fuego and S. Patagonia	Kerguelen	Arctica	Australia and New Zealand	Africa	S.America	N.America	Europe	Asia
N. kerguelensis Heid. & Kolbe							+							
N. kotschyi Grun.					+	+		+.		+			+	+
N. linearis (O.Müll.) Freng.						+		+				+	+	+
N. lucidula Grun.				+				+				+	+	+
N. megacuspidata Carlson					+									
N. minima Grun. var. atmoides (Grun.) Cleve							+			+			+	+
N. minutissima Cleve						+								
N. Murrayi W. & G. S. West	+													
N. — var. elegans W. & G. S. West	+	+												
N. mutica Kütz.	+		+		+	+		+	+	+	+	+	+	+
N. — var. Couhii (Hilse) Grun.						+		+	+	+	+		+	+
N. — var. geoppertiana (Bleich) Grun.						+		+		+	+	+	+	+
N. — var. producta Grun.	+				+	+		+					+	
N. — var. ventricosa (Kütz.) Grun.						+						+	+	+
N. muticopsiforme W. & G. S. West	+													
N. muticopsis Van Heurck	+	+	+						+					+
N. — forma capitata Carlson							+							
N. — forma lanceolata Freng.						+								
N. — var. linearis Freng.						+								
N. — forma reducta W. & G. S. West	+													
N. peraustralis W. & G. S. West	+													
N. peregrina (Ehrenb.) Kütz.						+		+			+	+	+	+
N. — var. perlonga Freng.						+								
N. perlepida Grun.	+						+							
N. placentula Ehrenb.						+			+		+	+	+	+
N. — var. lanceolata Grun.						+							+	
N. portomontana Cleve						+	+				+			
N. — var. fuegiana Freng.						+								
N. pseudocitrus Bourr. & Mang.							+							
N. pupula Kütz.						+		+	+		+	+	+	+
N. — var. bacillarioides Grun.						+		+				+		
N. — var. linearis M.Per.						+								
N. pusilla W.Sm. var. spetsbergensis Grun.						+		+						
N. pygmaea Kütz.						+		+			+	+	+	+
N. radiosa Kütz.	+			+		+		+	+	+	+	+	+	+
N. — var. acuta (W.Sm.) Grun.						+							+	+
N. rhynchocephala Kütz.	+					+		+	+	+	+	+	+	+
N. sculpta Ehrenb.						+								
N. seminulum Grun.	+					+	+	+		+	+	+	+	+
N. Schönfeldii Hust.							+			+			+	+
N. Shackletoni W. & G. S. West	+													
N. — var. pellucida W. & G. S. West	+													
N. spissata Bourr. & Mang.							+							
N. stauropteroides Fritsch	+													
N. suecorum Carlson						+								
N. undulatistriata Bourr. & Mang.							+							
N. viridula Kütz.						+		+	+	+	+	+	+	+
N. vulpina Kütz.						+		+	+	+		+	+	+

	Antarctica	S.Shetland	S.Orkney	S.Georgia	Falkland	Fuego and S. Patagonia	Kerguelen	Arctica	Australia and New Zealand	Africa	S.America	N.America	Europe	Asia
Pinnularia acrosphaeria (Bréb.) Kütz.						+						+	+	
P. alpina W.Sm. var. Kerguelensis (Heid. & Kolbe) Freng.							+							
P. appendiculata (Ag.) Kütz.						+			+	+	+	+	+	+
P. — var. irrorata Grun.						+						+		
P. Backebergii Bourr. & Mang.							+							
P. borealis Ehrenb.	+		+	+	+	+	+	+	+	+	+	+	+	+
P. — forma rectangularis Carlson					+	+								
P. — var. australis Bourr. & Mang.							+							
P. — var. cuneorostrata Bourr. & Mang.							+							
P. — var. linearis Hérib.						+								
P. braunii (Grun.) Cleve			+						+	+		+	+	+
P. Brébissonii (Kütz.) Cleve		+							+			+	+	+
P. — var. diminuta Grun.			+									+	+	+
P. circumducta Bourr. & Mang.								+						
P. divergens W.Sm.						+			+	+	+	+	+	
P. — var cuneata Grun.						+				+				
P. — var. elliptica Grun.						+			+	+	+			
P. — var. minor Temp. & Per.						+					+		+	
P. — var. parallela Brun.						+					+			
P. Doello-Juradoi Freng.						+					+			
P. esox Ehrenb.						+					+			
P. globiceps Greg.	+						+					+	+	+
P. — forma amphicephala Fritsch	+	+												
P. — forma elongata Fritsch	+	+												
P. interrupta W.Sm.						+	+			+		+	+	+
P. — forma stauroneiformis (Van Heurck) Cleve					+	+	+			+		+	+	
P. isostauron (Ehrenb.) Grun.						+	+					+	+	
P. Kolbei Bourr. & Mang.								+						
P. lata (Bréb.) W.Sm.						+	+		+	+	+	+	+	+
P. — var. curta Grun.						+	+						+	
P. — var. latestriata (Greg.) Cleve						+							+	
P. latevittata Cleve						+								
P. — var. spathulata Freng.						+								
P. macilenta Ehrenb.					+							+	+	+
P. major Kütz.						+							+	
P. — var. linearis Cleve						+							+	
P. — var. subacuta (Ehrenb.) Cleve						+							+	
P. — var. transversa (A.Schm.) Cleve						+							+	
P. microstauron (Ehrenb.) Cleve					+	+			+	+	+	+	+	+
P. — var. australis Bourr. & Mang.							+							
P. — var. elongata Bourr. & Mang.							+	+						
P. Peragillii Freng. var. gracilis Freng.						+								
P. quadratarea A.S. var. dulcicola Bourr. & Mang.								+						
P. stauroptera (Grun.) Rabenh. var. interrupta Cleve						+	+					+	+	
P. streptoraphe Cleve						+	+					+	+	

182

	Antarctica	S.Shetland	S.Orkney	S.Georgia	Falkland	Fuego and S. Patagonia	Kerguelen	Arctica	Australia and New Zealand	Africa	S.America	N.America	Europe	Asia
P. — var. gibbosa A. Cleve						+							+	
P. subcapitata Greg. var. hybrida (Grun.) Freng.						+								
P. subsolaris (Grun.) Cleve var. Kerguelensis Bourr. & Mang.							+							
P. viridis (Nitzsch) Ehrenb.			+	+		+	+	+	+	+	+	+	+	+
P. — var. commutata (A.Schm.) Cleve						+		+	+		+	+	+	+
P. — var. distinguenda Cleve						+		+	+		+	+	+	+
P. — var. lata Freng.						+								
P. — var. semicruciiata (Grun.) Cleve						+								
Trachyneis aspera (Ehrenb.) Cleve	+				+	+		+			+	+		
Amphora acutiuscula Kütz.						+				+	+			
A. coffaeiformis (Ag.) Kütz.						+			+			+	+	
A. — var. borealis (Kütz.) Cleve						+							+	
A. libyca Ehrenb.						+								
A. ovalis Kütz.		+	+			+			+	+	+	+	+	+
Cymbella americana A. Schm.						+						+		
C. — var. acuta Schm.						+								+
C. Aubertii Bourr. & Mang.							+							
C. cesati (Rabenh.) Grun.							+	+		+	+	+	+	+
C. cistula (Hempr.) Kirchn.	+		+			+		+	+		+	+	+	+
C. — forma minor Van Heurck						+					+	+		
C. — var. maculata (Kütz.) Van Heurck	+		+			+					+	+		
C. — var. Nordenskiöldii (O.Müll.) Carlson			+			+								
C. Clericu Freng.						+								
C. Ehrenbergii Kütz.						+			+		+	+	+	
C. — var. delecta (A.Schm.) Cleve						+								
C. gastroides Kütz. var. gigantea (Pant.) Freng.						+								
C. lanceolata (Ehrenb.) Kirchn.						+			+		+	+	+	
C. lacustris (Ag.) Cleve var. australis Bourr. & Mang.							+							
C. naviculiformis Auersw.						+			+	+	+	+	+	+
C. nodosa Bourr. & Mang.							+							
C. Nordenskiöldii O.Müll.						+								
C. parva (W.Sm.) Grun.						+			+		+		+	
C. pusilla Grun.	+								+		+		+	+
C. subantarctica Bourr. & Mang.							+	+						
C. tumida (Bréb.) Van Heurck						+			+		+	+	+	+
C. turgidula Grun.							+	+			+	+	+	+
C. ventricosa Kütz.	+			+	+	+			+	+	+	+	+	+
Gomphonema candelariae Freng.						+								
G. — var. elliptica Freng.						+								
G. — var. minor Freng.						+								
G. constrictum Ehrenb.						+			+	+	+	+	+	+
G. — var. capitata (Ehrenb.) Grun.						+			+	+	+	+	+	+
G. — var. clavata (Ehrenb.) Freng.						+							+	
G. gracile Ehrenb. var. dichotoma (Kütz.) Grun.						+			+		+	+	+	+

	Antarctica	S.Shetland	S.Orkney	S.Georgia	Falkland	Fuego and S. Patagonia	Kerguelen	Arctica	Australia and New Zealand	Africa	S.America	N.America	Europe	Asia
G. intricatum Kütz. var. dichotomum (Kütz.) Grun.				+		+			+	+	+	+	+	+
G. Kerguelensis Bourr. & Mang.							+							
G. — forma lanceolata Bourr. & Mang.							+							
G. — forma rhomboidea Bourr. & Mang.							+							
G. lanceolatum Ehrenb.						+			+	+	+	+	+	+
G. micropus Kütz.						+						+	+	
G. montanum Schum.			+					+		+			+	
G. olivaceum (Lyngb.) Kütz.						+		+		+			+	+
G. tenellum Kütz.					+							+	+	
Tropidoneis laevissima W. & G. S. West	+													
Denticula antarctica Castr.	+	+												
D. elegans Kütz. var. Kittoniana Grun.								+						
D. — var. robusta Bourr. & Mang.								+						
D. tenuis Kütz.			+		+	+		+		+		+	+	+
Epithemia zebra (Ehrenb.) Kütz.						+								
E. — var. elongata Grun.					+									
E. — var. porcellus Grun.	+					+	+			+		+	+	+
E. — var. proboscidea (Kütz.) Grun.						+		+				+		+
Rhopalodia gibba O.Müll.						+			+	+	+	+	+	+
R. — var. ventricosa (Ehrenb.) Grun.						+					+	+	+	+
R. gibberula (Ehrenb.) O.Müll.	+				+	+	+			+	+	+	+	+
Hantzschia amphioxys (Ehrenb.) Grun.	+			+		+		+	+	+	+	+	+	+
H. — var. arverna M. Per.						+								
H. — var. capitellata Grun.						+						+		+
H. — var. hyperborea Grun.						+								+
H. — var. minor Per.						+						+		
H. elongata (Hantzsch.) Grun.						+				+		+	+	+
H. — var. linearis O.Müll.						+							+	
Nitzschia acicularis (Kütz.) W.Sm.	+													
N. amphibia Grun.						+			+	+	+	+	+	+
N. — var. acutiuscula Grun.						+							+	
N. brevissima Grun.						+						+		
N. constricta Ralfs var. similis Grun.						+								
N. debilis (Arnott) Grun.	+				+			+					+	
N. denticula Grun.						+		+		+	+	+	+	+
N. frustulum (Kütz.) Grun. var. Kerguelensis Bourr. & Mang.							+							
N. gracilis Hantzsch							+			+	+	+	+	+
N. ignorata Krasske forma longissima Bourr. & Mang.							+							
N. linearis (Ag.) W.Sm.						+		+		+	+	+	+	+
N. obtusa W.Sm.												+	+	+
N. palea (Kütz.) W.Sm.						+		+	+	+	+	+	+	+
N. pandriformis Greg. var. parva Freng.						+								
N. sigma (Kütz.) W.Sm.						+			+	+	+	+	+	+
N. — var. rigidula Grun.						+							+	
N. subtilis (Kütz.) Grun.	+					+		+		+			+	+
N. — var. acicularis Freng.						+								

	Antarctica	S.Shetland	S.Orkney	S.Georgia	Falkland	Fuego and S. Patagonia	Kerguelen	Arctica	Australia and New Zealand	Africa	S.America	N.America	Europe	Asia
N. — var. paleacea Grun.						+						+	+	
N. vitrea Norm.						+				+	+	+	+	+
Pseudonitzschia migrans (Cleve) Van Heurck	+				+									
Surirella angusta Kütz.	+			+		+				+		+	+	+
S. — var. constricta Hustedt							+							+
S. apiculata W.Sm. var. pandriformis Freng.						+								
S. biseriata Bréb.						+		+	+	+	+	+	+	+
S. guatemalensis Ehrenb.						+					+	+		
S. Kerguelensis Grun.							+							
S. minuta Bréb.						+					+	+	+	
S. — var. pinnata Grun.						+		+					+	
S. ovalis Bréb.					+			+		+	+	+	+	+
S. striatula Turp.						+						+	+	
S. tuberosa O.Müll.						+								
S. — forma elongata Freng.						+								
S. — var. costata Freng.						+								
Chlamydomonas antarctica Wille	+													
C. caudata Wille		+												
C. Ehrenbergii Gorosch.	+	+					+							
C. intermedia Chodat forma antarctica W. & G. S. West	+	+												
C. nivalis (Bauer) Wille	+	+						+				+	+	
C. subcaudata Wille	+													
Chloromonas alpina Wille	+													
Pteromonas nivalis (Shuttlew) Chodat		+												
Eudorina elegans Ehrenb.						+		+	+	+	+	+	+	+
Pandorina morum Bory						+		+	+	+	+	+	+	+
Gonium pectorale Müller						+			+			+	+	+
Paulschulzia pseudovolvox (Schulz) Skuja						+							+	
Gloeocystis vesiculosa Näg.								+		+	+	+	+	+
Pseudotetraspora gainii Wille	+													
Elakatothrix gelatinosa Wille forma biplex Nygaard						+							+	
Apiocystis brauniana Näg.								+	+	+	+	+	+	+
Schizochlamys gelatinosa A.Br. var. minor Bernard							+						+	+
Stylosphaeridium stipitatum Geitler						+	+					+	+	
Chlorococcum infusionum Menegh.									+			+	+	+
C. — forma antarctica Wille	+													
Characium gracilipes Lampert						+					+		+	
C. limneticum Lemm.						+					+		+	+
C. obtusum A.Br.							+					+	+	
Codiolum gregarium A.Br. forma antarctica Wille							+							
Eremosphaera viridis De Bary			+							+	+	+	+	+
Mycacanthococcus antarcticus Wille	+													
M. cellaris Hansg.	+													

	Antarctica	S.Shetland	S.Orkney	S.Georgia	Falkland	Fuego and S. Patagonia	Kerguelen	Arctica	Australia and New Zealand	Africa	S.America	N.America	Europe	Asia
M. — forma antarctica Wille	+													
M. ovalis Wille	+	+												
Chlorella conglomerata (Artari) Oltm.	+												+	+
C. ellipsoidea Gorosch. forma arctica Wille								+						
C. — forma antarctica Wille	+	+												
C. koettlitzii (Fritsch) Wille	+	+												
C. tetraedrica Wille								+						
C. Werthii Wille								+						
C. vulgaris Beyer								+		+		+	+	+
Trochiscia antarctica Fritsch			+						+			+	+	+
T. aspera (Reinsch) Hansg.	+									+		+	+	
T. crassa Hansg.	+											+	+	
T. granulata (Reinsch) Hansg.						+						+	+	
T. hystrix (Reinsch) Hansg.	+	+	+	+								+	+	
T. nivalis Lagerh.		+											+	
T. pachyderma (Reinsch) Hansg.		+	+	+							+		+	
T. reticularis (Reinsch) Hansg.		+	+			+		+	+				+	
T. tuberculifera Gain	+					+				+		+	+	+
Tetraedron caudatum (Corda) Hansg. forma minor Wille							+							
T. punctulatum (Reinsch) Hansg. forma trigona Wille							+							
Oocystis crassa Wittr. var. marsonii (Lemm.) Printz.							+			+			+	
O. lacustris Chodat						+							+	+
O. — forma nivalis Fritsch			+											
O. natans (Lemm.) Wille						+							+	
O. solitaria Wittr.			+			+	+		+	+	+	+	+	+
Scotiella antarctica Fritsch	+		+				+							
S. polyptera Fritsch	+		+											
Chodatella brevispina Fritsch			+				+			+			+	
Botryococcus braunii Kütz.						+	+		+	+	+	+	+	+
Enallax costatus (Schmidle) Pascher						+	+							
Ankistrodesmus convolutus Corda						+								
A. falcatus (Corda) Ralfs						+	+	+	+	+	+	+	+	+
A. — var. spiralis W. & G. S. West						+	+		+			+	+	+
Kirchneriella obesa (West) Schmidle						+	+			+	+	+	+	+
K. — var. aperta (Teil.) Brunth.						+							+	
Dictyosphaerium Ehrenbergii Näg.						+				+			+	
D. pulchellum Wood						+	+		+	+	+	+	+	+
Pediastrum boryanum (Turp.) Menegh.						+	+		+	+	+	+	+	+
P. — var. campanulatum Wille							+							
P. — var. depauperatum Wille							+							
P. — var. granulatum (Kütz.) A.Br.						+		+		+	+	+	+	+
P. — var. longicorne forma mamillosum Krieger							+							
P. Braunii Wartm.							+	+				+	+	+
P. integrum Näg.							+	+	+	+		+	+	+
P. Kawraiskyi Schmidle						+	+					+	+	+

	Antarctica	S.Shetland	S.Orkney	S.Georgia	Falkland	Fuego and S. Patagonia	Kerguelen	Arctica	Australia and New Zealand	Africa	S.America	N.America	Europe	Asia
P. muticum Kütz. var. inermius Racib.							+						+	+
P. tetras (Ehrenb.) Ralfs						+	+		+	+	+	+	+	+
Soropediastrum kerguelensis Wille							+							
S. rotundatum Wille							+							
Sorastrum Hathoris (Cohn) Schmidle							+			+			+	
Scenedesmus acutiformis Schr.							+		+		+	+	+	
S. bijugatus (Turp.) Kütz.							+							
S. brasiliensis Bohlin							+				+	+	+	+
S. corallinus Chodat							+						+	
S. denticulatus Lagerh.							+	+		+	+	+	+	+
S. hystrix Lagerh. forma laevis Wille							+							
S. kerguelensis Wille							+						+	
S. obliquus (Turp.) Kütz.							+		+	+	+	+	+	+
S. pediastroides Wille							+							
S. quadricauda (Turp.) Bréb.							+	+	+	+	+	+	+	+
S. — var. rectangularis G. S. West						+							+	+
S. securiformis Wille							+							
Crucigenia antarctica Wille							+							
C. irregularis Wille							+				+	+	+	+
C. rectangularis (Näg.) Gay						+			+	+	+		+	
Coelastrum Bohlinii Schmidle & Senn							+						+	
C. cambricum Arch. var. intermedium (Bohlin) West							+			+		+	+	+
C. microsporum Näg.			+	+		+	+	+	+	+	+	+	+	+
C. — forma irregularis Fritsch			+							+				
C. proboscideum Bohlin							+	+		+	+	+	+	+
C. scabrum Reinsch							+			+	+	+	+	+
C. sphaericum Näg.			+		+		+		+		+	+	+	+
Chlorosphaera antarctica Fritsch			+											
C. kerguelensis Wille							+							
Ulothrix aequalis Kütz.	+		+	+				+		+			+	+
U. flacca (Dillw.) Thur.	+					+		+				+		+
U. flaccida Kütz.	+								+				+	
U. — var. fragilis (Hook & Harv.) Hansg.		+				+								
U. fragilis Kütz.						+								
U. imflexa Kütz.	+					+		+		+		+	+	
U. microspora Lagerh.						+								
U. moniliformis Kütz.								+	+		+	+	+	+
U. oscillarina Kütz.	+								+		+	+	+	+
U. stagnorum (Kütz.) Rabenh.							+							
U. subtilis Kütz.			+			+	+			+			+	+
U. — var. variabilis (Kütz.) Kirchn.	+								+	+				+
U. tenerrima Kütz. forma antarctica W. & G. S. West	+								+					
Hormidium parietinum Kütz.							+							
Stichococcus bacillaris Näg.	+								+	+	+	+	+	
S. — var. major (Näg.) Rabenh.	+												+	
S. — var. minor (Näg.) Rabenh.	+										+		+	

	Antarctica	S.Shetland	S.Orkney	S.Georgia	Falkland	Fuego and S. Patagonia	Kerguelen	Arctica	Australia and New Zealand	Africa	S.America	N.America	Europe	Asia
Raphidonema nivale Lagerh. forma minor Wille	+		+						+					
Binuclearia tatrana Wittr.	+													
Microspora flocculosa (Ag.) Thur.							+	+				+	+	+
M. pachyderma (Wille) Lagerh.							+	+				+	+	
M. willeana Lagerh.								+		+		+	+	+
Schizogonium murale Kütz.					+							+	+	
Prasiola antarctica Kütz.	+					+								
P. calophylla (Carmich) Menegh.	+												+	
P. crispa (Lightf.) Ag.	+												+	
P. fluviatilis (Sommerf.) Aresch.	+	+	+	+	+	+	+	+				+	+	+
forma antarctica Wille	+						+							
Stigeoclonium falklandicum Kütz.					+		+							
Draparnaldia glomerata (Vauch) Ag.							+			+				+
forma distans (Kütz.) Hansg.							+						+	
Protoderma Brownii Fritsch	+		+											
Microthamnion strictissimum Rabenh.							+			+		+	+	+
Trentepohlia aureus (L.) Mart.						+	+			+		+	+	
T. polycarpa Nees & Mont						+	+							+
Coleochaete javanica Wildem.							+							+
C. scutata Bréb.							+			+	+	+	+	+
Chaetosphaeridium globosum Klebs							+			+	+	+	+	+
Dicoleon Nordstedtii Klebs							+		+				+	
Pleurococcus antarcticus W. & G. S. West	+									+				
P. dissectus (Kütz.) Näg.	+									+			+	
P. frigidus W. & G. S. West	+													
P. koettlitzii Fritsch	+									+				
P. pachydermis Lagerh.	+												+	
P. — forma stipitata W. & G. S. West	+													
P. stercorarius Berk.					+									
P. tectorum Trvis forma														
P. — antarctica Wille	+													
P. vulgaris Menegh.	+	+	+		+			+			+	+	+	+
Cladophora glomerata (L.) Kütz.							+					+	+	+
Hormiscia penicilliformis (Roth) Fries	+	+			+		+	+						
Rhizoclonium angulatum (Hook & Harv.) Kütz.						+								
Urospora penicilliformis (Roth) Aresch.							+							
Pithophora aequalis Wittr.						+								
Bulbochaete mirabilis Wittr.							+					+		+
Oedogonium sp.			+											
Mesotaenium endlicherianum Näg.			+				+							
Cylindrocystis Brébissonii Menegh.			+				+	+		+	+	+	+	+
C. — var. minor W. & G. S. West							+				+		+	+
C. crassa DeBary	+		+					+		+			+	
Ancylonema Nordenskioldii Berggren.	+							+						
Netrium digitus var. lamellosum (Bréb.) Grönbl.							+	+	+		+	+	+	+
Closterium calosporum Wittr.							+	+					+	+

	Antarctica	S.Shetland	S.Orkney	S.Georgia	Falkland	Fuego and S. Patagonia	Kerguelen	Arctica	Australia and New Zealand	Africa	S.America	N.America	Europe	Asia
C. cornu Ehrenb.							+		+	+	+	+	+	+
C. cynthia DeNot.							+	+	+	+	+	+	+	+
C. dianae Ehrenb.							+		+	+	+	+	+	+
C. — var. arcuatum (Bréb.) Rabenh.							+	+	+		+	+	+	+
C. juncidum Ralfs							+							
C. — forma minor Wille							+							
C. Kützingii Bréb.						+	+	+	+		+	+	+	+
C. navicula (Bréb.) Lütkem.							+	+			+	+	+	+
C. parvulum Näg.						+	+	+	+	+	+	+	+	+
C. setaceum Ehrenb.							+		+	+			+	+
Pleurotaenium Ehrenbergii (Ralfs) Delp.							+	+	+		+	+	+	+
Tetmemorus granulatus (Bréb.) Ralfs.							+	+	+				+	+
T. laevis (Kütz.) Ralfs							+	+	+	+			+	+
Cosmarium abscissum Lütkem.													+	
C. angulosum Bréb. var. concinnum (Rabenh.) W. & G. S. West							+	+	+	+	+	+	+	+
C. antarcticum Gain	+													
C. botrytis (Bory) Ehrenb.						+	+	+	+	+	+	+	+	+
C. — var. tumidum Wolle							+					+	+	
C. contractum Kirchn. forma minor Wille							+							
C. crenatum Ralfs	+			+		+	+	+	+	+	+	+	+	+
C. cruciferum DeBary							+		+	+		+	+	
C. curtum (Bréb.) Ralfs	+							+				+	+	+
C. diplosporum (Lund.) Lütkem.							+	+	+		+	+	+	+
C. exiguum Arch. forma minor Wille							+							
C. — var. subrectangulum G. S. West							+						+	
C. gayanum DeToni							+							
C. — forma kerguelensis Wille							+							
C. granatum Bréb.							+	+	+	+	+	+	+	+
C. laeve Rabenh.							+	+	+	+	+	+	+	+
C. Meneghinii Bréb.							+	+	+			+	+	+
C. minimum W. & G. S. West							+		+			+	+	+
C. pseudoconnatum Nordst.	+								+	+	+	+	+	+
C. pseudopyramidatum Lund.							+	+	+	+	+	+	+	+
C. regnellii Wille							+	+		+	+	+	+	+
C. tinctum Ralfs							+					+	+	+
C. undulatum Corda	+							+	+	+	+	+	+	+
C. — var. minutum Wittr.	+							+				+	+	+
Arthrodesmus Ralfsii West							+	+					+	
Euastrum dubium Näg.							+	+		+		+	+	+
Staurastrum Aretasii Bourrelly & Manguin							+							
S. muticum Bréb.							+	+	+			+	+	+
Hyalotheca dissiliens (Sm.) Bréb.						+	+	+	+		+	+	+	+
Spondylosium planum (Wolle) W. & G. S. West							+					+	+	+
Sphaerozosma aubertianum West var. Archerii (Gutw.) West							+		+			+	+	+
Spirogyra stictica (Eng.Bot.) Wille							+					+	+	
S. tenuissima (Hass.) Kütz.							+					+	+	+

	Antarctica	S.Shetland	S.Orkney	S.Georgia	Falkland	Fuego and S. Patagonia	Kerguelen	Arctica	Australia and New Zealand	Africa	S.America	N.America	Europe	Asia
S. Weberi Kütz.												+	+	
S. varians (Hass.) Kütz.							+			+		+	+	+
Zygnema Vaucherii Ag.						+								
Mougeotia kerguelensis H.Krieger							+							
M. nummuloides Hass.							+					+	+	
M. parvula Hass.							+					+	+	
M. recurva (Hass.) DeToni							+					+	+	+
Chara longibracteata Wallm.					+									
Tolypella nidifica Leonh. forma antarctica (A.Br.) Nordst.							+							

LITERATURE CITED

BACHMANN, H. 1921. Beiträge zur Algenflora des Süsswassers von Westgrönland. *Mitt. naturf. Ges. Luzern* **8**.

BOHLIN, K. 1897. Die Algen der ersten Regnell'schen Expedition. I. Protococcoideen. *Bih. K. Vet. Akad. Handl.* **23**, Afd. III, 7.

BOLDT, R. Bidrag till kännedomen om Sibiriens Chlorophyllophycéer. *Öfvers K. Vet. Akad. Förh.* 1885 No. 2.

BOLDT, R. 1887. Grunddragen af Desmidieernas utbredning i Norden. *Bih. K. Sv. Vet. Akad. Handl.* **13**, Afd. III, No. 6.

BOLDT, R. 1887. Desmidieer fran Grönland. *Bih. K. Sv. Vet. Akad. Handl.* **13**, Afd. III, No. 5.

BORGE, O. 1893. Süsswasser-Chlorophyceen gesammelt von Dr. A. Osw. Kihlman im nördlichsten Russland, Gouvernement Archangel. *Bih. K. Sv. Vet. Akad. Handl.* **19**, Afd. III, No. 5.

BORGE, O. 1911. Die Süsswasseralgenflora Spitzbergens. *Vidensk. Skrifter I Math.-Nat. Kl.* 1911, No. 11.

BÖRGESEN, F. 1894. Ferskvandsalger fra Ostgrönland. *Medd. om Grönl.* **18**.

BÖRGESEN, F. 1910. Freshwater Algae from the "Danmark-Expedition" to north-east Greenland. *Medd. om Grönl.* **43**.

BOURRELLY, P. & E. MANGUIN, 1954. Contribution à la flora algale d'eau douce des îles Kerguelen. *Mém. Inst. Sci. Madag. B*, **V**.

BRUN, J. 1900. Diatomées de Jan Mayen et du Grönland. *Bih. K. Sv. Vet. Akad. Handl.* **26**, Afd. III, No. 18.

CHOLNOKY, B. J. 1954. Diatomeen und einige andere Algen aus dem 'de Hoek-Reservat in Nord-Transvaal. *Bot. Notis.* 1954.

CHOLNOKY, B. J. 1954. Diatomeen aus Süd-Rhodesien. *Portug. Acta biol.* B. **4**: 3-4.

CHOLNOKY, B. J. 1954. Neue und seltene Diatomeen aus Afrika. *Österr. bot. Z.* **101**: 4.

CHOLNOKY, B. J. 1955. Diatomeen aus salzhaltigen Binnengewässern der westlichen Kaap-Provinz in Südafrika. *Ber. dtsch. bot. Ges.* **68**: 1.

CHOLNOKY, B. J. 1955-1958. Hydrobiologische Untersuchungen in Transvaal I, II. *Hydrobiologia* **7**: 3, **11**: 3-4.

CHOLNOKY, B. J. 1956. Neue und seltene Diatomeen aus Afrika II. Diatomeen aus dem Tugela-Gebiete in Natal. *Österr. bot. Z.* **103**: 1.

190

CHOLNOKY, B. J. 1957. Beiträge zur Kenntnis der Südafrikanischen Diato-meenflora. *Portug. Acta biol.* B, **6**: 1.
CHOLNOKY, B. J. 1957. Über die Diatomeenflora einiger Gewässer in den Magalies-Bergen nahe Rustenburg.
CHOLNOKY, B. J. 1957. Neue und seltene Diatomeen aus Afrika III. Diato-meen aus dem Tugela-Flusssystem, hauptsächlich aus den Drakens-bergen in Natal. *Österr. bot. Z.* **104**: 1-2.
CHOLNOKY, B. J. 1958. Beiträge zur Kenntnis der Südafrikanischen Diato-meenflora II. Einige Gewässer im Walerberg-Gebiet, Transvaal. *Portug. Acta biol.* B, **6**: 2.
CHOLNOKY, B. J. 1959. Neue und seltene Diatomeen aus Afrika IV. Diato-meen aus der Kaap-Provinz. *Österr. bot. Z.* **106**: 1-2.
CHOLNOKY, B. J. 1960. Beiträge zur Kenntnis der Diatomeenflora von Natal. *Nova Hedwigia* **2**: 1-3.
CHOLNOKY, B. J. 1960. Beiträge zur Kenntnis der Ökologie der Diatomeen in dem Swartkops-Bache nahe Port Elizabeth. *Hydrobiologia* **16**: 3.
CLEVE, P. T. & A. GRUNOW, 1880. Beiträge zur Kenntnis der arktischen Diatomeen. *K. Sv. Vet. Akad. Handl.* **17**: 2.
CLEVE, P. T. 1894-1895. Synopsis of the naviculoid Diatoms. *K. Sv. Vet. Akad. Handl.* **26**: 2, **27**: 3.
CLEVE, P. T. 1898. Diatoms from Franz Josef Land. *Bih. K. Sv. Vet. Akad. Handl.* **24**, Afd. III, No. 2.
CLEVE, A. 1900. Beiträge zur Flora der Bären-Insel. *Bih. Sv. Vet. Akad. Handl.* **26**, Afd. III, No. 10.
CLEVE, A. 1951-1955. Die Diatomeen von Schweden und Finnland. *K. Sv. Vet. Akad. Handl.* Fjär. Ser.**2**: 1, **3**: 3, **4**: 1, **4**: 5, **5**: 4.
COMMITTEE ON POLAR RESEARCH 1961. Science in Antarctica. The Life Sciences in Antarctica. Publ. 839. Nat. Acad. Sci.-Nat. Research Council Wash. D.C.
FRENGUELLI, J. 1923-1924. Diatomeas de Tierra del Fuego. *Anal. Soc. Cient. Argent.* 96-98.
FRENGUELLI, J. 1943. Diatomeas de las Orcadas del Sur. *Rev. Museo La Plata***5**.
FRITSCH, F. E. 1910. Freshwater Algae collected in the South Orkneys by Mr. R. N. Rudmose Brown, B. Sc., of the Scottish National Antarctic Expedition, 1902-1904. *J. Linn. Soc. Bot.* **11**.
FRITSCH, F. E. 1912. Algae of the South Orkneys. Edinburgh. 1912 Report on the scient. results of the voy. of S.Y. "Scotia" III.
FRITSCH, F. E. 1912. Freshwater Algae in National Antarctic Expedition Nat. Hist. VI.
FRITSCH, F. E. 1917. Freshwater Algae. Brit Antarctic (Terra Nova) Expedi-tion, 1910. Nat. History Report Bot. 1.
FRITSCH, F. E. 1918. Contributions to our knowledge of the freshwater algae of Africa 2. First Report on the Freshwater Algae mostly from the Cape Peninsula, in the Herbarium of the South African Museum. *Ann. South Afr. Mus.* **9**.
FRITSCH, F. E. 1924. Contributions to our knowledge of the freshwater algae of Africa. 4. Freshwater and Subaerial Algae from Natal. *Trans. R. Soc. S. Afr.* **11**: 4.
FRITSCH, F. E. 1929. Contributions to our knowledge of the freshwater algae of Africa 7. Freshwater Algae from Griqualand West. *Trans. R. Soc. S. Afr.* **18**: 1.
FRITSCH, F. E. & RICH, F. 1929. Contributions to our knowledge of the fresh-water algae of Africa 8. Bacillariales from Griqualand West. *Trans. R. Soc. S. Afr.* **18**: 2.
FRITSCH, F. E. & STEPHENS, E. 1921. Contributions to our knowledge of the freshwater algae of Africa 3. Freshwater algae, mainly from the Transkei Territories, Cape Colony. *Trans. R. Soc. S. Afr.* **9**: 1.

FUKUSHIMA, H. 1959. General Report on Fauna and Flora of the Ongul Island, Antarctica, especially on Freshwater Algae. *J. Yokoh. Municip. Univ.* Ser. C-31, No. 112.

FUKUSHIMA, H. 1961. Algal Vegetation in the Ongul Islands, Antarctica. *Antarctic Record* No. 11.

FUKUSHIMA, H. 1962. Diatoms from the Shin-nan rock ice-free area, Prince Olav coast, the Antarctic Continent. *Antarctic Record* No. 14.

FUKUSHIMA, H. 1962. Notes on diatom vegetation of the Kasumi rock ice-free area, Prince Olav coast, Antarctica. *Antarctic Record* No. 15.

GAIN, L. 1911. Note sur la Flore algologique d'eau douce de l'Antarctide Sud-Américaine. *Bull. Mus. Hist. nat.*

GAIN, L. 1912. La Flore Algologique des régions antarctiques et subantarctiques. Paris 1912.

GEITLER, L. 1932. Cyanophyceae, in Rabenh. Krypt. Flor. 14.

GRÖNBLAD, R. 1945. De Algis Brasiliensibus, praecipue desmidiaceis, in regione inferiore fluminis Amazonas a professore August Ginzberger anno MCMXXVII collectis. *Acta Soc. Sci. Fenn.* Nov. Ser. B, **II**: 6.

HARIOT, J. 1908. Algues in Exp. antarct. franç. (1903-1905). Sciences Nat. Documents scient. Bot. Paris 1908.

HEURCK, H. VAN, 1909. Diatomées. Anvers. Exp. antarct. belg. Résultats du voy. du S.Y. Belgica Rapports scient. Bot.

HIRANO, M. 1959. Notes on some algae from the Antarctic, collected by the Japanese Antarctic Research Expedition. *Spec. Pub. Seto Marine Biol. Labor.* No. 3.

HIRANO, M. 1955-1960. Flora Desmidiarum Japonicarum I-VII. *Contr. Biol. Labor. Kyoto Univ.* No. 1, 2, 4, 5, 7, 9, 11.

HOLN, M. H. 1951. A Study of the Distribution of Diatoms in Western New York State. *Cornell Univ. Agr. Exp. Stat. Mem.* 308.

HOLMBOE, J. 1902. Navicula mutica Kütz. aus dem antarktischen Festlande. *Nyt. Mag. f. Naturv.* 40.

HOOKER, J. D. & HARVEY, W. H. 1847. Flora Antarctica I. Botany of the Antarctic Voyage.

HUSTEDT, F. 1921. Zellpflanzen Ostafrikas, gesammelt auf der Akademischen Studienfahrt 1910. *Hedw.* 63.

HUSTEDT, F. 1922. Bacillariales aus Innerasien, gesammelt von Dr. Sven Hedin. *South. Tibet* 6: 3, Bot.

HUSTEDT, F. 1927. Bacillariales aus dem Aokikosee in Japan. *Arch. Hydrobiol.* 18, 19.

HUSTEDT, F. 1927-1959. Die Kieselalgen, in Rabenh. Krypt. Flor. VII, Teil. 1, 2.

HUSTEDT, F. 1937. Süsswasserdiatomeen von Island, Spitzbergen und den Faeroes-Inseln. *Bot. Arch.* 38.

HUSTEDT, F. 1937-1938. Systematische und ökologische Untersuchungen über die Diatomeen-Flora von Java, Bali und Sumatra. *Arch. Hydrobiol.* Suppl. 15.

HUSTEDT, F. 1949. Süsswasser-diatomeen aus dem Albert-Nationalpark in Belgisch-Kongo. Expl. Parc. nat. Albert, Miss. H. Damas.

HUSTEDT, F. 1957. Die Diatomeenflora des Flusssystems der Weser im Gebiet der Hansestadt Bremen. *Abh. naturw. Ver. Bremen* 34: 3.

KOL, E. 1942. The snow and ice algae of Alaska. *Smiths. Misc. Coll.* 101: 16.

KOL, E. 1944. Vergleich der Kryovegetation der nördlichen und südlichen Hemisphäre. *Arch. Hydrobiol.* 40.

KRASSKE, G. 1938. Beiträge zur Kenntnis der Diatomeen-Vegetation von Island und Spitzbergen. *Arch. Hydrobiol.* 33.

KRIEGER, W. 1933-1937. Die Desmidiaceen, in Rabenh. Krypt. Flor. XIII, Abt. 1.

KRIEGER, W. 1938. Süsswasseralgen aus Spitzbergen. *Ber. dtsch. bot. Ges.* 56.

192

LAGERSTEDT, N.G.W. 1873. Sötvattens-Diatomaceer fran Spetsbergen och Beeren Eiland. *Bih. K. Sv. Vet. Akad. Handl.* **1**: 14.

LARSEN, E. 1904. The Freshwater Algae of East Greenland. *Medd. om Grönl.* **30.**

LARSEN, E. 1907. Ferskvandsalger fra Vest-Grönland. *Medd. om Grönl.* **33.**

MEISTER, F. 1912. Die Kieselalgen der Schweiz. *Beitr. z. Krypt. fl. d. Schweiz.*

MEISTER, F. 1932. Kieselalgen aus Asien.

NORDSTEDT, O. 1872. Desmidiaceae ex insulis Spetsbergensibus et Beeren Eiland in expeditionibus annorum 1868 et 1870 suecanis collectae. *Ofv. Kongl. Vet.-Akad. Förh.* 1872 No. 6.

NORDSTEDT, O. 1875. Desmidieae arctoae. *Ofv. K. Sv. Vet. Akad. Handl.* 1875.

NORDSTEDT, O. 1885. Desmidieer samlade af Sv. Berggren under Nordenskiöldska expeditionen till Grönland 1870. *Ofv. K. Vet. Akad. Handl.* 1885.

NORDSTEDT, O. 1888. Freshwater Algae collected by Dr. S. Berggren in New Zealand and Australia. *K. Sv. Vet. Akad. Handl.* **22.**

NYGAARD, G. 1932. Contributions to our knowledge of the freshwater algae of Africa. 9. Freshwater Algae and Phytoplankton from the Transvaal. *Trans. R. Soc. S. Afr.* **20**: 2.

OSTRUP, E. 1910. Diatoms from North-east Greenland collected by the Danmark-Expedition. *Medd. om Grönl.* **43.**

OSTRUP, E. 1920. Freshwater Diatoms from Iceland. *Bot. Icel.* **2**: 1.

PATRICK, R. 1936. A Taxonomic and Distributional Study of some Diatoms from Siam and the Federated Malay States. *Acad. Nat. Sci. Philad.* **78.**

PETERSEN, J. B. 1923. The Freshwater Cyanophyceae of Iceland. *Bot. Icel.* **II.**

PETERSEN, J. B. 1924. A Botanical trip to Jan Mayen. 2. Freshwater Diatomaceae. *Dansk. bot. Ark.* **4.**

PETERSEN, J. B. 1924. Freshwater Algae from the North-coast of Greenland collected by the late Dr. Th. Wulff. *Medd. om Grönl.* **64.**

PETERSEN, J. B. 1928. The aerial algae of Iceland. *Bot. Icel.* **II.**

PRESCOTT, G. E. 1951. Algae of the Western Great Lakes Area.

RICH, F. 1932. Contribution to our knowledge of the freshwater algae of Africa 10. Phytoplankton from South African Pans and Vleis. *Trans. R. Soc. S. Afr.* **20**: 2.

RICH, F. 1935. Contribution to our knowledge of the freshwater algae of Africa 11. Algae from a Pan in Southern Rhodesia. *Trans. R. Soc. S. Afr.* **23**: 2.

REINSCH, P. F. 1876. Species ac genera nova Algarum aquae dulcis quae sunt inventa in speciminibus in Expeditione "Vener" transit hieme 1874-1875, in insula Kerguelensi a clar. Eaton collectis. *J. Linn. Soc. Bot.* **15.**

REINSCH, P. F. 1878. Algae aquae dulcis Insulae Kerguelensis. *Philos. Trans. Roy. Soc. London* **168.**

REINSCH, P. F. 1883. Spec. et Gen. nov. Algarum ex insula Georg. Austri. *Ber. dtsch. bot. Ges.* **6.**

REINSCH, P. F. 1890. Die Süsswasseralgenflora von Süd-Georgien. Die intern. Polarforsch. 1882-1883. Die deutsch. Exp. und ihre Ergebn. II.

SCHMIDLE, W. 1898. Über einige von Knut Bohlin in Pite Lappmark und Vesterbotten gesammelte Süsswasseralgen. *Bih. K. Sv. Vet. Akad. Handl.* **24,** Afd. III, No. 8.

SKUJA, H. 1949. Zur Süsswasseralgenflora Burmas. *Nov. Acta Soc. Sci. Upsal.* **IV,** 14, No. 5.

SKUJA, H. 1955. Taxonomische und biologische Studien über das Phytoplankton schwedischer Binnengewässer. *Nov. Acta Reg. Soc. Sci. Upsal.* **IV,** 16, No. 3.

SKVORTZOW, B. W. 1936. Diatoms from Kizaki Lake, Honshu Island, Nippon. *Phil. J. Sci.* **61**: 1.

193

SKVORTZOW, B. W. 1936. Diatoms from Biwa Lake, Honshu Island. Nippon. *Phil. J. Sci.* **61**: 2.
SKVORTZOW, B. W. 1937. Diatoms from Ikeda Lake, Satsuma Province, Kiusiu Island, Nippon. *Philip. J. Sci.* **62**: 2.
SMITH, G. M. 1920-24. The Phytoplankton of the Inland Lakes of Wisconsin.
STOCKMAYER, S. 1906. Kleiner Beitrag zur Kenntnis der Süsswasseralgen- flora Spitzbergens. *Österr. bot. Z.* **56**.
STRÖM, K. M. 1923. The Alga-Flora of the Sarek Mountains. *Naturw. Unters. d. Sarekgeb.* **3**: 15.
STRÖM, K. M. 1926. Norwegian Mountain Algae. *Skr. Norske Vid.-Akad. Oslo Math.-Nat. Kl.* 1926. No. 6.
TAYLOR, W. R. 1934-35. The freshwater algae of Newfoundland. *Pap. Mich. Acad. Sci. Arts & Lett.* **19, 20**.
TAYLOR, W. R. 1954. Cryptogamic Flora of the Arctic. II. Algae: Non-planc- tonic. *Bot. Rev.* **20**.
THOMASSON, K. 1955. Studies on South American Freshwater Plankton. 3. Plankton from Tierra del Fuego and Valdivia. *Acta Horti Gotoburg.* **19**.
TILDEN, J. 1910. Minnesota Algae I.
WEST, W. & G. S. & CARTER, N. 1904-1923. A Monograph of the British Desmidiaceae I-V.
WEST, G. S. 1909. The Algae of the Yan Yean Reservoir, Victoria: a Biologi- cal and Oecological Study. *J. Linn. Soc. Bot.* **39**.
WEST, W. & G. S. 1911. Freshwater Algae. Brit. Antarct. Exp. 1907-09. Reports on the scient. invest. I: 7.
WHELDEN, R. M. 1947. Algae in N. Polunin, Bot. of the Canadian Eastern Arctic, II. Thallophyta and Bryophyta. *Nat. Mus, Canada Bull.* **97**.
WILDEMANN, E. DE. 1900. Expédition antarctique belge. Note prélim. sur les Algues rapportées par M. Racovitza. *Acad. roy. Belg.* Bull. classe d. sciences 1900.
WILDEMANN, E. DE. 1935. Observations sur des Algues. Exp. Antarct. Belge. Résultats du voyage de la Belgica en 1897-99. Rapports scient.
WILLE, N. 1879. Ferskvandsalger fra Novaja Semlja samlede af Dr. E. Kjellman paa Nordenskiölds Expedition 1875.
WILLE, N. 1902. Mitteilungen über einige von C. E. Borchgrevink auf dem antarktischen Festlande gesammelte Pflanzen III. Antarktische Algen. *Nyt. Mag. f. Naturvid.* **40**.
WILLE, N. 1924. Süsswasseralgen von der Deutschen Südpolarexpedition auf dem Schiff "Gauss". Dtsch. Südpolar-Exped. VIII. Bot.

LICHENS

BY

CARROLL W. DODGE
Department of Botany, University of Vermont, Burlington, Vermont

The climate of Antarctica except in the northern part of the Peninsula imposes severe limitations on the growth of land plants; only the lichens and a few mosses have been able to flourish under these limitations. Sunlight is abundant for nearly half the year and by heating the dark colored rocks, the lichen at the surface of the rock is probably growing at a higher temperature than the air temperature indicates. Except on rock surfaces moistened by water from melting snow and ice, the habitat is very dry and conservation of the water supply is a major problem for the lichen. Any mechanism which retards water loss, tends to reduce gas exchange in which the carbon dioxide is a necessity for photosynthesis. In exposed sites, any large foliose or fruticose lichen is subjected to wind strains and possible breakage. When a high wind drives snow crystals against a crustose thallus, it erodes the outer portion like a sand blast, removing part of the cortex and sometimes the algal layer. A few small species have solved this problem by growing in cracks of crystalline rocks under a translucent quartz crystal and sending to the surface a slender strand of hyphae bearing the reproductive organs.

Before considering the morphologic adaptations to a cold dry climate, we may sketch briefly the normal morphology of lichens. Most of the Antarctic lichens are crustose (31 genera), that is, a vegetative layer (thallus) closely adherent to the surface of a rock, less commonly to a low tuft of mosses. Fourteen genera are leaf-like (foliose) with dorsiventral differentiation into layers, often somewhat raised above the substrate and variously attached to it. The other twelve genera are fruticose, short, erect or recumbent, attached only at the base, more or less cylindric with an outer layer of cortex, a middle layer of algae and an inner medulla which may be variously differentiated to give additional strength for wind resistance.

The marine species of *Verrucaria* from the northern end of the Peninsula face the same physiologic problems as do the marine algae, thus differing from all other Antarctic lichens. One species grows only below the low tide level, several species grow between the tide levels and the rest just above the high tide level, where they are often wet by salt spray in stormy weather.

The rest of the Antarctic lichens show various adaptations to their environment. In some genera or species within other genera, the outer layer of the thallus, the cortex, is composed of intricately woven, predominantly periclinal hyphae, the outermost usually decomposed and disintegrating. These lichens are usually found in moist, sheltered habitats.

In *Alectoria*, the cortical hyphae are thickwalled, longitudinal and closely cemented together, resisting wind strain and water loss. The cortex is often pierced by small pores to allow gas exchange. These species are either recumbent or grow in small tufts, densely branched and not more than 5 mm tall, in contrast to north temperate species which may be pendent and 30 cm long.

In the remaining Antarctic lichens, the cortical hyphae are perpendicular to the surface of the thallus (fastigiate) and also cemented together in a layer nearly impervious to water vapor when dry, but easily permeated by water. Gas exchange is secured by cracks in this layer, rarely by definite pores. These lichens may grow in either sheltered or exposed sites. In regions of high winds on exposed sites, this layer is often partly, rarely completely, eroded away, exposing the algal layer to damage from drying.

In the Umbilicariaceae and in many species of various genera in other families, the outermost cells of the fastigiate cortex die and form a gelified amorphous layer, sometimes up to 100μ thick, being constantly renewed from below. This condition is reminiscent of the horny layer of the human skin which effectively protects the underlying layers of cells in the epidermis. Such a cortex is seldom seriously eroded by high winds.

In *Mastodia* and in lichens with Nostocaceous symbionts, the algae fill the whole thallus, the fungus hyphae growing in the gel between the algal cells and reproducing at the upper surface of the thallus. In the other lichens, the algae form a distinct layer under the upper cortex. The algal cells may occur singly or be united into small colonies, separated by fungus hyphae. *Trebouxia* of the Chlorphyceae is the usual symbiont in Antarctic lichens; *Trentepohlia*, so common in tropical and subtropical lichens and occasional in those of the temperate zone, is so far unknown in Antarctica, thus eliminating many families from consideration. *Mastodia* with a *Prasiola* symbiont, is confined to moist sites. Its fungus hyphae penetrate between the two layers of algal cells, stimulating them to further cell division, so that the resulting thallus is four or eight cells thick. Free-living *Prasiola* is found on rocks in pools formed by melting snow as far south as McMurdo, but the lichen *Mastodia* is confined to the northern portion of the Peninsula and to the coast of East Antarctica. Similarly free-living *Nostoc* is found at McMurdo, while *Collema* and *Leptogium* with *Nostoc* symbionts are confined to the same areas as *Mastodia*.

In the Heppiaceae with *Stigonema* symbionts, one species is occasional from the MacRobertson Coast to the George V Coast, not yet found at Cape Hallett, but occurring in the Edward VII Peninsula and in Marie Byrd Land. So far no species has been found in the Peninsula. In other parts of the world, the Heppiaceae are characteristic of very dry regions, either cold or hot.

In crustose species with a thin thallus, the medulla consists of closely woven, slender hyphae, but in thicker thalli and in the foliose and fruticose thalli, the upper portion of the medulla usually consists of very loosely woven hyphae with large air spaces, resulting in better aeration of the algal layer above. The lower portion may be very compactly woven, of thickwalled hyphae, resulting in greater mechanical strength in resisting wind strain. In the erect *Usnea* and *Stereocaulon*, the center of the thallus is occupied by a solid strand of sclerenchyma, while in *Cladina* and *Cladonia* the strand of sclerenchyma is a hollow cylinder.

A lower cortex is seldom differentiated in the crustose families; in the foliose families, it is present and usually of the same structure as the upper cortex, except that it lacks an amorphous layer. In *Physcia* and the subgenus *Physcioideae* of *Parmelia*, the upper cortex is fastigiate while the lower cortex is formed of thickwalled, periclinal hyphae, more or less cemented together. Without reproductive structures, it is practically impossible to separate these two taxa. As all of HUE's species of *Physcia* were based on sterile material, their systematic position remains in doubt until their apothecia are found.

In some species of the crustose families, a hypothallus of more loosely woven, brown or black hyphae is found between the rock and the rest of the thallus. This may serve for better aeration of the thallus along with the various systems of cracks in the upper cortex.

In general, Antarctic lichens are much smaller than those of the same genus farther north, for example, species of *Usnea* in Antarctica are seldom more than 5 cm tall, while some of those in the cooler part of the temperate zone may reach 100 cm or more. This tendency to reduce the size of the thallus and concentrate on production of a few, relatively large apothecia, often with dark discs, is especially noticeable in *Lecanora* subg. *Squamaria* and in *Rinodina fecunda*.

In 23 species of *Buellia*, the thallus consists of repeatedly branched rhizomorphs. Where a rhizomorph comes in contact with a suitable alga, it forms a small pulvinate assimilative areole. The apothecia are sometimes borne on the rhizomorphs, more often on or at the side of the assimilative areoles. A few species in other crustose genera show this tendency, but it is less highly developed.

There are also intriguing problems of symbiosis, parasymbiosis and parasitism. In *Rinodina fecunda*, the fungus hyphae and algae form a very small thallus on the side of a moss stem below the growing apex of the shoot, sheltered from the wind. The fungus hyphae grow downward in the moss stem as a true parasite, eventually killing the lower part of the stem, but never injuring the apex which continues growth and photosynthesis, providing nutrients for the relatively large apothecia on a very small thallus. In many species, the lichen clearly parasitizes the underlying moss and eventually kills it.

In some cases, e.g., *Buellia Johnstoni* starts growing as an epiphyte on *Lecanora exsulans*, but soon sends its hyphae down into the thallus below, making contact with its algae. Both lichens continue to grow and fruit normally. The upper *Buellia* thallus is reduced in size, but the apothecia are much larger than those of the same species when free-living on rock. In some species, colonies of bacteria, apparently the nitrogen-fixing *Azotobacter*, occur within the lichen thallus. Growth of such thalli is more luxuriant than those of the same species in which no bacterial colonies are found. Free-living *Azotobacter* has recently been found in Antarctica. This phenomenon has been previously reported by CENGIA-SAMBO in a few species from the Alps just below the snow line.

In *Thelidium Caloplacae*, growing on Caloplacaceous thalli, the thallus is extremely reduced, most of the nutrition coming from the host alga, with only a few algal cells in the *Thelidium* thallus, apparently producing just enough hormones to stimulate perithecial development; so far I have not found this species free-living. The fungus of *Thelidium parvum* apparently used *Trebouxia* and *Scytonema* algae indiscriminately as symbionts, a very rare occurrence in lichens.

Asexual multiplication is less common than in lichens of temperate and tropical climates. Propagula and isidia are very rare. Soredia are more common, but only in genera with relatively more sorediose species farther north.

Sexual reproduction is almost universal once the thallus is sufficiently mature. The spermogonia, or male organs, are often much larger with the wall folded, giving a greater area for the production of spermatiaphores, than in species of the same genera in tropical or temperate climates. This suggests a compensation for a far greater wastage of spermatia, the male gametes, than is found elsewhere. This phenomenon is also present in the subantarctic islands. I have not studied enough lichens from the Arctic to know if this phenomenon is common there.

Geographic distribution

After more than a century of exploration, many areas are still practically unknown, i.e. the south coast of the Weddell Sea, the coast of East Antarctica to 20° E., (and poorly known to 60° E.), the coast of Marie Byrd Land to Thurston Island and the base of the Peninsula (south of 69 °S.) as well as the mountains from 60— 144 °W. Although I have studied more than 8000 specimens, the sampling error is still high for many regions.

Isotherms show a greater correlation with species distribution than latitude alone, as they take into account, altitude, latitude and cloudiness during the growing season. Most of the rocks are volcanic, granites or highly metamorphosed sedimentary rocks, and show a richer flora than is found on similar rocks farther north. Calcareous outcrops are very rare and seem to have a much poorer flora than farther north.

The coast and adjacent islands of the Peninsula region is distinctly warmer north of 65 °30′S. and has a richer flora (93 species), including 20 genera not found elsewhere in Antarctica. Twenty-one species extend to the South Shetlands and South Orkneys and thirteen are reported to extend farther north to Fuegia and the Falkland Islands. I have seen very little material south of 69 °S. and the few species seen seem to be new. The few species seen from Thurston and Dustin Islands to the west, are either endemic or are found in the Peninsula.

The coast of East Antarctica from 20 °E. to 170 °E., mostly close to the Antarctic Circle, has a rich and fairly uniform flora, 70 species of which 34 are confined to the Coast and 36 species extend eastward to Marie Byrd Land. The sampling error is still large between 20° and 110° E. Twenty-three species seem endemic along the coast of Victoria Land; 20 species extend from Marie Byrd Land to Cape Adare but not beyond; 3 species are endemic to the Edward VII Peninsula; 16 species extend from the Edward VII Peninsula to adjacent Marie Byrd Land and 18 species seem endemic in Marie Byrd Land. Ross Island has much porous lava, a poor substrate for lichens and has a much poorer flora, although collections are abundant. No material has yet been seen from the Horlick Mountains. In the Queen Maud Mountains, 85 °30′S. to 86 °03′S., only seven species have been received, of which three are endemic, the others also occurring in Marie Byrd Land.

Origin of the lichen flora

As I see it, there are two alternatives: either a richer preglacial flora with the hardier genera and species surviving on nunataks and spreading slowly as the ice receded, or a colonization *de novo* by migration from Fuegia. In no case have I found any relationship to the flora of subantarctic Kerguelia or of Macquarie Island. If migration occurred from Fuegia, it was so long ago that the ancestry of the present species of Antarctica is scarcely recognizable. There are at least three genera endemic in the Peninsula and two others from Marie Byrd Land to the Queen Mary Coast. The other genera are also found in the south temperate zone. Several genera have more species in Antarctica than elsewhere, with one or two species extending northward along the Andes, sometimes as far as the subarctic.

Once migration occurred from Fuegia via the South Shetlands to the northern end of the Peninsula, the lichens may have spread southward along the Peninsula and adjacent islands to its base, thence eastward along the coast of East Antarctica in the direction of the prevailing winds. For example, *Acarospora Gwynni* from East Antarctica is closely related to a yellow species from Cockburn Island reported by MAGNUSSON. Except for a very different species from the northern end of the Peninsula, *Acarospora* is unkown in Antarctica. I have seen one species of *Physcia* from the Peninsula and another from the Knox Coast in East Antarctica. Other species may have spread from the base of the Peninsula westward along the mountain ranges and nunataks to Marie Byrd Land, Victoria Land and along the coast of East Antarctica (36 species).

If we consider the distribution of *Usnea*, subg. *Neuropogon*, agreeing with MOTYKA that the subg. *Protousnea* from southern South America is the most primitive, the species of the subg. *Neuropogon* from the subantarctic islands except Fuegia, are not closely related to those of the Antarctic flora. One or two species seem to have migrated northward along the Andes to the Arctic, although there are great gaps in their present distribution. The species of Antarctica show much diversity in the Peninsula. *Usnea antarctica* ranges from Marie Byrd Land to Cape Hallett and gives rise to other species along the coast of East Antarctica.

Species of *Buellia*, the largest genus in Antarctica (72 species), are found in practically every locality in Antarctica where collections have been made, but each species has a rather limited distribution. The group of species with a thallus of rhizomorphs and assimilative areoles seems endemic in Antarctica. *Lecanora* the next largest genus with 32 species has its species about equally divided between the Peninsula and from Marie Byrd Land to the Coast of East Antarctica, but each species has a limited distribution. In its subgenus *Squamaria* there are two species in the Peninsula, one in Marie Byrd Land and Victoria Land, two from Marie Byrd Land to northern Victoria Land and along the coast of East Antarctica and one restricted to the coast of East Antarctica, all the latter species are more closely related to each other than to the species in the Peninsula. *Lecidea* with 28 species has its subg. *Biatora* limited to the Peninsula and the rest of the species with quite limited distributions.

Rinodina frigida seems the most abundant lichen in East Antarctica. It is rather rare in Marie Byrd Land, increasingly common in Victoria Land and more so along the coast of East Antarctica. The species shows some variability but has not yet segregated into geographic races or micro-species.

The species of *Alectoria* in the Peninsula are closely related to the Andean and northern species, while *A. antarctica* from Marie Byrd Land to Ross Island, and *A. congesta* from the coast of East Antarctica, are very distinct and might be segregated as a distinct genus.

Altogether, migration from Fuegia seems the more attractive hypothesis, but concrete data in its favor are very few and many data can be interpreted by either hypothesis. If such a migration occurred, it happened so long ago that the evidence is quite obscure.

VASCULAR PLANTS

BY

N. M. WACE

Department of Geography, University of Adelaide, South Australia
(with 13 fig.)

Introduction

The Antarctic continent has for long occupied an important place in the theories of plant geographers. The presence of an Antarctic land mass formerly joined to other southern lands and with a warmer climate than at present, has been invoked to explain some of the widely disjunct distributions of vascular plants which today inhabit parts of South America, Southern Africa, Australia, New Zealand, and various islands in the Southern Ocean. Thus, in 1859

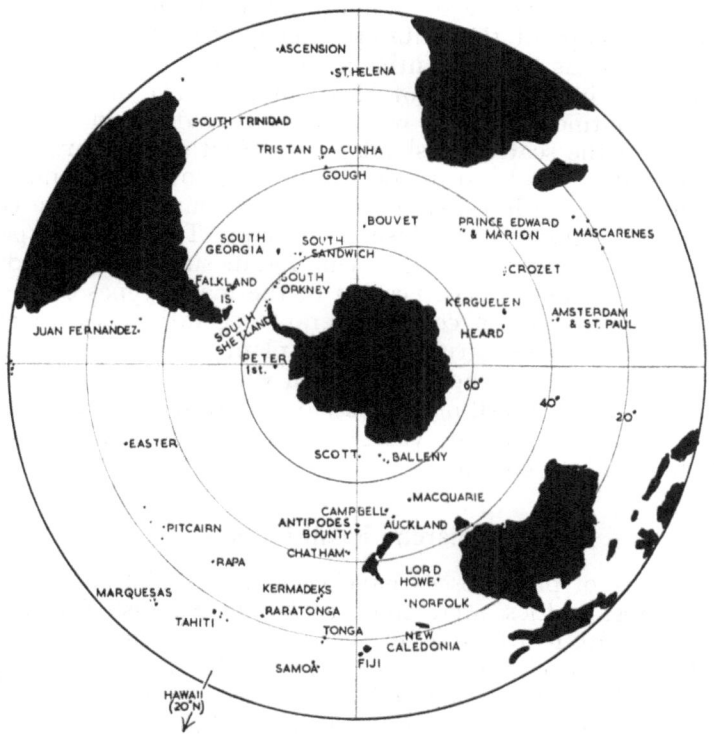

Fig. 1. The Southern Hemisphere, showing the positions of various islands mentioned in the text.

202

(long before the discovery of any plant fossils in the Antarctic),
DARWIN speculated upon:

"a former warmer period, before the commencement of the last glacial period, when
the Antarctic lands now covered with ice, supported a highly peculiar and isolated
flora."

of which

"a few forms had been... widely dispersed to various points in the Southern Hemisphere".

Similar views are expressed in the correspondence between DARWIN,
J. D. HOOKER, and Alfred Russell WALLACE (SKOTTSBERG, 1960a),
and the idea has been put forward many times since then (for a
recent review see SKOTTSBERG, 1956 pp. 389—393).

Antarctica today has a very meagre vascular flora. Any consideration of its former plant geographical relationships must therefore
depend almost entirely upon such fossils as have been discovered
there, and upon their relations to contemporary fossil floras, and to
present-day floras elsewhere. This account will therefore start with
an enumeration of the vascular plant fossils that are known from
the continent and its nearby islands, and will then discuss the present relationships of the Antarctic Cenophytic floras. (The fossil
floras and faunas of the continent are more fully dealt with by
HARRINGTON in Chapter I of this volume). The living circum-
Antarctic continental plants will next be considered in relation to
the floras of the scattered islands of the Southern Ocean; and an
attempt will be made to divide the present southern hemisphere
flowering plants with a disjunct distribution into "Tertiary continental" and "Quarternary insular" groups. The major types of
vegetation on the Antarctic continent and its surrounding lands are
then discussed, and a classification of the major types of circum-
Antarctic and southern cold temperate vegetation is attempted,
according to the gross structure of the plant communities. Finally,
some suggestions are made on research which could throw further
light upon the composition of the old Antarctic flora, and on the
history and biology of the vascular plants which now inhabit, or
have inhabited the continent.

Antarctic Fossil Floras

A number of plant fossils have been described from Antarctic
regions. Most of these were collected by the Swedish South Polar
Expedition (1901—04) and by parties from the British Falkland
Islands Dependencies bases (since 1945), in the Antarctic Peninsula
and its nearby islands; and also by the British Antarctic Expedition
(1910—13) and the United States Antarctic Expedition (1933—35)
in the Ross Sea area of the Trans-Antarctic Mountains. During and
since the International Geophysical Year (1957—58) numerous rock

collections bearing plant fossils have been made from many parts of the continent, but especially in the Trans-Antarctic Mountains and along the Indian Ocean coast. CROHN (1959), PAVLOV (1958), CRANWELL (1959) and CRANWELL, HARRINGTON & SPEDEN (1960) have described fossil pollens and spores, but most other collections of fossil plants have been examined only for macroscopic remains. BARGHOORN (1961) and PLUMSTEAD (1962) have recently reviewed the fossil floras of Antarctica, but they have been little concerned with the Cenophytic.

Several collections of fossil plants which have been made on some subantarctic islands will be considered here. Macroscopic fossils (SEWARD & CONWAY, 1934) and microspores (COOKSON, 1947) have been described from Kerguelen; fossil wood from South Georgia (DOUGLAS 1923, GORDON 1930); and lignite microspores from Macquarie Island (BUNT, 1956).

Paleozoic and Mesozoic Floras

All the Paleozoic terrestrial plant remains so far described from the Antarctic are from Beacon sediments or their equivalents (HARRINGTON, 1958) in Greater Antarctica (Fig. 2). A number of species of *Glossopteris* and *Vertebraria*, and unidentified fragments of wood were first described by SEWARD (1914) from collections made by E. A. WILSON on Buckley Mountain (85 °S.), a nunatak on the Beardmore Glacier. Further discoveries of *Glossopteris*-bearing sandstones, sometimes associated with coal seams containing gymnosperm wood, were described from Mt. Weaver (DARRAH, 1936), from the Horlick Mountains (SCHOPF, 1962), from erratics on the Priestley and Mackay Glaciers in Victoria Land (SEWARD, 1914), and from Horn Bluff (c.150 °E.) on the coast of Wilkes Land (MAWSON, 1940). Pollen and spores from this last site have been described by PAVLOV (1958), who compared them to Triassic deposits from the Soviet Arctic. Carbonaceous shales from the Amery Formation sediments in the Prydz Bay area (c.80 °E.) which are thought to be of Permian age, and regarded "as the equivalent of the lithologically and structurally similar beds of the Beacon Formation . . . from the Ross Sea area" (CROHN, 1959), were found to contain fossil microspores *(Florinites, Leucosporites, Pitysporites)*, and monocolpate pollen grains. Considerable controversy was aroused by SEWARD's (1914) description of what FLORIN (1940) subsequently accepted as "a possibly podocarpaceous pollen grain" at the Priestley Glacier site. In view of the present mainly southern distribution of the Podocarpaceae, this discovery and its tentative acceptance as podocarp pollen is of particular interest (see below). The Beacon sediments are widely distributed in Greater Antarctica (see chapter I of this volume), and the deposits have been placed from Carboniferous to early Jurassic (DARRAH, 1936) in age.

Fig. 2. Localities of pre-Tertiary vascular plant fossils known from the Antarctic regions.

Plant remains from Jurassic rocks at Hope Bay (c.63°S.) near the northern tip of the Antarctic Peninsula were described by HALLE (1913). The flora includes many gymnosperms (Bennetitales, Cycadales, araucarians and podocarps), but no lycopods, Gingkoales, or early flowering plants. Further collections of mid-Jurassic plant fossils have since been made from Mount Flora at Hope Bay in 1945 (ADIE, 1958), and a similar stratigraphic succession to that known from Mount Flora has been found in the Trinity Peninsula nearby. Jurassic plants which are probably similar to the Hope Bay fossils have also been found on James Ross Island, and as far south as Cape Longing and Cape Sobral (c.66°S.) on the eastern side of the Antarctic Peninsula; and also at King George Island (62°S.) in the South Shetland group, and in Alexander Island (71°—72°S.) on the western side of the Peninsula (FLEMING et al., 1938; ADIE, 1958). This Antarctic Peninsula Jurassic flora is the richest known from the Continent.

Petrified araucarian wood, thought to be of Paleozoic or Lower Mesozoic age has been found in South Georgia (DOUGLAS, 1923; CAMPBELL SMITH, 1930), but no remains of this age are known from any other subantarctic islands. Kerguelen, Heard and Macquarie Islands all contain Mesozoic rocks, but other circum-Antarctic islands which are situated away from the continental shelves are of more recent age.

These pre-angiosperm fossil remains can obviously give little information about the biogeographical relations of the present Antarctic and circum-Antarctic floras. The Antarctic occurrence of *Glossopteris* relates its Permo-Carboniferous flora to that of its contemporaries in all the southern continents and in India. The relationships of these *Glossopteris* floras have been discussed, amongst others, by SAHNI (1927), HILL (1929), JUST (1952) and PLUMSTEAD (1962); and ADIE (1952) has reviewed the representation of the Gondwana system in Antarctica. Paleogeographic problems of Gondwanaland have been reviewed by WALCOM (1949) and EDWARDS (1955). Recently, BALME (1962) has examined the world distribution of pollens which are associated with the *Glossopteris* floras, and has concluded that they give no support to the Gondwanaland concept since the *Glossopteris*-associated pollen types appear to be more-or-less universal in Permian times.

The fossil occurrence and distribution of the southern Mesozoic and Tertiary conifers has been discussed by FLORIN (1940), some of whose conclusions are considered later. The widespread range of the araucarians relates the late Paleozoic and Mesozoic Antarctic floras to their contemporaries in most other parts of the world; but the abundance of podocarps in the early Antarctic deposits relates the flora of the continent to contemporary and present gymnosperm floras in only the southern continents. These early genetic affinities amongst the conifers of the Antarctic with areas of land now widely separated from it, perhaps suggest that the Antarctic has had closer terrestrial connections with the other continents during the late Paleozoic and Mesozoic than it has at present. With the exception of a few small-seeded bird-dispersed genera *(Juniperus, Taxus)*, the living conifers are apparently incapable of crossing wide stretches of ocean today, and are not found on truly "oceanic" islands (GUPPY, 1919; LI, 1958). Whether any such connections have involved continental drifting, land bridges or island chains, the biological evidence for climatic change in the Antarctic since the Mesozoic can not be disputed. The very existence of the remains of large trees near the present position of the Pole probably implies some latitudinal movement of the rocks containing them, since it is difficult to envisage the survival of such plants through long periods of darkness in the winter months.

Tertiary Floras

Tertiary plant fossils have been described from the Antarctic Peninsula and its nearby islands, and some pollen-bearing morainic material from the McMurdo Sound area of the Ross Sea. No Tertiary plant fossils are known from most of Greater Antarctica, or from Byrd Land. Some Tertiary fossil plants have also been described from subantarctic Kerguelen and Macquarie Islands (Fig. 3).

Macroscopic plant remains were described from Seymour Island (c.64°S.) near the northern tip of the Antarctic Peninsula, by Dusen (1908). The flora includes ferns, gymnosperms and flowering plants, many of which were referred to living families, genera or species; but doubt has been cast on some of these identifications. Florin (1940) has re-identified some of the gymnosperms, and Cranwell (1959) has examined some of Dusen's material for microsfossils.

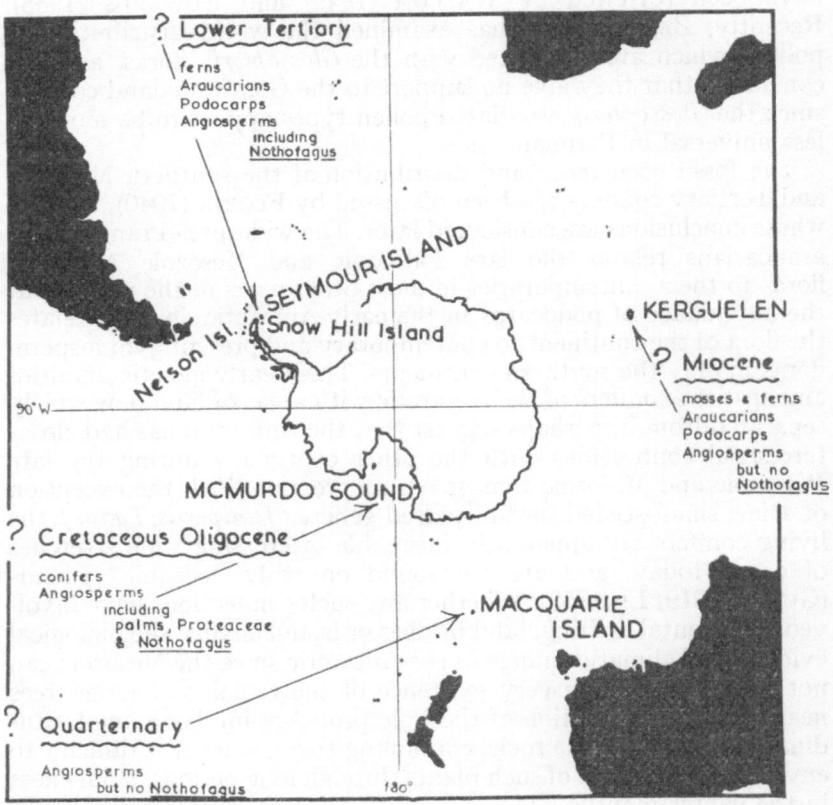

Fig. 3. Cretaceous and Tertiary plant fossils known from the Antarctic regions.

Table I.

Tertiary Fossil records from Seymour Island, Antarctic Peninsula (c. 64 °S., 56 °W.)

Compiled from Dusén (1908), Florin (1940) and Cranwell (1959). Pollen records are indicated by an asterisk, and genera with species now living outside Antarctica are in italics. The forms "Carpolites" (2 species) and "Phyllites" (26 species) have been omitted.

Ferns	* Cyatheaceae:	*Alsophila*
	Polypodiaceae	*Asplenium* (2 species)
		Dryopteris (2 species)
		Pecopteris (18 species)
		Polypodium (2 species)
		Sphaenopteris (10 species)
		Taenopteris (2 species, one resembling *Blechnum)*
Gymnosperms	* Schizaeaceae	
	Araucariaceae:	* *Agathis*
		* *Araucaria* subgenus *Colymbea*
	Podocarpaceae:	* *Acmopyle*
		* *Dacrydium*
		* *Phyllocladus*
		* *Podocarpus* section *Stachycarpus*
Angiosperms	Cyperaceae:	Scirpitis (cf. *Schoenoplectus)*
	Aquifoliaceae:	Iliciphyllum (2 species, cf. *Ilex)*
	* Cruciferae	
	Cunoniaceae	*Caldcluvia*
	* Fagaceae	* *Nothofagus* (pollen of *fusca* and *brassii* groups)
	Lauraceae:	Lauriphyllum
	Leguminosae:	Leguminosites
	* Loranthaceae	
	Melastomaceae:	Miconiiphyllum
	Monimiaceae:	*Laurelia*
		Mollinedia
	Myricaceae:	*Myrica*
	* Myrtaceae	
	* Onagraceae:	(pollen resembling *Fuchsia* sp.)
	* Proteaceae:	*Lomatia* (4 species)
		Knightia
	Winteraceae:	*Drimys*

The above list of plants (Table I) is a combination of Dusén's original list, Florin's amendements, and Cranwell's pollen and spore records. Cranwell states that "the dominants, as indicated by the pollen, are conifers and ... *Nothofagus*". She found that the total number of microspores or fragments per slide was very small, but that no lycopod spores or pollen of Compositae, Centrolepidaceae, Restionaceae, Pandanaceae, Gramineae or Cyperaceae were present in the sample that she examined. Gothan (1908) who described the fossil woods in the Seymour Island deposits, found an araucarian, two podocarps (one of which, according to Florin, may be *Acmopyle)*, two Lauraceous forms, and *Nothofagus ("Notho-*

fagoxylon"). The age of these important remains is in doubt: DUSEN considered that they were late Tertiary, but CRANWELL (1959) mentions the possibility that some microspores may be redeposited from the Cretaceous beds nearby, and that an upper age limit in the Tertiary can not be set for the main deposits.

Cretaceous, as well as Tertiary fossils from Snow Hill, Seymour, and Cockburn Islands were also collected in 1946; and plant fossils (probably Miocene) comparable to the Seymour Island deposits were discovered in 1949 on Nelson Island in the South Shetland group (ADIE, 1958). Work is now in progress on these Antarctic Peninsula and South Shetland Cenophytic floras by the British Antarctic Survey geologists at Birmingham University, England (ADIE, 1962a).

An important discovery of some plant and animal microsfossils in morainic material from Minna Bluff and White Island in McMurdo Sound, was described by CRANWELL, HARRINGTON & SPEDEN (1960). Pollens of conifers (unspecified), palms, Proteaceae and *Nothofagus* were found, together with marine microfauna which suggested a late Cretaceous to Oligocene age. No fungi, fern spores or Compositae pollens were found in the sample. The pollens were few, but the discovery is important because it contains the only flowering plant remains known from the whole of Greater Antarctica.

Some plant remains from Kerguelen Island (50°S., 70°E.) consisting of carbonised wood (EDWARDS, 1921), petrified wood and fragments of herbaceous plants (SEWARD & CONWAY, 1934), and plant microfossils (COOKSON, 1947) have revealed an extensive Tertiary flora there. The list of Kerguelen fossil plants (Table II) is compiled from the above three accounts, with FLORIN's (1940) assignment of the form-genera *Elatocladus* and *Cupressinoxylon* to Podocarpaceae, and *Dadoxylon* to *Araucaria*. The lignites and carbonaceous sandstones in which these plant remains were found, are interbedded with lava flows which are assumed to have been extruded in Tertiary fissure eruptions. The gymnosperm trunks were found throughout the flows near Christmas Harbour "so that it is impossible to reckon the period of time that must have elapsed between the origin, growth and destruction of the successive forests now buried" (EDWARDS, 1921). It is noteworthy that no remains of *Nothofagus* have yet been found on Kerguelen, and it seems from the pollen and spore frequencies that the flowering plants occupied only a subsidiary position in the Tertiary vegetation there (COOKSON, 1947).

BUNT (1956) in a preliminary report on the lignites of Macquarie Island (54°S., 159°E.) notes that their pollens are unlike those of the plants at present inhabiting the island, and that no gymnosperm or *Nothofagus* pollens were found. Dr. BUNT has very kindly shown me the figures and descriptions that he prepared, and more detailed work will be necessary before comparisons can be made to known

209

Table II.

Tertiary Fossils Recorded from Kerguelen (c. 50 °S., 70 °E.)

Compiled from EDWARDS (1921), SEWARD & CONWAY (1934), FLORIN (1940) and COOKSON (1947). Pollen records are indicated by an asterisk, and genera and species now living are in italics.

Fungi	Ascomycetes: Hemisphaerales (two fruit bodies resembling Australian Tertiary epiphyllous forms)
Algae	Chrysophyceae: Diatoms (6 genera, including the present sub-alpine *Fragilaria*)
Bryophytes	Mosses: "Dicranites", "Muscites" and other fragments of moss leaves.
Pteridophytes	Filicites: 3 leaf forms (resembling *Blechnum pennamarina, Gleichenia* and *Polystichum mohroides*)
	* Trilites: 9 spore forms (one resembling *Cyathea* and another *Ophioglossum*)
	* Monolites: 2 spore forms (undetermined)
Gymnosperms	*Araucaria*
	* Araucarites (2 pollen forms)
	Podocarpus, section *Eu-podocarpus*
	* Disaccites/Podocarpites (3 pollen forms)
	? *Dacrydium*
	* Polysaccites/Microcachryites (2 pollen forms)
	* Disaccites/Phyllocladites (3 pollen forms)
Angiosperms	* Monosporites: 2 pollen forms (one resembling Gramineae and the other *Typha* or *Sparganium*)
	* Tricolpites: 3 pollen forms (one probably Tubiflorae)
	* Tetracladites/Droseridites (pollen resembling *Drosera*)
	* Monosulcites: 3 pollen forms (all indeterminable, but one apparently resembling *Ginkgo*)
	Some leaf fragments, resembling *Ranunculus trullifolius* or *Azorella selago*

pollen types elsewhere. The Macquarie lignites whose pollen content was examined were covered with glacial debris, and it is possible that the remains are post-Tertiary and interglacial in age. More pollen-bearing material has recently been collected from Macquarie Island by Dr. DUIGAN of the Botany School, Melbourne University.

The Present Relationships of the Tertiary Floras of Antarctica

The relationships of the Seymour Island and other contemporary floras has been discussed by NANTHORST (1907), HILL (1929), and BERRY (1928 & 1938); and the literature concerning the Antarctic Peninsula, Patagonian and Chilean Tertiary floras has been reviewed by SKOTTSBERG (1956).

DUSEN (1908) considered that the Seymour Island flora was a mixture of species with different ecological preferences: one group related to the present temperate West Patagonian and South

Chilean flora, and typified in the deposits by *Caldcluvia, Drimys, Laurelia, Lomatia* and *Nothofagus;* and a subtropical group typified by *Miconiiphyllum, Mollinedia, Araucaria* and the ferns. He thought that the temperate group were washed into the subtropical deposits from montane communities. The view that the remains are "probably mainly from forest at different elevations", and that "some types may have been washed in from older deposits" was supported by CRANWELL (1959) from her studies of the pollens and spores. In view of the doubt that has been cast on the identification of the macro-fossils (DARRAH, 1960), and in view of their uncertain age, and that they may be derived from ecologically distinct communities laid down at different times, it is obviously hazardous to make any comparisons with contemporary vegetation elsewhere. The leaf sizes and morphology probably indicate that the climate at the time the deposits were laid down was moist and temperate to subtropical. The presence in the deposits of DUSEN's first (temperate) group of species, together with various podocarps, possibly suggests that the flora resembled some areas of the present Valdivian forest of Western Chile between 41° and 48°S. latitude (SKOTTSBERG, 1916; GODLEY, 1960). But SKOTTSBERG (1956) stated that he could see little reason for BERRY's assertion that the Seymour Island flora contained a large element of subtropical or warm temperate types mixed with forms resembling the present temperate flora of Southern Chile and Patagonia.

Information provided by the British Antarctic Survey (ADIE, 1962a) suggests that the Tertiary plant remains from King George Island (c.62°S.) in the South Shetland group are probably earlier than those from Seymour Island, and that climate during the time that the former deposits were laid down was probably the warmer. The meagre information so far available on the Antarctic Peninsula floras therefore suggests that there was a cooling of the climate there during the Tertiary, in common with that affecting northern floras of comparable latitudes. The palm pollens from the Lower Tertiary of the McMurdo Sound area may also indicate that this cooling can be taken to apply to Greater Antarctica. Any such cooling must have led to the extinction of much of the flora of Tertiary Antarctica, and the migration of those elements for which suitable land connections remained open, or which had the capacity to cross sea barriers to more northerly lands.

The occurrence of *Nothofagus* and Podocarpaceae in many southern Tertiary floras, and their importance in much southern cold temperate vegetation today, makes their movements during and since the Tertiary of great paleoclimatic and biogeographic interest — comparable to that of the northern Fagaceae and the Pinaceae in the northern hemisphere. COUPER (1960) drew attention to the fact that there were then no records of *Nothofagus* pollen from the north-

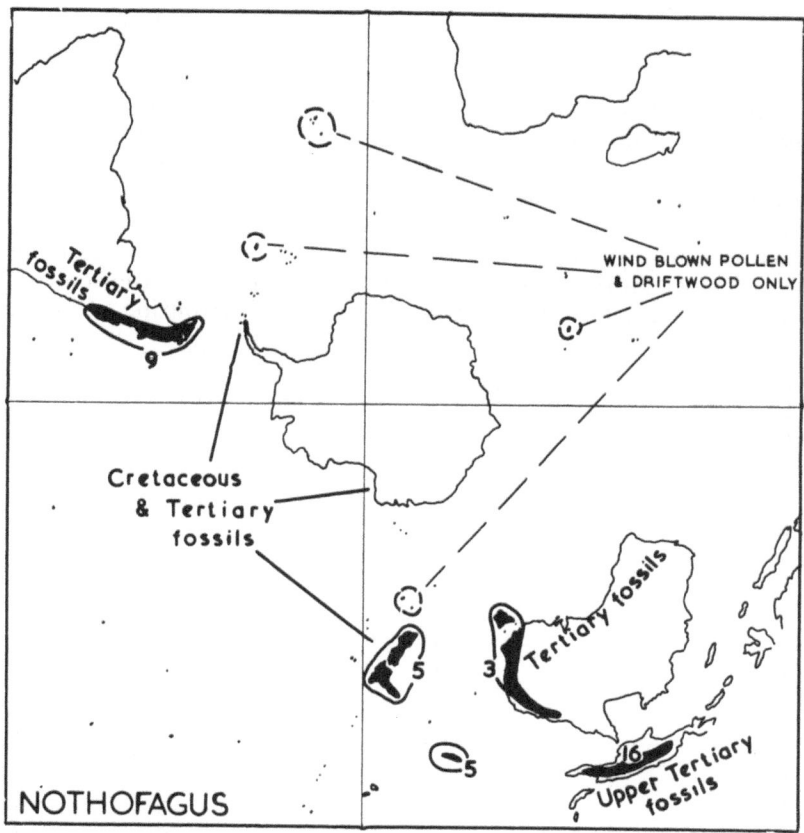

Fig. 4. The distribution of the "continental" genus *Nothofagus* Blume. Numbers of living species in various regions, and the known fossil range of the genus. (Wind blown pollens &c. from Tristan – Gough, Kerguelen, South Georgia and New Zealand shelf islands probably represent long range trans-oceanic dispersal of pollen only, and do not indicate that *Nothofagus* has ever inhabited these places). Doubtful fossil records from the Northern Hemisphere omitted. For references see Table III, and also Cranwell (1963).

ern hemisphere, the Asian mainland, or western Indonesia. He suggested that the movements of the three sections of the genus could most easily be explained by a northward migration from Lesser Antarctica (where remains of all three occur in the Seymour Island deposits), to their present areas of occurrence ranging from New Guinea to southern South America (Fig. 4). Ma Khin Sein's (1961) confirmation of *Nothofagus* pollen in the British Eocene perhaps refutes the idea that the genus has always been confined to the southern hemisphere (unless long range aerial transport can be invoked to account for the British grains). But whether Antarctica

212

has been the "cradle or the grave" in *Nothofagus* evolution (CRAN-WELL, 1962), it seems probable that the continent has played an important part in the spread of the genus.

COUPER (1960) also produced evidence from the known records and distribution of podocarp pollens of the genera *Dacrydium (franklinii), Microcachrys, Phyllocladus* and *Podocarpus* (section *Dacrycarpus*), to show that these may also have spread northwards from Lesser Antarctica, where the pollens of most of them are almost certainly present on Seymour Island or at Hope Bay. In his classic paper on the Mesozoic and Tertiary conifers of the southern hemisphere, FLORIN (1940) showed that the distribution of all the

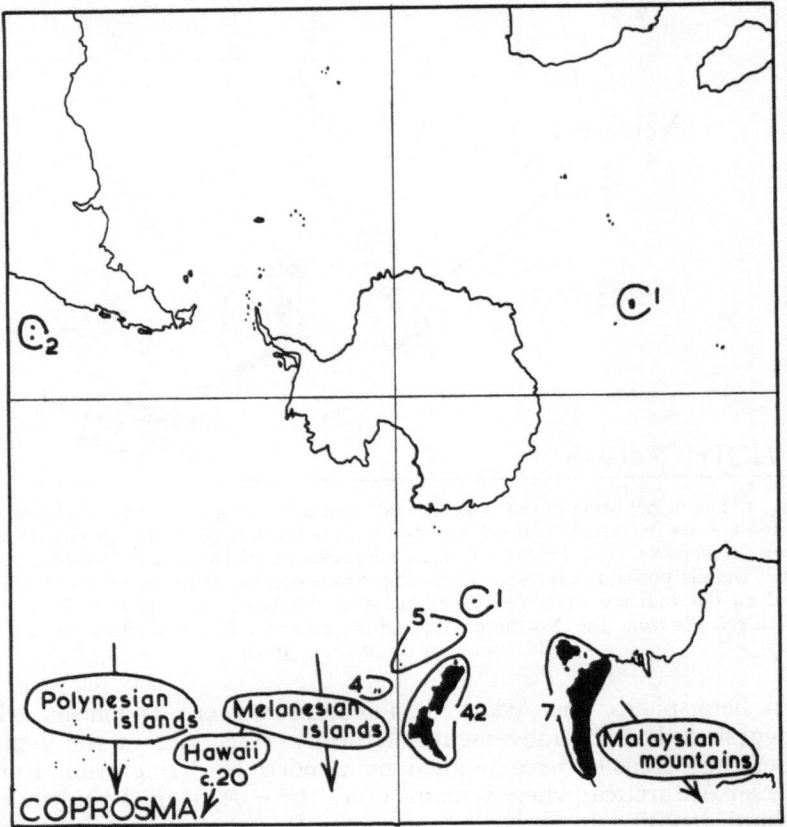

Fig. 5. The distribution of the species of *Coprosma* FORST. The genus is centred on New Zealand and Hawaii, with endemic representatives in many Pacific islands. A single "Quarternary insular" species *(C. pumila* HOOK. f.) inhabits temperate and subantarctic regions in the South West Pacific, and has recently been identified from Kerguelen (GREENE & GREENE, 1963). Data from ALLAN (1961), BURBIDGE (1963) and SKOTTSBERG (1956).

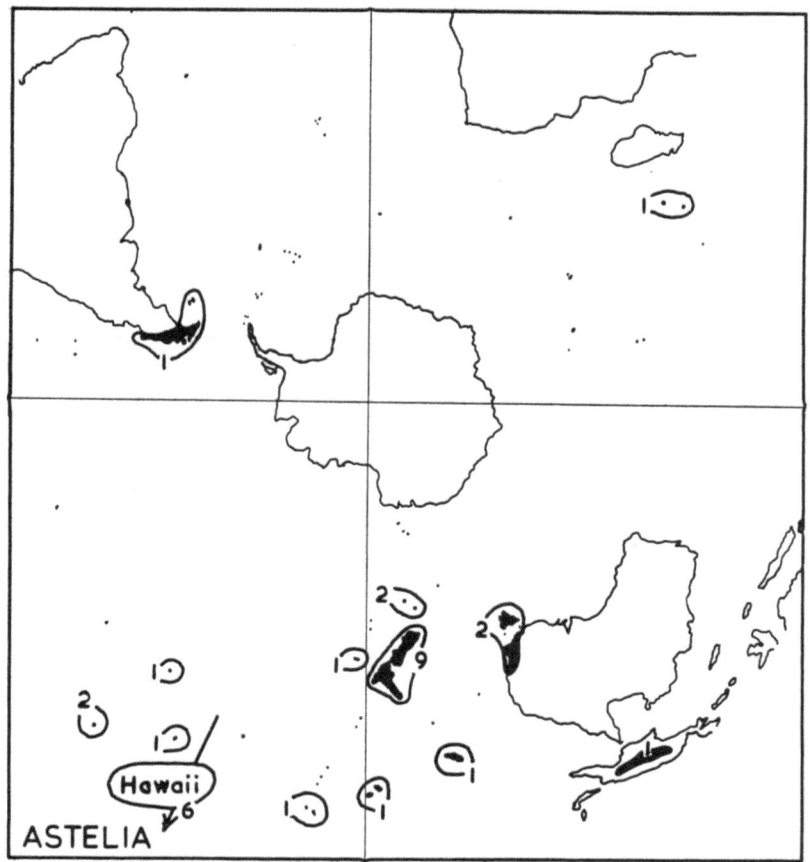

Fig. 6. The distribution of the species of *Astelia* BANKS et SOLANDER. This genus does not fall easily into either of the author's "continental" or "insular" categories, and perhaps owes its present range to a combination of old continental connections and more recent dispersals to remote islands. Data from SKOTTSBERG (1937 and 1960b).

conifers with a southern range today (with the exception of *Araucaria* and certain Cupressaceae) could best be explained on the basis of a northward migration from the Antarctic, for which was there fossil evidence. FLORIN stated that:

"Antarctica has played an important role in the development and distribution of the southern group of conifers. The data related to (conifer) distributions seem most readily explained by assuming land connections, or at least much closer proximity between Antarctica and the adjaceant southern ends of South America, Australia, New Zealand and South Africa".

CRANWELL (1962) has emphasised the affinity between the early Antarctic deposits and the (supposedly Tertiary) remains from

Kerguelen. In the presence of podocarp fossils, but no *Nothofagus* and few other flowering plants on Kerguelen, it is tempting to suppose that the southern conifers reached Kerguelen (and thence Africa?) by overland connections from the Antarctic during the Mesozoic, before many flowering plants had had time to spread.

The Tertiary plant remains that are known from the Antarctic continent are too few to provide any factual basis upon which a wholesale comparison with Tertiary floras elsewhere may be made. Although it seems possible that some genera may have spread northwards from Antarctica before, or possibly during the Tertiary, no con-

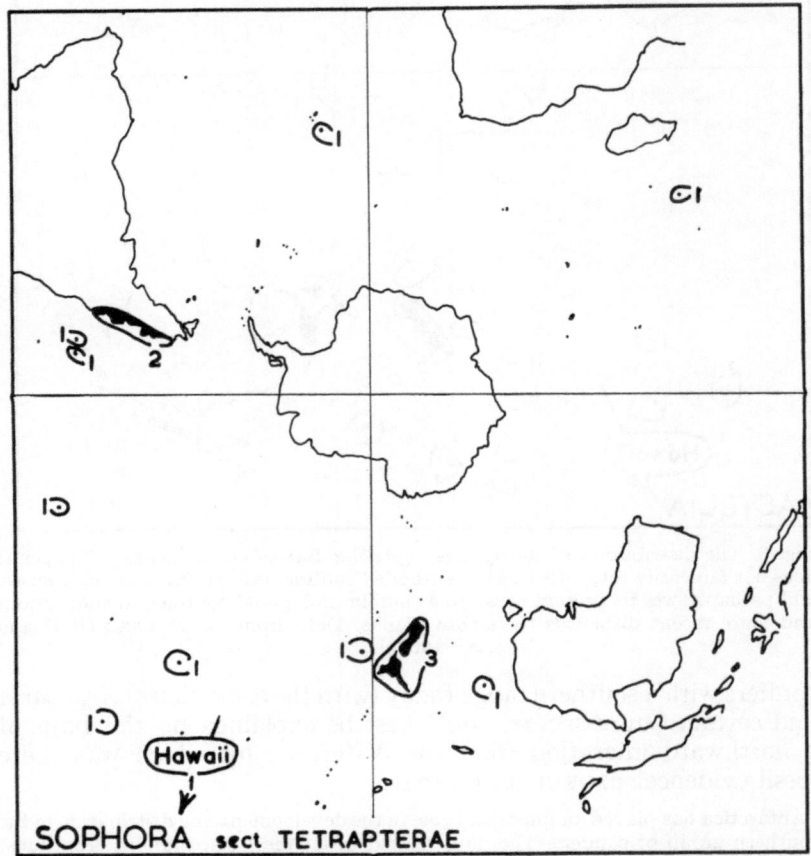

Fig. 7. The distribution of the species of *Sophora* L., section Tetrapterae (= *Edwardsia* SALISB.). An "insular" group of closely related species, mostly of coastal or streamside habitats, some members of which have a remarkable capacity for prolonged floatation of seeds in seawater (see WACE & DICKSON, 1964). Data from SKOTTSBERG, (1956).

clusions can be drawn regarding the ultimate origin of any taxonomic groups there, even though their presence in Tertiary Antarctica
may have been demonstrated, and their circum-Antarctic discontinuous distributions today suggest that they have spread from a
southern source. BERRY (1938) and SKOTTSBERG (1956) have both
drawn attention to the danger of assuming a southern origin for such
genera as *Araucaria, Drimys* or *Laurelia* simply because of their
fossil occurrence on Seymour Island and their present southern distributions. Many biogeographers would probably agree with SKOTTS-
BERG (1956):

"the little we know about the pre-glacial vegetation of Antarctica is sufficient to
prove that this large land mass . . . was inhabited by a large and varied flora, that it
may have been a primary centre of evolution, that in other instances it served as a
secondary centre, and that it was a much-trodden road between America and
Australia-New Zealand".

A land (or island chain) link between the Antarctic Peninsula and
the Southern Andean chain, by way of the Scotia Arc, during or
before the early Tertiary, seems to be well established geologically
(eg. HOLTEDAHL, 1929; HAWKES, 1962). The biological evidence
from the Tertiary floras of the Antarctic Peninsula and South
America supports the idea of such a land link. But land connections
between Antarctica and Australasia, and still more between Antarctica and Southern Africa during the Tertiary, are more problematical. Perhaps the greatest gap in our understanding of the past
biogeographic importance of Antarctica as a migration route or an
evolutionary centre lies in the almost complete ignorance of any
angiosperm remains throughout the whole of Greater Antarctica.
The very large structural and geological differences between the two
parts of the Continent (LOEWE, 1961; THIEL, 1961) may imply that
we are not justified in regarding Tertiary Antarctica as a single
biogeographic unit (TAYLOR, 1960, 1963). In view of this lack of
information on any Tertiary floras in Greater Antarctica (except
for the fragmentary but important finds from McMurdo Sound,
noted above), any suggested northward migrations of plants from
the Antarctic during the Tertiary to Australasia or Africa (eg.
CROIZAT, 1952) must remain largely speculative.

"Antarctic" Elements in the Present Floras of Southern Lands

J. D. HOOKER first drew attention to the floristic similarities that
are to be found between the widely separated continents of the
southern hemisphere, and between the remote islands of the Southern Ocean. He gathered the results of his botanical collections into
his magnificent *"Botany of the Antarctic Voyage . . ."* including the
suggestively entitled *"Flora Antarctica"* (1844—47). Although later

216

workers (eg. HEMSLEY, FLORIN, SKOTTSBERG) have in the main accepted HOOKER's contention that this scattered flora may have spread from the Antarctic, the flora itself has not recently been defined by a list of the plants that may be involved in all the southern lands.

Knowledge of the Tertiary flora of Antarctica is too fragmentary to be of much help in compiling such a list. The principal criterion for suggesting an Antarctic origin (or a sojourn in late Mesozoic or Tertiary Antarctica) for any taxonomic group, remains a widely discontinuous distribution in the southern hemisphere of the living plants today. In attempting to define any such "Antarctic" group from their present distributions as native plants, the following points must be borne in mind:

1. Many plant species which may have spread north from Antarctica, are likely to have reached (or survived in) only a single southern continent;

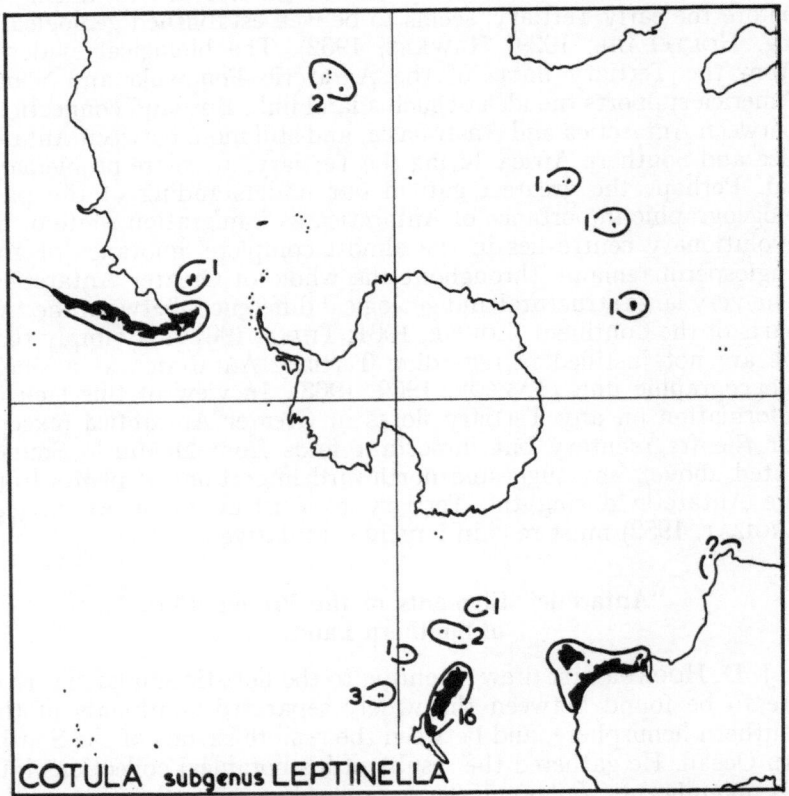

Fig. 8a.

217

2. Other species may have spread into northern latitudes, so that their southern derivation would be unsuspected from their present ranges;

3. Yet other species with a one-time cosmopolitan distribution may have achieved a southern discontinuous range today due to extinction in northern and tropical areas.

Some of the problems associated with defining any southern derived group of plants are discussed by DuRietz (1940), Gordon (1949), van Steenis (1953) and Skottsberg (1956). It is clear that a reliable definition of any such group involves the whole question of the phylogeny and area(s) of origin of the angiosperms, and will have to await a far wider range of knowledge of Cenophytic floras

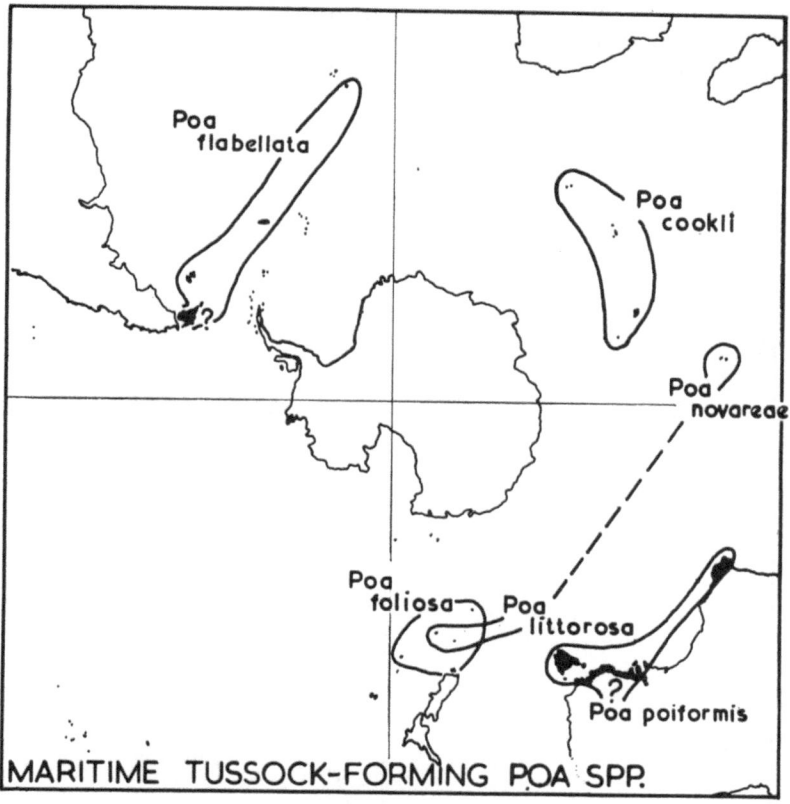

Fig. 8b.
Fig. 8. The distribution of *Cotula* L. subgenus *Leptinella*, and the large coastal tussock-forming species of *Poa* L. Both are examples of taxa of predominantly maritime habitats, with circum-Antarctic temperate to subantarctic "insular" ranges. Data for *Cotula* from Allan (1961), Greene & Greene (1963) and Wace & Dickson (1964); data for *Poa* from Petrie (1909), Wace (1960) and Willis (1962).

218

in both northern and southern hemispheres, and in the tropics.

In view of these limitations, and our meagre knowledge of the Tertiary fossil floras of the Antarctic, it may be impossible ever to compile a soundly based list of plants in whose evolution and dispersal Antarctica may have played a part. The list of "Antarctic" plants presented here (Table III) includes only:

a. Those whose fossil remains are known from the Antarctic continent and its offshore islands;

b. Those species and genera whose present ranges as native plants are discontinuous between two or more of the southern continents or islands groups, and which have not been found in the northern hemisphere as living plants.

Certain living taxa which have outliers in Malaysia, Polynesia or Central America are, however, included; as are some with fossil representatives in northern lands. Australia and New Zealand are considered as a single continental unit in assessing the discontinuous

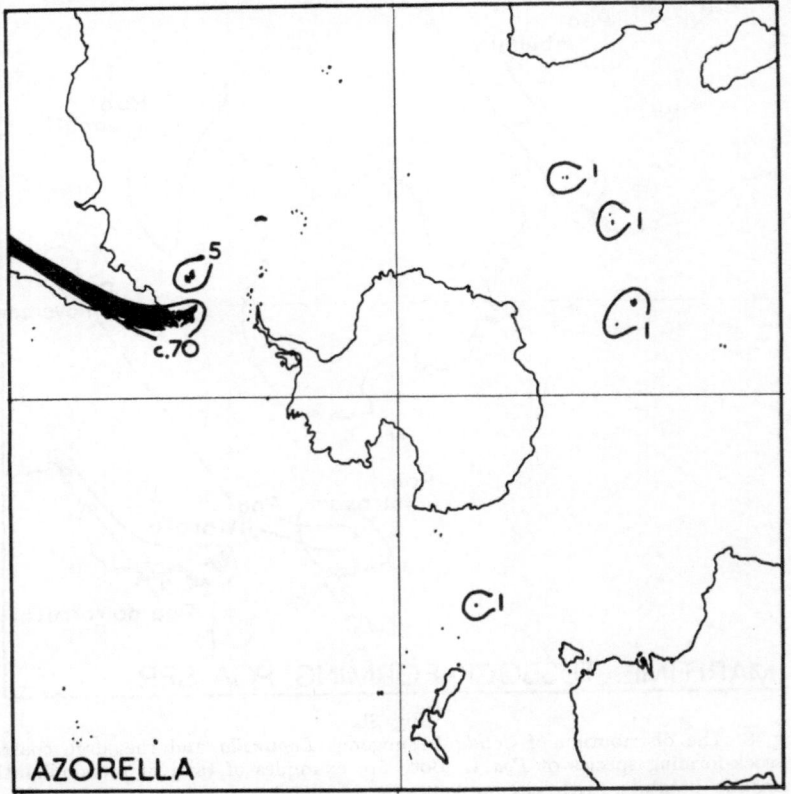

Fig. 9a.

ranges of the plants. The table does not include all of S\ᴋᴏᴛᴛsʙᴇʀɢ's (1960a) "Antarcto-Tertiary" plants, although it is based very largely on various papers by that author. Notwithstanding these difficulties, the table should include most of those seed plants which are now confined (or nearly confined) to the southern hemisphere, and whose present distribution is difficult to explain without invoking now disappeared land connections (probably involving Tertiary Antarctica), or long range trans-oceanic dispersal, or both of these phenomena.

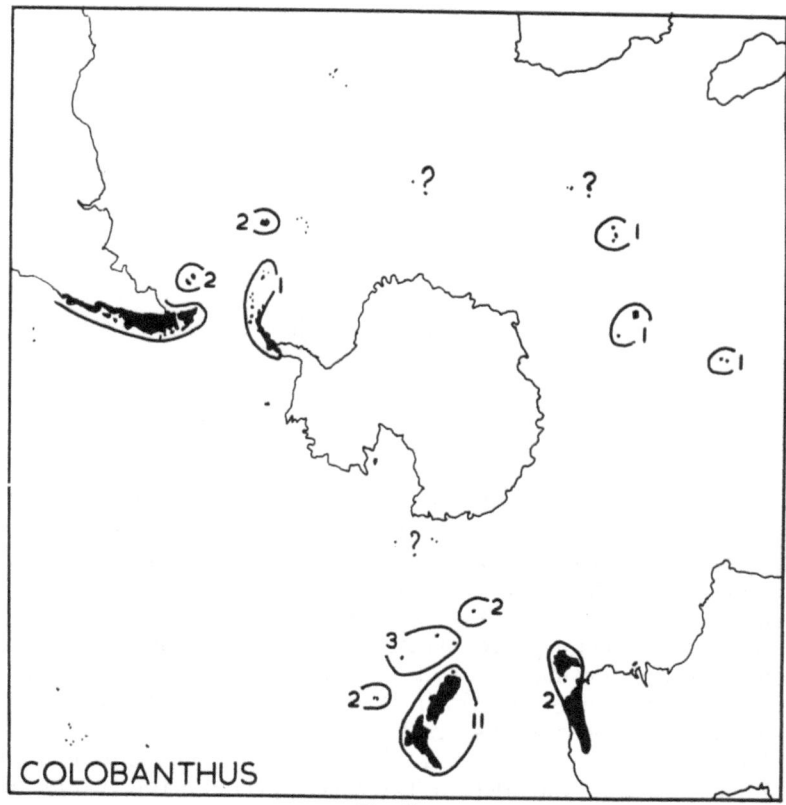

Fig. 9b.

Fig. 9. The distribution of *Azorella* Lᴀᴍ. and *Colobanthus* Bᴀʀᴛʟ. Both genera comprise cushion-forming feldmark species, and reach subantarctic and Antarctic provinces. *Azorella* is centred on southern America and the Andes, but a single "insular" species *(A. selago* Hooᴋ f.) inhabits all the subantarctic islands, although it has surprisingly not been discovered on South Georgia. Data from Aʟʟᴀɴ (1961), Gʀᴇᴇɴᴇ & Gʀᴇᴇɴᴇ (1963) and Sᴋᴏᴛᴛsʙᴇʀɢ (1960a).

Colobanthus is centred on New Zealand, but a single "insular" species *(C. crassifolius* (D'Uʀᴠ.) Hooᴋ. f. reaches the Scotia Arc islands and the Antarctic Peninsula, while related species inhabit the subantarctic islands in the Indian Ocean. Data from Sᴋᴏᴛᴛsʙᴇʀɢ (1954), Aʟʟᴀɴ (1961), Tᴀʏʟᴏʀ (1955a) and Gʀᴇᴇɴᴇ & Gʀᴇᴇɴᴇ (1963).

Since HOOKER's early work, much of the discussion on the origins of these southern disjunct plants has turned on this controversial question of long range trans-oceanic dispersal as against a formerly greater continuity of the southern land masses. GORDON (1949) listed the protagonists of the two schools of thought as follows:

Trans-oceanic dispersal	Greater land continuity
DARWIN	J. D. HOOKER
WALLACE	SKOTTSBERG
GUPPY	CAMPBELL
OLIVER	WULFF
GIBBS	CAIN
RIDLEY	GOOD
SETCHELL	

to which might be added:

	CROIZAT
	DU RIETZ
	FLORIN
	HEMSLEY
	VAN STEENIS

The considerable literature on this subject has been reviewed by TURRILL (1953) and SKOTTSBERG (1956, 1960a).

Both long range dispersal and greater land continuity in the past may be necessary to account for the present distributions of the southern disjunct plants. It seems that these "Antarctic" plants can be divided into two groups:

a. Southern plants with a more-or-less exclusively southern hemisphere continental range, which may have been driven out of the Antarctic during the Tertiary, and whose existing distribution and biology strongly suggests some previously closer land connections between their present areas of occurrence.

b. Southern hemisphere plants occurring in remote islands as well as on the continents, which may also have been driven out of Tertiary Antarctica, but which have achieved circumpolar ranges today by trans-oceanic dispersal during the Quaternary.

This division into "continental" and "insular" elements is noted against the plants included in Table III, although the two groups are not claimed to be rigidly exclusive and there are some taxa which do not fall clearly into either of them.

Some advocates of greater land continuity involving Antarctica, to account for the present range of the southern disjuncts, have also admitted the possibility of recent long range dispersal of insular species (eg. SKOTTSBERG, 1960 p. 455). But no previous attempt has been made to divide the southern plants as a whole into "insular" and "continental" groups, although VAN STEENIS' (1953) division of the Malaysian mountain plants into "circum-South Pacific", "old Oceanic" and "general subantarctic" categories to some extent recognises the division. DAWSON (1958) has also suggested that the floristic affinities between Australasia and South America have arisen "during two distinct periods; the first and longest before, and the second since the Pleistocene ice age". HOLDGATE (1960a) has also advanced the idea of older "continental" and younger "oceanic" elements amongst the southern circumpolar plants.

"Insular" and "Continental" plants

Of the hundred or so genera of seed plants listed in Table III, only about twenty (marked "I" in the left hand column) occur in more than one of the isolated island groups in the Southern Ocean. Most of the southern disjunct plants are confined to the continents (including New Zealand), and to such ancient continental islands as New Caledonia and the Falklands, which are not far removed from the main continental masses.

The "insular" plants listed in Table III are:—

> *Agrostis magellanica*
> *Deschampsia antarctica/elegantula*
> *Festuca erecta*
> *Poa cookii/foliosa/flabellata/litorosa*
> *Uncinia spp.*
> *Juncus pusillus/scheuzerioides*
> *Rostkovia spp.*
> *Colobanthus spp.*
> *Ranunculus biternatus/crassipes*
> *Cardamine corymbosa/glacialis*
> *Tilleaea moschata*
> *Acaena spp.*
> *Sophora sect. Tetrapterae*
> *Callitriche antarctica/christensenii*
> *Apium australe*
> *Azorella selago*
> *Coprosma pumila*
> *Galium antarcticum*
> *Nertera depressa/granadensis*
> *Cotula subgen. Leptinella*
> *Lagenophora spp.*

Certain taxa in *Scirpus, Carex* and *Hydrocotyle* should possibly be added to this list. Both species at present living on the Antarctic continent itself *(Colobanthus crassifolius* and *Deschampsia antarctica)* belong to this "insular" group. All the remote islands south of latitude 35°S. have native floras very largely composed of these circumpolar "insular" species: only a few otherwise "continental" species have spread to single islands or archipelagoes from their biologically parent (not necessarily their nearest) continents (Table IV). Although there has been some formation of insular endemics, most of the "insular" species are little differentiated on the various

Genera, sections of genera, and species of southern hemisphere gymnosperms and angiosperms which are known (either fossil or living) from Antarctica, or whose present range as native plants is discontinuous between two or more of the southern continents and island groups, are listed on the left of the Table. Numbers of species are entered in the appropriate columns across the table. Each entry is qualified by an index number referring to the source of information (see References).
The following symbols are also employed:—

+ : taxon present, but numbers of species uncertain
± : taxon closely allied to that listed, present in the area concerned
( ) : fossil records for Antarctica (and offlying islands), South Georgia and Kerguelen (see Tables I and II)
C : "Tertiary/Continental" plants ⎫
I : "Quarternary/Insular" plants ⎬ see text pp. 221 and 230.

	1	2	3	4	5	6	7	8
	ANTARCTICA, South Orkney, S. Shetland & S. Sandwich Is.	South Georgia	SOUTH AMERICA	Falkland Islands	Tristan – Gough Islands	SOUTHERN AFRICA	Madagascar & Mascarene Is.	Amsterdam & St. Paul Islands
C Agathis	$(+)^{57}$							
C Araucaria	$(+)^{57}$	$(+)^{67}$	3^{148}					
C sect. Eu-Libocedrus			2^{148}					
C Acmopyle	$(+)^{57}$							
C Dacrydium	$(+)^{57}$		1^{148}					
C Microcachrys	$(?)^{26}$							
C Phyllocladus	$(+)^{57}$							
C Podocarpus	$(+)^{57}$		$+^{97}$			$+^{169}$	$+^{181}$	
I Agrostis magellanica			1^{154}	1^{147}	$±^{174}$			$±^{78}$
I Deschampsia antarctica	1^{153}	1^{153}	1^{153}	1^{153}	$±^{174}$			
I Festuca erecta		1^{146}		1^{147}	$?^{174}$			
C Hierochloe redolens			1^{180}	$±^{147}$				
C Koeleria Sect. Dorsoaristatae			6^{148}	1^{147}				
I Poa cookii foliosa flabellata		1^{146}	$?^{90}$	1^{147}	1^{174}			
litorosa novareae								1^{78}
Carex trifida			1^{148}	1^{147}				
C Carpha			1^{148}			2^{169}	$+^{181}$	
Cladium Subgen. Machaerina			$+^{154}$				$+^{154}$	

III.

Southern Lands

This table has been compiled from the sources noted, and not (except in a few cases) from a personal knowledge of the plants in the field or the herbarium. It is impossible in tabular form to include the taxonomic discussion that many of the plants demand; and since it does not include taxa higher than genera, it omits some closely related genera found in different southern continents and islands (eg. *Diselma* and *Fitzroya; Hectorella* and *Lyallia; Lebentanthus* and *Prionotes; Metrosideros* and *Tepualia*).

Many of the taxa listed here are much in need of revision, and the floras and lists consulted are in many cases outdated: it is emphasised therefore that this table lacks taxonomic precision. It is included only to show in an easily comprehended form, the systematic variety of some of the seed plants in whose evolution and spread Antarctica may have played a part.

9	10	11	12	13	14	15	16	17	18	19	20
Marion & Prince Edward Is.	Crozet Islands	Kerguelen & Heard Island	AUSTRALIA (incl. Tasmania)	New Guinea, and the Indonesian Mountains	New Caledonia, Norfolk, and Lord Howe Islands	NEW ZEALAND	Macquarie Island	Antipodes, Auckland, Bounty, Campbell and Snares Islands.	Chatham Islands	Hawaiian Islands	Juan Fernandez Islands
			3[182]	+[158]	1[136]	1[4]					
		(+)[140]	2[182]	+[158]	+[97]	2[4]					
			3[158]		3[4]						
					1[26]						
		(+)[21]	1[37]	+[4]	+[4]	7[4]					
		(+)[26]	1[37]								
		(?)[21]	1[37]	+[65]		3[4]					
		(+)[140]	1[182]	2[158]	1[158]	7[4]					
1[70]	1[70]	1[27]	±[182]	±[65]		1[148]	1[164]	1[119]			±[154]
1[153]	1[70]	1[27]					±[164]	±[119]			
		1[27]					1[164]				
			1[180]	1[180]		1[180]					
			2[148]			3[148]					
1[78]	1[70]	1[27]									
							1[164]	1[119]			
								1[119]			
						1[148]	1[164]	1[19]			
						1[180]		1[19]			
			2[180]	1[158]							
			+[154]	+[154]	+[154]					+[154]	1[154]

Table

	1	2	3	4	5	6	7	8
	ANTARCTICA, South Orkney, S. Shetland & S. Sandwich Is.	South Georgia	SOUTH AMERICA	Falkland Islands	Tristan – Gough Islands	SOUTHERN AFRICA	Madagascar & Mascarene Is.	Amsterdam & St. Paul Islands
Oroebolus			1[154]	1[147]				2[78]
I Uncinia		1[120]	13[154]	1[147]	2[174]	9[184]		
C Hypolaena			+[148]			29[184]		
C Leptocarpus			1[148]			89[184]		
C Restio							+[181]	
C Gaimardia			1[148]	1[147]				
C Juncus planifolius			1[180]					
(pusillus								
I (scheuzerioides		1[146]		1[147]				
C Marsippospermum			1[148]	1[147]				
I Rostkovia		1[146]	+[155]	1[147]	1[174]			
Astelia			1[185]	1[147]			1[156]	
C Bulbine						c. 30[13]		
C Caesia						3[169]		
C Enargea			2[148]	1[147]				
Libertia			3[154]					
Myrica	(?)[51]							
C Nothofagus	(+)[30]		9[159]					
C Gevuina			1[16]					
C Knightia	(?)[51]							
C Lomatia	(?)[51]		3[148]					
C Oreocallis			2[65]					
C Orites			+[16]					
Mida								
C Muehlenbeckia			10[148]					
C Carpobrotus			+[13]			+[169]		
Tetragonia tetragonoides			1[154]			±[154]		
C Anacampseros						c. 50[13]		
I Colobanthus								
crassifolius	1[153]	1[146]	1[153]	1[153]				+[78]
C Caltha (Sect.			3[148]	2[147]				
Psychrophila)								
C Ranunculus acaulis			1[148]	1[147]				
(biternatus		1[146]	1[90]	1[147]	±[174]			1[78]
I (crassipes								
C Drimys	(?)[51]		3[154]					
C Laurelia	(?)[51]		1[148]					

III (continued)

9	10	11	12	13	14	15	16	17	18	19	20
Marion & Prince Edward Is.	Crozet Islands	Kerguelen & Heard Island	AUSTRALIA (incl. Tasmania)	New Guinea, and the Indonesian Mountains	New Caledonia, Norfolk, and Lord Howe Islands	NEW ZEALAND	Macquarie Island	Antipodes, Auckland, Bounty, Campbell and Snares Islands.	Chatham Islands	Hawaiian Islands	Juan Fernandez Islands
1^{70}		1^{27}	2^{180}	2^{154}		1^{154}		1^{19}		1^{150}	1^{154}
			4^{182}	2^{154}		14^{154}	1^{164}	1^{19}		1^{154}	5^{154}
			$+^{182}$			1^{20}					
			12^{182}	1^{65}		1^{20}			1^{20}		
			27^{184}								
			1^{180}	1^{65}		4^{20}		2^{116}			1^{154}
	1^{70}	1^{27}	1^{180}			1^{20}			1^{20}		
		1^{78}	1^{180}			1^{180}					
							1^{164}	1^{116}			
						1^{20}		1^{20}			
								1^{116}			
			2^{180}	1^{185}	1^{185}	9^{155}		2^{19}	1^{20}	6^{155}	
			2^{182}								
			6^{182}	1^{65}		1^{20}					
			$?^{182}$								
			2^{182}	1^{65}		2^{154}			1^{20}		1^{154}
			3^{159}	16^{159}	5^{159}	5^{4}					
			1^{182}	$+^{182}$		1^{4}					
					2^{4}						
			8^{182}								
			2^{182}	1^{65}							
			8^{182}								
			$\pm^{154}$	$\pm^{154}$		1^{4}				$\pm^{154}$	1^{4}
			10^{182}	1^{158}		5^{4}			1^{4}		
			$+^{182}$								
			1^{13}			1^{4}			$+^{4}$		1^{154}
			1^{182}								
	$+^{70}$	$+^{27}$	$\pm^{182}$			$\pm^{4}$	1^{164}	1^{116}	$\pm^{4}$		
			2^{182}			2^{4}					
			1^{37}			1^{4}		1^{4}	1^{4}		
1^{78}	1^{148}	1^{27}					1^{164}				
		1^{148}					1^{148}				
			6^{154}	29^{154}		$\pm^{4}$					1^{154}
						1^{4}					

Table

	1	2	3	4	5	6	7	8
	ANTARCTICA, South Orkney, S. Shetland & S. Sandwich Is.	South Georgia	SOUTH AMERICA	Falkland Islands	Tristan – Gough Islands	SOUTHERN AFRICA	Madagascar & Mascarene Is.	Amsterdam & St. Paul Islands
C Mollinedia	(?)[51]		+					
I Cardamine (corymbosa			1[4]					
(glacialis			1[148]	1[147]	1[174]			
C Drosera								
(Sect. Psychophila)			1[148]	1[147]				
I Tillaea moschata			1[148]	1[147]				
C Caldcluvia	(?)[51]		1[160]					
C Cunonia						1[64]		
C Weinmannia			+[160]				+[158]	
I Acaena		2[146]	c.100[154]	4[174]	2[147]	1[154]		1[78]
C Geum (Sect. Oncostylis)			2[150]					
I Sophora								
(Sect. Tetrapterae)			2[154]		1[174]		1[155]	
Geranium (microphyllum								
C (sessiliflorum			1[4]					
Pelargonium australe					1[174]			
Oxalis (lactea								
C (magellanica			1[148]	±[148]				
C Coriaria ("Lurida" agg.[4])			+[148]					
C Discaria			c. 11[154]					
C Aristotelia			2[148]					
C Eucryphia			3[154]					
C Drapetes			1[148]	1[147]				
Metrosideros						?[158]		
C Fuchsia	(?)[51]		c. 100[4]					
I Callitriche antarctica		1[146]	1[4]	1[147]	±[174]			
C Gunnera			15[154]	1[154]		1[154]	1[154]	
C Haloragis								
Myriophyllum elatinioides			1[148]	1[147]				
C Pseudopanax			2[148]					
I Apium australe			1[148]	1[147]	1[174]	1[18]		1[78]
I Azorella selago			1[90]	1[147]				
C Lilaeposis		?[70]	11[155]	1[147]				
C Oreomyrrhis			2[105]	1[105]				
C Schizeilema			2[148]	2[147]				
C Griselinia			4[148]					
C Pernettya			8[154]	1[147]				
C Tetrachondra			1[4]					

III (continued)

9	10	11	12	13	14	15	16	17	18	19	20
Marion & Prince Edward Is.	Crozet Islands	Kerguelen & Heard Island	AUSTRALIA (incl. Tasmania)	New Guinea, and the Indonesian Mountains	New Caledonia, Norfolk, and Lord Howe Islands	NEW ZEALAND	Macquarie Island	Antipodes, Auckland, Bounty, Campbell and Snares Islands.	Chatham Islands	Hawaiian Islands	Juan Fernandez Islands
						$?^4$	1^{164}	1^4			
						1^{148}		1^{148}			
1^{70}	1^{70}	1^{27}	1^{148}			2^{148}					
						1^4	1^{164}	1^4	1^4		
			$+^{160}$	$+^{160}$		$+^{160}$					
					12^{64}						
			$\pm^{182}$	$+^{65}$	2^{136}	2^4					
1^{78}	1^{78}	1^{27}	3^{154}	1^{154}		13^4	2^{164}	1^4		1^{154}	2^{154}
			1^{37}			$+^4$		$+^4$			
					1^{155}	3^{155}			1^{155}	1^{154}	2^{154}
			1^{37}	1^{158}		1^4		1^4			
			1^{37}			1^4					
			1^{37}			1^4			1^4		
			1^4			1^4					
				1^{158}							
				$+^{158}$		$+^4$			$?^{158}$		
			1^{37}			1^4			1^4		
			3^{148}			2^4					
			3^{182}								
			1^{148}	$+^{65}$		5^4					
			2^{182}	8^{158}		10^4		1^4		$+^{80}$	
						4^4		1^4			
1^{70}	1^{70}	1^{70}	1^{154}	2^{158}		10^4	1^{164}	1^{19}			
			59^{154}	8^{65}	$+^{158}$	7^4		1^4	1^4	2^{154}	3^{154}
			1^{180}			1^4	1^{164}		1^4		3^{154}
						8^4			1^4		
1^{78}	1^{70}	1^{27}	1^{13}		1^4	1^4	1^{164}	1^4	1^4		$\pm^{154}$
			3^{182}			3^4			1^4		
			7^{105}	6^{105}		3^{105}			1^4		
			1^4			10^4		1^4			
						2^4					
			2^{182}			2^4					1^{154}
						1^4					

Table

	1	2	3	4	5	6	7	8
	ANTARCTICA, South Orkney, S. Shetland & S. Sandwich Is.	South Georgia	SOUTH AMERICA	Falkland Islands	Tristan – Gough Islands	SOUTHERN AFRICA	Madagascar & Mascarene Is.	Amsterdam & St. Paul Islands
C Gratiola peruviana			1[148]					
Hebe			2[4]	1[147]				
C Jovanella			2[148]	?[147]				
C Ourisia			c. 12[4]					
I Coprosma pumila								
I Galium antarcticum		1[146]						
I Nertera depressa			1[148]	1[147]	1[174]			
C Hypsela			+[4]					
C Lobelia anceps s.l.			1[4]			1[4]	1[181]	
C Pratia (angulata				1[147]				
(repens								
C Selliera			+[4]					
C Donatia			1[4]					
C Forstera			1[160]					
C Phyllachne			1[148]					
Abrotanella			3[154]	2[147]			1[154]	
I Cotula								
subgen. Leptinella			+[148]	1[147]	2[174]			
I Lagenophora			3[154]	1[147]	1[174]			
C Microseris			+[16]					

Some genera which probably contain southern discontinuous taxa have been omitted from this table, due to taxonomic or other difficulties. These are:

Carex (Section Echinochlaenae) – see SKOTTSBERG, 1915.

Euphrasia – see DU RIETZ, 1940

Gentianella – see ALLAN, 1961 (p. 779)

Hydrocotyle – see ALLAN, 1961 (p. 448)

Myosotis (antarctica, and albiflora/rakiura) – see SKOTTSBERG, 1915

Samolus (repens) – see SKOTTSBERG, 1915

Senecio (Section Erechites) – see VAN STEENIS, 1934: and ALLAN, 1961.

Taraxacum (magellanicum) – see ALLAN, 1961 (p. 763)

Wahlenbergia (gracilis) – see ALLAN, 1961 (p. 788).

III (continued)

9	10	11	12	13	14	15	16	17	18	19	20
Marion & Prince Edward Is.	Crozet Islands	Kerguelen & Heard Island	AUSTRALIA (incl. Tasmania)	New Guinea, and the Indonesian Mountains	New Caledonia, Norfolk, and Lord Howe Islands	NEW ZEALAND	Macquarie Island	Antipodes, Auckland, Bounty, Campbell and Snares Islands.	Chatham Islands	Hawaiian Islands	Juan Fernandez Islands
			1[148]			1[4]		3[4]	4[4]		
			6[182]	2[158]		76[4]					
						2[4]					
			1[182]			10[4]					
	1[70]	1[70] / 1[27]	1[129]	±[158]		1[4]	1[164]	1[4]	±[4]	±[80]	±[154]
			1[4]	?[65]		1[4]		1[4]	1[154]	1[154]	1[154]
			2[182]			1[4]					
			1[4]	±[65]	1*	1[4]			1[4]		1[154]
			1[158]			1[4]					
			2[182]			1[4]					
			1[182]			1[4]					
			1[182]			4[4]		1[4]			
			1[36]			3[4]		2[4]			
			3[182]	1[154]		6[4]					1[154]
1[70]	1[70]	1[27]	+[4]			16[4]	1[164]	2[4]	3[4]		1[154]
			2[154]	+[65]		5[4]		2[4]	1[4]	3[154]	1[154]
			1[182]			1[4]					

* R. D. HOOGLAND, personal communication

islands, and are therefore likely to be recent immigrants (TAYLOR, 1955a; WACE & DICKSON, 1964).

In contrast to these "insular" plants, such families as Centrolepidaceae, Proteaceae and Restionaceae, and such genera as *Eucryphia*, *Fuchsia* and *Nothofagus* seem to be entirely continental in their present ranges, even though many islands appear to offer them suitable habitats. The absence of *Nothofagus* from all the islands on the New Zealand shelf, and from the Falklands*; or the absence of all Myrtaceae, Centrolepidaceae and Proteaceae from the subantarctic islands of the Indian Ocean; or the absence of the cushion-forming feldmark and montane bog species of *Abrotanella*, *Donatia*, *Gaimardia* and *Phyllachne* from the Tristan, Gough, Marion, Amsterdam and Kerguelen islands — these absences suggest that the plants concerned are incapable of crossing ocean barriers today. Apart from the lesser variety of habitats and the possible lack of open communities in which new arrivals could establish themselves on the islands, the reasons for the absence of these southern continental plants from the remote islands may be partly historical — that the islands in question have only very recently recovered from active vulcanism or glaciation or severe peri-glacial effects. The widespread southern continental plants may not have had sufficient time to colonise them. But the absence of any pollen of *Nothofagus* from the peats and lignites in the older islands such as Kerguelen and Macquarie (both of which contain pre-glacial plant remains), does suggest that *Nothofagus* at least has never inhabited them. Arguments such as these based on negative evidence (especially where the number of deposits investigated is so small) can not be conclusive, but in the case of such an abundant pollen producer as *Nothofagus* (LICITIS, 1953), negative evidence can be more than suggestive.

If the comparatively short oversea distances between the New Zealand mainland and the islands on the New Zealand continental shelf, or between West Patagonia and the Falkland Islands are sufficient to deter *Nothofagus* from spreading to the islands today, have its means of dispersal been more efficient in the past than they are now, or must we look for more continuous land formerly connecting its now widely disjunct continental areas of occurrence? The fact that it has carried its fungal parasite *Cyttaria* (which according to SANTESSON (1945) is not found on any other genera) and certain invertebrates associated with its moss epiphytes, to both South America and New Zealand, has been advanced also as evidence that *Nothofagus* increases its range only by marginal extensions, and not by long range oversea dispersal (RODWAY, 1914;

---

* VAN STEENIS (1953, p. 334) in his subdivision of *Nothofagus* and key of the species, notes the genus as present in the Falkland and South Pacific subantarctic islands. SKOTTSBERG'S (1913) and ALLAN'S (1961) floras do not substantiate this.

Table IV.

Geographical Elements in the Native Angiosperm Floras of some Southern Islands

Native species in the island floras are listed according to the areas in which their closest relatives are found. Taxonomic uncertainties make the positions of many of the entries uncertain, especially as between the "circumpolar" and "southern America" categories. Names used are those accepted by WACE & DICKSON (1964) for the Tristan-Gough group, and GREENE & GREENE (1963) for the subantarctic islands.

	South Georgia	Tristan-Gough Group	Kerguelen	Macquarie Island
Circumpolar subantarctic or southern cool temperate	Ranunculus biternatus Colobanthus crassifolius Acaena adscendens A. tenera Callitriche antarctica Galium antarcticum Juncus scheuzerioides Uncinia smithii Deschampsia elegantula Poa flabellata Festuca erecta	Ranunculus caroli Cardamine glacialis Sophora macnabiana Acaena sarmentosa A. stangii Callitriche christensnii Apium australe Nertera depressa Cotula goughensis C. moseleyi Lagenophora nudicaulis Uncinia brevicaulis U. compacta Agrostis ? magellanica Deschampsia ? kingii ? Festuca erecta Poa flabellata	Ranunculus biternatus R. trullifolius Colobanthus kerguelensis Acaena adscendens Callitriche antarctica Tillaea moschata Azorella selago Galium antarcticum Cotula plumosa Juncus pusillus Uncinia compacta Agrostis magellanica Deschampsia elegantula Poa cookii Festuca erecta	Ranunculus biternatus Cardamine corymbosa Colobanthus crassifolius C. muscoides Acaena adscendens A. anserifolia Tillaea moschata Callitriche antarctica Azorella selago Cotula plumosa Juncus scheuzerioides Cares trifida Uncinia riparia Agrostis magellanica Poa foliosa Festuca erecta
Southern America	Colobanthus subulatus Rostkovia magellanica	Hydrocotyle capitata Empetrum rubrum Chenopodium tomentosum Rumex frutescens Rostkovia tristanensis Tetroncium magellanicum Spartina arundinacea		

232

Table IV. (continued)

	South Georgia	Tristan-Gough Group	Kerguelen	Macquarie Island
Southern Africa		Pelargonium grossularioides Phylica arborea		
Australasian			Lyallia kerguelensis Coprosma pumila	Stellaria decipiens Epilobium linnaeoides E. nerteroides Stilbocarpa polaris Coprosma pumila Pleurophyllum hookerii
Other areas, or Unknown	Montia fontana Juncus inconspicuosus Phleum alpinum Alopecurus antarcticus	Nertera assurgens N. holmboei Gnaphalium pyramidale Peperomia tristanensis Carex insularis C. thouarsii Scirpus sulcatus S. thouarsianus Agrostis media A. carmichaelii A. nr. bergiana Glyceria sp. Polypogon sp.?	Ranunculus moseleyi Pringlea antiscorbutica Montia fontana Limosella lineate Poa kerguelensis	Montia fontana Myriophyllum elatinioides Hydrocotyle sp. Luzula campestris Scirpus aucklandicus Deschampsia chapmanii D. penicilliata Puccinellia macquariensis Poa hamiltonii
Sources of Information	Skottsberg, 1912 Greene & Greene, 1963	Wace & Dickson, 1964	Cour, 1958 Greene & Greene, 1963	Taylor, 1955a Greene & Greene, 1963

GORDON, 1949; DENNIS, 1960). Studies of living *Nothofagus* communities in New Zealand (HOLLOWAY, 1954) also support this idea of extension only by slow marginal spread. If therefore the genus has never been present in northern lands, and its capacities for dispersal have remained substantially unaltered, it must have spread to its present widely discontinuous areas of occurrence over land that formerly connected the southern continents, and Antarctica where its fossils are found. Arguments have already been mentioned to show that the present and fossil distribution of most of the southern hemisphere conifers demands some land connections in the south, probably involving the Antarctic continent. Similar arguments from their present distributions, and their failure to spread over short sea gaps could be presented for many of the "continental" plants listed in Table III.

Apart from these differences in distribution between the "continental" and "insular" species, and the inferred differences in their present-day dispersal capacities, many of the "insular" plants have obvious adaptations to aid dispersal. Examples are seen in the hooks attached to the fruit in *Uncinia*, the barbed spines on the achenes of *Acaena*, the adhesive cypselas in *Lagenophora*, the very thin and light cypselas that readily stick to wet surfaces in *Cotula (Leptinella)*, and the seeds that float unharmed for long periods in sea water in *Sophora* (WACE & DICKSON, 1964). Although some of the "continental" plants have adaptations to internal or external animal dispersal, such adaptations are not so apparent in the "continental" plants as a whole, and certain taxa (eg. gymnosperms, Proteaceae, *Nothofagus*) lack any obvious aids to very long range seed dispersal.

DAWSON (1958) has noted that the floristic elements in common to Australasia and South America which he regarded as arising in a common pre-glacial flora, and which are here regarded as "continental", show their relationships mainly at the generic level or above. Different species are generally found on the separate continents. The "insular" plants, however, consist of very closely related or single circumpolar species in the disjunct parts of their ranges. The greater morphological divergences found between the "continental" plants suggest that they have been isolated from one another for longer than the "insular" species, or possibly that divergence in morphology had already taken place in the Antarctic before they spread north to their present sites (SKOTTSBERG, 1960a).

Floristic and Faunistic Relationships between southern lands

The present distribution of those elements in the southern hemisphere floras in whose history Antarctica may have played a part — SKOTTSBERG's (1960a) "Antarcto-Tertiary" plants — is shown in

Fig. 10. The vascular floras of SKOTTSBERG'S Antarctic and subantarctic provinces are very largely composed of "insular" species,
but those of his Austral province and the outliers further north are

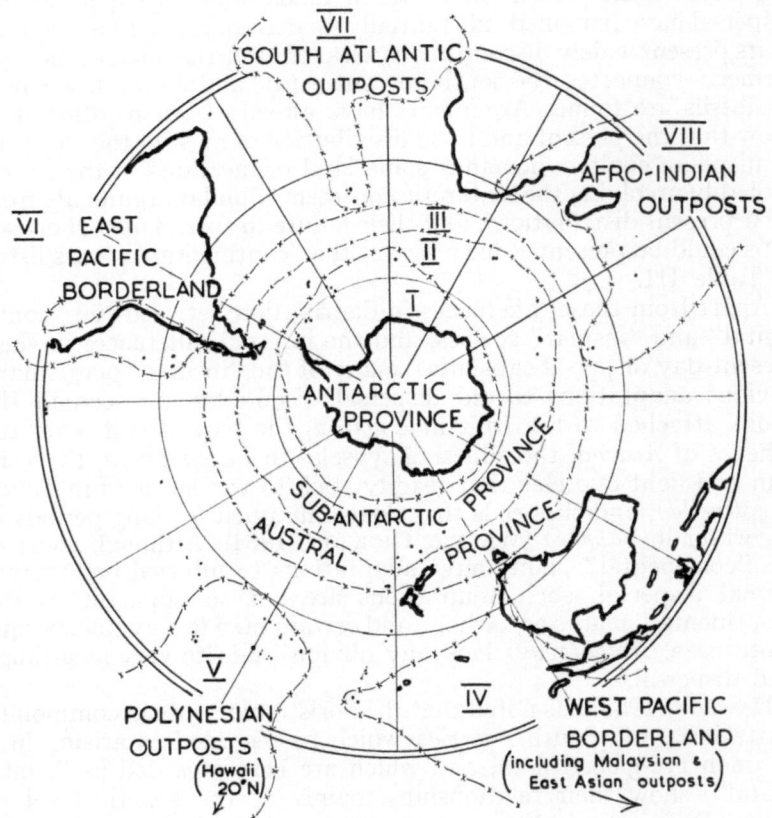

Fig. 10. The floristic regions of the "Antarcto-Tertiary" plants, according to
SKOTTSBERG (1960a).

mainly "continental" and they form only a small part of the total
vascular floras of the areas concerned. This map expresses what
HEMSLEY (1885, p. 51) remarked about the floristic relationships of
southern lands:

"It is quite clear that in the southern, as in the northern hemisphere, the only
admissable demarcation of the coldest floral region is a zonal one. Proceeding northward in the three great land masses, the differences in the vegetation (meaning the
floristic composition) soon become so pronounced that it is convenient to treat them
as distinct floral regions; yet apart from the Antarctic types, the relationships of the
distribution of other peculiar southern types is highly interesting, and seems to
point to a migration northward, and a former greater land connection in the southern hemisphere".

The floristic relations between southern Africa and the other southern continents are more distant than those between Australasia and South America. Not only are fewer taxa involved, but the relationships amongst those that are confined to the southern hemisphere are generally expressed only at the family level (eg. Cunoniaceae, Philesaiceae, Proteaceae, Restionaceae), although a few genera such as *Carpobrotus* and *Leptocarpus* are apparently native in all three southern continental areas (GOOD, 1953). The more distant relations of the African to the other southern continental floras may perhaps be taken to indicate a connection between them at a remoter period, in which Antarctica may have been involved.

Three levels of genetic relationship can therefore be recognised amongst the southern floras:

a. a distant relationship between a few of the continental plants of southern Africa, to those of the other southern continents;

b. a closer relationship between a larger proportion of the continental floras of temperate Australasia and South America;

c. a very close relationship between a small group of circum-Antarctic plants in both the continents and the remote islands of the Southern Ocean.

It is tempting to speculate that these relationships may have arisen in pre-Tertiary, Tertiary and Quaternary times respectively; and in the case of (a) and (b) possibly by way of Antarctica. It has been emphasised by many authors (eg. HOOKER, 1853; GOOD, 1933) that the floristic elements in common to the southern continents represent only a small — but not an insignificant — proportion of their total angiosperm floras. Some estimates of the size of this possibly southern-derived group in the floras of various southern lands are shown in Table V. Even if small in numbers of genera or species in comparison to their total indigenous floras, these can hardly be dismissed simply as a collection of "waifs and strays" (COCKAYNE, 1928 p. 413), whose range is attributable to the vagaries of long distance dispersal over "variable sweepstakes routes" (SIMPSON, 1953). The ecological preferences of its members suggest that the "Antarctic" plants are "part of a floristic element that has been maintained as a unit" (MATHIAS & CONSTANCE, 1955). Its existence continues to pose one of the major problems of historical phytogeography.

It is of interest to see whether elements similar to those of the "Antarctic" plants can be recognised amongst terrestrial animals; and if so whether their range can be most easily accounted for solely by oversea dispersal between the southern lands. KUSCHEL (1960) listed examples of various arthropod genera and families with southern disjunct ranges, and commented that most of the phytophagous insects in this group fed on ferns or gymnosperms, and that

236

Table V.

"Antarctic" Elements in Southern Continental Floras.

Estimates of the size of possibly southern-derived elements in the floras of various southern lands have been made by several authors. The estimates here quoted are based on numbers of genera, and therefore take no account of prolific speciation in some southern genera (eg. *Hebe* and *Coprosma* in New Zealand).

Region	A Approximate size of native Angiosperm flora	B "Antarctic", or southern-derived	B/A %	Sources
Temperate South America	1478	53	3½	GOOD, 1933
Temperate Southern Africa	1492	c. 20 ?	1	GOOD, 1933 & 1953
New Guinea	1094	34	3	VAN STEENIS, 1934
	1350	24	2	GOOD, 1960
Australia (including Tasmania)	1700	c. 35 ?	2	GOOD, 1953; BURBIDGE, 1960
Tasmania alone	401	c. 35 ?	9	BURBIDGE, 1960 & Table III
New Zealand	290	c. 50	17	COCKAYNE, 1928; ALLAN, 1961

"the nature of the vast majority of animals distributed over widely separated areas is such that transport either active or passive across the now existing geographical barriers under present climatic conditions would be extremely improbable".

HOLDGATE (1960b) recognised a southern circumpolar group of invertebrates in discussing the origins of the faunas of the South Atlantic and South Indian Ocean islands. Many of the groups mentioned as southern disjuncts by KUSCHEL (1960) seem to be absent from the islands, and it is possible that "insular" and "continental" categories could be recognised amongst the southern disjunct invertebrates, as with the vascular plants. Cf. EVANS 1959 (Ref. 196).

In contrast, DARLINGTON (1957, 1960) has pointed out that no terrestrial vertebrates have southern cold temperate zonal distributions similar to those of the "Antarctic" plants and invertebrates. The present distributions of such predominantly southern groups as marsupials, ratite birds, leptodactylid and hylid frogs, and galaxiad fish, can all be accounted for by spread from northern lands where they have subsequently become extinct, or by trans-oceanic dispersal, or by parallel evolution in the separate continents.

SIMPSON (1940) has also argued cogently against the acceptance of Antarctica as a faunal migration route, at least for the vertebrates. The reasons for these differences in distribution pattern between the plants and invertebrates on the one hand, and the vertebrates on the other, remain obscure. DARLINGTON (1960) states that the differences could be due to an earlier spread, or easier dispersal on the part of the plants and the invertebrates; but he thinks that the reasons are more likely to be connected with the general impoverishment of both floras and faunas at the southern extremities of Australasia and South America, and to the differing abilities of plants and animals to survive climatic changes under conditions of isolation through the Tertiary, and in later competition with species invading from the north. The discovery of any remains of terrestrial vertebrates on the Antarctic continent, which would be an event of outstanding biogeographic importance, might also help to resolve this apparent difference in present distribution patterns between the southern plants and animals.

Antarctic and Circum-Antarctic Vegetation

Few accounts have been published of terrestrial vegetation on the Antarctic continent, but more or less detailed vegetational descriptions exist for many of the remote subantarctic and cold temperate islands in the Southern Ocean (references in WACE, 1960), and for parts of the southern cold temperate areas on the continents (see especially REICHE, 1907; COCKAYNE, 1928; GODLEY, 1960; GIBBS, 1920; SUTTON, 1928; COSTIN, 1954 & 1959; and the references listed in the more recent papers). Most workers on the southern floras have

238

been concerned mainly with the taxonomic affinities between the plants of southern lands, but in addition to these genetic bonds between them, structurally similar types of plant community can also be recognise in the widely separated land masses of the southern hemisphere surrounding the Antarctic continent. It is important in the study of Antarctic ecology and biogeography to arrive at a classification of the circum-Antarctic terrestrial vegetation types independently of the supposed origins of the constituent plants. Such a classification must be based on the structure of the vegetation, and the growth forms of the plants comprising it.

Many writers have agreed upon the presence of three major zones — of climate, of surface waters, and of marine life, around the Antarctic continent south of about 30°S. latitude. These have generally been called Antarctic, subantarctic, and subtropical, although DEACON (1960) has suggested that a temperate zone might be recognised between the last two. The life zones of those organisms dependent entirely upon the sea for their existence seem more-or-less to follow these hydrological zones bounded by the Antarctic and subtropical convergences (MURPHY, 1936; KNOX, 1960; MACKINTOSH, 1960), but the limits of the major types of terrestrial vegetation do not coincide with them so closely.

SKOTTSBERG (1904) defined the zones of terrestrial vegetation, when he recognised "Antarctic" and "Austral" zones in Fuegia, the Scotia Arc islands, and the Antarctic Peninsula (Table VI). Subsequently, SKOTTSBERG (1905) discussed the earlier attempts of J. D. HOOKER, DE CANDOLLE, ENGLER, GRISEBACH, DRUDE, and others to delimit the zones of terrestrial vegetation around the Antarctic, and himself recognised four "dominions". These domini-

Table VI.

The Vegetation zones of the American Quadrant of Antarctic Regions, (according to SKOTTSBERG, 1904)

Category	Major Characteristics	Places
Austral	1. Closed forest or grassland vegetation 2. Littoral or sublittoral algae with large floating fronds (mostly *Durvillea* and *Macrocystis* spp.)	Tierra del Fuego Staten Island Falkland Islands South Georgia South Sandwich Islands
Antarctic	1. "Cold Desert" with scattered phanerograms only 2. No coastal algae with large floating fronds	South Orkney Islands South Shetland Islands Antarctic Peninsula and its offlying islands.

239

ons, and his later Antarctic, subantarctic and Austral regions in southern lands (SKOTTSBERG, 1953 & 1960a) were defined largely on geographic and floristic criteria. Thus, he defined his Austral zone "south of approximately 40°" as having "a conspicuous antarctic element . . . often forming distinct communities in a flora and vegetation of different origin" (SKOTTSBERG, 1960a p. 449). In order to arrive at a vegetational classification, the major features of the structure of the plant communities should be used exclusively to delimit the different zones. It is here proposed to use the limit of tree or woody shrub growth to separate the temperate from the subantarctic, and the limit of closed phanerogamic communities to separate the subantarctic from the Antarctic vegetation. This classification, which extends that used by the author (1960) in discussing the vegetation of the southern oceanic islands, attempts to be independent of climatic, geographic, or other environmental

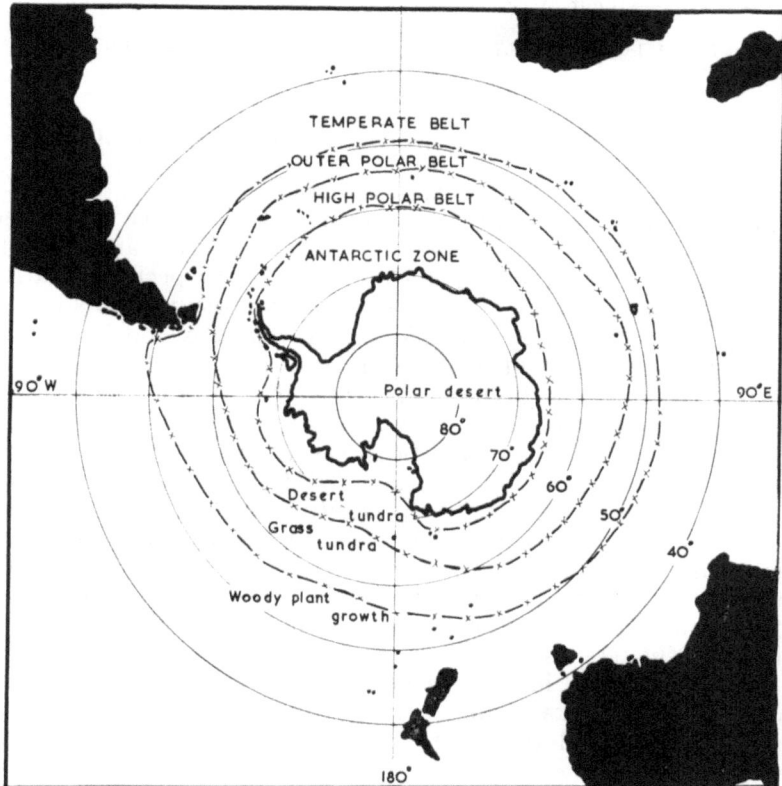

Fig. 11. The Natural Provinces of the Antarctic regions, according to NORDEN-SKJÖLD (1928, pp. 61–65 & 75).

240

data, or of the supposed floristic origins of the plants involved. It
follows, with some divergences, the zonation proposed by NORDEN-
SKJÖLD (1928), but with a different nomenclature. The equivalent
areas are shown in Table VII and Fig. 11 and 12.

Table VII.

The Nomenclature and Distinguishing Characteristics of the zones of terrestrial
vegetation in Southern Latitudes, according to NORDENSKJÖLD and the Author.

NORDENSKJÖLD, 1928	WACE, 1960 and 1964
TEMPERATE	TEMPERATE
——————————— limit of tree or woody shrub growth ———————————	
OUTER POLAR BELT	SUBANTARCTIC
——————————— limit of closed phanerogamic communities ———————————	
HIGH POLAR BELT	LOW ANTARCTIC
——————————— limit of closed cryptogamic communities ———————————	
ANTARCTIC	HIGH ANTARCTIC

Antarctic Vegetation

The present land vegetation of continental Antarctica is confined
to a few peripheral and mountainous ice-free areas, and consists
almost entirely of lower cryptogams. Only two species of native
vascular plants are known to grow on the continent and its immediately adjoining islands — *Colobanthus crassifolius* and *Deschampsia antarctica* (SKOTTSBERG 1954, LLANO 1956 & 1962)*. On the
continent, these are both confined to the Antarctic Peninsula
(BERTRAM, in FLEMING et al. 1938; LLANO 1962), and probably do
not extend far inside the Antarctic Circle. Their southernmost recorded locality is near Stonington Island (68°12′S.) on the western side
of the Peninsula (BRYANT, 1945). *Deschampsia antarctica* has been
noted covering patches "as much as a yard or two square" on Signy
Island in the South Orkney group (SKOTTSBERG, 1954), but the
flowering plants on the Antarctic mainland probably nowhere form
closed communities, and are found only in rock crevices and other
relatively sheltered situations.

Several observers have remarked upon the size and depth of the
mats and hummocks of moss in the Antarctic Peninsula and its
nearby islands, and the Antarctic vascular plants are frequently

* LLANO (1962) recognises two species of *Deschampsia* in the Antarctic in addition to
Colobanthus; but SKOTTSBERG (1954) recognises only *Deschampsia antarctica* and
Colobanthus crassifolius as native to the continent and its offshore islands.

found in association with such communities. BERTRAM (in FLEMING et al. 1938) described patches of moss up to three feet deep and nearly an acre in extent in the Argentine Islands (65 °S.). HOLDGATE (1962b) notes moss banks up to 175 cm deep on the sheltered northerly slopes of Signy Island (61 °S.), the surface layers of which are quickly warmed by the sun and may reach a temperature of more than 10°C when the air temperature is near freezing point. Both authors note that the moss mats are permanently frozen beneath the surface, even in summer. SKOTTSBERG (1905) contrasts the luxuriance of the moss vegetation on the west of the Antarctic Peninsula with its sparsity in the same latitudes on the east coast. HOLDGATE (1963) mentions luxuriant moss growth on ground warmed by volcanic heat in the South Sandwich Islands (56°— 60 °S.), where *Deschampsia antarctica* also occurs. RUDOLPH (1963) described the terrestrial vegetation at Hallet Station (c.72 °S.) in the Ross Sea area of Greater Antarctica, where no vascular plants are present, and algae *(Prasiola crispa)*, various lichens and a moss *(Bryum argentium)* form open communities with about 15% ground cover. A rise in air temperatures immediately above the mosses is noted, similar to that recorded from the Antarctic Peninsula.

The extreme paucity of the present phanerogamic vegetation in the Antarctic, as compared to that of similar northern latitudes was considered by RUDMOSE BROWN (1928) to be due to the shortness of the Antarctic summer, and to the low temperatures during the few months of the year when the ground is free of snow. Higher plants thus have to contend with a very short growing season, and may be unable to complete their life cycles. Both *Colobanthus* and *Deschampsia* are perennial species, and were said by RUDMOSE BROWN (1928) to reproduce only vegetatively. SKOTTSBERG (1954 and 1956 p. 337 footnote) maintains that both species flower and set seed in the Antarctic Peninsula; *Deschampsia antarctica* having cleistogamous flowers. Other factors preventing the growth or establishment of higher plants on Antarctic coasts were said to include competition from penguins on possible sites that might otherwise be suitable for occupation, and the very cold dry winds from the interior of the Continent (SKOTTSBERG, 1904; RUDMOSE BROWN, 1928). Ice free areas near the coasts of Greater Antarctica also lack vascular plants, and have only a very meagre bryophyte cover, probably due to the excessively dry conditions there (AVSUYK et al., 1956).

In addition to these environmental influences limiting plant growth in the Antarctic, it is probably that one of the most important factors contributing to the poverty of its flora (and hence the scantiness of its vegetation), is its present isolation from other lands. No other continent is so isolated. The investment of the Antarctic by both eastward and westward flowing winds and ocean currents

242

(MECKING, 1928), and the latitudinal zonation of most of the birds of Antarctic regions (DELL, 1960; MURPHY, 1960) would make the natural spread of seed or other plant disseminules to the continent very difficult under present conditions. It will be of interest to see whether the breakdown of this isolation by man, with the establishment on the continent of a considerable number of permanently inhabited bases during and since the International Geophysical Year (1957—58), leads in turn to establishment of any alien plants in the Antarctic. Only three species are known so far as weeds there: a single plant of *Poa annua* found in 1953 on Deception Island (SKOTTSBERG, 1954), and *Poa pratensis* and *Stellaria media* recorded from the Antarctic Peninsula and Signy Island respectively (GREENE & GREENE, 1963).

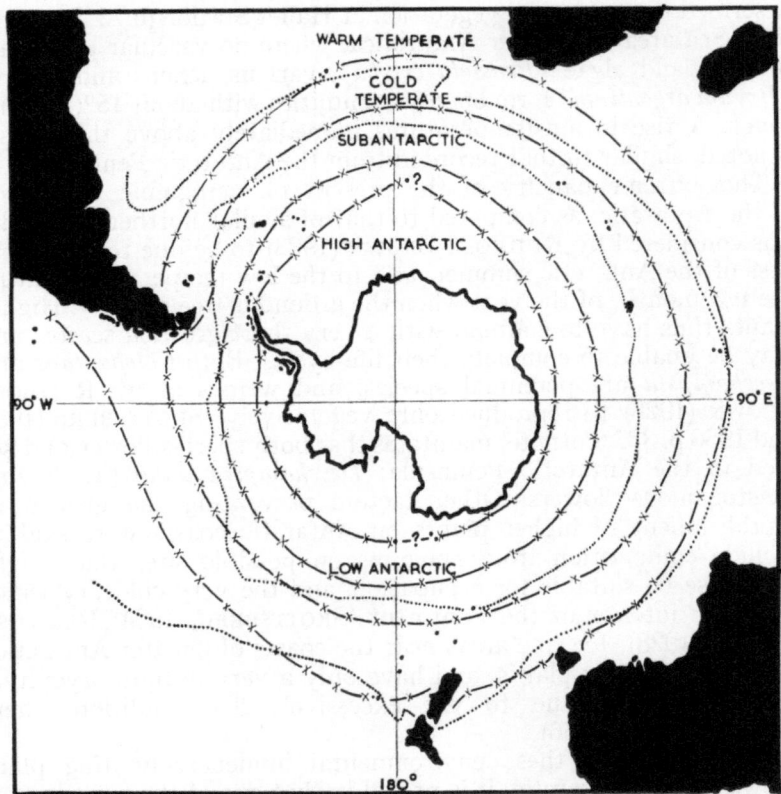

Fig. 12. The terrestrial vegetation zones of Antarctic regions, according to the author. The dotted lines indicate the approximate positions of the Antarctic and subtropical convergences. Accounts of vegetation (or lack of it) on Bouvet, Scott and Balleney Islands are too vague to enable their placing in the Low or High Antarctic zones to be made with certainty.

Dr. MARTIN HOLDGATE has suggested *(in litt.)* that it may be possible to recognise two distinct vegetation provinces within the Antarctic zone: a High Antarctic province in which there are no closed plant communities, and in which lichens, algae, and small tufted mosses are the sole land vegetation; and a Low Antarctic province in which extensive and deep moss mats occur in favourable sites, but with only scattered vascular plants. The High Antarctic province would include the whole of continental Antarctica south of the Antarctic Circle, together with its offlying islands and probably the more remote Peter I, Scott and Balleny Islands also. It is the area of extremely cold continental climate, with easterly winds on the coasts and much sea ice throughout the year. The Low Antarctic province includes parts of the west coast of the Antarctic Peninsula to the north of the Antarctic Circle together with the offlying islands, and the South Shetland, South Orkney and South Sandwich archipelagoes, and possibly Bouvet Island also (Fig. 12). It is the area subjected to some maritime influences, on the fringe of the circum-Antarctic low pressure belt.

Subantarctic vegetation

The term "subantarctic" has been used by oceanographers to delimit the zone of waters between the Antarctic and subtropical convergences (DEACON, 1960); but it has also been used in purely geographic (MECKING, 1928), and in floristic senses (SKOTTSBERG, 1905 & 1953). GODLEY (1960) has defined the region partly on climatic and partly on plant physiognomic criteria. Following SKOTTSBERG (1905), RUDMOSE BROWN (1928) observed that

"the term 'sub-Antarctic' is justified rather by proximity to the Antarctic than by any real approximation to Antarctic conditions. The true Antarctic climate is typically continental in contrast to the climate of the sub-Antarctic islands, which is essentially oceanic, and in most respects cool temperate rather than polar."

In view of these different uses, and the fact that the zone is of differing extent according to the distinguishing criteria used, it may be better eventually to employ another term (? antiboreal) to distinguish those lands around the Antarctic which have an indigenous terrestrial vegetation with closed communities of vascular plants, but without shrubs or trees and without ombrogenous bogs. As understood here, the subantarctic vegetation zone is characterised by terrestrial plant communities with the following characteristics:

1. Soligenous mires in which the important peat-forming plants are Bryales and Juncaceae (not sphagna);
2. Feldmark communities composed of flowering plants with very close-growing and compact mat and cushion forms;

3. Closed herbfield communities in which large-leaved perennial herbs are conspicuous;

4. Communities of large trunk-forming tussock grasses around the coasts.

Some or all of these types of vegetation are found in New Zealand and South America, as well as in the Falkland Islands (SKOTTSBERG, 1913) and the islands on the New Zealand continental shelf (CHILTON, 1909; COCKAYNE, 1928). These areas are here considered as temperate, however, due to the presence on them of forest or shrubs, or of bogs formed by phanerogamic cushion plants or sphagna. As defined here, the subantarctic zone includes no continental areas, and consists only of islands which are themselves remote from the continents (Fig. 1). It is a zone of cool, wet hyper-oceanic climates.

Peat forming communities are present on all the subantarctic islands, except possibly Heard Island. The mires are probably all soligenous "paludification bogs" or else purely topogenous (CRANWELL, 1953). It is doubtful whether truly ombrogenous convex-surfaced peat deposits are formed either on open slopes or in valleys in the subantarctic region. The dominant plants of the peat-forming communities are Bryales (especially *Breutelia)* and Hepaticae, usually associated with Juncaceae which may themselves be important in peat formation. SKOTTSBERG (1912) notes *Rostkovia magellanica* as a dominant species in the South Georgia bogs; COUR (1959) lists *Juncus pusillus* as dominant and exclusive to the bog communities on Kerguelen; and TAYLOR (1959) notes that *Juncus scheuzerioides* is common in bogs on Macquarie Island. The *Distichia* bogs of the high Peruvian Andes (WEBERBAUER, 1911) may also resemble the subantarctic communities in structure. Sphagna are either absent or of only minor importance in the subantarctic bog communities. The genus, not recorded by STEERE (1961) has recently been reported from S. Georgia by GREEN, in CARRICK (Ref. 195), and on Macquarie Island, *Sphagnum* occurs only in small patches in the bogs, and is not common (TAYLOR, 1955a). Discussing the mossbogs of Kerguelen, COUR (1959) remarks how few phanerogams are present in them, and comments:

"De nombreuses mousses et hépatiques (surtout *Marchantia polymorpha)* sont également visibles. Il n'existe pas tourbières à sphaignes, le genre *Sphagnum* n'était pas représenté dans l'archipel".

It is not known whether any sphagna are present in the Marion or Crozet Island groups, but the genus is not recorded by HEMSLEY (1885) or WERTH (1928), so that *Sphagnum* can scarcely be a prominent feature of their vegetation. The distribution of the southern bog types is further discussed below, and mapped in Fig. 13.

The importance of large cushion-forming flowering plants has long been considered a characteristic feature of subantarctic vegetation (WARMING, 1909; SCHROTER & HANRI, 1914). Large hummocks of

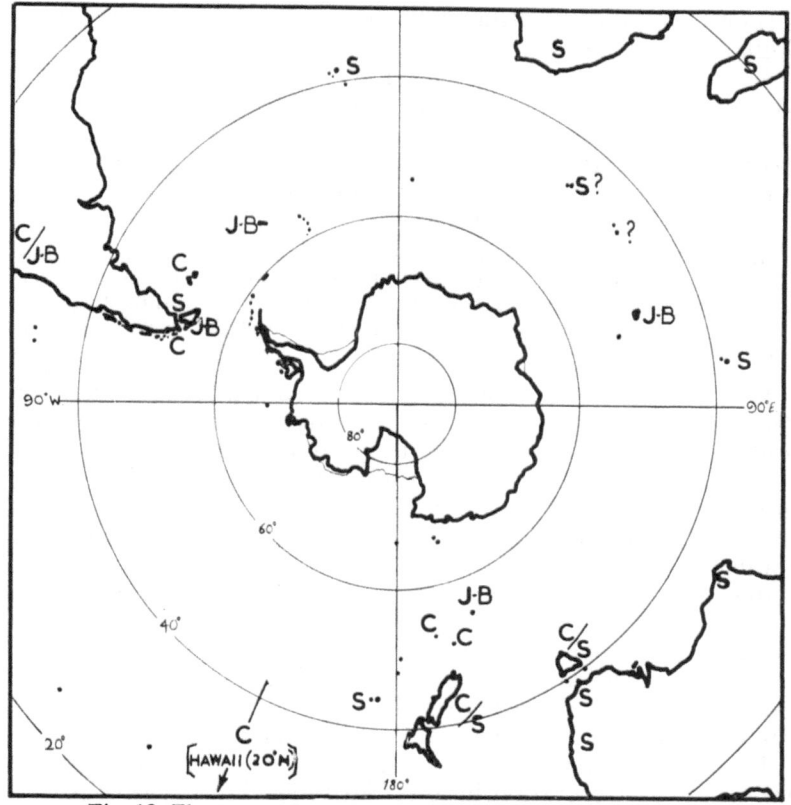

Fig. 13. The types of ombrogenous mires in southern regions.
J–B: Juncaceae & Bryales dominated bogs
C   : Flowering plant cushion bogs
S   : *Sphagnum* bogs.
For discussion, see text. Data from CRANWELL (1953), COCKAYNE (1928), TAYLOR, (1955a), GODLEY (1960), WACE (1960), COUR (1959), MILLINGTON (1954) and COSTIN (1957).

Azorella selago are the chief components of the "desert" and "tundra" formations on Kerguelen (WERTH, 1928), and appear to form pure communities in places, with the hummocks coalescing to make an undulating carpet. Although such closed communities approach the cushion bogs in structure, which are discussed later, it seems that most of the communities dominated by flowering plant cushions in the subantarctic are open, and do not form a continuous layer of peat and are therefore best included under the heading of feldmark. *Azorella selago* also dominates feldmark communities at altitudes over 200 m on Macquaire Island, where it is important in the formation of gravel terraces (TAYLOR, 1955b). *Azorella* is unknown from South Georgia, where there appear to be no strictly comparable feldmark cushion communities, although some *Colobanthus* spp. there have a similar habit.

Several investigators (eg. MOSELEY, 1892; ANDERSSON quoted in NORDENSKJÖLD, 1928) have noted the higher temperatures within the plant cushions compared to that of the ambient air. Thus, MOSELEY found that *Azorella* cushions on Marion Island had internal temperatures of 50 °F. (10 °C) while that of the outside air was 45 °F. (7.2 °C). He commented:

"It is evident that these mounds retain and store up a considerable quantity of the sun's heat: and this fact probably yields a partial explanation of their peculiar form, which is like that of so many otherwise widely different Antarctic plants... No doubt power gained of resistance to wind is one of the chief causes of their assumption of this form".

WERTH (1928) also considered that much of the Kerguelen vegetation was a "wind desert" whose physiognomy was determined primarily by exposure. In addition to this exposure factor, the adoption of cushion and mat forms by flowering plants in the subantarctic, and by bryophytes in higher latitudes, is no doubt connected with their ability to absorb and store solar radiation. It is notable that only bryophytes form cushions in the most severe climates, where the soil (or the lower layers of the cushions themselves) becomes permanently frozen. The ability of mosses to absorb water through their stems and leaves may make it possible for growth to take place on the surface of the bryophyte cushions, even when the lower stems are dead; whereas the low temperatures would kill or render ineffective flowering plant roots and conducting systems.

The subantarctic plant communities share many characteristics with montane vegetation in lower latitudes, but it seems probable that the predominance of compact and mat-forming chamaephytes is a biological reflection of their cold isothermal climates. TROLL (1958) has stressed the similarity in life forms and climates between the subantarctic regions and certain altitudinal zones in the high tropical Andes. RAUNKIAER (1934), although working with inadequate data for many of the subantarctica island floras, found that their biological spectra differed from those of the cool temperate islands near New Zealand and in the Indian and South Atlantic Oceans in having more than 20% chamaephytes.

Closed herbfield communities in which large-leaved (mesophyll to macrophyll) species are a conspicuous element, occur in both Macquarie Island and the Kerguelen, Heard and Crozet and Marion Island groups. The dominant species *(Stillbocarpa polaris* and *Pleurophyllum hookerii* on Macquarie, and *Pringlea antiscorbutica* on the Indian Ocean islands) have no counterpart of the same growth form in South Georgia. The herbfield communities throughout the subantarctic are composed mainly of smaller species of grasses and sedges (notably *Agrostis magellanica, Festuca erecta, Carex* spp.,

Uncinia spp.) together with *Ranunculus biternatus, Colobanthus* spp.,
Acaena spp., *Callitriche* spp., and the fern *Blechnum penna-marina*.
As in montane herbfield communities elsewhere, the majority of the
subantarctic herbfield species are hemicryptophytes. Floristically,
the large herbs of the Macquarie and Indian Ocean Island communities give the appearance of being relict "epibiotics" (CAIN,
1944) in the subantarctic, and they are more narrowly distributed
than the other components of the herbfield communities most of
which are "insular" species and circumpolar in range.

Communities of large tussock grasses, sometimes forming stout
erect trunks with foliage borne up to a metre or more above the
ground, are characteristic of subantarctic coasts. They are often
growing on peat which has been enriched with animal droppings,
and are partly a biotic climax associated with bird and sea mammal
breeding grounds (GILLHAM, 1960). The communities frequently
consist of pure stands of the tussock grass, whose life form is well
adapted to survive seal movements and penguin trampling (WACE,
1960). Unlike the other subantarctic vegetation types, counterparts
of which occur in lower latitudes, the northward extensions of the
large tussock communities are not also accompanied by a rise in
their altitudinal limits. The dominant species are probably halophy-
tes (SKOTTSBERG, 1942), and the maritime tussock communities of
the subantarctic and southern cool temperate regions are structurally unlike the sod-tussock and tall tussock communities of the
Australian and New Zealand mountains (COSTIN, 1954; COCKAYNE,
1928). They are now almost exclusively an insular type of vegetation,
whose present range is complementary to that of native herbivorous
land mammals (WACE, 1960) The importation by man of herbivores
such as sheep and rabbits to subantarctic islands has led to the destruction of their coastal tussock grassland communities (COSTIN &
MOORE, 1960; HOLDGATE & WACE, 1961).

As in the Antarctic zone, it is probable that the extreme isolation
of the subantarctic islands has an important influence on the
nature of their vegetation. The survival of any flowering plants on
Macquarie Island or South Georgia through the Pleistocene glaciations is thought to be very unlikely (TAYLOR, 1954; SKOTTSBERG,
1956), although some may have survived in the Indian Ocean islands (? *Pringlea, Lyallia)*. Most of the vacular plants of the subantarctic belong to the "insular" group of species (Table III), and
are probably postglacial arrivals on the islands. How far this isolated
and species-poor subantarctic vegetation is in equilibrium with its
present climate is unknown, but it is likely that many niches are
untenanted and that its ecosystems are inherently unstable (ELTON,
1958). WERTH (1928) noted that on Kerguelen the different vegetational formations were characterised by changes in the proportions and the ecological preferences of the constituent species, rather

than by their presence in some and absence from others. The accelerated arrival of new species of plants, both accidentally and intentionally introduced by man, is having a profound effect on the vegetation of the subantarctic islands, especially when preceeded by the importation of herbivorous mammals (HOLDGATE & WACE, 1961).

Cold temperate circum-Antarctic vegetation

It is beyond the scope of the present account to review the characteristics of all the southern cold temperate vegetation types; but certain structurally distinct communities (i.e. formations in the sense of WARMING and later authors) to the north of the subantarctic zone have a particular importance to the present discussion because they contain a high proportion of "Antarctic" plants, including the dominant species. Bog and rain forest communities will be considered here.

Plant communities growing on peat which may be as much as several metres deep, and dominated by very hard and compact cushion-forming flowering plants, are one of the characteristic features of the wet montane regions of New Zealand (COCKAYNE, 1928; CRANWELL, 1953), and of the very wet coastal strip of extreme south-west Chile and Tierra del Fuego (AUER, 1933). The cushion communities of these two areas, and of the New Zealand shelf islands, have been compared by GODLEY (1960); and similar communities are also found in the Tasmanian mountains (GIBBS, 1920; SUTTON, 1928) and in the Falkland Islands (SKOTTSBERG, 1913), with outliers in Hawaii (SKOTTSBERG, 1940), and possibly also in the New Guinea mountains and the equatorial high Andes (Fig. 13). Cushion plant communities are absent from mainland Australia (COSTIN, 1959). Lists of the important species concerned in the formation of cushion communities are given in Table VIII.

Many authors, starting with COCKAYNE (1928) have described the cushion plant communities on deep peat as "cushion bogs". The term will be used here, although it is doubtful whether they are all purely ombrogenous in nature. The cushion bog communities are obviously related structurally to the subantarctic feldmark formation already referred to, and a succession from fell field to swamp has been described in the Tasmanian montane cushion communities (CURTIS & SOMERVILLE, 1949). Apart from the striking growth forms of the dominant plants which give them their distinctive character, the cushion bogs are often distinguished by mats of densely tufted plants (notably *Astelia*) between the cushions, and by small rosette epiphytes on the surface of the cushions (especially *Drosera* and *Plantago* species). Ferns are generally inconspicuous, and mosses (including sphagna) are unimportant or altogether absent (CRANWELL, 1953; SKOTTSBERG, 1940). Floristically, the cushion bogs are characterised by having a very high proportion of "Antarctic"

Table VIII.

Cushion-Forming Flowering Plants in Southern Lands

Cushion plant genera are listed according to the major habitats they occupy in the various regions. Associated species of dwarf conifers, and mat-forming tufted plants are listed below. Compiled from the sources noted.

	Southern Chile	Peruvian Andes	Falkland Islands	Subantarctic Islands	Tasmania	New Zealand & shelf islands
Feldmark communities (open)	Azorella	Azorella Opuntia Pycnophyllum.	Azorella Bolax	Azorella Colobanthus Lyallia	Dracophyllum Pterygopappus	Haastia Raoulia
Bog communities (closed)	Donatia Gaimardia Oreobolus Phyllachne	Distichia	Astelia Gaimardia Oreobolus	—	Abrotanella Donatia Dracophyllum Phyllachne	Astelia Giarmardia Oreobolus Phyllachne
Mat-forming tufted plants	Astelia pumila				Astelia alpina Milligania densiflora	
Dwarf conifers	Dacrydium fonckii		—	—	Microcachrys tetragona Podocarpus alpina	Dacrydium laxifolium Podocarpus nivalis
Sources of information	GODLEY, 1960	WEBERBAUER, 1911 TROLL, 1958	SKOTTSBERG, 1913	TAYLOR, 1955a SKOTTSBERG, 1912 WERTH, 1928	GIBBS, 1920 SUTTON, 1928	COCKAYNE, 1928 CHILTON, 1919 GODLEY, 1960

250

species, and most of these (including the dominants such as *Donatia*, *Gaimardia, Oreobolus* and *Phyllachne)* are "continental" in their range.

Bogs dominated by sphagna appear to be complementary in their distribution in the southern temperate regions to the cushion bogs. *Sphagnum* bogs have been described from southern and eastern Australia (CROCKER & EARDLEY, 1939; COSTIN, 1954 and 1957; MILLINGTON, 1954), from New Zealand and the Chatham Islands (COCKAYNE, 1928), from western Chile and northern Tierra del Fuego (AUER, 1933), and from Tristan da Cunha (WACE & HOLD-GATE, 1958) and Gough Island (WACE, 1961). They also occur in Tasmania, and are probably present on Amsterdam Island (DE LA RUE, 1954). Truly ombrogenous bogs with domed profiles, marginal drainage and low pH are probably present in all these areas, and "regeneration complex" features have been described from the Australian bogs (COSTIN, 1954). Floristically the southern *Sphagnum* bogs contain a lower proportion of "Antarctic" species than the cushion bogs: *Astelia, Oreobolus* and various Restionaceae (notably *Calophorus/Hypolaena)* are probably always associated with them in continental regions, but plants of the autochthonous floristic elements are also important associates. Thus various epacrids (especially *Richea)* in Tasmania, myrtaceous and epacrid shrubs in Australia and New Zealand, and *Empetrum* in South America are often associated or co-dominant with the sphagna, and play an essential part in stabilising the moss hummocks.

Nowhere do sphagna and cushion plants form mixed bog associations, although some associates of the cushion plants are often present in *Sphagnum* bogs, and small patches of sphagna are occasionally present in cushion communities. The two peat-forming communities appear to be more or less distinct ecologically, although it is difficult to relate their present ranges to any simple climatic factors of precipitation or temperature. While the cushion bogs are not found in any remote oceanic islands in the southern hemisphere and are essentially south Pacific in range, *Sphagnum* bogs occur in both islands and continents all round the Antarctic. In Tasmania and New Zealand the two bog communities appear to be altitudinally zoned. COCKAYNE (1928) includes *Sphagnum* bogs as a lowland vegetation type and cushion bogs as montane in New Zealand. In the Tasmanian central mountains, the *Sphagnum-Richea-Gleichenia* bogs are confined to the more sheltered valleys below 1000 m, while the cushion bogs are developed in more exposed situations above this altitude. In South America and the Falklands, the cushion bogs are apparently found under more oceanic climates, while the *Sphagnum* bogs are associated with more continental conditions typical of the deciduous forests (AUER, 1933). CRANWELL (1953) has found evidence of *Sphagnum* peat beneath cushion communities in New Zealand, and AUER (1933 and 1958) records many peat profiles in which *Sphagnum* peat overlies that formed by cushion plants. A better understanding of the present ecological relationships and the past distribution of the two peat-forming communities, and of the Juncaceae and moss-bogs of the subantarctic, would add greatly to our knowledge of the recent movements of "Antarctic" plants and of possible climatic changes in the southern hemisphere.

The temperate rain forests of New Zealand and southern Chile, dominated by evergreen species of *Nothofagus,* have long been recognised as broadly similar to one another in structure as well as containing taxonomically related species (eg. SCHIMPER 1903), although some workers have denied any physiognomic resemblances between them (OLIVER, 1925). The evergreen forests of these two areas are too diverse to enable any adequate summary of their main ecological characteristics to be attempted here, but they resemble one another chiefly in their close-growing nanophyll to microphyll dominants with a deep canopy, giving rise to forests of a dark and gloomy aspect in which ferns and mosses are the principal associates, chiefly as epiphytes. New Zealand *Nothofagus* forests have been described by COCKAYNE (1928), and the Chilean forests by REICHE (1907). GODLEY (1960) has listed the floristic elements in common to these forests, and has found that the New Zealand forests are floristically more diverse than the Chilean. Tasmanian rain forests have been described by GILBERT (1959), who notes the extreme floristic poverty of the *Nothofagus — Atherosperma* communities, which may lack any phanerogamic associates. The temperate *Nothofagus* rain forests are confined to continental regions, and neither *Nothofagus* itself, nor its associated trees with a southern disjunct range *(Laurelia/Atherosperma, Eucryphia, Drimys, Weinmannia)* occur on the southern islands. Like the cushion bogs, the southern temperate rain forests contain a high proportion of "Antarctic" plants with a "continental" range (see above), so that it seems probable that the formation has migrated as a unit to its present areas. The ecological relations between the *Nothofagus*-dominated "Antarctic" vegetation, and that dominated by the autochthonous floristic elements has been described from Tasmania (GILBERT, 1959), New Zealand (HOLLOWAY, 1954), and South America (AUER, 1958). Studies of the interrelationships of these forests, and of forest histories in the southern continents, could add considerably to an understanding of recent climatic changes, and the interaction between the differently derived elements in their vegetation (ROBBINS, 1962).

The woody vegetation of the southern temperate islands differs from that of the continents in its structure. The dominant species generally form thickets amongst the smaller herbaceous communities, in which ferns are often dominant. Such communities have been described as "fern bush" (WACE, 1960 and 1961). They occur in the islands on the New Zealand continental shelf, the Falklands, the Tristan — Gough group, and Amsterdam Island; and their southern limit marks the poleward extent of the temperate islands as recognised here. The dominant shrubs and small trees are semi-procumbent heath-like leptophyll to microphyll evergreen species, which are generally interspersed with wet heath or bog communities. Fern bush is structurally similar to communities found near the tree line on the tropical mountains (TROLL, 1958). Floristically, the fern bush communities consist mostly of "insular" southern disjunct species, especially among the pteridophytes; but in marked contrast to the continental temperate rain forests the dominant woody species do not generally belong to "Antarctic" taxa. Thus *Phylica* of the Atlantic and Indian Ocean islands is African, and *Metrosideros* of the islands on the New Zealand continental shelf is Malaysian and Polynesian, and neither belong to southern disjunct groups.

In addition to these distinct formations which are of interest because of the presence of so many "Antarctic" plants, the montane bog and rain forest fringes in the southern continents also appear to be the habitat of many of the relict endemic conifers, such as *Arthrotaxis, Diselma, Microcachrys* and *Phaerosphaera* of Tasmania, and *Pilgerodendron* and *Saxegothea* of Chile. Both the bog and rain forest communities are typical of areas with constantly wet oceanic climates. Most of the "Antarctic" species of "continental" range which are listed in Table III grow in these communities, or in other vegetation in which local conditions lead to abundant groundwater (streamsides, swamps & c.). It is largely this preference for humid microthermal habitats which gives an ecological unity to the taxonomically diverse assemblage of "Antarctic" plants and many of their associated invertebrates. Because the related taxa are found in comparable communities, with similar types of association between their members, it is difficult to escape the conclusion that the formations have migrated as entities (EVANS, Ref. 196). Only paleoecological work in the Antarctic itself can determine whether that continent has played a direct part in any such migrations.

Conclusions and Problems for Research in the Biogeography of Antarctic Vascular Plants

In writing the foregoing accounts, the author has adopted a partisan attitude to the problem of the southern hemisphere plants and vegetation with a disjunct range. He has favoured the view that an "Antarctic" flora and vegetation exits, and that its remains now to be found in southern lands can best be explained by closer connections between the southern continents during or since the Cretaceous. These are subjects of dispute: other explanations of present distributions in the southern hemisphere may suffice

(SIMPSON, 1940). But it is striking how little we have progressed towards a solution of this — one of the major problems of historical biogeography — since the problem was first recognised and defined over a century ago. Modern methods of taxonomic and ecological analysis are only just beginning to be employed on a large scale in researches on Antarctic and paleo-Antarctic plants and vegetation. Following the ideas put foward by various authors in three recent symposia*, some suggestions are included here on topics of research and areas of special interest in biogeographic studies on Antarctic vascular plants.

Paleontology

"An understanding of plant distribution in time is an essential precursor to an understanding of its present distribution in space" (PLUMSTEAD, 1962). However suggestive modern distributions may be, only fossil remains can provide incontrovertible evidence of the part that Antarctica may have played in the evolution and spread of floras. SIMPSON (1953) has suggested that where the fossil record of a group of organisms is poor or lacking, attempts to give historical explanations of present distributions are of doubtful value, and it must be admitted that the fossil evidence in favour of Antarctica as a floral migration route is slender. As has been stressed by BARG-HOORN (1961), there is a special need for further information on Tertiary floras in the Antarctic. This applies particularly to Greater Antarctica and Byrd Land. CROHN (1959) estimated that in the segment of Greater Antarctica between the meridians of 50° and 80°E., less than one percent of the area consists of rock outcrops, and this is probably true for most of the rest of the continent. Much reliance may therefore have to be placed upon erratics or samples dredged from offshore; but if these contain microspores, they could give valuable information from areas where it may be impossible to obtain macroscopic remains.

The subantarctic islands of Macquarie, South Georgia, and especially Kerguelen, are of critical importance in the study of circum-Antarctic Tertiary floras, and the presence of permanently inhabited bases on all of them should make it possible to obtain rock samples with relative ease. The confirmation of the presence or absence of podocarp and *Nothofagus* remains on Macquarie and Kerguelen (the only isolated southern islands which are known to contain Tertiary plant-bearing deposits), would add greatly to our

---

* Discussion on the Biology of the Southern Cold Temperate Zone, under the leadership of C. F. A. PANTIN, F.R.S. – *Proc. Roy. Soc. Lond.* B, **152**. "The Life Sciences in Antarctica": Report by the Committee on Polar Research part **1** – Nat. Acad. Sciences, Washington (pub. no. 839). "Symposium on Antarctic Biology", held in Paris, September 1962 (abstracts of many papers in *Polar Record*, **11** (72) October 1962; and see Ref. 195).

knowledge of the time during which the Antarctic flora may have spread northwards. Further analyses of the pollen and spore content of the Kerguelen lignites, combined with isotope dating of the intercalated lavas, would be particularly useful.

The role of the Antarctic in the evolution and spread of floras can not be viewed in isolation from the history of the angiosperms in other lands. Antarctica is considered by some workers to hold the key to the origin of the angiosperms themselves (CROIZAT, 1952; PLUMSTEAD, 1962). A systematic reassessment of all the northern hemisphere plant fossils belonging to taxa which now have a southern range, would be a useful contribution towards an understanding of the possible importance of Antarctica in historical biogeography.

Taxonomy

Taxonomic units must be precisely defined before their ranges can be mapped or tabulated (TURRILL, 1953 p. 226). The only excuse for the use of such makeshift methods as have been employed, for example, in Table III of this account, is that taxonomic revisions are needed of so many of the plants involved, that to wait for them all to be published would be to wait indefinitely. Several points seem immediately relevant to taxonomic work on "Antarctic" plants:

a. Duplicate specimens of plant collections made in southern cold temperate and subantarctic regions should be circulated to herbaria in all the southern continents, as well as the larger northern herbaria. Only thus can southern hemisphere workers who are familiar with the plants in the field, see the variation in "Antarctic" taxa. For historical reasons it is inevitable that most type specimens and early collections should be in northern herbaria, but it is absurd that Tasmanian or New Zealand botanists, for example, should still have to travel to Europe to see specimens of South American plants similar to their own.

b. Modern taxonomic methods, especially those using the biometric, experimental and cytological criteria of the "biosystematist", should be employed in comparing the disjunct populations of "Antarctic" taxa. Apart from the considerable theoretical interest of experimental taxonomic studies and chromosome counts in insular floras, (eg. MOORE, 1960 and 1962; HAIR & BUEZENBERG, 1961), the incidence of polyploidy may give a guide to the direction of plant migrations amongst floristic elements in continental regions (CAIN, 1944; STEBBINS, 1950; RATTENBURY, 1962).

c. Botanical "inventories" of all the remote southern islands should be compiled before human interference and the spread of exotics destroys or distorts the evidence of natural distributions. The aim should be to produce for southern regions, distribution maps as detailed and taxonomically well founded as those published for the circumpolar Arctic plants (HULTÉN, 1958). There is a very great need for the taxonomy of the southern disjunct plants to be worked through critically, with the eventual aim of producing a modern Flora Antarctica.

Dispersal

The efficiency of plant dispersal across wide stretches of ocean is still largely a matter of conjecture, yet it is essential to an understanding of the present ranges of many of the "Antarctic" plants. Although a considerable number of observations on the subject have been made (eg. RIDLEY, 1930), and correlations pointed out between the ranges of particular plants and the ranges or directions of movement of their potential dispersing agents (eg. TAYLOR, 1954; FALLA, 1960; HOLDGATE, 1960c), there is almost no experimental evidence available, and the subject is left wide open to speculation.

A systematic approach to the problem of pre-human dispersal of plants between the southern lands, could include the following observations and experiments:

Sea dispersal:
a. The examination of beach drift, including the identification of driftwood by anatomical means, and germination tests on any plant propagules found (eg. BARBER et al., 1959).
b. Floatation experiments in sea water, followed by germination tests, on the diaspores of southern disjunct (and especially insular) plants.

Air dispersal:
Atmospheric and ice-core studies similar to those carried out in the Arctic by POLUNIN (1951 and 1958) and on Antarctic and subantarctic insects by GRESSITT (see GRESSITT et al. 1960, 1961).

Animal dispersal:
Examination of the down, gut content and faeces of birds breeding in or visiting southern islands and continents with attempts to germinate any seeds & c. found. Birds caught at sea, whose movements might account for circumpolar plant ranges, would be particularly interesting in this work.

Such experiments and observations as these might yield entirely negative results, but they are still worth making because of the almost total lack of reliable data on the subject of long range dispersal by natural agencies in southern regions.

Ecology

Ecological studies in most of the Antarctic, and many circum-Antarctic lands have not progressed beyond the stage of simple and often cursory description. In a few areas there are no published

256

descriptions of vegetation (eg. Marion and Prince Edward Islands), or of the probable lack of it (eg. Bouvet, Ballenny, Scott and Peter I Islands). In others, notably the Tasmanian mountains, much of Fuegia, and some of the subantarctic islands in the Indian Ocean, the existing accounts are much in need of revision. The close study of plant communities in the Antarctic, and in the subantarctic islands, could give some indications on whether the poverty of the flora in high southern latitudes is due to hostile environments or to the difficulties of dispersal. In the southern cold temperate regions, the inter-relationships of the *Nothofagus*-dominated and other forests, and the relationships of the cushion-bogs, *Sphagnum*-bogs, and Juncaceae-Bryales bogs are of particular interest, as noted above.

Most "Antarctic" vascular plants inhabit regions with oceanic climates in which peat is formed or in which open water allows the accumulation of organic remains in conditions suitable for the preservation of pollens. Although pollen analyses of deposits in New Zealand (MOAR, WARDLE) and Fuego-Patagonia (AUER) have enabled a start to be made there in studies of vegetation history, no Quarternary vegetation studies have yet been made in Tasmania, the Falklands or any of the subantarctic islands (except Macquarie). Such work could lead to a far better understanding of Pleistocene climatic changes and consequent plant movements in the southern hemisphere in the recent past. Pollen studies and carbon dating of moss peats in the Low Antarctic, could also yield valuable evidence of the age of the present vegetation on the continent.

The early voyagers who sought the Terra Australis Incognita, regarded their quest as part of a major problem of global geography. Now that man has established himself on the Antarctic, and the great distances between the southern islands and continents are less of a barrier than they were in the past, the investigation of the ecology and biogeography of the Antarctic and circum-Antarctic flora as a unit is becoming easier. A synthetic viewpoint, following DARWIN, HOOKER and SKOTTSBERG, which seeks to uncover the importance of the Antarctic, both to historical biogeography and in its effects on present environments in surrounding regions, is the only one which can lead to a better understanding of the derivation and the ecology of plant life in the higher southern latitudes.

Addendum

Since writing this account, several papers concerning the biogeography of Antarctic vascular plants have been published, notably in the symposia of the 10th Pacific Science Congress held at Honolulu in 1961. These papers are here referred to briefly, and cited in full at the end of the bibliography.

ADIE (1962b) outlined the geological history of Antarctica, and emphasised the total dis-similarity of Greater and Lesser Antarctica "each of which has evolved independently, but in adjacent positions in similar climatic environments". In another paper ADIE (1963) reviewed the geological evidence for possible Antarctic land connections, and stated that "the problems concerning the distribution of faunas and floras of the Southern Hemisphere fall into two clear-cut time categories: pre-Cretaceous and post-Cretaceous..." Land connections involving Antarctica are perhaps acceptable in the former period, but "present day land distribution is essential for the post-Cretaceous faunal and floral distribution patterns". FLEMING (1963) advanced the idea that the southern disjunct plants should be considered in two groups, which are broadly equivalent to the "continental" and "insular" plants of the present author (Table III). He has called these groups the Paleoaustral (or Paleonotian) and Neoaustral (or Neonotian) respectively. FLEMING also considered the timing of the first appearances in the New Zealand pollen record of the now southern disjunct plants, and was inclined to give "wavering support" to the view that the dispersal of both groups had been across the sea because the process of spread to New Zealand appeared to be intermittent throughout the Cenophytic. CRANWELL (1963) discussed the living and fossil range of *Nothofagus* (including the problematical records from the northern hemisphere), and concluded that land connections involving Antarctica were necessary to account for both the past and present ranges of the genus. VAN STEENIS (1963) considered a southern Pacific land bridge probably involving Antarctica as absolutely necessary to account for the range of the southern disjunct plants. His ideas are expanded in another paper (VAN STEENIS, 1962), in which the disjunct "continental" taxa are listed, and the problems of their range placed in a world-wide setting in relation to the past dispersals of vascular plants and possible land bridges between the continents.

PREEST (1963) examined experimentally the present dispersal capacities of the New Zealand podocarps and *Nothofagus*, and concluded that "the chances of the effective dispersal of viable seed of either group on or over the sea, and its survival entry and subsequent growth to maturity, and the naturalisation of the species in a new land hundreds or thousands of kilometres distant, are of an extremely low order of probability". He therefore favoured the idea of former land connections between the now separated parts of their fossil and living ranges. HAIR (1963) also favoured the closer proximity in the past of the "Gondwana" continents, all of which contain Mesozoic and Tertiary podocarps. He proposes a rearrangement based on chromosome numbers and characteristics of all the podocarps, by the use of which the northward spread of different

258

groups from the Antarctic is easier to envisage than with the existing taxonomic arrangements based on gross morphology.

KUSCHEL (1963) proposed the establishment of an Austral biogeographic region centring on Antarctica, and embracing the whole of Australia and New Guinea as well as Chile and Patagonia and the Cape region of Southern Africa, and all the subantarctic islands. Although possibly useful as a concept for the study of certain taxonomic groups, the size and biotic diversity of such a region (and especially the inclusion of the Cape) makes it of doubtful use as a biogeographic realm in which much of the biota has had a common history.

KNOX (1963) in summarising the symposium on "Antarctic Relationships" pointed to the value of comparative ecological and evolutionary (as well as purely systematic and regional) studies amongst disjunct taxa. In the present author's opinion, the most striking fact that emerges from these recent papers is that although our geological and paleontological knowledge of Antarctica and its surrounding lands is more complete than it was, and although the range of the "Neoaustral"or "oceanic" elements is now conceded to be due largely to long range trans-oceanic dispersal, the division of opinion between the "diffusionists" and the adherents of past land connections remains as acute as ever in respect of most of the plants in whose history Antarctica may have been involved.

REFERENCES

1. ADIE, R. J., 1952. Representatives of the Gondwana System in Antarctica. *Symposium sur la série de Gondwana, publié par le XIXe Congrès Géologique International, Algiers (393—399)*.
2. ADIE, R. J. 1958. Geological investigations in the Falkland Islands Dependencies since 1940. *Polar Rec.* 9 (58), *3—17*.
3. ADIE, R. J. 1962a. Personal Communication.
4. ALLAN, H. H. 1961. Flora of New Zealand. Wellington, Government Printer. (Vol. I)
5. AUER, V. 1933. Verschiebungen der Wald- und Steppengebiete Feuerlands in Postglazialer Zeit. *Acta geogr.* 5 (2), *1—313*.
6. AUER, V. 1958. The Pleistocene of Fuego-Patagonia: II, The History of the Vegetation and Flora. *Ann. Acad. Sci. Fenn.* A, 50.
7. AVSYUK, G. A., MARKOV, P. A. & SHUMSKII, L. 1956. Cold wastelands in the Antarctic Continent; and: Geographical Observations in an Antarctic Oasis. Both in English translation published for National Science Foundation & Dept. Interior, Washington by Israel Program for Scientific Translations.
8. BALME, B. E. 1962. Some Palynological evidence bearing on the development of the Glossopteris-flora. Chapter 25 (pp. *269—280)* in: The Evolution of Living Organisms (Ed. G. W. LEEPER). Melbourne University Press.
9. BARBER, H. N., DADSWELL, H. E. & INGLE, H. D. 1959. Transport of driftwood from South America to Tasmania and Macquarie Island. *Nature, Lond.* 184 (4681), *203—204*.

10. BARGHOORN, E. S. 1961. A brief review of fossil plants of Antarctica, and their geological implications. Chapter 1 (pp. 5—9) in: The Life Sciences in Antarctica. National Academy of Sciences, Washington: publication no. 839.

11. BERRY, E. W. 1928. Tertiary fossil plants from the Argentine Republic. *Proc. U.S. Nat. Mus.* **73.**

12. BERRY, E. W. 1938. Tertiary flora from the Rio Pichileufu, Argentina. *Geol. Soc. America* special paper no. **12.**

13. BLACK, J. M. 1943—60. Flora of South Australia, second edition. Adelaide, Government Printer.

14. BRYANT, H. M. 1945. Biology at East Base, Palmer Peninsula, Antarctica. *Proc. Amer. phil. Soc.,* **89** (1), *256—269.*

15. BUNT, J. 1956. Living and fossil pollen from Macquarie Island. *Nature, Lond.* **177** (4530) *337.*

16. BURBIDGE, N. T. 1960. The phytogeography of the Australian Region. *Aust. J. Bot.,* **8** (2), *75—211.*

17. CAIN, S. A. 1944. Foundations of Plant Geography. New York, Harper.

18. CAMPBELL SMITH, W. 1930. Report on the geological collections made during the voyage of the "Quest" on the Shackleton-Rowett Expedition to the South Atlantic and Weddell Sea in 1921—22. London, British Museum.

19. CHILTON, C. (Ed.) 1909. The Subantarctic Islands of New Zealand. Wellington, Canterbury Philosophical Institute.

20. COCKAYNE, L. 1928. The Vegetation of New Zealand. *Veg. Erde,* **14** (Ed. ENGLER & DRUDE).

21. COOKSON, I. C. 1947. Plant microfossils from the lignites of Kerguelen Archipelago. *British-Australian-New Zealand Antarctic Res. Exp. Rep. A.* **2** (8), *129—142.*

22. COSTIN, A. B., 1954. A study of the ecosystems of the Monaro Region of New South Wales. Sydney, Government Printer.

23. COSTIN, A. B. 1957. The high mountain vegetation of Australia. *Aust. J. Bot.* **5** (2), *173—189.*

24. COSTIN, A. B. 1959. Vegetation of high mountains in Australia in relation to land use. Chapter 26 (pp. *427—451*) in Biogeography and Ecology in Australia (Ed. A. KEAST). The Hague, Dr. W. Junk.

25. COSTIN, A. B. & MOORE, D. M. 1960. The effects of rabbit grazing on the grasslands of Macquarie Island. *J. Ecol.,* **48,** *727—739.*

26. COUPER, R. A. 1960. Southern hemisphere Mesozoic and Tertiary Podocarpaceae and Fagaceae, and their paleogeographic significance. *Proc. Roy. Soc. London, B.,* **152,** *491—500.*

27. COUR, P. 1958. A propos de la flore de l'archipel de Kerguelen. *Terres Australes et Antarctiques Franç.,* **4** et **5,** *10—32.*

28. COUR, P. 1959. Flore et végétation de l'archipel de Kerguelen. *Terres Australes et Antarctiques Franç.,* **8** et **9,** *3—40.*

29. CRANWELL, L. M. 1953. An outline of New Zealand peat deposits, with notes on the condition of the rain-fed cushion bogs. *Proc. 7th. Pacific Sci Congr. (1949),* **5,** *186—208.*

30. CRANWELL, L. M. 1959. Fossil pollen from Seymour Island, Antarctica. *Nature, Lond.* **184** (4701), *1782—1785.*

31. CRANWELL, L. M. 1962. Antarctica: cradle or grave for *Nothofagus? Pollen et Spores,* **4** (1), *190—192.*

32. CRANWELL, L. M., HARRINGTON, H. J. & SPEDEN, I. G. 1960. Lower Tertiary microfossils from McMurdo Sound, Antarctica. *Nature, Lond.* **186** (4726), *700—702.*

33. CROCKER, R. L. & EARDLEY, C. 1939. A South Australian *Sphagnum* bog. *Trans. Roy. Soc. S. Aust.* **63** (2), *210—214.*

260

34. CROHN, P. W. 1959. A contribution to the geology and glaciology of the western part of Australian Antarctic Territory. *Aust. Nat. Antarctic Res. Exp. Rep., A, 3, 1—103.*

35. CROIZAT, L. 1952. Manual of Phytogeography, or an account of plant dispersal throughout the world. The Hague, Dr. W. Junk.

36. CURTIS, W. M. 1946. *Phyllachne colensoi* Berggren, an addition to the list of subantarctic plants in the Tasmanian flora. *Proc. Roy. Soc. Tasmania for 1946, 31—34.*

37. CURTIS, W. M. 1956. The student's flora of Tasmania. I. Hobart, Government Printer.

38. CURTIS, W. M. & SOMERVILLE, J. 1949. The vegetation (of Tasmania). Aust. N-Z Assoc. Adv. Sci. Handbook for Hobart meeting, *51—57.*

39. DARLINGTON, P. J. 1957. Zoogeography. New York, Wiley.

40. DARLINGTON, P. J. 1960. The zoogeography of the southern cold temperate zone. *Proc. Roy. Soc. London*, B, **152**, *658—668.*

41. DARRAH, W. C. 1936. Antarctic fossil plants. *Science*, **82**, *390—391.*

42. DARRAH, W. C. 1960. Principles of paleobotany 2nd ed. New York, Ronald Press.

43. DARWIN, C. 1859. On the origin of species by means of natural selection . . . London, Murray.

44. DAWSON, J. W. 1958. Inter-relationships of the Australasian and South American floras. *Tuatara*, **7** (1), *1—6.*

45. DEACON, G. E. R. 1960. The southern cold temperate zone. *Proc. Roy. Soc. London*, B, **152**, *441—447.*

46. DE LA RUE, E. A. 1954. Les îles Saint Paul et Nouvelle Amsterdam. *Geographia, 38—43.*

47. DELL, R. K. 1960. Sea bird logs between New Zealand and the Ross Sea. *Rec. Dominion Mus.* **3**, (4), *293—305.*

48. DENNIS, R. W. G. 1960. Discussion on cryptogamic botany. *Proc. Roy. Soc. London*, B, **152**, *539.*

49. DOUGLAS, G. V. 1923. Geological results of the Shackleton-Rowett (Quest) Expedition. *Proc. geol. Soc. London*, **79**, *x-xi.*

50. DU RIETZ, G. E. 1940. Problems of bi-polar plant distribution. *Acta phytogeogr. suecica*, **13**, *215—282.*

51. DUSEN, P. 1908. Über die Tertiäre Flora der Seymour-Insel. *Wiss. Erg. Schwed. Südpolar Exp. 1901—03*, **3** (3), *1—27.*

52. EDWARDS, W. N. 1921. Fossil coniferous wood from Kerguelen Island. *Ann. Bot.* **35** (140), *609—617.*

53. EDWARDS, W. N. 1955. The geographical distribution of past floras. *Adv. Science*, **12** (46), *165—176.*

54. ELTON, C. S. 1958. The ecology of invasions by animals and plants. London, Methuen.

55. FALLA, R. A. 1960. Oceanic birds as dispersal agents. *Proc. Roy. Soc. London*, B, **152**, *655—659.*

56. FLEMING, W. L. S., STEVENSON, A., ROBERTS, B. B. & BERTRAM, G. C. L. 1938. Notes on the scientific work of the British Graham Land Expedition, 1934—1937. *Geogr. J.* **91**, *508.*

57. FLORIN, R. 1940. The Tertiary fossil conifers of southern Chile and their phytogeographical significance, with a review of the fossil conifers of southern lands. *Kungl. Svensk. Vet. Akad. Handl.* **19** (2),

58. GARDNER, C. A. 1931. Enumeratio plantarum Australiae Occidentalis. Perth, Government Printer.

59. GIBBS, L. S. 1920. Notes on the phytogeography and flora of the Mountain summit plateau of Tasmania. *J. Ecol.* **8** (1), *1—17* and *89—117.*

60. GILBERT, J. M. 1959. Forest succession in the Florentine Valley, Tasmania. *Proc. Roy. Soc. Tasmania*, **93**, *129—151.*

61. GILLHAM, M. E. 1960. Modification of subantarctic flora of Macquarie Island by sea birds and sea elephants. *Proc. Roy. Soc. Vict.* **74** (1),
62. GODLEY, E. J. 1960. The botany of southern Chile in relation to New Zealand and the subantarctic. *Proc. Roy. Soc. London* B, **152**, *457— 475.*
63. GOOD, R. D'O. 1933. A geographical survey of the flora of temperate South America. *Ann. Bot.* **47** (188), *691—725.*
64. GOOD, R. D'O. 1953. The geography of the flowering plants. 2nd. Edition. London, Longmans.
65. GOOD, R. D'O. 1960. On the geographical relationships of the angiosperm flora of New Guinea. *Bull. Brit. Mus. Nat. Hist., Botany,* **2** (8).
66. GORDON, H. D. 1949. The problem of subantarctic plant distribution. *Rep. Aust. N-Z Ass. Adv. Sci.* **27**, *142—149.*
67. GORDON, W. T. 1930. A note on *Dadoxylon (Araucaroxylon)* from the Bay of Isles. Report on the geological collections made during the voyage of the "Quest" . . . in 1921—22 London, British Museum.
68. GOTHAN, W. 1908. Die fossilen Hölzer von den Seymour und Snow-Inseln. *Wiss. Erg. Schwed. Südpolar Exp. 1901—03,* **3** (8), *1—33.*
69. GREENE, S. W. (in the press) The vascular flora of South Georgia. *Brit. Antarctic Survey Sci. Rep.,* **45**.
70. GREENE, S. W. & GREENE, D. M. 1963. Check list of the subantarctic and Antarctic vascular flora. *Polar Rec.* **11** (73), *411—418.*
71. GRESSITT, J. L., LEECH, R. E., LEECH, T. S., SEDLACEK, J. & WISE, K. A. J. 1960. Trapping airborne insects in the Antarctic area, I. *Pacific Insects,* **2**, *345.*
72. GRESSITT, J. L., LEECH, R. E., LEECH, T. S., SEDLACEK, J. & WISE, K. A. J. 1961. Trapping airborne insects in the Antarctic area, II. *Pacific Insects,* **3**, *559.*
73. GUPPY, H. B. 1919. The island and the continent. *J. Ecol.* **7**, *1.*
74. HAIR, J. B. & BEUZENBERG, E. J. 1961. High polyploidy in a New Zealand *Poa. Nature, Lond.* **189** (4759), *160.*
75. HALLE, T. G. 1913. The Mesozoic flora of Graham Land. *Wiss. Erg. Schwed. Südpolar Exp. 1901—03,* **3** (14), *1—123.*
76. HARRINGTON, H. J. 1958. Nomenclature of rock units in the Ross Sea region, Antarctica. *Nature, Lond.* **182** (4631), *290.*
77. HAWKES, D. D. 1962. The structure of the Scotia Arc. *Geol. Mag.* **99** (1), *85—91.*
78. HEMSLEY, W. B. 1885. Report on the present state of knowledge of various insular floras . . .; and Report on the botany of the Bermudas, and various islands in the Atlantic and Southern Oceans I & II. *Sci. Res. voyage Challenger, Botany,* **1**.
79. HILL, A. W. 1929. Antarctica and the problems of geographical distribution. *Proc. 5th. int. bot. Congr.* (1926),
80. HILLEBRAND, W. 1888. Flora of the Hawaiian Islands. Heidelberg.
81. HOLDGATE, M. W. 1960a. The Royal Society Expedition to southern Chile. *Proc. Roy. Soc. London,* B, **152**, *434—441.*
82. HOLDGATE, M. W. 1960b. The fauna of the mid-Atlantic islands. *Proc. Roy. Soc. London,* B, **152**, *550—576.*
83. HOLDGATE, M. W. 1960c. Future research in the southern cold temperate zone (discussion). *Proc. Roy. Soc. London,* B, **152**, *674—675.*
84. HOLDGATE, M. W. 1962a. Ecology on the Antarctic fringe. *Polar Rec.,* **11** (72), *335—336.*
85. HOLDGATE, M. W. 1962b. Personal communication, dated 6.4.62.
86. HOLDGATE, M. W. 1963. Observations in the South Sandwich Islands, 1962. *Polar Rec.* **11** (73), *394—405.*
87. HOLDGATE, M. W. & WACE, N. M. 1961. The influence of man on the floras and faunas of southern islands. *Polar Rec.,* **10** (68), *475—493.*

262

88. HOLLOWAY, J. T. 1954. Forests and climates in the South Island of New Zealand. *Trans. Roy. Soc. N-Z.* **82** (2), *329—410.*

89 HOLTEDAHL, O. 1929. On the geology and physiography of some Antarctic and subantarctic islands. *Sci. Res. Norweg. Antarctic Exp. 1927—28,* **3.**

90. HOOKER, J. D. 1844—47. I: Flora Antarctica (2 vols).

91. HOOKER, J. D. 1853—55. II: Flora Novae-zelandiae(2 vols).

92. HOOKER, J. D. 1855—60. III: Flora Tasmaniae (2 vols).
All included in: The botany of the Antarctic Voyage of H.M. Discovery ships Erebus and Terror in the years 1839—1843. London, Reeve.

93. HULTÉN, E. 1958. The amphi-Atlantic plants and their phytogeographical connections. *Kungl. Svenska Vet. Akad. Handl.* **7** (1), *1—340.*

94. JUST, T. 1952. Fossil floras of the southern hemisphere, and their phytogeographical significance. *Bull. Amer. Mus. Nat. Hist.* **99** (3), *189—203.*

95. KNOX, G. A. 1960. Littoral ecology and biogeography of the southern oceans. *Proc. Roy. Soc. London,* B, **152,** *577—624.*

96. KUSCHEL, G. 1960. Terrestrial zoology in southern Chile. *Proc. Roy. Soc. London,* B, **152,** *540—550.*

97. LI, H. L. 1953. Present distribution and habitats of conifers and taxads. *Evolution,* **7,** *245—261.*

98. LICITIS, R. 1953. Air-borne pollen and spores sampled at five New Zealand stations, 1951—2. *N-Z. J. Sci. Technol.,* B, **34** (4), *291—316.*

99. LLANO, G. A. 1956. Botanical research essential to a knowledge of Antarctica. In 'Antarctica in the I.G.Y.' (Ed. A. P. CRARY *(Geophysical Monograph no. 1,* American Geophysical Union, Washington DC.

100. LLANO, G. A. 1962. The terrestrial life of the Antarctic. *Sci. Amer.,* **207** (3), *212—230.*

101. LOEWE, FR. 1961. Fortschritte in der physikalisch-geographischen Kenntnis der Antarktis. *Erdkunde,* **15** (2), *81—92.*

102. LOURTEIG, A. & COUR, P. 1963. Essai sur la distribution géographique des plantes vasculaires de l'archipel de Kerguelen. *Comité Nat. franç. Rech. Antarctiques,* **3,** *63—70.*

103. MACKINTOSH, N. A. 1960. The pattern of distribution of the Antarctic fauna. *Proc. Roy. Soc. London,* B, **152,** *624—631.*

104. MA KHIN SEIN, 1961. *Nothofagus* pollen in the London clay. *Nature, Lond.* **190,** (4780), *1030—1031.*

105. MATHIAS, M. E. & CONSTANCE, L. 1955. The genus *Oreomyrrhis* (Umbelliferae): a problem in South Pacific distribution. *Univ. California Publ. in Bot.,* **27** (6), *347—416.*

106. MAWSON, D. 1940. Sedimentary rocks. *Sci. Rep. Aust. Antarctic Exped. 1911—14,* A, **4** (11), *347—367.*

107. MECKING, L. 1928. A regional geography of the Arctic and the Antarctic. Pp. 93—338 in "The Geography of the Polar Regions" (Ed. W. L. G. JOERG): *Amer. geogr. Soc., special Publ. no.* **8.**

108. MILLINGTON, R. J. 1954. *Sphagnum* bogs on the New England plateau, New South Wales. *J. Ecol.* **42** (2), *328—344.*

109. MOORE, D. M. 1960. Chromosome numbers of flowering plants from Macquarie Island. *Bot. Not.* **113** (2), *185—191.*

110. MOORE, D. M. 1962. Experimental taxonomic studies in Antarctic floras. *Polar Rec.* **11** (72), *323.*

111. MOSELEY, H. N. 1892. Notes by a naturalist in H.M.S. Challenger. London, Murray.

112. MURPHY, R. C. 1936. Oceanic birds of South America (2 vols). New York, McMillan.

113. MURPHY, R. C. 1960. Oceanic birds. *Proc. Roy. Soc. London*, B, **152**, *642—654.*

114. NANTHORST, A. G. 1907. On the Upper Jurassic flora of Hope Bay, Graham Land. *C.R. Xme. Congr. Geol. Mexico (1906)*, **10** (2), *1269.*

115. NORDENSKJÖLD, O. 1928. A general characterisation of polar nature. Pp. 3—90 in "The Geography of Polar Regions" (Ed. W. L. G. JOERG): *Amer. geogr. Soc., special Publ.*, **8.**

116. OLIVER, R. L. & SORENSEN, J. H. 1951. Botanical investigation on Campbell Island. *Cape Exp. Series*, **7.**

117. OLIVER, W. R. B. 1925. Biogeographical relations of the New Zealand flora. *J. Linn. Soc. Bot. London*, **47**, *99.*

118. PAVLOV, V. V. 1958. Rezul'taty palinologicheskogo analiza obraztsov iz otlozheniv osadochnovulcan-icheskoy serii bikon (Antarkida, Zemlya Korolya Georga V Mys Blaff. *Nauchno-Issledovatel'skiy Institut Geologii Arktik Spornik Statey po Paleontologii i Biostratigrafi. Vypusk* **12**, *77—79.* (English summary in *Polar Rec.* **10**, *322).*

119. PETRIE, D. 1909. Gramina of the subantarctic islands of New Zealand. In "The subantarctic islands of New Zealand" (Ed. C. CHILTON). Vol. **2**, *472—481.* Wellington, Canterbury Philosophical Institute.

120. PHILCOX, D. 1961. An *Uncinia* from South Georgia. *Kew Bull.* **15** (2), *229.*

121. PLUMSTEAD, E. P. 1962. Fossil floras of Antarctica. *Sci. Rep. Trans-Antarctic Exp.* **9.**

122. POLUNIN, N. 1951. Seeking airborne botanical particles about the North Pole. *Svensk. bot. Tidskr.* **45** (2), *320—354.*

123. POLUNIN, N. 1958. The botany of ice island T-3. *J. Ecol.* **46** (2), *323—347.*

124. RATTENBURY, J. A. 1962. Origins of the New Zealand flora: cytogeobotanical observations on the Malayan element. Chapter 34 (pp. *417—425)* in "The Evolution of Living Organisms" (Ed. G. W. LEEPER) Melbourne University Press.

125. RAUNKIAER, C. 1934. The Arctic and Antarctic chamaephyte climate. Chapter 7 in "The Life Forms of Plants and statistical plant geography". Oxford, Clarendon.

126. REICHE, K. 1907. Grundzüge der Pflanzenverbreitung in Chile. *Veg. Erde*, **8**, *1—374.*

127. RIDLEY, H. N. 1930. The dispersal of plants throughout the world. Ashford, Reeve.

128. ROBBINS, R. G. 1962. The Podocarp-broadleaf forests of New Zealand. *Trans. Roy. Soc. N-Z.* **1** (5), *33—75.*

129. RODWAY, L. 1903. The Tasmanian flora. Hobart, Govt. Print.

130. RODWAY, L. 1914. Botanical evidence in favour of land connection between Fuegia and Tasmania during the present floristic epoch. *Proc. Roy. Soc. Tasmania* for 1914, *32—34.*

131. RUDMOSE BROWN, R. N. 1928. Antarctic and subantarctic plant life and some of its problems. *Amer. geogr. Soc. special Publ.*, **7**, *343—352.*

132. RUDOLPH, E. D. 1963. Vegetation of Hallett Station area, Victoria Land, Antarctica. *Ecology*, **44** (3), *585—586.*

133. SAHNI, B. 1927. The southern floras: a study in the plant geography of the past. *Proc. 13th. Indian Sci. Congr.*, *229—254.*

134. SANTESSON, R. 1945. *Cyttaria*, a genus of inoperculate Discomycetes. *Svensk bot. Tidskr.* **39** (4), *319—345.*

135. SCHIMPER, A. F. W. 1903. Plant geography upon a physiological basis. (English translation) Oxford, Clarendon.

136. SCHLECHTER, R. 1906. Beiträge zur Kenntnis der Flora von Neu-Kaledonien. *Engl. bot. Jb.* **39.**

264

137. SCHOPF, J. M. 1962. Preliminary report on plant remains and coal of the sedimentary section in the central range of the Horlick Mountains, Antarctica. *Polar Studies Inst. Rep.* **2** (Ohio State University Research Foundation).

138. SCHRÖTER, C. & HANRI, O. B. 1914. Versuch einer Übersicht der siphonogamen Polsterpflanzen. *Engl. bot. Jb.* **50** (suppl.)

139. SEWARD, A. C. 1914. Antarctic fossil plants. *British Antarctic Exp. 1910: Natural History Reports, Geology.* **1** (1), *1—49.*

140. SEWARD, A. C. & CONWAY, V. 1934. A phytogeographical problem: fossil plants from the Kerguelen Archipelago. *Ann. Bot.* **48** (141), *715—742.*

141. SIMPSON, G. G. 1940. Antarctica as a faunal migration route. *Proc. 6th. Pacific Sci. Congr. (1939)* **2**, *755—768.*

142. SIMPSON, G. G. 1953. Evolution and geography: an essay on historical biogeography, with special reference to the mammals. Condon Lectures, **8** (Eugene, Oregon, U.S.A.).

143. SKOTTSBERG, C. 1904. On the zonal distribution of South Atlantic and Antarctic vegetation. *Geogr. J.* **24**, *655—663.*

144. SKOTTSBERG, C. 1905. Some remarks upon the geographical distribution of vegetation in the colder southern hemisphere. *Ymer*, **25**, *402—427.*

145. SKOTTSBERG, C. 1910. Übersicht über die wichtigsten Pflanzenformationen Südamerikas S. von 41°. *Svensk Vet. Akad. Handl.* **46** (3).

146. SKOTTSBERG, C. 1912. The Vegetation of South Georgia. *Wiss. Erg. Schwed. Südpolar Exp. 1901—03,* **4** (13).

147. SKOTTSBERG, C. 1913. A botanical survey of the Falkland Islands. *Bot. Erg. Schwed. Exp. Patagonien und Feuerlands, 1907—09,* **3** (50).

148. SKOTTSBERG, C. 1915. Notes on the relations between the floras of subantarctic America and New Zealand. *Plant World*, **18** (5), *129—142.*

149. SKOTTSBERG, C. 1916. Die Vegetationsverhältnisse langs der Cordillera de Los Andes S. von 41°. *Svensk Vet. Akad. Handl.* **56** (5).

150. SKOTTSBERG, C. 1940. Report on Hawaiian bogs. *Proc. 6th Pacific Sci. Congr. (1939),* **4**, *659.*

151. SKOTTSBERG, C. 1942. The Falkland Islands. *Chron. Bot.* **7**, *23—26.*

152. SKOTTSBERG, C. 1953. Influence of the Antarctic Continent on the vegetation of southern lands. *Proc. 7th. Pacific Sci. Congr.* (1949), **5**, *92—99.*

153. SKOTTSBERG, C. 1954. Antarctic flowering plants. *Bot. Tidsskr.* **51**, *330—338.*

154. SKOTTSBERG, C. 1956. Derivation of the Flora and Fauna of Juan Fernandez and Easter Island. *Nat. Hist. Juan Fernandez & Easter Island,* **1** (3), 5, *193—438.*

155. SKOTTSBERG, C. 1960a. Remarks on the plant geography of the southern cold temperate zone. *Proc. Roy. Soc. London,* B, **152**, *447—457.*

156. SKOTTSBERG, C. 1960b. *Astelia* on Mauritius. *Svensk bot. Tidskr.* **54** (3), *477—482.*

157. STEBBINS, G. L. 1950. Variation and evolution in plants. New York, Columbia Univ. Press.

158. STEENIS, C. G. G. J. VAN, 1935. On the origin of the Malaysian Mountain flora, part 2. *Bull. Jard. bot. Buitenzorg, series III,* **13** (3), *129—257.*

159. STEENIS, C. G. G. J. VAN, 1953. Results of the Archbold Expedition's Papuan *Nothofagus. J. Arnold Arboretum*, **34** (4), *300—374.*

160. STEENIS, C. G. G. J. VAN, 1962. The land bridge theory in botany, with particular reference to tropical plants. *Blumea*, **11** (2), *235—542.*

161. STEERE, W. C. 1961. The bryophytes of South Georgia. Chapter 5 (pp. *34—48*) in "The Life Sciences in Antarctica". National Academy of Sciences, Washington; publication no. **839**.

162. SUTTON, C. S. 1928. A sketch of the vegetation of the Cradle Mountains, Tasmania; and a census of the plants. *Proc. Roy. Soc. Tasmania for 1928, 132—159.*

163. TAYLOR, B. W. 1954. An example of long distance dispersal. *Ecology,* **35** (4),

164. TAYLOR, B. W. 1955a. The flora, vegetation and soils of Macquarie Island. *Aust. Nat. Antarctic Res. Exp. Rep.* B, **2,**

165. TAYLOR, B. W. 1955b. Terrace formation on Macquarie Island. *J. Ecol.* **43** (1), *133—137.*

166. TAYLOR, G. 1960. Structure of Antarctica. *Aust. J. Sci.,* **22** (7), *299—300.*

167. TAYLOR, G. 1963. Probable disintegration of Antarctica. *Geogr. J.* **129** (2), *190—191.*

168. THIEL, E. 1961. Antarctica: one continent or two? *Polar Rec.,* **10** (67), *335—348.*

169. THONNER, F. 1915. The flowering plants of Africa. London, Dulau & Son.

170. TROLL, C. 1958. Tropical mountain vegetation. *Proc. 9th. Pacific Sci. Congr.* **20**, *37—46.*

171. TURRILL, W. B. 1953. Pioneer plant geography: the phytogeographical researches of Sir Joseph Dalton Hooker. The Hague, M. Nijhoff.

172. WACE, N. M. 1960. The botany of the southern oceanic islands. *Proc. Roy. Soc. London.* B, **152**, *475—490.*

173. WACE, N. M. 1961. The vegetation of Gough Island. *Ecol. Monogr.* **31** (4), *337—367.*

174. WACE, N. M. & DICKSON, J. (in the press). The terrestrial botany of the Tristan da Cunha Islands. *Phil. Trans. Roy. Soc. London,* B.

175. WACE, N. M. & HOLDGATE, M. W. 1958. The vegetation of Tristan da Cunha. *J. Ecol.* **46** (3), *593—620.*

176. WALKOM, A. B. 1949. Gondwanaland: a problem of palaeogeography. *Rep. Aust. & N-Z. Ass. Adv. Sci.* **27**, *1—17.*

177. WARMING, E. 1909. Oecology of Plants (English translation) Oxford, Clarendon.

178. WEBERBAUER, A. 1911. Die Pflanzenwelt der Peruanischen Anden. *Veg. Erde,* **12.**

179. WERTH, E. 1928. Die Vegetation der Subantarktischen Inseln Kerguelen, Posession und Heard Eiland. *Dtsch. Südpol. Exp. 1901—03,* **8** (1) no. 6, *125—176.*

180. WILLIS, J. H. 1962. A handbook to plants in Victoria, I Melbourne University Press.

181. STRAKA, H. 1960. Über Moore und Torf auf Madagaskar und den Maskarenen. *Erdkunde,* **14** (2), *81—98.*

182. BURBIDGE, N. T. 1963. Dictionary of Australian plant genera (Gymnosperms and Angiosperms). Sydney, Angus & Robertson.

183. FLORIN, R. 1962. The distribution of conifer and taxad genera in space and time. *Act. Hort. Berg.,* **20** (4).

184. WEIMARCK, H. 1941. Phytogeographical groups, centres and intervals within the Cape Flora. *Lunds Univ. Årsskr.* NF 37 (5), *1—143.*

185. SKOTTSBERG, C. 1937. Recent researches in *Astelia. Trans. Roy. Soc. N.Z.,* **67**, *218—226.*

186. ADIE, R. J. 1962b. The geology of Antarctica. *Amer. geophys. Union Monogr.* **7**, *26—39* (Ed. H. WEXLER, M. J. RUBIN & J. E. CASKEY).

187. ADIE, R. J. 1963. Geological evidence on possible Antarctic land connections. Pp. *455—463* in "Pacific Basin Biogeography" (Ed. J. L. GRESSITT) – Bishop Museum Press, Honolulu.

188. CRANWELL, L. M. 1963. Nothofagus: living and fossil. Pp. *387—400* in "Pacific Basin Biogeography" (Ed. J. L. GRESSITT) – Bishop Museum Press, Honolulu.

266

189. FLEMING, C. A. 1963. Paleontology and southern biogeography. Pp. *369—385* in "Pacific Basin Biogeography" (Ed. J. L. GRESSITT) – Bishop Museum Press, Honolulu.
190. HAIR, J. B. 1963. Cytogeographical relationships of the southern podocarps. Pp. *401—414* in "Pacific Basin Biogeography" (Ed. J. L. GRESSITT) – Bishop Museum Press, Honolulu.
191. KNOX, G. A. 1963. Antarctic relationships in Pacific Biogeography. Pp. *465—474* in "Pacific Basin Biogeography" (Ed. J. L. GRESSITT) – Bishop Museum Press, Honolulu.
192. KUSCHEL, G. 1963. Problems concerning an Austral region. Pp. *443—450* in "Pacific Basin Biogeography" (Ed. J. L. GRESSITT) – Bishop Museum Press, Honolulu.
193. PREEST, D. S. 1963. A note on the dispersal characteristics of the seed of the New Zealand podocarps and beeches and their biogeographical significance. Pp. *415—424* in "Pacific Basin Biogeography" (Ed. J. L. GRESSITT) – Bishop Museum Press, Honolulu.
194. STEENIS, C. G. G. J. VAN, 1963. Transpacific floristic affinities, particularly in the tropical zone. Pp. *219—232* in "Pacific Basin Biogeography" (Ed. J. L. GRESSITT) – Bishop Museum Press, Honolulu.
195. CARRICK, R., HOLDGATE, M. W. & PREVOST, J., (Eds.), 1964. Biologie Antarctique. *Proc. 1st. Symposium Antarctic Biology (Paris, Sept. 1962)* – Paris, Hermann.
196. EVANS, J. W. 1959. The Zoogeography of some Australian insects. Chapter 9 (pp. *150—163)* in "Biogeography and Ecology in Australia" (Ed. A. KEAST) – The Hague, Dr. W. Junk.

MICROBIOLOGY OF ANTARCTICA

BY

JOHN MCNEILL SIEBURTH

Associate Professor of Oceanography, Narragansett Marine Laboratory of the Graduate School of Oceanography, University of Rhode Island, Kingston, Rhode Island, U.S.A.

Introduction

The association of microorganisms with chemical changes in organic materials was ably demonstrated by LOUIS PASTEUR in the late 1850's. These observations aroused the curiosity of naturalists and physicians who accompanied exploring vessels into the high latitudes. At a time when heterogenesis still had proponents, NYSTRÖM (1868) aboard the Sofia repeated PASTEUR's classical "sterile broth" experiments at Spitzbergen. This demonstration of the presence of bacteria in Arctic air was apparently the first polar bacteriological study.

NANSEN (1897) made frequent microscopic examinations of melt-pools on Arctic ice-flows during the summer. Algae and diatoms bloomed at the bottom of these pools and provided food for protozoans. Bacteria were occasionally observed. An exogenous source of microorganisms was suggested by the presence of dust sprinkled on the surface of snow. LEVIN (1899) was apparently the first to attempt the cultivation and isolation of bacteria in the Arctic during the Nathorst Expedition to Spitzbergen. *Escherichia coli* was obtained from the intestines of a polar bear, seals and gulls, while other mammals and birds were "bacteriologically sterile". Sea water was found to have less than one organism per ml by LEVIN's procedures.

The Sixth International Geographical Congress held in London in 1895 drew attention to the lull in Antarctic exploration during the previous 50 years. It emphasized the importance of Antarctic exploration and urged scientific societies to promote expeditions. Although the numerous expeditions which followed up to World War I concentrated on geographic exploration, some microbiological studies were conducted under all but impossible conditions. By 1900 bacteriological techniques had been developed which permitted the enumeration of heterotrophic bacteria, the classification of bacteria by cultural and morphological characteristics, and the detection of certain biochemical activities. GAZERT, the bacteriologist for the German South Polar Expedition, planned a study which was outlined before the expedition in both German (GAZERT, 1901a) and English (GAZERT, 1901b). These accounts may have stimulated the medical officers of the other expeditions, thereby precipitating the Antarctic bacteriological activity that occurred during the next decade.

The German South Polar Expedition of 1901—1903, led by DRY-GALSKI in the Gauss, visited Kerguelen and Heard Islands and the mainland at Gaussberg (90° E. Long.). GAZERT (1912) studied the microflora of the air, sea water and intestinal contents of birds, seals and fish. During the concurrent Swedish South Polar Expedition of 1901—1903, led by NORDENSKJÖLD in the Antarctic, EKELÖF (1907, 1908) conducted bacteriological studies at Snow Hill Island (east coast of the Antarctic Peninsula) on soil, air and surface sea water. The Scottish National Antarctic Expedition of 1902—1904, led by BRUCE in the Scotia, wintered over at Laurie Island in the South Orkneys (60° S, 44° W) and crossed the Weddell Sea to Coats Land. PIRIE (1912) made air exposure plates, examined sea water both in the Weddell Sea and at Laurie Island, and cultured birds and seals and a fish at Laurie Island. ROUX (1907), the noted bacteriologist, prepared a bacteriological program for the French Antarctic Expedition of 1903—1905 which explored the western coast of the Antarctic Peninsula. Its leader, CHARCOT (1908), examined intestinal material from birds, seals and fish and brought back mixed cultures and two soil samples. These materials were used by Mlle. TSIKLINSKY and BELLIAEFF under the direction of Professor METCHNIKOFF to isolate and describe their bacterial types (TSIKLINSKY, 1908). E. L. ATKINSON, the surgeon with SCOTT's British Antarctic Expedition of 1910—1913, made snow cultures (SCOTT, 1914) but apparently did not publish his findings. The Australasian Antarctic Expedition of 1911—1914, led by MAWSON in the Aurora, established a base at Commonwealth Bay (King George V Land), from which the surgeon, McLEAN (1918, 1919) conducted studies on seaweed, glacier ice, air, birds, seals, and the nose and throat of man.

The only bacteriological study between World Wars I and II was conducted during the Byrd Antarctic Expedition II of 1933—1934. Samples and broth enrichment cultures of snow as well as samples of plant debris and soil brought back by SIPLE were used by DARLING to isolate and identify bacteria (DARLING & SIPLE, 1941).

SLADEN (1961), a medical officer with the Falkland Island Dependencies Survey from 1948—1951 at Hope Bay and other bases, followed up McLEAN's (1919) observations on the Staphylococcal flora of the nasopharynx of men in Antarctica and examined penguins for fungi (SLADEN, 1954). As a member of the 1951—1952 Australian National Antarctic Research Expedition to Macquarie Island, BUNT studied the microflora of birds and a sea elephant (BUNT, 1955a); sea water and kelp (BUNT, 1955b) and soil (BUNT & ROVIRA, 1955a, 1955b). McBEE (1960) made cultures from a few birds and seals at McMurdo Sound during the U.S. Expedition

Deep Freeze II of 1956—1957. In connection with the International Geophysical Year and the Argentine Antarctic Expedition of 1957—1958 in the South American Quadrant of Antarctica, SIEBURTH (1959a, 1959b, 1959c) examined the ecological relationships between Antarctic birds and their diet in an attempt to explain the occurrence of "bacteriologically sterile" animals found by most of the earlier workers. The respiratory flora and diseases of birds were also studied (SIEBURTH, 1958b). Field studies conducted during the 1958—1959 Argentine Expedition included observations of the microflora of soil and guano at penguin rookeries (SIEBURTH, 1962) and led to the isolation and identification of the algal antibacterial detected the previous year (SIEBURTH, 1960a, 1961b). The continuance of the IGY of 1957—1958 for an unspecified time has maintained much activity in Antarctica as is shown by an increasing number of notes which will be described later in detail. Unfortunately there is a growing tendency of these workers to use undescribed frozen samples randomly collected by others as a source for study.

The purpose of this chapter is to attempt an appraisal of our knowledge in this field by drawing together the heterogeneous fragments obtained over the last 60 years. It is hoped that the great voids in our knowledge will stimulate others to conduct comprehensive ecological studies on the numerous habitats in the Antarctic environment.

Air, snow and ice habitats

The most obvious microbial link between the Antarctic Continent and other land masses is the air with its dust particles and associated microflora. There are two outstanding questions. To what degree do these air-borne organisms survive their trip? How able are they to adapt to the meager substrates and the extremes of temperature, pressure and light?

The expeditions between 1901 and 1904 attempted to cultivate air-borne bacteria. While the Gauss was frozen in the pack-ice GAZERT (1912) attempted to determine if bacteria were present in the air. His cultures of freshly fallen snow were sterile in every instance. EKELÖF (1908) made air exposure plates at Snow Hill Island during the summer and obtained bacterial growth on more than half of them. He estimated that the settling rate on the plates was two hours per bacterium. The source of these organisms was believed to be the microflora from the surrounding soil which was stirred up by the unusually strong winds. PIRIE (1912) exposed plates and tubes in the crow's nest of the Scotia's main mast for periods up to 20 hours during the voyage into the Weddell Sea. These cultures and those exposed during the winter at Laurie Island were negative. The only positive plates were those exposed on the deck, which were believed to be contaminated from the ship and the sea spray. These three studies indicated that if the air were not sterile then the content was lower than could be detected by the methods used.

However, affirmative results were obtained by two investigations between 1910 and 1914 in the Australian Quadrant of Antarctica. ATKINSON (SCOTT, 1914) examined snow and isolated a "very motile bacterium". It was believed to be air-borne and to have been brought down by the snow although bacteria were not found in the air itself. McLEAN (1918, 1919) observed by direct microscopic and cultural means that falling snow, glacier-ice and thaw-water consistently contained certain bacterial and yeast forms. On the basis of these observations and current views of ice formation he developed an interesting hypothesis on the development of an autochthonous ice microflora from air-borne microorganisms. This theory was that bacteria adhering to dust particles ascend by the rising equatorial air and are brought pole-ward by winds at high altitudes where they are frozen to snow flakes and drop to the Antarctic plateau. During the transformation of snow to ice crystals, impurities such as inorganic salts are concentrated to form an inter-crystalline "brine," with a lower freezing point (as low as —13° C), which at low temperatures binds the interlocking grains. The glaciers so formed approach the coast with the passage of time and undergo a super-ficial surface thaw during the four summer months. During this period the temperatures approach or exceed 0° C permitting the "brine" to thaw forming a menstrum in the inter-crystalline channels where an indigenous microflora could develop and thrive.

During the second Byrd Antarctic Expedition, 1933—1934, SIPLE (DARLING & SIPLE, 1941) made field observations on the microflora of snow at Little America. Direct platings of one to two ml aliquots incubated at 37° C yielded no growth. However, the addition of 50—200 ml aliquots of melted snow to reconstitute concentrated sterile broth yielded growth in many instances. Pint jars of broth exposed to air atop mountain peaks for several days also yielded growth. DARLING used dessicated cultures, broth enrichment cultures, and melted snow samples brought back by SIPLE to isolate as many different species as possible that would grow on nutrient agar after incubation for two days at 35° C and five days at room temperature. The organisms which were described in detail according to source consisted of various *Bacillus, Achromobacter, Flavobacterium, Proactinomyces* and *Micrococcus* species.

BUNT (1955b) exposed plates of nutrient agar for ten minute periods at various altitudes from sea-level to 1200 feet at Mac-quarie Island. The plate counts and the wind speed were used to

271

estimate the number of bacteria per 100 cubic yards of air*. The value of 300 organisms/100 cubic yards of air at sea level fell off rapidly to 13 at 100 feet, 4 at 200 feet, and none at 300 and 500 feet. BUNT believed that these air-borne organisms, which were mainly Gram-negative rods, were of marine origin. Unfortunately a sea water medium was not used in addition to the fresh water medium.

SIEBURTH (1962) discussed some preliminary observations made on the bacterial content of air during the December 1957 crossing of the Drake Passage. Samples of air ranging from 25—100 liters were obtained from the foredeck of the ship with a membrane filter aerosol sampler. The membranes which were cultured on Eugonagar (BBL) at 37° C for one week, failed to detect growth. While guano and soil samples were being obtained from penguin rookeries at three islands in the South Shetland group in January, 1959, air exposure plates were also made. Despite the presence of nesting penguins and winds, heart infusion agar plates (Difco) exposed on the ground between nests for periods up to eight hours also failed to yield detectable organisms.

A frozen sample of glacier ice examined by STRAKA & STOKES (1960) contained no detectable organisms when plated at a 10^{-2} dilution. Broth enrichment yielded an undescribed yeast with a temperature optimum between 5 and 10° C and a maximum growth temperature of 20° C.

The important conclusion that can be drawn from these studies is that the viable microflora of air surrounding the Continent is quite sparse, but that sufficient organisms survive to at least contaminate snow and glacier ice in detectable quantities. These survivors may even give rise to an autochthonous snow-ice flora. During the IGY aerial sampling was conducted during flights to and over the Continent. This material is being examined under the direction of DR. ORVILLE WYSS of the University of Texas, but unfortunately the data have not been made available. Such studies when compared with similar observations over temperate, equatorial, and arctic areas would help clarify the potential microbial inoculum of Antarctic air. Ice cores from various glaciers were also collected and sent to the Snow, Ice and Permafrost Research Establishment (now the Cold Regions Research Engineering Laboratory) of the U.S. Army Corps of Engineers. Apparently the anticipated bacteriological studies were not conducted.

Terrestrial habitats

Soil

The origin of the air-snow-ice microflora in Antarctica has been traced to air-borne microorganisms arising from the soils in lower

* cubic yard = 0.7645 m³.

272

latitudes. But what of the soil microflora in Antarctica? Except for the mountain ranges and the nunatuks which jut out from the snow-ice cover the only other exposed regions are the rocky and volcanic areas bordering the Continent and innumerable islands. Moraines are the principal soil accumulations. In many areas soil is quite scarce. The "soil" is often of laval origin or consists of a thin layer rich in humus from penguin feces and the debris from mosses and lichens.

The sinking of the Antarctic and the enforced additional year of the Swedish Expedition gave EKELÖF (1908) an opportunity to continue his studies on the soil microflora of Snow Hill Island. Ninety-three of the 105 soil samples examined contained bacteria. Samples obtained from depths greater than 20 cm were free of detectable bacteria. The counts obtained in summer were ten times those of the winter samples. The bacterial content averaged 2×10^4 organisms/cc of soil while the maximum count was 1.4×10^5 organisms/cc. At the optimum temperature of $17.5°$ C as long as six to eight days incubation was required for detectable growth on gelatin media. The cultural characteristics were given for 29 different bacillary, coccal, spiral and filamentous forms. EKELÖF is apparently the only Antarctic bacteriologist who has conducted an ecological study on a seasonal basis.

TSIKLINSKY (1908) examined two soil samples brought back by CHARCOT. One sample was from a moraine frequented by penguins and other birds and the second was from the top of a rock covered with vegetation. Although the bacterial content was low compared to French soil, sufficient growth was obtained to select eight different isolates including *Bacillus* species, a filamentous organism *(Streptothrix)*, a coccus and a red yeast. There were no obligate anaerobes present. Attempts by OMELIANSKY to obtain nitrifiers by enrichment culture were negative, presumably as a result of the age of the two year old samples. The culture plates always gave rise to fungi including *Aspergillus glaucus*, *Penicillum glaucum* and a species of *Mucor*.

McLEAN (1919) examined morainic mud from Cape Denison. A wet mount exhibited numerous rod and coccal forms which were cultured and briefly described. This same material supported the growth of wheat and barley to a height of eight to ten inches. Smears of a number of samples of morainic mud and granitic sand which were not examined until four years later indicated that the numbers of bacteria were greatly reduced in comparison to the fresh material. Organisms recovered by broth enrichment culture, unlike the organisms cultured in the Antarctic, grew better at $37°$ C than at 18—$20°$ C. The cultural characteristics of these isolates were given in detail.

Only one of three samples of alluvial soil from Byrd Land examined by DARLING & SIPLE (1941) yielded detectable growth. *Bacillus fusiformis, B. mesentericus* and *B. subtilus* were obtained.

When the Aurora visited sub-Antarctic Macquarie Island, McLEAN (1919) was impressed with the bacteriological opportunities to study the soil and their plant associations. Although these materials were described at length, only one sample was examined. BUNT & ROVIRA (1955a) studied the bacterial flora, and other characteristics of relatively fresh soil samples representative of certain plant associations on Macquarie Island. The bacterial content of the roots and rhizosphere was greater than that of the adjacent soils. The few morphological and cultural properties studied failed to indicate a qualitative difference between the rhizosphere and non-rhizosphere isolates. Some isolates grew better at 10° than at 25° C, but the majority did better at the higher temperature. Respiration studies indicated that activity was dependent upon the total carbon content. The effect of temperature and heat treatment on respiration of Macquarie soils was determined by BUNT & ROBIRA (1955b). Respiration increased between 10° and 37° C and decreased between 37° and 50° C. The immediate respiration which followed the wetting of air dried soils was attributed to the oxidative enzymes of "dead" microorganisms.

Nine frozen "soil" samples of an undescribed nature from the McMurdo Sound region were examined by STRAKA & STOKES (1960) for the presence of "psychrophiles". Growth was detected in the 30°C platings from seven samples but only in four samples incubated at 0° C for a fortnight. The 0° C counts ranged up to 2×10^4/g but never exceeded the 30° C counts which were as high as 4.1×10^5. Although organisms were present which could grow at Antarctic temperatures, the higher 30° C counts and the temperature optima above 20° C suggest that these organisms may have been of animal origin. Unfortunately none of the isolates were keyed to genera.

FLINT & STOUT (1960) described the generic nature of the microfauna and microflora of soil samples from the McMurdo Sound area. Terrestrial algae and ciliate protozoans typical of temperate soils were found. In contrast the monera were more distinctive. Fungi were uncommon and only one site yielded *Streptomyces*. There were more chromogenic bacteria than are normally found in temperate soils, spore-forming bacteria were rare, and there was a large proportion of Gram-positive cocci. The bacterial isolates grew at both 4° and 25° C. Attempts to isolate aerobic nitrogen fixing bacteria were unsuccessful. In contrast, BOYD & BOYD (1962) were able to isolate *Azotobacter chroococcum* and *A. indicus* from enrichment cultures of a frozen sample of beach material from the Windmill Islands, Wilkes Station (110° E. Long.).

Frozen soil samples obtained from the McMurdo Sound area by

the New Zealand group during the IGY were examined for the numbers and species of yeasts by DI MENNA (1960). From a series of seven samples taken in 1958, less than 100 yeasts/g were obtained from one out of three samples plated on acid glucose peptone agar (AGPA) and from four out of the seven samples plated on casein agar. An exception was one sample plated on AGPA which yielded 5×10^3 colonies/g. Eighty-one of the 84 isolates were *Candida scottii*. From a series of five samples obtained in 1959, four yielded yeasts when plated on AGPA with or without aureomycin. *C. scottii* accounted for two of the 14 isolates obtained from plates incubated at room temperature (20° C) and 18 of the 26 isolates incubated at 4° C. The other isolates were *Sporobolomyces odorus* (5), *Cryptococcus laurentii* (5), *C. albidis* (2), *C. luteolus* (1), *C. diffluens* (2), *Rhodotorula minuta* (4), and *R. graminus* (1). Enrichment cultures in nitrogen depleted broth selected a somewhat different flora. *Candida scottii*, the dominant yeast species obtained, was an obligate psychrophile. All 18 *C. scottii* isolates tested grew after four days at 0° C. However, growth was only detected in half of the isolates at 15–16° C and in none at 20° C after ten days incubation.

An admirable step taken by DI MENNA was to deposit cultures of *C. scottii* in collections at other laboratories.

Forty-eigth soil samples from East and West Ongul Islands and the Continent (Syowa, 40° E. Long.) collected during the third and fourth Japanese Antarctic Research Expeditions (1959—1960) were kept in a frozen state aboard the Soya until examined. They were used to inoculate acid malt agar plates for the isolation of fungi (TUBAKI, 1961) and yeasts (SONEDA, 1961) at 15°, 20° and 25° C. Fungi and yeasts were not detected in 25 of the 48 samples. Thirteen isolates of five yeast species were obtained from these samples. *Cryptococcus laurentii* and *Rhodotorula mucilaginosa* (five isolates each) were considered by SONEDA (1961) to exist in Antarctic soil due to their dominance and ability to grow at lower temperatures (5° C) than the other isolates. Singles isolates of *Cryptococcus albidis*, *Torulopsis famata* and *Trichosporon cutaneum* were also obtained. The 19 fungal isolates belonged to four species. All grew at 5° C but none at 30° C. Fourteen isolates of *Blodgettia borneti* were obtained from two locations. This apparently algicolous fungus was considered by the author to be a member of the indigenous flora. A single isolate of *Chrysosporium pannorum* grew well at 5° C and was considered to be a temperate contaminant adapted to the Antarctic environment. Two isolates of *Chrysosporium verrucosum* sp. nov. and a single isolate of *Cylindrium griseum* were not considered of any significance.

HARDER & PERSEIL (1962) used five refrigerated soil samples from three areas in the South American Quadrant of Antarctica to isolate lower soil *Phycomycetes*. Approximately ten sea water and ten fresh water enrichment cultures were incubated at both 10—15° and 22° C. The six *Phycomycete* isolates were obtained from the fresh water enrichments and belonged to species (*Rhizophydium utriculare, Rhizophydium sphaerotheca,* and *Hyphochytrium* cf. *Catenoides*) which were considered ubiquitous. The presence of Actinomycetes, Pythiaceae, Saprolegniaceae, Zygomycetes and fungi with septate-hyphae was noted.

Fecal material in various stages of decomposition at penguin rookeries on three islands in the South Shetland group were examined for differences in pH and the bacterial flora by SIEBURTH (1962). The volcanic ash on Deception Island apparently prevented the usual accumulation of guano, which did not exceed a quarter inch in depth. In contrast to the other rookeries studied the alkaline samples were devoid of antibacterial activity, and the total bacterial counts of the different samples varied considerably. A pooled sample of chick and adult feces (4.0×10^7 organisms/g) was dominated by *E. coli* while a vacant nest contained a mixed flora (1.5×10^5/g) of *Sarcina, Micrococcus, Mycococcus* and *Bacillus* species. The "soil" on a steep incline of a ridge high above the rookery (free of direct penguin contamination) contained a dominant *Mycococcus* flora (10^5 organisms/g) and a minor *E. coli* (6×10^2/g) population. However the "soil" between nests had only a small population (10^3/g) of fastidious organisms. The "soil" in a ravine between nesting areas contained a mixed "soil" and "fecal" flora of 10^5 organisms/g. (*Mycococcus* spp., *E. coli, Streptococcus faecalis* and *Sarcina ureae*). An adjacent pond in the ravine which was green with a bloom of filamentous algae was free of detectable bacteria. The variation between fresh samples from one small area illustrates the folly of describing the microflora of the Antarctic from the study of a few frozen samples of unspecified origin.

The ecology of soil microorganisms must be studied over a certain period of time on location. The study underway at the McMurdo Biological Laboratory by Dr. WILLIAM L. BOYD of Ohio State University is a necessary step in the right direction.

Cryptogamic flora

One of the most overlooked habitats in the Antarctic is the cryptogamic flora. These plants are one of the few non-fecal sources of organic matter. The known antibacterial properties of lichens suggests some very interesting ecological relationships.

Moss from the McMurdo Sound region has been studied by the New Zealand workers. FLINT & STOUT (1960) noted that in contrast to the distinctive soil flora, the bacterial flora of moss was quite similar to epiphytes found in temperate climates. This consisted mainly of chromogenic Gram-negative organisms. A week-old moss sample *(Bryum antarcticum)* flown out in 1957 was studied by DI

MENNA (1960). Cultures incubated at both 4° C and room temperature yielded similar populations (7—7.5 × 10^4/g) and species distributions (*Cryptococcus albidis*, 72 and 75%; *C. laurentii*, 18 and 20%; and *Rhodotorula minuta*, 7 and 8.5% of the 95 isolates). Although these results were similar to her observations on comparable New Zealand material, they differed greatly from frozen soil samples which yielded the obligate psychrophile, *Candida scottii*, as the dominant yeast.

BURKHOLDER (1961) discussed his unpublished observations made during January, 1959, on the antibiotic activity of some 100 samples of lichens examined. About half of the samples tested exhibited activity and one, a species of *Ramalina*, possessed the ability unique among lichens to inhibit Gram-negative bacteria. The possible effect of the anti-bacterial substance produced by these organisms on their epiphytic flora would be of interest.

Ponds

The numerous ponds and lakes which thaw during the summer and possess simplified and unique microbiotas are of great interest. Unfortunately they have been virtually overlooked. An exception is the note by BARGHOORN & NICHOLS (1961) who reported the presence of a rather complex ecosystem involving the sulfur cycle under high salinity and great temperature variation (—51° to 4.5° C). They studied the pyritic sediments of kettle holes in the Wright Dry Valley in the McMurdo Sound region. A 25 feet by 10 feet by 1 foot deep saline pond (7.6 × 3.0 × 0.3 m) containing 132‰ of dissolved solids, filamentous algae and diatoms, had sediments whose black sub-surface color was attributed to hydrogen sulfide. The presence of *Desulfovibrio* was indicated by the precipitation of iron sulfide in broth media. This reaction was more rapid at room temperature than at 5° C.

Marine habitats

Sea water and sediments

The highly productive marine basins surrounding the Continent are an apparent paradox since their rich organic content seems to maintain such a negligible bacterial content. Equally rich water in temperate zones at comparable temperatures ($\pm 2°$ C) maintain levels of 10^4 organisms/ml. Examination of Antarctic marine waters for their bacterial content in the years 1901—1904 are only antedated by a few classical marine bacteriological studies of temperate waters (ZoBELL 1946).

GAZERT (1912) used the methods of FISCHER, 1894, (see ZoBELL, 1946) to enumerate bacteria in the surface waters of the Atlantic between Kiel and Capetown. His count of less than one organism/ml was low compared to Fischer's average of 10^3 organisms/ml for 175 samples of Atlantic water. In the Antarctic Gazert found 0—10 organisms/ml except around attached algae where 10—30 organisms/ml were found. Appreciable numbers were found in sediments at depths as great as 5,320 meters. Attempts to detect nitrifying organisms were negative. Denitrifying bacteria were found in small numbers, and although activity was observed at 5°—10° C, it was greater between 20° and 25° C. Inshore surface waters were also examined by Ekelöf (1908) near Snow Hill Island. Although never more than 21 organisms/ml were observed, his average count was 4.4 organisms/ml. Five different spiral and rod forms were isolated.

Broth and gelatin slants inoculated with several drops of surface water from the Weddell Sea by Pirie (1912) yielded growth in seven out of ten cultures incubated at 7°—15° C. He observed denitrifiers quite commonly but only at temperatures above 0° C. Surface water counts obtained by Pirie on denitrification media yielded 35, 170 and 334 colonies/ml while sea water gelatin yielded 112 colonies/ml in a fourth sample. In contrast to his average count of these four samples (1.63×10^2/ml) water from 2,000 and 2,500 fathoms (3,700—4,700 m) yielded two and one colony, respectively, from 5.2 ml aliquots. Cultures from 13 sediment, bottom water and intermediate depth samples (5—10 ml) were free of detectable bacteria. Five attempts to detect nitrifiers in two enrichment media were negative. Two samples of bottom sediments from the Bay of Whales, McMurdo Sound, yielded only one unidentified culture (Darling & Siple, 1941).

Bunt (1955b) inoculated fresh water nutrient broth with aliquots of plankton hauls from the bays at Macquarie Island. The enrichment broths were plated on fresh water nutrient agar and 20 isolates of "marine bacteria" were keyed to seven genera.

A systematic study of the heterotrophic bacterial content of Antarctic and sub-Antarctic waters was included in an investigation of the Indian Ocean during the second Antarctic voyage of the Soviet vessel Ob, 1956—1957 during the IGY. The procedures and general results obtained from the 1,117 samples from 61 stations taken along the 20° and 40° E. longitudes between Antarctica and Africa and along 97° and 90° E. longitude between Antarctica and the Ganges River were outlined by Lebedeva (1958). Forty ml aliquots were used to inoculate membrane filters which were incubated on fish peptone agar at 18° to 25° C in the Antarctic and sub-tropics and 28° to 35° C in the tropics. Forty-three per cent of 693 samples of Antarctic, sub-Antarctic, and sub-tropical waters were "sterile" while 13% contained one colony. Bacterial populations increased with proximity to each of the continents, including Antarctica.

278

The data of LEBEDEVA (1958) have formed the basis of a number of papers on the use of the horizontal and vertical distribution of cultivable heterotrophs as hydrological indicators of ocean circulation (KRISS, 1959; KRISS et al., 1960a, 1960b). This use of cultural rather than their direct microscopic procedures is a radical departure from the work of KRISS and colleagues over the past decade (SIE-BURTH, 1960b). The rationale behind this new approach is that bacteria are more sensitive indicators of organic matter than present chemical methods and that the bacterial content is an index of readily assimilable organic matter which characterizes a water mass. On this basis the "sterile" and "low" content layers are assumed to be Antarctic water masses which are distinguished from the "high" content equatorial-tropical water masses. The penetration of different water masses into other areas was indicated by layers of different bacterial density. The cultural differences of representative isolates were apparently characteristic of the water mass. Some thousand isolates are being studied to elaborate on the specificity of bacterial types with specific water layers.

Membrane filters inoculated with surface waters from the Drake Passage in 1957 were cultivated on broth media made with varying proportions of eugonbroth (BBL) and sea water (SIEBURTH, 1962). A few samples south of the Antarctic Convergence contained as many as a hundred orange pigmented mesophilic heterotrophs per ml, but most of the samples contained fewer than ten organisms/ml. Sea water agar pour plates (heart infusion agar base, Difco) prepared by PAUL R. BURKHOLDER during the 1958—1959 phytoplankton studies in the Bransfield and Gerlache Straits (BURK-HOLDER & SIEBURTH, 1961) never yielded more than a few organisms per ml in surface water samples. However, BURKHOLDER (1961) found that low counts of bottom sediments from Melchior Archipelago were markedly increased when the medium was supplemented with partially decomposed phytoplankton net concentrates before autoclaving.

A better understanding of nutrient-temperature interrelationships of psychotrophic bacteria may reveal a functional bacterial microflora in Antarctic waters which is not only associated with the rich phytoplankton blooms but sustains it by the *in situ* recycling of limited or marginal nutrients. However, another factor which may limit bacterial populations is the liberation of antibacterial substances by marine algae.

STEVENS aboard the USS Staten Island during the 1960—1961 Antarctic program in McMurdo Sound determined the "most probable numbers" of aerobic and anaerobic bacteria as well as the soluble organic carbon content of sea water, bottom sediments and pack ice (unpublished data, STEVENS & OPPENHEIMER, Institute of Marine Science, University of Texas). This preliminary work involving chemical characterization of the environment in conjunction with bacterial studies is very encouraging.

More meaningful studies on the ecology of marine bacteria cannot be conducted until more useful apparatus and analytical techniques are developed for shipboard use. Waters must be characterized by the types and amounts of their organic compounds and their microfloras must be described in regard to their thermic, physiologic and taxonomic types. Zero time enzyme procedures and nuclear methods would be helpful in determining the rates of specific bacterial processes. Only with such techniques can one attempt to assess the ecological significance of marine bacteria in certain processes and in the economy of the sea in general.

Phytoplankton and seaweed

The apparent sparsity of bacteria in Antarctic marine waters has been discussed in the foregoing section. In addition to producing organic matter suitable as bacterial substrates, the phytoplankton may also produce materials inhibitory to bacterial growth. Although it seems unreasonable to expect that this is the main factor accounting for the sparsity of bacteria, the very presence of such compounds makes it necessary to consider them when studying this problem.

The well recognized bactericidal property of sea water has been reviewed by CARLUCCI & PRAMER (1960). LUCAS (1947) has drawn attention to the possibility that antibiotics are produced in the sea by marine organisms. While studying the gastrointestinal microflora of Antarctic birds, SIEBURTH (1959a) observed the absence of a detectable gastrointestinal microflora in the anterior segments of certain pygoscelid penguins and the absence of lactose-fermenting strains of *Escherichia coli* in all pygoscelid penguins studied. Bacterial inhibition in their stomach contents was due to their euphausid diet. The antibacterial activity of the euphausids *(Euphausia superba)* was traced to their stomach contents and then to phytoplankton blooms dominated by a green mucilaginous colonial alga (SIEBURTH, 1959b). These observations were confirmed and the alga was identified as *Phaeocystis pouchetii* during a second field study when algal concentrates were collected for laboratory studies (SIEBURTH & BURKHOLDER, 1959).

The antibiotic principle of *Phaeocystis pouchetii* was found to be a volatile acid which was isolated as the sodium salt and identified chemically, physically and biologically as acrylic acid, $CH_2 = CH—COOH$ (SIEBURTH, 1960a). The antibiotic properties of acrylic acid *in vitro* such as its acid potentiation as well as its *in vivo* activity in the acidic avian gut have been described (SIEBURTH, 1961b). This naturally occurring antibacterial is an inexpensive industrial chemical and its potential use as a feed additive and prophylactic has been recognized (SIEBURTH, 1961b; WHITE-STEVENS, et al., 1961). The observations by the early workers that the gastrointestinal microflora of certain birds was suppressed led to this demonstration of an antibacterial in marine phytoplankton and its application to the feed industry. This work suggests that other basic research in Antarctica may lead to concrete applications in our everyday life.

An immediate objection to attributing the sparsity of marine bacteria in Antarctic waters to the high acrylic acid content (8% dry weight) of the *Phaeocystis* blooms (SIEBURTH, 1960a) is that an acidic medium such as the avian gut enhances the antibacterial activity while sea water is highly buffered and has an alkaline pH (8.0 ± 0.5). Observations on the effect of free acrylic acid upon natural populations of marine bacteria (SIEBURTH, 1962) indicate that concentrations of 1—100 μg/liter enhanced growth while those above 10 mg/liter which decreased the pH also suppressed growth. It is possible that the microinterface between the algal cell (pH 6.05) and the sea water is also acidic and antibacterial as a result of acid excretion by actively metabolizing cells. It seems plausible that algal acids may provide a nutrient source for marine bacteria as well as serve as a protective mechanism against bacterial decay of the metabolizing algal cells.

BUNT (1955b) discussed the possible importance of the large brown seaweeds in the ecology of the narrow littoral zone of Macquarie Island. He conducted an experiment on the decomposition of the giant kelp, *Durvillea antarctica*. Bacterial development was restricted mainly to the water surface and the parts of the kelp that were exposed to the air. Kelp flies accelerated decomposition and offset the effects of low temperature, presumably by spreading bacteria. Yeasts were numerous within the kelp tissues where they were believed to break down the algal carbohydrates. Large numbers of protozoa were present in the liquid.

Fish and invertebrates

Studies are being conducted on the occurrence, distribution and taxonomy of fish during the expeditions of a number of countries. Unfortunately, allied bacteriological investigations like those of SHEWAN et al. (1960) are not being undertaken in conjunction with these observations. A few miscellaneous notes on fish were included in the earlier studies.

GAZERT (1912) obtained bacteria from the skin slime and gastrointestinal contents of single specimens of notothenid and lycodid fish. Nitrifying and denitrifying bacteria were not detected. TSIKLINSKY (1908) found that all the fish isolates from CHARCOT's cultures were marine psychrophiles which grew well in sea water media. The organisms included pleomorphic Gram-negative nonsporing rods, micrococci and yeasts. PIRIE (1912) failed to obtain growth in cultures from the intestines of *Notothenia coriiceps*.

McLEAN (1919) remarked that smears from fish "showed an interesting assortment of organisms, their number being fairly high in the intestine". Growth was obtained from rectal cultures of *Notothenia coriiceps* (bacilli and coccus forms) and a specimen of *Chalinura ferrieri* (Gram-negative spore-forming bacillus) taken from a depth of 1,700 fathoms in a deep sea trawl.

Except for two attempts (McLEAN, 1919) to culture holothurians, which are quite common in Antarctic waters, the invertebrates have apparently not been studied. Freshly caught euphausids from the stomach contents of penguins contained a considerable bacterial load as shown by Gram stained smears. Studies of marine waters should include the microflora associated with both the phyto- and zooplankton.

Avian and mammalian habitats

Gastrointestinal microflora and fecal accumulations

Very little is known of the microflora of wild mammals and birds in their natural habitat even in temperate zones. Such factors as the random capture of food, seasonal changes in the nature and type of food materials and its bacterial content, as well as the effect of forced dieting especially during the nesting and breeding periods, must have a profound effect on the type and extent of the enteric microflora. Such investigations on warm blooded animals have been an attractive field of study in the Antarctic (SIEBURTH, 1959d, 1961a). This is probably due to the relative abundance of seals, penguins and flying birds compared to the barren landscape and to the fact that this is one of the few rich sources of organic matter upon which readily cultivated bacteria can develop. The early studies concluded that gastrointestinal populations were smaller than those in temperate zone animals and that "bacteriologically sterile" animals were not uncommon.

GAZERT (1912) made both aerobic and anaerobic cultures of the stomach and intestinal contents of birds and seals. Bacterial growth was obtained from one tern and one Adelie penguin while none was detected in other Adelie penguins and terns, king penguins, Antarctic petrels and snowy petrels. Bacteria were seldom observed in the stomach and small intestine of crabeater, Weddell and leopard seals, but organisms were always present in the large intestine. EKELÖF (1908) failed to detect bacteria in the intestinal contents of Adelie and gentoo penguins, terns or cormorants. Bacilli were obtained from only two of a number of skua gulls examined. PIRIE (1912) made agar and gelatin stab cultures from stomach and intestinal materials. Growth was obtained from nine out of 21 birds and three out of seven seals cultures. CHARCOT (1908) examined intestinal material from gulls, penguins, petrels and seals and found organisms of various forms. Although bacteria were present in smaller numbers than in animals from temperate regions, he did not report "bacteriologically-sterile" specimens. TSIKLINSKY (1908) used mixed cultures from these materials to describe fifteen isolates including rod forms, micrococci and sarcinae. Observations from the field notebook of McLEAN (1919) indicate that bacterial growth was obtained from two out of three seals and five out of eight birds cultured. The primary objective of these early studies was to observe the presence or absence of bacterial growth and to note some of the cultural and morphological characteristics of the mixed and isolated organisms.

BUNT (1955a) examined qualitatively the feces of single specimens of fifteen species of birds and one seal. Two birds were "bacteriologically sterile." This was ascribed to a possible microbial antagonism within the gut. Since many of the non-coliform cultures did not survive, the presence of certain morphological types was tabulated. The penguins presented a more simplified picture; a king contained only *Escherichia coli*, while a royal also yielded *E. coli* and *Bacillus* forms. A rockhopper had only *Bacillus* species, while a gentoo contained non-sporing rods in addition to *Bacillus* species. At least two morphological types were distinguished in each of the nine flying birds which contained bacteria. Seven contained non-sporing rods, five had micrococci while *E. coli* and *Bacillus* species occurred in four specimens. *E. coli* was the only organism isolated from a sea elephant *(Morunga elephantine)*.

During a biological survey in the Ross Sea area, McBEE (1960) cultured the large intestine, rectum and feces of two Adelie penguins, two Weddell seals and one skua gull. Nutrient agar alone or with various additives failed to yield aerobic growth when incubated at either 37° or 55° C. Although growth was obtained in glucose agar shake cultures, neither population estimates nor isolates were made. A frozen sample of seal intestinal material was examined for clostridia by a colleague, L. DS. SMITH, who isolated four strains of *Clostidium perfringens* (type A) and one strain each of *C. sordellii* and *C. difficile*.

In an attempt to determine why the above studies recorded either a partial or total suppression of organisms in the gastrointestinal tract of Antarctic animals, SIEBURTH (1959a, 1960a) conducted a quantitative and qualitative study on sixteen Antarctic birds representing both the plankton and predatory-scavenging feeding types. In addition to the enumeration of certain bacterial types and the classification of the dominant organisms, the antibacterial activity of the samples was tested against intestinal isolates. Four birds were free of an aerobic flora, although anaerobes were detected in varying numbers. By the methods employed in earlier studies, in which anaerobic procedures were crude or were not used, these birds might have been considered "bacteriologically sterile". Other birds were free of detectable bacteria in the anterior segments of their gastrointestinal tracts. Such samples exhibited antibacterial activity against coliform isolates. This phenomenon was most marked in the plankton eating pygoscelid penguins. Typical lactose-fermenting strains of *Escherichia coli* were not present in the isolates of the dominant microflora of these penguins, although they were obtained from all the other birds possessing an aerobic flora. The antibacterial substance has been shown to be acrylic acid, which is quite active against lactose-fermenting strains of *E. coli* in the acidic avian gut (SIEBURTH, 1961b). The antibiotic properties of acrylic acid and the mechanical transfer of the "antibiotic" containing algae from the water to the penguins by their euphausid diet has been discussed on p. 279.

Suppression of the gastrointestinal microflora of a predatory giant petrel which had fed on penguin chicks was also observed (SIEBURTH, 1959a). The antibacterial activity of blood serum from a number of penguins and flying birds was of sufficient concentration to explain the antibacterial activity and inhibition observed in this predatory bird. These observations offer several explanations for the apparent "bacteriological sterility" in animals representing various feeding types reported by the previous workers.

The classification of enteric organisms is of more than academic interest since ecological studies of other habitats such as soil must depend upon a recognition of fecal types in order to detect contamination from this source. ISACHENKO & SIMAKOVA (1934) discussed the role that birds may play in distributing bacterial types in Arctic soils which have a low indigenous bacterial content. Attempts at classification of the enteric organisms other than *Escherichia coli* obtained during the Antarctic studies have been limited to descriptions based upon colonial characteristics and Gram-stained smears. The problems of classification in this area are the usual problems of all taxonomy such as insufficient description of accepted species, artificial criteria and discrepancies in usage. Problems unique to the study of natural populations involve the maintenance of subcultures of wild isolates with exacting cultural requirements long enough to learn how to grow them before they die out. Of the some 200 isolates studied by SIEBURTH (1959a) the obligate anaerobic rods and streptococci from four birds and fastidious Gram-negative rods from three other birds were impossible to maintain in a viable state aboard ship. Two of the latter birds also contained in smaller numbers an organism similar to *Streptococcus equinus*. The remaining nine birds had an aerobic, facultatively anaerobic flora in which members of the family Enterobacteriaceae were the dominant forms. Delayed or negative lactose-fermenting types were

284

present instead of or in addition to the rapid lactose-fermenting types. The enterobacteria consisted of typical and atypical *Escherichia coli* strains, *Aerobacter-Hafnia* species, a Klebsiella-like organism and members of the Providence and Bethesda groups. In addition to these dominant organisms Coryneforms, *Streptococcus* species, *Achromobacter eurydice* and *Micrococcus ureae* were obtained in lesser numbers.

The bacterial flora of fecal accumulations have been examined for a number of reasons. Since MᴄLᴇᴀɴ (1919) was unsuccessful in bringing back intestinal cultures, six samples of penguin guano gathered from the Commonwealth Bay area were used by the Bureau of Microbiology in Sydney to obtain isolates. Gram-positive sporing and non-sporing bacilli, Gram-negative sporing bacilli, Gram-positive cocci and a yeast were obtained. MᴄLᴇᴀɴ pointed out that *Escherichia coli* did not survive while sporing bacilli dominated. He emphasized the point that guano samples in addition to their indigenous microflora are also probably contaminated by bacteria from "... the scanty granite sand, lichen soil, moss soil, morainic mud, algae, snow, ice and sea water (frozen spray)."

Frozen samples of animal feces from the McMurdo Sound region were examined by Sᴛʀᴀᴋᴀ & Sᴛᴏᴋᴇs (1960) for the presence of "psychrophiles" (organisms capable of giving detectable growth after fourteen days incubation at 0° C). A sample of skua gull feces yielded 3×10^3 organisms/g at 0° C and 3×10^4 at 30° C while a sample of seal feces yielded 9×10^4 organisms in plates incubated at both 0° and 30° C. Pony debris and horse feces of an undescribed origin and age were also studied. This material was probably from Hut Point, where both Sʜᴀᴄᴋʟᴇᴛᴏɴ (1908—1909) and Sᴄᴏᴛᴛ (1911—1912) quartered ponies for their attempts to reach the South Pole (Sᴜʟʟɪᴠᴀɴ, 1957). In either event the normally mesophilic flora of these materials was subjected to temperatures for some 50 years which would undoubtedly permit a flora to develop which could grow at 0° C. Counts of 10^7 organisms/g were obtained at both 0° and 30° C. Although the isolates were capable of growth at either 0° or —7° C, their optimum was between 20° and 30° C and their maximum, between 28° and 37° C.

Penguin feces which form guano deposits at rookeries apparently undergo bacterial decomposition at ambient temperatures ($< 5°$ C) to yield a "humus" or "soil" that supports a bryophytic flora. In an attempt to characterize changes which occur in penguin fecal material under natural conditions, penguin guano in various stages of decomposition as well as soil samples (see p. 275) were studied at penguin rookeries at three islands in the South Shetland group (Sɪᴇʙᴜʀᴛʜ, 1962). The freshly collected samples were examined for pH and antibacterial activity against *Staphylococcus aureus* and *Sarcina lutea*, and were used to inoculate general and selective media for the enumeration of bacterial types. The cultures were incubated at 25° C and isolates from the countable cultures were used to determine the taxonomic and physiological types in the dominant flora. The acidic feces (pH 4.5—6.5) contained a variable fecal flora and apparently underwent rapid bacterial decomposition as indicated by the alkaline reaction (pH 8.5—9.9) and antibacterial activity (presumably as a result of ammonia formation) of fresh guano in occupied nests. Nest guano had a "fecal", mixed "fecal and soil", or "soil" flora depending upon age and nesting activity in the rookery. Various organisms such as *Sarcina* and *Micrococcus* species which were minor components of the fecal flora became dominant. Old nest guano, "soil" and "soil humus" which was acidic in nature contained organisms similar to *Mycococcus albus* subsp. *lactis* and proteolytic *Bacillus* species including *B. tinakiensis* (KUROCHKIN, 1958)*). Many of the isolates required incubation temperatures below 20° C after the initial subculture. Most of the isolates grew well at the salinity of sea water and a few apparently grew better. The contribution of marine bacteria to enteric, guano and soil floras merits investigation. A transition from a "fecal" to a "soil" flora in penguin guano of increasing age and state of decay was accompanied by changes in pH and anti-bacterial activity. This indicated that at least certain components of the bacterial flora were active and able to decompose this material at the near freezing summer temperatures.

Respiratory flora and medical observations

The respiratory flora is probably a reflection of the amount of aerial contamination and the health of the individual. Due to the large sea masses surrounding Antarctica, the microbial population of Antarctic air is so slight that it is almost impossible to detect by simple procedures (see p. 269). In Antarctic animals nasal, pharyngeal and tracheal floras have been demonstrated. They appeared to be associated with mild respiratory infections or wounds.

Bacterial growth was obtained from both the nose and throat of a crabeater and a Weddell seal by MCLEAN (1919). The mucous and nasal secretions from a sea elephant with a nose wound were found by BUNT (1955a) to contain a mixed flora. Tracheal swab cultures indicated that the pygoscelid penguins were free of a detectable respiratory flora while the flying birds, such as the skua gull, have a flora of Gram-negative organisms associated with a tracheitis (SIEBURTH, 1958b). Four *Escherichia coli* and *Proteus vulgaris* isolates from these flying birds were pathogenic and caused lesions in the air sacs of chickens. SLADEN (1954) attempted to isolate

---

* Subsequently shown to be strains of *Bacillus licheniformis* apparently producing the red pigment pulcherrimin *(Arch. f. Mikrobiol.* **46,** *414—427, 1963).*

Aspergillus fumigatus from penguins. None of the few fungi obtained were considered pathogenic.

McLEAN (1919) was also interested in whether the nasal and pharyngeal microflora of man persisted or was altered during his isolation in the Antarctic. Cultures were obtained from two members of the expedition over a nine month period. A potential pathogen, *Staphylococcus aureus*, was present only during the first three months while the non-pathogen, *S. albus* persisted over the entire observation period. *Streptococcus* and *Bacillus* species appeared to decrease in occurrence. SLADEN (1961) summarized his unpublished observations on the persistence of potential pathogens during man's over-winter isolation. Contrary to McLEAN's observations, *S. aureus* as well as alpha-hemolytic streptococci persisted throughout the two years of study. Despite the crowded conditions in the huts and sledging tents, the carriers of coagulase-positive strains of *S. aureus* tended to maintain their own phage type. This is not true in temperate zones. In recent years, one of the most vexing problems in hospitals has been the rapid rate of spread of antibiotic resistant *S. aureus* between patients and hospital personnel by nasal carriers.

The absence of acute upper respiratory infections in the Antarctic such as the common cold, and the severe infections resulting from the bringing in of new personnel or the return of over-wintering personnel to civilization have often been observed (SLADEN 1961; FRAZIER 1945). A long term investigation of this problem "Operation Snuffles", was initiated during Operation "Deep Freeze IV", 1958—1959 (SLADEN & GOLDSMITH, 1960). A laboratory was set up on the USS Staten Island. Blood samples and swabs for bacterial and virus isolations were obtained from four types of "communities" which differed in their degree of isolation. These materials are being used in an attempt to isolate the causal agents and to match them with antibodies developing in the blood. In this way the spread of upper respiratory infections from relief to over-wintered personnel can be traced by their pre- and post-exposure histories.

Wounds and skin abrasions and their associated infections which are often slow to heal are one of the hazards recorded in polar literature. The necessity for using such affected limbs plus one's enforced isolation and concern for his required health may magnify the importance of these minor injuries. A deep cut on SIPLE's hand which became infected from seal blubber spread to other abrasions (DARLING & SIPLE, 1941). *Bacillus megatherium*, one of the most common organisms found by them in the habitat of the Byrd Expedition was isolated from the pus. The wounds resulting from the seals fighting among themselves have been found by both McLEAN (1919) and SIPLE (1941) to be commonly infected. SIEBURTH (personal observations) found during the 1957—1958 expedition that skin abrasions arising while hunting and handling birds were often infected. However, the greater number of abrasions caused by handling oceanographic gear during the 1958—1959 expedition such as laceration of the fingers and knuckles which lay open for months with proud flesh were not infected despite minimal water for sanitation. The almost daily handling of phytoplankton net concentrates with their bacteriostatic activity may have been a factor. KNOEDLER & STANMEYER (1958) found that alveolar osteitis was a problem once after every four or five tooth extractions in the Antarctic, despite adequate precautions. The usual rate in a dental school was one out of a hundred. They believed that this high incidence was incited by exposure of the patients to severe cold during the healing process. Cold induced dental problems were the chief medical concern during the U.S. Antarctic Service Expedition (FRAZIER, 1945).

Whales which are commercially harvested and internationally inspected, offer a fertile field for bacteriological study aboard the laboratory equipped factory ships. Apparently the normal microflora of these squid and euphausid feeding mammals has not been studied. ROBERTSON (1954) described a single blue whale whose flesh, blubber and bones were studded every inch or two with cysts which were suggestive of advanced miliary tuberculosis in children. Unfortunately the specimens disappeared before adequate examinations could be made. COCKRILL (1951) served as the veterinary surgeon in charge of examinations and standards aboard the S.S. Southern Venturer during her 1948—1949 Antarctic whaling season to procure whale meat for human consumption. No evidence of specific disease was noted during the inspection of some 2,000 whales. Only one carcass was rejected *in toto* due to an extensive peritonitis and diseased vertebrae of unknown cause. Wound lesions were localized. The commonest pathological lesions were single and multiple nodules and encapsulated bodies which showed evidence of caseation and calcification. They occurred in the muscles, organs and small intestine. Staphylococci were recovered from these lesions but acid-fast organisms were never detected. COCKRILL believed that these nodules and cysts were the end stage of multiple small abcesses formed during the life cycle of internal parasites.

No comprehensive studies of bacterial and viral diseases of animals have been conducted in the Antarctic. It is likely that chronic diseases and low mortality syndromes have apparently gone unnoticed. Mass mortalities of penguins and seals have been observed in close proximity to established Antarctic stations, but these reports are verbal only and they have not been documented. Most natural mortalities have been ascribed by the inquiring medical officers to the ever present cestode and trematode infestation. The role of bacteria in these mass mortalities is unknown. Members of the genus

Salmonella are ubiquitous pathogens which have low host specificity and would be an ideal subject for study. Their occurrence in oceanic birds such as the herring gull (STEINIGER & HAHN, 1953) undoubtedly provides a means for their transmission to remote areas. Penguins are known to be susceptible (COCKBURN, 1947). A preliminary survey using serologic means has indicated that both salmonellosis and ornithosis do occur in Antarctic birds (SIEBURTH, 1958b). An investigation of a mass mortality of both adult and nesting gentoo penguins near Port Lockroy (1959) indicated an acute enteritis (personal observation). Unfortunately the "salmonella-like" isolates from a hundred moribund and morbid penguins were destroyed during a storm at sea. The occurrence of animal diseases transmissable to man has considerable importance from the human health standpoint. Conversely, the proximity of human habitation to animal breeding grounds and the accumulation of waste and refuse from Antarctic bases and supply ships may pose a serious threat to the wildlife of Antarctica. A University of Ohio team is currently searching for pathogens in animal fecal accumulations in the McMurdo Sound region. Domestic animals such as dogs, pigs, cows, sheep and chickens are being introduced and maintained at some Antarctic bases as a source of work animals, pets, milk, eggs and fresh meat. The presence of disease in such animals has already been noted (SIEBURTH, 1958a). Such practices unless accompanied by adequate quarantine, especially under veterinary supervision, also pose a threat to both man and the natural fauna of Antarctica.

Perspectives

Our present knowledge of Antarctic microbiology is based largely upon random observations. Purposely-planned and well executed programs involving teams of scientists are beginning to appear. Such studies will undoubtedly do much to clarify specific problems, especially those that lend themselves to routine analyses. The most profitable approach will probably involve groups of physical and biological scientists studying their own disciplines but working on a common subject. In this way the individual works can stand alone but the overall results can be used to create a more complete bio-physical picture. The establishment of laboratories on land and at sea and the facilities to fly all or most of the way has done much to facilitate studies. The formation of more polar institutes as university departments might attract competent scientists into this field who otherwise would be prevented from getting away from their academic duties during the Antarctic summer. There is an increasing use of frozen samples during the present intensive era of Antarctic research. Surely the months of laboratory study spent would warrant a month or two in the field which would greatly increase the accuracy and significance of the data and broaden the scope of the research.

Antarctic microbiology is essentially the microbial ecology of the various habitats in the area encompassed by waters of the Antarctic convergence. Aside from the limitations of the source material available and the hardships imposed by the environment, the problems of methodology and experimental approach are the same as for microbial ecology in general. These are the characterization of the physical, chemical, and biological environments as well as the indigenous microflora and its biochemical activities. Developments in the field of biophysical instrumentation, particularly in the area of solid state physics and its electronic circuitry, are providing versatile instruments for detecting and recording subtle changes in the physical parameters necessary for ecological studies. The availability of instruments which continuously record oxygen and carbon dioxide content, pH and redox values, pressure, temperature, salinity, humidity, solar radiation, etc., now permits the microbiologist to monitor these factors more closely and yet concentrate on his microorganisms.

Characterization of the chemical environment was once restricted to the determination of certain mineral elements and proximate analyses such as protein nitrogen and fat. Thin layer chromatography now permits the rapid separation of generic entities such as phospholipids, diglycerides and steroids, while gas chromatography and such instruments as amino acid analyzers permit the separation and quantitative analysis of microquantities of compounds on the species level. Trace elements are now being determined in biological materials by neutron activation analyses.

The weakest phase in studying the microbial ecology of natural habitats, at least those poor in nutrients such as ice and sea water, is the lack of methods for studying the autochthonous microflora. Levels of nutrients present in these habitats are sufficient to support an active flora but are insufficient to detect growth visually by standard cultural methods. Increasing the nutrients to concentrations sufficient to permit the development of detectable growth undoubtedly selects a microflora to fit the medium. The nutrient requirements of microorganisms appear to be controlled to an unexpected degree by temperature.

The concepts of cardinal growth temperatures (minimum, optimum, and maximum) and the classical growth temperature curve are useful in teaching bacteriology but they restrict enquiry into the effect of temperature on bacterial processes. The recently recognized phenomenon of multiple temperature optima (OPPENHEIMER & DROST-HANSEN, 1960) challenges our concepts of cardinal temperatures and growth temperature curves. The use of thermal gradient blocks for incubation permits the observation of the effect of small temperature differences on both growth and enzymatic processes. Present studies by this writer indicate that multiple temperature optima are quite common, at least in recent isolates, and that growth peaks do occur within limits as narrow as 2° C. This phenomenon is so marked that it is not masked by the overlapping populations of natural inocula at the extremes of growth temperature. It seems possible that each temperature optimum represents a single or multiple enzyme system. If this is the case, then there must be enzymes which prefer 6°—12° C and which are still operative at 0 °C thereby permitting the development of psychrotrophic (EDDY, 1960) or psychrophilic (INGRAHAM & STOKES, 1959) microfloras. Although the dehydrogenase activity (INGRAHAM & BAILEY, 1959) and proteolytic enzymes (PETERSON & GUNDERSON, 1960) of "psychrophiles" could not be differentiated from the cold-sensitive enzymes of mesophiles, enzymatic changes must accompany the development of bacteria in natural products at low temperatures. Marked changes in the amount and nature of various protein fractions of skim milk were shown to accompany an increase in the population of psychrophilic *Pseudomonas* species (SKEAN & OVERCAST, 1960). Lipolysis of milk fat and the formation of free fatty acids accompanied the development of lipolytic organisms in milk at 4° C (OVERCAST & SKEAN, 1959).

The effect of temperature on vital and non-vital enzyme systems has been studied with mesophilic organisms, although its importance to psychrotrophic organisms is highly suggestive. Elevation of the incubation temperature of a temperature-sensitive mutant of *Escherichia coli* above 30° C inhibited growth (MAAS & DAVIS, 1952), which was restored at 38° C by the addition of pantothenate. This phenomenon was ascribed to an unusually heat sensitive pantothenate synthesizing enzyme. BOREK & WAELSCH (1951) observed that *Lactobacillus arabinosus* required phenylalanine at 37° C but not at 35° C. The effect of incubation temperature on biochemical tests in the genera *Pseudomonas* and *Achromobacter* was studied by ALFORD (1960). A few degrees increase in temperature inhibited non-vital enzyme activity upon which biochemical tests for taxonomic studies are based. The implication of the effect of multiple temperature optima and excessive temperature on the nutrition and taxonomy of psychrotrophic organisms is obvious. Procedures used in recent attempts to clarify the taxonomy of marine bacteria (COLWELL & LISTON, 1960; SHEWAN, et al., 1960a) would not take the temperature requirements of the more psychrotrophic bacteria into consideration. Definitive studies on the nutrition and metabolism of these organisms are greatly needed. The psychrotrophs of the naturally cold habitats in Antarctica undoubtedly offer a fertile field for the investigation of bacterial processes and enzymatic reactions which occur at low temperatures.

The most formidable task of microbial ecologists who are not taxonomists is the classification of enormous numbers of widely-differing, poorly-growing wild strains of microorganisms. The only apparent relief on the horizon is the rapid acceptance and use of the Adansonian principle of equally weighted characteristics and the application of computer techniques suggested by SNEATH (1957a, 1957b). Statistical models have been constructed to show the taxonomic-relationships of microorganisms within genera (LYSENKO & SNEATH, 1958). Other applications of computer techniques have permitted the division and comparison of taxonomically and physiologically related microorganisms (SNEATH & COWAN, 1958; RHODES, 1961; CHEESEMAN & BERRIDGE, 1959; LISTON, 1960). A simplified key to marine bacteria and a logical redefinition of the troublesome genus *Achromobacter* has been suggested by SHEWAN et al. (1960a, 1960b). The further development of such an approach will permit the ready identification of the main isolates at least to genera and eventually lead to major subdivisions or biotypes which would be more meaningful than many species as they now stand.

The original International Geophysical Year program included biological aspects; however, it has taken several years for these researches to be implemented. The initiation of new microbiological studies each year is indeed encouraging. As greater interest develops in this field many of the voids will be filled and much that has been written here will only be of historical interest.

REFERENCES

ALFORD, J. A. 1960. Effect of incubation temperature on biochemical tests in the genera *Pseudomonas* and *Achromobacter*. *J. Bact.*, 79, 591—593.

BARGHOORN, E. S. & NICHOLS, R. L. 1961. Sulfate-reducing bacteria and pyritic sediments in Antarctica. *Science*, 134, 190.

BOREK, E. & WAELSCH, H. 1951. The effect of temperature on the nutritional requirements of microorganisms. *J. biol. Chem.*, 190, 191—196.

BOYD, W. L., & BOYD, J. W. 1962. Presence of *Azotobacter* species in polar regions. *J. Bact.*, 83, 429—430.

BUNT, J. S. 1955a. A note on the faecal flora of some Antarctic birds and mammals at Macquarie Island. *Proc. Linn. Soc. N. S. Wales*, 80, 44—46.

BUNT, J. S., 1955b. The importance of bacteria and other microorganisms in the seawater at Macquarie Island. *Austr. J. mar. Freshw. Res.*, 6, 60—65.

BUNT, J. S. & ROVIRA, A. S. 1955a. Microbiological studies of some subantarctic soils. *J. Soil Sci.*, 6, 119—129.

BUNT, J. S. & ROVIRA, A. S. 1955b. The effect of temperature and heat treatment on soil metabolism. *J. Soil Sci.*, 6, 129—136.

BURKHOLDER, P. R. 1961. General microbiology of Antarctica, in *Science in Antarctica*, Part 1 *The Life Science in Antarctica*. Nat. Acad. Sci. — Nat. Res. Coun. Pub. No. 839, Chap. XV, 129—137.

BURKHOLDER, P. R. & SIEBURTH, J. McN. 1961. Phytoplankton and chlorophyll in the Gerlache and Bransfield Straits of Antarctica. *Limnol. and Oceanogr.* 6, 45—52.

292

CARLUCCI, A. F. & PRAMER, D. 1960. An evaluation of factors affecting the survival of *Escherichia coli* in sea water. 1. Experimental procedures. *Appl. Microbiol.*, **8**: *243—247*.

CHARCOT, J. B. 1908. Journal de l'expédition. Expédition Antarctique Française 1903—1905. Masson et Cie, Paris, 120 pp. (p. 54).

CHEESEMAN, G. C. & BERRIDGE, N. J. 1959. The differentiation of bacterial species by paper chromatography VII. The use of electronic computation for the objective assessment of chromatographic results. *J. appl. Bact.*, **22**, *307—316*.

COCKBURN, T. A. 1947. Salmonella typhi-murium in penguins. *J. comp. Path.*, **57**, *77—78*.

COCKRILL, W. R. 1951. Antarctic pelagic whaling: the role of the veterinary surgeon in the whaling industry, with special reference to standards of inspection in the production of whalemeat for human consumption, and some notes on the pathology of the baleen whales. *Vet. Rec.*, **63**, *111—125*.

COLWELL, R. R. & LISTON, J. 1960. Microbiology of shellfish. Bacteriological study of the natural flora of Pacific Oysters *(Crassostrea gigas)*. *Appl. Microbiol.*, **8**, *104—109*.

DARLING, C. A. & SIPLE, P. A. 1941. Bacteria of Antarctica. *J. Bact.*, **42**, *83—98*.

EDDY, B. P. 1960. The use and meaning of the term "psychrophilic". *J. appl. Bact.*, **23**, *189—190*.

EKELÖF, E. 1907. Studien über den Bakteriengehalt der Luft und des Erdbodens der Antarktischen Gegenden, ausgeführt während der Schwedischen Südpolar-Expedition 1901—1903. *Z. Hyg. Infek.*, **56**, *344—370*.

EKELÖF, E. 1908 Bakteriologische Studien während der Schwedischen Südpolar-Expedition. Wiss. Ergeb. d. Schwed. Südpolar-Exped. 1901—1903, Stockholm, Band VII, Lief. 7, 120 pp.

FLINT, E. A. & STOUT, J. D. 1960. Microbiology of some soils from Antarctica. *Nature, Lond.*, **188**, *767—768*.

FRAZIER, R. G. 1945. Acclimatization and the effects of cold on the human body as observed at Little America III, on the United States Antarctic Service Expedition 1939—1941. *Amer. philos. Soc. (Philadelphia)*, **89**, *249—255*.

GAZERT, H. 1901a. Bakteriologische Aufgaben der Deutschen Südpolar-Expedition. *Petermann's M. Gotha*, **47**, *153—155*.

GAZERT, H. 1901b. The bacteriological work of the German South Polar Expedition. *Scott. Geogr. Mag. Edinburgh*, **17**, *470—473*.

GAZERT, H. 1912. Untersuchungen über Meeresbakterien und ihren Einfluss auf den Stoffwechsel im Meere. Deutsche Südpolar-Expedition 1901—1903, Georg Reimer, Berlin 7 (3), *1—296*.

HARDER, R. & PERSIEL, I. 1962. Notiz über das Vorkommen niederer Erdphycomyceten in der Antarktis. *Arch. Mikrobiol.*, **41**, *44—50*.

INGRAHAM, J. L. & BAILEY, G. F. 1959. Comparative study of effect of temperature on metabolism of psychrophilic and mesophilic bacteria. *J. Bact.*, **77**, *609—613*.

INGRAHAM, J. L. & STOKES, J. L. 1959. Psychrophilic bacteria. *Bact. Rev.*, **23**, *97—108*.

ISACHENKO, B. L. & T. L. SIMAKOVA, 1934. Bacteriological investigation of Arctic soils. *Leningrad, Vses. Arkt. Inst. Trudy*, **9**, *107—124*.

KNOEDLER, D. & STANMEYER, W. 1958. Dental observations made while wintering in Antarctica, 1956—57. *J. dent. Res.*, **37**, *614—622*.

KRISS, A. E. 1959. Distribution of water masses in the Indian Ocean and in the central Pacific Ocean according to microbiological data. Int. Oceanogr. Congress, Preprints, Amer. Assoc. Adv. Sci., Wash., pp. *555—560*.

293

KRISS, A. E., ABYZOV, S. S., LEBEDEVA, M. N., MISHUSTINA, I. E. & MITS-KEVICH, I. N. 1960a. Geographic regularities in microbe population (Heterotroph) distribution in the world ocean. *J. Bact.*, **80,** *731—736*.
KRISS, A. E., LEBEDEVA, M. N. & MITZKEVICH, I. N. 1960b. Micro-organisms as indicators of hydrological phenomena in seas and oceans. II Investigation of the deep circulation of the Indian Ocean using microbiological methods. *Deep-Sea Res.*, **6,** *173—183*.
KUROCHKIN, B. N. 1958. A new species of sporogenous bacteria from the soil of several lakes. *Mikrobiologiia*, **27,** *221—225* (AIBS trans. *217—221*).
LEBEDEVA, M. N. 1958. Quantitative distribution of heterotrophic micro-organisms in the Indian Ocean and in adjoining Antarctic Seas. *Doklady Akad. Nauk SSSR*, **121,** *650—653* (AIBS trans. *217—221*).
LEVIN, E. I. 1899. Les microbes dans les régions Arctiques. *Ann. Inst. Pasteur, Paris*, **13,** *558—567*.
LISTON, J. 1960. Some results of a computer analysis of strains of *Pseudomonas* and *Achromobacter*, and other organisms. *J. Bact.*, **23,** *391—394*.
LUCAS, C. E. 1947. The ecological effects of external metabolites. *Biol. Rev.*, **22,** *270—295*.
LYSENKO, O. & SNEATH, P. H. A. 1958. The use of models in bacterial classification. *J. gen. Microbiol.*, **20,** *284—290*.
MAAS, W. K. & DAVIS, B. D. 1952. Production of an altered pantothenate synthesizing enzyme by a temperature sensitive mutant of *Echerichia coli*. *Proc. Nat. Acad. Sci. U.S.*, **38,** *785—797*.
McBEE, R. H. 1960. Intestinal flora of some Antarctic birds and mammals. *J. Bact.* **79,** *311—312*.
McLEAN, A. L. 1918. Bacteria of ice and snow in Antarctica. *Nature, Lond.* **102,** *35—39*.
McLEAN, A. L. 1919. Bacteriological and other researches. Australasian Antarctic Expedition, 1911—1914, *Sci. Repts.*, Sidney Series C 7 (4), *1—130*.
MENNA, M. E. DI, 1960. Yeasts from Antarctica. *J. gen. Microbiol.*, **23,** *295—300*.
NANSEN, F. 1897. *Farthest North*. Harper & Bros. Pub. New York. 2 Vol. 714 pp.
NYSTRÖM, C. 1868. Om fasnings och forruttnelsprocesserna pa Spetsbergen. Upsala Lakareforen. Forh.
OPPENHEIMER, C. H. & DROST-HANSEN, W. 1960. A relationship between multiple temperature optima for biological systems and the properties of water. *J. Bact.*, **80,** *21—24*.
OVERCAST, W. W. & SKEAN, J. D. 1959. Growth of certain lipolytic microorganisms at 4° C and their influence on free fat acidity and flavour of pasteurized milk. *J. Dairy Sci.*, **42,** *1479—1485*.
PETERSON, A. C. & GUNDERSON, M. F. 1960. Some characteristics of proteolytic enzymes from *Pseudomonas fluorescens*. *Appl. Microbiol.*, **8,** *98—103*.
PIRIE, J. E. H. 1912. Notes on Antarctic bacteriology. Scottish National Antarctic Expedition. Report of the Scientific Results of the S. Y. Scotia, Vol. 3 Botany, No. 10, *137—148*.
RHODES, M. E. 1961. The characterization of *Pseudomonas fluorescens* with the aid of an electronic computer. *J. gen. Microbiol.*, **25,** *331—345*.
ROBERTSON, R. B. 1954. *Of Whales and Men*. Alfred A. Knoff, Inc., New York. 300 pp. (pp. 206—209).
ROUX, E. 1907. Recherches microbiologiques, in *Instructions pour l' expédition Antarctique organisée par le Dr. Jean Charcot*. Gauthier-Villars, *39—43*.
SCOTT, R. F. 1914. Scott's Last Expedition. London, Vol. 1, p. 211.

294

SHEWAN, J. M., HOBBS, G. & HODGKISS, W. 1960a. A determinative scheme for the identification of certain genera of gram-negative bacteria, with special reference to the *Pseudomonadaceae*. *J. appl. Bact.*, **23**, 379—390.

SHEWAN, J. M., HOBBS, G. & HODGKISS, W. 1950b. The *Pseudomonas* and *Achromobacter* groups of bacteria in the spoilage of marine white fish. *J. appl. Bact.*, **23**, 463—468.

SIEBURTH, J. McN. 1958a. Antarctic microbiology. *Amer. Inst. Biol. Sci. Bull.*, **8**, 10—12.

SIEBURTH, J. McN. 1958b. Respiratory flora and diseases of Antarctic birds. *Avian Dis.*, **2**, 402—408.

SIEBURTH, J. McN. 1959a. Gastrointestinal microflora of Antarctic birds. *J. Bact.*, **77**, 521—531.

SIEBURTH, J. McN. 1959b. Antibacterial activity of Antarctic marine phytoplankton. *Limnol. and Oceanogr.* **4**, 419—424.

SIEBURTH, J. McN 1959c. Estudios microbiológicos en aves y fitoplancton marino Antárcticos. Contr. Inst. Antarctico Argentino, Bs. As., No. 35, 41 pp.

SIEBURTH, J. McN 1959d. Estudios microbiológicos sobre animales Antárcticos. Contr. Inst. Antarctico Argentino, Bs. As., No. 45, 19 pp.

SIEBURTH, J. McN. 1960a. Acrylic acid, an "antibiotic" principle in Phaeocystis blooms in Antarctic waters. *Science*, **132**, 676—677.

SIEBURTH, J. McN. 1960b. Soviet aquatic bacteriology: A review of the past decade. *Quart. Rev. Biol.*, **35** (3), 179—205.

SIEBURTH, J. McN. 1961a. Antarctic animal bacteriology, in *Science in Antarctica*. Part 1 *The Life of Science in Antarctica*. Nat. Acad. Sci. — Nat. Res. Coun. Pub. No. 839, Chap. XVI, pp. 138—146.

SIEBURTH, J. McN. 1961b. Antibiotic properties of acrylic acid, a factor in the gastrointestinal antibiosis of polar marine animals. *J. Bact.*, **82**, 72—79.

SIEBURTH, J. McN. 1962. *Bacterial habitats in the Antarctic environment.* In *Symposium on Marine Microbiology* sponsored by Amer. Soc. for Microbiol., C. H. Oppenheimer, Ed., C. C. Thomas Pub. House, Springfield, Ill. (in press).

SIEBURTH, J. McN, & BURKHOLDER, P. R. 1959. Antibiotic activity of Antarctic phytoplankton. Internat. Oceanogr. Congress, Preprints. Amer. Ass. Adv. Sci., Wash., 933—934.

SKEANS, J. D. & OVERCAST, W. W. 1960. Changes in the paper electrophoretic protein patterns of refrigerated skim milk accompanying growth of three pseudomonas species. *Appl. Microbiol.*, **8**, 335—338.

SLADEN, W. J. L. 1954. Penguins in the wild and in captivity. *Avicultural Mag.*, **60**, 132—142.

SLADEN, W. J. L. 1961. Medical microbiology, in *Science in Antarctica* Part 1 *The Life Sciences in Antarctica*. Nat. Acad. Sci. — Nat. Res. Coun. Pub. No. X 839, Chap. XVIII, pp. 151—155.

SLADEN, W. J. L. & GOLDSMITH, R. 1960. Biological and medical research based on U.S.S. Staten Island, Antarctic, 1958—59. *Polar Rec.*, **10**, 146—148.

SNEATH, P. H. A. 1957a. Some thoughts on bacterial classification. *J. gen. Microbiol.*, **17**, 184—200.

SNEATH, P. H. A. 1957b. The application of computers to taxonomy. *J. gen. Microbiol.*, **17**, 201—226.

SNEATH, P. H. A. & COWAN, S. T. 1958. An electro-taxonomic survey of bacteria. *J. gen. Microbiol.*, **19**, 551—565.

SONEDA, M. 1961. On some yeasts from the Antarctic region. Spec. Pub. Seto Marine Biol. Lab., Sirahama, Wakayama-ken Japan, "Biol. Results Jap. Antarctic Res. Exped". No. 15, 10 pp.

STEINIGER, F., & E. HAHN. 1953. Über den Nachweis von Keimen der Typhus-Paratyphus-Enteritis-Gruppe aus Vogelkot von der Stora Karlsö. *Acta. path. microbiol. scand.* **33**, *401—408*.

STRAKA, R. P. & STOKES, J. L. 1960. Psychrophilic bacteria from Antarctica. *J. Bact.*, **80**, *622—625*.

SULLIVAN, W. 1957. *Quest for a Continent.* McGraw-Hill Book Company, Inc., New York. 372 pp.

TSIKLINSKY, 1908. La flore microbienne dans les régions du Pôle sud. Expédition Antarctique Française 1903—1905. Masson et Cie, Paris, 36 pp.

TUBAKI, K. 1961. On some fungi isolated from the Antarctic materials. Spec. Pub. Seto Marine Biol. Lab., Sirahama, Wakayama-Ken Japan. "Biol. Results Jap. Antarctic Res. Exped.", No. 14, 9 pp.

WHITE-STEVENS, R. H., PENSACK, J. M., STOKSTAD, E. L. R. & SIEBURTH, J. McN. 1961. The effect of acrylic acid salts on growth of chicks. *Poultry Sci.*, **40** (5), *1469*.

ZOBELL, C. E. 1946. *Marine Microbiology.* Chronica Botanica Co., Waltham, Mass., 240 pp.

Addendum

This manuscript was completed December, 1961. Some pertinent publications which have appeared since are listed below.

BOYD, W. L. & BOYD, J. W. 1962. Viability of thermophiles and coliform bacteria in arctic soils and water. *Canad. J. Microbiol.* **8**, *189—192*.

BOYD, W. L. & BOYD, J. W. 1963a. Viability of coliform bacteria in antarctic soil. *J. Bact.* **85**(5), *1121—1123*.

BOYD, W. L. & BOYD, J. W. 1963b. Soil microorganisms of the McMurdo Sound area, Antarctica. *Appl. Microbiol.* **11**, *116—121*.

BOYD, W. L. & BOYD, J. W. 1963c. Enumeration of marine bacteria of the Chukchi Sea. *Limnol. & Oceanogr.* **8**(3), *343—348*.

HOLM-HANSEN, O. 1963. Algae: Nitrogen fixation by Antarctic species. *Science* **139**, *1059—1060*.

MEYER, G. H., MORROW, M. B., & WYSS, O. 1962. Viable microorganisms in a fifty-year-old yeast preparation in Antarctica. *Nature, Lond.* **196**, *598*.

MEYER, G. H., MORROW, M. B., WYSS, O., BERG, T. E., & LITTLEPAGE, J. L. 1962. Antarctica: The microbiology of an unfrozen saline pond. *Science* **138**, *1103—1104*.

SIEBURTH, J.McN. 1964a. Polymorphism of a marine bacterium (Arthrobacter) as a function of multiple temperature optima and nutrition. In *Symposium on Experimental Marine Ecology*, Misc. Pub. No. 2, Grad. Sch. Oceanogr., Univ. of R.I., Kingston, pp. *11—16*.

SIEBURTH, J.McN. 1964b. Antibacterial substances produced by marine algae. Developments in Indust. Microbiol., **5**, (in Press).

Contribution No. 84 from the Graduate School of Oceanography, University of Rhode Island, Kingston, R.I.

THE CHAETOGNATHA
OF THE SOUTHERN OCEAN

BY

P. M. DAVID

National Institute of Oceanography, Wormley, Godalming, Surrey
(with 16 figs.)

Introduction

The Chaetognatha are a widespread almost wholly planktonic group occurring, often in large numbers, in the open ocean. They are hermaphrodite but almost certainly not self fertilising animals and exhibit relatively little anatomical variation throughout the group.

The species of this group have, because of the apparently close adaptation of some of them to characteristic water masses, been used as "indicators" of hydrographical conditions. It seems that the association of a species with a particular water mass is not as obvious in oceanic conditions as it is in the proximity of neritic waters: but nevertheless it can be seen from a study of the chaetognaths of the Antarctic regions that although a species may be recorded from several distinct water masses it is to a greater or lesser extent characteristic of one water mass.

The numerous expeditions to the Antarctic Continent have passed through the Southern Ocean and some of these have collected plankton samples on the way. In addition purely oceanographic expeditions have operated in far Southern waters. The commoner elements of the chaetognath fauna have therefore been known for many years, and recent work has shifted the emphasis from systematic to distributional and other studies of more direct interest to the ecology of the Southern Ocean.

So far the life history of only one Antarctic species has been examined in detail (DAVID 1955), but it is probable that others will be so treated in years to come.

All the species seem to be basically alike in their mode of life, that is they are actively swimming carnivores feeding upon the smaller, mainly crustacean plankton, and perhaps the life cycle of one of the species may well be an indication of the life cycle of all of them.

Historical

The first plankton hauls to be made in the Southern Ocean were those made by Sir JAMES CLARKE ROSS 1839—43, but unfortunately these were never reported upon and were not to be found after his death.

The first systematic oceanographic observations were performed by the Challenger Expedition in 1873 but the chaetognath collections were not reported upon until 1907 when FOWLER who had examined the material from the British Antarctic Expedition 1901—04 added a short note on the Antarctic collections taken by HMS Challenger. Although this had priority by date of collection it was by no means the first Antarctic material to be reported. LEVINSON (1885) had recorded *S. hexaptera* (= *S. gazellae* RITTER ZAHONY, probably) from 42°53′S 46°38′W. PARKER (1896) recorded *Spadella hamata* (= *E. hamata*) from the Auckland Islands and other localities, and STEINHAUS (1900) had recorded *Krohnia* (= *Eukrohnia*) *hamata* and *S. hexaptera* (= *S. gazellae?*) from the material collected by the Hamburg Magellan expedition of 1892—93. The early confusion of *S. hexaptera* and *S. gazellae* was due to the use of alcohol as a preservative which made the characteristic fin differences unrecognisable; though it must be added that RITTER ZAHONY (1909) described the type of *S. gazellae* from alcohol preserved material.

It seems likely that FOWLER (1907) was the first to recognise the species now known as *S. marri* DAVID, from the Challenger's Collections though he described it as *S. zetesios*; the appearance of this species is not greatly affected by preservation in alcohol because of the strong body musculature.

In 1909 RITTER ZAHONY described *S. gazellae* as a new species from the material obtained by the Gazelle expedition of 1874—76.

In 1911 RITTER ZAHONY reported upon the extensive material collected by the German South Polar Expedition (Gauss 1901—03) and, while he may not in some cases have recognised the fact, had all the Antarctic species now known before him. His "Revision of the Chaetognaths" stands today as a milestone in the study of this group, and so far as the Antarctic is concerned subsequent workers have, where progress has been made at all, done little but dot the i's and cross the t's in the systematic study of the Chaetognatha.

GERMAIN (1913) added some most useful information about *S. gazellae* and *S. maxima* but the most surprising thing about his report is that the collections of the Pourquoi pas? did not seem to contain *E. hamata*, the most numerous of all Antarctic chaetognaths.

JAMESON (1914) found the usual common forms in the Scotia Collections and also reported *Heterokrohnia mirabilis* RITTER ZAHONY though it seems more probable now that this specimen was *E. hamata*.

JOHNSTON & TAYLOR (1921) with a totally inadequate collection of material attempted to show that *S. gazellae* RITTER ZAHONY was synonymous with *S. lyra* KROHN, an opinion which confused systematists for many years. BURFIELD (1930) and BOLLMAN (1934) reporting on the "Terra Nova" (1910—13) and Deutschland (1910—12) collections described the common Antarctic forms but in addition

BURFIELD reported *S. serratodentata* from the Ross Sea: what the explanation for this curious anomaly is, is not known, but the species has not been reported since from the Antarctic area proper. It is perhaps worth noting that both authors regarded *S. gazellae* as a species rather than as a synonym of *S. lyra* despite JOHNSTON & TAYLOR's (1921) report. In HARDY & GUNTHER (1936) the Chaetognaths were described by Dr. H. E. BARGMANN who also regarded *S. gazellae* as a species in its own right.

THIEL (1938) who worked on the Meteor collection considered that the anatomical similarities between *S. lyra, gazellae* and *maxima* justified the distributional treatment of the three species as one group; whether he in fact thought that they would one day be regarded as a single species is not now clear, but subsequent work has shown them to be distributionally and anatomically quite distinct.

It is, however, most unfortunate that at least *S. maxima* was not clearly distinguished as the Meteor collections have perhaps been the most important so far from the tropical Atlantic and might have supplied the answer to the theory of suggested tropical submergence in that species.

DAVID (1958) working on the extensive Discovery collections (1925—1951) reported the usual Antarctic species, confirming much of RITTER-ZAHONY's earlier work, and added a number of adventitious subantarctic species.

STADEL (1958) reporting on the Schwabenland (1938—39) collections was able to confirm RITTER-ZAHONY's and DAVID's result. FAGETTI (1959) has also reported *S. tasmanica, E. hamata* and *S. gazellae* from Antarctic waters.

Environment

The Southern Ocean is an area of relatively simple hydrographical conditions, which have been most adequately summarised by DEACON (1963). Fig. 1 is a diagram drawn by Mr. R. I. CURRIE showing water masses and water movements in the Southern Ocean and is reproduced with his permission. For convenience the starting point may be taken in the region of the Antarctic continental shelf where under the influence of the extreme conditions of cooling and mixing, a very dense water is produced which sinks down the continental slope of Antarctica principally in the Weddell sea area and forms a bottom current flowing out northwards and eastwards into the ocean basins. This water, the Antarctic Bottom Water, flowing away from the continent is compensated for by a southerly flow of the main body of deep water which rises towards the surface in the Antarctic region. The surface circulation above these deep movements is mainly influenced by the wind and in the vicinity of the continent a westerly flowing current (the East Wind Drift) is induced by the easterly winds of these high latitudes. Over the greater part of the Antarctic zone, however, the predominantly

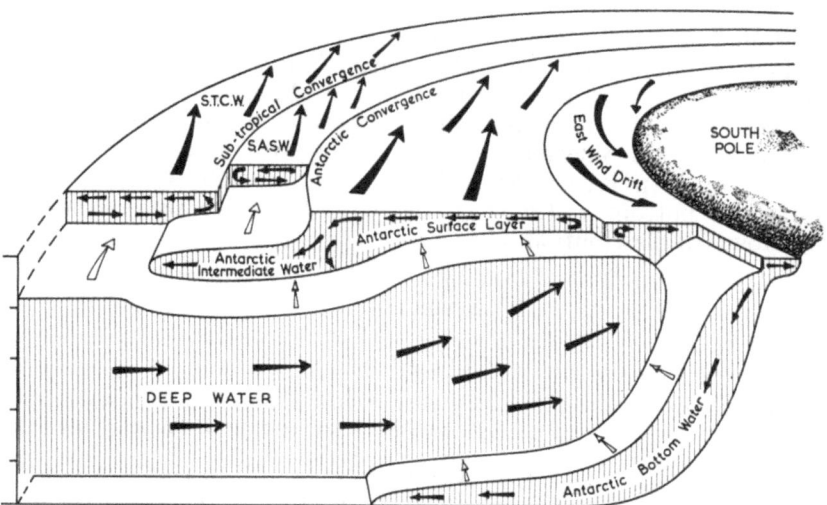

Fig. 1. A three dimensional diagrammatic representation of the water masses and circulation of the Southern Ocean. S.A.S.W. = Subantarctic surface water. S.T.C.W. = Subtropical central water. (previously unpublished diagram by R. I. Currie).

westerly winds carry the surface waters to the east, also imparting to them a northerly component (the West Wind Drift). Between these two surface movements there is a fairly well defined region of divergence.

As the bulk of the Antarctic surface water moves to the north it reaches a point where its density is such that it sinks from the surface over the rising Deep Water. The position where this sinking takes place, the Antarctic Convergence, shows surprisingly little variation and it encircles the Antarctic Continent, as a well defined "front" lying in about 50 °S in the Atlantic sector and somewhat farther south in the Pacific Sector. The surface waters sinking at the Antarctic convergence descend steeply at first and then more gradually, and flow northwards at a depth of some 800 m into the major ocean basins, forming the Antarctic Intermediate layer.

To the north of the Antarctic Convergence the surface waters, still under the influence of the westerly winds are carried to the east within the sub-antarctic region, evidently with some northward component at the surface and a southerly component at about 100 m depth. This flow seems to be an almost self contained circulation and along its northern limit it flows adjacent to the warmer eastwardly flowing subtropical waters. The boundary between these eastward flowing currents of cold and warm water is clearly a region of active convection; this region is called the Subtropical Convergence.

From the distribution of chaetognaths it will be seen that the surface boundary of the Subtropical Convergence is the most important hydrographic feature followed closely by the Antarctic Convergence. The division of the Southern Ocean into two distinct surface zones is reflected by the Chaetognath fauna, but neither the warm deep layer, the Antarctic Intermediate Water or the Antarctic bottom water seem particularly important factors in chaetognath distribution.

Systematic

Sagitta gazellae

Sagitta gazellae	RITTER-ZAHONY, 1909. Die Chaetognathen der Gazelle-expedition. *Zool. Anz.* **XXXIV**, 787—93.
S. gazellae	RITTER-ZAHONY 1911: GERMAIN, 1913: JAMESON, 1914: BURFIELD, 1930; BOLLMANN, 1934; HARDY & GUNTHER, 1936; DAVID, 1955. 1958. 1959; STADEL 1958: FAGETTI 1959.
S. hexaptera	STEINHAUS, 1900; FOWLER, 1907; GERMAIN, 1913.
Sagitta innom.	FOWLER 1908.
S. lyra	JOHNSON & TAYLOR, 1921; BOLLMANN, 1934 (part); THOMSON, 1947 (part).
S. maxima group	THIEL, 1938 (part).

The recent description of *S. scrippsae* ALVARINO 1962 has helped to explain the anomalous N. Pacific forms reported by TCHINDONOVA (1955). There seems now little doubt that *S. gazellae* is a valid species, FRASER 1957, STADEL 1958, FAGETTI 1959, ALVARINO 1962.

Sagitta marri

Sagitta marri	DAVID 1956, *Sagitta planctonis* and related forms: *Bull. Brit. Mus. (Nat. Hist)*, Zool, **IV**, 8 435—51.
S. marri	STADEL 1958.
S. zetesios	FOWLER, 1907.
S. planctonis	(non STEINHAUS) RITTER-ZAHONY, 1911 (part); JAMESON, 1914; JOHNSTON & TAYLOR, 1921; BURFIELD, 1930 (part); BOLLMANN 1934 (part); MACKINTOSH, 1937; THIEL, 1938 (part).
S. planktonis,	HARDY & GUNTHER, 1936.

Eukrohnia bathyantarctica

Eukrohnia bathyantarctica	DAVID, 1958. A new species of *Eukrohnia* from the Southern Ocean with a note on fertilization. *Proc. Zool. Soc. Lond.* **131**, part 4. *597—606.*
E. fowleri	RITTER-ZAHONY 1911 (part).
Eukrohnia sp.	DAVID, 1958b.

Sagitta maxima

Sagitta maxima	(CONANT) 1896.
Spadella maxima	CONANT, 1896, *Johns Hopkins Univ. Circ.* **15,** *84.*
Sagitta whartoni	FOWLER 1898.
S. gigantea	BROCH, 1906.
S. maxima	RITTER-ZAHONY, 1911; GERMAIN, 1913; JAMESON, 1914; HUNTSMAN, 1919; BURFIELD, 1930; BOLLMANN, 1934; THIEL 1938 (part); FRASER, 1952.
S. lyra	MICHAEL 1911 (part)

Eukrohnia hamata

Eukrohnia hamata	(MÖBIUS), 1875.
Sagitta hamata	MÖBIUS 1875, Vermes, in: Die Expedition zur physikalisch-chemischen und biologischen Untersuchung der Nordsee im Sommer 1872. *Wiss. Meeresunters. Kiel,* **II**.
Eukrohnia hamata	RITTER-ZAHONY 1909, 1911; JAMESON, 1914; JOHNSTON & TAYLOR, 1921; BURFIELD, 1930; BOLLMANN, 1934; HARDY & GUNTHER, 1936; MACKINTOSH, 1937; THIEL, 1938; GHIRARDELLI, 1953; DAVID, 1958, STADEL, 1958; FAGETTI, 1959.
Eukrohnia hamata var. antarctica	JOHNSTON & TAYLOR 1921.
Spadella hamata	PARKER 1896.

Distribution

The Chaetognath fauna of the Southern Ocean is composed of three elements (a) three endemic species (fig. 2) (b) four species common to other regions but which maintain populations within the Southern Ocean (c) eight species which probably do not maintain populations in the Southern Ocean but which are carried into the area by water movements and are able to live there as adults though probably unable to breed there.

The three endemic species are *S. gazellae* RITTER-ZAHONY, *S. marri* DAVID, and *E. bathyantarctica* DAVID.

S. gazellae. Horizontal distribution. Circumpolar; restricted to the North by the region of the Subtropical Convergence and to the south by the Antarctic neritic zone. Figure 3 shows the distribution of this species in the uppermost 100 m layer. The species is found in

302

largest numbers in the Subantarctic region. It would appear that it is found in very reduced numbers in the region of the Antarctic Convergence, see fig. 4. There are two races of the species, one, the

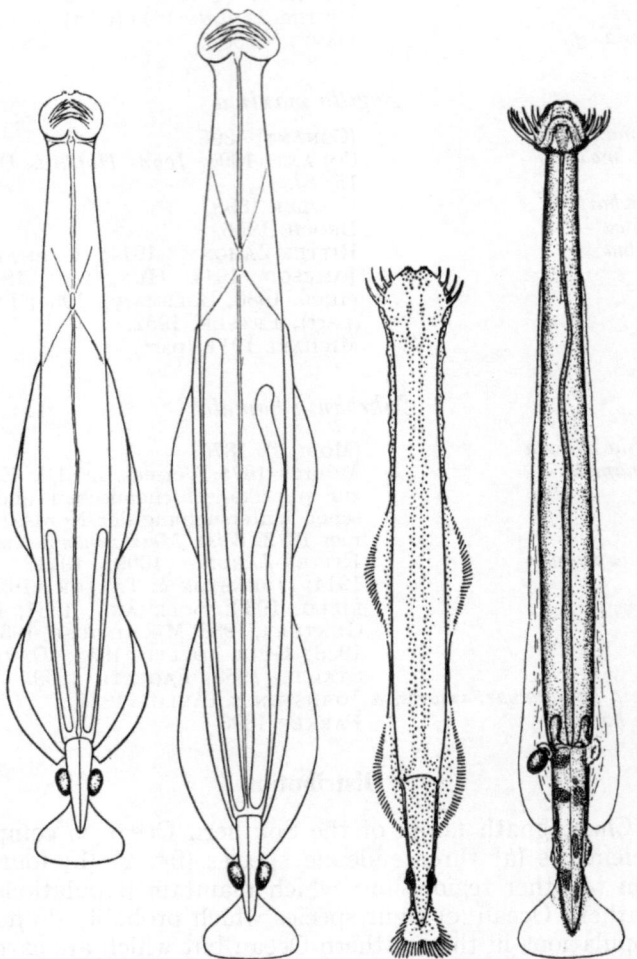

Fig. 2. The three endemic species from the Southern Ocean. a. *S. gazellae* small northern race; b. *S. gazellae* large southern race; c. *S. marri*; d. *E. bathyantarctica*.

southern form is larger at maturity than the northern subantarctic form and there are small anatomical differences between them (DAVID 1955 p. 254 et seq).

Vertical distribution. The vertical distribution of this species on three lines of stations in the Atlantic, Indian and Pacific Ocean

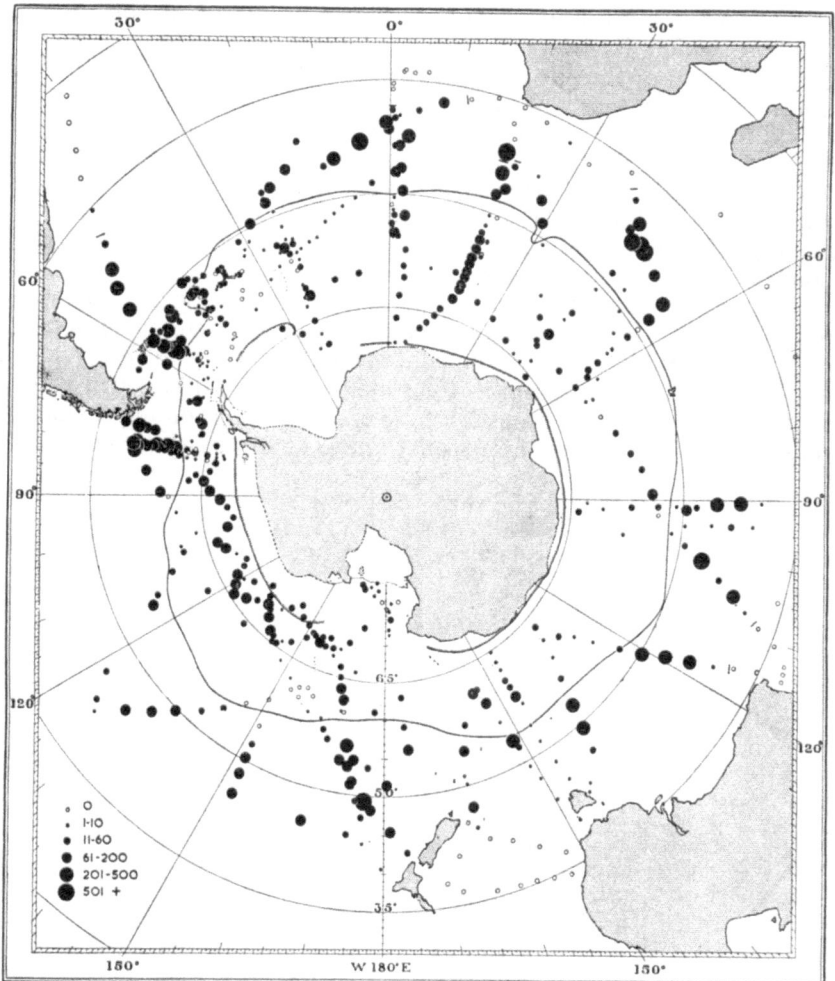

Fig. 3. The horizontal distribution of *S. gazellae* in the top 100 m of water. The chart shows the Antarctic Convergence, and, closer to the continent, the normal position of the pack-ice edge March. The Subtropical Convergence is shown as a short line at right angles to each line of stations which crossed it, and is the contemporary position found by reference to a continuous thermograph and not a mean position.

Sectors of the Southern Ocean is shown in fig. 5, there is some slight extension north of the Subtropical Convergence, but this is evidently not very marked and to the south numbers fall off as the Antarctic shelf is approached. Fig. 6 shows the seasonal variation of the species in the meridian of 0° in the Atlantic sector. This indicates that there is a seasonal migration out of the surface 100 m in the winter months, which is most marked in August when the maximum concentration of the species is in the 250—100 m layer. There does not seem to be any diurnal vertical migration in the species as fig. 7 shows. The average vertical distribution is shown in fig. 8 which indicates that it is most numerous in the 100—50 m layer. The species has two clearly marked centres of abundance, one in the Antarctic and the other in the subantarctic. These correspond to each of the two races. The life history of the species involves a considerable amount of seasonal vertical movement.

The eggs probably hatch at about 250 m and develop into juvenile animals in the upper 100 m layer, and the growing animals migrate to the 250—100 m layer in the winter months, rising to the surface again in spring. As they reach their maximum length, about twelve months after hatching they sink to deeper layers becoming sexually mature and undergoing certain anatomical changes (see DAVID 1955

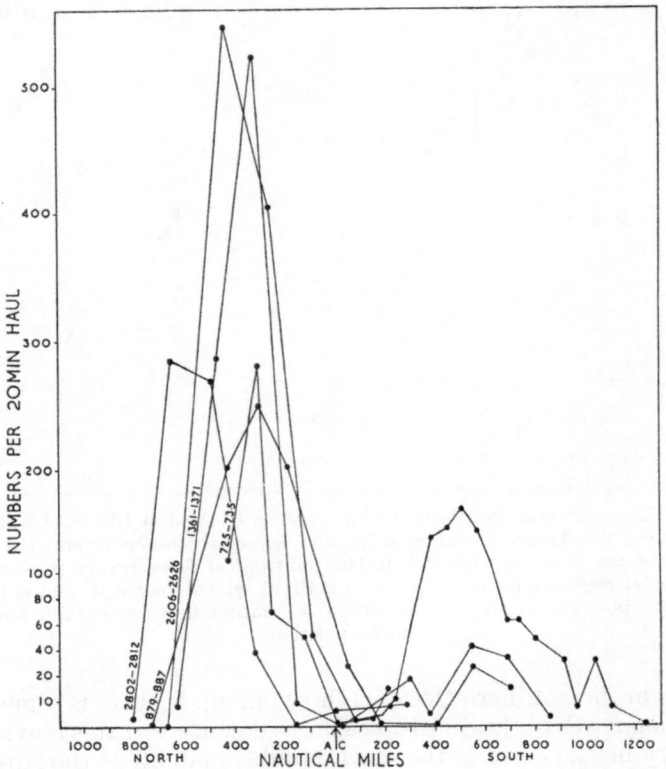

Fig. 4. A series of hauls from North-South lines of stations plotted by distance in miles from the Antarctic convergence.

305

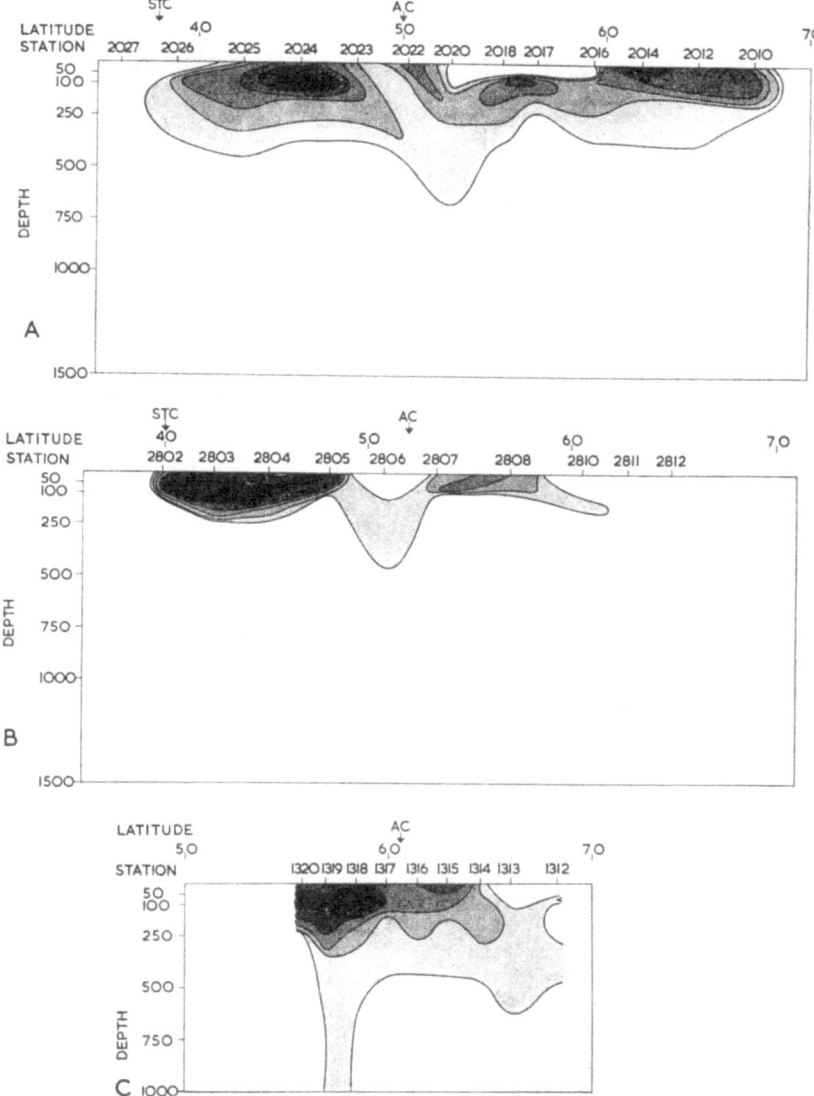

Fig. 5. The vertical distribution in summer of *S. gazellae* in 0°(A) 90°E (B) and 80°W (C).

p. 271 et seq). At a depth of about 1000 m they shed their eggs which float towards the surface hatching at about 250 m to recommence the life cycle. This cycle has only been demonstrated for *S. gazellae*, but deep breeding is a feature of several other species and the life cycle of these may be very similar.

306

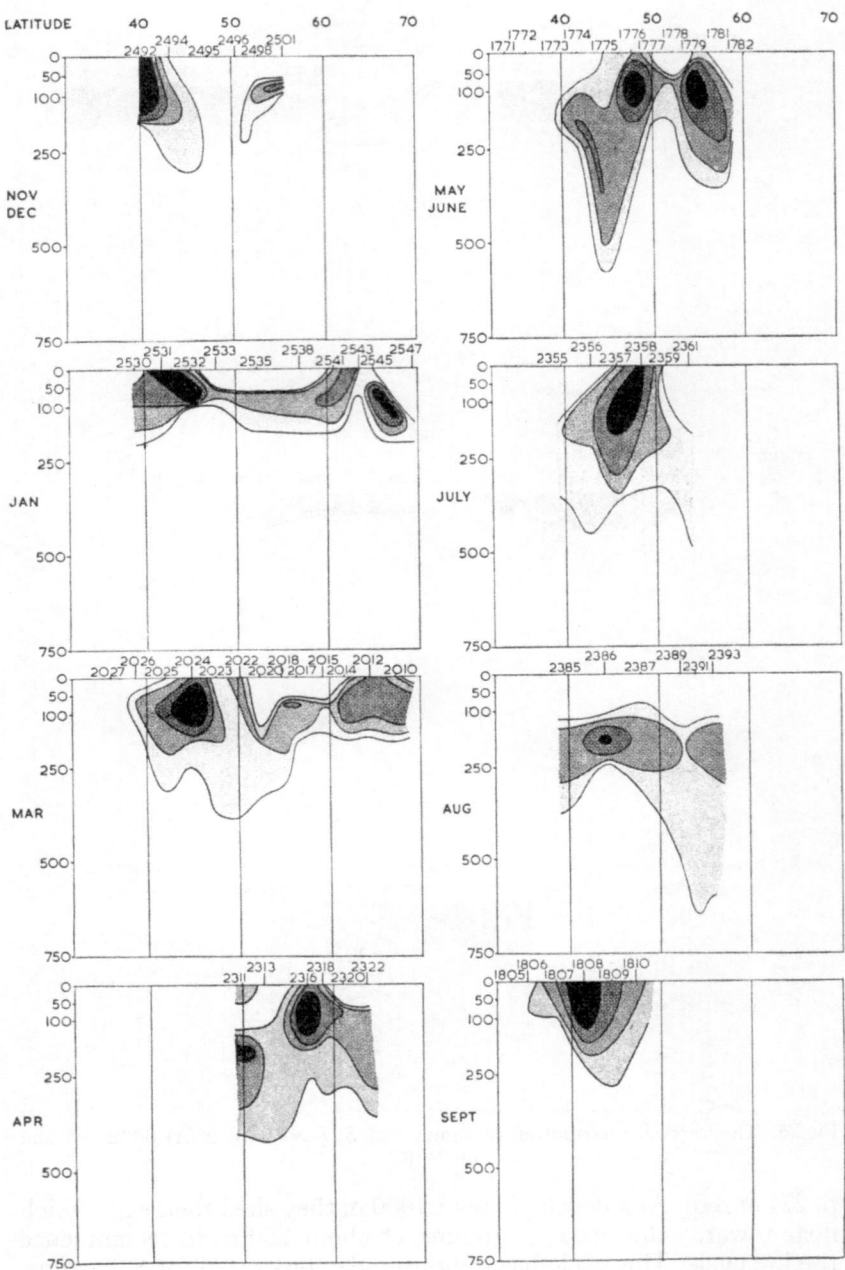

Fig. 6. The vertical distribution of *S. gazellae* for 8 months of the year on the meridian of 0°. The shading represents density per 250 m haul: (i) (the darkest) > 41 individuals, (ii) 21—40 individuals (iii) 11—20 individuals, (iv) (the lightest) 5—10 individuals. Hauls containing less than 5 individuals have been disregarded. The mean position of the Antarctic Convergence on this meridian is 51°S.

S. marri. Horizontal distribution. Circumpolar (BAKER 1954). It is most abundant in the Antarctic zone, and the specimens taken in the subantarctic are usually found in the Antarctic Intermediate water.

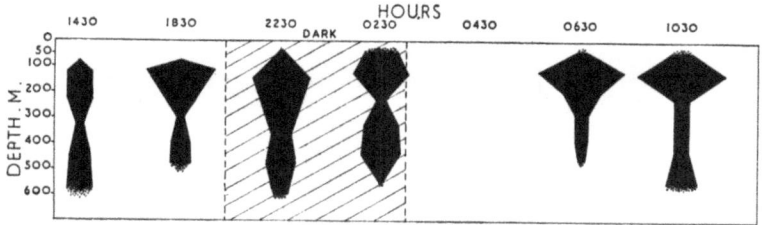

Fig. 7. The vertical distribution of *S. gazellae* at a 24 hour station (Discovery Station 461) showing absence of diurnal migration.

Vertical distribution. The main concentrations of this species are found well to the south of the Antarctic Convergence at depths of about 500—250 m, see fig. 9. The species does occur in the uppermost 100 m but usually only in small numbers. Occasional specimens may be found between 1500 and 1000 m and although the species matures in deep water the specimens found in deep nets are often immature.

The species is characteristic of the Warm Deep Water in the Antarctic zone, and fig. 10 shows how closely defined are the limits of the conservative properties of the water in which its maximum concentrations are found.

Nothing is known of its life history except that it matures in the deeper water layers between about 1000 and 750 m. MACKINTOSH (1937) has suggested that it migrates into deeper water during the winter months, but due to its southerly distribution and the prevalence of ice cover this has not yet been thoroughly demonstrated.

E. bathyantarctica. The limits of the horizontal distribution of this species are not yet known. It is probably circumpolar though this has yet to be proved.

Its vertical distribution seems to be from 1000—2400 m but it is very uncommon in hauls of 1500—1000 m while the few hauls at depths below this have yielded quite considerable numbers, suggesting that the main concentration of the species is between 2000 and 1500 m.

The four species common to other regions are *S. maxima, S. macrocephala, E. hamata* and *H. mirabilis.*

S. maxima. This species is the subantarctic counterpart of *S. marri.* Although its northern limit appears to be just north of the Subtropical Convergence, it has been found in the Peru and Benguela current regions.

308

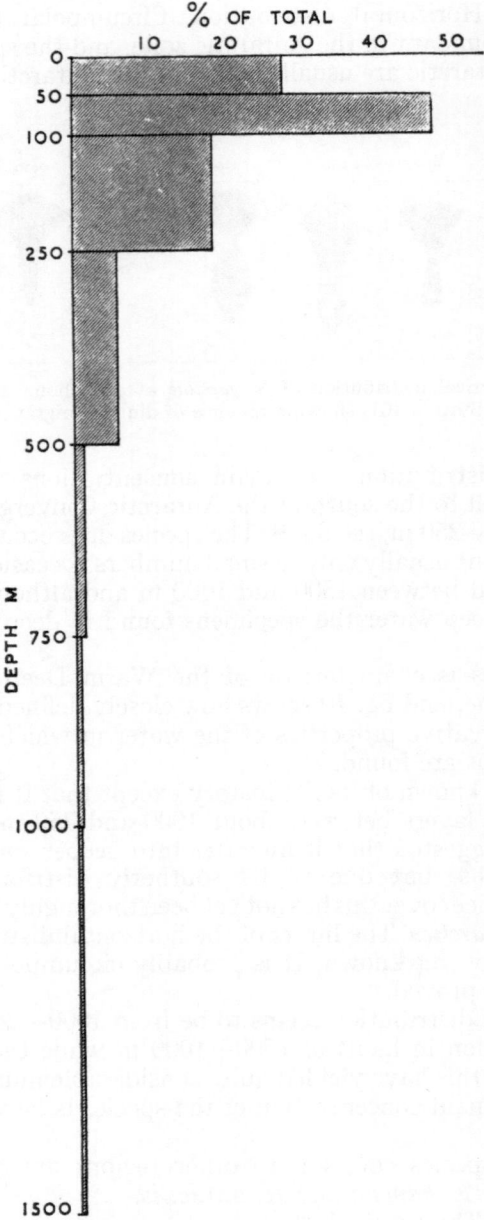

Fig. 8. The vertical distribution of *S. gazellae* at 129 vertical stations: the numbers at each depth are represented as a percentage of the total for all depths.

309

The southern limit is not usually far south of the Antarctic Convergence though it does occur well into antarctic water as is shown by the catches of the BANZAR Expedition (DAVID 1959.)

The species is most numerous between 150 and 500 m but is found in small numbers to 1500 m, fig. 11. It is sometimes found in subantarctic surface waters but not commonly. In other regions the

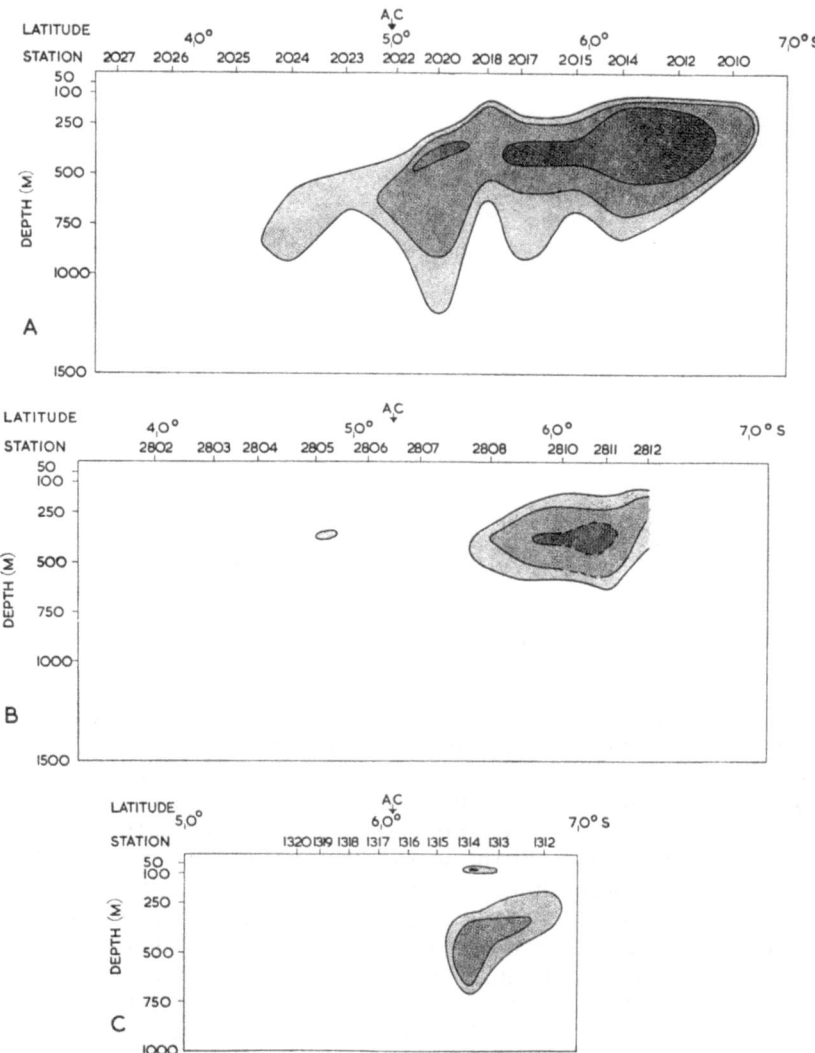

Fig. 9. The vertical distribution of *S. marri* in 0° (A), 90°E (B), and 80°W (C).

species is found at depths of 750 m or more, though it also occurs commonly at the surface in subarctic waters.

In the Southern Ocean the species breeds at depths around 1000 m.

S. macrocephala. This species is not common in the Southern

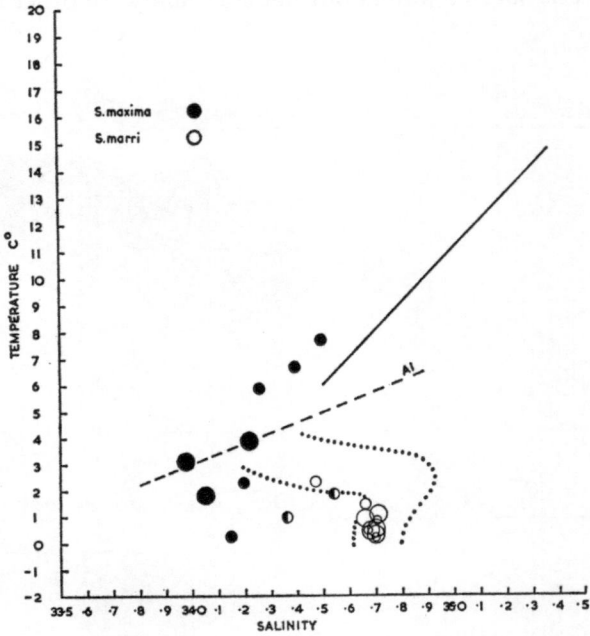

Fig. 10. The two highest orders of abundance of *S. marri* and *S. maxima* on the 0° meridian in March plotted again temperature and salinity.

Ocean but occurs sporadically at depths between 750 and 1500 m. It is slightly commoner in the subantarctic than the antarctic, and may perhaps be a straggler from the warmer seas. Nothing is known of its breeding nor have the ripe seminal vesicles been described.

E. hamata. This is the commonest species in the Southern Ocean and occurs in almost all the world's oceans; one cannot therefore define its northern limits, though its northern limit in surface waters is within the subantarctic zone. To the south it has been found even in the shelf waters of the Antarctic continent.

While its maximum numbers are found in deep waters in the other oceans, in the Southern Ocean the maximum concentration is in the top 500 m, fig. 12. The species is common right at the surface.

Like *S. gazellae* there are two races of this species, a large form in

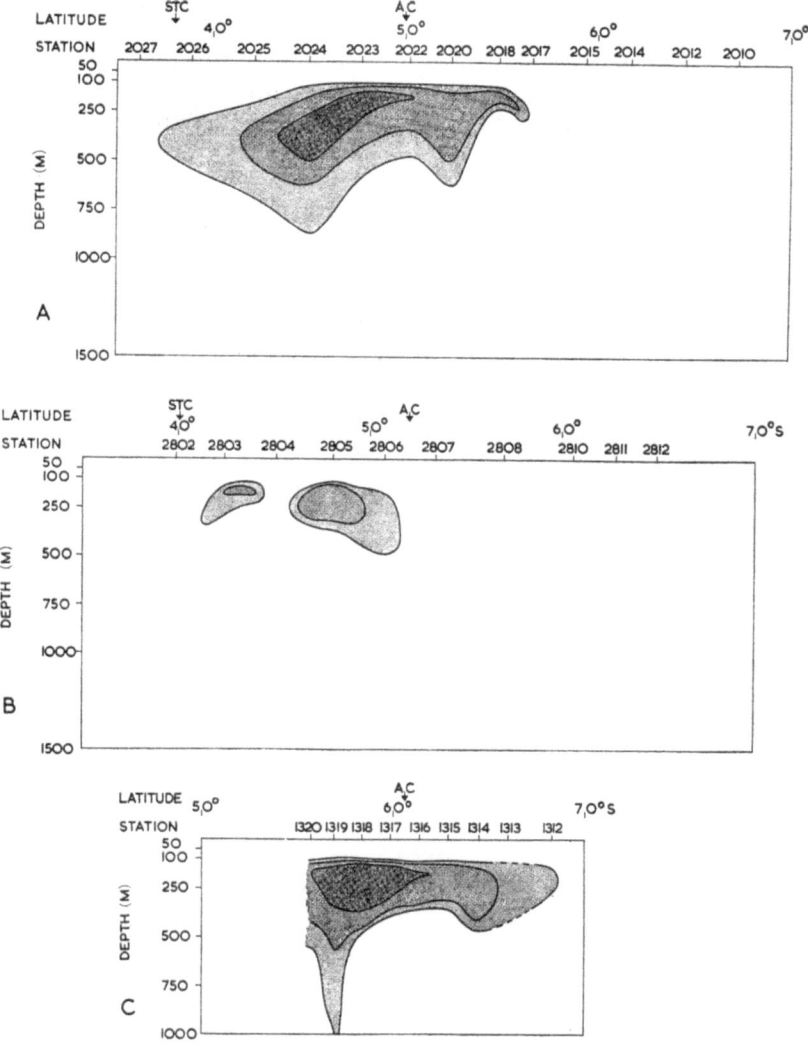

Fig. 11. Vertical distribution of *S. maxima* in 0° (A), 90°E (B), and 80°W (C).

cold antarctic waters, and a smaller form in the subantarctic. Where the two forms mix they can be distinguished, and there do not seem to be any intermediate forms.

The mature specimens are found in deeper waters from 750—1500 m, but the spent from *E. hamata* var. *antarctica* may also be found at the surface; it seems unlikely that this stage survives long

312

after spawning as the gut becomes but a thin strand and the body becomes very flaccid and transparent.

The eight species which probably do not maintain populations in the Southern Ocean but which are carried into the area by water movements from warmer seas are: *S. hexaptera* D'ORBIGNY (1843), *S. lyra* KROHN (1853), *S. zetesios* FOWLER (1905), *K. subtilis* (GRASSI

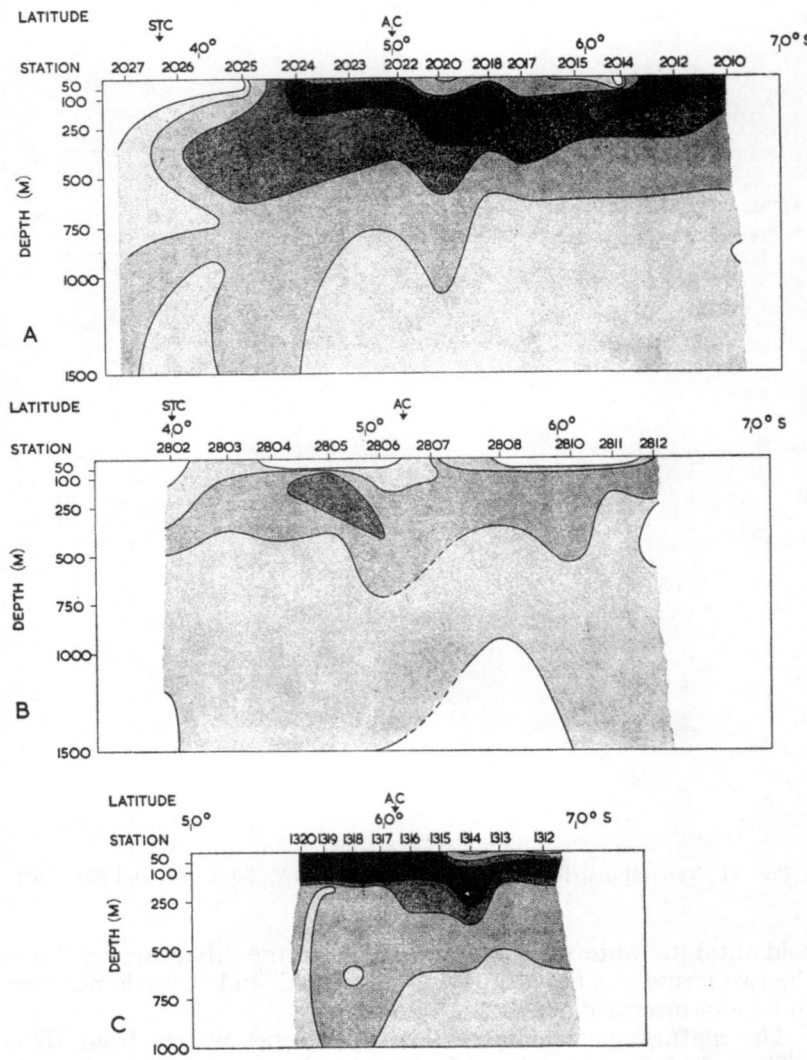

Fig. 12. Vertical distribution of *E. hamata* in 0° (A), 90°E (B), and 80°W (C)

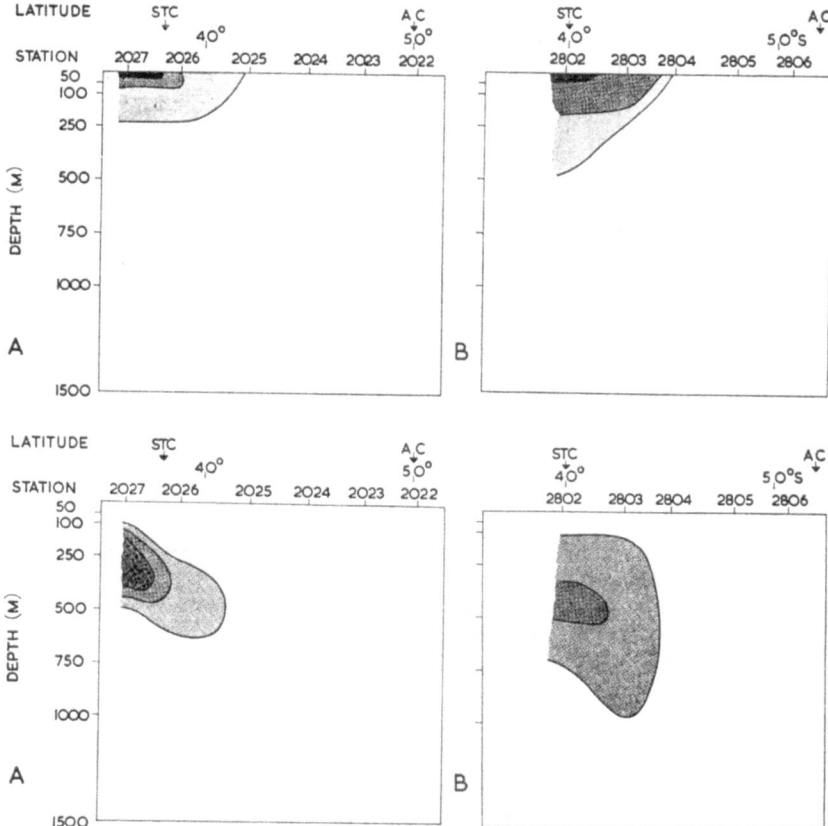

Fig. 13. Vertical distribution of *S. decipiens* and *S. serratodentata* in 0° (A) and 90°E (B).

1883), and *P. draco* (KROHN 1853). Of these only *S. decipiens* and *S. serratodentata* occur frequently enough to justify description of their distribution.

S. serratodentata and *S. decipiens*. These two widespread subtropical species are found in the northern part of the subantarctic. Fig. 13. The usual form of *S. serratodentata* is *tasmanica*, but these are not always typical.

H. mirabilis. This is only found below 2000 m. There is no evidence for a circumpolar distribution, but since the species has been found in other oceans it is probable that it is widespread in very deep water.

Table I.

Reports of species from the Southern Ocean. Records shown thus (v) were recorded as some other species but are now thought to be as shown in table on pp. 300—301.

	S. gazellae	E. hamata	S. marri	S. maxima	E. bathyantarctica	H. mirabilis	S. hexaptera	S. lyra	S. serratodentata	S. decipiens	S. zetesios	S. minima	P. draco	S. macrocephala
Levinson 1885	(v)	v					?							
Parker 1895	(v)	v	(v)				?		v					
Steinhaus 1900	(v)	v							v					
Fowler 1907	v		(v)	v		v	(v)							
R. Zahony 1909	v		(v)	v	(v)		v (v)							
R. Zahony 1911	v	v	(v)	v			(v)							
Germain 1913	v		(v)											
Jameson 1914	(v)		(v)											
Johnson & Taylor 1921	v	v	(v)	v					v					
Burfield 1930	v	v	(v)											
Bollman 1934	v	v												
Hardy & Gunther 1936	(v)	v	(v)											
Thiel 1938	v	v	(v)	?			?	?	v	v	?	?		
David 1958		v	v	v	v	v	v	v	v	v	v	v	v	v

Additional records: + S. bipunctata, + K. subtilis, + S. neglecta

The general pattern of distribution

Although much work, especially in European waters (MEEK 1928, RUSSELL 1935, 1936, 1939, FRASER 1937, 1952, 1961) has shown that chaetognaths appear to be very sensitive to hydrographical conditions, so much in fact that they may be used to indicate otherwise hardly detectable hydrographical changes, work in the open ocean does not seem to be corroborative. To consider the Southern Ocean species for example, only *S. marri* shows to any degree a narrow tolerance to hydrographical factors (see p. 307) and that only in so far as its maximum concentrations are concerned, (whereas the presence of but a few specimens may be regarded as "indications" in coastal waters). It is obvious, however, that in a broad sense the species are confined to particular zones and depth horizons, but by vertical migrations and as a result of the currents and water movements they may be found outside their normal habitats, and the presence (for example) of *S. maxima*, in the warm deep water far south in the Antarctic zone cannot be taken to indicate the presence of the Antarctic Intermediate water where it is usually to be found.

It seems more likely that to a great extent the pattern of distribution is what HUTCHINSON (1953) has called a "coactive" one and that each species is limited by competition with another species. Each is indeed closely adapted to a particular set of physical and chemical factors, though not so closely that it cannot exist in other conditions, and it is the degree of success against competitors under particular conditions which limits the distribution of a species. This point of view necessarily argues a limited number of ecological niches in any one part of the ocean. In the Southern Ocean for example there are evidently niches for two surface species, *E. hamata* and *S. gazellae*. These coexist throughout most of the region but *E. hamata* is more tolerant to physical and chemical conditions than *S. gazellae* as is evinced by its wide vertical distribution. It is possible that there is some difference in vertical distribution between these two species which cannot be detected in our rather crude sampling but more probably the anatomical differences which distinguish the two genera may be reflected in differences in behaviour which insulate them from direct competition where they share the same habitat.

There are niches for several deep living species which are filled by *E. hamata* down to about 1500 m, by *E. bathyantarctica* from 1500 to about 2500 m and *H. mirabilis* at depths beyond this. While *E. hamata* may compete to some extent with various deep breeding forms such as *S. gazellae*, *S. marri* and *S. maxima* and with occasional specimens of *S. macrocephala*, it would appear that *E. bathyantarctica* and *H. mirabilis* are virtually without competitors within their individual depth zone.

316

The pattern of distribution is summarised in fig. 14 which has been compiled from contour diagrams of the vertical distribution of six species. For clarity only the two contours of highest abundance for each species have been shown.

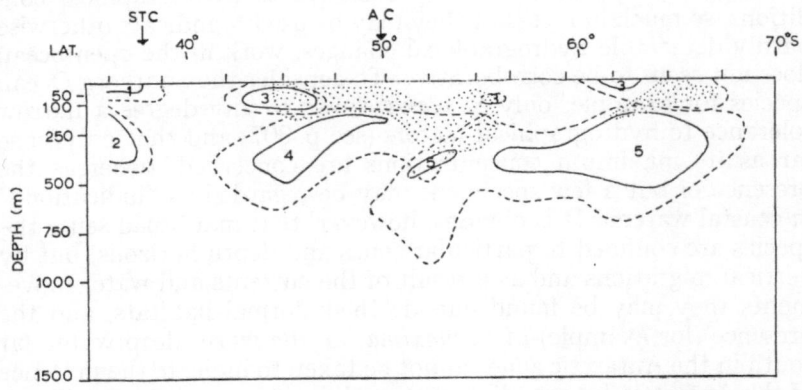

Fig. 14. Contour diagram showing the vertical distribution of sex species in the 0° meridian. For the five species of *Sagitta* the two highest orders of abundance have been plotted, for *E. hamata* only the highest order is shown. 1. *S. serratodentata;* 2. *S. decipiens;* 3. *S. gazellae;* 4. *S. maxima;* Stippled, *E. hamata;* S.T.C. Subtropical Convergence; A.C. Antarctic Convergence.

It will be seen that apart from the extensive overlapping of *E. hamata* with the other species, they are remarkably clearly segregated and the figure emphasises the geographic separateness of the centres of abundance of each species. It can also be seen that *S. decipiens* is distributionally analogous to *S. maxima* in the sub-antarctic and *S. marri* in the Antarctic. *S. decipiens* is, in the northern-most part of the subantarctic, at the southern limit of its range and although it occupies much the same depth horizon in the subtropics, there are a number of other species at the same depths there, though a diagram constructed in the same way as fig. 14 for the warmer parts of the Indian Ocean shows a similar segregation for the commoner species (DAVID 1963).

The situation within the Southern Ocean may be taken to be simple in comparison with other parts of the oceans and one can expect, and will find, more niches available for chaetognaths in warmer waters. It has for a long time been recognised that the cold waters of the ocean support far fewer species than the warm and HEINRICH (1962) has collected comparative numbers from the Pacific Arctic, tropics and Antarctic, for several groups. The seven chaetognath species which maintain large populations in the Ant-arctic can be compared with the sixteen or more species in the sub-tropics and the twenty six or so in tropical waters. It is not alto-

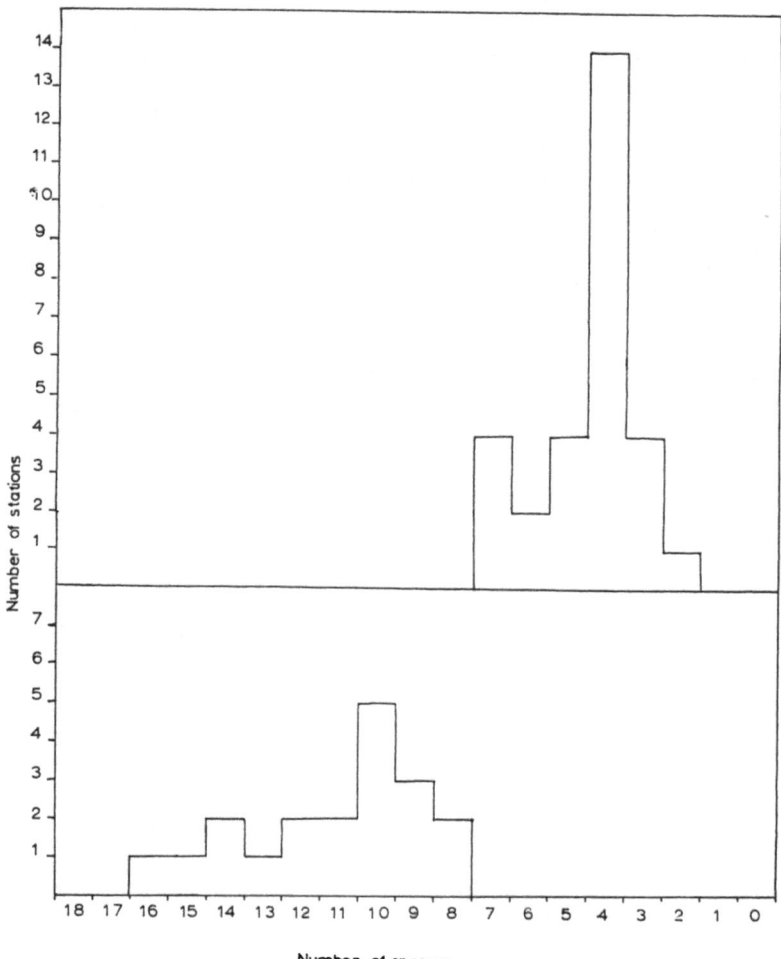

Fig. 15. The frequency of occurrence of a particular number of species at 29 stations in the Southern Ocean (top diagram) and 19 stations in the tropics and subtropics of the Indian Ocean (Lower diagram).

gether because the warmer seas present a greater selection of habitats than the Southern Ocean due to geographic features and more self contained circulatory systems, though this is probably important also, but because at any one spot the diversity of species is greater; as many as sixteen species of chaetognath may be taken between 1000 m and the surface in the warm parts of the Indian Ocean though ten is the most frequent number (DAVID 1963), yet in the Antarctic, the largest number to be found is seven, a total

318

which includes several exotic forms at the northernmost part of the subantarctic and the most frequent number is four, see fig. 15. The diversity of species in the tropics makes for great stability in the ecosystem, and it may well be that prolonged study of the antarctic populations will reveal large scale fluctuations in numbers resulting in part at least from the inherent ecological instability. DUNBAR (1960) has suggested that selection is operating in the Arctic in such a way that it favours the evolution of increased stability, and by analogy the same should be happening in the Antarctic though no evidence for it has yet come to hand.

Growth rate

The growth rates of the Southern Ocean chaetognaths are unknown except in the case of the small northern race of S. *gazellae*.

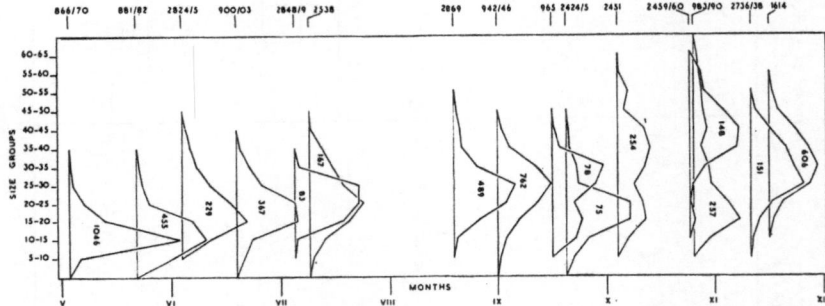

Fig. 16. The growth of S. *gazellae* based on counts and measurements of specimens in the 100-0m layer for the months of May to November.

In this form due to a peculiarity of the life history one particular brood remains within the 100—0 m layer during the winter months while the remainder of the population is deeper in the water column, it is therefore possible to follow the growth of this brood without difficulty. Mature specimens of this species are taken in all months of the year and it appears that some spawning takes place practically all through the year though spring and summer are evidently the peak periods.

In late March and early April the brood whose growth can be followed is spawned and fig. 16 shows the progress of the young until mid-November when the spring-spawned animals rise to the surface, mix with the autumn brood and make it difficult to follow the further progress of growth. The variation in upper size limit in this figure is probably due to variation in the depths at which the nets (nominally 100—0 m) fished which causes older animals deeper down to be sampled. It would seem likely that other broods from spawnings later than the March/April one appear occasionally as for example in mid-July.

The growth rate is seen to be about 5 mm a month for the winter months. More food is available in the spring and summer and the rate then is presumably higher.

The full size at maturity of this race of *S. gazellae* is 55—60 mm which would be reached in 12 months at 5 mm per month. However the full size is reached in the surface water before the breeding migration takes place; there must therefore be a month or so in which this migration and the maturation of the gonads takes place and if, as seems most probable, the life cycle is one year then an increased growth rate in the summer months would be essential if the animal is to acheive 60 mm in 10 or 11 months.

The growth rate of the Large Southern form of *S. gazellae* is not known; it matures at 75—80 mm in length and at 5 mm a month (by analogy with the SN race) maturity would be reached in 15—16 months. The evidence for some Antarctic animals suggests a slow growth and a two year life cycle e.g. *Euphausia superba* (MARR 1962), nor is a two year cycle uncommon in the Arctic, but it may be that the large amount of food available would permit a greater rate of increase. It is my own opinion that a greatly increased growth rate in summer results in a 12 month life cycle but the evidence is not sufficient to be certain.

It is not possible to express an opinion as to the life cycle of the other Southern Ocean chaetognaths, though as in *S. gazellae* mature individuals of the common species have been taken in enough months as to suggest that spawning takes place to a greater or lesser extent throughout the year.

Colonisation and speciation

At a time when *H. mirabilis* was considered to be an endemic Antarctic species it was reasonable to suppose that the Chaetognath fauna of the Southern Ocean was one of considerable antiquity, for the aberrant characters which define the genus and species would suggest a prolonged period of isolation from its closest relatives. Now that the species has been recorded elsewhere it becomes necessary to reconsider the relationships of the other endemic forms. These do not in fact differ much from their relatives in warmer seas; *S. gazellae* for example was for many years regarded as a synonym of *S. lyra*, and although the original data adduced for this by JOHNSON & TAYLOR (1921) was not satisfactory, the fact that the synonymy was generally accepted shows that there is great anatomical similarity between the species. It seems probable that *S. gazellae* has diverged from the same stock as *S. lyra* at a fairly recent date as an adaptation to cold water conditions.

S. marri was originally described as *S. zetesios* by FOWLER (1906) and later as *S. planctonis* STEINHAUS by RITTER-ZAHONY 1911 and others (see p. 300). The form of the mature seminal vesicles is entirely diagnostic in chaetognaths and when that of the *S. marri* was compared with that of *S. planctonis* it was obvious that the two were separate species.

E. bathyantarctic is evidently closely related both to *E. fowleri* and also *E. hamata* having many characters intermediate between those species, differentiation depends in fact upon a few minor characters.

It can be seen then that the endemic Antarctic Chaetognath species have diverged very little from the chaetognaths of the subtropics and one may suppose that those of Southern Ocean have not been isolated tor long. It seems most probable that the colonisation of the Southern Ocean has taken place from the north: though ALVARINO (1962) has suggested that *S. lyra* evolved from isolated populations of *S. gazellae* and the general surface circulation would tend to favour this view, it seems more likely that *S. gazellae* is fairly recently evolved from a subtropical *S. lyra* like ancestor. The distribution of *S. decipiens* and *S. serratodentata* (see p. 313) suggests the first stage in colonisation of the Southern Ocean when normally warm water species are transported below the surface disturbances of a convergence as in the case of *S. decipiens* or by adaptation to colder than normal conditions in *S. serratodentata* which is a species capable of thriving in mixed water conditions. After the colonisation of the subantarctic, another convergence must be passed and the Antarctic can be colonised; the two forms of *S. gazellae*, the Small Northern and Large Southern, may be examples of a slightly later stage.

S. marri may have reached the Antarctic via the Warm Deep Water which has a southward moving component and the hypothetical subantarctic part of this species has been lost, perhaps in competition with *S. maxima* or perhaps because at its optimum depth the water transport is to the north.

The surface living forms must somehow have evolved a means of remaining within a water mass which though its movement is predominantly circumpolar has a northerly component. MACKINTOSH (1937) has suggested that a seasonal migration to a deeper level where the water moves southwards may be the mechanism. It is possible also that in *S. gazellae* the main part of the population in the subantarctic lives at a layer of little or no motion between the northward and southward moving parts of the surface water.

The effect of high total wind in mixing the surface layers and maintaining genetic continuity throughout each species (DAVID 1961) has probably done more to reduce the numbers of species than the difficulties of colonisation.

Conclusions

The chaetognath fauna of the Southern Ocean is made up of seven species three of which are endemic forms.

The basic pattern of distribution is circumpolar and the Antarctic and Subtropical Convergences act to a greater or lesser extent as boundary regions for the fauna.

The life histories of most of the species involve a considerable amount of vertical seasonal and breeding movements.

It seems likely that colonisation of the region has been from the North and relatively recent.

LITERATURE

ALVARINO, A. 1962. Two new Pacific chaetognaths. *Bull. Scripps Inst. Oceanog.* **8**, 1, *1—50*.

BAKER, A. DE C., 1954. The circumpolar continuity of antarctic plankton species. *Discovery Rept.*, **XXVII**, *201—18*.

BOLLMAN, A., 1934. Die Chaetognathen der deutschen antarktischen Exped. auf der 'Deutschland' 1911—1912. *Int. Rev. Hydrobiol.* (Leipzig) **XXX**, *251—305*.

BURFIELD, S. T., 1930. Chaetognatha. Brit. Antarct. 'Terra Nova' Exped. (Zool.), **VII** (4), *203—28*.

CONANT, F. S., 1896. Notes on the Chaetognatha. *John Hopkins Univ. Circ.* **15**, *82—5*.

CURRIE, R. I., 1964. Environmental features in the ecology of Antarctic seas. Proc. of the SCAR Symposium on Antarctic Biology. Biologie Antarctique. Paris: Hermann.

DAVID, P. M. 1955. The Distribution of *Sagitta gazellae* Ritter-Zahony. *Discovery Rept.* **XXVII**, *235—78*.

DAVID, P. M., 1956. *Sagitta planctonis* and related forms. *Bull. Brit. Mus. (Nat. Hist.)*, **4**, 8, *435—51*.

DAVID, P. M. 1958a. The distribution of the Chaetognatha of the Southern Ocean. *Discovery Rept.* **XXIX**, *201—208*.

DAVID, P. M. 1958b. A new species of *Eukrohnia* from the Southern Ocean, with a note on fertilisation. *Proc. zool. Soc. London*, **131** (4) *597—606*.

DAVID, P. M. 1959. Chaetognatha. *Banzare Rep.* Ser. B, **VIII**, 2, *73—79*.

DAVID, P. M. 1961. The influence of vertical migration on Speciation in the oceanic plankton. *Syst. Zool.* **10** (1), *10—16*.

DAVID, P. M. In Press. Some aspects of speciation in the Chaetognatha Systematic Assoc. Publ. No 5, Speciation in the Sea.

DEACON, G. E. R., 1964. Antarctic Oceanography; The Physical Environment. Proceedings of the S.C.A.R. Symposium on Antarctic Biology. Biologie Antarctique. Paris: Hermann.

DUNBAR, M. J. 1960. The evolution of stability in marine environments; Natural selection at the level of the ecosystem. *Amer. Naturalist.* **875**, 94, *129—136*.

FAGETTI, E. 1959. Quetognatos presentes en muestras Antarcticas y subantarticas. *Rev. Biol. Mar.* **IX**, *251—255*.

FOWLER, 1907. Chaetognatha. Nat. Antarctic Exped. 1901—1904. III Zoology and Botany, *1—6*.

322

FOWLER, 1908. Notes on a small collection of plankton from New Zealand, I, *Ann. Mag. Nat. Hist.* ser. 8, **1**, *240—1.*

FRASER, J. H. 1937. The distribution of Chaetognatha in Scottish waters during 1936, with notes on the Scottish indicator species. *J. Cons. int. Explor. Mer*, **12** (3), *311—320.*

FRASER, J. H. 1952. The Chaetognatha and other zooplankton of the Scottish Area and their value as biological indications of hydrological conditions. Scottish Home Dep., Marine Research, **2**, *1—52.*

FRASER, J. H. 1961. The oceanic and bathypelagic plankton of the North-East Atlantic. Mar. Res. Scot. 1961, **4**, *1—48.*

FRASER, J. H. 1957. Chaetognatha. Fich. Ident. Zool. 1 (Revision).

GERMAIN, L., 1913. Chaetognathes. Deuxième expédition antarctique française (1908—10). Sciences naturelles: Documents scientifiques. Zoologie, *87—104.*

GHIRARDELLI, E., 1953. Chetognati. Echantillons rapportés par les Docteurs J. Sapin-Jaloustre et G. Cendron, Médécins Biologistes des Deux Expéditions en Terre Adelie: 1949—1951. *Boll. Zool.*, **XX**, 1-2-3, *39—43.*

HARDY, A. C. & GUNTHER, E. R., 1935. The plankton of the South Georgia whaling grounds. *Discovery Rept.* **XI**, *1—456.*

HEINRICH, A. K. 1962. Features of the main plankton communities in the Pacific. *Trudy Inst. Oceanol.* **58**, *114—134.*

HUNTSMAN, A. G., 1919. A special study of the Canadian Chaetognaths, their distribution etc., in the waters of the Eastern coast. Canad. Fish. Exped. 1914—1915. Dept of Naval Service, Ottawa, *421—85.*

HUTCHINSON, G. E. 1953. The concept of pattern in ecology. *Proc. Acad. Sci., Philad.* **105**, *1—12.*

JAMESON, A. P., 1914. The Chaetognatha of the Scottish National Antarctic Expedition 1902—04. *Trans. Roy. Soc. Edinb.* **49**, *978—89.*

JOHNSTON, T. H. & TAYLOR, B. B., 1921. The Chaetognatha. Aust. Antarct. Exped., secr. C, **VI** (2), *1—17.*

LEVINSON, 1885. Spolia Atlantica. Om nogle pelagiske Annulata. *Videnskab. Selskab. Skrifter* (6) III. 2 Kjobenhavn 1885. *341—343.*

MARR, J. W. S., 1962. The natural history and geography of the Antarctic Krill *(Euphausia superba* Dana). *Discovery Rept.* **XXXII**, *33—464.*

MACKINTOSH, N. A., 1937. The seasonal circulation of the antarctic macroplankton. *Discovery Rept.* **XVI**, *365—412.*

MEEK, A., 1928. On *Sagitta elegans* and *Sagitta setosa* from the Northumbrian plankton, with a note on a trematode parasite. *Proc. zool. Soc. Lond.* 1928, *67—84.*

MICHAEL, E. L., 1911. Classification and vertical distribution of the Chaetognatha of the San Diego Region. *Univ. Calif. Publ. Zool.* **VIII**, *21—186.*

MICHAEL, E. L., 1919. Report on the Chaetognatha of the 'Albatross' expedition to the Philippines. *Bull. U.S. Nat. Mus.* **100**, 1, 4, *235—77.*

MÖBIUS, K., 1875. Vermes. Die Expedition zur phys.-chem. und biol. Untersuch. der Nordsee im Sommer 1872. *Jb. Comm. Wiss. Untersuch. d. deutsch. Meere in Kiel.* **II—III**, *153—70.*

PARKER, F. J. 1896. Remarks upon an arrow worm *(Spadella)* from New Zealand waters. *Trans. Proc. N.Z. Inst.* 1895, **28** Wellington 1896, *756.*

RITTER-ZAHONY, R. VON. 1909. Die Chaetognathen der Gazelle Expedition. *Zool. Anz.* **XXXIV**, *787—793.*

RITTER-ZAHONY, R. VON, 1911. Revision der Chaetognathen. Deutsche Südpolar Exped. **XIII** (5), *1—71.*

RUSSELL, F. S., 1935. On the value of certain plankton animals as indicators of water movements in the English Channel and North Sea. *J. mar. Biol. Ass. U.K.* **20** (2), *309—332.*

RUSSELL, F. S., 1936. The importance of certain plankton animals as indicators of water movements in the western end of the English Channel. *Rapp. Cons. Explor. Mer*, **100** (3), *7—10*.

RUSSELL, F. S., 1939. Hydrographical and biological conditions in the North Sea as indicated by plankton organisms. *J. Cons. int. Explor. Mer*, **14** (2), *171—192*.

STADEL, O., 1958. Die Chaetognathen Ausbeute Deutsche Antarktische Expedition 1938—39. **II**, *208—244*.

STEINHAUS, O., 1900. Chätognathen. Ergebn. d. Hamb. Magalhaens. Sammelreise **III**, 10 pp.

STEINHAUS, O., 1896. Die Verbreitung der Chätognathen im südatlantischen und indischen Ozean. Inaug. Diss. Kiel., *1—49*.

TCHINDONOVA, Y. G., 1955. Chaetognatha of the Kurile-Kamchatka trench. *Trudy Inst. Okeanol.*, **12**, *298—310*.

THIEL, M. E., 1938. Die Chaetognathen-Bevölkerung des südatlantischen Ozean. Wiss. Ergebn. dtsch. atlant. Exped. "Meteor" **XIII**, 1.

THOMSON, J. M. 1947. The Chaetognatha of South-eastern Australia. *Bull. Coun. Sci. industr. Res. Aust.* **222**, *1—43*.

ANTARCTIC AND SUBANTARCTIC DECAPOD CRUSTACEA

BY

J. C. YALDWYN

Australian Museum, Sydney, N.S.W.

I. Antarctic Decapoda

One of the remarkable aspects of Antarctic benthic fauna is the almost complete absence of the crustacean order Decapoda. Many of the characteristic groups within this order are apparently completely absent, or at least not yet recorded. As one author has recently put it (BROCH, 1961:27), "everybody who has worked in Antarctic Waters has been struck by the peculiar absence of crabs, lobsters, shrimps and prawns etc. in shallow waters". Although this exemplifies the situation, it is not strictly correct as shrimps and prawns (Natantia) of at least two circumpolar species form a distinct feature of the shallow-water benthic fauna of such low-Antarctic areas (in the sense of EKMAN, 1953) as South Georgia and such an extremely high-Antarctic area as McMurdo Sound. Thus it is the S.O. Reptantia, comprising the crayfish, lobsters, anomurans, hermit crabs and true crabs, that is apparently absent in this region though represented in Polar Arctic seas. The malacostracan order Stomatopoda, comprising the mantis shrimps, is similarly absent in the Antarctic, and in this case also absent in the Arctic.

The first Antarctic Natantia, *Chorismus antarcticus* and *Notocrangon antarcticus*, were taken at South Georgia by the expedition of the German Polar Commission in 1882—83, and described by PFEFFER in 1887. These two species have since been recorded, often associated together, by most of the major Antarctic expeditions to date and are now known to be circum-Antarctic (off South Georgia, Enderby Land, MacRobertson Land, Kaiser Wilhelm II Land, Queen Mary Land, George V Land, Victoria Land, Ross Sea and Bellingshausen Sea) from seal stomachs and from recorded depths of 7 to 500 fathoms (12.8—915 m). The only other identified benthic species is *Spirontocaris antarcticus* taken by the British Australian N.Z. Antarctic Research Expedition off Adelie Land in 320 fathoms (585 m) and described by HALE in 1941. An historically interesting, unidentified prawn is that taken from a seal stomach off Cape Adare, Ross Sea, by the ill-fated young biologist NICOLAI HANSON and illustrated by his figures in BERNACCHI's account (1901) of BORCHGREVINK's expedition. Although HODGSON (1902) identified this as a "large member of the Palaemonidae", it appears to be more like a hippolytid and was probably *Chorismus* or *Spirontocaris*, the rostral profile possibly indicating the latter.

In the checklists and discussion that follow the Antarctic region is regarded as being bounded by the Antarctic Convergence to the north, and thus including South Georgia, South Sandwich Islands and the other groups completing the arc to Graham Land as well as the Antarctic continent itself. Bouvet and Heard Islands are south of this convergence but will not be considered further as no Decapoda have been recorded from their vicinity. Kerguelen, lying on the convergence, and Macquarie Island, lying just beyond, are regarded as extreme Subantarctic islands from the point of view of their decapod fauna and will be considered in the second part of this account. The Subantarctic Zone (the "Antiboreal" of EKMAN, 1953) is bounded by the Antarctic Convergence to the south and the Subtropical Convergence (the "Antiboreal Convergence" of EKMAN) to the north. It thus includes, in addition to Kerguelen and Macquarie, the southernmost part of South America, the Falkland Islands, Gough Island, Marion and Prince Edward Islands, Crozet Islands, Auckland and Campbell Islands, as well as Antipodes and Bounty Islands. Southern New Zealand is here not regarded as subantarctic, but Tristan da Cunha in the South Atlantic, New Amsterdam and St. Paul in the Indian Ocean and the Chatham Islands off New Zealand, all within the zone of fluctuation of the Subtropical Convergence, will be mentioned in the discussion of the Subantarctic Decapoda.

Checklist of benthic Antarctic Decapoda

S.O. NATANTIA

Fam. Hippolytidae
Chorismus antarcticus (PFEFFER, 1887)
Spirontocaris antarcticus HALE, 1951[1]
Fam. Crangonidae
Notocrangon antarcticus (PFEFFER, 1887)[2]

Checklist of pelagic, bathypelagic and abyssal Decapoda recorded from Antarctic waters

S.O. NATANTIA

Fam. Sergestidae
Petalidium foliaceum BATE, 1888
Fam. Pasiphaeidae
Pasiphaea longispina LENZ & STRUNK, 1914[3]
Pasiphaea scotiae (STEBBING, 1914)[3]

---

[1] *Spirontocaris* is used here in *sensu lato*. HOLTHUIS (1947) has discussed the subdivision of this unwieldy genus and it is clear from HALE's description that this species must be placed in the genus *Lebbeus* WHITE, but I hesitate to make this change in a general account.
[2] A slender geographic race *N. antarcticus gracilis* BORRADAILE, 1916, has been described and will be discussed below.
[3] These two species are possibly synonymous. They have never been critically compared.

326

Fam. Nematocarcinidae
   Nematocarcinus lanceopes BATE, 1888[4]
Fam. Oplophoridae
   Hymenodora gracilis SMITH, 1886

Records of the S.O. Reptantia from the Antarctic Region

Two references to reptants from Antarctic waters should be discussed here. STEBBING (1914) records the hymenosomid crab *Halicarcinus planatus* from Macdougal Bay, South Orkney Islands, supposedly collected during the Scottish National Antarctic Expedition in 1903. *H. planatus* is an abundant circum-Subantarctic species, recorded from the shores of all land areas and from all major islands in the Subantarctic Zone, but other than this one record, completely unknown from the Antarctic as defined above. Although well known from the Falkland Islands, it has not been recorded from South Georgia, an island much more fully investigated and more northerly situated than the often pack-surrounded South Orkneys, where the mean summer surface temperature of the sea is 0 °C. Until this record of *Halicarcinus* can be reliably confirmed, I feel it should be disregarded and the presence of brachyuran crabs in the Antarctic should remain unproved.

In 1958, the news bulletin *Antarctic*, published by the New Zealand Antarctic Society, carried an account (1(11):293) of a Russian Expedition on the oceanographic vessel Ob during the summer of 1957—58. The report stated that in "the area of Scott Island, north of the Ross Sea, a crab was caught very similar to Kamchatka crabs." If this last statement is correct, this would indicate a lithodid anomuran rather than a true crab and this might even be *Lithodes murrayi*, or a similar species. *L. murrayi* is known from Subantarctic deep water—off Marion Island in the South Indian Ocean and off Macquarie Island and southern New Zealand (unpublished record) in the South Pacific Ocean. This Russian record then is probably the first anomuran to be recorded from Antarctic waters if it can be substantiated.

Annotated bibliography of Antarctic Decapoda[5]

BAGE, F., 1938.   Crustacea Decapoda (in part). *Aust. Ant. Exped. 1911—14.* C, II (6): 1—13, pl. 4. *C. antarcticus* (from 25—354 fms and from seal stomachs); *N. antarcticus gracilis* (240—358

---

[4] The genus *Nematocarcinus* is regarded by some authors (see BARNARD, 1950: 671) as most probably bathybenthic rather than bathypelagic in habit. This species is abyssal.
[5] A number of other Antarctic expeditions have reports published on Decapoda collected during the expedition, but not containing reference to specimens collected in Antarctic waters. These, of course, are not included in this bibliography. 1 fathom = 1.829 m.

fms) and *Nematocarcinus lanceopes* (described as *Acanthephyra antarctica* n. sp.) 870 fms, recorded off the coasts of Australian Antarctic Territory.

BATE, C. S., 1888.

Report on the Crustacea Macrura collected by H.M.S. *Challenger* during the years 1873—76. *Rept. Voy. Challenger* Zool. 24: 1—942, 150 pls. *Nematocarcinus lanceopes* n. sp. and *Hymenodora gracilis* (as *H. mollicutis* n. sp.) from deep water collections off Australian Antarctic Territory.

BORRADAILE, L. A., 1916.

Crustacea Part. I. Decapoda. *British Ant. (Terra Nova) Exped. 1910* Zool. III (2): 75—110, 16 figs.
Pasiphaea longispina (from surface or midwater); *C. antarcticus* (50—300 fms) and *N. antarcticus* var. *gracilis* n. var. (50—300 fms) recorded from the Ross Sea. Knowledge of Antarctic Decapoda summarised to date.

CALMAN, W. T., 1907.

Crustacea. I. Decapoda. *National Ant. Exped. 1901—04.* Nat. Hist. II, 7 pp. *C. antarcticus* (from 20—500 fms and from seal stomachs) and *N. antarcticus* (from 100—500 fms and from seal stomachs) recorded from the Ross Sea.

COUTIÈRE, H., 1900.

Note préliminaire sur quelques Crustacés décapodes recueillis par l'expédition antarctique belge. *Bull. Mus. Hist. Nat. Paris* 1900 (5): *238—241* and also in *C.R. Acad. Sci.* 130: *1540—1543.*
N. antarcticus (from 200—250 fms) Bellingshausen Sea.

COUTIÈRE, H., 1917.

Crustacés Schizopodes et Décapodes. *Deux. Expéd. Ant. Franç. 1908—10*, 8 pp. *C. antarcticus* and *N. antarcticus* recorded from the Bellingshausen Sea.

HALE, H. M., 1941.

Decapod Crustacea. *British Aust. N.Z. Ant. Res. Exped.* B, IV (9): 258—285, pl. 3.
Pasiphaea longispina, Nematocarcinus lanceopes, Petalidium foliaceum and *Hymenodora gracilis* from pelagic or bathypelagic collections off the coasts of Australian Antarctic Territory. *C. antarcticus* (80—150 fms); *N. antarcticus* (100—270 fms) and *Spirontocaris antarcticus* n. sp. (320 fms) from bottom in same area.

HODGSON, T. V., 1902.

Crustacea. *Rept. Coll. Nat. Hist. "Southern Cross"*: 228—261, pls. 29—40. "A large member of the Palaemonidae" from seal stomach, Ross Sea.

LENZ, H. & STRUNK, K., 1914.

Die Dekapoden der Deutschen Südpolar-Expedition 1901—1903.
I. Brachyuren und Macruren mit Ausschluss der Sergestiden. *Deutsch. Südpolar-Exped. 1901—1903*, XV Zool. VII (3): 261—345, pls. 12—22. *Pasiphaea longispina* n. sp. (pelagic); *C. antarcticus* and *N. antarcticus* from off the coast of Australian Antarctic Territory.

328

PFEFFER, G., 1887. Die Krebse von Süd-Georgien nach der Ausbeute der Deutschen Station 1882—83. I. Teil. *Jb. Hamburg wiss. Anst.* 4: *41—150*, pls. 1—7.

C. antarcticus (as *Hippolyte antarcticus* n. sp.) and *N. antarcticus* (as *Crangon antarcticus* n. sp) from South Georgia (7—9 fms).

STEBBING, T. R. R., 1914. Stalk-eyed Crustacea Malacostraca of the Scottish National Antarctic Expedition. *Trans. Roy. Soc. Edinburg* 50: *253—307*, pls. 23—32.

Phye (= Pasiphaea) scotiae n. sp. and *Nematocarcinus lanceopes* from bathypelagic collection in the Weddell Sea.

Distribution and Relationships

With the exception of *Chorismus antarcticus*, which is also found in the Magellan region of South America, the benthic species listed above are restricted to the Antarctic, though both belong to genera, or groups of genera, strongly represented in the Northern Hemisphere. Of the pelagic and deep-water natants recorded from the Antarctic only the two species of *Pasiphaea* appear to be restricted to this region. The sergestid *Petalidium foliaceum* is a widespread southern deep-water form, found in both the Atlantic and Pacific sectors; *Nematocarcinus lanceopes* is a southern abyssal species also recorded with some doubt from off South Africa, while *Hymenodora gracilis* is one of the few truly cosmopolitan (from the point of view of temperature rather than geographic distribution) bathypelagic species (EKMAN, 1953:365; SIVERTSEN & HOLTHUIS, 1956:17).

The monotypic genus *Notocrangon* is most closely related to *Sclerocrangon* and like it appears to have developed from *Crangon*-like ancestors. *Sclerocrangon* is a bathymetrically wide-ranging genus of about 22 species, mainly restricted to arctic and temperate waters of the Northern Hemisphere. The exceptions are one species recorded from South African waters and two recently described from the Chatham Rise archibenthal between southern New Zealand and the Chatham Islands (YALDWYN, 1960). A geographic race of *Notocrangon antarcticus*, from Australian Antarctic Territory and the Ross Sea, with a more slender body form and more elongated appendages has been distinguished as *N. a. gracilis*. The typical race is found in South Georgia and the Weddell and Bellingshausen Seas. This is the only evidence the Decapoda can supply for the division of the Antarctic into the subregions of West Antarctic, with the Weddell Sea and Graham Land at its centre, and East Antarctic, centred on the Ross Sea and the Australian Territories (EKMAN, 1953; 227).

Ecology of Antarctic Decapoda

Little detailed information is available on the ecology of Antarctic shelf animals. In the case of the Natantia the most dramatic is the extensive underwater photographs taken in the Ross Sea by the New Zealand Oceanographic Institute in 1958—60. For example, Station A 538, at which both *Chorismus antarcticus* and *Notocrangon antarcticus* were taken (unpublished report by author), is well covered by underwater photographs (FELL, 1961, pls. 1—3) showing the extensive bottom life of this area, and by colour photographs (FELL, 1961, colour pl. 2) of the associated animals. BULLIVANT (1959) gives monchrome and colour underwater photographs of other Ross Sea stations from which natants were taken. The extensive collection of shrimps taken during the Commonwealth Trans-Antarctic Expedition 1957—58 and the N.Z. Oceanographic Institute Antarctic cruises 1957—61 is now being worked up by the author, though no additional Antarctic species appear to be present.

A photograph showing an unidentified shrimp (probably *C. antarcticus*) and associated invertebrates from 164 fathoms in the Weddell Sea is published in *Sea Frontiers* 6 (4):223 (1960).

No reason is put forward here for the small number of Antarctic Decapoda. However, it should be pointed out that in contrast there are only 21 species in the Polar Arctic (including such genera as *Spirontocaris*, *Pandalus* and *Sclerocrangon*), and that this reverses the usual invertebrate situation, where groups are often much better represented in the Antarctic than the Polar Arctic, but is still a low figure (there are more species in subantarctic South America) for such a successful group.

II. Subantarctic Decapoda

There is little in common between the Antarctic and Subantarctic Decapoda. As mentioned above, the Antarctic species *Chorismus antarcticus* occurs also in the Magellanic area of Subantarctic South America. The latter region is the only land mass or island in the Subantarctic Zone to have direct and continuous, coastal and shelf connections with the Northern Hemisphere. In consequence of this the decapod fauna is relatively rich and varied, the Natantia for instance being represented by about 8 genera (including such forms as *Nauticaris*, *Austropandalus* and *Campylonotus*), the Anomura by at least 5 genera (including *Munida*, *Lithodes* and *Petrolisthes*) and the Brachyura or true crabs by 23 genera (including *Halicarcinus*, *Hemigrapsus* and *Cancer*). Many of these are in reality temperate zone forms extending into the Subantarctic, for instance 19 of the above bracyuran genera also occur in the Peruvian-North Chilean region.

The Falkland Islands have, as might be expected from their relationship to the Patagonian shelf, a relatively rich decapod fauna for a Subantarctic island, but their total of about 10 genera (representing 11 species) contrasts dramatically with the 36 genera (representing at least 45 species) from Subantarctic South America. All Falkland species are known from the Magellanic area of South America, the only natant being the hippolytid *Nauticaris magellanica;* the Anomura being represented by 2 species of *Munida* *(M. subrugosa* and *M. gregaria), Lithodes antarcticus, Paralomis* and *Pagurus*, while the Brachyura include the circum-Subantarctic *Halicarcinus planatus* in addition to *Acanthocyclus, Peltarion, Eurypodius* and the pelagic grapsid *Planes*. Three normally characteristic intertidal families, Porcellanidae, Grapsidae (excluding the pelagic *Planes)* and Xanthidae, present in Subantarctic South America *(Petrolisthes, Hemigrapsus, Gaudichaudia, Pilumnoides* etc.) are absent from the Falklands and indeed from all the other Subantarctic islands from Gough through to the Auckland and Campbell Islands. In the New Zealand area there are no porcellanids or grapsids even at the Chatham Islands (KNOX, 1954; DELL, 1960) astride the Subtropical Convergence.

Nauticaris is recorded from isolated Gough Island in the South Atlantic and the commercial palinurid crayfish *Jasus lalandii* is taken in large numbers in the shallow waters surrounding Gough and the Tristan da Cunha group. The majid *Achaeopsis* and the goneplacid *Pilumnoplax* both occur in the deeper water at the edge of the Gough Island shelf and are the only crabs reported from this Subtropical Convergence area.

Marion and Prince Edward Islands in the South Indian Ocean have *Halicarcinus planatus* in the intertidal and *Nauticaris marionis* in shallow water, as well as an interesting deep water fauna of 7 genera taken by the Challenger Expedition. These include the natants *Chorismus tuberculatus* (the only other species in this otherwise Antarctic genus) and *Campylonotus*, as well as *Lithodes murrayi* and *Paralomis* among the anomurans.

No Decapoda are recorded from the Crozet Islands, though *Halicarcinus planatus* could well be there, and in fact, is the only littoral or shallow-water decapod known from Kerguelen and Macquarie (for recent work on the latter see KENNY & HAYSON, 1962), thus apparently confirming the extreme Subantarctic nature of these two islands. The anomurans *Lithodes murrayi* and *Sympagurus arcuatus* subsp. were taken on the Macquarie Island shelf by the B.A.N.Z.A.R. Expedition in 1930. As explained above the former species is known also from off southern New Zealand as well as off Marion Island, while the latter occurs in eastern Australian waters.

Auckland and Campbell Islands on the Campbell Plateau to the south of New Zealand have at least 14 genera (17 species including unpublished records) represented in their littoral and shallow-water zones. *Nauticaris marionis* and *Halicarcinus planatus* are both abundant species at these islands and both are found off the New Zealand mainland proper. All genera recorded are known from the New Zealand fauna, though the monotypic majid *Jacquinotia* comes close to being the only endemic decapod genus in the Subantarctic. This spider crab is an abundant, shallow water scavenger, and grows to a large size (carapace length up to 200 mm) in Auckland and Campbell Island waters, but small specimens do occur in the archibenthal off southern New Zealand. The natants *Pontophilus* and *Tozeuma*, the anomurans *Callianassa, Galathea, Munida (M. subrugosa* and *M. gregaria), Porcellanopagurus* and *Pagurus*, as well as the crabs *Leptomithrax, Cancer* and *Nectocarcinus* make up the remainder of the species recorded. The palinurid *Jasus lalandii* has been recorded once in seal vomit from the Auckland Islands (YALDWYN, 1958), but is substantiated from the more northerly Snares Islands (to the south of Stewart Island) where it is the most southerly record for the entire family of spiny crayfish (Palinuridae). The few species now known (unpublished information) from the Antipodes and Bounty Islands represent a greatly reduced sampling of the above fauna.

In conclusion it appears that the nearest approach to a typical Subantarctic species is the widespread *Halicarcinus planatus*, though the two species of the genus *Nauticaris* occur extensively throughout this zone. *Lithodes*, represented by more than one species, has a disjointed distribution from South America to southern New Zealand, and probably will be found to occur abundantly in Subantarctic waters when more is known of the shelf and slope faunas of this zone. The two characteristic species of *Munida* form an important seasonal vertebrate food source in both the Patagonian-Falkland area and the Auckland-Campbell-Southern New Zealand area, though known elsewhere from Tasmanian waters. On the other hand, *Jasus lalandii*, in its various geographical forms, has an interesting distribution in relation to the Subantarctic Zone. This crayfish is restricted to the coasts of the southern continental land masses, excluding South America, and to the circumpolar oceanic islands on or near the Subtropical Convergence (Tristan da Cunha and Gough, New Amsterdam and St. Paul, Chatham Islands and Juan Fernandez in the eastern Pacific). Many of the important commercial grounds of this species are in areas of mixed waters, astride the fluctuating convergence. Other than the one Auckland Island record discussed above, *Jasus* does not occur at the true Subantarctic islands discussed above.

LITERATURE CITED

(other than references listed in the annotated bibliography)

BARNARD, K. H., 1950. Descriptive Catalogue of South African Decapod Crustacea. *Ann. S. Afr. Mus.* **XXXVIII**: *1—837*, 154 figs.

BERNACCHI, L., 1901. *To the South Polar Regions. Expedition of 1898—1900.* Hurst & Blackett, London. 348 pp. Numerous pls.

BROCH, H., 1961. Benthonic Problems in Antarctic and Arctic Waters. *Sci. Res. Norwegian Ant. Exped. 1927—1928.* **38**: *1—32*, 6 figs.

BULLIVANT, J. S., 1959. Photographs of the Bottom Fauna in the Ross Sea. *N. Z. J. Sci.* **2** (4): *485—597*, pls.

DELL, R. K., 1960. Crabs (Decapoda, Brachyura) of the Chatham Islands 1954 Expedition. *N. Z. Dept. Sci. Indust. Res. Bull.* **139** (1): *1—8*, 2 pls.

EKMAN, S., 1953. *Zoogeography of the Sea.* Sidgwick & Jackson, London. 417 pp., 121 figs.

FELL, H. B., 1961. The Fauna of the Ross Sea. Part 1. Ophiuroidea. *N.Z. Dept. Sci. Indust. Res. Bull.* **142**: *1—79*, 19 pls.

HOLTHUIS, L. B., 1947. The Decapoda of the Siboga Expedition. Part. IX. The Hippolytidae and Rhynchocinetidae collected by the Siboga and Snellius Expeditions with remarks on other species. *Siboga Exped.* **39a 8**: *1—100*, 15 figs.

KENNY, R., & HAYSON, N., 1962. Ecology of Rocky Shore Organisms at Macquarie Island. *Pacific Sci.* **XVI** (3): *245—263*, 12 figs.

KNOX, G. A., 1954. The Intertidal Flora and Fauna of the Chatham Islands. *Nature* **174**: *871.*

SIVERTSEN, E. & HOLTHUIS, L. B., 1956. Crustacea Decapoda (The Penaeidea and Stenopodidea excepted). *Michael Sars N. Atlant. Exped. 1910* **V** (12): *1—54*, 4 pls.

YALDWYN, J. C., 1958. Decapod Crustacea from Subantarctic Seal and Shag Stomachs. *Rec. Dom. Mus.* **3** (2): *121—127.*

YALDWYN, J. C., 1960. Crustacea Decapoda Natantia from the Chatham Rise: A Deep Water Bottom Fauna from New Zealand. *N.Z. Dept. Sci. Indust. Res. Bull.* **139** (1): *13—53*, 10 figs.

MOLLUSCA OF ANTARCTIC
AND SUBANTARCTIC SEAS

BY

A. W. B. POWELL

Assistant Director, Auckland Institute and Museum
(with 6 figs.)

In evaluating the molluscan faunas of Antarctic and Subantarctic seas many factors both past and present require to be considered.

It is desirable to treat the Antarctic and Subantarctic faunas together since there is difficulty in drawing a hard and fast line between them. Many of the species range over both zones especially at locations such as Kerguelen and Macquarie Islands where the Antarctic Convergence runs very close to them and also in the case of South Georgia where faunas are telescoped to a certain extent owing to geographical location, the presence of a sizeable land mass and special associated ecological conditions.

The Antarctic environment

Firstly must be stressed the remoteness and distinctiveness of the Antarctic Continent compared with the other land masses of the world.

Antarctica is larger than Europe, being about 3400 miles in length from Joinville Land to Adelie Land, about 2800 miles from Charcot Land to Gaussberg and 2500 miles from Coats Land to Oates Land.

The average height of this vast South Polar Continent is, including the ice cap of generally indeterminate thickness, approximately 6500 feet. No terrestrial molluscs are known from this wilderness of ice and snow.

The extensive areas of barrier of shelf ice formations that surround the actual land area of Antarctica not only greatly increase its size in varying degree according to season but also result in conditions totally unfavourable to the establishment of an intertidal fauna.

For the greater part of the year the coastline is so surrounded by thick ice that the shoreline down to a depth of several fathoms* is swept clean and rendered almost devoid of macrofauna. Below this level, however, out of range of the scouring action of moving masses of ice, a wealth of marine life exists. Masses of algae provide shelter and food for a wide variety of animals, sponges in particular frequently taking on a dominant role. Such conditions appear to extend down to about twenty-five fathoms after which terrigenous

---

* 1 fathom = 6 ft = 1.829 m.

and pelagic deposits are encountered with increasing depth and distance from land.

The waters of the Southern Ocean present remarkable uniformity and this has made possible the mapping of broad clearly defined distributional patterns for surface organisms that are shown to be more or less linked with the physical factors.

As a general observation it is evident that in these southern surface waters in particular the number of species of organisms is few, but each occurs in countless numbers, whereas in tropical waters the reverse is the case.

The work of the Discovery II has shown that concentrations of phosphates and nitrates associated with immense quantities of phytoplankton are far greater than elsewhere, appearing always to be in excess of requirements.

The Subantarctic environment

The terms Antarctic and Subantarctic have often been loosely applied for whilst the former is clearly from that area extending from the Antarctic Continent to the Antarctic Convergence, the latter, which in theory occupies the zone between the Antarctic and Subtropical convergences, is not so clear cut, especially where it encounters the southern portions of the continental and other land masses.

The Antarctic Convergence marks a more or less definite boundary where ice diluted cold dense water of the Antarctic, 300 to 800 feet in depth meets the warmer but more saline therefore lighter water of the Subantarctic Zone.

At this seasonally fluctuating line of contact the Antarctic water sinks abruptly beneath the warmer Subantarctic water and continues to flow northwards as a deeper intermediate layer. Further north at approximately 40° S. latitude occurs the Subtropical Convergence, a similar but less pronounced phenomenon.

Whereas the Antarctic Zone is largely covered by pack-ice for most of the year the Subantarctic is relatively free of ice but is dominated by the vigorous and persistent West Wind Drift which sweeps little impeded by land masses around the whole extent of the Southern Ocean.

The West Wind Drift is undoubtedly the chief distributing agent in the circum-Subantarctic spread of marine molluscs, particularly those species associated with algae. Drifting masses of algae are undoubtedly the means by which the herbivorous Trochoids *Margarella* and certain Patellids, *Nacella* and *Patinigera* of the *fuegiensis* group as well as the byssiferous attached bivalves such as *Hochstetteria* have achieved their wide lateral dispersal.

The sources of molluscan material

Although vast areas of the Southern Ocean and long stretches of the coastline of Antarctica are still more or less unexplored biologically, nevertheless oceanographic investigation of these seas has been more intensive than in many more accessible regions of the world.

Exploration of the Antarctic with expedition ships equipped for oceanographic work has resulted in a wealth of information and the acquisition of much useful material.

Following the purely explorative period the first major contribution to our knowledge of Antarctic and Subantarctic molluscs resulted from the famous voyage of the English oceanographic expedition of H.M.S. Challenger, 1872—1874. Then at about the turn of the century onward rivalry among European countries resulted in a number of expeditions each fully equipped for scientific research.

The more important of these from the viewpoint of subsequent molluscan research were the Belgian Antarctic Expedition, "Belgica", 1897—1899, the English "Southern Cross" Expedition, 1898—1900, the German Deutsche Südpolar Expedition, 1901—1903, the Swedish Antarctic Expedition led by Dr. OTTO NORDENSKJÖLD, 1902, the Scottish National Antarctic Expedition, "Scotia", 1903—1904, Captain SCOTT's first National (Discovery) Expedition, 1901—1904, the French Expeditions under Dr. JEAN CHARCOT, 1903—1905 and again 1908—1910 and Captain SCOTT's last "Terra Nova" Expedition, 1910—1913.

Then came Sir DOUGLAS MAWSON's two Antarctic Expeditions, the Australasian Antarctic Expedition of 1911—1914 followed by the British Australian New Zealand Antarctic Research Expeditions of 1929—1931.

From 1925 when the Discovery Committee was set up in England there followed an intensive biological investigation of Antarctic and Subantarctic seas. At first these investigations were undertaken primarily in the interests of whaling but the need for complete relevant data was soon apparent with the result that our knowledge of the fauna as well as the physical data for these southern seas is now entering its final stage.

Although the work of the Discovery Investigations tends to center around life in the oceanic waters, nevertheless a great deal of bottom sampling by means of several kinds of dredges and nets was undertaken.

Reviewing the respective results of these expeditions so far as molluscs are concerned, varying degrees of efficiency in the processing of material becomes apparent. With some of the expeditions it is obvious that only the macrofauna was hand picked and the sediments discarded. In others the entire material including the substratum was meticulously examined with the result that a considerable microfauna is now known from some stations, but not from the entire area collectively covered by the expeditions.

A further factor that will eventually alter the picture again is in the use of large nets and trawls. When the Discovery Committee's "William Scoresby" was employed on an extensive fisheries investigation over the East Patagonian Shelf a number of large sized molluscs that had previously escaped smaller dredging equipment were secured in numbers.

Age of the Antarctic molluscan fauna

The high hopes of early investigators of Antarctic deep-water faunas that these remote waters would likely reveal living representatives of archaic organisms were soon dashed. So much so in fact that a totally different concept arose out of the initial studies of the material obtained — that of bipolarity, the supposed near identity between Arctic and Antarctic faunas.

SOOT-RYEN (1951), in reviewing the Antarctic pelecypods considers that there is evidence of an early Antarctic fauna from a time of more genial climatic conditions that still constitutes an element in the Recent fauna. He instances three bivalve molluscs, the Pectinid, *Adamussium colbecki*, *Laternula elliptica* and *Thracia meridionalis*.

The genus *Laternula* is a relatively large thin shelled bivalve widely distributed in the tropical Indo-Pacific and it ranges southward to southern Australia and Tasmania but there are no living members of the genus in South America, South Africa or New Zealand.

It is the habit of this animal to burrow deeply into soft mud and this insulating factor is probably the main reason why an otherwise warm-water mollusc has managed to adjust itself to a cold-water habitat. Similarly *Thracia* is an infaunal animal but *Adamussium* is a free swimmer.

Among the gastropods, *Perissodonta* stands out as a constituent of this relict fauna, shown by the Cretaceo-Tertiary ancestry of the Struthiolarids in both Patagonia and New Zealand. A living species *Perissodonta mirabilis* (SMITH) occurs at both South Georgia and Kerguelen Islands. A comparative study of the dentition reveals that the triangular-shaped laterals of *Perissodonta* approach those of *Aporrhais* and this feature, coupled with the deep labial sinus, makes the derivation of the Struthiolariidae from the *Aporrhaidae* a very reasonable assumption. The Volutes *Guivillea* and *Provocator* are probably further members of the older fauna.

Large Marginellids are usually considered characteristic of the tropics but species from the Magellan Region, *dozei* and *warrenii* have the size, bright coloration and enamel of their warm water relatives.

Ancestors of these could have entered the area during former warmer climatic periods by a gradual extension of their normal range down the continuous eastern coastline of South America. Another moderately large Marginellid, *ealesae*, although thin-shelled and colourless, comes from Enderby Land to the Ross Sea, again suggesting entry during a past period considerably warmer than at present.

SOOT-RYEN (1951) also points out that nearly all the shallow-water Antarctic bivalves that may be considered descendants of the older fauna of that region are ovoviviparous, retaining their young within the mantle cavity. This applies to the *Gaimardiidae, Cyamiidae, Philobryidae,* many of the *Leptonacea,* the *Carditidae* and *Laternula.*

THORSON (1936) has shown with Arctic bivalves that the eggs are large with a great amount of yolk and that the embryos develop without or with a very short pelagic larval stage, which enables them to survive unfavourable climatic conditions.

Summarized the present Antarctic molluscan fauna appears to be in an active stage of colonization, for the most part a replacement of an earlier fauna that must have been largely exterminated during the rigours of the comparatively recent ice-ages.

A significant observation in this connection is the inferred adaptive radiation of a few of the more vigorous and plastic families of the gastropoda, the *Trochidae, Cominellidae* and *Buccinulidae* in particular.

The gastropod families *Patellidae, Trochidae, Littorinidae, Naticidae, Muricidae, Cominellidae, Buccinulidae, Volutidae* and *Turridae* make up approximately 28% of the Antarctic-Subantarctic fauna — i.e. 75 genera comprising 245 species, out of a total of 875 nominate species.

The richest area of the Antarctic-Subantarctic in molluscan species is, as one would expect, the American Quadrant since that area has long been the most effective colonizing route. This region also has the greater extent of bays, channels and sheltered waterways, contrasted with the vast stretches of the Antarctic Continent associated with the Great Ice Barrier.

The continuity of the combined western coastlines of the Americas extending through the Scotia Arc, that great eastern "stepping stone", sweep by way of the Burdwood Bank, Shag Rocks, South Georgia, South Sandwich group, South Orkneys and South Shetlands to the vicinity of Graham Land on the Antarctic Continent, forms the, at present, chief colonizing route to Antarctica, whilst the Atlantic-Indian Ocean Cross Ridge, which runs from the Argentine Basin almost to the Kerguelen-Gaussberg (radial) Ridge, is effective in the eastward spread of Subantarctic benthic species. Possibly other radial extensions from Antarctica in former times may have operated also in distributing southern fauna to the New Zealand and Australian areas. The Subantarctic cross-ridges also may have been much shallower during some former period than at the present time and thus more effective in distributing organisms more or less restricted to the Continental Shelf.

Bipolarity

The term bipolarity was employed by the biogeographers of last century in the belief that large numbers of species of molluscs were common both to the Arctic and to the Antarctic but did not occur in the seas between.

However much of the evidence that formed the basis of this concept has since been discredited. For the most part the supposed resemblance between these high latitude northern and southern molluscs is ecologic rather than morphologic. Their thin shells are characteristic of the quiet depths at which many of them live, the lack of colour is due to restricted light penetration of the water and the general acquirement of a thick periostracum is resultant from the more erosive quality of ice-diluted water.

The three main hypotheses that have been advanced to explain the phenomenon of bipolarity have been summarized as follows: (1) Bipolar animals are relics of a former cosmopolitan fauna, the tropical portion of which is now extinct (examples, the Naticoid genus *Amauropsis*, the Trichotropid gastropods and the pelecypod genus *Astarte*). (2) Animals that may have migrated from one polar region to the other through the cold deep water regions of the tropics (examples, the gastropod genera *Aforia*, *Fusitriton* and *Puncturella*). (3) Parallel independant development in the two areas that has resulted in apparent identity (examples, gastropods of the families *Trochidae*, *Muricidae* and the *Buccinidae-Cominellidae-Buccinulidae* complex.

A fourth category could well be added to account for wide areas of dispersal over the floors of the more or less interconnected deep ocean basins (examples, the Turrid genera *Pleurotomella*, *Leucosyrinx* and *Pontiothauma*).

In the second category in which genera have achieved a bipolar distribution through going deep over the tropics, the western coastline of the Americas, as already mentioned, is the significant area not only because they afford an uninterrupted north to south land mass almost linked to Antarctica through submerged ridges and the island groups comprising the Scotia Arc but also from the phenomenon of upwelling cold water which is known to

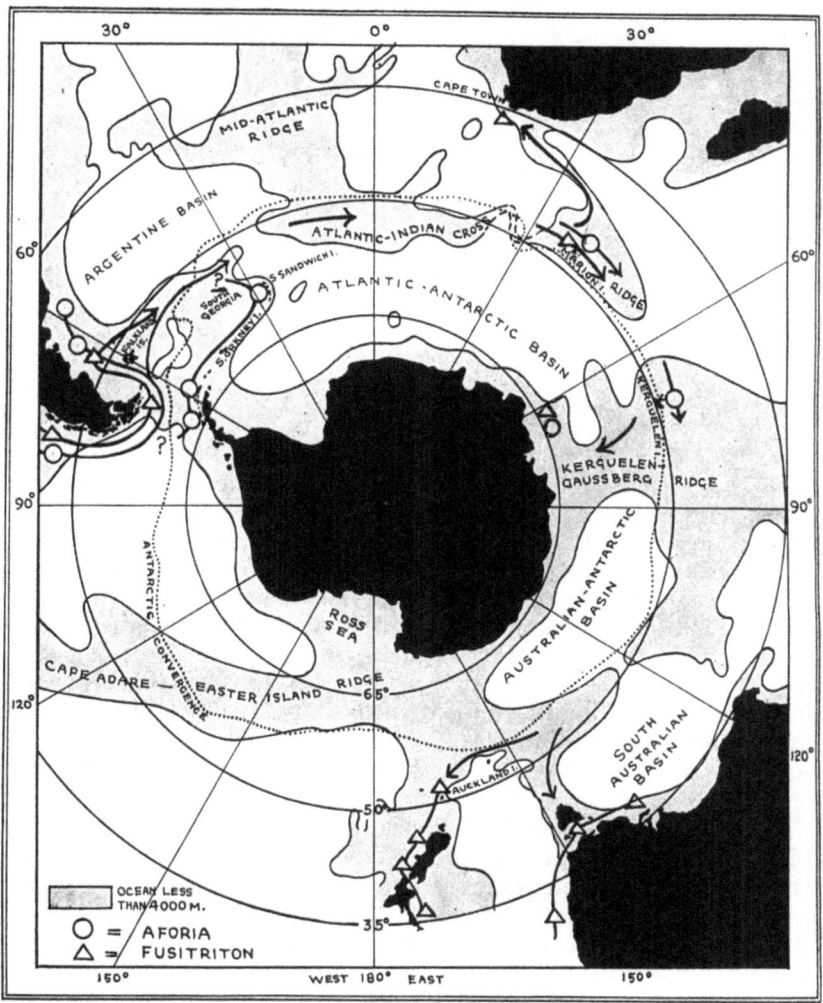

Fig. 1.

occur extensively along the west coast of both North and South America.

Figure 2 shows the distribution of three "bipolar" genera, *Acanthina* (shallow water) and *Fusitriton* and *Aforia* (continental shelf and deeper).

Species of *Acanthina* range along the western coastline of the Americas from North West America to the Magellan Region inclusive of the Falklands but are not known from elsewhere. They are shallow-water molluscs that have sufficient temperature toler-

340

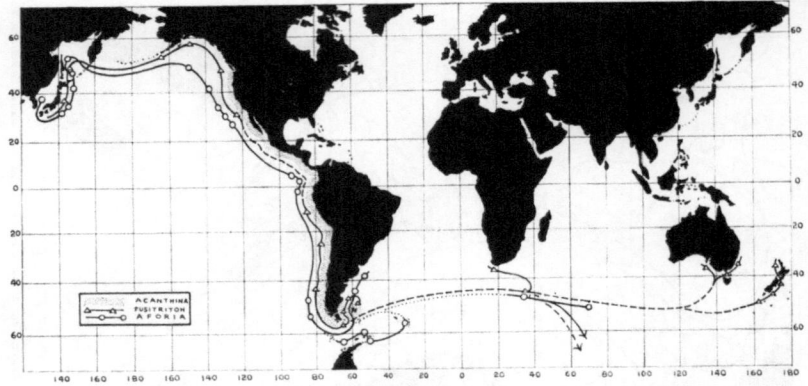

Fig. 2.

ance to enable them to occupy and continue across the Central American tropical belt.

The Cymatid genus *Fusitriton* is stenothermic, with a continental shelf to Archibenthal station, reaching a depth of 1800 fathoms in one instance and although widespread has a pattern of distribution by which it follows more or less the continental margins of the Pacific, ranging from northern Japan through the Aleutian Chain down the west coast of the Americas to the Magellan Region, including the Falklands and thence eastward including South Africa, southward to Enderby Land in the Antarctic, New Zealand, including its Southern Islands, New South Wales and Southern Australia.

The Turrid genus *Aforia* is the best example of a stenothermic "bipolar" mollusc that goes deep over the tropics. The following table lists a number of species of *Aforia* which are distributed more or less connectedly off the western coasts of the Americas, have followed through the Drake Strait into the Argentine Basin and thence eastward via the Atlantic-Indian Ocean Cross Ridge to the vicinity of Kerguelen as well as penetrating southward through the Scotia Arc chain to Antarctica where a species has spread along the continental edge to at least as far as MacRobertson Land.

The temperatures cited are associated with specimens dredged at various seasons nevertheless the lowest temperature record, 30.9° F (—0.6° C), and the highest, 41.8° F (5.4° C), show a surprisingly small range, with a maximum difference of only 10.9° F (6.0° C). That these stenothermic animals go deep over the tropics in order to keep within their temperature requirements is clearly shown by the *persimilis* records, especially that of the Ecuador occurrence, which at a depth of 741 fathoms and a temperature of 38.4° F, is situated only 1.03° N. of the equator.

Distribution of Aforia

Species and location	Depth in fathoms	Temp. F°	Date
insignis (JEFFREYS), "Ice Sea", Siberia	—	—	—
okhotskensis BARTSCH, Okhotsk Sea, 50°—60°N.	73	30.9°	—
kincaidi (DALL), Shelikoff Strait, Alaska	—	—	—
sakhalinensis BARTSCH, Off Sakhalin Id.	399	38.1°	—
diomedea BARTSCH, Hokkaido, Japan	266—399	32.1°—38.1°	—
chosensis BARTSCH, Japan Sea, 40°N.	122	34.1°	—
japonica (DALL), Honshu, Japan, 35 N.	369	41.8°	—
hondoana (DALL). Hondo, Japan	614	37.5°	—
crebristriata (DALL), S.W. of Sitka, Alaska	1569	34.9°	—
persimilis leonis (DALL), Washington	877	36.6°	—
persimilis blanca (DALL), Cape Blanco, Oregon,	1064	35.9°	—
amycus (DALL), off Monterey, California	871	38.0°	—
persimilis (DALL), Cortez Bank, California	984	38.0°	—
Gulf of Panama	1020	36.8°	Mar. 1891
Off Panama	1067	37.0°	Feb. 1891
Off Ecuador, 1 03 N.	741	38.4°	—
S.W. coast of Chile	677	38.0°	—
goodei (DALL), N.W. Patagonia, 45°35'S.	1050	36.9°	—
goniodes (WATSON), S.E. of Rio de Plata, Argentina	600	37.2°	Feb. 1876
between Falklands and Patagonia	82	—	Mar. 1932
staminea (WATSON), Marion and Prince Edward Islands	1375	35.6°	Dec. 1873
Kerguelen Island	105	—	—
magnifica (STREBEL), S.W. of Snow Hill Island, 64°S.	82	—	—
Palmer Archipelago	152—273	32.38°	Mar. 1927
South Shetlands	214	32.07°	Jan. 1937
South Sandwich Island	180	32.7°	Feb. 1930
Princess Elizabeth Land	437	—	—
MacRobertson Land	177	—	—

Regarding the fourth category which concerns molluscs of the deep ocean basins it would appear from the scant evidence available that certain genera have an extensive range through inter-connected basins from Antarctica to the equatorial region and even into the deep waters of the Northern Hemisphere, the process being most likely a gradual infiltration from deep equatorial basins both northward and southward following the deep cold waters.

Oceanographers have shown that the northward flowing cold Antarctic waters plunge deeply at the Antarctic Convergence, proceeding thence beneath the warmer Subantarctic waters and then again at the Subtropical Convergence, eventually reaching the abyssal depths even in the equatorial zone.

The Turrid genera *Leucosyrinx* and *Pleurotomella*, both well

342

represented in the deep waters of the North Atlantic and Caribbean, extend via the Subantarctic of the American Quadrant to the Antarctic Continent.

Another wide ranging Turrid genus *Pontiothauma*, described originally from the Bay of Bengal in 753—1250 fathoms, is now known from a number of stations in Indonesian waters and occurs also off Enderby land in Antarctica. However, a Japanese deep water species, recently described and ascribed to *Pontiothauma*, is not that genus but an allied one, *Spergo*, the type of which is from deep water off Hawaii.

When more is known concerning the molluscan faunas of the deep ocean basins, some very extensive distributional patterns should become evident.

Characteristics of the faunal areas of Antarctic and Subantarctic waters

Falkland Islands

The Falklands are a group of two large and many small islands lying upon the very extensive East Patagonian Continental Shelf, but almost severed from it by the transverse Falkland Trough which lies at a depth of from 150—200 metres, between two tongues of deep water which impinge both from the north and from the south. The Falklands lie within the Subantarctic Zone of surface waters (isotherms between 6° and 12° C), and are outside the northern limit of pack-ice. They are strongly influenced by the Cape Horn Current, composed largely of water of Pacific origin, which is swept through Drake Strait by the West Wind Drift and then turns northwards to the Falklands and resolves into the Falkland Current, which continues northward between the Falklands and Patagonia. The West Wind Drift proper passes well to the south of the Falklands. Owing to its position upon the Patagonian Shelf the Falkland marine molluscan fauna is predominantly Magellanic, and the scant terrestrial molluscan fauna, notably the presence of the freshwater genus *Chilina*, points to a former land link with the Patagonian mainland.

North of the Falklands at the limit of the Continental Shelf, the bottom descends steeply to the Argentine Basin, which comes within the influence of the warm Brazilian Current, which there has a seasonal temperature range of from 11.5° to 14.5° C.

Another important factor influencing the Falkland fauna is the presence of the Atlantic-Indian Ocean Cross Ridge which runs from the Argentine Basin almost to the Kerguelen-Gaussberg (radial) ridge and forms the northern boundary of the Atlantic-Antarctic Basin.

Deep water to the south effectively separates the Falklands from the Burdwood Bank and the rest of the extensive Scotia Arc.

The Falkland marine molluscan fauna is Subantarctic, predominantly Magellanic but with a strong admixture of continental temperate extralimital forms which have been induced to extend far south of their normal range through the continuity of the Patagonian land mass with its wealth of sheltered inlets, bays and channels.

The following instance temperate forms induced far south of their normal range due to these special conditions: *Fissurella (Balboaina)*, *Calliostoma*, *Polinices*, *Trochita*, *Nassarius*, *Acanthina*, *Typhis*, *Adelomelon*, and *Marginella*.

Restricted Magellanic genera are the two Calliostomid derivatives *Photinula* and *Photinastoma*, the Trophonids *Xymenopsis*, *Fuegotrophon* and *Stramonitrophon* and the Cominellid genera *Savatieria*, *Anomacme* and *Meteuthria*.

The Atlantic-Indian Ocean Cross Ridge has allowed an interchange of several genera and species between the Falklands and the Marion Island-Kerguelen area, notably *Provocator*, *Trophon declinans* and *Philine kerguelensis* as well as facilitating the eastward spread of both *Aforia* and *Fusitriton*.

The characteristic Subantarctic genera represented at the Falklands are *Gaimardia*, *Cyamium*, *Nacella*, *Patinigera*, *Margarella*, *Laevilitorina*, *Pareuthria* and *Kerguelenella*.

Apart from *Prosipho* which is abundantly represented both in the Antarctic as well as in the Subantarctic other Antarctic ranging genera are not strongly represented at the Falklands.

Burdwood Bank

The Burdwood Bank is a large shoal of from 80—150 metres in depth situated south of the Falklands and separated from them by deep water, 500—2000 metres. The shoal lies east of Tierra del Fuego, and it is now generally accepted that the line ot folding represented by the Andes and their former continuity in what is now termed the Scotia Arc passed through the Burdwood Bank and not the Falkland Islands. A trough of moderately deep water, 250—500 metres severs the bank from Tierra del Fuego, and a ridge connecting it to the eastward with Shag Rocks and South Georgia varies between 1000 and 2000 metres, severed in several places by deeper water of between 3000 and 4000 metres.

From his studies on fossil foraminifera dredged from the Burdwood Bank, MACFADYEN (1933) states in summarizing these fossil occurrences that the "beds are clearly shown to be the continuation of those exposed on Tierra del Fuego and Staten Island, and a part of the (renamed) Scotia Arc of folding, which is continued on a trend precisely determined by soundings to lie on the line of Shag

Rocks, South Georgia, Clerke Rocks, South Sandwich Islands, South Orkney Islands to the South Shetlands and Graham Land".

Only a small list of mollusca is available for this area but at least 13 species out of a total of 55 are apparently restricted to the locality. Of the remainder, 15 are found at the Falklands, but only three of them, *Davisia cobbi*, *Brookula calypso* and *Colpospirella algida*, are not generally distributed in the Magellanic Province.

The bulk of the fauna is Magellanic, but 4 Antarctic species, *Schizotrochus euglyptus*, *Pellilitorina pellita*, *Balcis antarctica* and *Paradmete fragillima*, here apparently reach their northern limit for the American Quadrant.

South Georgia and Shag Rocks

This very distinctive biogeographic unit lies from 12° to 20° east of the Burdwood Bank and is part of the Scotia Arc, although it is now surrounded by deep water to a depth of 3000 metres. As pointed out by EARLAND (1933), South Georgia is located in only slightly higher latitudes (54—55 °S.) than the Falklands (51°—52°30′S.), yet the contrast in both their physical conditions and their respective faunas is profound.

The Falklands are situated on the Patagonian Shelf, not the Scotia Arc, and the surrounding waters are ice-free, being out of the influence of the cold West Wind Drift. South Georgia, on the other hand, is an isolated area in a region of deep water, entirely within the influence of the cold West Wind Drift and even land conditions are glacial.

Partially resolving upon these conditions the bottom sediments are mainly tenacious blue muds in contrast to the sandy deposits of the Falkland area and the coarse sandy and often volcanic debris of the South Sandwich — South Shetland section of the Scotia Arc.

Owing to the far southward extension of the South American land mass the Antarctic Convergence is forced below its average latitude, with the result that the Falklands lie in the Subantarctic and South Georgia, since it is well to the eastward, comes within the Antarctic zone of surface waters.

Regarding the Foraminifera, EARLAND (1934) remarked that "In its isolation, South Georgia has either preserved or developed species which are almost confined to the island".

The molluscan fauna has scarcely any species common to the Falklands. On the other hand, the Discovery II collections revealed a number of genera and species previously considered characteristic of the Kerguelen and the Ross Sea areas. Their significance, however, is partly ecological, in that the blue muds of South Georgia are more comparable with the Ross Sea bottom than they are with the coarse sandy bottom of the shallower areas of much of the Scotia

Arc. This would account for the presence in South Georgia of the Antarctic species *Trichoconcha mirabilis*, but its apparent absence from other Scotia Sea localities.

The South Georgian fauna must now be fairly completely known. It is very distinctive, and as already noted has little in common with that of the Falklands, which are almost of the same latitude.

There are several endemic genera: *Promargarita*, *Pfefferia*, *Chlanidotella* and *Cavineptunea*. Antarctic genera and species are strongly represented; *Patinigera polaris*, *Venustatrochus*, *Laevilacunaria*, *Trichoconcha*, *Prosipho*, *Chlanidota*, *Probuccinum* and *Neactaeonina*. The widely distributed Antarctic — Subantarctic genera *Gaimardia* and *Margarella* are represented almost entirely by endemic species.

The characteristic Magellanic genera *Nacella*, *Photinula*, *Photinastoma*, *Xymenella* and *Adelomelon* are not represented. Evidence, however, that the Scotia Arc was formerly a more effective route than at present for the southward spread of Magellanic fauna is shown by the presence of a *Trochita* and a derivative of *Calliostoma*, *Venustatrochus georgianus*.

Deep water severing of the arc has culminated by isolation in the development of the South Georgian fauna as a distinctive unit.

On the other hand, there is a strong representation of both eastern Subantarctic and Antarctic forms characteristic of the Victoria and Enderby Quadrants; *Pallilitorina setosa*, *Amauropsis (Kerguelenatica) grisea*, *Perissodonta georgiana*, *Sinuber*, *Falsimohnia*, *Proneptunea*, *Probuccinum* and *Prosipho hunteri*.

South Sandwich Islands and Remainder of Scotia Arc

On the basis of foraminiferal studies, EARLAND (1934) has divided this section of the Antarctic into the following faunal areas: (1) Weddell Sea; (2) South Sandwich Islands; (3) South Orkney Islands to Clarence Island on the Continental Shelf and Slope; (4) Scotia Sea from 29°15′ to 60°W; (5) Drake Strait from 60°W; (6) Bransfield Strait and South Shetland Islands; (7) Palmer Archipelago and (8) Bellingshausen Sea.

The whole area is within Antarctic waters, and most of it is either on the Antarctic Continental Shelf or tied to it through the Scotia Arc, which links with Graham Land.

Our knowledge of the molluscan fauna of this area is too incomplete to gauge whether EARLAND's foraminiferal faunules apply to molluscs as well except to note that a number of the molluscs are common to two or more of his eight areas.

The molluscan list is made up as follows: South Sandwich Islands, 12 species; South Orkneys, 48 species; South Shetlands, 35 species;

Palmer Archipelago, 20 species; with a total of 93 species for the four areas; obviously an incomplete list.

Regarding the above four localities, provisionally as a whole, there are three marked influences apart from an apparently endemic faunule; (1) South Georgian, (2) Kerguelenian and (3) eastern Antarctic.

The South Georgian influence is represented by *Tropidomarga biangulata, Leptocollonia thielei, Pellilitorina pellita, Eatoniella kerguelenensis major, Amauropsis aureolutea, Sinuber sculpta scotiana, Prosipho astrolabiensis, Trophon minutus, T. shackletoni paucilamellatus* and *Neactaeonina cingulata*, although perhaps some of the above mentioned species could be more correctly regarded as a Weddell influence in the South Georgian fauna.

The Kerguelen influence is shown by *Pellilitorina setosa, Amauropsis (Kerguelenatica) grisea, Marseniopsis pacifica, Paradmete fragillima* and *Neactaeonina edentula*.

Finally, the eastern Antarctic (Enderby and Victoria Quadrants) influence is provided by *Falsimargarita gemma, Antimargarita dulcis, Subonoba fraudulenta, Balcis solitaria, Neobuccinum eatoni, Prosipho madigani, Acteon antarcticus, Philine alata* and *Toledonia major*.

Many of these correlatives are of deep water occurrence and in consequence are not restricted by the deep-water breaks in the Scotia Arc. The occurrence of the Magellanic genus *Tromina* may be accounted for in this manner.

Localization in *Chlanidota*, at least, is shown by the occurrence of three species not recorded from elsewhere.

Antarctic continent

The continental Antarctic molluscan fauna is incompletely known as instanced by the fact that many species are still on record only from the type locality. The writer's recently published results of the B.A.N.Z.A.R.E. mollusca (POWELL, 1958), largely from Enderby Land to Adelie Land, has greatly extended the range both geographic and bathymetric of some of the species previously known only from either the German Gauss Station, Davis Sea or the Ross Sea.

Large sections of the Antarctic Continent are still unknown from a faunistic standpoint, especially the coast between the Bellingshausen and Ross Seas.

Indications are that when this continental fauna is more completely known, a much larger percentage of the species will be found to be of circum-Antarctic range.

Worthy of note is the fact that the whole of the shoreline of Antarctica is ice-bound for most of the year and that only in a few places is a littoral fauna present.

Regarding the Ross Sea area considerable collections made by New Zealand Expeditions during recent years have not as yet been reported upon, but the indications are that molluscs rich in numbers if not particularly rich in species characterize the results. The published records for the area amount to 108 species exclusive of pelagic forms and of this number 20 are Nudibranchiata.

Even among species of known circum-Antarctic range certain ones achieve an optimum development at one or another of the established Antarctic stations. The Ross Sea area for instance seems best suited to the requirements of the truly Antarctic Pectinid, *Adamussium colbecki* and the chiton *Notochiton mirandus*.

Although the large *Limopsis jousseaumi grandis* ranges in deep water off the coast from Enderby Land to the Ross Sea and is doubtfully distinct from the typical species from the Western Antarctic, it reaches its optimum development and its greatest abundance apparently off Enderby Land. The largest example taken measured 80 mm × 65 mm.

Some spectacular new additions to the Antarctic fauna from the vicinity of Enderby Land are a large Fissurellid, *Tugali mawsoni*, a second species of *Venustatrochus*, *Fusitriton antarcticus*, *Chlanidota smithi*, *Chlanificula thielei*, the Turrids *Leucosyrinx mawsoni* and *L. macrobertsoni* and *Limopsis enderbyensis* (all POWELL, 1958).

Previously known ranges are extended considerably by the finding of a *Fusitriton* in a truly Antarctic locality and also in the recording of *Aforia magnifica* from Princess Elizabeth Land and MacRobertson Land.

Kerguelen Island

This island, situated in the Enderby Quadrant, lies on the extensive Kerguelen-Gaussberg Ridge which connects radially with Antarctica, separating the Atlantic-Antarctic Basin from the Australian-Antarctic Basin. The Antarctic Convergence is just south of the island but Heard Island to the south east is within the Antarctic Zone. To the west, on the Atlantic-Indian Ocean Cross Ridge, the Crozets, Marion and Prince Edward Islands, which have some faunal characteristics in common with Kerguelen, are situated within the Subantarctic.

The Kerguelen molluscan fauna is a rich one of approximately 145 species, which results from several factors, the sheltered waters of Royal Sound with its extensive islets and channels, the situation between two zones of surface waters, the underwater continuity with the Antarctic Continent and the lateral proximity of cross ridges extending westward to the vicinity of the Scotia Arc and eastwards towards Macquarie Island.

The shallow water molluscs are mostly of Subantarctic character, but the deeper water ones are mainly of Antarctic affinity which is to be expected with a location almost at the Antarctic Convergence.

The soft muds of Kerguelen, particularly in the inland waterways provide ecological conditions of the sea floor somewhat similar to those at South Georgia and this accounts for some similarities between the respective faunas.

The mud dwelling Struthiolarid, *Perissodonta mirabilis* and the deep burrowing bivalve, *Laternula elliptica* are common to both areas along with a number of other genera and species.

Among the bivalves mussels are represented by *Mytilus edulis desolationis* and *Aulacomya ater regia*, a giant subspecies allied to others that range from Patagonia to South Africa, southern New Zealand and the Antipodean.

Other very interesting inclusions in the Kerguelen fauna are two species of *Yoldia (Aequiyoldia)*, *Malletia gigantea*, the largest known of its genus, suggesting South American origin and *Astarte longirostris* otherwise known only from the Magellan Region. Another bivalve, *Cyclocardia astartoides*, has also a circum-Antarctic range.

Gastropods, too, are well represented, particularly in the families *Trochidae, Rissoidae, Naticidae, Cominellidae, Buccinulidae, Muricidae* and *Turridae*.

Macquarie Island

Macquarie Island is situated just within the Subantarctic, to the south of the southern islands of New Zealand (Auckland, Campbell, Antipodes and Bounty Islands), for which the writer designated the faunal province Antipodean (POWELL, 1955).

The molluscan fauna is a sparse one of only 54 known species, including one land snail. The shores of the island are rugged without inlets or shelter.

An evaluation of the Macquarie molluscan fauna does not substantiate TOMLIN's (1948) claim that there is a closer faunal relationship with the Antarctic Continent than with any other Southern area, for it has been found necessary (POWELL, 1955) to reject the following of TOMLIN's claimed Antarctic records — *Harpovoluta vanhoffeni, Toledonia globosa, Prosipho madigani, Probuccinum tenerum* and *Limatula hodgsoni*, and to replace these erroneous records with species that tend to swing the relationship of the fauna towards Kerguelen rather than to the Antarctic Continent.

Endemism is high and the Subantarctic element shows reasonably close relationship with the Kerguelen fauna, to which province the writer provisionally assigned it (POWELL, 1955).

There is, however, a well marked relict New Zealand mainland influence shown by the presence of the gastropod genera *Notoacmea, Actinoleuca, Plumbelenchus* and *Maurea*, the pelecypods *Pronucula*,

Tawera, and *Chlamys (Zygochlamys)* and the land snail *Phrix-gnathus.*

The most spectacular inclusion in the Macquarie fauna is the large abundantly represented Pectinid, *Chlamys (Zygopecten) delicatula* HUTTON (= *subantarctica* HEDLEY = *campbellicus* ODHNER).

This Pectinid belongs to a Southern Ocean group which ranges from Patagonia, with Miocene, Pleistocene and Recent species there, Graham Land, with a Pliocene or Pleistocene species, Kerguelen and Heard Island, upper Tertiary species, to the New Zealand area where the genus is known from the Miocene of Chatham Islands as well as the mainland of New Zealand, where it reached the North Island during the lower Pleistocene and remains a living constituent of the mainland New Zealand fauna along the eastern Otago shelf (Forsterian).

A surprise inclusion in the Macquarie fauna is *Cymatona tomlini* POWELL, 1955, representative of a deep water genus now known from the Chatham Rise as well as from the edge of the East Australian Continental Shelf.

Superimposed upon the impoverished basic Macquarie fauna is a Subantarctic drift element represented by *Gaimardia, Kidderia, Margarella, Laevilitorina* and *Kerguelenella.*

Links with the Kerguelen fauna other than the drift element just mentioned are shown by the presence of two species of the limpet genus *Patinigera, Puncturella,* two species of true *Trophon,* the Naticoid genus *Falsilunatia,* and the genera *Pareuthria, Prosipho* and *Admete.*

A genus of small Littorinids, *Macquariella,* a seaweed dweller, has one species, *hamiltoni* restricted to Macquarie Island but with relatives at Auckland and Campbell Islands, the Antipodes, Chathams and Stewart Island. The chitons are represented by an endemic species of *Hemiarthrum* which is closely allied to the wide ranging Subantarctic *setosum,* common at the Strait of Magellan, South Georgia and Kerguelen.

The large sized *Plaxiphora aurata campbelli* from Macquarie, Auckland and Campbell Islands is very closely allied to the typical species from Falkland Islands.

Subantarctic Islands of New Zealand

The molluscan faunas of the southern islands of New Zealand are obviously relict. The apparently high percentage of extinctions is presumed to have occurred following segregation of the islands. Dwindling of areas to a mere fraction of their former size, and increased exposure in a generally unsatisfactory environment of prevailing high winds and low temperatures would in time result in a reduction of species.

It is worthy of note that Auckland and Campbell Islands, the only two of these southern islands with sheltered waterways, have more species of littoral molluscs than have the exposed Bounty and Antipodes groups.

In 1951 the writer proposed the Antipodean, an adaptation of WAITE's Antipodes District, equalling in part the Rossian of FINLAY (1925) and in (1955) the writer restricted the use of the Antipodean for the New Zealand southern islands of Auckland, Campbell, Antipodes and Bounty, but excluded the Snares, which are considered Forsterian and Macquarie, which was referred to the Kerguelenian.

The Antipodes Province has a recorded total of 225 species distributed as follows; Auckland Islands, 169(13), Campbell Island, 70(6), Antipodes Islands, 52(4) and Bounty Islands, 75(0). The figures in brackets refer to terrestrial species.

The basic Antipodean fauna is predominantly relict New Zealand with both a superimposed Subantarctic drift element and at least three Subantarctic genera, *Plaxiphora*, *Patinigera* and *Pareuthria* that are unlikely to have reached this area by drift under present conditions.

The characteristic Subantarctic genera are *Margarella*, *Laevilitorina*, *Pareuthria*, *Kerguelenella*, *Gaimardia*, *Kidderia* and *Plaxiphora*. Endemic species belonging to typical mainland genera are representative of *Cellana*, *Maurea*, *Trichosirius*, *Tawera* and *Austrovenus*.

New Zealand polytypic species represented by Antipodean sub-species are *Notoacmea pileopsis subantarctica*, *Haliotis virginea huttoni*, *Thoristella chathamensis aucklandica*, *Microlenchus caelatus mortenseni* and *Eucominia nassoides nodicincta*.

Some wide ranging common New Zealand mainland species also occur over most of the Antipodean; these are *Zediloma digna*, *Lepsithais lacunosus*, *Mytilus aoteanus* and *Paphirus largillierti*.

A few forms, notably *Macquariella aucklandica*, *Maurea spectabilis* and *Chlamys (Zygochlamys) delicatulus* are either restricted to the Antipodean or extend only to the southern part of New Zealand in the Forsterian. Another characteristic Antipodean form is *Plumbelenchus*, a Subantarctic derivative of the northern Neozelanic and southern Australian genus *Cantharidus*.

Land and freshwater molluscs of the Subantarctic

Land and Freshwater molluscs are represented by a few species in the Subantarctic only, at the Falklands, Kerguelen Island and Macquarie Island.

The Falklands have 6 species of freshwater gastropods, one freshwater pelecypod and one small land snail, Kerguelen Island and Macquarie Island each has one indigenous land snail.

The Falkland terrestrial fauna is closely akin to that of Patagonia, which is what one would expect since these islands are contained within the Patagonian Shelf. The Kerguelen Island solitary land snail, *Notodiscus hookeri*, is probably a remnant from a former widespread Subantarctic snail distribution with which the Patagonian *Amphidoxa* is probably allied. The Macquarie Island *Phrixgnathus hamiltoni*, a very small snail, was probably derived by chance means from the New Zealand mainland at some time in the not very remote past.

The Antipodean Auckland and Campbell Islands have a more extensive terrestrial molluscan fauna, 11 land snails and 2 freshwater species in the case of the former and 6 land species for the latter.

All the species, however, are either identical with widespread New Zealand mainland species or are endemic derivations from mainland species. Both of these Antipodean faunas could have been derived during the past by chance means rather than by direct land connection.

Biogeographical provinces for the Antarctic and Subantarctic

It is not desirable at this stage of our knowledge of southern high-latitude molluscs to formulate a comprehensive scheme of biogeographical provinces.

A set of quadrant names proposed by MARKHAM (1912) for the Antarctic and by WAITE (1916) for the Subantarctic extensions of these quadrants may be usefully employed for indicating positions and in recording distribution but they have no biogeographical significance.

The Antarctic quadrants named by MARKHAM were:

Victoria Quadrant:	90° to 180° E.
Ross Quadrant:	180° to 90° W.
Weddell Quadrant:	90° to 0° W.
Enderby Quadrant:	0° to 90° E.

The Subantarctic extended quadrants of WAITE were:

Australian Zonal Quadrant.	An extension of Victoria Quadrant.
Pacific Zonal Quadrant.	An extension of Ross Quadrant.
American Zonal Quadrant.	An extension of Weddell Quadrant.
African Zonal Quadrant.	An extension of Enderby Quadrant.

To these arbitrarily defined areas WAITE proposed the following faunal districts:

A. Antipodes District. Subantarctic Islands of New Zealand, including Macquarie Island, Stewart Island and southern Otago of the South Island of New Zealand.

M. Magellan District.

K. Kerguelen District. Including Marion Island, Crozets and Heard Island as well as Kerguelen.

G. Glacial District. The whole of the Antarctic within the Antarctic Convergence.

FINLAY (1926) proposed five New Zealand faunal provinces, one of which, the Rossian, was intended to cover the Subantarctic Islands of New Zealand including both the Snares and Macquarie Islands.

The name Rossian was an unfortunate choice since it implied the Ross Sea area but was not intended to do so. For this reason the writer has reverted to an adaptation of WAITE's earlier designation Antipodes District to cover most of the area intended by FINLAY's Rossian.

WAITE's Glacial District is pointless, since it merely substitutes a term for Antarctic and doubtless, the restricted Antarctic circumpolar area will later require subdivision on a faunal basis.

In 1951, the writer proposed a tentative subdivision of the Antarctic and Subantarctic faunas, later modified (1955 and 1960) into the following biogeographical provinces.

Subantarctic

Magellan: Patagonia from below Chiloe Island (west coast) and Cape Blanco (east coast), Tierra del Fuego and the East Patagonian Continental Shelf, including the Falkland Islands as well as the Burdwood Bank, to the south of the East Patagonian Shelf.

Kerguelenian: Kerguelen Island, Crozets, Marion and Prince Edward Islands, Macquarie Island and possibly Heard and Bouvet Islands.

Antipodean: The southern islands of New Zealand, Auckland, Campbell, Antipodes and Bounty Islands but excluding Macquarie Island and also the Snares Islands which belong to the New Zealand mainland province, the Forsterian.

Antarctic

Georgian: South Georgia and Shag Rocks.

Indications are that the continental fauna of Antarctica is remarkably uniform, for a number of species have proved to be of circumpolar range. However, since large sections of Antarctica are still unexplored biologically, it is inadvisable at the present stage of our knowledge to suggest subdivisions other than to note the special character of the South Georgian fauna which is predominantly Antarctic although situated far to the north of the Antarctic Continent, almost as far north as the Subantarctic Falklands Group.

The molluscan families represented in Antarctic and Subantarctic seas

Gastropoda

Scissurellidae: This family is represented by 7 species, two of which are of Antarctic occurrence, *timora* MELVILL & STANDEN, from the South Orkneys and *petermannensis* LAMY from Petermann Island; 2 species of *Schizotrochus, amoenus* (THIELE) from Gauss Station, Davis Sea and *euglyptus* (PELSENEER), which is widespread in both Antarctic and Subantarctic areas. Finally a New Zealand genus, *Sinezona* is represented by one species, *subantarctica* (HEDLEY), at Macquarie Island.

Fissurellidae: Four species of typical key-hole limpets, *Fissurella (Balboaina)* and 2 not definitely attributable to the subgenus as well as a *Megatebennus* have extended their range from the Magellan area to the Falklands but are not found elsewhere in either the Subantarctic or the Antarctic. The genus *Puncturella* is represented by 4 species, one or another of which is found in most Antarctic and Subantarctic localities. The one most widely distributed, *conica* (ORBIGNY), is very closely similar to the Boreal type of the genus, *noachina* (LINNAEUS). The *"Scutus"*-like genus *Parmaphorella* has an Antarctic and a Subantarctic member, the former, *mawsoni* POWELL, from off Enderby Land, the latter, *melvilli* (THIELE) from the Falklands area. Two other species are known, one from the Strait of Magellan and the other from South Africa. The origin of this genus is uncertain, that is, whether it originated in the Magellanic area or whether it is an outlier from the Indo-Pacific-Neozelanic ranging *Tugali*.

Acmaeidae: Limpets of this family occur, with the genera *Patelloida* and *Scurria* both in the Magellanic Area, including the Falklands, but are not found in the Antarctic or elsewhere in the Subantarctic. At Macquarie Island, the Neozelanic genera *Actinoleuca* and *Notoacmea* reach their extreme southern range.

Patellidae: The southern limpets of the genera *Nacella* and *Patinigera* differ from the temperate and tropical *Patella, Cellana* and *Scutellastra* in having the gill cordon complete, not interrupted by the head.

The genus *Nacella*, which is a seaweed dweller with a thin shell and the apex at the extreme anterior end, ranges from the Magellan Province to Kerguelen. The species of *Patinigera* are normally Patellid-shaped but are at once recognized by the bronzy lustre of the interior of the shell. Some of them are rock dwellers, but *fuegiensis* (REEVE) and its subspecies *edgari* POWELL have thin shells of very low profile adapted for life on seaweeds. One species, *polaris* (HOMBRON & JACQUINOT) is wide ranging in the Western Antarctic but also extends to South Georgia and Bouvet Island.

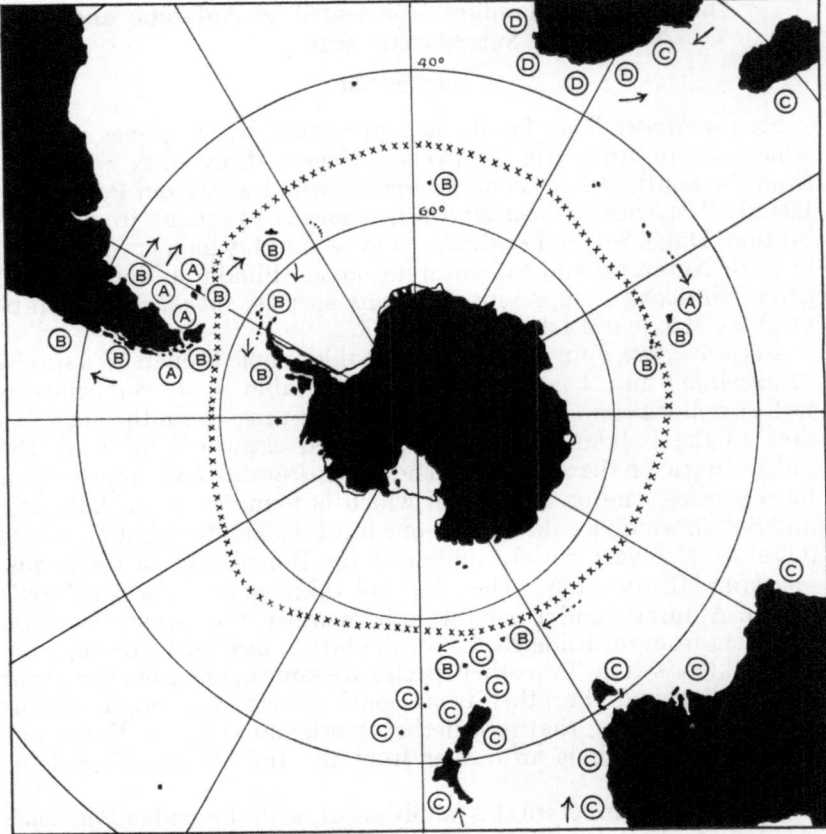

Fig. 3. Distribution of Southern Ocean Patellidae (Limpets).
A. *Nacella* — Lives on giant kelp — distributed by the West-Wind Drift from the Magellan Region to as far east as Kerguelen Island. B. *Patinigera* — Mostly associated with kelp, also — distributed eastward from the Magellan Region over all Subantarctic groups — reaches Antarctica at Graham Land and extends to Macquarie Island and Campbell Island in the New Zealand area. C. *Cellana* — A warm-water Indo-Pacific genus which extends southward down the east coast of Africa to Natal and to Auckland and Campbell Islands in the New Zealand area. D. *Patella* — An Atlantic genus which extends down to and around the Cape of Good Hope.

Two species of *Patinigera* occur at Macquarie Island, the Kerguelen species, *kerguelenensis* (SMITH) and an endemic species *macquariensis* FINLAY. The New Zealand Antipodean has one species of *Patinigera*, *veneris* (FILHOL) at Campbell Island but otherwise the New Zealand Subantarctic limpets are derivations of *Cellana* which is the common limpet of New Zealand, Australia and throughout much of the Indo-Pacific.

Lepetidae: The small white limpet, *Lepeta coppingeri* SMITH is probably circum-Antarctic and ranges from the Magellanic area to Kerguelen Island in the Subantarctic. Another Antarctic member is *Propilidium pelseneeri* THIELE, from 3397 metres north-west of Gauss Station.

Trochidae: The most characteristic southern Trochoid is *Margarella*, which may be considered the southern equivalent of the Boreal

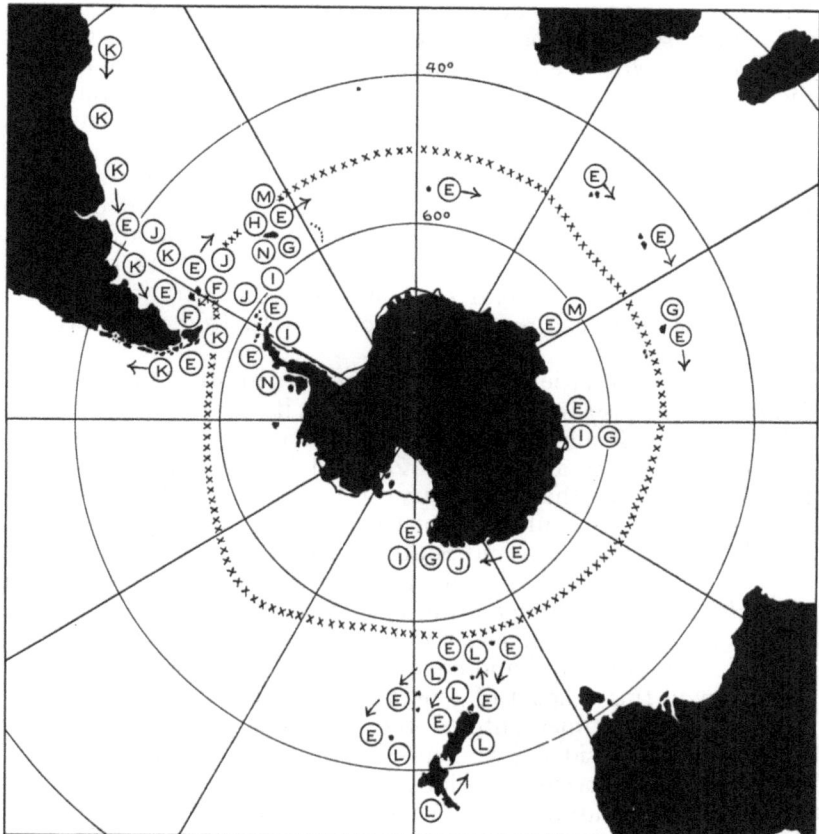

Fig. 4. Distribution of Southern Ocean Trochidae (Top-shells).
E. *Margarella* — Lives mostly associated with kelp, and widely distributed by agency of the West-Wind Drift — extends eastward to all the Subantarctic islands and in the New Zealand area to about the middle of the South Island — also at South Georgia and around much of the Antarctic continent. F. *Photinula & Photinastoma* Magellanic Region only. G. *Submargarita* — South Georgia, Antarctica and Kerguelen. H. *Promargarita* — South Georgia only. I. *Antimargarita* — Antarctic. J. *Falsimargarita* — Magellan Region, South Shetlands and Oates Land. K. *Calliostoma*, Magellanic Region. L. *Maurea* — a calliostonid genus restricted to the New Zealand area, southward to Macquarie Island. M. *Venustatrochus* — South Georgia and Mackenzie Sea. N. *Tropidomarga* — South Georgia and South Shetlands.

Margarita. Most species of *Margarella* live on algae and some, the New Zealand species in particular, inhabit cavities in the holdfast of *d'Urvillea.* The genus is probably circum-Antarctic and is well represented also in all the Subantarctic Islands, including the Antipodean. In New Zealand a species, *rosea,* occurs as far north as Banks Peninsula, inside the southern boundary of the Cookian Province.

South Georgia has in addition to *Margarella* a subgenus *Promargarita* which seems to be restricted to that province. Two other Trochoids, *Photinula* and *Photinastoma,* which are similar in appearance to *Margarella* but are larger, more solid and attractively spirally colour lined, are confined to the Magellan Province, including the Falklands.

True *Calliostoma* is well represented in the Magellan-Falklands area but has not spread to the eastward or southward either in Subantarctic or Antarctic waters. However a moderately large-sized Calliostomid genus, *Venustatrochus,* which is characterized by the great number of lateral teeth in the radula, occurs at South Georgia and there is a related Antarctic species from 540 metres in the Mackenzie Sea.

From off Macquarie Island in 120 metres another Calliostomid, *Maurea megaloprepes* (TOMLIN), clearly has its relationship with this otherwise New Zealand Antipodean and mainland genus.

Macquarie Island and the New Zealand Antipodean Subantarctic Islands constitute the entire range of another Trochoid, *Plumbelenchus,* which is a subgenus of *Cantharidus,* well represented in New Zealand, Tasmania and Southern Australia.

Other Trochoids of both Subantarctic and Antarctic distribution are *Submargarita, Antimargarita, Tropidomarga* and 4 species plus a subspecies of *Solariella.* Deep waters off Marion Island, Bellingshausen Sea and Kemp Land have produced 2 species of *Calliotropis.* Probably all this bracket of genera originated from ancestral stock that at various times entered the American Sector. *Calliotropis* could well be considered a member of the earlier Antarctic fauna.

Liotiidae: 13 species of Liotids, members of 6 genera inhabit Antarctic-Subantarctic Seas. They are *Cirsonella,* represented by one Antarctic species, *extrema* THIELE and one Subantarctic species *kerguelensis* THIELE; *Brookula,* with 4 species ranging from the Falklands to South Georgia and the western Antarctic, and a *Crosseola, pseudocollonia* POWELL, from 120 metres off Kerguelen. The remaining species have been somewhat doubtfully ascribed to the genera *Tharsiella, Circulus* and *Cyclostrema.*

Turbinidae: This family is represented only by 2 small species of an exclusively Antarctic genus *Leptocollonia, innocens* (THIELE) from Gauss Station and Enderby Land and *thielei* POWELL from South Georgia and Palmer Archipelago.

Skeneopsidae: To this family, THIELE (1912) proposed the genus *Microdiscula*, with 2 species of small size, *subcanaliculata* (SMITH) from Kerguelen and *vanhoffeni* THIELE, from Gauss Station, Davis Sea.

Omalogyridae: The north European genus *Omalogyra* is represented by the wide ranging minute *atomus*, with a southern range covering Marion Island, Prince Edward Island and Gauss Station. A subspecies *atomus burdwoodianus* STREBEL comes from the Burdwood Bank in 137—150 metres.

Trachysmatidae: THIELE ascribed 2 Gauss Station species, *ignobile* and *tenue* THIELE to *Trachysma*, this otherwise Boreal and North Atlantic genus.

Rissoellidae: This is represented by *Jeffreysiella* THIELE, with 2 species, *notabilis* THIELE, from Kerguelen and *edwardiensis* WATSON, from off Prince Edward Island in 310 fathoms.

Littorinidae: The Littorinids are represented by 10 species of *Laevilitorina*, the most widespread being *caliginosa* (SMITH), which ranges from Tierra del Fuego to the Falklands, South Georgia, South Shetlands, South Orkneys, Kerguelen and Macquarie Island. Other allied genera and subgenera are *Corneolitorina*, *Laevilacunaria* and *Pellilacunella*, South Orkneys, South Shetlands and Palmer Archipelago. Two larger, thin-shelled globular periwinkles, with an epidermis studded with short hair-like processes, *Pellilitorina*, occur at South Georgia, *pellita* (MARTENS) and *setosa* (SMITH), the latter ranging also to South Orkneys, Cape Adare and Kerguelen Island. At Macquarie Island, a further genus, *Macquariella*, reaches its southern limit, the other members of the genus being from the Antipodean, Moriorian (Chatham Islands) and Stewart Island (Forsterian). Not one of these southern species shows much external resemblance to the Northern Hemisphere Littorinids.

Lacunidae: Two species of *Lacuna* were recorded for southern seas by MELVILL & STANDEN, one, *abyssicola* MELVILL & STANDEN is doubtfully that genus and the other record, that of the North Atlantic — Boreal *divaricata* (FABRICIUS) from Falkland Islands probably represents an accidental introduction.

Hydrobiidae: Represented by 2 species referred to *Potamopyrgus* with some doubt. They are *georgiana* (PFEFFER), from South Georgia and *melvilli* (HEDLEY) from Macquarie Island. It is worthy of note that in New Zealand, *Potamopyrgus antipodum* (GRAY) is often found on marine mud flats where there is a brackish influence, although the genus is otherwise freshwater.

Rissoidae: The most characteristic Rissoids belong to the genera *Ovirissoa*, *Eatoniella* and *Eatoniopsis*. *Ovirissoa* is a restricted Antarctic genus with 4 species occurring at South Georgia and at a number of localities from Western Antarctica to the Ross Sea. *Eatoniella*, with 5 species and 2 subspecies has a wide Antarctic and Subantarctic range and the minute seaweed dwelling *Skenella georgiana* PFEFFER has a New Zealand allied species in *pfefferi* SUTER. Nineteen other species of featureless Rissoids have been described from southern seas but since most are unknown anatomically, no inferences can be drawn except to note that *Subonoba* is otherwise a New Zealand genus, strongly represented in the Antipodean.

Litiopidae: This family has 2 representatives, *Alaba incolorata* THIELE from Gauss Station, Davis Sea and *Diala limnaeiformis* (WATSON) from off Prince Edward Island in 50—150 fathoms.

Tornidae: A doubtful member of the type genus is *Tornus antarcticus* THIELE from the Gauss Station.

Trochaclidae: The monotypic *Trochaclis* is a small featureless turbinate shell of uncertain systematic position and the species *antarctica* also comes from the Gauss Station.

Cerithiidae — Cerithiopsidae: A species of *Ataxocerithium, pullum* (PHILIPPI) is common in the Magellanic region including the Falklands. Eight species ascribed to the Boreal genus *Cerithiella* are widely distributed at Subantarctic and Antarctic deep water stations. Two other Cerithiopsid genera apparently restricted to the Subantarctic, including Macquarie Island, and Antarctica are *Cerithiopsilla*, 9 species and *Eumetula*, 6 species.

Triphoridae: A single species, *Triphora delicatula* THIELE, comes from the Gauss Station, Davis Sea.

Turritellidae: Two Turritellid genera, *Banzarecolpus* and *Colpospirella* are of Subantarctic range, neither being very closely allied to the warmer water members of the family. The former, represented by 2 species, *austrina* (WATSON) and *frigida* (THIELE) is from Kerguelen Island and the latter, *algida* (MELVILL & STANDEN) is from the Falklands area.

Three Antarctic species are ascribed to the North Atlantic genus *Turritellopsis* and 3 Magellanic species to the widely distributed genus *Mathilda*.

Vermetidae: The only recorded Vermetid is *Serpulorbis murrayi* (HEDLEY) from Cape Royds, Antarctica in 60—80 fathoms.

Epitoniidae: This family comprises 3 species of *Coroniscala*, two of which are Magellanic, including the Falklands and the third is from South Georgia. Two species, one from the Falklands and the other from the Ross Sea area and Kerguelen are ascribed to the northern genus *Acirsa*.

Eulimidae: Represented by 6 species of *Balcis*, 5 of them Antarctic, 2 *Melanella*, also Antarctic, and *Diacolax cucumariae* MANDAHL-BARTH, a monotypic genus from between the Falklands and Patagonia.

Stiliferidae: A single species, *Stilifer polaris* HEDLEY, comes from off the Shackleton Ice-Shelf, Antarctica.

Trichotropidae: This family is represented by 4 very distinctive

Antarctic genera, *Antitrichotropis*, *Trichoconcha*, *Neoconcha* and *Discotrichoconcha*. All are almost entirely chitinous and *Trichoconcha mirabilis* SMITH, in particular will shrink proportionately to about one third of its original size, when dried out. In *Discotrichoconcha* the shell is an almost flat compressed helicoid spiral.

All the above genera are very different from the genus *Trichotropis* of Boreal seas.

The related family *Lippistidae* is represented by an Antarctic species, *exilis* POWELL, from off Enderby Land.

Calyptraeidae: This includes 3 species of *Trochita*, a typically South American genus but with 2 from the Magellanic area, including the Falklands and one from South Georgia. Another South American genus, *Crepipatella* has one species extending from the Magellan area to the Falklands.

Capulidae: The sole Antarctic species, *Capulus subcompressus* PELSENEER, has a conical shell, greatly compressed laterally. It is found in deep water from the Bellingshausen Sea to McMurdo Sound.

Struthiolariidae: A species, *Perissodonta mirabilis* (SMITH) occurs both at South Georgia and at Kerguelen Island. It is a living link between the Patagonian and New Zealand Cretaceo-Tertiary Struthiolarids. Other Recent Struthiolarids are *Struthiolaria* and *Pelicaria* from New Zealand and *Tylospira* from New South Wales and Southern Australia. The genus *Perissodonta* is doubtless a surviving member of the earlier Antarctic fauna.

Naticidae: This family is represented by 8 genera and 29 species. Ten of them have been ascribed to the Boreal genus *Amauropsis* and are mostly of Antarctic range. *Falsilunatia*, 6 species, is mostly from the Falklands area but one is found in the vicinity of Kerguelen and another from off Macquarie Island. *Prolacuna*, with 4 species, seems to be exclusively Antarctic and *Sinuber* has 2 species and a subspecies ranging from the Falklands to South Georgia, South Orkneys, Kerguelen and the Ross Sea area. The widespread genus *Polinices* has 2 Magellanic species, *magellanica* and *patagonicus* (PHILIPPI) which reach the Falkland and South Georgia areas. Another widespread Naticoid genus, *Tectonatica* is represented by *impervia* (PHILIPPI) from the Magellan area, Falklands and South Georgia and a subspecies *major* (STREBEL) from Paulet Island.

Lamellariidae: Three genera, *Lamellaria*, *Marseniopsis* and *Lamellariopsis* have 10 species in all. The 2 latter genera appear to be restricted to Antarctic and Subantarctic Seas.

Cymatiidae: The genus *Fusitriton* is represented by 2 species, a Magellanic one, *cancellatum* (LAMARCK) and *antarcticus* POWELL from 603 metres off Kemp Land, Antarctica.

This genus, already referred to under the heading "Bipolarity" is mostly archibenthal, distributed from Japan across the Aleutian Chain, the west coast of both Americas, Patagonia, Argentina,

360

Uruguay, Falkland Islands, South Africa, Marion Island, Kemp Land, Antarctica, New Zealand and its Southern Islands, off Southern Australia and New South Wales.

Another genus, *Argobuccinum*, is of circum-Southern Ocean distribution but except for one species in the Strait of Magellan it otherwise ranges in the zone of mixed waters just north of the Subtropical Convergence.

A third genus, *Cymatona*, has one species, *tomlini* POWELL, from 69 metres off Macquarie Island. Related species indicate its range to be associated with the East Australian Continental Shelf.

Cominellidae: The most characteristic southern member of this family is *Pareuthria* with 15 species, mainly Subantarctic, extending from the Strait of Magellan to Kerguelen and with one species at Campbell Island, New Zealand (Antipodean). Also, there are 2 Antarctic members, *innocens* (SMITH) and *plicatula* THIELE, Gauss Station to Enderby Land and Ross Sea. *Pareuthria* resembles the Mediterranean *Euthria* in shell form but the dentition shows the latter to belong to the *Buccinulidae*.

Nineteen other species, more or less related to *Pareuthria* in a broad sense, and mostly from the Magellanic Region belong to the following genera, endemic to that area: *Glypteuthria*, *Meteuthria*, *Tromina*, *Parficulina*, *Antistreptus*, *Anomacme* and *Savatieria*. Another related genus, *Falsimohnia*, ranges from South Georgia to Kerguelen Island.

Buccinulidae: This family is abundantly represented in the Subantarctic and Antarctic by shells resembling the northern whelks *(Buccinidae)* but with different radular and opercular characters. The most conspicuous of these is *Neobuccinum eatoni* (SMITH) from the Subantarctic of Kerguelen and Heard Islands but with a circum-Antarctic distribution as well, from shallow water down to at least 600 metres.

Another group of whelk-like shells, *Chlanidota*, is exclusively of Subantarctic and Antarctic occurrence and is represented by 8 species. They are thin shelled, mostly ovate to globular and have a pile-like epidermis. A third genus, *Pfefferia*, has a heavier shell, and distinctive opercular and radular characters. The 4 known species are restricted to deep water off South Georgia.

The genus *Probuccinum* comprises thin, more or less hyaline, fusiform shells with a distinctive D — shaped aperture. There are 8 species, all deep water and of Antarctic range except for 2 from Kerguelen Island and vicinity.

Thirty two species of another genus, *Prosipho*, are restricted to Subantarctic and Antarctic waters. These shells are of small size, mostly of fusiform shape and several are sinistral. The most distinctive feature is in the form of the lateral teeth of the radula which have the base produced downwards into handle-like appendages.

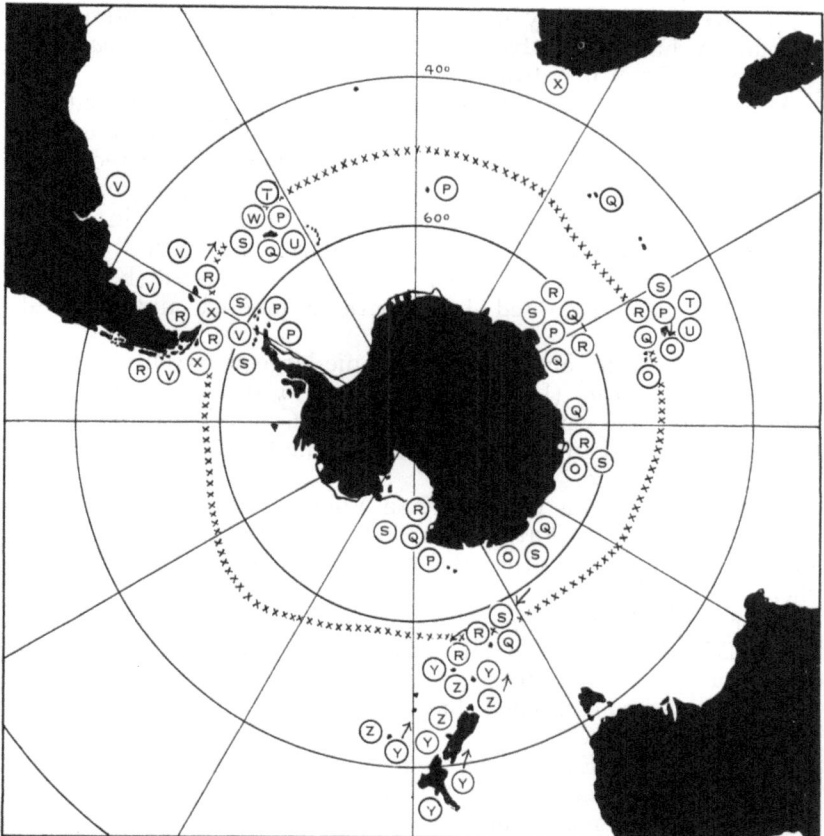

Fig. 5. Distribution of Southern Ocean Cominellidae-Buccinulidae (Whelks).
O. *Neobuccinum* — Circum-Antarctic, Kerguelen and Heard Islands. P. *Chlanidota* —
Circum-Antarctic, South Georgia, Bouvet Island and Kerguelen. Q. *Probuccinum* —
South Georgia, Kerguelen and Antarctica. R. *Pareuthria* — Magellan Region,
Kerguelen, Antarctica, Macquarie Island and Campbell Island in New Zealand
Region. S. *Prosipho* — Circum-Antarctic, Kerguelen and Macquarie Islands.
T. *Proneptunea* and U. *Falsimohnia,* — South Georgia and Kerguelen. V.
Tromina — Magellan Region and South Shetlands. W. *Pfefferia* — South Georgia
only. X. *Glypteuthria* — Magellan Region and South Africa. Y. *Buccinulum* and
Z. *Eucominia* — New Zealand to Auckland and Campbell Islands.

The remaining genera making up this family, all of which are exclusively Subantarctic and Antarctic ranging, are *Notoficula, Chlanificula, Chlanidotella, Bathydomus, Fusinella, Cavineptunea* and *Proneptunea*.

The combined whelk families, *Cominellidae* and *Buccinulidae* exhibit vigorous radiation and have representatives throughout the Antarctic and Subantarctic from shallow water to a depth of at least 1600 fathoms.

Nassariidae: A solitary member of this prolific tropical and temperate family has one Subantarctic member in *Nassarius vallentini* (MELVILL & STANDEN) from the Falkland Islands.

Muricidae: This equally prolific family is strongly represented in Antarctic and Subantarctic seas by 28 species of *Trophon* and 11 of the closely allied *Xymenopsis*. The latter genus is restricted to the Magellan Region, including the Falklands. Typical *Trophon* occurs strongly all around the Antarctic Continent and in all the Subantarctic Islands including Macquarie. A species reminiscent of warmer seas is *Typhis (Typhina) belcheri* (BRODERIP) which has been recorded from both Patagonia and the Falklands.

Thaisidae: A characteristic Magellanic large strong littoral shell, *Acanthina imbricata* (LAMARCK) occurs also at the Falklands but another well known Magellanic genus, *Concholepas concholepas* (BRUGUIERE) has not as yet extended to the Falklands or elsewhere in the Subantarctic.

Volutidae: A genus of large-sized Volutes, *Adelomelon*, which is restricted to the East Patagonian Shelf, extends with 3 species to the Falklands, which are situated within the eastern boundary of this shelf. Also, *Miomelon*, a related Volute with a Patagonian Tertiary ancestry has a living member in *scoresbyana* POWELL, from between the Falklands and the Strait of Magellan. Otherwise these genera have not spread in southern seas. Four other distinctive Volutid genera, however, combined, have a wide distribution in Antarctic and Subantarctic waters.

The Challenger Expedition dredged a large fragile shelled Volute, *Guivillea alabastrina* (WATSON) from 1600 fathoms, between Marion Island and the Crozets, and this species closely comparable with no other Volute has remained monotypic. It also could well be a member of the earlier Antarctic fauna which largely disappeared during the Pleistocene ice-ages. Another distinctive Volutid genus, *Provocator*, has 2 species ranging in moderately deep water from between Kerguelen Island and the vicinity of the Falklands. Still another Volutid genus restricted to the Antarctic is *Harpovoluta* with 2 species and a subspecies. In appearance and in the thinness of the shell this genus resembles the northern Buccinid genus *Volutharpa*. Finally a group of 6 Mitridlike Volutes, *Paradmete*, are strongly represented both in Antarctica and South Georgia to Kerguelen, while one species, *crymochara* ROCHEBRUNE & MABILLE, extends to Tierra del Fuego.

Cancellariidae: Eight species attributed to the Boreal genus *Admete* have a wide Antarctic and Subantarctic range from the Magellan area to Macquarie Island.

Olividae: This tropical to warm temperate family has one representative in *Baryspira longispira* (STREBEL), taken from between

South Georgia and the Falklands in 2675 metres. It probably represents a new genus.

Marginellidae: Again this family is mainly of warm water range yet there are 2 Antarctic and 3 Subantarctic species. Two species, *Marginella warrenii* MARRAT and *M. dozei* ROCHEBRUNE & MABILLE, are common on the East Patagonian Shelf, including the Falklands and they are remarkable in that their relatively large, polished and colourful shells are comparable with those of the tropical members of the family.

They probably extended their range down the eastern coastline of South America during the past at a time when Magellanic conditions were. much warmer than at present.

The two Antarctic Marginellids occur in deep water from Enderby Land to the Ross Sea.

Turridae: Forty one species of Turrids, belonging to 12 genera inhabit the Subantarctic-Antarctic Zones plus a further series of at least 16 species that are known only from East Patagonia and the Magellan Region.

The more significant genera are *Aforia, Leucosyrinx, Pleurotomella* and *Pontiothauma.* These have already been referred to under the heading of "Bipolarity", where it is pointed out that *Aforia* has an almost continuous distribution from the Arctic to the Antarctic, its continuity having been achieved by going deep over the tropical zone, assisted in some measure by the upwelling of cold water along its West American route. *Aforia* has one Antarctic member, *magnifica* (STREBEL), ranging from the Palmer Archipelago to Enderby Land and the Mackenzie Sea but the 4 Subantarctic species extend from the Patagonian Shelf to Kerguelen Island.

Two species ascribed to *Leucosyrinx* inhabit the Patagonian Shelf in the vicinity of the Falklands but 3 others range around the Antarctic Continent from Palmer Archipelago to Enderby Land and the Mackenzie Sea. It is suggested that this genus has gradually extended its range from the North Atlantic — Caribbean deeps along the floors of the ocean basins of the South Atlantic and thence eastward by the same means through the more or less interconnected deeps of the Southern Ocean.

Another characteristically North Atlantic genus, *Pleurotomella* has 7 members, 4 of them of deep water Antarctic occurrence, Davis Sea to Ross Sea, and 3 of Subantarctic range, Falklands to Kerguelen.

The occurrence of a *Pontiothauma* in Antarctica suggests another colonizing route, for that genus is otherwise known only from deepwater off India and Indonesia. It has probably gradually extended its range southward along the floor of the ocean deeps assisted by the northward bottom flowing cold water from the melting Antarctic ice.

A group of genera bearing some resemblance to the Boreal *Oeno-*

364

potinae ("Bela" — "Lora" complex), but probably not closely related are *Belaturricula, Conorbela, Lorabela* and *Belalora*, which bracket of genera the writer proposed in 1951 to emphasize the distinctiveness of these shells which are mostly from the American Quadrant of Antarctica.

A North Atlantic genus, *Spirotropis*, has an Antarctic member in *remota* POWELL from off Enderby Land and there is also a Kerguelen species, *studeriana* (MARTENS). To still another North Atlantic genus, *Typhomangelia*, a Kerguelen species, *fluctuosa* (WATSON) and a Davis Sea species, *principalis* THIELE, are tentatively referred.

A *Drillia*-like shell, *fuegiensis* (SMITH) from the East Patagonian Shelf, including the Falklands is considered to belong to *Eumetadrillia*, a Jamaican Miocene genus. Finally there is a distinctive Daphnellid genus, *Typhlodaphne*, with 4 southern members, occurring at Kerguelen Island, South Georgia and Enderby Land in moderately deep water.

Acteonidae: This family is represented by the type genus *Acteon* with one Magellanic species, *bullatus* GOULD, found also off the Falklands and one Antarctic species, *antarcticus* THIELE, ranging from the South Shetlands to the Davis Sea. A restricted Antarctic genus, *Neactaeonina*, has 3 species ranging from South Georgia and the South Shetlands to the Ross Sea and another strictly southern genus, *Toledonia*, has 10 species ranging from the Magellan Region to Kerguelen Island and around most of the Antarctic Continent. The genus also extends northward to the New Zealand Subantarctic Islands with one species from off the Auckland Islands in 95 fathoms.

Scaphandridae: Three species of the cosmopolitan genus *Cylichnina* range from South Georgia to the Ross Sea and another, which the writer named *Kaitoa scaphandroides* POWELL, 1951 seems to be related to this otherwise New Zealand Miocene genus.

Retusidae: There are 3 species of the cosmopolitan type genus, *Retusa*, ranging from South Georgia to the Ross Sea.

Diaphanidae: This is represented by 6 species of *Diaphana*, four of them of Antarctic occurrence, South Orkneys, South Georgia and Davis Sea, as well as 2 subantarctic species, one Magellanic and the other from Kerguelen.

To this family also belongs the restricted Antarctic genus *Newnesia* of 2 species, one from off Cape Adare and the other from near Paulet Island.

Aplustridae: This family is represented by the monotypic *Parvaplustrum tenerum* POWELL, 1951, so far known only from off the Falklands in 104—300 metres.

Philinidae: Seven species of the cosmopolitan genus *Philine* occur, four of them of Antarctic range and three from the Subantarctic, Falklands to Kerguelen.

Pyramidellidae: Ten species are recorded representing the cosmopolitan genera *Odostomia, Turbonilla* and *Chemnitzia,* and they range from the Magellan Region to South Georgia, McMurdo Sound and Macquarie Island. The apparently monotypic *Streptocionella singularis* MARTENS & PFEFFER is known only from South Georgia.

Pteropoda

Thirteen species of the order Pteropoda have been recorded from Antarctic and Subantarctic waters. Three characteristics of the Southern Ocean are *Thilea procera* STREBEL, *Clione antarctica* SMITH and *Spongiobranchaea australis* ORBIGNY. Most of the pelagic cosmopolitan species of *Cavolina, Embolus* and *Spiratella* occur in vast numbers in the surface waters.

Nudibranchiata

The nudibranchs are abundantly represented in both Antarctic and Subantarctic waters. Fifty eight species have been described or recorded from these seas and 12 families with 28 genera are represented.

Among the *Dorididae,* the genus *Cadlina* is represented by one Antarctic species, *affinis* OHDNER, from McMurdo Sound, a Falkland species, *falklandica* ODHNER and one from Kerguelen Island, *kerguelensis* THIELE. This genus has a remarkable "bipolar" distribution, for in the Northern Hemisphere it occurs widely but with its apparent southern limits in California, Gulf of Mexico and Cape Verde Islands.

Another Dorid genus, *Austrodoris,* seems to be confined to southern seas, with one species ranging from the Magellan Region to South Georgia and 5 in the Antarctic proper.

An exclusive Antarctic family, the *Charcotidae,* consists of 2 monotypic genera, *Charcotia* and *Telarma.*

In the *Eolidacea* an exclusively Antarctic genus, *Notaeolidia,* is represented by 5 species ranging from the South Orkneys to McMurdo Sound, and another endemic inclusion in this family is the monotypic *Pseudotritonia quadrangularis* THIELE from the Gauss Station, Davis Sea.

In the *Pleurobranchidae,* 2 species of *Bouvieria* occur in the Magellan Province, including the Falklands. In the *Duvauceliidae* there are 4 species of *Duvaucelia,* two of them extending to Antarctic waters, as well as two exclusively Antarctic members of the genus *Tritoniella.* The family *Bathydorididae* has 5 exclusively Antarctic members of the type genus *Bathydoris.*

Other families represented are the *Notodorididae,* with 2 Antarctic species of *Aegires (Anaegires), Onchidoridae* with a species of

Acanthodoris at the Falklands and one of *Prodoridunculus*, monotypic, from Gauss Station, Davis Sea, *Dorididae*, with, in addition to *Cadlina* and *Austrodoris*, mentioned above, the genera *Doridigitata*, one species from the Falklands, *Archidoris*, with an Antarctic species, *nivalis* THIELE, and two from Kerguelen Island. Also the genera *Geitodoris*, *Diaulula* and *Gargamella*, each with a single species from the Falklands area, *Iduliidae* with a species, *Idulia antarctica* ELIOT from Cape Adare to the Ross Sea, *Eubranchidae*, with 3 genera, *Eubranchus*, 2 species, Falklands and Cape Adare, *Galvinella*, 2 Antarctic and one Magellanic species and a *Coryphella* ranging from the Falklands to South Georgia, the *Tergipedidae* with one species of *Tergipes* from West Antarctica, the monotypic *Guyvalvoria francaisi* VAYSSIÈRE, from Wandel Island, 3 Ross Sea members of *Cuthonella*, 3 species of *Cuthona*, all from South Georgia and 2 Subantarctic members of *Cratena*, one from Kerguelen and the other from the Falklands.

To complete the list *Aeolidia serotina* BERGH was described from the Falklands and the following 3 Magellanic Nudibranchs are on record — *Ancula fuegiensis* ODHNER, *Holoplocamus papposus* ODHNER and *Trippa hispida* ORBIGNY.

Siphonariidae: The pulmonate limpets are represented by 2 genera, *Pachysiphonaria* and *Kerguelenella*. The former is of Magellanic occurrence, including the Falklands but the latter is circum-Subantarctic, from the Magellan Region to Macquarie Island, extending southward also to South Georgia and northward from Macquarie Island through the Antipodean to Stewart Island, southern New Zealand.

The genus *Kerguelenella* closely resembles the Aleutian genus *Liriola*, and in some way they may have had a common origin in the past when conditions were such, that shallow-water continuity was possible along the western coastline of the Americas.

Scaphopoda

The genus *Dentalium* has 5 species and a subspecies, the large Subantarctic *aegeum* WATSON, being common in dredgings from around Kerguelen Island. The other species are from either the Magellan Region or from deep water around much of the Antarctic Continent. More study of the material will probably result in a redtution of the number of species recorded.

In the *Siphonodentaliidae* there are 2 species of *Cadulus (Polyschides)*, both of Antarctic occurrence and one of *Siphonodentalium* from off the Gauss Station, Davis Sea in 3423 metres.

Pelecypoda

Solemyidae: Two species of the subgenus *Acharax* occur in the

Magellan area but not at the Falklands, and the family is unknown from elsewhere in Subantarctic or Antarctic waters.

Nuculidae: The type genus *Nucula* has 3 Subantarctic species, *falklandica* PRESTON and *pisum* SOWERBY, both from the Falklands, although the type locality for the latter is Valparaiso, and *kerguelensis* THIELE, from Kerguelen Island. An Antarctic species, *notobenthalis* THIELE, comes from north west of Gauss Station in 2725 metres. In the New Zealand Antipodean the mainland species *hartvigiana* PFEIFFER extends down to the Auckland and Campbell Islands and a restricted Antipodean species *rossiana* FINLAY occurs only at Auckland and Antipodes Islands.

At Macquarie Island *Pronucula* achieves its most southern range in the species *mesembrina* HEDLEY. This genus occurs also at Auckland and Campbell Islands with the species *bollonsi* POWELL, and elsewhere the genus is known from mainland New Zealand, the Kermadec Islands, Tasmania and South Australia.

Nuculanidae: The genus *Nuculana*, in the broad sense, is represented by one Western Antarctic species, *inaequisculpta* (LAMY), from South Orkneys, South Shetlands and Palmer Archipelago in 75—150 metres.

Five species of Antarctic Nuculanids have been attributed to the North Atlantic genus *Yoldiella; antarctica* (THIELE) from Gauss Station and Enderby Land, *ecaudata* and *oblonga* (PELSENEER), West Antarctica, to Gauss Station and off Shackleton Ice-shelf, *valetii* LAMY from South Orkneys and *profundorum* (MELVILL & STANDEN) from 1410 fathoms, Weddell Sea.

The rostrate *Propeleda longicaudata* (THIELE) is probably circum-Antarctic. Three species are attributed to *Yoldia* and for these, SOOT-RYEN, 1951, provided a new subgeneric name, *Aequiyoldia*, since these shells have points of difference in the anatomy and in the presence of a few hinge teeth that distinguish them from the typical Boreal genus. The species of *Aequiyoldia* are *isonota* MARTENS and *subaequilateralis* SMITH from Kerguelen Island and *woodwardi* HANLEY, with a wide southern distribution ranging from the Falklands to Magellan area to South Georgia and Western Antarctic to the Ross Sea. The little known North Atlantic genus *Silicula* has an Antarctic member in *rouchi* LAMY, and the genus is known also from Patagonia.

Malletiidae: Five species of *Malletia* occur, *cumingi* (HANLEY), Falklands, and Magellan area to Rio de Janeiro, *concentrica* and *pellucida* THIELE from the Gauss Station, *sabrina* HEDLEY, from off the Shackleton Ice-shelf to MacRobertson Land and finally *gigantea* (SMITH), the largest member of the genus, reaching 62 mm in length, which is common in shallow water at Royal Sound, Kerguelen Island.

Arcidae: This family is represented in the Western Antarctic by

2 small species of *Bathyarca, sinuata* PELSENEER, 1903 and *strebeli* MELVILL & STANDEN. A New Zealand species, however, *cybaea* HEDLEY, extends south to Macquarie Island.

Limopsidae: The genus *Limopsis* is abundantly represented by 10 species and a subspecies and in these seas reaches its optimum Recent development, with one species, *grandis* (SMITH) attaining a size of 80 × 65 mm.

The Subantarctic species are *hardingi* MELVILL & STANDEN, from the Falklands, *hirtella* ROCHEBRUNE & MARBILLE, from Patagonia and the Falklands, *straminea* SMITH, from between Kerguelen and Heard Islands and *marionensis* SMITH, from off Marion and Prince Edward Islands. Typical *jousseaumi* ROCHEBRUNE & MABILLE comes from Tierra del Fuego but the species extends to the Antarctic in the Bellingshausen Sea area. The strictly Antarctic species are *enderbyensis* POWELL, from off Enderby Land, *jousseaumi grandis* (SMITH), from Ross Sea to Enderby Land, *laeviuscula* and *longipilosa* PELSENEER from Western Antarctica, *scabra* THIELE, from Gauss Station and *lilliei* SMITH from McMurdo Sound to Enderby Land.

The large *jousseaumi grandis* is a very spectacular and distinctive species, not only because of its large size but also owing to the thick epidermal cover of the shell with long dense hair-like filaments.

Philobryidae: The Southern Ocean Philobryas belong to four widely distributed genera, *Hochstetteria*, which has 13 strictly Antarctic and Subantarctic species and some that range north of the Subtropical Convergence at the islands of St. Paul and Amsterdam, as well as the whole coastline of New Zealand, *Lissarca*, with 6 species, widespread in both Antarctic and Subantarctic seas, including 2 species from the New Zealand Antipodean at Auckland Islands and Bounty Islands, the strictly circum-Antarctic monotypic *Adacnarca nitens* PELSENEER and finally *Hochstetterina limopsoides* THIELE, from Gauss Station and off Enderby Land.

All the Hochstetterias have a considerable epidermal cover, often in the form of lapped thatch-like processes and filaments that extend beyond the perimeter of the shell. They are attached by a byssus, often to seaweeds and this factor accounts for their wide Southern Ocean distribution, doubtless achieved by the eastward drift of detached masses of seaweed.

The small *Adacnarca nitens* is not only circum-Antarctic but it also has a vertical range down to at least 600 metres. This species develops the brood within the gill-cavity of the shell, a provision that undoubtedly helps greatly in the survival of the species in these frigid rigorous waters.

The genus *Lissarca*, formerly considered to belong to the *Limopsidae* is now referred to the *Philobryidae*. These small ovate shells showing a weakly developed taxodont hinge have mostly a nestling habit, the Antarctic *notorcadensis* MELVILL & STANDEN and the Kerguelen *rubrofusca* SMITH often occurring in great numbers on and among the spines of sea-urchins. A more strongly developed taxodont hinge is found in the small *Hochstetterina limopsoides* from Gauss Station and off Enderby Land in deep water.

Mytilidae: Large sea mussels are common in most Subantarctic waters but are conspicuously absent from Antarctic Seas. In fact the only members of the family so far recorded for Antarctica are 2 minute species, ascribed, probably incorrectly to the Boreal genus *Dacrydium*.

In the Magellanic Province at least 5 species of mussels extend to the Falklands. They are *Mytilus edulis chilensis* HUPE (GAY), *Perna perna* (LINNAEUS), *Choromytilus chorus* (MOLINA), *Aulacomya ater* (MOLINA), better known as "*magellanicus*", and 2 species of *Hormomya*, *blakeanus* MELVILL & STANDEN and doubtfully, *ovalis* (LAMARCK). In the Subantarctic to the eastward, the cosmopolitan *edulis* has a local subspecies, *desolationis* LAMY from Kerguelen, where it is known also from late Tertiary beds. Another subspecies of the *edulis* complex is *aoteanus* POWELL, which is found throughout New Zealand and the Antipodean but not so far at Macquarie Island.

The ribbed mussels, *Aulacomya*, have a wide lateral range in Southern seas but are found mostly in the zone of mixed waters north of the Subantarctic Zone. The genus extends from the Magellan area to South Africa, Kerguelen, the southern islands of New Zealand, excluding Macquarie and also the whole of the South Island of New Zealand as well as the southern part of the North Island and the Chathams. The Kerguelen subspecies is a giant form reaching a maximum length of 145 mm.

Pinnidae: The genus *Pinna* is known from mainland Magellanic localities but has not extended its range either to the Falklands or elsewhere in Subantarctic or Antarctic areas.

Pectinidae: The most striking of the southern Pectens is the circum-polar, large, thin-shelled, biconvex species, *Adamussium colbecki* (SMITH), formerly considered to belong to the *Amussiidae*, but SOOT-RYEN (1951) showed it to be a very distinct member of the true *Pectinidae*. It is a free swimmer and occurs at depths of from 4 to 700 metres.

Another Pectinid group of species, *Chlamys (Zygochlamys)*, once had a circum-Subantarctic range but now the group survives only at the extremities of this range, the Magellan area to the west and Macquarie Island and southern New Zealand to the east. Dr. C. A. FLEMING (1957), in describing a new species of this subgenus from upper Tertiary, probably Pliocene rocks at Heard Island, reviewed the fossil and Recent occurrences of *Zygochlamys* as follows: Patagonia: *geminata* (SOWERBY), Miocene and *actinodes* (SOWERBY),

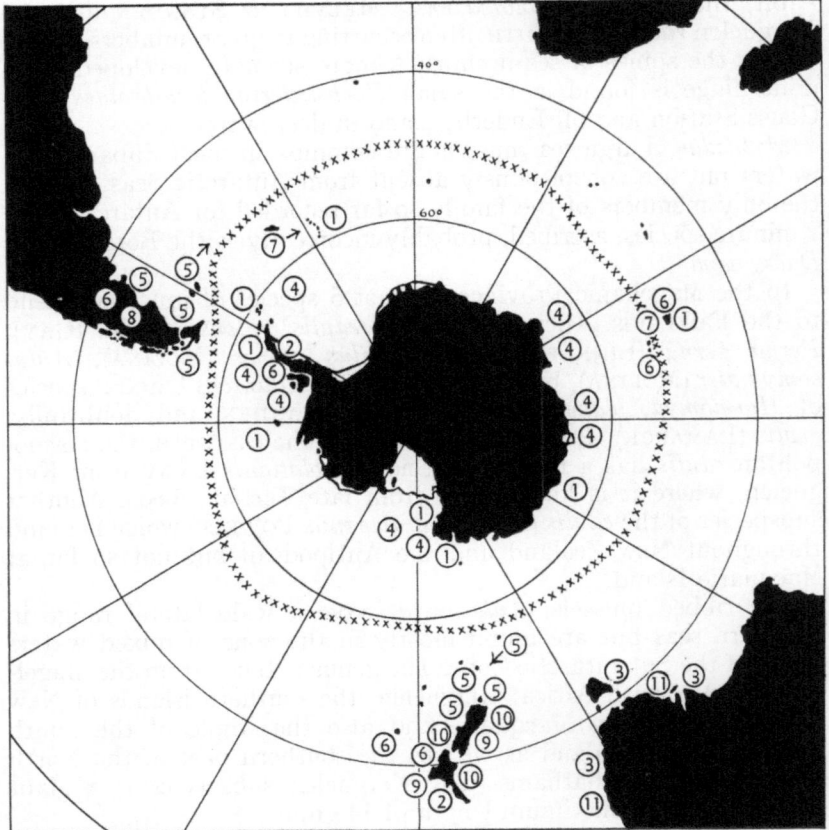

Fig. 6. Distribution of the older Antarctic Molluscan Fauna: Laternulidae, Pectinidae and Struthiolariidae.

1. *Laternula elliptica* (Recent) circum-Antarctic and Kerguelen. 2. *Laternula* (Tertiary-Pleistocene) — Graham Land and Auckland, New Zealand. 3. *Laternula* (Recent sps.) — Tropical East Indies to Tasmania and Southern Australia. 4. *Adamussium Colbecki* — Circum-Antarctic. 5. *Chlamys (Zygochlamys)* (Recent) — Magellan Region, Macquarie Island and New Zealand to middle of South Island. 6. *Chlamys (Zygochlamys)* (Tertiary-Pleistocene) — Patagonia, Kerguelen, Heard Island, Chatham Islands and middle of North Island of New Zealand. 7. *Perisso-donta* (Recent) — South Georgia and Kerguelen. 8. *Perissodonta* (Tertiary) — Pata-gonia. 9. and 10. *Struthiolaria* and *Pelicaria* — Recent and Tertiary, New Zealand. 11. *Tylospira* — Recent and Tertiary, South East Australia.

Lower Pleistocene; *patagonica* (KING) and *patriae* DOELLA-JURADO, living: Grahamsland: *anderssoni* HENNIG, Cockburn Island, Plio-cene and Pleistocene; Heard Island, *heardensis* FLEMING, Pliocene?; Kerguelen Island, *mawsoni* FLETCHER, upper Tertiary; Macquarie Island *(subantarctica)* HEDLEY, which probably equals *delicatula* (HUTTON), living; New Zealand: *delicatula* (HUTTON), Campbell

Island, Lower Pliocene, southern part of the North Island, lower Pleistocene and living to as far north as North Otago, South Island of New Zealand; Chatham Islands, *seymouri* (MARWICK), uppermost Miocene or lower Pliocene.

The remaining Antarctic — Subantarctic Pectinids belong to four genera each of wide deep sea occurrence in other areas. They are *Cyclopecten*, 4 species, *Palliolum*, 5 species, *Hyalopecten*, 1 species and *Variamussium*, 2 species. The Antipodean Pectinids include besides *Zygochlamys*, allies of 2 species of *Chlamys* and one *Cyclopecten* which have extended southward from the New Zealand mainland.

Limidae: Six Antarctic ranging species of the widespread genus *Limatula* are known and two of them extend to Subantarctic waters, *pygmaea* (PHILIPPI) to the Strait of Magellan, Falklands, Marion and Prince Edward, Kerguelen and Macquarie Islands, and *hodgsoni* (SMITH), which is now known to extend to Bouvet Island (SOOT-RYEN, 1951) and probably Macquarie Island (TOMLIN, 1948). The New Zealand species *Escalima regularis* POWELL, extends southward in the Antipodean to near the Auckland Islands.

Astartidae: This family, which is so much in evidence in all northern and Arctic waters is poorly represented in the south, which is not surprising since it is a stenothermic animal for which the equatorial waters, except in the great depths, present a formidable barrier. Three species are recorded, *antarctica* THIELE, from off Gauss Station and Enderby Land, and two Magellanic species, *magellanica* SMITH, from Strait of Magellan and Burdwood Bank and the allied *longirostris* (ORBIGNY) from both the Falklands and Kerguelen.

Carditidae: Six species of the northern genus *Cyclocardia* represent this family in Antarctic-Subantarctic waters, 2 species, *malvinae* and *thouarsii* (ORBIGNY) were described from the Falklands, *congelascens* MELVILL & STANDEN from the Burdwood Bank, 2 exclusively Antarctic species, *antarctica* (SMITH) and *intermedia* (THIELE) from Davis Sea and Mackenzie Sea and finally *astartoides* (MARTENS), described from Kerguelen but widely distributed around the Antarctic Continent, as well as at Bouvet Island and South Georgia. In the Antipodean, New Zealand species of *Cardita*, *Venericardia*, *Pleuromeris* and *Verticipronus* occur but not to as far south as Macquarie Island. The latter, a small seaweed nestling shell, long considered monotypic, is now known to occur at Tristan da Cunha in *Verticipronus tristanensis* SOOT-RYEN, 1952.

Condylocardiidae: This family is represented by *Carditella pallida duodecimcostata* MELVILL & STANDEN from the Burdwood Bank and three other species of *Carditella* from the Magellan area but none elsewhere in either the Antarctic or Subantarctic.

Vesicomyidae: A minute bivalve from near Alexander Ist. Land, West Antarctica is named *Vesicomya laevis* (PELSENEER), and it is reputed to be closely allied to a North Atlantic species, *atlantica* (SMITH). An Antarctic genus, *Ptychocardia*, with 2 species, *rudis* HEDLEY and *vanhoffeni* THIELE, is also included in this family.

Cyamiidae: The small smooth bivalves, *Cyamium*, which are mostly thin-shelled and of elongate shape are represented by ten species, eight of them from the Falklands, one from Kerguelen and one from South Georgia. Two other related genera have Antarctic representatives, the exclusively Antarctic *Pseudokellya* with 3 species and *Cyamiomactra*, with one Western Antarctic species, *laminifera* LAMY, but the type of the genus is from New Zealand, where it extends south to the Antipodean.

Perrierinidae: This family is well represented in New Zealand, including the Antipodean by several species each of *Perrierina* and *Legrandina*, small Leptonid-like shells with a weak taxodont hinge. The characteristic Antarctic and Subantarctic members of this family are *Cyamiocardium*, small thin-shelled bivalves with weak radiate sculpture and circular outline. The species are *denticulatum* (SMITH), Kerguelen and Burdwood Bank to Ross Sea, *rotundatum* (THIELE), Gauss Station to Adelie Land and *fragillimum* (THIELE) from Kerguelen Island.

Gaimardiidae: These are the most abundant bivalves of the Subantarctic. They belong to two genera, *Gaimardia*, 5 species and 2 subspecies and *Kidderia*, with 7 species, ranging from the Magellan area to New Zealand. They are thin-shelled and of shallow-water occurrence, mostly on seaweeds. *Gaimardia* grows to 30 mm or more, is often brightly coloured, yellowish, red or purple-brown, and has a weakly to strongly rostrate anterior end. The species of widest distribution is *trapesina* with a range for the typical species from the Magellan area to Kerguelen, but subspecies occur at Macquarie Island, *coccinea* HEDLEY and at Auckland Islands, *flemingi* POWELL.

The other genus, *Kidderia* is even wider in distribution for it extends from the Magellan area up through the Antipodean even to the extreme north of New Zealand. The shell of *Kidderia* is small and elongate-ovate, with the umbonal area close to the anterior end. A third genus *Costokidderia*, of Antipodean-Southern New Zealand range resembles *Kidderia* but has a radiate sculpture of strong ribs.

Lucinidae: The widespread genus *Lucinoma* has a single representative in the Patagonian *lamellata* (SMITH) which reaches the Burdwood Bank.

Thyasiridae: This wide ranging family found in all cold seas is represented by 3 southern species of *Thyasira*, *bongraini* LAMY from Western Antarctica, *marionensis* (SMITH) from between Prince Edward and Marion Islands and *falklandica* (SMITH), which extends from the Falklands southward to South Georgia and the South

Orkneys. A second member of the family, ascribed to the Boreal genus *Axinopsis*, comes from deep water at Gauss Station and off the Shackleton Ice-shelf. In the Antipodean a New Zealand genus *Maorithyas* extends to Auckland and Campbell Islands, *flemingi* POWELL.

Leptonidae: A New Zealand genus, *Notolepton*, with species common throughout the Antipodean is represented at Macquarie Island and at Kerguelen by *umbonatum* (SMITH). Another Kerguelen species of the genus, *parasiticum* (DALL) is commensal upon the spatangid echinoid *Abatus cavernosus*. Still another small Leptonid genus, *Davisia*, of three species has been found so far, only at the Falklands and on the Burdwood Bank.

Erycinidae: The cosmopolitan seaweed nestling *Lasaea* has 3 species in southern waters, *consanguinea* (SMITH), Kerguelen, Falklands and South Orkneys, a Magellan species, probably erroneously identified as the Mediterranean *miliaris* (PHILIPPI) and *rossiana* FINLAY of Antipodean range only. The equally widespread genus *Kellia*, larger shells of circular outline, fragile and usually living buried in mud, is represented by 4 named species, *nimrodiana* HEDLEY and *simulans* SMITH of Antarctic range, a Kerguelen species, *nuculina* MARTENS and *magellanica* SMITH from the Magellan area, including the Falklands. Another minute member of the family is *Scacchia plenilunium* MELVILL & STANDEN, from the Falklands.

Montacutidae: The genus *Mysella*, abundantly represented in both New Zealand and Australian waters is also common throughout Antarctic and Subantarctic waters. Ten species are recorded and one of them, *charcoti* (LAMY) extends from West Antarctica to South Georgia, Kerguelen and Macquarie Island.

Galeommatidae: A single member of this family, *Solecardia antarctica* HEDLEY, comes from 10—20 fathoms, Cape Royds, Antarctica. The genus is more typical of warm seas.

Cardiidae: The true cockles are represented by *Trachycardium delicatulum* (SMITH) from west of the Falklands in 125 fathoms. The New Zealand *Nemocardium (Pratulum) pulchellum* (GRAY), however, extends south over most of the Antipodean.

Veneridae: This widespread family is not represented in the Antarctic, but in the Subantarctic Recent fauna infiltration occurs at both the Falklands and at Macquarie Island. At the former, the Patagonian *Eurhomalea exalbida*, a large heavy bivalve, is common accompanied by an apparently endemic, much smaller Venerid, *Gomphina (Jukesena) foveolata* (COOPER & PRESTON). The family is well represented in the New Zealand Antipodean by large Venerids of the genera *Paphirus*, *Chione (Austrovenus)*, *Plurigens* and *Tawera* and the last mentioned is represented very abundantly at Macquarie Island in the species *mawsoni* (HEDLEY).

Mactridae: The large Magellanic Mactrid *Darina solenoides* KING is recorded from the Falklands but the family is not known from elsewhere in Subantarctic — Antarctic waters.

The *Garidae* and the *Tellinidae* are both absent from the Falklands, the other Subantarctic islands and the Antarctic, but the Semelid genus *Leptomya* extends to the Auckland Islands from the New Zealand mainland.

Solenidae: The Magellanic razor-shell, *Solen macha* MOLINA has reached the Falklands but is unknown from elsewhere in the Subantarctic — Antarctic Zones.

Myidae: The single Falklands member ascribed to this family, *"Mya" antarctica* (MELVILL & STANDEN), described as a *Thracia*, then thought to be the young of *Laternula elliptica*, is still in doubt regarding its taxonomic status.

Hiatellidae: Two nominate species of *Hiatella* range throughout Antarctic and Subantarctic seas but valid differences between them are not clearly evident. They are *antarctica* (PHILIPPI) and *subantarctica* (PRESTON). A third species, *bisulcata* (SMITH), described from Kerguelen, has not since been collected.

Pholadidae: The family is absent except for species of *Barnea*, *Xylophaga* and *Nettastomella* from mainland Magellanic areas only.

Teredinidae: An apparently endemic ship-worm, *Bankia odhneri* ROCH, was described from the Falklands.

Lyonsiidae: One member of this family, *Entodesma arcaeformis* (MARTENS) extends from the Magellan area to the Falklands and the other *Mytilimeria falklandica* PRESTON, seems to be endemic there.

Pandoridae: The characteristic West American and Patagonian genus *Kennerlia* has not been recorded from the Falklands or elsewhere in Southern Ocean localities.

Pholadomyidae: This ancient family is recorded for the Eastern Antarctic by 3 species described by HEDLEY, *adelaidis*, *antarctica* and *mawsoni*.

Thraciidae: This cosmopolitan family, more typical of cold waters, is abundantly and widely distributed in both Antarctic and Subantarctic seas by *Thracia meridionalis* (SMITH), originally described from Kerguelen.

Laternulidae: It is anomalous to find a large sized flourishing member of this otherwise warm-water genus, *Laternula*, in Antarctic and Subantarctic waters. This shell, *Laternula elliptica* (KING & BRODERIP) is a member of the earlier Antarctic fauna, as is shown by a fossil occurrence in the Pleistocene deposits at Cockburn Island, and it is now abundantly distributed from the Falklands to Kerguelen and around the whole of the Antarctic Continent, wherever the ecological requirement of soft mud is present. It is apparently this burrowing habit in the deep muds of the ocean floor that has afforded insulation for the species during extreme variations in temperature.

Verticordiidae: An Antarctic member of this family is *Lyonsiella planulata* THIELE, known from deep water at the Gauss Station and also off Enderby Land.

Poromyidae: A member of this deep water family is *Poromya spinosula* THIELE, also from the Gauss Station.

Cuspidariidae: This cosmopolitan family is abundantly represented in Antarctic and Subantarctic waters with 5 species of *Cuspidaria* and one each of *Myonera* and *Cardiomya*.

Cephalopoda

A very useful summary of the Southern Ocean Cephalopods was published recently by DELL (1958) in the Reports of the British Australian New Zealand Antarctic Research Expedition. Dr. DELL lists 20 species for the Antarctic proper, 18 for the Magellanic Province, of which 5 reach the Falklands, 5 for South Georgia, 3 for the Kerguelen Province and 5 for the New Zealand Subantarctic (Antipodean).

Collections are at present inadequate for providing more than an elementary knowledge of the extent and distribution of the Southern Ocean Cephalopods but a few generalisations can be made.

Amongst the Octopoda, for instance, *Octopus*-like forms with biserial suckers occur in the Subantarctic at Tierra del Fuego, Kerguelen and Campbell Island, but in the Antarctic proper the dominant Octopoda are the Eledonids, one of which, *Pareledone charcoti* (JOUBIN) is certainly circum-Antarctic ranging and it is likely that *Pareledone turqueti* (JOUBIN) also will prove to be similarly generally distributed in Antarctic Seas.

Three species of the 8 recorded Decapod Cephalopods, *Alluroteuthis antarcticus* ODHNER, *Crystalloteuthis glacialis* CHUN and *Psychroteuthis glacialis* THIELE are endemic to either the Antarctic or adjacent seas.

The remaining Cephalopods so far recorded from Antarctic waters all belong to wide ranging pelagic genera and four of them are members of the family *Cranchiidae*.

REFERENCES

BERGENHAYN, J. R. M., 1937. Antarktische und subantarktische Polyplacophoren. *Sci. Res. Norweg. Antarct. Exped.* 17, *1—12.*

BERGH, R. 1884. Report on the Nudibranchiata. *Zool. Challenger.* 10, *1—154.*

BERGH, R. 1886. Report on the Marseniadae collected by H. M.S. Challenger. *Zool. Challenger,* 15 (41), *1—24.*

BERRY, S. S., 1917. Cephalopoda. *Austr. Antarct. Exped. (1911—1914), Ser. C., Zool. & Bot.* 4 (2), *1—38.*

376

BURNE, R. H., 1920. Mollusca, Pt. 4, Anatomy of Pelecypoda. *Brit. Antarct. ("Terra Nova") Exped.*, 1910, *Zool.* **2** (10).

CARCELLES, A. R. & WILLIAMSON, Susana I., 1951. Moluscos Marinos de la Provincia Magellanica. *Rev. Inst. Nac. Invest. Ciencias Nat. Buenos Aires*, **2**, 5, *225—383*.

COOPER, J. E. & PRESTON, H. B., 1910. Diagnoses of new species of marine and freshwater shells from the Falkland Islands, including description. of two new genera of marine Pelecypoda. *Ann. Mag. Nat. Hist.* (Ser. 8), **5**, *110—114*.

COTTON, B. C., 1937. Loricata *B.A.N.Z.A.R.E. Rep.* **4B**, *9—19*.

CUNNINGHAM, R. O. 1871. Mollusca of Fuegia and the Magellan Straits. *Trans. Linn. Soc., London*, **27**, *474—488*.

DALL, W. H., 1876. Mollusks. Contributions to the Natural History of Kerguelen Island. *Bull. U.S. Nat. Mus.* **3**, *42—48*.

DAVID, L., 1934. Zoologische Ergebnisse der Reisen von Dr. Kohl-Larsen nach den subantarktischen Inseln bei Neuseeland und nach Südgeorgien, Pt. 9, *Senckenbergiana*, **16**, *126—137*.

DELL, R. K. 1952. Marine Biology, in "The Antarctic Today", Wellington, New Zealand.

DELL, R. K., 1964. Marine Mollusca from Macquarie and Heard Islands. *Rec. Domin. Mus., Wellington, New Zealand*, 4(20), *264—301*.

EALES, N. B., 1923. Mollusca. Pt. 5. Anatomy of Gastropoda (except the Nudibranchiata). *Brit. Antarct. ("Terra Nova") Exped.*, 1910, *Zool.* **7** (1), *1—46*.

EARLAND, A., 1933. Foraminifera, Part. 2, South Georgia. *Discovery Rep.* **7**, *27—138*.

EARLAND, A., 1934. Foraminifera, Part. 3. The Falkland sector of the Antarctic (excluding South Georgia), *Discovery Rep.*, **10**, *1—208*.

ELIOT, C. N. E., 1905. The Nudibranchiata of the Scottish National Antarctic Expedition. *Trans. Roy. Soc. Edinb.*, **41**, *519—532*.

ELIOT, Sir Charles, 1907. Nudibranchs from New Zealand and the Falkland Islands. *Proc. Mal. Soc., London*, **7**, *327—361*.

ELIOT, Sir Charles, 1907. Mollusca. 4. Nudibranchiata in National Antarctic Expedition (1901—1904), Nat. Hist. **2**, *1—28*.

ELIOT, Sir Charles, 1907. Mollusca. 6. Pteropoda in National Antarctic Expedition (1901—1904), Nat. Hist. **3**, *1—15*.

ELIOT, Sir Charles, 1910. The Nudibranchiata of the Scottish National Antarctic Expedition. *Rep. Sci. Res. Voy. S.Y. "Scotia"*, **5**, *11—24*.

FLEMING, C. A. 1957. A new species of Fossil Chlamys from the Drygalski Agglomerate of Heard Island, Indian Ocean. *J. Geol. Soc. Austr.* **4** (1), *13—19*.

FLETCHER, H. O., 1938. Marine tertiary fossils and a description of a recent Mytilus from Kerguelen Island. *B.A.N.Z.A.R.E. Rep. Ser.* A. **2** (6), *103—116*.

HADDON, A. C., 1886. Report on the Polyplacophora collected by H.M.S. Challenger. *Zool. Challenger*, **15**, *1—50*.

HEDLEY, C., 1911. Mollusca. British Antarctic Expedition (1907—9), Biol. **2** (1), *1—8*.

HEDLEY. C., 1916. Mollusca. Australasian Antarctic Expedition, Ser. C., Zool. & Botany, **4** (1), *1—80*.

HEDLEY, C., 1916. Report on Mollusca from Elevated Marine Beds "Raised beaches" of McMurdo Sound. *Brit. Antarct. Exped. 1907—9, Geology* **2** (5), *85—88*.

HOMBRON & JACQUINOT, 1841. Description de quelques Mollusques provenant de la campagne de l'Astrolabe et de la Zelee. *Ann. Sci. Nat.* (Zool.) **16**, *62—64* and *190—192*.

HOYLE, W. E., 1885. Preliminary Report on the Cephalopoda collected during the cruise of H.M.S. Challenger. *Proc. Roy. Soc. Edinb.* **13**, *94—114*.

HOYLE, W. E., 1907. Mollusca, Pt. 1, Cephalopoda. Nat. Antarct. Exped.
Zool. 2.

HOYLE, W. E., 1912. Cephalopoda of the Scottish National Antarctic Expedition. *Trans. Roy. Soc. Edinb.* **48**, *273—283*.

HUBENDICK, B., 1946. Systematic Monograph of the Patelliformia. *K. svenska
Vetensk. Akad. Handl.* **23** (5), *1—94*.

HUBENDICK, B., 1951. Pteropoda, with a New Genus. *Further Zoological
Results of the Swedish Antarctic Expedition, 1901—1903*, **4** (6), *1—10*.

HUPE, L. H. (GAY, C.), 1854. Historia física y política de Chile, 8, Mollusca.

IHERING, H. VON, 1902. Die Photinula-Arten der Magellan-Strasse. *Nachr. Bl.
dtsch. malakozool. Ges.*, *97—104*.

JOUBIN, L. 1903. Cephalopodes. Resultats du voyage S.Y. Belgica, Zool. 1—4.

JOUBIN, L. 1905. Description de deux Eledones provenant de l'expédition
du Dr. Charcot dans l'Antarctique. *Mém. Soc. Zool. Fr.* **18**, *22—31*.

JOUBIN, L. 1907. Expédition Antarctique Française, 1903—1905. Cephalopodes.

JOUBIN, L. 1914. Némertines, Céphalopodes, Brachiopodes. *Deux. Expéd.
Antarct. Franç.* (1908—1910).

LAMY, ED., 1905. Gastropodes prosobranches recueillis par l'expédition
antarctique française du Dr. Charcot. *Bull. Mus., Paris, 475—483*.

LAMY, ED., 1906. Sur quelques Mollusques des Orcades du Sud. *Bull. Mus.
Paris, 121—126*.

LAMY, ED., 1906. Lamellibranches recueillis par l'expédition antarctique du
Dr. Charcot. *Bull. Mus., Paris, 44—52*.

LAMY, ED., 1907. Gastropodes Prosobranches et Pélécypodes. *Expéd. Antarct.
Franç.* (1903—5), *1—20*.

LAMY, ED., 1910. Mollusques recueillis par M. Rallier du Baty aux Iles
Kerguelen. *Bull. Mus., Paris, 198*.

LAMY, ED., 1910. Mission dans l'Antarctique dirigée par le Dr. Charcot:
Gastropodes (pp. *318—324*), Pélécypodes (pp. *388—394*). *Bull. Mus.
Paris*, **16**.

LAMY, ED., 1911. Gastropodes, Prosobranches, Scaphopodes et Pélécypodes.
Deux. Exped. Antarct. Franç. (1908—1910), *1—32*.

LAMY, ED., 1911. Sur quelques Mollusques de la Géorgie du Sud, et des îles
Sandwich du Sud. *Bull. Mus., Paris, 22—27*.

LAMY, ED., 1915. Mollusques recueillis aux îles Kerguelen par M. Loranchet.
Bull. Mus., Paris, 68—76.

LONNBERG, E., 1899. On the Cephalopods collected by the Swedish Expedition to Tierra del Fuego (1895—96). *Svenska Exped. Magellanslandern*
2, *49—64*.

MACFADYEN, W. A., 1933. Fossil Foraminifera from the Burdwood Bank.
Discovery Rep., **7**, *1—16*.

MARTENS, E. VON, 1881. Über mehrere von Sr. Maj. Schiff Gazelle von der
Magellaen-strasse der Ostküste Patagoniens und der Kerguelen-Insel
etc. *S.B. Gesell. naturf. Fr.*, *75—80*.

MARTENS, E. VON, 1885. Vorläufige Mitteilung über die Mollusken-fauna von
Süd-Georgien. *S.B. Ges. naturf. Fr., Berlin, 89—94*.

MARTENS, E. VON & PFEFFER, G., 1886. Die Mollusken von Süd-Georgien.
J. Hamburg, Wiss. Anst., **3**, *65—135*.

MARTENS, E. VON & THIELE, J. 1903. Die beschalten Gastropoden der deutschen Tiefsee-Expedition, 1898—99. *Wiss. Erg. "Valdivia"*, **7**,
1—146.

MASSY, ANNE L., 1916. Mollusca, Pt. 2. Cephalopoda. *British Antarct. ("Terra
Nova") Exped. (1910), Zool.* **2**, *141—175*.

MASSY, ANNE L., 1920. Mollusca. Pt. 3. Eupteropoda (Pteropoda Thecosomata) and Pterota (Pteropoda Gymnosomata). *British Antarct. ("Terra
Nova") Exped. (1910), Zool.* **2**, *203—232*.

378

MASSY, ANNE L., 1932. Mollusca. Gastropoda. Thecosomata and Gymnosomata. *Discovery Rep.*, **3**, *267—296.*

MELVILL, J. C. & STANDEN, R., 1898. Notes on a collection of marine shells from Lively Island, Falklands, with list of species. *J. Conch.*, **9**, *97—105.*

MELVILL, J. C. & STANDEN, R., 1901. Mollusca collected by Mr. Rupert Vallentin at Stanley Harbour, Falkland Islands. *J. Conch.*, **10**, *43—47.*

MELVILL, J. C. & STANDEN, R., 1907. The marine mollusca of the Scottish National Antarctic Expedition. *Trans. Roy. Soc. Edinb.* **46**, *119—157.*

MELVILL, J. C. & STANDEN, R., 1912. The marine mollusca of the Scottish National Antarctic Expedition, Pt. 2. *Trans. Roy. Soc. Edinb.* **48**, *333—366.*

MELVILL, J. C. & STANDEN, R., 1914. Notes on Mollusca collected in the northwest Falklands by Mr. Rupert Vallentin, F.L.S., with descriptions of six new species. *Ann. Mag. Nat. Hist.* (Ser. 8) **13**, *109—136.*

ODHNER, N. H., 1923. Die Chitonen. *Wiss. Erg. Schwed. Südpolar-Exped.*, **1** (3), *1—4.*

ODHNER, N. H., 1923. Die Cephalopoden. *Further Zoological Results of the Swedish Antarctic Expedition*, 1901—1903, **1**, (4), *1—7.*

ODHNER, N. H., 1926. Die Opisthobranchien. *Further Zoological Results of the Swedish Antarctic Expedition*, 1901—1903, **2** (1), *1—100.*

ODHNER, N. H., 1931. Die Scaphopoden. *Further Zoological Results of the Swedish Antarctic Expedition*, 1901—1903, *1—8.*

ODHNER, N. H., 1934. The Nudibranchiata. British Antarct. ("Terra Nova") Exped. (1910), Zool, **7**, *229—310.*

PELSENEER, P., 1903. Résultats du Voyage du S.Y. 'Belgica'. Zoologie, Mollusques (Amphineures, Gastropodes et Lamellibranches) Anvers, *1—85.*

PLATE, L., 1908. Scaphopoden. Résultats du Voy. du S.Y. 'Belgica' en 1895—9, *1—4.*

PLATE, L., 1908. Die Scaphopoden der Deutschen Südpolar Exped. (1901—3), *Wiss. Erg. dtsch. Südpol. Exped.* **10**, *1—6.*

POWELL, A. W. B., 1951. Antarctic and Subantarctic Mollusca: Pelecypoda and Gastropoda. *Discovery Rep.*, **26**, *47—196.*

POWELL, A. W. B., 1955. Mollusca of the Southern Islands of New Zealand. *Cape Exped. Ser. Bull.* **15**, *D.S.I.R.*, Wellington, N.Z., *1—152.*

POWELL, A. W. B., 1957. Mollusca of Kerguelen and Macquarie Islands. *B.A.N.Z.A.R.E. Rep. (Ser. B.)*, **6** (7), *107—150.*

POWELL, A. W. B., 1958. Mollusca from the Victoria-Ross Quadrants of Antarctica. *B.A.N.Z.A.R.E. Rep. (Ser. B.)*, **6** (9), *165—216.*

POWELL, A. W. B., 1960. Antarctic and Subantarctic Mollusca (catalogue), *Rec. Auck. Inst. Mus.*, **5** (3—4), *117—193.*

PRESTON, H. B., 1912. Characters of six new Pelecypods and two new Gastropods from the Falkland Islands. *Ann. Mag. Nat. Hist.* (Ser. 8) **9**, *636—640.*

PRESTON, H. B., 1913. Descriptions of fifteen new species and varieties of marine shells from the Falkland Islands. *Ann. Mag. Nat. Hist.*, **11**, *218—223.*

PRESTON, H. B., 1916. Descriptions of eight new species of marine Mollusca from the South Shetland Islands. *Ann. Mag. Nat. Hist.* **18**, *269—272.*

ROBSON, G. C., 1930. Cephalopoda 1. Octopoda. *Discovery Rep.*, **11**, *373—402.*

ROCHEBRUNE, A. T. DE & MABILLE, J., 1889. Cephalopoda, Gastropoda et Lamellibranchiata. *Mission Scientifique du Cap Horn* (1882—3), **6**, Zool., *1—126.*

SMITH, E. A. 1875. Descriptions of some new shells from Kerguelen's Island. *Ann. Mag. Nat. Hist.* (4) **16**, *67—73.*

SMITH, E. A. 1877. Mollusca in Zoology of the Transit of Venus Expedition. *Phil. Trans. Roy. Soc. London,* **168** (extra vol.), *167—192.*

SMITH, E. A. 1881. Account of the zoological collections made during the survey of H.M.S. Alert in the Straits of Magellan and on the coasts of Patagonia. Mollusca. *Proc. Zool. Soc., London, 22—44.*

SMITH, E. A. 1885. Report on the Lamellibranchiata collected during the voyage of H.M.S. Challenger. *Zool. Challenger,* **13** (35), *1—341.*

SMITH, E. A. 1902. On the supposed similarity of the mollusca of the Arctic and Antarctic regions. *Proc. Mal. Soc., London,* **5,** *162—166.*

SMITH, E. A. 1902. Report on the collections of Mollusca made in the Antarctic during the voyage of the 'Southern Cross', Pt. **7,** *201—213.*

SMITH, E. A. 1907. Mollusca and Brachiopoda. *National Antarctic Exped. ('Discovery') Nat. Hist.* **2,** *1—12.*

SMITH, E. A. 1907. Lamellibranchiata. *National Antarctic Exped. ('Discovery') Nat. Hist.* **2,** *1—7.*

SMITH, E. A. 1915. Mollusca, Pt. 1. Gastropoda Prosobranchia, Scaphopoda and Pelecypoda. Brit. Antarct. ('Terra Nova') Exped. (1910), Zool. **2,** *61—112.*

SOOT-RYEN, T., 1951. Antarctic Pelecypods. *Sci. Res. Norw. Antarct. Expeds.,* 1927—1928, No. 32, *1—46.*

STREBEL, H., 1904—1907. Beiträge zur Kenntnis der Mollusken Fauna der Magalhaen-Provinz. *Zool. Jb. Abt. Syst.* Jena. Pt. 1, 1904, **21,** *171—248,* Pt. 2, 1905, Suppl. 8, *121—166,* Pt. 3, 1905, **22,** *575—666,* Pt. 4, 1906, **24,** *91—174* & Pt. 7, 1907, **25,** *79—196.*

STREBEL, H., 1908. Die Gastropoden. *Wiss. Erg. schwed. Südpolar-Exped.* (1901—1903), **6** (1), *1—112.*

STUDER, T. 1879. Die Fauna von Kerguelensland. *Arch. Naturgesch.* **45,** *104—141.*

THIELE, J. 1903. Die beschalten Gastropoden der deutschen Tiefsee-Expedition, 1898—9 (B). *Wiss. Erg. 'Valdivia',* **7,** *147—174.*

THIELE J., 1906. Über die Chitonen der deutschen Tiefsee-Expedition. *Wiss. Erg. 'Valdivia'* **9,** *325—336.*

THIELE, J. 1906. Note sur les Chitons de L'Expédition Antarctique du Dr. Charcot. *Bull. Mus. Paris,* **12,** *549—550.*

THIELE, J. 1907. Amphineures. *Expéd. Antarct. Franç.* (1903—5), *1—3.*

THIELE, J. 1908. Die antarktischen und subantarktischen Chitonen. *Dtsch. Südpolar-Exped.* (1901—3), *7—23.*

THIELE, J. 1911. Amphineures. Deuxième Expéd. Antarct. Franç. (1908—10), *33—34.*

THIELE, J. 1912. Die antarktischen Schnecken und Muscheln. *Dtsch. Südpolar-Exped.* (1901—3), **13,** *183—285.*

THIELE, J. 1920. Die Cephalopoden der Deutschen Südpolar-Exped. *Dtsch. Südpol. Exped.* (1901—3), **16,** *431—465.*

THIELE, J. 1925. Gastropoda der deutschen Tiefsee-Expedition, 2, *Wiss. Erg. 'Valdivia',* **17** (2), *36—382.*

THORSON, GUNNAR, 1936. The Larval Development, Growth and Metabolism of Arctic Marine Bottom Invertebrates compared with those of other seas. *Meddelelser om Gronland,* **100.**

TOMLIN, J. R. LE B., 1948. The Mollusca of Macquarie Island, B.A.N.Z.A.R.E. Rep. (Ser. B) **5,** 221—232.

VAYSSIERE, A. 1906. Diagnoses génériques des Mollusques gastropodes nouveaux rapportés par L'Expédition antarctique du Dr. Charcot. *Bull. Mus. Paris,* *147—149.*

VAYSSIERE, A. 1906. Sur les Gastéropodes Nudibranches de l'Expédition antarctique du Dr. Charcot. *C.R. Acad. Sci. Paris,* **142,** *718—719.*

VAYSSIERE, A. 1907. Mollusques Nudibranches et Marséniadés. Exped. Antarct. Franç. (1903—5), *1—51.*

VAYSSIERE, A. 1916. Sur un Amphineure et sur quelques Gastropodes Opis-
thobranches et prosobranches de la Deuxième Exped. du Dr. Charcot.
C.R. Acad. Sci. Paris, **162**, *271—273.*

VAYSSIERE, A. 1917. Recherches zoologiques et anatomiques sur les mollusques amphineures et gastéropodes. *Deuxième Expéd. Antarct. Franç.*
(1908—10), *1—50.*

WATSON, R. B. 1886. Report on Scaphopoda and Gasteropoda collected by
'Challenger'. *Zool. Chall. Exped.*, **15**, *1—756.*

NOTES ON THE BIOGEOGRAPHY AND ECOLOGY OF FREE-LIVING, MARINE COPEPODA

W. VERVOORT

Rijksmuseum van Natuurlijke Historie, Leiden

(with 6 figs.)

Both marine and freshwater Copepods have been described from the Antarctic and sub-Antarctic region. Though a variety of brackish water habitats undoubtedly occur in the area under observation, true Antarctic (or sub-Antarctic) brackish water Copepoda so far have remained unobserved, though they will almost certainly be present.

The freshwater Copepoda will not be discussed here. The marine Copepoda may be arbitrarily divided into three ill-defined groups, the truly planktonic species, the littoral or neritic forms and the parasites. The parasites too will remain unobserved here.

Though the life history of only very few species is known in some detail, our knowledge of the planktonic forms is far superior to that of the neritic forms. This curious fact has been brought about by the routine usually carried out on board of Antarctic research vessels. The working of stations, where hydrographical data along with plankton samples are being collected, forms an integral part of the program of nearly all Antarctic research expeditions, whereas only very few expeditions established bases in the Antarctic zone which permitted continuous work on biological and hydrographic scale in inlets favourable for the study of neritic life. Parasitic copepods have so far only incidentally been collected.

Copepods of the Antarctic and sub-Antarctic marine plankton

The life history of planktonic marine Copepods is so intimately linked with the hydrological conditions prevailing in the Antarctic and sub-Antarctic, that it can only be understood if considered in connection with some basic facts of water movements, and temperature and salinity distributions in the South. The following account enumerates only those facts that are necessary to understand the biology of such species of Copepods that have been sufficiently studied; it forms a much simplified condensation of DEACON's (1933, 1937) papers and at many places it is considerably more complicated or even completely useless on account of local conditions.

The surface of the Southern Ocean is covered with a 100 to 250 m thick layer of cold water of comparatively low salinity, which rests on a very thick layer of warmer water of a higher salinity (the water of the warm, deep Antarctic current). Below this layer of warm

382

water there is a layer of Antarctic bottom water of low temperature
and fairly low salinity. In the surface layer the temperature varies
between —1°.5 C and 0° C, the salinity between ± 32.85 and 34.5‰;
in the deep, warm current temperatures range from 2°.0 C to 0°.5 C
and salinity varies from 34.70 to 37.90‰. In the Antarctic bottom
layer temperatures are somewhere between —0.2 and —0°.7 C

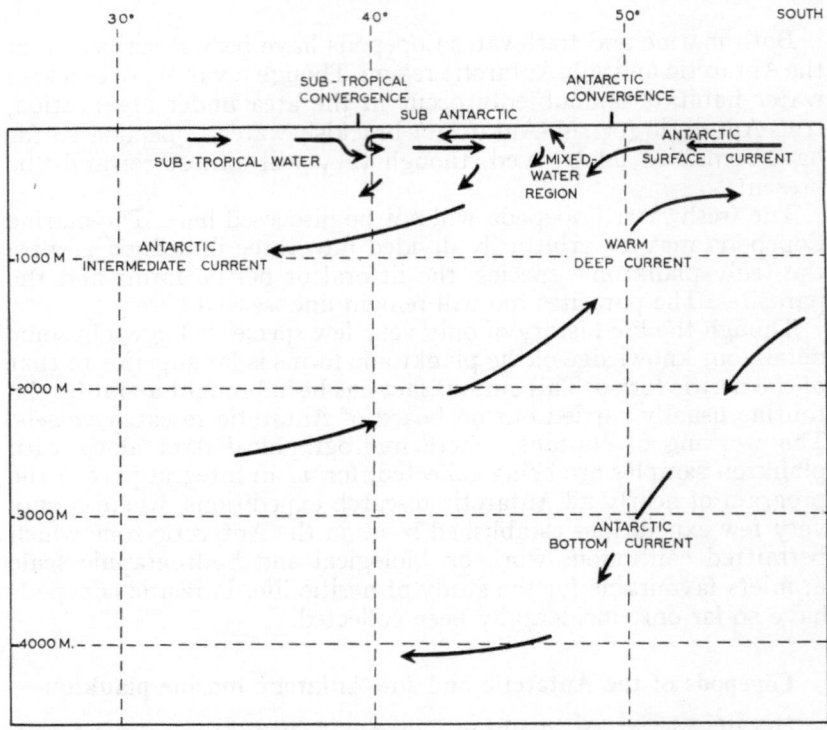

Fig. 1. Diagrammatic representation of the water movements in the southern
Oceans. After DEACON, 1937, modified.

whilst the salinity is generally less than 34.64‰. There is a distinct
discontinuity layer separating the Antarctic surface water from the
intermediate water layer, within which both temperature and
salinity rise rapidly with increasing depth.

The whole body of water, from surface to bottom, is in movement,
moving generally from west to east, except at the highest latitudes.
There is, however, a distinct, northerly directed component in the
movement of the Antarctic surface and bottom layers, and a southerly component in the movement of the warm, deep current. A freely
floating object, therefore, would be carried in a northerly direction

in the Antarctic surface water, to be carried southward again if it sank into the belt of warm water, whilst its general movement again would be northerly if it reached the Antarctic bottom layer.

The northerly movement of the Antarctic surface water can be followed to a region roughly between 55° and 50° S, where it sinks beneath a warmer, sub-Antarctic water layer of higher salinity, though still maintaining its northerly direction. This area, where the cold, Antarctic surface water meets the much warmer sub-Antarctic water, is marked by rapidly rising surface temperatures when moving from south to north and is usually indicated as the Antarctic Convergence. Though this Antarctic Convergence forms a circular belt round the Antarctic continent, its geographical position is different in the various Antarctic sectors (about 50° S in the Atlantic sector, about 60° S in the Pacific sector).

The warm sub-Antarctic water reaches the region of the Antarctic Convergence as southerly directed subsurface current, whose southerly movement is checked in the region of the Antarctic Convergence by the downwardly moving Antarctic surface water. Part of the sub-Antarctic sub-surface water flows off in a generally northerly directed sub-Antarctic surface current, traceable until the region of the sub-tropical convergence, part of it moves downwards and northwards along with the original Antarctic surface water, now moving north as a deeper, Antarctic intermediate current.

It stands to reason that a continual northern drift in the surface and bottomlayers as well as a southward drift in the deep, warm layer would ultimately result in the depletion of pelagic animal life in the Southern Ocean. And yet the development of pelagic animal life in the Southern Ocean as a result of the tremendous development of the phytoplankton is so overwhelming, that the Antarctic region is amongst the very richest. There must, consequently, be some mechanism which ensures the continual return of eggs or larvae to waters threatened with depletion. Though only in some isolated cases the very first details of this mechanism have been clarified, we do know that for the bulk of the inhabitants of the surface zone this mechanism consists of seasonal vertical migrations coupled with reproductory phenomena. Some of the characteristic surface inhabitants do not show such seasonal migrations and we are still at a loss how they maintain their usually large numbers. The same, to some extent, pertains to the characteristic mid-water dwellers: though their number is never so overwhelming as that of surface species they do form a regular component of the mid-water plankton. Their cycle of life is still practically unknown. We are just as poorly informed about the abyssal inhabitants of the bottom layer.

Copepods play a very important role in the Antarctic and sub-Antarctic zooplankton and their importance and abundance is only rivaled by two other animals: *Euphausia superba* DANA en *Eukroh-*

nia hamata (MÖBIUS). Some peculiarities of the standing crop of Antarctic and sub-Antarctic zooplankton can therefore, with some prudence, be taken as starting point for our consideration of the Copepods. Data on the distribution and quantity of the standing crops of zooplankton supplied by FOXTON (1956) show that the vertical distribution of this standing crop is similar in the Antarctic and sub-Antarctic in any one month, i.e., the greatest quantity of the zooplankton in Antarctic and sub-Antarctic is at a more or less uniform depth in a given month. During the summer months this highest concentration of zooplankton occurs at the surface; during the winter months there is a migration into deeper water and the largest concentrations are to be found in relatively warm intermediate water. Though there are local variations in the amount of zooplankton the quantity as a whole is remarkably uniform over the whole Antarctic region, the region of the West Wind Drift being slightly richer than the remaining areas.

BAKER (1954) has demonstrated that some of the more important constituents of this zooplankton, in the region south of the Antarctic Convergence, show circumpolar continuity, i.e. though their relative number, in the top 100 m water layer, fluctuates, they are constantly represented throughout the Southern Ocean. The species studied by BAKER include *Calanus propinquus, C. simillimus, Calanoides acutus, Rhincalanus gigas, Euchirella rostromagna, Euchaeta antarctica, Metridia gerlachei, Pleuromamma robusta, Heterorhabdus austrinus, Haloptilus ocellatus* and *H. oxycephalus*. Some peculiarities of the Antarctic and sub-Antarctic surface, bathypelagic and abyssal copepods will now be discussed below.

A. Surface plankton (epiplankton)

Calanus simillimus GIESBRECHT, *Clausocalanus laticeps* FARRAN and *Eucalanus longiceps* MATTHEWS can be taken as representing characteristic sub-Antarctic surface species. Of these three *Calanus simillimus* undoubtedly plays the most important role, a role which can to some extent be compared with that of the northern *Calanus helgolandicus* CLAUS and *C. finmarchicus* (GUNNERUS). Very little, unfortunately, is known of its life history but we know that it is capable of rapid vertical migrations. The occurrence both north and south of the Antarctic Convergence can be explained in two ways. HARDY & GUNTHER (1935) think *C. simillimus* to be an Antarctic form which is not so "purely" Antarctic as *C. propinquus*, with whom it is at times found together in large concentrations. An Antarctic form, therefore, which finds its optimal living conditions near the northern boundaries of the Antarctic region and which is able to survive in the sub-Antarctic. Personally I believe the species to live and breed in the sub-Antarctic, where large quantities, including mature females and males as well as develop-

385

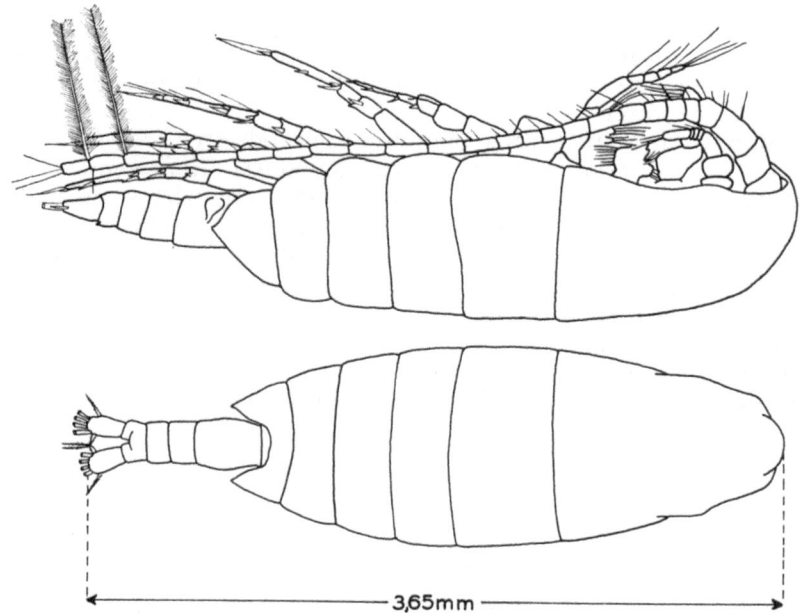

Fig. 2. *Calanus simillimus* GIESBRECHT, adult female, lateral and dorsal. After
VERVOORT, 1951.

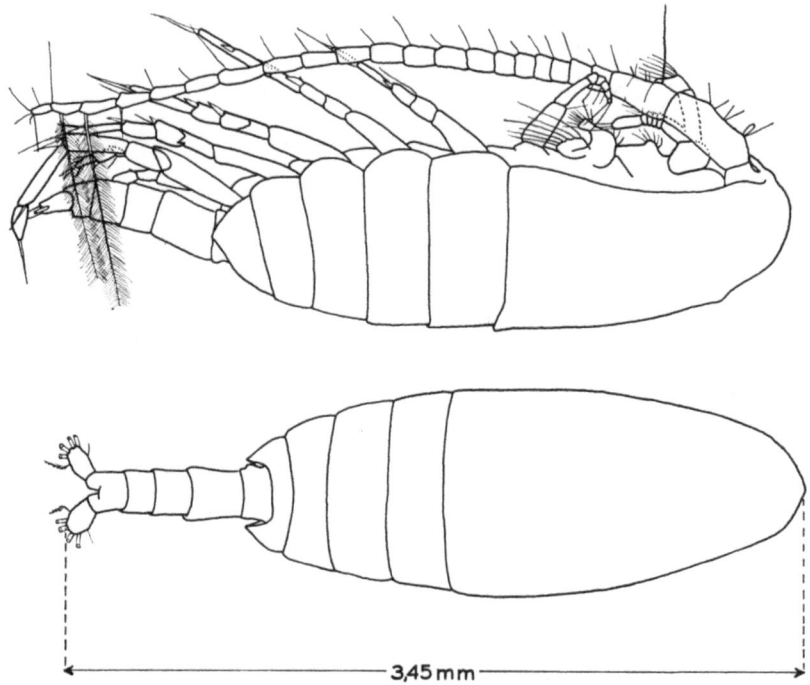

Fig. 3. *Calanus simillimus* GIESBRECHT, adult male, lateral and dorsal. After
VERVOORT, 1951.

mental stages, can be sampled in the surface 100 m layer. Its occurrence in some quantities in areas south of the Antarctic convergence
might therefore be indicative of the influx or mixing of sub-Antarctic surface water.

Purely Antarctic surface plankton Copepods are *Calanus propinquus* BRADY, *Calanoides acutus* (GIESBRECHT), *Rhincalanus gigas*
BRADY and *Metridia gerlachei* GIESBRECHT along with such smaller
forms as *Scolecithricella glacialis* (GIESBRECHT) and *Oithona frigida*
GIESBRECHT. We know next to nothing of *Scolecithricella glacialis*
and *Oithona frigida*, a little of *Calanus propinquus* and some aspects
of the life history of the remaining species.

Calanus propinquus is a surface copepod with shows a very marked
diurnal migration: the species is almost absent from the surface
during the day (it then dwells in deeper water), to move to the

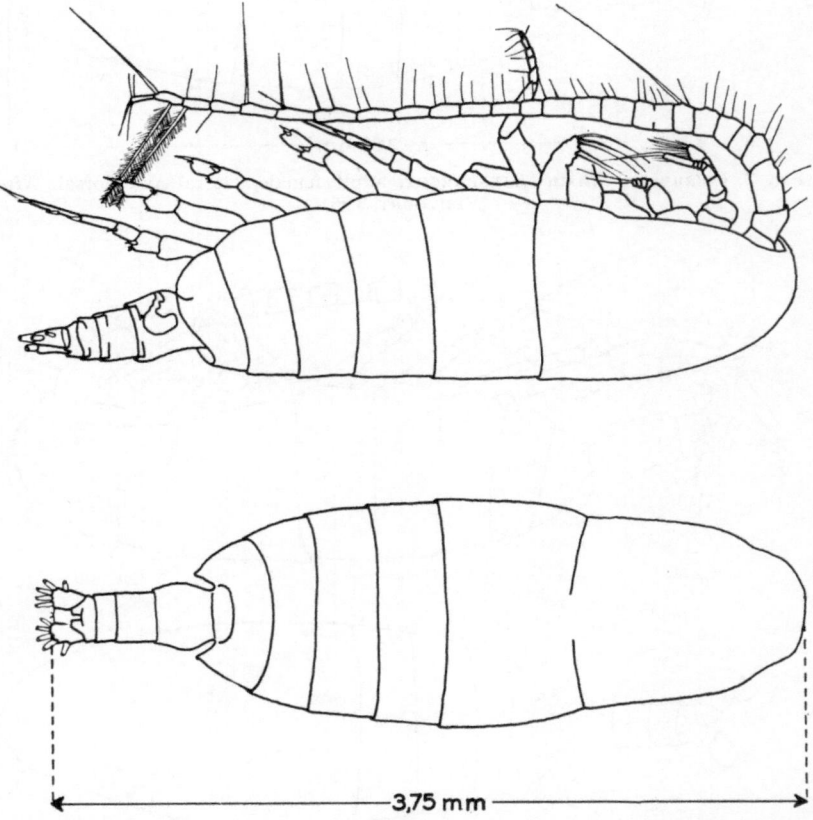

Fig. 4. *Calanus tonsus* BRADY, adult female, lateral and dorsal. After VERVOORT,
1957.

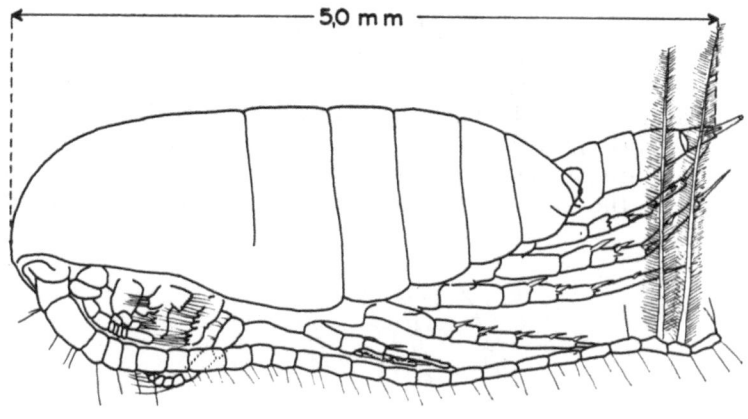

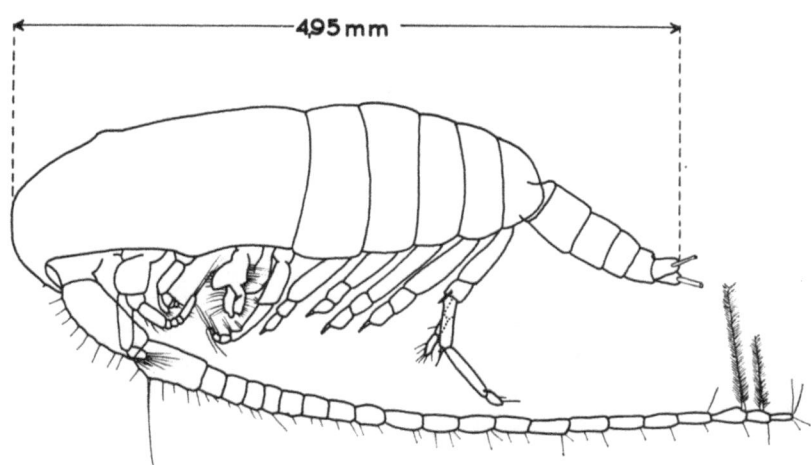

Fig. 5. *Calanus propinquus* BRADY, adult male (lower) and adult female (upper), both in lateral view. After VERVOORT, 1951.

surface at the onset of darkness. This diurnal migration does not prevent a gradual transportation of the stock in northerly direction; there is, moreover, a distinct increase in the number of copepodite stages in localities near the convergence. This seems to suggest reproduction in the Antarctic surface layer; a number of the copepodites of this species is being carried farther north in the Antarctic intermediate current as the species is occasionally met with far to the north of the convergence. How the stock is replenished we do not know for certain but it seems likely that eggs or larvae are carried south in the warm deep current.

Some features of the biology of *Calanoides acutus* (GIESBRECHT) and *Rhincalanus gigas* BRADY may as well be discussed together. Both species are of enormous importance as they form a large percentage of the zooplankton of the Southern Ocean. Though

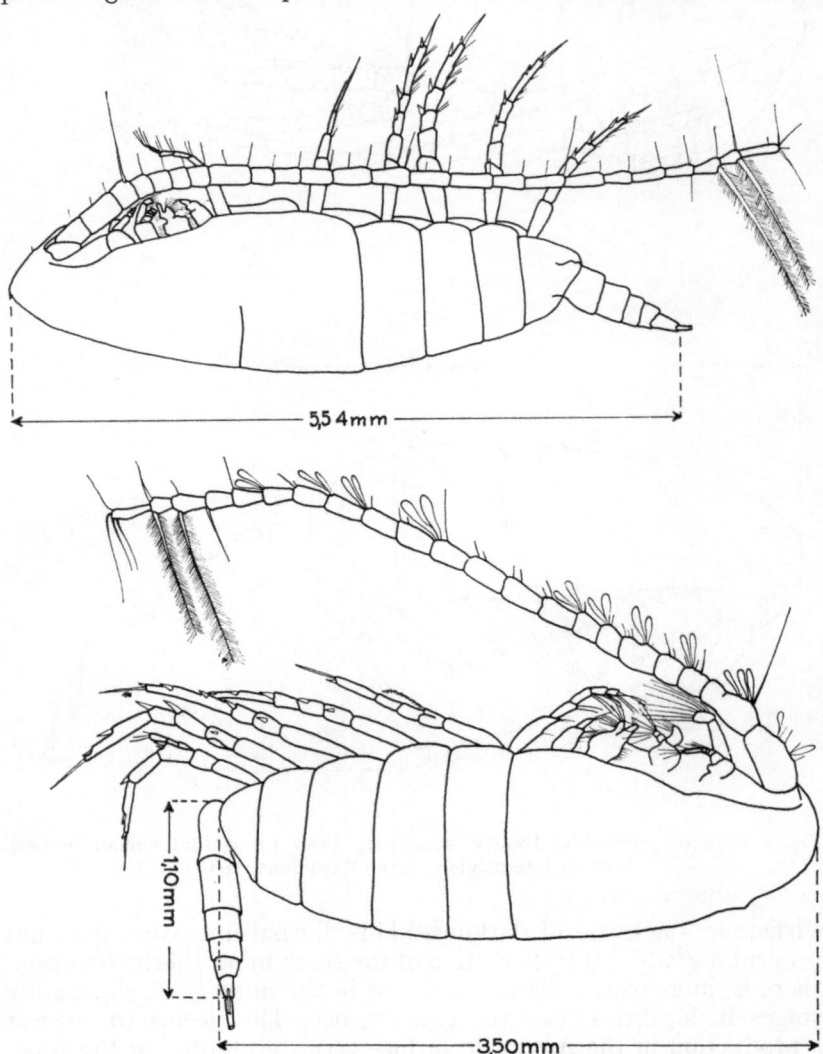

Fig. 6. *Calanoides acutus* (GIESBRECHT), adult female (upper) and adult male (lower), both in lateral view. After VERVOORT, 1951.

their distribution is decidedly circumpolar and they may be captured in large numbers at almost any locality south of the Antarctic Convergence, a certain fluctuation of their number in horizontal hauls suggests patchiness in their distribution over a smaller area, whilst certain bodies of water may be fairly poor in one or in both species. The Weddell Sea, for instance, seems to be characterized by a remarkable scarcity of *Rhincalanus gigas* (OMMANNEY, 1936). Seasonal vertical migrations of both species, coupled with the northerly drift in the surface water and the southerly movement in the deeper water, have been irrefutably proved by MACKINTOSH (1937) and form the keystones to the biology of both species.

The life history of *Rhincalanus gigas* has been investigated by OTTESTAD (1932, 1936), HARDY & GUNTHER (1935), MACKINTOSH (1934) and OMMANNEY (1936). Concerning *Calanoides acutus* we are less well informed, but both species apparently have much in common. Both are subjected to a very distinct seasonal vertical migration which, during the months of the southern summer, brings the species in the Antarctic surface water, in which they are carried northward. The Antarctic Convergence marks the northern limit of distribution of both species. In *Rhincalanus gigas* at least this northern distributional limit is fairly distinct in the southern winter but much less clear in the southern summer, when specimens of *Rhincalanus gigas* are carried far to the north and they mix with the closely related *Rhincalanus nasutus* GIESBRECHT. *Calanoides acutus* does not appear to survive the conditions north of the Antarctic Convergence and though stray specimens do at times occur in sub-Antarctic midwater hauls, their state of preservation suggests that they were dead when captured.

In the Antarctic surface waters both adults and developmental stages of *Rhincalanus gigas* and *Calanoides acutus* do occur, but the developmental (copepodite) stages predominate and the sex ratio of the adults suggests that most of the males have died. In fact, while adult males of *Rhincalanus gigas* can be found in surface hauls, especially from localities in high latitudes, those of *Calanoides acutus* appear to be extremely rare (only four adult males have so far been found, one by VERVOORT (1951), four by ANDREWS in material from the Discovery Investigations (ANDREWS, in litt.)). Having regard to the large quantities of sexually mature females in the surface hauls the scarcity of males in my opinion can only point to a very short mating season and short-lividity of the adult male phase. The last fact is also supported by the great reduction of the oral appendages, particularly the cutting edge of the mandibular praecoxa, in the sexually mature male.

As a working hypothesis it can be assumed that at or near the Antarctic Convergence, the copepodites of *Calanoides acutus* with or without surviving adult females, descent in the deep, warm current of the Antarctic at the onset of the southern winter. During this winter *Calanoides acutus* is carried southward in relatively warm water. In the female the final moulting, which brings them from the Vth copepodite into the adult stage, and the ripening of the ovaries may set in long before the southern journey is finished. There are reasons to suggest that in the male, still in the Vth copepodite stage, the development of the gonads also proceeds during the journey south, but the spermatophore only becomes available for fecundation after the final ecdysis. This final moulting, followed by copulation and spawning, takes place at the beginning of the southern summer along the edge of the sea ice which fringes the Antarctic continent, but it is unknown whether copulation takes place in the warm intermediate water (which would involve only the vertical migration of the fecundated females and account for the scarcity of males in the surface layer) or in surface waters after upward movement of both sexes. Anyhow, the eggs are pelagic and likely to drift northward in the Antarctic surface current, along with laying and spent females. It must be born in mind, however, that only part of the picture drawn up above is supported by anatomical facts.

In *Rhincalanus gigas* the picture in so far is more complicated, that there are, at least in the Falkland sector, two distinct spawning periods during the year. The onset and duration of these spawning periods are largely governed by the temperature of the water, but roughly the picture is as follows. There is a summer spawning period, usually starting in the middle of December, during which spawning takes place in the surface waters and which produces a summer generation. This summer generation, as early copepodites, descends into the deep, warm water, where it reaches sexual maturity late May or early June and produces the over-wintering generation, which spends the months of July, August and September in deep water, mostly as copepodites in the IVth and Vth stage, that reappear at the surface in October, where maturation, dependent upon the water temperature, proceeds and spawning sets in in the middle of December (OMMANNEY, 1936). The scarcity of *Rhincalanus gigas* in the Weddell Sea can probably be explained from the fact that no spawning takes place there and the stock of *Rhincalanus* in that area is dependent upon regular introduction of copepodites in bodies of water from the South Atlantic or the Southern Indian Ocean.

Only very few facts concerning the biology of *Metridia gerlachei* are known. The species has a distinct preference for the deeper water, particularly the adults, so that during the day it is almost completely absent in the top 50 m water layer; the largest number of individuals, both during the day and the night, occurs between 250 and 100 m depth. The species, however, is a powerful swimmer, so that the number of specimens captured at the surface during the night may be considerable as a result of rapid migration from deeper waters. The species is particularly abundant in the region of the West Wind drift and in the Weddell Sea. Spawning seems to take place at the end of the southern summer, but the number of generations is unknown. Evidence published by BEKLEMISHEV (1958) suggests that the bathymetrical and horizontal distribution of this species in high latitudes is largely governed by the —1 ° C isotherm.

B. *Plankton of the warm, deep current (bathypelagic species)*

DEACON (1937, p. 3) has made it clear that in all sectors of the Southern Ocean there is a southwardly directed, deep, warm current. The general direction of this current is horizontal until the region of the Antarctic Convergence is reached, where it climbs steeply over a northwardly directed current of Antarctic bottom water. The deep, warm current, consequently, in subtropical waters a stratum found at a considerable depth (some 1500—3500 m depth), in the Southern Ocean climbs till the subsurface (250—500 m depth) is reached; moreover the layer is less thick, the Antarctic bottom current may be found at 1500—2000 m depth. The steep slope of the deep, warm current determines the position of the Antarctic convergence and is itself dependent upon the northwardly directed bottom current. This bottom current and the influence which it exercises upon the slope of the warm, deep current are governed amongst other factors by the configuration and topography of the sea bottom and the distance from the source of origin of the Antarctic bottom current.

The deep, warm current is inhabited by a number of Copepod species that prefer intermediate depths. The separation from the surface zone is fairly distinct, though it has already been pointed out that several surface inhabitants *(Calanoides acutus, Rhincalanus gigas)* descent in the deep, warm water during the winter to be transported in a southerly direction. *Metridia gerlachei*, on the other hand, has a distinct preference for the deeper water without being a characteristic component of the deep, warm current: it seems to occur both in the deeper strata of the surface water and the upper strata of the deep, warm current.

The remaining Copepods that are regularly found in the deep warm current can be divided into two groups, of which the first represents characteristic inhabitants of the Southern Ocean (*Aetideopsis minor* (WOLFENDEN), *A. antarcticus* (WOLFENDEN), *Chiridius polaris* WOLFENDEN, *Euchirella rostromagna* WOLFENDEN, *E. latirostris* FARRAN, *Pseudochirella elongata* (WOLFENDEN), *Euchaeta antarctica* GIESBRECHT, *E. austrina* GIESBRECHT, *E. similis* WOLFENDEN, *E. erebi* FARRAN, *Onchocalanus magnus* WOLFENDEN, *O. wolfendeni* VERVOORT, *Cornucalanus robustus* VERVOORT, *Amallophora altera* FARRAN, *Racovitzanus antarcticus* GIESBRECHT, *R.*

erraticus VERVOORT, *Scolecithircella dentipes* VERVOORT, *S. polaris* WOLFENDEN, *Scaphocalanus subbrevicornis* (WOLFENDEN), *Heterorhabdus austrinus* GIESBRECHT, *H. farrani* BRADY, *Haloptilus ocellatus* WOLFENDEN, *Candacia falcifera* FARRAN, *Oncaea curvata* GIESBRECHT). The second group, which is much larger, includes the species that also occur in the deep water of Atlantic, Indian and Pacific Oceans and of which the area of distribution includes the Antarctic deep, warm current. It should be born in mind that this stratum approaches the surface much more closely than in latitudes north of the Convergence, so that the number obtained in mid-water oblique and vertical hauls in the Antarctic is usually much larger than elsewhere. The majority of inhabitants of the deep, warm layer, moreover, consists of powerful swimmers that are able to make extensive diurnal vertical movements, which at nighttime brings them in the cold, Antarctic surface water. This not only applies to the "real" Antarctic species such as *Aetideopsis minor*, *Euchirella rostromagna*, *Eucheata antarctica*, *Scaphocalanus subbrevicornis* and many others, but also to such species as *Gaidius tenuispinus* (G. O. SARS), *Ctenocalanus vanus* GIESBRECHT, *Scaphocalanus brevicornis* (G. O. SARS), *Heterostylites major* (F. DAHL), e.a., that have a very extended area of distribution and apparently tolerate the low temperatures at the surface quite well.

There is no distinct separation between the inhabitants of the deep, warm current and those of the Antarctic bottom layer, that is to say there is a large group of species that prefer either the deep, warm current or the Antarctic bottom layer and quite a number that seems to have no preference (*Undeucheata major* GIESBRECHT, *Scaphocalanus affinis* (G. O. SARS), *S. echinatus* (FARRAN), *Metridia princeps* GIESBRECHT, *Metridia curticauda* GIESBRECHT, *Pleuromamma robusta* (F. DAHL), *Lucicutia frigida* WOLFENDEN, e.a.). It seems likely that a number at least of these carries out vertical migrations that brings them from one stratum into the other.

No details concerning the life history or biology of the inhabitants of the warm, deep current are known but it does not appear unlikely that a mechanism, comparable with that of the inhabitants of the surface zone and dependent upon vertical migrations and water currents, ensures the renewal of the stock.

C. *Plankton of the Antarctic bottom water (abyssal species)*

The Copepods inhabiting this zone are all abyssal species that have been found in the deep water of all great oceans; there is apparently no characteristic Antarctic Copepod fauna of the bottom water layer as far as pelagic species are concerned. The most constant and characteristic feature of the abyssal Copepods is their scarcity in numbers in vertical or oblique hauls, which only to some extent can be explained by their ability to avoid the net. Their density of distribution, apparently, is much less than that of species in intermediate water layers. The males, usually, are only occasionally met with and in many species one sex only, the female, is known. The very distinct prevalence of females over males in the bathypelagic and abyssal plankton has recently received the attention of MEDNIKOV (1961). MEDNIKOV has distinctly demonstrated that this prevalence of females is certainly due to a shift in the sex ratio. The biological value of the prevalence of the females in deep water, according to this author, is the increased fecundity of the species under consideration, because of the very limited supply of food. The extremely low density of the males in deep water plankton presents another important biological problem: that of the chance of encounters of both sexes of a certain species. Clearly there is no parthenogenesis in deep water Calanoids and there must be biological mechanisms (luminescence, chemical perception, e.g.) that safeguard the reproduction of the species.

Littoral Antarctic Copepods

Though the number of characteristic Antarctic littoral Copepods will undoubtedly prove to be larger, only one species, *Drepanopus pectinatus* BRADY, has to some extent been investigated. This species occurs in enormous numbers in the coastal waters around South Georgia (HARDY & GUNTHER, 1935) and around Kerguelen (VERVOORT, 1957). I have also seen samples from the waters around Heard Island that are almost exclusively composed of this species. The species occurs plentifully in the waters above the continental shelf surrounding these islands and in the many inlets of the coast; the number decreases very rapidly in localities beyond the edge of this continental shelf. The species shows marked diurnal migration and is most numerous in the top 150 m stratum; it is rarely observed deeper than 250 m. In its occurrence, however, it is distinctly patchy, whilst no information concerning its seasonal distribution and biology is available.

Table I.

List of the Antarctic and sub-Antarctic pelagic Copepoda[1]

Species	Type	Neritic	sub-Antarctic	Antarctic	Abyssal
Calanoida					
Calanus australis BRODSKY	S				
Calanus propinquus BRADY	S—B*		FARRAN, 1929 VERVOORT, 1957 BRODSKY, 1959	VERVOORT, 1957	
Calanus tonsus BRADY	S*		VERVOORT, 1957	VERVOORT, 1957	
Calanus similimus GIESBRECHT	S*		VERVOORT, 1957		
Calanoides acutus GIESBRECHT	V*			VERVOORT, 1957	
Megacalanus princeps WOLFENDEN	A				VERVOORT, 1957
Bathycalanus bradyi (WOLFENDEN)	A				VERVOORT, 1957
Eucalanus elongatus (DANA)	A		FARRAN, 1929		
Eucalanus longiceps MATTHEWS	S—B		VERVOORT, 1957	HARDY & GUNTHER, 1935	
Rhincalanus gigas BRADY	S—B			VERVOORT, 1957	
Rhincalanus nasutus GIESBRECHT	V			VERVOORT, 1957	
Microcalanus pygmaeus (G. O. SARS)	S—B		VERVOORT, 1957	VERVOORT, 1957	
Gaidius affinis G. O. SARS	B				

[1] This list does not claim completeness. The references are to papers giving more complete information. No references to BRADY's (1910) paper have been included here. There are many unrecognizable species amongst the many new forms mentioned by that author, whilst the references to known species must be very carefully considered. The species listed by HARDY & GUNTHER (1936) have been identified by A. SCOTT; some of these at least need confirmation. The abbreviations used under the heading "type" are: S = surface species; B = bathypelagic species; A = abyssal species; V = species that shows seasonal vertical migrations. Species characteristic of the Antarctic are marked *.

Species		(a)	(b)	(c)	(d)
Gaidius intermedius WOLFENDEN	A				VERVOORT, 1957
Gaidius tenuispinus (G. O. SARS)	B—A				VERVOORT, 1957
Gaetanus antarcticus WOLFENDEN	B—A			VERVOORT, 1957	VERVOORT, 1957
Gaetanus latifrons G. O. SARS WOLFENDEN	A		VERVOORT, 1957	VERVOORT, 1957	VERVOORT, 1957
Gaetanus minor FARRAN	B		VERVOORT, 1957		
Euchirella lativostris FARRAN	B*		VERVOORT, 1957		
Euchirella rostrata (CLAUS)	B				
Euchirella rostromagna WOLFENDEN	B*				
Pseudochirella elongata (WOLFENDEN)	B*			VERVOORT, 1957	VERVOORT, 1957
Pseudochirella hirsuta (WOLFENDEN)	A			VERVOORT, 1957	VERVOORT, 1957
Pseudochirella mawsoni VERVOORT	B—A		VERVOORT, 1957		VERVOORT, 1957
Pseudochirella notacantha (G. O. SARS)	A			VERVOORT, 1957	VERVOORT, 1957
Pseudochirella pustulifera (G. O. SARS)	A				VERVOORT, 1957; HARDY & GUNTHER, 1935
Pseudeuchaeta brevicauda (G. O. SARS)	A				
Undeuchaeta major GIESBRECHT	B—A		VERVOORT, 1957		VERVOORT, 1957
Ctenocalanus vanus GIESBRECHT	B				VERVOORT, 1957
Clausocalanus arcuicornis (DANA)	S			VERVOORT, 1957	
Clausocalanus laticeps FARRAN	S*		FARRAN, 1929		
Farrania frigida (WOLFENDEN)	A		VERVOORT, 1957		
Drenanopus pectinatus BRADY	S	VERVOORT, 1957			VERVOORT, 1957
Spinocalanus abyssalis GIESBRECHT	A				VERVOORT, 1957
Spinocalanus magnus WOLFENDEN	A				VERVOORT, 1957
Spinocalanus spinosus FARRAN	A				FARRAN, 1929

Table I. (continued)

Species	Type	Neritic	sub-Antarctic	Antarctic	Abyssal
Mimocalanus cultrifer FARRAN	A			TANAKA, 1960	VERVOORT, 1957
Stephus longipes GIESBRECHT	S		VERVOORT, 1957		
Aetideus armatus (BOECK)	B		VERVOORT, 1957		
Euaetideus australis VERVOORT	B				
Aetideopsis antarcticus (WOLFENDEN)	B*			FARRAN, 1929	
Aetideopsis minor (WOLFENDEN)	B*			VERVOORT, 1957	
Chiridius polaris WOLFENDEN	B*			VERVOORT, 1957	
Undeuchaeta minor GIESBRECHT	S—B		FARRAN, 1929	VERVOORT, 1957	
Euchaeta antarctica GIESBRECHT	B*			VERVOORT, 1957	
Euchaeta austrina GIESBRECHT	B*			VERVOORT, 1957	
Euchaeta biloba (FARRAN)	S—B		VERVOORT, 1957	VERVOORT, 1957	
Euchaeta erebi (FARRAN)	B*			FARRAN, 1929	
Euchaeta exigua WOLFENDEN	A				VERVOORT, 1957
Euchaeta farrani WITH	A				VERVOORT, 1957
Euchaeta rasa (FARRAN)	B—A			VERVOORT, 1957	VERVOORT, 1957
Euchaeta scotti FARRAN	B—A			HARDY & GUNTHER, 1935	HARDY & GUNTHER, 1935
Euchaeta similis WOLFENDEN	B*			VERVOORT, 1957	VERVOORT, 1957
Valdiviella insignis FARRAN	A				
Onchocalanus magnus (WOLFENDEN)	B*			VERVOORT, 1957	
Onchocalanus wolfendeni VERVOORT	B*			VERVOORT, 1957	
Cornucalanus robustus VERVOORT	B—A			VERVOORT, 1957	VERVOORT, 1957
Cephalophanes frigidus	A				VERVOORT, 1957
Amallophora altera	B*			VERVOORT, 1957	
Undinella brevipes FARRAN	A				VERVOORT, 1957

Species	Code	Reference 1	Reference 2
Racovitzanus antarcticus GIESBRECHT	B*		VERVOORT, 1957
Racovitzanus erraticus VERVOORT	B*		VERVOORT, 1957
Scolecithricella glacialis (GIESBRECHT)	S*		VERVOORT, 1957
Scolecithricella dentipes VERVOORT	B*		VERVOORT, 1957
Scolecithricella emarginata (FARRAN)	B		HARDY & GUNTHER, 1936
Scolecithricella incisa (FARRAN)	S—B*		FARRAN, 1929
Scolecithricella minor (BRADY)	B		HARDY & GUNTHER, 1936
Scolecithricella ovata (FARRAN)	B	VERVOORT, 1957	
Scolecithricella polaris WOLFENDEN	B*		VERVOORT, 1957
Scolecithricella robusta (T. SCOTT)	B		VERVOORT, 1957
Scolecithricella valida (FARRAN)	A		VERVOORT, 1957
Scaphocalanus affinis (G. O. SARS)	B—A		VERVOORT, 1957
Scaphocalanus brevicornis (G. O. SARS)	B		HARDY & GUNTHER, 1935
Scaphocalanus echinatus (FARRAN)	B—A	FARRAN, 1929	
Scaphocalanus magnus (T. SCOTT)	A		
Scaphocalanus subbrevicornis (WOLFENDEN)	B*		VERVOORT, 1957
Temorites brevis G. O. SARS	A		
Metridia curticauda GIESBRECHT	B—A		VERVOORT, 1957
Metridia gerlachei GIESBRECHT	S—B*		VERVOORT, 1957
Metridia lucens BOECK	S—B	VERVOORT, 1957	VERVOORT, 1957

Table I. (continued)

Species	Type	Neritic	sub-Antarctic	Antarctic	Abyssal
Metridia princeps GIESBRECHT	B—A				VERVOORT, 1957
Pleuromamma borealis (F. DAHL)	B		FARRAN, 1929		
Pleuromamma gracilis (CLAUS)	B		FARRAN, 1929		
Pleuromamma robusta F. DAHL) f. *antarctica* STEUER	B—A		VERVOORT, 1957		
Lucicutia curta FARRAN	A			VERVOORT, 1957	VERVOORT, 1957
Lucicutia frigida WOLFENDEN	B—A			VERVOORT, 1957	VERVOORT, 1957
Lucicutia grandis GIESBRECHT	A				VERVOORT, 1957
Lucicutia macrocera G. O. SARS	A				VERVOORT, 1957
Lucicutia magna WOLFENDEN	A				FARRAN, 1929
Lucicutia maxima STEUER	B			HARDY & GUNTHER, 1935	
Lucicutia wolfendeni SEWELL	A				VERVOORT, 1957
Disseta palumboi GIESBRECHT	A				HARDY & GUNTHER, 1935
Heterorhabdus austrinus GIESBRECHT	B*			VERVOORT, 1957	
Heterorhabdus compactus G. O. SARS	A			HARDY & GUNTHER, 1935	FARRAN, 1929; HARDY & GUNTHER, 1935
Heterorhabdus farrani BRADY	B*			VERVOORT, 1957	
Heterorhabdus pustulifer FARRAN	B*			VERVOORT, 1957	
Heterostylites major (F. DAHL)	A			HARDY & GUNTHER, 1935	VERVOORT, 1957
Haloptilus fons FARRAN	B				
Haloptilus ocellatus WOLFENDEN	B*			VERVOORT, 1957	
Haloptilus oxycephalus (GIESBRECHT)	B			VERVOORT, 1957	
Augaptilus glacialis G. O. SARS	B			VERVOORT, 1957	

Species				
Augaptilus megalurus Giesbrecht	B		Hardy & Gunther, 1935	
Euaugaptilus laticeps (G. O. Sars)	A			Vervoort, 1957
Euaugaptilus magnus (Wolfenden)	A			Vervoort, 1957
Centraugaptilus rattrayi T. Scott	B		Hardy & Gunther, 1935	
Pseudaugaptilus longiremis G. O. Sars	A			Vervoort, 1957
Pontoptilus ovalis G. O. Sars	A			Vervoort, 1957
Pachyptilus eurygnathus G. O. Sars	A			Vervoort, 1957
Arietellus simplex G. O. Sars	A			Vervoort, 1957
Phyllopus bidentatus Brady	B	Vervoort, 1957	Vervoort, 1957	
Candacia cheirura Cleve	S—B	Vervoort, 1957	Vervoort, 1957	
Candacia falcifera Farran	B		Vervoort, 1957	
Candacia maxima Vervoort	B	Vervoort, 1957		
Paralabidocera antarctica (I. C. Thompson)	S		Vervoort, 1957	
Cyclopoida				
Pseudocyclopina belgicae (Giesbrecht)	S		Giesbrecht, 1902	
Mormonilla phasma Giesbrecht	B—A		Hardy & Gunther, 1935	Hardy & Gunther, 1935
Oithona frigida Giesbrecht	S	Vervoort, 1957	Vervoort, 1957	
Oithona similis Claus	S—B	Vervoort, 1957	Vervoort, 1957	
Ratania atlantica Farran	S—B		Vervoort, 1957	
Oncaea conifera Giesbrecht	S—B		Vervoort, 1957	
Oncaea curvata Giesbrecht	B		Vervoort, 1957	
Oncaea mediterranea Giesbrecht	S—B	Vervoort, 1957		
Oncaea notopus Giesbrecht	B		Vervoort, 1957	
Oncaea venusta Philippi	B		Tanaka, 1960	
Conea rapax Giesbrecht	B—A		Hardy & Gunther, 1935	
Lubbockia aculeata Giesbrecht	B—A		Vervoort, 1957	

400

REFERENCES *

BAKER, A. DE C., 1954. The circumpolar continuity of Antarctic plankton species. *Discovery Repts.*, **27**, *201—218*, figs. 1—5, tabs. 1—3.

BEKLEMISHEV, K.V., 1958. (The biogeographical characteristic of some Antarctic zooplankton species). *Doklady Akad. Nauk SSSR*, **120**, 3, *507—509*, figs. 1—3.

BRODSKY, K. A., 1958. (Plankton Research of the Soviet Antarctic Expedition, 1955—1958). *Bjull. Sov. Antarkt. Eksped.*, no. **3**, *25—30*.

BRODSKY, K. A., 1959. (On phylogenetic relations of some Calanus (Copepoda) species of Northern and Southern Hemispheres). *Zool. Zh.*, **33**, *1537—1553*, figs. 1—5.

DEACON, G. E. R., 1933. A general account of the hydrology of the South Atlantic Ocean. *Discovery Repts.*, **7**, *171—238*.

DEACON, G. E. R., 1937. Hydrology of the Southern Ocean. *Discovery Repts.*, **15**, *1—124*, figs. 1—22, pls. 1—44.

FARRAN, G. P., 1929. Copepoda. British Antarctic ("Terra Nova") Expedition, 1910. Nat. Hist. Report, Zool., **8**, 3, *203—306*, pls. 1—4, figs. 1—37.

FOXTON, P., 1956. The distribution of the standing crop of Zooplankton in the Southern Ocean. *Discovery Repts.*, **28**, *191—236*, figs. 1—19, tabs. 1—11.

GIESBRECHT, W., 1902. Copepoden. In: Résult. Voy. Belgica, Rapp. Sci., Zool., *1—49*, pls. 1—13.

HARDY, A. C. & E. R. GUNTHER, 1935. The Plankton of the South Georgia Whaling grounds and adjacent waters, 1926—1927. *Discovery Repts.*, **11**, *1—456*, figs. 1—193.

MACKINTOSH, N. A., 1934. Distribution of the Macroplankton in the Atlantic sector of the Antarctic. *Discovery Repts.*, **9**, *65—160*, figs. 1—48.

MACKINTOSH, N. A., 1937. The Seasonal Circulation of the Antarctic macroplankton. *Discovery Repts.*, **16**, *365—412*, figs. 1—9, tabs. 1—21.

MEDNIKOV, B. M., 1961. On the Sex Ratio in Deep water Calanoida. *Crustaceana*, **3**, *105—109*, figs. 1—3.

OMMANNEY, F. D., 1936. Rhincalanus gigas (Brady), a Copepod of the Southern Macroplankton. *Discovery Repts.*, **13**, *277—384*, figs. 1—29.

OTTESTAD, P., 1932. On the Biology of some Southern Copepoda. *Hvalrådets Skr.*, no. **5**, *1—61*.

OTTESTAD, P., 1936. On Antarctic Copepods from the "Norvegia" Expedition 1930—1931. Sci. Res. Norwegian Antarctic Exped. of Consul Lars Christensen, no. 15, *1—44*, figs. 1—11.

TANAKA, O., 1960. Pelagic Copepoda. In: Biological Results of the Japanese Antarctic Research Expedition, no. 10. *Spec. Publ. Seto Mar. Biol. Lab.*, *1—95*, pls. 1—40.

VERVOORT, W., 1951. Plankton Copepods from the Atlantic sector of the Antarctic. *Verh. Kon. Ned. Akad. Wet.* Afd. Natuurkunde, sectie 2, **47**, 4, *1—156*, figs. 1—82.

VERVOORT, W., 1957. Copepods from Antarctic and sub-Antarctic plankton samples. Rep. B.A.N.Z. A. R. Exped., 1929—1931, (B), **3**, *1—160*, figs. 1—138.

---

* Only the principal references are cited here. A fairly complete bibliography of Antarctic (Calanoid) Copepoda can be found in VERVOORT, 1957.

BRYOZOA OF THE ANTARCTIC

BY

MARY D. ROGICK

College of New Rochelle, New Rochelle, New York

Extensive collections of Bryozoa have been made in the Antarctic but some still remain to be written up. Table I lists the bryozoologists and the expeditions which collected their Bryozoa. A key letter precedes each bryozoologist's name, as *F* for BORG, B for BUSK, *H* for HASTINGS, etc. The key letters will also be used in Tables II and III.

Table I.

Bryozoologists and the sources of their collections.

F — BORG 1944. — Nordenskjöld's Swedish Antarctic Expedition of 1901—1903.

B — BUSK 1881, 1884, 1886. — "Challenger" Expedition of 1873—1876.

C — CALVET 1904a, 1904b, 1904c. — Hamburger Magalhaensische Sammelreise.

CC — CALVET 1909. — Charcot's Expédition Antarctique Française 1903—1905.

N — HASENBANK 1932. — "Valdivia" Deutsche Tiefsee Expedition 1898—1899.

H — HASTINGS 1943. — "Discovery" Reports for 1925—1933. "William Sowerby" Expeditions of 1926—1932. "Terra Nova" British Antarctic Expedition of 1910, "Discovery" National Antarctic Expedition of 1901—1904, Shackleton-Rowett "Quest" Expedition of 1922.

J — JOHNSTON & ANGEL 1940. — British, Australian and New Zealand Antarctic Expedition of 1929—1931.

K — KIRKPATRICK 1902. — "Southern Cross" Voyage.

G — KLUGE 1914. — Drygalski's Deutsche Südpolar-Expedition 1901—1903.

L — LIVINGSTONE 1928. — MAWSON's Australasian Antarctic Expedition of 1911—1914.

R — ROGICK 1955—1962. — United States Navy's 1947—1948 Antarctic Expedition.

T — THORNELY 1924. — Mawson's Australasian Antarctic Expedition of 1911—1914.

V — VIGELAND 1952. — Norwegian Antarctic Expedition of 1927—1928.

W — WATERS 1888. — "Challenger" Expedition of 1873—1876.

WW — WATERS 1904. — "Belgica" Voyage of 1897—1899.

The "Antarctic" waters in which bryozoans have been collected include not only those immediately around the Antarctic continent but also the waters about the islands of Bouvet, Clarence, Elephant, Grass, Heard, King George, Montague, Penguin, South Georgia, South Orkney, South Sandwich, South Shetland, Palmer Peninsula and Shag Rocks.

The most complete account of the correlation between bryozoan species distribution and the hydrological conditions of the Antarctic and sub-Antarctic is that of HASTINGS 1943 (pp. 477—501). The latitudinal range of Antarctic waters is very extensive in some areas. She notes (p. 491) "the presence of a fauna with Antarctic affinities in a zone off the Patagonian shelf," and also that "the fauna of Heard Island is similar to that of the other islands in the South Indian Ocean despite its different hydrological conditions."

The total number of bryozoan species reported from the Antarctic waters, as defined above, is about 321. Of these about 179 species are known exclusively from the Antarctic, up to the present time. The remaining 142 species have been reported both from the Antarctic and from warmer, more northerly waters about Australia, New Zealand, South Africa, South America and elsewhere. How many of these 321 species are synonyms, valid species or possibly involve more than a single species, is not certain. Bryozoan species determinations are sometimes very unreliable because they may be based on characters that vary too greatly, or because there may be a change of emphasis on what constitutes a proper diagnostic feature. For example, in the early days, the colony's growth habit was the important criterion for identifying a species, while now the emphasis is on the nature of the individual zooids of a colony. Moreover, some characteristics, such as size of structures, tentacle number, spine number or nature, presence or absence of structures and the degree of calcification of the zooids may be very stable in some species but are extremely variable in other closely related forms. The extent to which ecology influences the variability has not been fully investigated in this group. Therefore, geographical distribution records of old bryozoan species are unreliable simply because in some cases it is impossible to determine to what species a given name actually applied.

It is highly debatable whether any of the Antarctic bryozoan species also occur in the Arctic, or vice versa. WATERS 1904, pp. 5—6, does not believe that the Bryozoa offer any support for the bipolar theory.

However, there are some very cosmopolitan bryozoans which range from high northern latitudes through intervening regions and down to the sub-Antarctic, or Antarctic in a few instances. WATERS 1904, p. 9, cites *Micropora coreacea* ESPER, *Hippothoa divaricata* LAMOUROUX, *Hippothoa distans* MACGILLIVRAY, *Smittina reticulata* MACGILLIVRAY, *Idmonea atlantica* FORBES and *Entalophora proboscidea* MILNE-EDWARDS as examples of such widely distributed species.

The 179 species to date known only from the Antarctic are classified thus: 1 Entoproct, 2 Ctenostomes, 18 Cyclostomes or Stenolaemates, 89 Anascan Cheilostomes and 69 Ascophoran Cheilostomes. The following account will deal with these 179 exclusively Antarctic species rather than with those whose geographic range extends beyond the Antarctic.

Table II lists these 179 species by number. The same number will be used to identify the species in Table III. Table II also gives the depth range for most species and the bibliographic source (see Key Letters of Table I) for each record.

Table II.

The 179 exclusively Antarctic Bryozoa and their depth range.

Species number and name	Depth range in meters, and reporting author	
	Least depth	Greatest depth
Entoprocta		
1. *Barentsia capitata* CALVET 1904b		
Ctenostomata		
2. *Alcyonidium antarcticum* WATERS 1904		459 WW
3. *Alcyonidium flabelliforme* KIRKPATRICK 1902		
Cyclostomata or Stenolaemata or Stenostomata		
4. *Defrancia sarsi* BORG 1944	75 F	195 F
5. *Diastopora gracilis* BORG 1944		104 F
6. *Diastopora* sp. WATERS 1904		480 WW
7. *Diastopora solida* WATERS 1904		435 WW
8. *Domopora antarctica* BORG 1944	100 F	150 F
9. *Filicrisia* sp. BORG 1944		95 F
10. *Hastingsia irregularis* BORG 1944	104 F	920 F
11. *Hastingsia pygmaea* BORG 1944	252 F	310 F
12. *Hornera smitti* BORG 1944	125 F	920 F
13. *Idmidronea curvata* BORG 1944	104 F	150 F
14. *Idmidronea hula* BORG 1944		104 F
15. *Idmidronea obtecta* BORG 1944	95 F	920 F
16. *Lichenopora tubicen* BORG 1944		95 F
17. *Stomatopora antarctica* WATERS 1904	435 WW	569 WW
18. *Stomatopora divergens* WATERS 1904	480 WW	500? WW
19. *Stomatopora major* JOHNSTON var., WATERS 1904		480 WW
20. *Tubulipora gracillima* BORG 1944	95 F	104 F
21. *Tubulipora spatiosa* BORG 1944	100 F	150 F
Cheilostomata Anasca		
22. *Amastigia abyssicola* (KLUGE) 1914	2450 G	2800 WW
23. *Amastigia antarctica* (KLUGE) 1914	329 H	480 WW
24. *Amastigia cabereoides* (KLUGE) 1914	329 H	385 G
25. *Amastigia solida* (KLUGE) 1914	70 G	385 G
26. *Aspidostoma coronata* (THORNELY) 1924		220 T
27. *Beania erecta* WATERS 1904	20 CC	480 WW
28. *Beania erecta* var. *livingstonei* HASTINGS 1943	20 V	342 H
29. *Beania scotti* HASTINGS 1943	329 H	366 H
30. *Brettia triplex* HASTINGS 1943	329 H	366 H
31. *Callopora onychocelloides* (CALVET) 1909	6.4 L	40 CC
32. *Camptoplites angustus* (KLUGE) 1914	75 V	385 G
33. *Camptoplites areolatus* (KLUGE) 1914	25 CC	500? WW

Table II. (continued)

Species number and name	Least depth	Greatest depth
34. *Camptoplites bicornis* var. *compacta* (KLUGE) 1914	259 H	594 H
35. *Camptoplites bicornis* var. *elatior* (KLUGE) 1914	70 G	610 H
36. *Camptoplites bicornis* var. *magna* (KLUGE) 1914	24 H	1080 H
37. *Camptoplites giganteus* (KLUGE) 1914	130 H	385 G
38. *Camptoplites latus* (KLUGE) 1914	385 G	567 H
39. *Camptoplites latus* var. *aspera* HASTINGS 1943	122 H	270 H
40. *Camptoplites latus* var. *striata* HASTINGS 1943	203 H	315 H
41. *Camptoplites lewaldi* (KLUGE) 1914	351 H	567 H
42. *Camptoplites multispinosus* (KLUGE) 1914		385 G
43. *Camptoplites rectilinearis* HASTINGS 1943	256 H	441 H
44. ?*Camptoplites reticulatus* var. *spinosa* (WATERS) 1904	500 WW	569 WW
45. *Camptoplites retiformis* (KLUGE) 1914	60 H	610 H
46. *Camptoplites retiformis* var. *tenuispina* HASTINGS 1943	99 H	391 H
47. *Camptoplites tricornis* (WATERS) 1904	106 H	287 L
48. *Cellaria aurorae* LIVINGSTONE 1928	201 L	287 L
49. *Cellaria diversa* LIVINGSTONE 1928	82 L	421 L
50. *Cellaria lata* WATERS 1904		480 WW
51. *Cellaria mawsoni* LIVINGSTONE 1928	201 L	582 L
52. *Cellaria moniliorata* ROGICK 1956c	106 R	183 R
53. *Cellaria vitrimuralis* ROGICK 1956c.		106 R
54. *Cellaria wandeli* CALVET 1909	40 CC	220 L
55. *Cellariaeforma coronata* ROGICK 1956c		106 R
56. *Cellariaeforma extentamuralis* ROGICK 1956c		183 R
57. *Cellariaeforma parvimuralis* ROGICK 1956c		106 R
58. *Chaperia cylindracea* var. *protecta* WATERS 1904		480 WW
59. *Chaperia dichotoma* KLUGE 1914		385 G
60. *Chaperia gaussi* KLUGE 1914		370 G
61. *Chaperia lepralioides* KLUGE 1914	350 G	385 G
62. *Chaperia patulosa* WATERS 1904	350 G	500? WW
63. *Chaperia quadrispinosa* KLUGE 1914	380 G	385 G
64. *Chaperia simplicissima* KLUGE 1914	350 G	385 G
65. *Chaperia spinosissima* CALVET 1904b	30 C	40 C
66. *Cornucopina angulata* (KLUGE) 1914		2450 G
67. *Cornucopina dubitata* (CALVET) 1909		110 CC
68. *Cornucopina lata* (KLUGE) 1914	329 H	385 G
69. *Cornucopina ovalis* HASTINGS 1943		177 H
70. *Cornucopina polymorpha* (KLUGE) 1914	106 H	567 H
71. *Cribrilina projecta* WATERS 1904	201 T	480 WW
72. *Erymophora klugei* HASTINGS 1943	385 G	610 H
73. *Farciminellum antarcticum* HASTINGS 1943	77 H	567 H
74. *Farciminellum lineare* (KLUGE) 1914		2910 G
75. *Figularia spatulata* (CALVET) 1909	30 CC	201 L
76. *Flustra angusta* KLUGE 1914	120 V	457 N
77. *Flustra curva* KLUGE 1914	90 V	421 L

Table II. (continued)

Species number and name	Depth range in meters, and reporting author	
	Least depth	Greatest depth
78. *Flustra drygalski* KLUGE 1914	380 G	385 G
79. *Flustra flagellata* WATERS 1904	30 CC	500? WW
80. *Flustra renilla* (PFEFFER) 1889		
81. *Flustra tenuis* KLUGE 1914	40 G	400 G
82. *Flustra vanhöffeni* KLUGE 1914	80 V	385 G
83. *Flustra vulgaris* KLUGE 1914	350 G	567 N
84. *Himantozoum antarcticum* (CALVET) 1909	0 H	594 H
85. *Himantozoum sinuosum* var. *variabilis* (KLUGE) 1914		3397 G
86. *Klugella echinata* (KLUGE) 1914	329 H	457 H
87. *Labioporella adeliensis* LIVINGSTONE 1928	110 L	220 L
88. *Mawsonia extensalata* ROGICK 1956c	64 R	183 R
89. *Mawsonia membranacea* (THORNELY) 1924	183 R	642 L
90. *Megapora hyalina* WATERS 1904	436 WW	460 WW
91. *Melicerita latilaminata* ROGICK 1956c		106 R
92. *Melicerita obliqua* (THORNELY) 1924	183 R	582 L
93. *Membranipora constantia* KLUGE 1914	350 G	385 E
94. *Membranipora gigantea* KLUGE 1914	350 G	385 G
95. *Membranipora inconstantia* KLUGE 1914		385 G
96. *Membranipora perlucida* KLUGE 1914		350 G
97. *Membranipora strigosa* WATERS 1904		459 WW
98. *Membranipora uniserialis* WATERS 1904	435 WW	480 WW
99. *Membraniporella antarctica* KLUGE 1914	350 G	385 G
100. *Notoplites antarcticus* (WATERS) 1904	120 H	610 H
101. *Notoplites crassiscutus* HASTINGS 1943	120 H	204 H
102. *Notoplites drygalskii* (KLUGE) 1914	5.5 L	1080 H
103. *Notoplites klugei* (HASENBANK) 1932	77 H	594 H
104. *Notoplites perditus* (KLUGE) 1914		2450 G
105. *Notoplites tenuis* (KLUGE) 1914	46 G	594 H
106. *Notoplites tenuis* var. *uniserialis* HASTINGS 1943	77 H	594 H
107. *Notoplites vanhöffeni* (KLUGE) 1914	183 H	385 G
108. *Notoplites watersi* (KLUGE) 1914	238 H	480 WW
109. *Ogivalina lata* (KLUGE) 1914	46 L	385 G
110. *Thalamoporella steganoporoides* var. *granulata* (BUSK)		137 B
Cheilostomata Ascophora		
111. *Catenicella frigida* WATERS 1904		500? WW
112. *Cellarinella foveolata* WATERS 1904	459 WW	582 L
113. *Cellarinella laytoni* ROGICK 1956c		106 R
114. *Cellarinella margueritae* ROGICK 1956c		73 R
115. *Cellarinella njegovanae* ROGICK 1956c	106 R	183 R
116. *Cellarinella nodulata* WATERS 1904	64 T	582 T
117. *Cellarinella nutti* ROGICK 1956c	73 R	183 R
118. *Cellarinella rossi* ROGICK 1956c	106 R	183 R
119. *Cellarinella roydsi* ROGICK 1956c	64 R	106 R
120. *Cellarinella watersi* CALVET 1909	46 L	287 T
121. *Cellepora horneroides* WATERS 1904	410 WW	500? WW

Table II. (continued)

Species number and name	Depth range in meters, and reporting author	
	Least depth	Greatest depth
122. *?Cellepora setosa* Thornely 1924	90 V	750 V
123. *Clithriellum inclusum* (Waters) 1904	106 R	569 WW
124. *Dakaria dabrowni* Rogick 1962		
125. *Emballotheca contortuplicata* (Calvet) 1909	64 R	648 L
126. *Emballotheca phylactelloides* (Calvet) 1909	40 CC	220 L
127. *Escharoides barica* Rogick 1955b	64 R	183 R
128. *Escharoides biformata* Waters 1904		500? WW
129. *Escharoides bubeccata* Rogick 1955b	64 R	106 R
130. *Escharoides praestita* (Waters) 1904	73 R	480 WW
131. *Escharoides tridens* (Calvet) 1909	30 CC	110 CC
132. *Euthryrisella carthagensis* (Calvet) 1909		40 CC
133. *Fenestrulina exigua* (Waters) 1904	46 L	500 WW
134. *Hippadenella carsonae* Rogick 1957		106 R
135. *Hippellozoon gelidum* (Waters) 1904	110 L	500? WW
136. *Hippellozoon hippocrepis* (Waters) 1904	220 L	500? WW
137. *Hippellozoon lepralioides* (Waters) 1904	287 L	459 WW
138. *Isoschizoporella tricuspis* (Calvet) 1909	30 CC	648 L
139. *Kymella polaris* (Waters) 1904	40 CC	500? WW
140. *Lepralia frigida* Waters 1904	435 WW	500? WW
141. *Microporella proxima* Waters 1904	6 T	569 WW
142. *?Microporella trinervis* Waters 1904	183 R	582 L
143. *Orthoporidra compacta* (Waters) 1904	435 WW	500? WW
144. *Osthimosia clavata* Waters 1904	435 WW	500? WW
145. *Osthimosia milleporoides* (Calvet) 1909	20 CC	106 R
146. *Parasmittina hymanae* Rogick 1956c	156 R	183 R
147. *Phylactellipora lyrulata* (Calvet) 1909	46 L	648 L
148. *Porella clivosa* var. *inerma* Calvet 1909	30 CC	40 CC
149. *Porella marginata* (Calvet) 1909	73 R	287 L
150. *Retepora antarctica* Waters 1904	220 L	580 WW
151. *Retepora laevigata* Waters 1904	436 WW	460 WW
152. *Retepora protecta* Waters 1904	410 WW	500? WW
153. *Retepora protecta* var. *crassa* Waters 1904		460 WW
154. *Reteporella flabellata* Busk 1884		137 B
155. *Rhamphostomella bassleri* Rogick 1956c	64 R	183 R
156. *Schizoporella eatoni* var. *areolata* Calvet 1909		
157. *Schizoporella gelida* Waters 1904		435 WW
158. *Schizoporella pellucidula* Calvet 1904b		
159. *Sertella frigida* (Waters) 1904	6 T	435 WW
160. *Smittina abditavicularis* Rogick 1956c	183 R	287 L
161. *Smittina alticollarita* Rogick 1956c		183 R
162. *Smittina antarctica* (Waters) 1904	201 T	582 L
163. *Smittina canui* Rogick 1956c	73 R	183 R
164. *Smittina conspicua* (Waters) 1904	201 L	582 T
165. *Smittina crassatina* (Waters) 1904	435 WW	500? WW
166. *Smittina directa* (Waters) 1904	82 L	500? WW
167. *Smittina excertaviculata* Rogick 1956c		183 R
168. *Smittina gelida* (Waters) 1904		500 WW
169. *Smittina obicullata* Rogick 1956c	64 R	73 R
170. *Smittina oblongata* Rogick 1956c		183 R

Table II. (continued)

Species number and name	Depth range in meters, and reporting author	
	Least depth	Greatest depth
171. *Smittina pileata* (WATERS) 1904		500 WW
172. *Smittina reptans* (WATERS) 1904	436 WW	500? WW
173. *Smittina tripora* (WATERS) 1904	110 T	500? WW
174. *Smittinella rubrilingulata* ROGICK 1956c		64 R
175. *Smittoidea ornatipectoralis* ROGICK 1956c		183 R
176. *Smittoidea ornatipectoralis brevior* ROGICK 1956c	73 R	183 R
177. *Systenopora contracta* WATERS 1904	220 L	500? WW
178. *Toretocheilum absidatum* ROGICK 1960		73 R
179. *Umbonula dentata* (WATERS) 1904	156 R	500? WW

Bryozoa occur from shoreline to great depths. Some are abyssal: *Amastigia abyssicola* 2450—2800 m, *Cornucopina angulata* and *Notoplites perditus* from 2450 m, *Farciminellum linearis* from 2910 m and *Himantozoum sinuosum variabilis* from 3397 m. Other species have a considerable depth range: *Camptoplites bicornis magna* 24—1080 m, *Himantozoum antarcticum* 0—594 m, *Isoschizoporella tricuspis* 30—648 m, *Microporella proxima* 6—569 m, *Notoplites drygalskii* 5½—1080 m, *Phylactellipora lyrulata* 46—648 m and *Sertella frigida* 6—435 m.

SILÉN (1951, pp. 63, 65) reported *Levinsenella magna* (BUSK) 1884 and *Cornucopina rotundata* (KLUGE) 1914 from depths of 4540—4600 m from the North Atlantic Station 387 of the Swedish Deep-Sea Expedition of 1947—1948 (N. Lat. 40°33' to 34',W. Long. 35°24' to 52'). Both species also occur in Antarctic waters. KLUGE (1914, pp. 640, 650) reported *C. (Bicellaria) rotundata* from 3397 and 3423 m and *L. (Farciminaria) magna* from 3397 m from the Kaiser Wilhelm II Sector. HASTINGS (1943, 407) remarked that *C. rotundata* also occurred in the BUSK 1884 Challenger material from 4026 m at 35°39'S. Lat. and 50°47'W. Long.

For convenience, the Antarctic waters have been divided into 9 sectors of approximately 40° each, and several groups of islands. The distribution of the exclusively Antarctic 179 species is indicated on Table III, except for Species # 124 *(Dakaria dabrowni)* whose type locality is unknown. The region around the Palmer Archipelago yielded the greatest number of species. The Kaiser Wilhelm and Victoria and Adelie Land Sectors also yielded a goodly number while the remaining sectors had relatively few species.

The 9 sectors and the number of their endemic (exclusively Antarctic) bryozoan species are as follows.

The New Schwabenland Sector, from 20°W. to 20°E. Long., had two species: No. 51, *Cellaria mawsoni*, and No. 147, *Phylactellipora lyrulata*.

The Enderby Land Sector, from 20°E. to 60°E. Long. had no endemic bryozoans.

The Kaiser Wilhelm II Land Sector, from 60°E. to 100°E. Long., had 77 species.

The Wilkes Land Sector, from 100°E. to 140°E. Long., had 28 species.

The Victoria and Adelie Land Sector, from 140°E. to 180° Long., includes the eastern part of the Ross Sea. It had 81 species.

The West Ross Sea Sector, 180° to 140°W. Long., had 13 species.

The Amundsen Sea Sector, 140°W. to 100°W. Long., had one endemic species, No. 22, *Amastigia abyssicola*.

The Bellingshausen Sector includes part of the Palmer Peninsula and Archipelago and Bellingshausen Sea, an area extending from 100°W. to approximately 60°W. Long. It had 104 species.

The Weddell Sea Sector which includes Weddell Sea and part of the Palmer Peninsula and Archipelago extends from approximately 60°W. to 20°W. Long. and had 20 species.

In addition to the above described 9 sectors several far-removed island groups produced an Antarctic fauna. These islands are as follows: Heard Island, with two species, the South Orkneys with three species, Bouvet Island with five, Shag Rocks with seven, the South Sandwich and South Shetlands each with eight species and the South Georgia Islands with nineteen species.

Table III lists these sectors and island groups, the particular species found in each and the author reporting each species.

Table III.

Geographic distribution of the 179 numbered endemic species.

Antarctic sector or island group	Species reported therefrom and by whom. See Table I for key letters representing authors and Table II for species names which correspond to these species numbers.
Heard Island	Species No. 110, Author B (Busk); 154 B
South Orkneys	32 H, 36 H, 84 H
Bouvet Island	49 V, 76 N, 83 N, 84 H, 103 N
Shag Rocks	4 F, 27 H, 47 H, 69 H, 84 H, 100 H, 101 H
South Sandwich Islands	27 H, 46 H, 70 H, 73 H, 84 H, 102 H, 103 H, 105 H
South Shetlands	28 H, 32 H-V, 33 H, 36 H, 46 H, 70 H, 73 H, 102 H

Table III. (continued)

Antarctic sector or island group	Species reported therefrom and by whom. See Table I for key letters representing authors and Table II for species names which correspond to these species numbers.
South Georgia Islands	1 C, 4 F, 28 H, 36 H, 37 H, 39 H, 45 H, 47 H, 65 C, 70 H, 73 H, 80 C, 84 H 100 H, 101 H, 102 H, 103, H, 105 H, 158 C
Kaiser Wilhelm Sector, 60°E—100°E. Long.	22 G, 23 G, 24 G, 25 G, 26 T, 27 G, 32 G, 33 G, 34 G, 35 G, 36 G, 37 G, 38 G, 41 G, 42 G, 45 G-L, 47 G, 48 L, 49 L, 51 L, 54 T-L, 59 G, 60 G, 61 G, 62 G, 63 G, 64 G, 66 G, 68 G, 70 G, 72 G, 73 G, 74 G, 75 T-L, 76 G-L, 77 G-L, 78 G, 79 G-L, 81 G-L, 82 G-L, 83 G, 84 G, 85 G, 86 G, 87 L, 89 T-L, 92 T-L, 93 G, 94 G, 95 G, 96 G, 99 G, 100 G, 102 G-L, 103 G, 104 G, 105 G, 107 G, 108 G, 109 G, 116 T-L, 120 T-L, 122 T-L, 125 T-L, 126 T-L, 135 T-L, 136 L, 139 L, 142 T-L, 147 T-L, 149 T-L-R, 150 T-L, 159 T-L, 162 T-L, 164 L, 173 T-L, 177 L
Wilkes Land Sector, 100°E.—140° E. Long.	49 L, 52 R, 54 R, 56 R, 76 L, 77 L, 79 L, 84 L, 88 R, 89 R, 92 R, 102 L, 115 R, 117 R, 118 R, 123 R, 126 R, 127 R, 130 R, 142 R, 155 R, 160 R, 161 R, 163 R, 167 R, 170 R, 175 R, 176 R
Victoria and Adelie Land Sector, 140°E.—180° Long.	3 K, 23 H, 24 H, 25 H, 27 H, 28 L-H, 29 H, 30 H, 31 L, 32 H, 33 H, 34 H, 35 H, 36 H, 37 H, 38 H, 41 H, 43 H, 45 H, 47 T-L-H, 48 L, 49 L, 51 L, 52 R, 53 R, 54 T, 55 R, 57 R, 68 H, 70 H, 71 T, 72 H, 73 L-H, 75 T-L, 76 L, 79 L, 81 L, 84 T-L-H, 86 H, 89 T-L, 91 R, 92 T-L, 100 H, 102 L-H, 103 H, 105 H, 106 H, 107 H, 108 H, 109 L, 112 T-L, 113 R, 115 R, 116 T-L, 117 R, 118 R, 119 R, 120 T-L, 122 T-L, 123 R, 125 T-L, 129 R, 131 L, 133 L, 134 R, 137 T-L, 138 T-L, 139 T-L, 141 T-L, 142 T-L, 145 R, 147 T-L-R, 149 T-L, 150 T, 159 T-L, 160 L, 162 T-L, 164 T-L, 166 L, 173 T-L, 177 T-L
West Ross Sea Sector, 180°—140°W. Long.	27 H, 32 H, 34 H, 36 H, 38 H, 41 H, 45 H, 47 H, 70 H, 73 H, 84 H, 102 H, 105 H
Amundsen Sea Sector, 140°W.—100°W. Long.	22 WW
Bellingshausen Sea Sector, including the western part of Palmer Peninsula, or parts falling within approximately 100°W.—60°W. Long.	2 WW, 6 WW, 7 WW, 10 F, 12 F, 17 WW, 18 WW, 19 WW, 23 WW, 27 WW-CC-V, 28 H-V, 31 CC, 32 WW-H, 33 WW-CC, 34 WW-H, 35 H, 36 H, 40 H, 44 WW, 45 H, 47 WW, 50 WW, 54 CC, 58 WW, 62 WW, 65 C, 67 CC, 71 WW, 72 WW, 75 CC-V, 76 V, 77 V, 79 WW-CC-V, 81 V, 82 V, 84 CC-H-V, 88 R, 90 WW, 97 WW, 98 WW, 100 WW-H, 102 H, 103 H, 105 H, 107 H, 108 WW, 109 V, 111 WW, 112 WW, 114 R, 116 WW, 117 R, 119 R, 120 CC, 121 WW, 122 V, 123 WW, 125 CC-R,

410

Table III. (continued)

Antarctic sector or island group	Species reported therefrom and by whom. See Table I for key letters representing authors and Table II for species names which correspond to these species numbers.
	126 CC-R, 127 R, 128 WW, 129 R, 130 WW-R, 131 CC-V-R, 132 CC, 133 WW, 135 WW, 136 WW, 137 WW, 138 CC-R, 139 WW-CC, 140 WW, 141 WW, 142 WW, 143 WW, 144 WW, 145 CC-R, 146 R, 147 CC-V-R, 148 CC, 149 CC-V-R, 150 WW, 151 WW, 152 WW, 153 WW, 155 R, 156 CC, 157 WW, 159 WW-CC, 162 WW, 163 R, 164 WW, 165 WW, 166 WW, 168 WW, 169 R, 171 WW, 172 WW, 173 WW, 174 R, 176 R, 177 WW, 178 R, 179 WW-R
Weddell Sector, includes Weddell Sea, and the part of Palmer Peninsula area, extending approximately 60°W.—20°W. Long.	4 F, 5 F, 8 F, 9 F, 10 F, 11 F, 12 F, 13 F, 14 F, 15 F, 20 F, 21 F, 27 H, 28 H, 33 H, 35 H, 45 H, 72 H, 100 H, 102 H
New Schwabenland Sector, 20°W.—20°E. Long.	51 V, 147 V

Of these 179 endemic species *Himantozoum antarcticum* was reported from ten different regions (i.e. sectors or island groups), *Notoplites drygalskii* occurred in nine regions, *Camptoplites bicornis magna* was in seven regions. Nine species, Nos. 27, 32, 45, 47, 70, 73, 100, 103 and 105, were reported from six regions. *Beania erecta livingstonei, Camptoplites areolatus* and *Flustra angusta* occurred in five regions; eight species occurred in four regions, twenty-nine species were found in three regions, thirty-eight species were from two regions and eighty-nine species occurred in only one sector or island group. These eighty-nine are Species Numbers 1 to 3, 5 to 9, 11, 13 to 21, 26, 29, 30, 39, 40, 42 to 44, 50, 53, 55 to 61, 63, 64, 66, 67, 69, 74, 78, 80, 83, 85, 87, 90, 91, 93 to 99, 104, 106, 110, 111, 113, 114, 121, 128, 132, 134, 140, 143, 144, 146, 148, 151 to 154, 156 to 158, 161, 165, 167 to 172, 174, 175, 178 and 179.

Upon completion of further studies of collections the above distribution lists will be augmented by the addition of new species and more extensive distribution records for the present species.

It is quite probable that examination of existing collections of mollusks, crabs, echinoderms, worm tubes, corals, hydroids, alcyonarians, brachiopods, tunicates and algae would produce additional new species and new distribution records because bryozoans grow readily on the shells or exoskeletons of such animal groups and on many substrates that offer room for attachment.

With very few exceptions bryozoans are sessile, colonial, tentaculate, ciliary feeders that attach permanently to such substrates as mentioned above, and to rocks, pebbles, spines and even to the carapace of some vertebrates, like turtles.

As to general appearance and growth habit, some bryozoans are dendritic or foliaceous, others are encrusting. Some may be rubbery in texture, others may be membranous, gelatinous, chitinous, calcareous or silicious. Some species form soft mats, others form thin crusts or even nodules. The Retepores resemble a handful of stiffly starched coarse lace. The colonies are sometimes quite sizeable but the individual zooids are microscopic. Some calcareous colonies have a relatively clean appearance, permitting no other sessile organism to attach while the colony is still alive. Whether this is because of avicularial activity, ciliary currents caused by tentacle cilia or because of toxicity is uncertain. Some non-Antarctic bryozoans produce toxic substances which when released into the surrounding water are toxic to such organisms as fish.

Some areas of a bryozoan colony may die while other areas (as at the edges) of the same colony continue growing. Sometimes dying areas regenerate new zooids to replace the degenerated individuals, thus repopulating the same zooecia (exoskeleton).

Little is known of the embryology of the Antarctic bryozoans. HASTINGS studied and illustrated some of the ancestrulae. The U.S. Navy collections made during the months of December, January and February, 1947—1948, had not only mature colonies but in some species there were also embryos, ancestrulae and beginning colonies, indicating that sexual reproduction (as well as the more common asexual or budding type of reproduction) was occurring.

As expected, Antarctic zooids were comparatively more robust and larger than were zooids of similar bryozoan species from warmer waters.

Most bryozoologists who worked on the Antarctic forms concerned themselves primarily with taxonomy rather than with the ecology of the forms so little ecological data other than depth and geographic distribution is available.

BIBLIOGRAPHY

BORG, F. 1944. The Stenolaematous Bryozoa. Further Zoological Results of the Swedish Antarctic Expedition 1901—1903. Stockholm. III (5): *1—276*.

BUSK, G. 1881. Descriptive Catalog of the species of *Cellepora* collected on the "Challenger" Expedition. *J. Linn. Soc. Zool.*, **XV**: *341—356*.

412

Busk, G. 1884. Report on the Polyzoa collected by H.M.S. *Challenger...* Part I. The Cheilostomata, Report on the Scientific Results of the Voyage of H.M.S. *Challenger...* 1873—1876, Zoology, X (V): i-xx and *1—216.*

Busk, G. 1886. Part II. The Cyclostomata, Ctenostomata and Pedicellinea. Report on the Scientific Results of the Voyage of H.M.S. *Challenger...* 1873—1876, XVII (L): i-viii and *1—47.*

Calvet, L. 1904a. La distribution géographique des Bryozoaires marins et la théorie de la bipolarité. *C.R. Acad. Sci. Paris,* **CXXXVIII:** *384—387.*

Calvet, L. 1904b. Diagnoses de quelques espèces de Bryozoaires nouvelles ou incomplètement décrites de la Région sub-antarctique de l'Océan Atlantique. *Bull. Soc. zool. France,* **XXIX** (3): *50—59.*

Calvet, L. 1904c. Bryozoen. Hamburger Magalhaensische Sammelreise. Hamburg. 45 pp.

Calvet, L. 1909. Bryozoaires. Expédition Antarctique Française (1903—1905) commandée par Dr. Jean Charcot. Sciences Naturelles: Documents scientifiques. 50 pp.

Hasenbank, W. 1932. Bryozoa der Deutschen Tiefsee-Expedition. I. Teil. Wiss. Ergebn. Deutsch. Tiefsee-Exped. XXI (2): *318—380.*

Hastings, A. B. 1943. Polyzoa (Bryozoa). I. Scrupocellariidae... *Discovery Reports.* **XXII:** *301—510.*

Johnston, T. H. & L. M. Angel. 1940. Endoprocta. B.A.N.Z. Antarctic Research Expedition 1929—1931. Reports. Series B (Zool. and Botany) IV (7): *215—231.*

Kirkpatrick, R. 1902. Polyzoa. Report on Collections of Nat. Hist. made in the Antarctic Regions during the voyage of the *"Southern Cross".* XVI: *286—289.*

Kluge, G. 1914. Die Bryozoen der Deutschen Südpolar-Expedition. I. Die Familien Aetidae, Cellularidae... Deutsche Südpolar-Exped. 1901—1903 von Drygalski, Band XV; Zool. Band VII: *599—678.*

Livingstone, A. 1928. The Bryozoa. *In:* Sci. Reports Mawson's Australasian Antarctic Exped. 1911—1914. Ser. C, Zool. Bot., IX (1): *5—94.*

Rogick, M. D. 1955a. Genus *Emballotheca* L. 1909. *Trans. Amer. microsc. Soc.* **74** (2): *103—112.*

Rogick, M. D. 1955b. Studies on marine Bryozoa, VI. Antarctic *Escharoides.* *Biol. Bull.* **109** (3): *437—452.*

Rogick, M. 1956a. Studies on marine Bryozoa, V. *Clithriellum...* *Trans. Amer. microsc. Soc.* **75** (1): *70—74.*

Rogick, M. 1956b. Studies on marine Bryozoa. VII. *Hippothoa. Ohio J. Sci.* **56** (3): *183—191.*

Rogick, M. 1956c. Bryozoa of the U.S. Navy's 1947—48 Antarctic Expedition. *Proc. U.S. Nat. Mus.* **105** (3358): *221—317.*

Rogick, M. 1957a. Studies on marine Bryozoa, IX. *Phylactellipora. Ohio J. Sci.* **57** (1): *1—9.*

Rogick, M. 1957b. Studies on marine Bryozoa, X. *Hippadenella...* *Biol. Bull.* **112** (1): *120—131.*

Rogick, M. 1959a. Studies on marine Bryozoa, XI. Antarctic *Osthimosiae.* *Ann. N.Y. Acad. Sci.* **79** (2): *9—42.*

Rogick, M. 1959b. Studies on marine Bryozoa, XII. *Porella. Ohio J. Sci.* **59** (4): *233—240.*

Rogick, M. 1960. Studies on marine Bryozoa, XIII... *Biol. Bull.* **119** (3): *479—493.*

Rogick, M. 1962. Studies on marine Bryozoa, XIV. *Dakaria. Trans. Amer. microsc. Soc.* **81** (1): *84—89.*

Silén, L. 1951. Bryozoa. Reports of the Swedish Deep-Sea Expedition, vol. II. Zool. No. 5: *63—69.*

THORNELY, L. 1924. Polyzoa. *In:* Sci. Reports Mawson's Australasian Antarctic Exped. 1911—1914. Ser. C, Zool. Bot., VI (6): *1—23*.

VIGELAND, I. 1952. Antarctic Bryozoa. Det Norske Vid-Akad. Oslo. Sci. Results Norweg. Antarctic Exped. 1927—1928. No. 34: *1—16*.

WATERS, A. W. 1888. Supplementary Report on the Polyzoa . . . Report on the Scientific Results of the Voyage of H.M.S. *Challenger* . . . 1873—1876, XXXI (LXXIX): *1—41*.

WATERS, A. W. 1904. Bryozoa. Expéd. Antarct. Belge, Résult. Voy. S.Y. *Belgica* 1897—1899 . . . de Gomery, Rapp. Sci. Zool. 114 pp.

Editor's Note

The Editors regret that Dr. Rogick died in New Rochelle, N.Y., on October 25, 1964.

THE ACAROLOGY OF THE ANTARCTIC REGIONS

BY

PER DALENIUS

Zoological Institute Lund, Sweden

(with 2 figs.)

Introduction

In more than one sense both the Antarctic continent and the Sub-Antarctic islands are white spots on the map. The existence of such areas has always inspired men to enlarge our knowledge, and several expeditions have been equipped to explore the unknown lands of Antarctica.

The interest seems, however, to have been chiefly "geographical", the biological research was directly or indirectly influenced by seas rich in fair game. Besides, the expeditions were chiefly interested in the geology, glaciology, and meteorology of the white continent. It is true that the organizers and members of the expeditions did not lack in interest in biology, but for various reasons the tangible results of one expedition could not be very considerable. From the biologist's viewpoint the value of these expeditions lies in the fact that the separate observations form together a valuable ground for future studies.

As for the field work, biological research demands a great variety of methods and instruments for collecting and studying the material. The equipment required for the study of marine subjects is quite different from that which is used by a student of terrestrial animals or plants. A botanist does not work in the same way as a zoologist. But they all need an equipment which is more or less voluminous.

In my view, this explains why biological research in the Antarctic regions has mostly got its results by chance, even if it is easy to point out exceptions. Thanks to its typically German thoroughness the German South Polar Expedition of 1901—03, for example, brought home a collection of great biological interest. During one of his excursions inland WILSON, the physician of NSBX 1949—52, discovered areas with mites as well as lichens and mosses. A recent map of Antarctica reveals the fact that there are such spots of plant life all around the continent (fig. 1). Indeed, the regions round the South Pole afford possibilities for the study of problems concerning fundamental ecology, life at its border, geographical distribution, and the development of species.

A collocation published in Scientific American (September, 1962) shows that all the nine countries performing scientific research 1962 in Antarctica included in their programmes "biology and medicine".

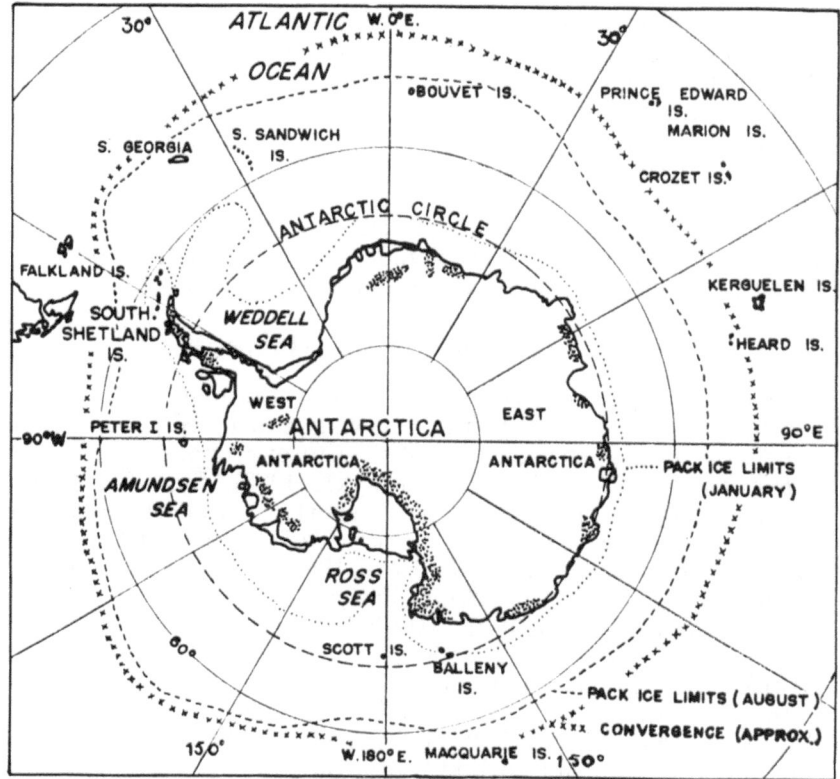

Fig. 1. Finds of Mosses and Lichens in Antarctica.

To no other branch of science is such great interest devoted, except meteorology.

If we only consider the terrestrial life of Antarctica, we can state that insects and arachnids together are the highest forms of native life. There are springtails, a wingless fly, lice, and mites, about 50 species in all. If we take also the Sub-Antarctic mites, marine as well as terrestrial, into consideration, the list of species is much more comprehensive. The number of at least 111 species is imposing, but many of them are parasites and transported to the area by birds or marine mammals. The *Halacaridae* form another large group of mites. Many of them have a wide distribution in the seas from the north to the south. And only comparatively few species are native to the Antarctic and Sub-Antarctic areas.

List of species

In this list of species found in the Antarctic and the Sub-Antarctic regions, I have followed the systematical order used by VITZTHUM in Brohmer-Ehrmann-Ulmer: Die Tierwelt Mitteleuropas, III. Band. Within the different groups, however, the species are put in alphabetic order. As acarology comprises a great many species of each of the three groups Parasitiformes, Trombidiformes, and Sarcoptiformes, it is impossible to be a master of them all. Consequently this means that it might be advisable not to accept the names of critical species without reserve. In this field, viz. acarology of Antarctica, there is indeed much work to be done by future specialists of the different groups.

Parasitiformes/Mesostigmata (16 species)

Asca crozetensis RICHTERS, 1908. The German South Polar Expedition of 1901—03 (in the following called GSP) found it in mosses on the Possession Island of the Crozet group.

Digamasellus Racovitzai (TROUESSART, 1903). Syn. *Gamasellus Racovitzai* TRST. TROUESSART (l.c.) reports it from the Gerlache Strait, where it was found in mosses. TRÄGÅRDH (1907) received specimens from Hovgaard Island and Wandel Island collected by Première Expédition Antarctique Française (PEAF), The Swedish South Polar Expedition 1901—03 (SSE) found this mite species in mosses and lichens on the S. Shetland Islands (TRÄGÅRDH 1907a). BERLESE (1917) reports it from Graham Land.

D. Racovitzai ssp. *neo-orcadensis* (TROUESSART, 1912). The Scottish National Antarctic Expedition with SY Scotia (SYS) found the mite in mosses on a rock in Scotia Bay.

Eulaelaps mawsoni WOMERSLEY, 1937. The Australasian Antarctic Expedition of 1911—14 (AAE) found this mite in the tide water zone at the Macquarie Islands.

Gamasellus crozetensis RICHTERS, 1908 is collected by GSP on Possession Island (Crozet), where it was found in mosses.

G. Jeanneli ANDRE, 1947 was found on Kerguelen Island during the Bougainville expedition (CB). It was found on the rocks at the sea.

Gamasiphis crozetensis (RICHTERS, 1908). Syn. *Heydeniella crozetensis* RICHT. GSP found it in mosses on Possession Island (Crozet).

Hydrogamasus antarcticus TRÄGÅRDH, 1907. It has been collected by SSE (on Paulet Island of the Antarctic region) as well as by AAE, who took it in the tide water zone at the Macquarie Islands.

Laelaps grahamensis TRÄGÅRDH, 1907. SSE found it in Graham Land, where a single female was found under a stone.

Neoparasitus crozetensis RICHTERS, 1908. GSP collected this mite in mosses on Possession Island (Crozet).

Opisthope crozetensis RICHTERS, 1908. It is reported from Possession Island, where it was found in mosses.

Pachylaelaps macquariensis WOMERSLEY, 1937. AAE found it in the tide water zone at the Macquarie Islands.

Pachyseius adeliensis WOMERSLEY, 1937. This species is a mite of the deep sea. AAE found it in Commonwealth Bay at the depth of 20—30 fathoms (37—55 m).

Piracarus crozetensis RICHTERS, 1908 was found by GSP in the same locality as the species mentioned above.

Sejulus plumosus OUDEMANNS, 1905 is reported from Possession Island of the Crozet group (RICHTERS, 1908), where it was found in mosses. Possibly this mite is an European species which has been transported by the members of GSP to the Sub-Antarctic region.

Zercon tuberculatus TRÄGÅRDH, 1907. The species was described on one nympha which SSE found in mosses in Graham Land, as well on specimens collected by PEAF on Wandel Island.

Parasitiformes/Ixodides (3 species)

Ceratixodes uriae (WHITE, 1852). Syn. *Hyalomma puta* CAMBRIDGE, *Ixodes borealis* KRAMER & NEUMANN, *I. fimbriatus* KRAMER & NEUMANN. This mite has been found on birds, *Pygosceles taeniatus*, *Chionis minor*, *Uria* sp., *Larus* sp., and others which gives it a very wide distribution in both the Northern and Southern Hemispheres.

Ixodes auritulus NEUMANN, 1904. Syn. *I. thoracicus* NEUM., *I. percavatus* NEUM. It has been judged to be a fairly common tick on sea birds, and thus it has a wide distribution in the Southern Hemisphere. ZUMPT (1952) reports it from Heard Island, but this mite has also been recorded from the Macquarie Islands, Kerguelen Island, and Tristan da Cunha, as well as from S. and N. America.

I. kerguelenensis (ANDRÉ & COLAS-BELCOUR, 1942). Syn. *I. canisuga* JOHNSTON v. *kerguelenensis* A. & B. This ixodid mite has been recorded from Kerguelen (ANDRÉ, 1947). Australian National Antarctic Research Expedition (ANARE) took a collection on Heard Island. No ecological data are given.

Trombidiformes/Tarsonemini (1 species)

Disparipes antarcticus RICHTERS, 1908. GSP found it in mosses on Heard Island as well as on Possession Island (Crozet).

Trombidiformes/Prostigmata, Bdellidae (2 species)

Bdella antarctica TRÄGÅRDH 1907. SSE collected several specimens on the rocks, S. Georgia.

B. pallida (CAMBRIDGE, 1876). Syn. *Scirus pallidus* CAMBR. It has been recorded from Kerguelen and Possession Island (Crozet) (CAMBRIDGE, 1876; ENDERLEIN, 1903; ANDRÉ, 1947). STUDER (1879) received several specimens collected in different dry localities.

418

Trombidiformes/Prostigmata, Tetranychidae (1 species)

Bryobia praetiosa C. L. KOCH, 1836. Syn. *Torrynophora serrata*
CAMBRIDGE 1876, *Bryobia speciosa* C.L.K./KRAMER & NEUMANN
1883. The species is found on Kerguelen and St. Paul Island (ENDER-
LEIN, 1903; ANDRÉ, 1947), where it lives under stones, and — as
well — in the Arctic regions (see TRÄGÅRDH, 1910).

Trombidiformes/Prostigmata, Halacaridae (49 species)

Agaue agauoides (LOHMANN, 1908) Syn. *Halacarus (Polymela)
agauoides* LOHM. GSP found it in the Antarctic Sea (—385 m).
WOMERSLEY (1937) reports this halacarid mite from Commonwealth
Bay, King George V Land.
  A. affinis SOKOLOW 1962. Found by a Russian expedition, 1956
at a depth of 197 m, 65°52'S, 111°40'E.
  A. consobrina (ANDRÉ, 1933) is reported from St. Paul Island.
  A. drygalski (LOHMANN, 1908). Syn. *Halacarus (Polymela)
drygalski* LOHM., *Leptospathis d.* (LOHM.), cf. VIETS 1950. It has been
found on Kerguelen Island in the littoral zone and in the Antarctic
Sea, —385 m (GSP). AAE collected it in Commonwealth Bay
(—25 fathoms) and at the Macquarie Islands, in the tide water
zone. VIETS (1950) reports it from the Falkland Islands.
  A. Hamiltoni WOMERSLEY, 1937. AAE found it at the Macquarie
Islands.
  A. longissima SOKOLOW 1962. Found by a Russian expedition
1956 at a depth of 300 m, 66°12,2' S, 91°37,2' E.
  A. parva (CHILTON, 1883). Syn. *Leptospathis occultus* (LOHM.)
in TROUESSART 1914, *L. Bouvieri* TRST. 1907, *Halacarus (Polymela)
occultus* LOHM. It has been found in the sea at a depth of 20 m to
385 m by GSP, PEAF, and AAE.
  A. parva (CHILT.) v. *Womersley* WOMERSLEY/VIETS, 1950. Syn.
A. occultus (LOHM.) v. *setifera* WOMERSLEY, 1937. AAE found it in
Commonwealth Bay, at the depth of 25 fathoms (46 m).
  A. pilosa (GIMBEL, 1919). This mite was collected by GSP at
their Gauss station.
  A. tenuirostris (LOHMANN, 1908). Syn. *Halacarus (Polymela)
tenuirostris* LOHM. GSP found it in the Antarctic Sea at the depth
of —385 m. TROUESSART (1914) received specimens found by PEAF
on Bryozoans.
  A. veles TROUESSART, 1907. PEAF found it in the sea, —20 m, at
Victoria Land.
Agauopsis antarctica (LOHMANN, 1908). Syn. *Agaue antarctica*
LOHM. GSP found it at Kerguelen Island in the littoral zone where
it lived among algae, and — as well — in the Antarctic Sea at a
depth of between 46 m and 385 m. WOMERSLEY (1937) reports this

mite from Commonwealth Bay, where it was found at the depth of 20 to 25 fathoms (37—46 m).

A. microrhynca TROUESSART v. *paulensis* LOHMANN, 1908 was taken by GSP on algae in shallow water at St. Paul Island. It is also to be found at the shores of Kerguelen Island (ANDRÉ, 1933).

Copidognathus acanthophorus VIETS, 1950. SSE found it at S. Georgia.

C. floridus TROUESSART, 1914. Deuxième Expédition Antarctique Française (DEAF) found it on bryozoans at Port Lockroy, Antarctica.

C. kerguelensis LOHMANN, 1908. LOHMANN (l.c.) as well as ANDRÉ (1933) report this mite from Kerguelen Island, where it has been found on algae in shallow water.

C. liouvillei TROUESSART, 1914. DEAF found it on ascidians in the Antarctic Sea.

C. Marc-Andréi VIETS, 1950 is reported from S. Georgia, where it was found by SSE.

C. oculatus (HODGE, 1863). Syn. *Halacarus oculatus* HODGE. LOHMANN (1908) as well as WOMERSLEY (1937) report the species from the Antarctic region. In Commonwealth Bay it was found at the depth of 25 fathoms (46 m).

C. simonis LOHMANN, 1908. GSP found it in algae in the littoral zone, at Cape Town, while TROUESSART (1914) reports it from the Antarctic region, where DEAF collected this halacarid mite on ascidians.

C. vanhöffeni LOHMANN, 1908 is recorded from the German station of GSP, found at a depth of 46 to 385 m. It was judged to be more common, however, in the upper parts of these depths.

Halacarus actenos TRST. v. *robustus* LOHMANN, 1908. GSP found it on algae in the littoral zone at Kerguelen Island. VIETS (1950) reports it from the Falkland Islands.

H. excellens LOHMANN, 1908 is living in the Antarctic Sea, in deep water. It was found by GSP.

H. gracile-unguiculatus LOHMANN, 1908. This species was found at Kerguelen Island, where it lived on algae in the littoral zone. AAE collected it below the tide water zone at the Macquarie Islands.

H. harioti TROUESSART, 1889. AAE found it on a rock below the tide water zone at the Macquarie Islands. It is also recorded from the Falkland Islands, by SSE.

H. harioti TRST. v. *kerguelensis* LOHMANN, 1908 was found in samples taken among algae from the littoral zone at Kerguelen Island.

H. latirostris BIMBEL, 1919. GSP collected it at their Gauss station.

H. minor LOHMANN, 1908 was found by GSP at their Antarctic stations at the depths of 45 to 385 m. AAE recorded it from Commonwealth Bay.

420

H. nanus GIMBEL, 1919 is recorded from GSP's Gauss station.

H. novior LOHMANN, 1908. GSP collected this mite species on algae in shallow water at Kerguelen Island.

H. novus LOHMANN, 1908. The first collections of this mite species were taken at Kerguelen and St. Paul Island, where it lived on algae in the littoral zone. It is also breeding in the sea east of Graham Land (TROUESSART, 1914), at S. Georgia (VIETS, 1950), and at the Macquarie Islands (TROUESSART, 1914; WOMERSLEY, 1937). It seems to prefer shallow water.

H. nudipelliger ANDRÉ, 1933 is reported from St. Paul Island.

H. (Halacarellus) uschakovi SOKOLOW 1962. Was found in the Antarctic Sea by a Russian expedition 1956.

H. validus GIMBEL, 1919. GSP collected it at their Gauss station.

H. villosus LOHMANN, 1908 has been found once in a sample taken by GSP from the bottom of the Antarctic Sea.

H. werthi LOHMANN, 1908 is collected on algae in shallow water, Kerguelen Island.

Leptospathis scriptor TROUESSART, 1907. PEAF collected this halacarid species at the depth of 20 m at Queen Victoria Land.

Lohmannella falcata (HODGE, 1863). Syn. *Leptognathus falcata* HODGE, *Raphignathus falcatus* HODGE/BRADY, *Trouessartella falcata* HODGE/LOHM. GSP collected it on algae in the uppermost littoral zone at Kerguelen Island as well as in the Antarctic Sea at the depth of 385 m. TROUESSART (1914) reports it from Port Lockroy, Petermann Island, where it was collected on bryozoans and sponges.

L. gaussi LOHMANN, 1908. GSP collected it in the same locality as the species mentioned above.

L. gaussi LOHM. v. *kerguelensis* LOHMANN, 1908 was also found at Kerguelen Island.

Rhombognathus apsteini LOHMANN, 1908. GSP collected it on algae of the littoral zone at Kerguelen Island.

R. magnirostris TRST. v. *lionyx* TROUESSART, 1914. It was described from specimens collected on bryozoans and algae. AAE collected this subspecies in sea-weed in the tide water zone at the Macquarie Islands. It is also recorded from Tierra del Fuego.

R. magnus LOHMANN, 1908. LOHMANN reports this halacarid species from Kerguelen Island, where it was collected on algae in the littoral zone.

Simognathus sculptus (BRADY, 1875). Syn. *Pachygnathus sculptus* BRD. AAE found it on a rock at the Macquarie Islands.

Werthella bouvieri TROUESSART, 1914. DEAF collected this species on sponges at Petermann Island.

W. Johnstoni WOMERSLEY, 1937. The species was collected among cestodes from the intestines of a Weddell seal.

W. parvirostris (TROUESSART, 1889). Syn. *Halacarus parvirostris* TRST., *Agaue parvirostris* TRST. It was first found at New Zealand, but

GSP collected it also at Kerguelen Island, where it lived between littoral algae.

Trombidiformes/Prostigmata, Eupodidae (9 species)

Gainia nivalis TROUESSART, 1914. It was found in samples taken from algae on snow, Petermann Island.

Lorryia leptonychotes WOMERSLEY, 1937. AAE reported it from samples taken among cestodes in a Weddell seal.

L. polaris WOMERSLEY, 1937 was collected by AAE at the depth of 15 to 20 fathoms (27—37 m).

Nanorchestes amphibius TOPSENT et TROUESSART, 1890. This mite species is described from samples taken at the West-European coasts. AAE found it, however, at the Macquarie Islands, where it was collected in sea-weed.

Rhagidia gerlachei (TROUESSART, 1903). Syn. *Nörneria gigas* (CANESTRINI) ssp. *gerlachei* TRST., *R. gigas* ssp. *gerlachei* TRST. It has been found in sea water (BERLESE, 1914) as well as in mosses on rocks near the Antarctic Peninsula (TROUESSART, 1903; TRÄGÅRDH, 1907).

R. kerguelenensis (CAMBRIDGE, 1876). Syn. *Poecilophysis kerguelenensis* CAMBR. It has been found on several islands in the Sub-Antarctic region (Kerguelen I., Macq. I., Possession I., and St. Paul I.), where it seems to live mainly under stones (CAMBRIDGE, 1876; WOMERSLEY, 1937; ANDRÉ, 1947).

R. Johnstoni WOMERSLEY, 1937. AAE found it at King George V Land on cestodes from a Weddell seal.

Stereotydaeus villosus (TROUESSART, 1903). Syn. *Penthaleus (Tectopenthalodes) villosus* TRST. It seems to be one of the most numerous mites in mosses and lichens. It has been recorded by TROUESSART 1903 and 1912, TRÄGÅRDH 1907 and 1907a, BERLESE 1917, EWING 1945, and is to be found in both Antarctica and the Sub-Antarctic regions.

Tydaeus antipodus WOMERSLEY, 1937. AAE collected this eupodid mite on sea-weed at the tide water zone of Macquarie Islands.

Sarcoptiformes/Acaridiae (14/17 species)

Alloptes aschizurus GAUD, 1952 is a parasite of birds which has been found at Kerguelen Island on *Chionis minor*.

A. bisetatus HALLER, 1881 is found on *Sterna virgata*, Kerguelen Island (GAUD 1952).

A. crassipes (CANESTRINIS, 1878). DEAF collected this bird parasite on *Chionis alba*.

Carpoglyphus neglectus (CAMBRIDGE, 1876). See ANDRÉ 1947. ENDERLEIN (1903) reports the species from Kerguelen Island.

422

Glycyphagus domesticus (DE GEER, 1778). AAE found it in the Macquarie Islands.

Hyadesia kerguelensis LOHMANN, 1908 was found by GSP on algae, Kerguelen Island.

H. uncinifer MEGNIN 1889. The prototype specimens were collected on Patagonia. AAE found it in green sea-weeds from between tide marks.

Megninia antarctica GAUD, 1952 was found on *Pelecanoides georgicus*, Kerguelen Island.

M. forcipata (HALLER, 1878). Syn. *M. centropodus* v. *forcipata* BERL./TROUESSART 1914. It is a parasite of birds, in Antarctica it has been collected on *Chionis alba*.

Thecarthra incerta GAUD, 1952 is parasitic on *Pelecanoides georgicus*, Kerguelen Island.

Torynophora saxorum (STUDER, 1879). Syn. *Acarus saxorum*. Has been collected on Kerguelen Island.

Trichotarsus antarcticus TRÄGÅRDH, 1907. This mite species was collected by PEAF on algae, Wandel Island.

Tyroglyphus farinae (LINNÉ, 1758). AAE found it in fresh water lakes on Macquarie Islands. WOMERSLEY (1937) supposes that the existence of this species in the Sub-Antarctic region is due to "an infestation of the container from the ship's stores". It is perhaps a cosmopolitan species.

T. longior (GERVAIS, 1844). Syn. *T. dimidiatus* (HERM.) v. *longior* (GERV.) Cf. ANDRÉ 1947. This mite species has been collected in Commonwealth Bay at the depth of 20 to 39 fathoms (37—54 m) as well at St. Paul Island. "Cette forme, absolument cosmopolite, se rencontre généralement sur toutes les substances végétales en voie de décomposition" (ANDRÉ, 1947, p. 100).

Zachvatkinia puffini (Buchholzia, 1869). Syn. *Dermaleichus puffini, Buchholzia 1869, Bonnetella (Buchholzia) p.* It is a cosmopolitan parasite of birds. GAUD (1952) received it from the Kerguelen Island, where it was taken on *Catharacta skua lönnbergi*. (To these species of *Acaridiae* may be added *Glycyphagus spinipes* (KOCH) and *Aleurobius farinae* (DE GEER). BERLESE (1917) supposes, however, that they have come from the ship's stores, and thus they do not belong to the fauna of the Antarctic regions.)

Sarcoptiformes/Oribatei (21 species)

Alaskozetes antarctica (MICHAEL, 1903). Syn. *Notaspis antarctica* MICH. Cf. WALLWORK, 1962b. It has a circumpolar distribution (DALENIUS & WILSON, 1958) and lives in mosses, lichens, grass as well as under stones or on rocks.

A. antarctica (MICH.) v. *Grandjeani* DALENIUS, 1958. ANARE collected it on the Macquarie Islands on a fresh water pool and on cliffs, as well as on Heard Island in grass.

Anarea macquariensis DALENIUS, 1958. It was collected by ANARE on the Macquarie Islands, where it was found living among algae and *Ulva* sp. in the littoral zone.

A. marina (LOHMANN, 1908). Syn. *Halozetes marina* LOHM., *Notaspis marina* LOHM. GSP found it on Kerguelen Island and St. Paul Island. WOMERSLEY (1937) reports this oribatid mite from the Macquarie Islands.

Antarctozetes crozetensis (RICHTERS, 1908). Syn. *Oribata crozetensis* RICHT., *Jeannelia c.* in D. & W. 1958, cf. BALOGH, 1961. This oribatid species was found on Kerguelen Island and on Possession Island (Crozet).

A. Gaussi (RICHTERS, 1908). Syn. *Oribata Gaussi* RICHT. GSP found it on Possession Island, Crozet group.

Ceratozetes antarctica (MICHAEL, 1895). Syn. *Oribata antarctica*. It lives on rocks and is found on S. Georgia.

Galumna alata (HERMANN, 1804) ssp. in MICHAEL 1895. It is doubtful, whether this subspecies, reported by •MICHAEL from S. Georgia, belongs to *G. alata*. Probably it is the same oribatid species as was found by SSE under stones in a pool on S. Georgia (TRÄGÅRDH, 1907a).

Liebstadia Anareensis DALENIUS, 1958. It was collected by ANARE in a *Colobanthus* cushion, on a rock, Macquarie Islands.

L. Nordenskjöldi (TRÄGÅRDH, 1907). Syn. *Oribatula Nordenskjöldi* TGDH. SSE found it under stones, on the S. Shetland Islands as well as on Tierra del Fuego.

Maudheimia petronia WALLWORK, 1962. Five specimens were collected from lichens on rock 300 m.a.S. at Hallett Glacier (WALLWORK, 1962a).

M. Wilsoni DALENIUS, 1958. O. WILSON, NSBX, collected this oribatid mite on Passat and Ekberget, two nunataks far inland Queen Maud Land. It is more numerous under stones than in mosses and lichens.

Notaspis Scotiae TROUESSART, 1912. The data given by TROUESSART do not allow any correct systematical location. It was found on the S. Orkney Islands.

Oppia crozetensis (RICHTERS, 1908). Syn. *Notaspis crozetensis* RICHT. GSP collected it on Possession Island, Crozet.

O. crozetensis (RICHT.) ssp. *Anareensis* DALENIUS, 1958. This subspecies was collected by ANARE from *Colobanthus* on rocks, the Macquarie Islands.

O. nitens (MICH.) ssp. *brachytrichinus* DALENIUS, 1958. Syn. *Damaeus nitens* MICH. in RICHTERS 1908. GSP found it on Possession Island, Crozet.

Pertorgunia Belgicae (MICHAEL, 1903). Syn. *Notaspis Belgicae* MICH. MICHAEL reports it from Graham Land, where it also was found by other expeditions (DEAF, PEAF, SSE) in the same area.

United States Antarctic Service Expedition of 1939—41 found this oribatid mite as far south as 68°S. ANARE collected it on the Macquarie and Heard Islands. WALLWORK (1962c) reports it from Base Gonzales Videla and Deception I. It lives among lichens and mosses but also under stones.

P. colobanthi DALENIUS, 1958. ANARE collected it in *Colobanthus* sp. on the Macquarie Islands. WALLWORK (1962c) gives a description of the immature stages.

P. crozetensis (RICHTERS, 1908). Syn. *Scutovertex crozetensis* RICHT. It seems to be very tolerant to the humidity of the environment. GSP found it on Possession Island, Heard Island, and Kerguelen Island, from where also ANDRÉ (1947) has reported this species.

P. macquariensis DALENIUS, 1958 was collected by ANARE on the Macquarie Islands, where it lives on the shore of fresh water pools.

Podacarus Auberti GRANDJEAN, 1955. The prototype specimens were collected on Kerguelen Island. ANARE found it on the Macquarie and Heard Islands, where it lives in grass soils with *Poa annua* and *Agrostis magellanica* (DALENIUS & WILSON, 1958).

In the collection of STUDER (1879) there is a species called *Acarus riparius* which — owing to ANDRÉ 1947 — is an oribatid mite. Its systematical position, however, is quite impossible to establish.

Table I.

The number of mite species from the Antarctic regions. (Those species which are collected in terrestrial as well as in marine environments have been counted twice.)

Group of mites	Number of						
	terrestrial species		marine species		parasitic species		total
	Abs.	%	Abs.	%	Abs.	%	
Parasitiformes	11	57.9	5	26.3	3	15.8	19
Trombidiformes	8	13.6	49	83.0	2	3.4	59
Sarcoptiformes	23	62.2	5	13.9	9	25.0	37
	42	36.5	59	51.8	14	12.4	115

Discussion

If we compare the land communities of Antarctica and of the Sub-Antarctic regions with those of Arctica, we can easily state the faunistical poorness of the South Polar lands. As for the sea, however, the situation is rather the contrary. At least we can establish the fact that the sea environment is extremely rich in various organisms. This is also true of the acarids, of which hitherto 59 species are found in the sea, i.e. 52% of the total acarid fauna known up to date (table I). The dominance of the sea water mites grows still stronger, if we consider the fact that the parasitic species live on sea birds, seals or on other sea animals. The table shows that the terrestrial mites are mainly recruited from Parasitiformes and Sarcoptiformes, of which the gamasid and oribatid mites live more or less far from sea water. Among the Trombidiformes the halacarid mites create the highest per cent of marine species (83.0%).

On the whole, however, it is easy to see that the land communities are more or less closely dependent on sea. The terrestrial mites feed on algae, mosses or lichens, which in turn get their need of nitrogen and other substances covered by excreta of sea birds. This was observed by WILSON (DALENIUS & WILSON, 1958) on Passat and Ekberget on Queen Maud Land.

It is an interesting fact that the Antarctic soil contains rather a rich flora of micro-organisms, such as bacteria, which are important for the development of soil. And this condition is the real base

Fig. 2. Mites. Note eggs at upper left. Antarctic Peninsula. Photo: R. E. LEECH. Jan. 1961.

426

for the whole community, plants as well as soil animals, which is hold together by a food-chain. In Antarctica, however, this is remarkably short: excreta and inorganic substances — bacteria — lichens and mosses — springtails and mites feeding on vegetables — carnivorous mites. But here the chain ends. Consequently the community of Antarctica is not comparable with one of a region with a more highly developed food chain, the balance between the four constituents of the ecosystem: abiotic substances, producers, consumers and decomposers, is different because of the quantitatively poor development of the community.

This condition makes the ecosystem less complicated than in other regions of the world. But that is precisely why, as I see it, Antarctica might be the land where the fundamentals of ecology should be studied.

In a certain area there are three factors especially important to the soil fauna, viz. nourishment, temperature and humidity. Of course their importance changes from one area to another. Thus, in Antarctica nourishment does not seem to be a limiting factor, judging from field studies upon *Maudheimia Wilsoni*. It lives under stones, only exceptionally it has been collected in mosses or lichens. Evidently it prefers the poorer stones to the richer plant environment. There seem to be similar conditions for the acarids collected by recent expeditions. Probably in Antarctica the influence of temperature is more important to the soil mites than is nourishment. Yet it is a remarkable fact that the soil fauna is able to survive the harsh environment of an Antarctic mountain. Unfortunately it was not possible to make any observations on the thermoclimate of the habitat, where *Maudheimia Wilsoni* was found. It was, however, presumed that *M. Wilsoni* is able to sustain very low temperatures, which was, in part, also proved at the Zoological Institute of Lund University, Sweden. But it must also be presumed that this oribatid species can profit by the few occasions of favourable temperature, even if it can move and perhaps breed in a temperature between 0° and —10° C.

Because of the clean air of the Antarctic continent an intense radiation of heat can be expected. The dark rock will also absorb a great amount of heat. There is too a considerable reflexion back to the air. Consequently there are occasionally high temperatures to be found in the exposed parts of the ground, and because of the physical properties of the substrate the temperature grows higher in and just below the stones than in the mosses and lichens. On the other hand the daily amplitude of the temperature is greater under stones than in mosses. But owing to experiences from the Arctic regions of the Swedish mountains it seems, as if the maxima of temperature is more important to the soil fauna than is the mean temperature.

There are recent observations of the temperature of rocks in the Antarctic mountains. PAUL A. SIPLE (see LLANO, 1962) studied the climate of inland nunataks of the Horlick Mts. four degrees from the geographical South Pole. It was then observed that within three hours the temperature of the rock rose from —15° C to such high values as +27.8° C. This condition explains, why life is possible in the coldest area of the world, where the winter temperature of the air falls to values lower than —65° C. The organisms are extremely resistant against low temperatures and can rapidly use the short moments of favourable situations. Here there is an open field for interesting physiological studies.

But, as was hinted above, living organisms of Antarctica, plants as well as springtails and mites, are strongly influenced also by lack of free water. The nunataks are naked even in winter, because the snow blows away. In summer there is practically no rain. Water from melting snow evaporates rapidly because of the extreme radiation of heat. It has been observed, however, that sometimes ice crystals are frequently transported by wind from snowfields nearby to the areas with mosses or lichens, where because of inso-lated heat they melt into water. Such a minute amount of water is sufficient only to those organisms which have become adapted to the extremely dry environment of Antarctica. This fact that in Antarctica drought is a limiting factor seems like a paradox. But the expression "frozen desert" illustrates the curious situation of the organisms living there. The ability to resist low temperatures is undoubtedly connected with an ability of water regulation in the body tissues. The cold death is caused by the freezing of water in the body. To prevent this the animals consequently have to transfer the free water of the tissues into chemically bound water.

Probably there is also a correlation between this condition of the tissues and their ability to resist drought. There might be an inter-action between the resistibility to cold as well as to drought. In both cases it is necessary to bind water. This is a condition which I have observed on Swedish oribatid mites of the Torneträsk area (North Sweden). When autumn comes, the organisms are well adapted to the falling temperature of the ground, but then there seems also to have developed an ability to resist the low humidity of the environment during winter. It seems to be quite a natural condition, because in winter the soil is dry, just as it is in Antarctica. This is, however, a complexity of different problems, to be solved by future zoologists.

There seems also to be a correlation between this adaptability of the terrestrial organisms and their geographical distribution. This statement is not easy to prove at the present stage of our knowledge, because too few observations have been made in the Antarctic regions. And because not much news has appeared, the following will be only a very short résumé of what is known concerning the geographical distribution of some Antarctic terrestrial mites.

It is evident that some of the oribatid mites are distributed almost all round the continent. DALENIUS & WILSON (l.c.) give an account of the distribution of the oribatids known up to that date. Two of them are circumpolar, viz. *Alaskozetes antarctica* and *Pertorgunia Belgicae*, which are both adapted to an environment in some cases dry, in others wet. That may be the reason, why they are to be found in Antarctica as well as on the islands all round the continent. *Podacarus Auberti* may also belong to the same group. Among the species of other groups, the eupodid mite *Rhagidia kerguelenensis* can be mentioned in this connection. It has a wide distribution, at least from the Crozet to the Macquarie islands. *Stereotydaeus villosus* is also widely distributed in Antarctica as well as in the Sub-Antarctic islands.

The distributional agents have been discussed by both ANDRÉ (1947) and DALENIUS & WILSON (l.c.), who point out the importance of both winds and birds for the Antarctic mites and of sea currents for the mites living at the shores of the islands all round the continent.

Also to the most resistant species there are obstacles to their distribution. It is probable that spores of the organisms, such as mites, are more or less evenly dispersed all over the Antarctic continent. But nevertheless it is impossible for them to colonize everyone of the new areas, where they are dropped and to form there a population able to survive the harsh climate. Thus in the different areas, where plants have been found in Antarctica (fig. 1) the mite populations form isolated islands. Such areas are well adapted for the development of species. This has been suggested by DALENIUS & WILSON (l.c.), and the new locality of a species belonging to the genus *Maudheimia* (see above) gives material to the discussion of species development. But both as to the geographical distribution and the development of species at the present stage only some more or less vague tendencies can be discussed. Here much more studies in the field are wanted, which also indicates some of the objects of future biological research in Antarctica. The problems have recently been discussed by WALLWORK (1963).

The words of P. PAULIAN (1952, p. 138) are worth being considered: "La vie aux Iles Kerguelen, connue dans ses grandes lignes, offre encore pour de longues années des surprises au naturaliste. La travail n'est qu'ébauché mais il se révèle déjà passionnant et grâce aux installations maintenant construites des chercheurs peuvent se donner entièrement à leur tâche." Indeed Antarctica has many surprises to offer the students interested in the different biological problems of the white continent.

429

ACKNOWLEDGMENT

Fil. dr., lektor AINA RUBENIUS, Uppsala, Sweden has kindly helped me by correcting my English manuscript. I want to express my gratitude to her.

CONSULTED LITERATURE

ANDRÉ, M., 1947: Acariens. Croisière du Bougainville aux îles australes françaises. *Mém. Muséum Nat. Hist. nat., Paris* N. S. **XX,** *65*.

BALOGH, J., 1961: Identification Keys of world oribatid (Acari) families and genera. *Acta Zool. Acad. Sci. Hung.* VII, 3—4, Budapest.

BERLESE, A., 1917: Acariens. Deuxième Expédition Antarctique Française (1908—1910). Doc. sci. Paris. p. 1.

CAMBRIDGE, O. P., 1876: On a new order and some new genera of Arachnida from Kerguelen's Land. *Proc. Zool. Soc. London.*

DALENIUS, P. & WILSON, O., 1958: On the soil fauna of the Antarctic and of the Sub-Antarctic Islands. The Oribatidae (Acari). *Arkiv f. Zool.* Ser. 2, 11, **23,** *393*.

ENDERLEIN, G., 1903: Die Landarthropoden der von der Tiefsee — Expedition besuchten antarktischen Inseln. I. Die Insekten und Arachnoideen der Kerguelen. Wiss. Ergebn. der deutschen Tiefsee-Expedition auf dem Dampfer "Valdivia" 1898—1899. Bd. III. Jena. p. *199*.

EWING, H. E., 1945: Mites of the United States Antarctic Service Expedition 1939—1941. *Proc. Amer. phil. Soc.* **89,** *296*.

GAUD, J., 1952: Acariens plumicoles (Analgesidae) de quelques oiseaux des Îles Kerguelen (Récolte P. Paulian). *Mém. Inst. Sci. Madagascar.* Série A, **VII,** 2, *161*.

LLANO, G. A., 1962: The terrestrial life of the Antarctic. *Sci. American,* September 1962, *213*.

LOHMANN, H., 1908: Die Meeresmilben der Deutschen Südpolar-Expedition 1901—1903. Deutsche Südpolar-Expedition 1901—1903. Bd IX, Zool. **I,** *361*.

MICHAEL, A. D., 1903: Acarida (Oribatidae). Résultats du voyage du S.Y. Belgica en 1897—1898—1899. *Rapp. Sci. Zool. Acariens libres* R. 17. Anvers. p. 1.

PAULIAN, P., 1952: La vie animale aux îles Kerguelen. *Terre et la Vie,* **99,** *129*.

RICHTERS, F., 1908: Die Faunen der Moosrasen des Gaussbergs und einer südlicher Insel. Deutsche Südpolar-Expedition 1901—1903. Bd. IX Zool. 1, *259*.

SOKOLOW, I., 1962: Biological results of the Soviet Antarctic Expedition (1955—1958), I. Academy of Science of the USSR, Zoological Institute. Explorations of the fauna of the seas I (IX). Moskva. In Russian.

SPEISER, P., 1909: Milben (Acarina). Deutsche Südpolar-Expedition 1901—1903. Bd. X Zool. 2.

STUDER, TH., 1879: Die Fauna von Kerguelensland. *Arch. Naturgesch.* **45,** erster Band.

TRÄGÅRDH, I., 1907: Acariens terrestres. Expédition Antarctique Française (1903—1905). Sci. natur.: doc. sci. Paris. p. 11.

TRÄGÅRDH, I., 1907a: The Acari of the Swedish South Polar Expedition. Wiss. Ergebn. d. Schwedischen Südpolar-Expedition 1901—03. Bd. V, Lief. 11, *1*.

TRÄGÅRDH, I., 1910: Acariden aus dem Sarekgebirge. Naturw. Unters. d. Sarekgeb. in Schwedish-Lappland. Bd. IV, Zool., Lief. 4, *375*.

TROUESSART, E.-L., 1907: Acariens marins. Expédition Antarctique Française (1903—1905). Sci natur.: doc. sci. Paris.

430

TROUESSART, E.-L., 1912: Acariens de l'Expédition Antarctique Nationale Ecossaise. Scottish National Antarctic Expedition. Report on the scient. results of the voyage of S.Y. "Scotia" during the years 1902, 1903 and 1904. Vol. VI, Zool, *81*.

TROUESSART, E.-L., 1914: Acariens. Deuxième Expédition Antarctique Française (1908—1910). Sci. natur.: doc. sci. Paris. *1*.

VIETS, K., 1950: Die Meeresmilben (Halacaridae, Acari) der Fauna Antarctica. Further Zoological Results of the Swedish Antarctic Expedition 1901—1903. Vol. IV, No. 3, *1*.

WALLWORK, J. A., 1962a: Maudheimia petronia n.sp. (Acari: Oribatei), an oribatid mite from Antarctica. *Pacific Insects* 4(4), 15, 1962, Honolulu.

WALLWORK, J. A., 1962b: A redescription of Notapsis antarctica Michael, 1903 (Acari: Oribatei). *Pacific Insects* 4(4), 15, 1962, Honolulu.

WALLWORK, J. A., 1962c: Notes on the genus Pertorgunua Dalenius, 1958 from Antarctica and Macquarie (Acari: Oribatei). *Pacific Insects* 4(4), 15, 1962, Honolulu.

WALLWORK, J. A., 1963: Phylogenetic relationships and geographical distribution of some Oribatei (Acari) from Antarctica. *Proc. XVI. int. Congr. Zool., Washington.* Vol. 1. Washington.

WOMERSLEY, H., 1937: Acarina. Australasian Antarctic Expedition 1911—1914. Sci. Rep. Ser. C, Zool. and Bot. Vol. 10, part 6, *1*.

ZUMPT, F., 1952: The ticks of sea birds. ANARE Reports. Ser. B, vol. 1 Zool., *12*.

BIOGEOGRAPHY AND ECOLOGY OF LAND ARTHROPODS OF ANTARCTICA*

BY

J. L. GRESSITT

Bishop Museum, Honolulu, Hawaii

(with 29 figs.)

Arthropods (certain insects and mites) appear at present to have the southernmost known records for resident animals (nearly 84° S. Lat.). Possibly rotifers, tardigrades or protozoans may occur farther south. Arthropods are the dominant strictly land animals of Antarctica, as they are also of oceanic islands and of unfavorable environments in general, such as high mountain ridges or other cold areas. It is true that arthropods are the dominant land animals throughout the world, but in the less favorable environments and on land separated from continental areas by wide sea barriers, the land arthropods hold a much greater position of dominance. Land vertebrates, particularly mammals, reptiles, amphibians and fresh-water fish, are generally lacking in such situations, and land birds are generally few or lacking so that the only vertebrates are often sea birds, marine mammals and fish. This is likewise the situation on the Antarctic continent, and to a large degree on the subantarctic islands also.

It is of considerable interest that certain free-living insects and mites are able to exist in Antarctica. Parasitic insects benefit by the body temperatures of their host birds or mammals, but the free-living forms must contend with the harsh environment. The sea water around the continent is richly endowed with life, as its temperature may not fall below —2° C. But the terrestrial environments present much more rigorous conditions, with temperatures falling even below —70° C in some areas.

Only a few phyla and classes of land animals occur on the continent. Of the arthropods, only two classes appear to be present, Acarina and Insecta, besides a very few freshwater Crustacea, and

* Partial results of research under the United States Antarctic Research Program, in part supported by a grant from the National Science Foundation.

In connection with the preparation of this paper, I am indebted to G. E. DUNNET, R. R. FORSTER, M. W. HOLDGATE, J. ILLIES, H. JANETSCHEK, G. KUSCHEL, R. E. LEECH, T. S. LEECH, C. J. MITCHELL, M. D. MURRAY, C. W. O'BRIEN, M. E. PRYOR, K. RENNELL, J. T. SALMON, C. N. SMITHERS, W. O. STEEL, R. W. STRANDTMANN, K. WATSON, N. WILSON, and K. A. J. WISE. The illustrations and maps were largely prepared by PHYLLIS HABECK. Other assistance was provided by CAROL NAKASHIGE, SETSUKO NAKATA and CLARA UCHIDA.

A number of statements or references to zoogeographical studies not documented here are documented bibliographically in GRESSITT & WEBER, 1960; GRESSITT & PRYOR, 1961; and GRESSITT, 1961.

432

Tardigrada. Of the Insecta, five orders are present, and of the Acarina, five orders also. Of the five orders of Insecta, two are free-living
(Collembola and Diptera); two are permanently parasitic (Anoplura,
Mallophaga); and one (Siphonaptera) spends part of its life-cycle in
temporarily abandoned nests of birds.

The Collembola are represented by the families Hypogastruridae,
Onychiuridae and Isotomidae and the Diptera by the Chironomidae
(subfamilies Clunioninae and Podonominae). Anoplura are represented by the family Echinophthiriidae. Mallophaga are represented
by the families Menoponidae and Philopteridae. Of the Acarina,
the Mesostigmata are represented by Cercomegistidae, Neoparasiti-
dae, Laelaptidae, Ameroseiidae, Parasitidae, Veigaiadae, Rhoda-
caridae, Uropodidae, Ascaidae, Zerconidae (mostly free-living),
Rhinonyssidae (respir. tracts of penguins), Halarachnidae (nasal
mites of seals, fig. 6); the Metastigmata by Ixodidae (ticks); the
Prostigmata (Trombidiformes) by Penthalodidae, Eupodidae, Rhagi-
diidae, Bdellidae, Erythraeidae, Pachygnathidae (all free-living);
the Astigmata (Sarcoptiformes) by Proctophyllodidae, Analgesidae
(feather mites), Glycyphagidae (cheese mites); the Cryptostigmata
(Oribatei) by Oribatulidae (fig. 7), Eremaeidae, Ceratozetidae, all
free-living).

The presence of mites (Acarina), springtails (Collembola) and a
wingless midge (Clunioninae: *Belgica)* on the Antarctic continent
has been known for some 60 years. However, there is still relatively
little known about the extent of occurrence, and the ecology, of the
probably several dozen species of free-living terrestrial arthropod
species of Antarctica proper. In addition to these are perhaps two
dozen species of biting lice on the antarctic birds, and a half dozen
species of sucking lice on the various antarctic seals, as well as some
parasitic mites and two species of ticks (GRESSITT & WEBER, 1960).
Only recently a species of flea (fig. 21) was found in nests of petrels at
the edge of the continent (SMIT & DUNNET, 1962).

A few representatives of groups of insects not known for certain
from the Antarctic proper have been taken on the Antarctic mainland. These include three beetles of the family Lathridiidae taken
in air-borne insect nets at McMurdo installation, Ross Island,
77°48′ S. Lat. It is possible that these got into the nets accidentally
in New Zealand, or were blown to McMurdo Sound from New Zealand. At Marble Point (77°26′), across McMurdo Sound from Ross
Island, a mature male spider in good condition, probably of the
Micryphantidae, was also taken in a net for air-borne insects. Though
this might have been blown from New Zealand or elsewhere, it
might even represent a resident species. At Cape Hallett (72°19′)
another mature male spider, belonging to the family Attidae, was
actually found in moss and lichens (FORSTER, *in litt.).* At the Dailey
Islands (77°51′), H. JANETSCHEK found remains in soil which seemed

to be those of a thysanuran. Again, they might have been brought by air currents, or might represent local fauna. Further investigations may prove some of these, or other groups, to be actually represented in the true Antarctic fauna.

Environment

A few isolated or incomplete studies, and limited field surveys at widely spaced sites provide us with an imperfect picture of land arthropod life on the Antarctic continent. As representative of different degrees of severity of the environment as regards to terrestrial arthropods, the following samples are selected, involving the principal areas studied, progressing from north to south. Degrees of latitude are indicated for each, and also longitude:

1.	63° S.	Southern South Shetland Islands (58°—63° W).
2.	64°—65°	Danco coast, Antarctic Peninsula (Palmer Pen., Graham Land; 62°—63° W).
3. (a, b).	66°—68°	Fringes of East Antarctica (a, Wilkes, 66°S, 110°E; Mirny, 66°S, 93°E; b, Davis, 68°S, 78°E).
4.	72°	Cape Hallett, N. Victoria Land coast (170°E; also Cape Adare, 71°S).
5.	72°±	Dronning Maud Land, nunataks, up to 1,250 m (0°—4°W).
6.	75°—76°	Middle Victoria Land coast (currently being investigated by Bishop Museum team; 162°±E).
7.	77°+	S. Victoria Land, coastal areas and northern capes of Ross Island (164°—169°E).
8.	77°+	S. Victoria Land, peaks and ridges to west of coastal area, up to 2,000 m; 161°E).
9.	78°+	Foot of Mt. Discovery, S. Victoria Land, edge of Ross Ice Shelf (165°E). (Also Cape Armitage, southern end of Ross Island; 78°—S; 167°E).
10.	84°—	Hood Glacier, near foot of Beardmore Glacier, to 600 m (172°—173°E).
	(85°+	Plunket Point, upper end of Beardmore Glacier; 168°E.; Negative).

These sites represent very limited sections of the continent. Most of them are situated in two narrow sectors between 55°—66° W. Long. and between 160°—173° E. Long. The others are isolated spots on the coast between 78°—111° E. Long., and one (no. 5) at 0°—4° W. Long. slightly inland from the coast. These sites are indicated on the map (fig. 1) by the above numbers. In fig. 2, the respective numbers of known species of the various arthropod groups (exclusive of those permanently associated with the bodies of their hosts) are shown for these same localities.

Ecology

In the following paragraphs, some condensed information is presented on the numbered areas of the preceding list, and on Fig. 1.

434

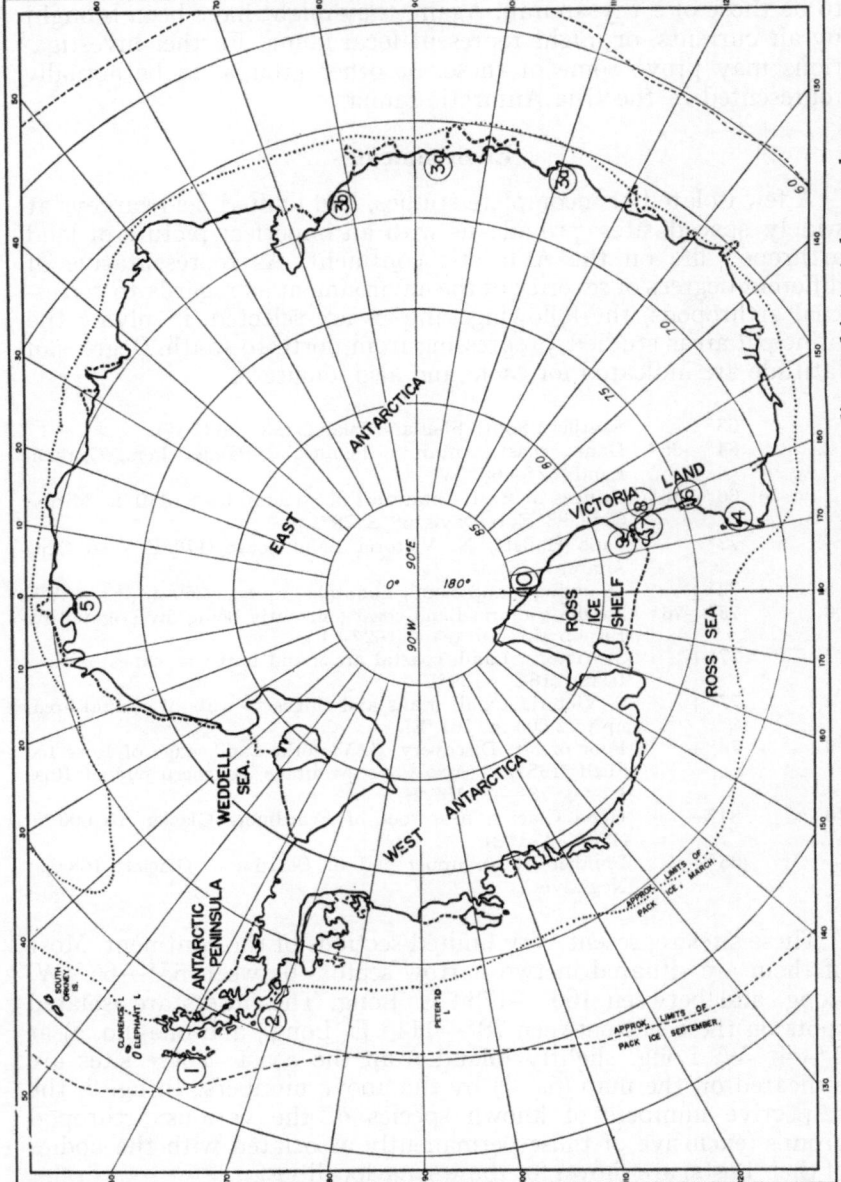

Fig. 1. Map of Antarctica indicating arthropod collecting areas discussed in text, by numbers.

435

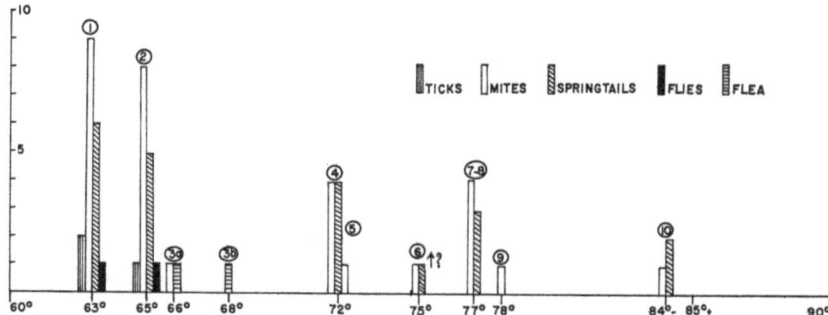

Fig. 2. Graph of arthropod representation at antarctic collecting sites, by degrees of South Latitude (numbers as in fig. 1 and text).

1. Southern South Shetland Islands (62°—63° S. Lat.; 58°—63° W. Long.)

This environment is one of the most favorable in the Antarctic. It is of the same latitude as the northern tip of the Antarctic Peninsula (Palmer Pen. or Graham Land), and is just to its west. Here the climate is relatively mild, the temperature being above freezing for considerable periods. The vegetation is somewhat varied, and besides the usual algae, lichens, micro-fungi and mosses (fig. 3) found in many of the ice-free areas of the continent, it includes liverworts, a larger fungus, a grass *(Deschampsia)* and an herb *(Colobanthus)*.

The arthropod fauna here is probably as varied as in any part of Antarctica proper. It has not been exhaustively sampled and reported for any one island, or for the archipelago in general. No detailed ecological studies have been made in the area. TORRES (1959) studied the podonomine midge. From occasional brief samplings (including some by R. E. LEECH, 1960 and R. E. and T. S. LEECH, 1960—61) there would appear to be a free-living fauna of about 20 species, including over 10 species of free-living mites (oribatids, penthalodids, eupodids, etc.); about six species of Collembola (including *Friesea grisea, Cryptopygus antarcticus, Tullbergia mixta* and *Hypogastrura antarctica;* SALMON, 1962b); and the midge *Parochlus steineni.* Ticks occur around the rookeries, and on penguins or other birds, along with other parasites, including mites and lice.

The free-living mites and springtails occur primarily in the areas of rich moss and lichen growth in snow-melt drainage areas, as well as in association with the grass, herb and liverworts. The winged podonomine midge *(P. steineni,* fig. 13), lives in brackish to fresh water. Larvae were found in a low crater-pool with partial tidal interchange through the rocks. In early summer, the water is fresher as a result of snow-melt. Adults occur on the rocks adjacent to the water, and fly on the splashing of water onto the rocks.

436

Fig. 3. Deception Island, South Shetland Is. 62°57′ S.; rich moss growth, collecting site for mites and springtails; T. S. LEECH at right; Jan. 1961 (R. E. LEECH).

2. Danco Coast, middle of west side of Antarctic Peninsula (64°—65° S. Lat.; 60°—63° W. Long.)

This area, part of the Palmer Peninsula or Graham Land, is of historic interest biologically. It is the general area of some of the earliest investigations in Antarctica, and the type locality of the wingless clunionine midge *Belgica antarctica* and several of the mites, springtails and others from the Belgian expedition of 1897—99.

On the rather limited ice-free coastal slopes, in areas of appropriate moisture, there are rich banks of moss where free-living mites and springtails abound (mostly the same as those mentioned under 1). *Belgica* breeds in small rocky snow-melt pools which may be influenced by salt spray in summer (fig. 4). These rock crevices are likely to contain debris blown by wind from nearby penguin rookeries. The adults (figs. 14, 15) may crawl about and breed on snow. They may be bothered by springtails crawling over them.

Grass, herb, fungi, liverworts and other plants found in the South Shetland Islands occur also in this general area, offering one of the richest environments of the antarctic, and probably the richest on the continent itself. Springtail populations are often quite dense (fig. 8).

437

Fig. 4. Danco Coast, Antarctic Peninsula, 64°51′ S.: site of mites, springtails and apterous midge *(Belgica)*, Base Gonzales Videla; 6 Jan. 1961 (R. E. LEECH).

3. Fringes of East Antarctica (66°—68° S. Lat.; 78°—110° E. Long.)

The great arc of the continental coast opposite Africa, the Indian Ocean, Australia and southern New Zealand, aside from the barren ice cap and ice shelves, biologically appears to be the poorest part of the continent. However, it has been little investigated for arthropods. This coastal area is much farther north than many populated areas in Victoria Land, and much of the coast is as far north as parts of the Antarctic Peninsula. There is relatively little exposed land, and most of it consists of bare rocky coastal outcrops or off-shore islets. Many of these spots are used by Adelie penguins for rookeries. Also, there are some slightly inland "oases", with ponds and rich algal growth. Possibly no mites or Collembola have been recorded except on the shores of the Ross Sea and from near 0° Long. (see following sections). Recently, M. E. PRYOR had taken at least free-living mites near Myrny (3a), but they have not yet been studied. Recently also, fleas (fig. 21) representing a new genus and species (SMIT & DUNNET, 1962) were taken in nests of petrels *(Fulmarus glacialoides, Pagodroma nivea)* near Wilkes (3a) and Davis (3b). The

438

fleas apparently over-winter in the nests which are abandoned by the birds for most of the year and are covered with a meter or so of snow during most of this period. No midges have been recorded from these coasts.

If it is true that there is so little arthropod life on this relatively northern portion of Antarctica, it may be related to the proximity to areas of minimum temperature near the center of the wider portion of the ice-cap, and also to the relative scarcity of mountains with their wind protection, exposed rock and solar radiation effects. None of the higher plants have been reported from East Antarctica. Mosses are relatively rare; so lichens, algae and perhaps microscopic fungi form the plant foods available for arthropods.

4. Cape Hallett, northern Victoria Land (72° S. Lat.; 170° E. Long.)

This locality on the coast of the Ross Sea was recently studied ecologically (PRYOR, 1962). The coast is mountainous with extensive rock outcrops and scree slopes, as well as ice-free beach and other areas which are bare of snow in mid-summer. Mosses occur in protected damp environments, and lichens and algae are widespread on rocky slopes. Four species each of free-living mites and springtails were observed at Hallett. One of the springtails, *Isotoma klovstadi*, was studied in detail, and PRYOR'S report on its ecology is summarized in the next section. No higher plants have been found here or southward, and no oribatid mites have as yet been found south of Cape Hallett.

About 100 km north of Cape Hallett is Cape Adare, site of some earlier records, including at least one mite *"Stereotydeus"*, and the above springtail *Isotoma*.

5. Dronning Maud Land, nunataks (71°—72° S. Lat.; 0°—4° W. Long.)

This area, at the border of West and East Antarctica, not far from the Weddell Sea, is the site of the only ecological study that has been made of a mite in Antarctica (DALENIUS & WILSON, 1958). Two nunataks, Passat at alt. 150 m, and Ekberget at 1650 m, were found to be inhabited by an oribatid, *Maudheimia wilsoni*, described by DALENIUS from the two localities. Ekberget, just west of the meridian of Greenwich, is some distance inland. Vegetation is primarily of lichens, and some debris from petrel nesting was found in the inhabited areas.

6. Middle Victoria Land (75°—76° S. Lat.; 160°—163° E. Long.)

This area, in the neighborhood of Terra Nova Bay, is currently being investigated by Bishop Museum's field team, under charge of K. A. J. WISE. Mites and springtails have been collected at several localities, but the number of species is not yet known.

439

7. Ross Island and South Victoria Land: low altitudes (77° + S. Lat.; 164°—169° E. Long.).

Several localities, primarily Cape Crozier, Cape Royds and Cape Barne on Ross Island, and Marble Point and Granite Harbor on the opposite coast of the mainland, have been the sites of surveys by several members of Bishop Museum teams, 1959—64. The lower ridges leading down to the capes, particularly in somewhat protected areas, or low areas, are often inhabited by abundant individuals of the springtail *Gomphiocephalus hodgsoni*, which is discussed in the next section. Also, a trombidiform mite (penthalodid) is widespread and abundant; and a few other kinds of mites, mostly trombidiform, have been found in the area. The springtails are abundant on crustaceous lichens and the mites are generally found on the damp undersides of stones where there is snow-melt or soil moisture. Some mites and springtails, as well as tardigrades, rotifers, nematodes and others are found among mosses.

8. South Victoria Land: peaks and ridges (77° + S. Lat.; 161° E. Long.)

On ridges, peaks and nunataks (fig. 5) inland and west of the coastal areas of area 7, both mites and springtails have been found at altitudes of about 2000 meters; the highest records for insects in

Fig. 5. South Victoria Land, 77° S.: Upper Mackay Glacier, looking west, northern tip of Willett Range at left; white springtails taken in center, mites at right (K. A. J. WISE, Dec. 1960).

440

Antarctica. Mosses are rare, lichens are the dominant visible plants, but microscopic fungi and other organisms may supply much of the primary food. One locality was on the south side of Taylor Dry Valley, below a glacier. Some aspects of ecology in this area and the preceding area are discussed in the next section.

9. Foot of Mt. Discovery, South Victoria Land (78° 24' S. Lat.; 165° E. Long.)

The next to the southernmost record is for a species of trombidiform mite, *Stereotydeus mollis*, at the foot of Mt. Discovery, toward Minna Bluff, southwest of Ross Island. The exact locality was on moraine separated from the foot of the mountain by a largely iced-over stream; the moraine being pushed northward by the end of a small glacier coming around the lower slopes of the mountain. Limited algal growth was present. The mites were running from one rock to another in the sunshine on a north-facing protected gravel slope.

On Observation Hill, between McMurdo Station and Cape Armitage, on the southwestern tip of Ross Island, a small trombidiform mite was found, but no springtails.

10. Hood Glacier, near Beardmore Glacier (83°55' S. Lat.; 172°— 173° E. Long.)

This is the southernmost known area of occurrence of free-living permanent resident animals. Two species of springtails, *Biscoia sudpolaris*, genus and species known only from this area, and *Anurophorus subpolaris*, endemic species (SALMON, 1962a), and one small mite, representing the family Pachygnathidae (Trombidi-

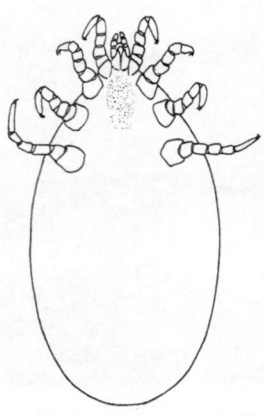

Fig. 6. Nasal mite of elephant seal; *Halarachne miroungae* FERRIS (Halarachnidae), Danco Coast (after DOMROW, 1962, Pacific Ins. vol. 4, no. 4).

formes), occur here. They were found by H. TYNDALE-BISCOE and colleagues at several localities on ridges and nunataks at altitudes from 150 to 600 meters (TYNDALE-BISCOE, 1960). A color photograph of the area is shown in GRESSITT, 1961 (fig. 23, f).

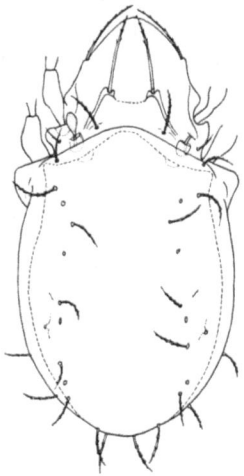

Fig. 7. Free-living oribatid mite, *Maudheimia petronia* WALLWORK (Oribatulidae), Cape Hallett, 72° S. (after WALLWORK, 1962, Pacific Ins. vol. 4, no. 4).

A more southern area, Plunket Point, at the upper end of the Beardmore Glacier (85°5′ S. Lat.; 168° E. Long.) was carefully searched by Dr. H. JANETSCHEK during the 1961—62 season. It was found to be completely negative for living organisms, the area being both very cold and too dry to support life. It is possible that arthropods may occur in mountains this far south, such as the Queen Maud Range to the east.

Provided sufficient moisture is present, certain types of environments are frequently inhabited by arthropods. In general, these are in protected situations, facing northward or at least away from the prevailing winds from the ice-cap. Amphitheatres providing reflection of solar radiation from rock or snow banks warm the rock surfaces. Moisture from snow-melt or subsurface moisture is essential (GRESSITT & LEECH, 1961).

Another type of environment is provided by freshwater ponds. These ponds, generally rich in algae, rotifers and protozoans, are sometimes inhabited by mites or springtails. The mites may be in the algae, generally at the margins, and the springtails may be floating on the surface. More often, springtails are seen floating on small pools in melt streams.

442

Insect ecology on the fringe of Antarctica

HOLDGATE (1964) has given an interesting account of the general
terrestrial ecology of the Antarctic fringe, based on the preliminary
results of an ecological primary survey in progress on Signy Island
of the South Orkney Islands. The South Orkneys are just south of
60° S. Lat., and thus barely within the antarctic zone or region.
HOLDGATE presents an interesting classification of far southern
zones:

	South Temperate	Subantarctic	Outer Antarctic	Inner Antarctic
Temp. regime (mean temp., months)	2 above 8.5° 7 + above 5.4°	7 — above 5.4° 6 + above 0°	1 + above 0°	None above 0°
(Winter, mean monthly temp.)			Rarely below —10°	Well below —20°
Vegetation	Woody veg. in lowlands	Vascular: tussock	2 flowering plants, liverworts	No liverworts or flowering plants

Signy Island's vegetation is typical of the "Outer Antarctic"
zone. It does not cover all of the nearly one-half ice-free surface of
the island. Some bare slopes are greatly disturbed by solifluxion, and
have only isolated patches of moss. Three formations were distinguished: a) *Andreaea — Usnea* on knolls and ridges; b) *Polytrichum
— Dicranum* on well-drained situations; and c) *Drepanocladus —
Acrocladium — Brachythecium* of runnels, snowpatches and some
coastal areas. The two flowering plants grow only at low altitudes
on more or less northward facing slopes, in favorable radiation traps.

On Signy, summer arthropod activity is correlated with warming
of the ground by direct radiation. On sunny days the temperature in
the moss mats increases relatively much more than the air temperature. Also, moss and soil differ in speed with which they absorb
radiation, and in the temperatures they reach. Even mats of different species of mosses attain different temperatures. For instance,
on one sunny afternoon, three moss mats composed of different
genera of mosses gave temperature readings of 3°, 5.25° and 7.5° C,
respectively, when air temperature was 2°. The mosses consistently
reaching higher temperatures have a much higher percentage of air
space and a lower percentage of water-content.

On Signy, these moss mats have a rich microfauna, particularly
of mites, springtails and nematodes. Preliminary studies indicate

that abundance of individuals differs at different levels. On 13 December 1961, a patch of *Dicranum — Polytrichum* mat 20 × 8 cm yielded 264 mites and 232 springtails from the upper 2.5 cm, but only 51 mites and 3 springtails between depths of 2.5 cm and 5 cm. This type of mat, which has a high air-content, supports more individuals than the wetter types of bryophyte mats.

The South Sandwich Islands are northeast of the South Orkney Islands, and just north of the 60° meridion of latitude, an arbitrary boundary of the Antarctic region. However, these more isolated, largely ice-covered small islands have an environment similar to those on the "Antarctic fringe", like Signy, just discussed, or even a more rigorous environment. Very few arthropods are known from the South Sandwich Islands. HOLDGATE found, however, that around fumaroles and the main crater on Bellingshausen Island, there were rich moss mats and hepatics. Mites and springtails were also abundant in these mats. Since fumaroles are likely to be short-lived phenomena, HOLDGATE suggested that this wealth of local development might prove to be an indication that ecological

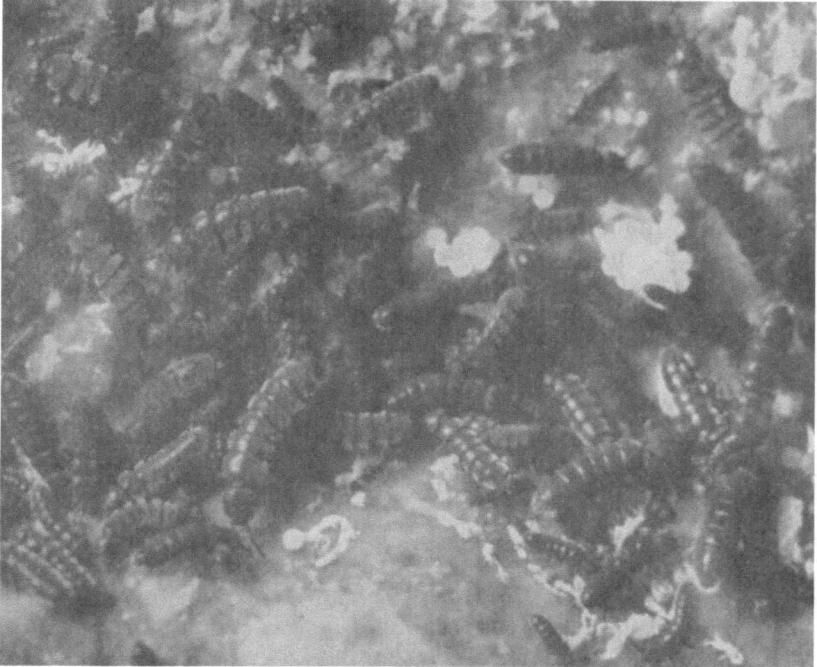

Fig. 8. Springtail, *Cryptopygus antarcticus* WILLEM (Isotomidae) with eggs, Base Gonzales Videla, Danco Coast, Antarctic Peninsula, 64°51′ S.; Jan. 1961 (R. E. LEECH).

444

poverty may prove to be more important in the populating of far southern areas than the obstacles to dispersal.

For the winged fly *Parochlus steineni* (GERCKE) of the South

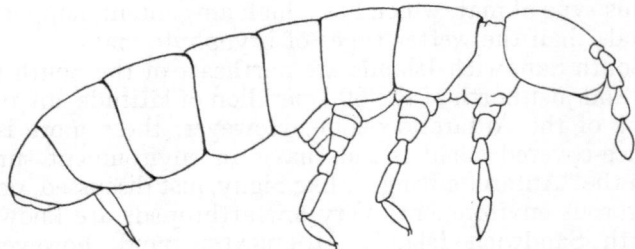

Fig. 9. Springtail, *Cryptopygus antarcticus* WILLEM (Isotomidae), same locality as in fig. 8.

Shetland Islands and South Georgia (fig. 13), TORRES (1959) has given a few ecological notes. It was observed flying close to the surface over snow-melt ponds on King George Island, in January. Males were more abundant than females. In South Georgia, it has been observed in numbers on the snow surface. Air temperature recorded was 0.25°—1.5° C.

For the wingless fly, *Belgica antarctica* JACOBS on the Antarctic Peninsula (fig. 14), TORRES (1953) summarizes biological information. Larvae were found in small ponds with green algae and guano from nearby rookeries. Adults were found in moss, on ground, and on damp rocks. At certain times, during the mating season, adults were seen floating in masses on ponds. These were as many as 5000—6000 individuals in a mass. Of 2000 individuals, 82% proved to be males. As mentioned above, LEECH & LEECH found mating individuals on snow (fig. 15), and larvae in the water, in cavities of rocks, in a penguin nesting area.

Insect ecology in outer middle Antarctica

The springtail *Isotoma klovstadi* CARPENTER (fig. 10) was studied by PRYOR (1962) at Cape Hallett, northern Victoria Land, during the summer seasons 1958—59 and 1959—60. It is also known from Giekie Land (Cape Adare), Tierra del Fuego and Macquarie Island. PRYOR found this springtail on most slopes near Hallett Station except those facing west toward the source of prevailing winds. It was at all elevations where snow-melt was available, including drainage channels, but not on flooded flats. It was abundant among mosses, but scarce among lichens, and absent from rookeries. Individuals were found in dry depressions on undersides of stones. At temperatures below freezing, they were found in clusters in frost fractures of stones. Preferred niches were large flat rocks anchored on the windward side. When temperatures were high and humidity low, they moved downward to cooler, damper areas, sometimes clustering in branches of moss gametophytes. They were often found with nematodes, protozoans and tardigrades, but rarely with mites. They were found with feather boluses and egg shells, as well.

PRYOR found that these springtails move considerably when circumstances are not favorable. When temperatures are high on talus slopes, they may move into mosses where the temperature may be 11° C cooler. They become immobile at temperatures below freezing. The temperature becomes lethal for this species between —50° C and —60° C. This species, like all Antarctic Collembola, has cutaneous respiration. It is active in the morning and afternoon when humidity is higher. During mid-day, the humidity may fall to 15%, which causes the springtails to become inactive. Low humidity has the least effect on the eggs, but is most harmful to young springtails. Eggs are not viable after exposure for 24 hours at less than 5% relative humidity, but adults died in 15 minutes at 5—10% relative humidity. Adults were still alive after being submerged in water for 5 days.

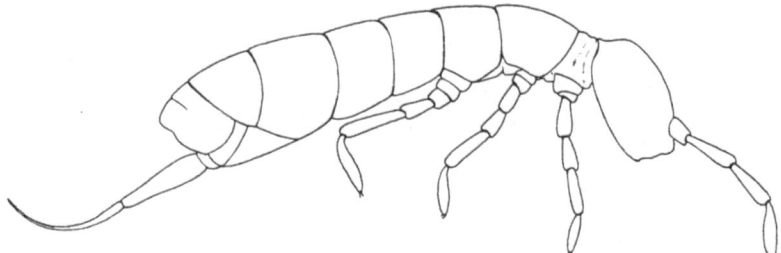

Fig. 10. Springtail, *Isotoma klovstadi* CARPENTER (Isotomidae), Cape Hallett, 72° S.

The reaction to light on the part of the springtails is of minor significance according to PRYOR, and much less important than reaction to temperature and humidity. He found that there was no response to light at low temperatures, although there was response to heat in darkness. He concluded that *Isotoma* is neither positively nor negatively phototropic, and that the tendency to move to the other side of an overturned stone is a reaction to temperature and humidity rather than to light.

This species overwinters in egg and adult stages. The adults are resistant to low temperatures, but may die of desiccation. The eggs, laid singly in mosses or soil, are more resistant to desiccation. The adults may aestivate in the middle of warm summer days when humidity drops.

Food consists primarily of mosses and algae, and adults may be

446

crowded together when feeding. Spores of algae and mosses were found in digestive tracts. Sometimes they feed on dead birds. Experimentally, springtails were reared on lichens for short periods.

Climate dominates the environment, and temperature, moisture, wind and soil are critical factors. Daily temperature fluctuation may be as great in summer as seasonal fluctuation in temperate climates.

Cape Hallett has the richest fauna so far surveyed in Antarctica except for the Antarctic Peninsula.

Some of the different interpretations regarding reaction to light between the conclusions of DALENIUS & WILSON (for the mite *Maudheimia*) and PRYOR (for the springtail *Isotoma*) may relate partly to greater dependence on moisture in the case of the springtail.

Insect ecology in inner Antarctica

As a result of recent field work in southern Victoria Land and on Ross Island by LEECH, GRESSITT (GRESSITT & LEECH), WISE (GRESSITT, 1961; GRESSITT, LEECH & WISE 1963), TYNDALE-BISCOE (1960), and particularly by JANETSCHEK (1963), some preliminary information is available on the ecology of land arthropods in the most southern known inhabited areas. The areas investigated are between 77° and 85° S. Lat., and both mites and springtails (fig. 12) were found to 83°55′ S., as well as at considerable altitudes. These include areas 7, 8, 9 and 10 in the preceding list.

The only proven free-living land arthropods to be found in the area are a few species of mites and springtails, aside from tardigrades. Some of these have just been named, and some are still without names. In some particular environments, some of the species are abundant. But often they appear to be absent in niches that outwardly seem appropriate. Thus, the geographical occurrence of the species appears to be limited and spotty. A few of the species are known only from a single area, and others occur in widely separated localities. No doubt further field work, which will become more possible with longer range helicopters, will reduce wide gaps, as well as bring to light additional species. The nature of the spotty occurrence of insects in this far southern area suggests that even though the environment is inhospitable, there are many niches which are empty because of insufficient time to populate them in the face of the many obstacles. This spotty occurrence strongly suggests that the primary means of spread of species in Antarctica is by air dispersal. As shown below, air dispersal has no doubt also been important in bringing forms of life to the continent, although some of the residents may represent remnants of the once rich fauna which existed in Antarctica in late Paleozoic and perhaps much of the Mesozoic and later. It would appear that some of the species have evolved in the area, and others are recent immigrants. More of the former appear to be endemics restricted in distribution, and more of the latter to be wide-spread, interestingly enough.

In this far southern area, occurrence of arthropods is naturally limited to ice-free areas, although inhabited niches may be snow-covered in winter and presumably reach very low temperatures. Adequate studies of winter conditions tolerated by these species remain to be made in the future. In general, the arthropods occur in special types of environments where protection from the elements is afforded. Although many of the ice-free areas are thus because almost constant winds carry the falling snow elsewhere; they are at the same time relatively warm areas in summer, as the exposed rock absorbs solar energy and the substrate, particularly the under-sides of thin rocks, reaches temperatures far higher than does the air a short distance above the rocks. A most important requirement for life is moisture, and some of the ice-free areas lack sufficient moisture, and are barren deserts. Others, such as many scree slopes, provide too ephemeral an environment for establishment of colonies of plants or animals.

Two terrestrial ecosystems (and biocenoses) exist in this southern area, according to JANETSCHEK (1963):

a) *Chalikosystem, or bare gravel system:* system of bare gravel consisting of weathering products of bedrock and/or moraine, without visible (macrophytic) vegetation. It is inhabited by an hemiedaphic mesofauna of 3—4 species of Collembola and 1—2 species of trombidiform mites. The simple food-chain apparently begins with the Collembola feeding on minute soil fungi. Soil microphytes appear to be in all places inhabited by arthropods which lack visible vegetation, up to altitudes of 2000 meters (at 77° + S.) which seems to be the upper limit of the Chalikosystem. At this possible absolute limit for animal life in these latitudes, the fauna is very impoverished, with 1—2 species of Collembola only; and sometimes one species of mite. This corresponds with the situation six degrees farther south at one-third or less the altitude, at the Hood Glacier. The species here (not yet identified), as compared with *Gomphiocephalus*, are far fewer in numbers of individuals, are more active at lower temperatures, and probably cold-stenothermic. The springtails and mites of this system are soft skinned, requiring high humidity. Thus, the most important limiting factor is not temperature, but moisture. The humidity of the soil air and the surface is closely correlated with the occurrence of soil on clay (from local rocks or brought in moraine).

b) *Bryosystem:* The other local terrestrial ecosystem is more or less open, rarely closed, and consists of macrophytic vegetation made up of mosses, lichens and rarely thin algal cover. Lichens are dominant, but mosses provide the richer environment and thus house more of the characteristic fauna. This is the climax vegetation

448

of the area. Its fauna is similar to the bryofauna elsewhere, and probably some of the species are widespread. Besides just a few springtails and mites, the fauna includes Protozoa, bdelloid rotifers, nematodes and tardigrades. The vertical distribution of the bryo-system corresponds with the upper limits of scattered mosses and soil lichens, up to about 1300 meters in the areas of Mt. Seuss and Mt. England (77° + S.)

The Chalikosystem and the Bryosystem often occur side by side, or inter-mixed, depending on local conditions, often forming a mosaic. This may be largely the result of microrelief and its direct influence on insolation and wind, and indirect influence on tempera-ture, snow, snowborne humidity, or others, thus controlling humidi-ty and temperature.

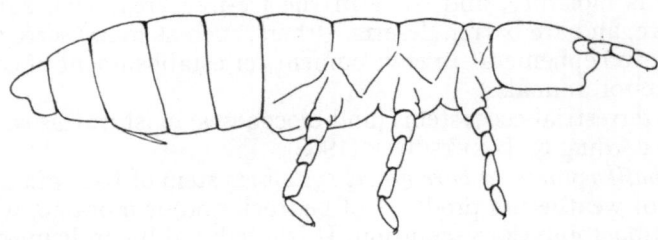

Fig. 11. Springtail, *Gomphiocephalus hodgsoni* CARPENTER (Hypogastruridae),. Marble Point, 77°27′ S.

Actually, JANETSCHEK found plant and animal life much higher, at 3600 meters on Mt. Erebus (Ross I.), but it included only fungi, bacteria, Cyanophyta, rhizopod Protozoa, a small bdelloid rotifer and a tardigrade cyst. This may be attributed to the heat supplied by the active volcano, together with constant moisture supply.

JANETSCHEK did some laboratory and field tests on tolerances of the common lowland springtail of the McMurdo Sound area, *Gomphiocephalus hodgsoni* CARPENTER (fig. 11). Unexpectedly, its temperature preference is +11.32 ±0.55° C. The highest commence-ment of reversible cold stupor was observed at +6.5 °C, and commen-cement of anabiosis with complete cessation of movement at —18°. Verified frost-resistance extends to —20 ±2°, while temperatures below —28° appear to be lethal. The lowest commencement of temporary heat stupor was observed at about +20° C, but supra-optimal increased activity commences at +17.5°, with failure to accomplish normal feeding. Temperatures above +33° were lethal. There are phobic reactions to heat, and thermoreceptor organs are probably on the antennae, as deduced from the behavior. There seem to be no similar protective mechanisms against cold. This sensory reaction apparently accounts for an asymmetrical frequency

curve (JANETSCHEK, 1963, fig. 1) of activity in relation to temperature, by which activity decreases much more rapidly with increased
temperature than with decreased temperature. But depending on a
complex of factors, the ranges of normal and disturbed activity and
reversible stupor more or less overlap. There seems to be some correlation between the time period of recovery from cold stupor and the
velocity of temperature increase of the environment. Much further
research on this and related subjects is needed. Temperature change
gradients are probably of ecological importance, besides the maxima,
minima, averages and sums.

It is somewhat surprizing to find such a high temperature preference for one of the southernmost terrestrial animals. This suggests
that the species is not in very close harmony with its environment
and not peculiarly adapted to the severe climate of this area. It
may have immigrated to this area in fairly recent times. However,
it has not been positively recorded from as far north as Cape Hallett.
Its recent entry into the area could explain its absence from many
niches in the region which appears quite appropriate. It is also relevant that this species has been taken on several occasions in air-
borne insect trappings. In spite of the abundance of the species, it is
not difficult to understand that the environment offers many obstacles to frequent successful colonizations, even where niches are

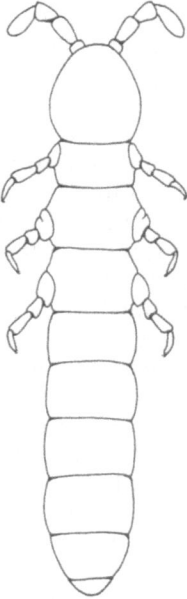

Fig. 12. Springtail, *Anurophorus subpolaris* SALMON (Isotomidae), Hood Glacier,
near Beardmore Glacier, 83°55′ S.

empty. One important factor, of course, involves the limited periods of activity above ground, which are less for this species than for others more adapted to the low temperatures. On the other hand, the range of activity under laboratory experiments, correlated with the microclimatic conditions of the natural environment, showed that during much of the 24 hour cycle some soil temperatures are high enough to permit feeding and reproduction. With absorption of solar energy by gravel and sand, it is possible for the subsurface environment to become too warm for the springtail. Areas normally thus overheated are not occupied by arthropods. However, in many places, insufficient humidity is the reason for lack of life. The dry soils here reach higher temperatures than the temperate alpine soils because of less cooling by evaporation.

Among areas studied, Cape Crozier, and some small spots between Cape Royds and Cape Barn, all on Ross Island, seemed to provide the most favorable environments, and the largest populations. These are at low altitudes. As many as 100 *Gomphiocephalus* were found in about 5 cm^2 on the underside of a small rock encrusted with yellow lichens near Cape Barne. Populations are highly concentrated in the most favorable spots, but the general production of animal life here is comparable to the scanty production of pioneer phases of animal population development in primary soils very poor in humus on higher mountains in temperate zones.

Of great ecological importance here are the eustatic (stable) or astatic (unstable) conditions of the environment. The microclimate of the Chalikosystem (bare gravel) showed the greatest stability in the fine gravelly and sandy soil near the surface. Thus this appears to be the optimal ecotope. Thin snow patches remaining in summer do not contribute to a satisfactory environment, as in the dry air the snow may sublimate directly into the air, without providing soil moisture. The most stable humidity source seems to be the subsurface clay, at least where the permafrost level is depressed sufficiently to release the moisture from the clay. Absence of life in many localities may be explained by the lack of clay. Another explanation may be failure of recently de-glaciated areas to become populated because of insufficient time to overcome the odds against successful dispersal and colonization.

Two other species of springtails present in this area of southern Victoria Land, at higher altitudes, appear to be autochthonous. They probably evolved here, and are better adapted to the environment. They occur in the more impoverished facies of the Chalikosystem, under extreme conditions of tolerance, up to altitudes of 2000 m. They might be the same as those just described from the Hood Glacier area. The Chalikosystem is undoubtedly the oldest system here, and probably that in which the endemic forms developed.

One bit of possibly contradictory evidence was the observation on

Mt. Suess of large rock areas entirely covered with lichens which were almost entirely dead. This suggests that there was a warmer period after the period of maximum glaciation of the continent.

Soil development in these ecosystems usually only reaches a poor primary soil with only a little primary humus in the Chalikosystem, but in locally favorable places may reach a stage like a kind of "Protoranker" and "Ranker" (KUBIËNA, 1948) in the Bryosystem. Considering together both pedological and coenological characteristics of the local systems, and ranging them from poorer to better conditions, JANETSCHEK (1963) proposed the following schematic arrangement:

Chalikosystem → mosaic of both → Bryosystem
Primary soils → "Protoranker" → "Ranker"

JANETSCHEK observed the best development of the first on Mt. Seuss at altitudes of 800 to 1200 m, and the best development of the last on quite young moraines of the Blue Glacier near the outlet at the coast, at 60 m altitude. At that point there are more or less closed stocks of mosses with an extent of 50 to 60 m². In the mosaic, the Bryosystem occurs in the more protected concavities in the terrain, and the Chalikosystem on the more exposed rock areas.

The influence of birds on the development of these biocoenoses

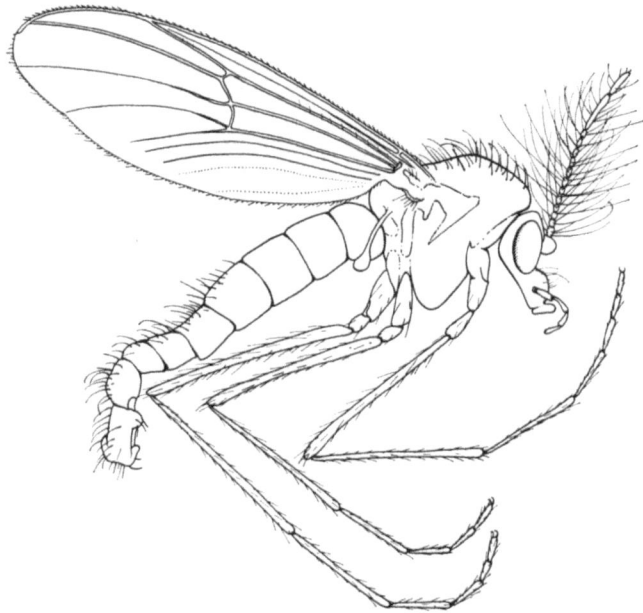

Fig. 13. Podonomine midge, *Parochlus steineni* (GERCKE), Chironomidae, Penguin Island, South Shetland Is., 62°04′ S.

452

and soils seems to be of rather little importance, and limited to small areas near penguin rookeries and skua territories. Within the penguin rookeries no arthropods were found except the ectoparasitic Mallophaga.

Ecological features

A characteristic of both Antarctic and subantarctic terrestrial faunae is a general scarcity of predators and parasites of insects. Most of the species are primary feeders on plants, animals, carrion or decaying vegetation. Some of the mites may be predators on insects (particularly on their eggs) but their habits are not well known. In the subantarctic areas, spiders are probably the principal predators. On Campbell Island, the red-billed gull was seen to catch syrphid flies hovering in the air. The scarcity of entomophagus insects may be related to the improbability that a parasitic insect which is dependent upon a particular host may succeed in reaching and establishing in the same locality as its host, after the host has become established. Larvae of some of the coelopid flies tend to be carnivorous, feeding on other larvae in the rotting kelp on beaches.

Food cycles are thus very simple in general, with often only a single direct step of plant and plant-feeder, debris and scavenger, or vertebrate and ectoparasite. The lack of entomophagous insects, again, is often reflected in superabundance of the insect species

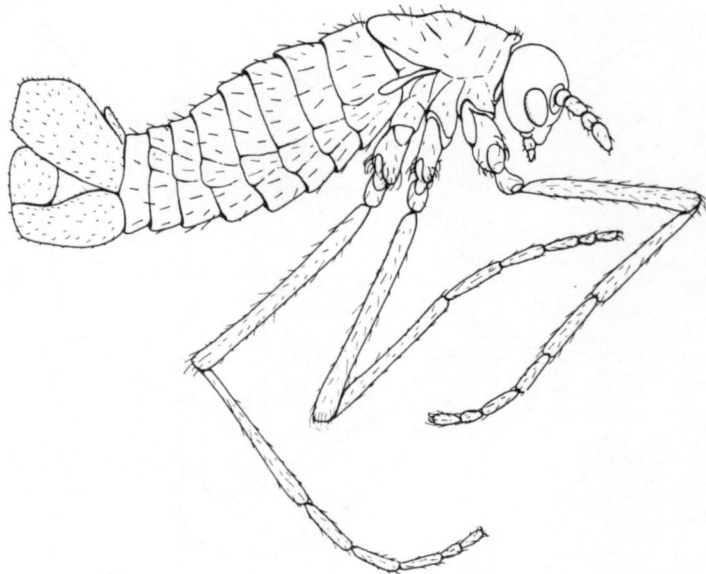

Fig. 14. Clunionine midge, *Belgica antarctica* Jacobs (Chironomidae), Base Gonzales Videla, Danco Coast, Antarctic Peninsula, 64°51′ S.

453

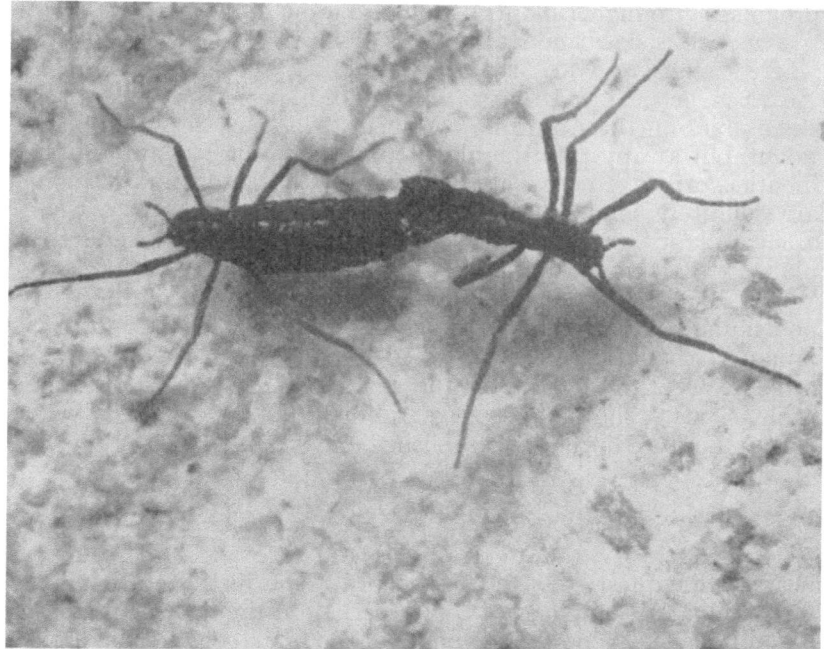

Fig. 15. Clunionine midge, *Belgica antarctica* Jacobs (Chironomidae), Base Gonzales
Videla, Danco Coast, 64°51′ S.; March 1961 (R. E. Leech).

present, where the environment is sufficiently favorable. This is
particularly true of Collembola, both on the continent and on the
islands. This has also been noted in very high mountain areas, as in
the high Himalaya (Mani, 1962). This must surely be in part attributable to lack of competitors and natural enemies in the extreme
environments.

On the subantarctic islands, and to some extent on the rare ice-
free coasts of northernmost parts of the continent, seals, particularly the elephant seal, greatly modify the environment. They
flatten vegetation, and in areas of sand, soil or bogs, make wallows
which are polluted with their excrement. Aside from disturbing
plants and arthropods, a very rich environment is created, which
supports a number of kinds of insects. On Macquarie mites, springtails, psychodid flies and others were found in this environment.
On Campbell quite a varied fauna was evident in such environments.
In the rotting carcasses of elephant seals, another rich environment
is presented, and with progression to more northern islands, the
variety of inhabitants is increased.

In general, large segments of the fauna spend much or all of their
existence underneath the mat plants, including mosses, lichens,

progressing to higher plants which also grow in low mats, and on to the tussock, sedges and leaf-mold beneath the larger herbs or scrub where such occur (GRESSITT, 1962).

It has been stated that no introduced species have established themselves on the Antarctic continent. This may prove to be mistaken, but at any rate the likelihood of many successful establishments is rather remote. On the subantarctic islands, however, there are definitely some establishments among arthropods, as well as among plants and domestic animals. Great changes have taken place in the vegetation and ecology of some islands (HOLDGATE & WACE, 1961). How much affect this may have had upon the arthropod fauna may be difficult to assess because of insufficient earlier faunal investigations, poor knowledge of arthropod food habits and other aspects of ecology on the islands. Feral domestic animals, such as cattle, sheep, goats, pigs, rabbits, rats and others as well as certain kinds of birds, are established on various islands.

Dispersal

The question of the origin of the land arthropods of Antarctica bears on the question of the history of the continent and the history of its climate. If the earlier colder period on the continent was too severe for the survival of remnants of the fauna which existed there when the continent's climate was temperate, then it is likely that all or much of the present fauna has resulted from immigration in fairly recent time. Thus it suggests transport in air currents (GRESSITT, 1961) or by the agency of birds (FALLA, 1960; TAYLOR, 1954).

There has been little positive evidence presented that birds carry insects from place to place, although the points made by TAYLOR and FALLA will bear further attention. It seems quite reasonable that birds may play a part in the spread of arthropods. DALENIUS & WILSON (1958) suggested that most mite dispersal has been effected by birds to Antarctica, and that winds are responsible only for local dispersal within limited areas on the continent. Some indirect evidence cited was that debris of bird remains was found in the places where free-living mites were found living in Dronning Maud Land, showing that petrels had nested in the area. These petrels regularly migrate to subantarctic islands and return to the continent in summer to nest. It has also been shown that in some areas of the continent plants are more abundant where birds nest. PRYOR (1962) pointed out that insect populations were greater near rookeries (but not within penguin rookeries). This is in part correlated with food supply and soil production, for bird rookeries are the main source of organic debris. However, there may be some correlation with dispersal by birds. JANETSCHEK (1963) minimizes direct relationships between rookeries and insect populations.

Table I.

Summary of arthropods trapped from the air in Antarctic and Subantarctic regions.

Group	No. of specimens	Range of latitude	Maximum Distance from probable source, in km	Group occurs in Antarctica	Group occurs in Subantarctic Is.
Acarina	4	41—62°	1600	(X)	X
Araneida	3	52—59°	500	?	X
	1	77°	3000?		
Collembola	5	63—77°	1000±	X, (X)	X
Blattaria	2	42—72°	2600?	—	(X)
Thysanoptera	14	39—58°	1200	—	(X)
Psocoptera	5	42—58°	1200	—	X
Homoptera	6	39—58°	1200	—	X
Heteroptera	9	39—66°	2000	—	(X)
Neuroptera	7	42—48°	400	—	X
Lepidoptera	6	40—71°	3500±	—	(X)
Diptera	113	38—75°	2000	(X)	X
Coleoptera	8	39—77°?	3000?	—	X
Hymenoptera	41	38—58°	1200	—	X

X = group represented by species or families trapped.
(X) = group represented, but not, or probably not, by types trapped.

In regard to transport in air currents, there is mounting evidence to indicate that this method of transport may be important in the populating of new areas. It seems to be particularly applicable to arthropods. Trapping experiments carried on during recent years in the Pacific and in antarctic areas (GRESSITT, 1961; GRESSITT & YOSHIMOTO, 1963), has shown that many insects are carried great distances across ocean in air currents. This work has involved the screening of large volumes of air by means of large nets and suction traps on ships, a trap in a superconstellation airplane, and traps on smaller planes. In the work done in the Pacific (primarily tropical) over a longer period, interesting results have been obtained which indicate considerable correlation between the types of insects trapped and those established naturally on the more isolated and smaller islands. This is a strong indication that natural dispersal is largely by air currents to land which is greatly isolated from continental areas, and has been for longer periods or since elevation above sea. Most of the groups of insects occurring naturally on small isolated islands are small insects of low specific gravity, which might have been so transported. This is also true of most of the groups represented in Antarctica and on the subantarctic islands (see Tables I and II). Of course many types of insects which might be blown to Antarctica would not succeed in establishing themselves because of the inhos-

Table II.

Groups of insects trapped south of 50° South Latitude.

	No. of specimens	Range of degrees Lat.	Long.	Distance from probable source in km
Acarina	3	55—62	159—172 E	1600
Araneida	1	59	175 E	1300
?Lycosidae	1	52	175 W	900
Salticidae	1	52	175 W	900
?Micryphantidae	1	77	162 E	3200?
Collembola				
Poduridae	4	63—77	62 W / 164—171 E	1—400
Tomoceridae	1	70	85 W	500+
Blattaria	1	72	170 E	2700?
Thysanoptera				
Thripidae	1	58	175 E	1200
Psocoptera				
Liposcelidae	3	58	175 E	1200
Homoptera				
Aphididae	2	54	175 E	800
Coccidae	1	58	175± E	1200
Jassidae	1	53	175± W	1000
Heteroptera				
Lygaeidae: *Nysius huttoni*	1	60±	171 E	1400
Lepidoptera				
Nymphalidae: *Vanessa gomerilla*	1	71	97 W	3500
Gelechiidae	1	68	80 W	1400
Microlepidoptera	1	54	175 E	800
Diptera				
Mycetophilidae:				
Mycetophila ?*marshalli*	1	54	169 E	800
Piophilidae	1	55	159 E	1200±
Coelopidae	1	55	159 E	3 (Macquarie)
Sphaeroceridae	1	51	170 E	600
Coleoptera	1	75	166	3000
Lathridiidae	3	77	166 E	3200?
Hymenoptera				
Eulophidae	1	53	175 W	1000
Ichneumonidae	2	53	175 W	1000

pitable environment. Another important obstacle is that prevailing wind directions on the continent are from the south, or at least from the ice-cap and thus unfavorable for immigration. Probably the air masses cooling and sinking at the south pole have come great distances at high altitudes from south temperate areas. Furthermore, around the continent there is a fairly regular trend of winds from the west, in regular rotation in the areas between the continent and the other southern continents. Between the "forties" and the "sixties" (to about 66° S. Lat.) there is practically no land except for the few small subantarctic islands, the southern tip of South America, and the tip of the Antarctic Peninsula. Thus, insects cannot easily be blown very directly to the antarctic continent. Again, the habits of individual species have a great bearing on the likelihood of their becoming air-borne. In addition to the trapping results shown in the tables, PRYOR (1962) trapped springtails by various means at Cape Hallett.

It has been debated whether insects carried great distances in air currents would survive the adverse factors, primarily of desiccation. GISLEN (1948) stated that small organisms can survive many hours in the air. Of course some types of insects are better adapted than others for withstanding desiccation. Cloudy weather is more favorable than sunny weather, and air humidity is a critical factor.

Wing reduction

There is a distinct tendency toward reduction of wing size or complete loss of wings among insects of normally winged groups. This is characteristic of populations on very small islands, but is more strikingly so on far southern islands. There appears to be a clear association with non-use of wings in connection with the strong winds which are normal in these far southern latitudes. Thus there appears to be a rapid selection in the direction of wing reduction. It may well be argued that insects which fly may be carried out to sea, and also that insects with fully developed wings, even if not flying, are more apt to be picked up by strong gusts of wind and carried away than those with small wings or no wings.

On Campbell Island, with the help of K. RENNELL, I carried out trapping experiments to attempt to determine both the transport of insects to the island from elsewhere, and the blowing of insects off the island. Batteries of nets were placed both on windward cliff-tops and on a leeward shore. The results have not been analyzed, but there was a very obvious correlation in the leeward shore trap battery between flight activity and wind speed. When wind was strong, very few insects were flying, but when wind was only moderate, many more were on the wing. Large numbers of insects were trapped. Presumably many of these would have been blown into the sea by

458

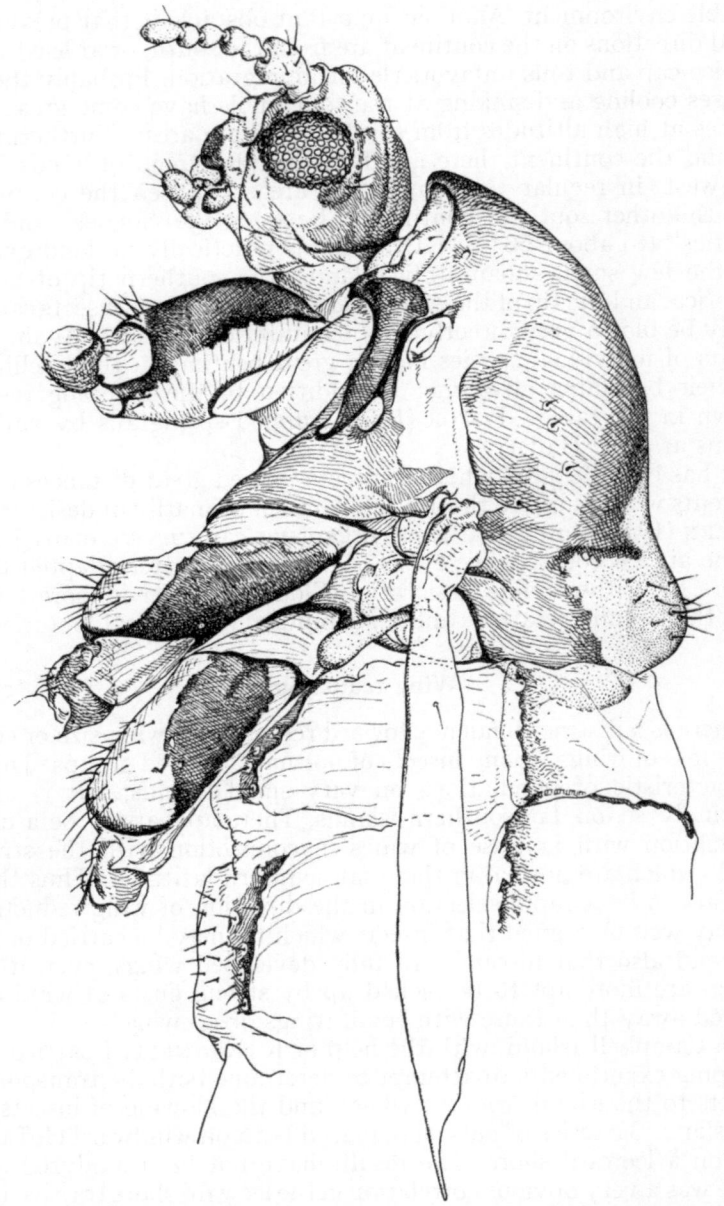

Fig. 16. Clunionine midge, *Halirytis macquariensis* BRUNDIN, female (Chironomidae), Macquarie Island, 54°30′ S. (after BRUNDIN, 1962, Pacific Ins., vol. 4, no. 4).

459

sudden gusts, down drafts just off-shore, or general increase in wind speed, had they not been trapped. On very windy days, at the same location, outgoing waves and spray were seen to arise just off the shore, indicating that the wind closely followed the land contour into the water. Often weather balloons were blown directly down into the water on release.

On these more southern islands there appears to be a general tendency for some of the wingless or short-winged species to be more active than some of the winged species. In connection with the various brachypterous moths of Campbell Island, their tendency to jump and sham death has been noted (SALMON & BRADLEY, 1956).

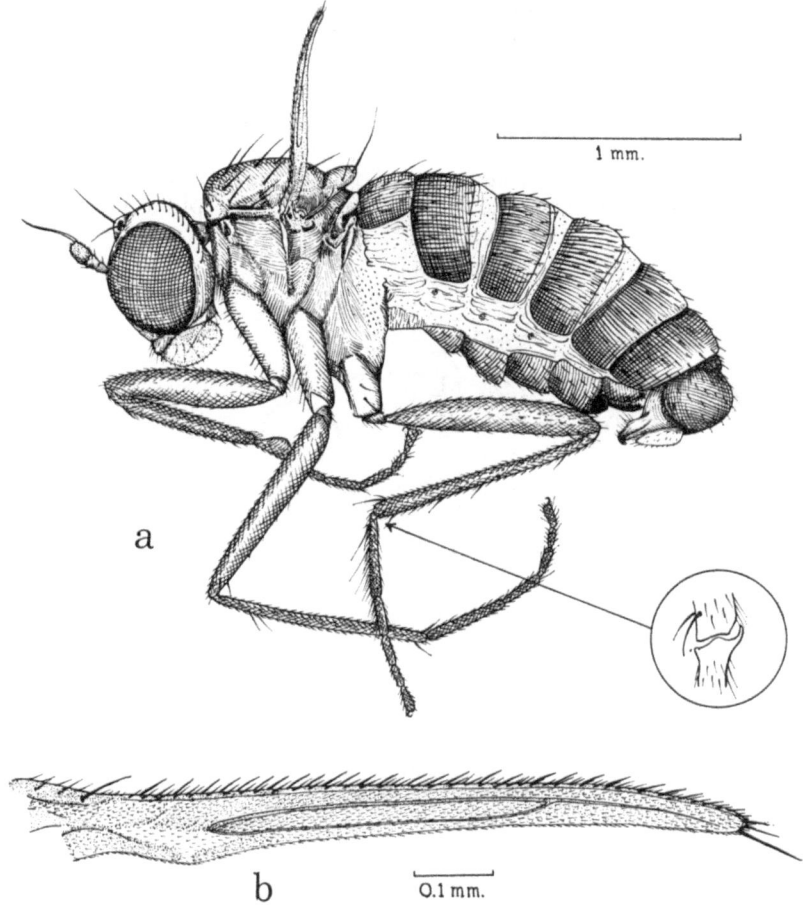

Fig. 17. Brachypterous fly, *Schoenophilus pedestris* LAMB (Dolichopodidae), Macquarie Island (after KOHN, 1962, Pacific Ins. vol. 4, no. 4).

460

Some of them run rapidly along the ground in grassy areas, or jump from one sedge or tussock to the next. One species lives on exposed rock on the shores.

On Macquarie Island, *Halirytis* (fig. 16) lives on rocky shores, and *Schoenophilus* (fig. 17) and *Apetaenus* (fig. 18) run on mat plants, mosses, bare rock or sand.

Some of the wingless flies on Campbell are extremely active. *Baeopterus* (fig. 19) and *Icaridion* (fig. 20) breed in rotting *Durvillea* kelp on the beaches. When disturbed they run down under the rocks and are very difficult to find. The winged coelopids, also on kelp, often tended to drop to the ground rather than fly when kelp on which they were resting was picked up. Some of the wingless Dolichopodidae and Anthomyiidae on Campbell occur on the flowers of *Bulbinella*, and run rapidly down or jump when approached, and may be as hard to catch as winged species.

Among normally winged groups which have undergone wing reduction on subantarctic islands are the following. Heteroptera: Enicocephalidae; Psocoptera; Lepidoptera: Tineidae, Yponomeutidae, Cosmopterygidae, Elachistidae, Tortricidae, Crambidae; Diptera: Tipulidae, Chironomidae, Dolichopodidae, Coelopidae, Cypselidae,

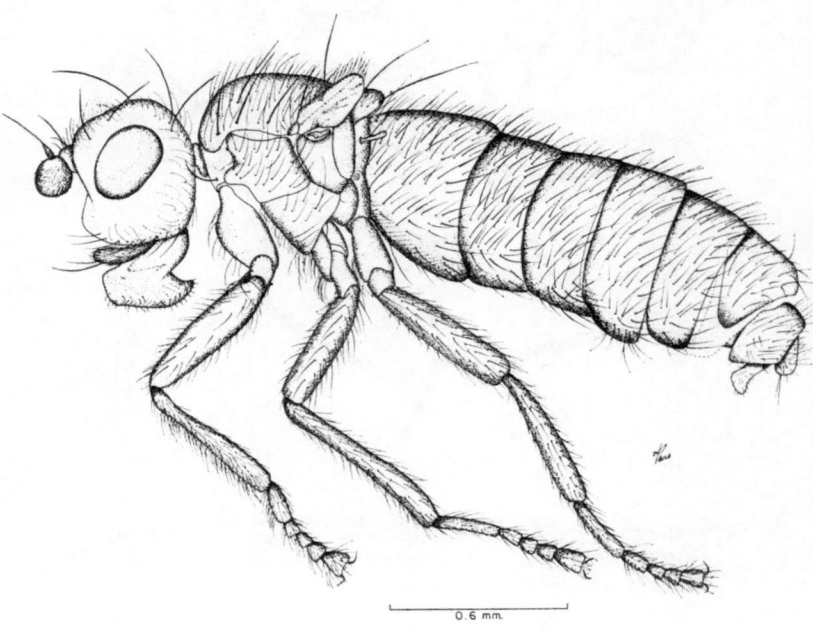

0.6 mm.

Fig. 18. Subapterous fly, *Apetaenus watsoni* HARDY (Coelopidae), Macquarie Island (after HARDY, 1962, Pacific Ins. vol. 4, no. 4).

Ephydridae, Anthomyiidae; Hymenoptera: Diapriidae, Encyrtidae, Cynipidae, Ichneumonidae.

The approximate percentages of species with reduced wings, by islands and orders, are shown in table III.

Heard Island has no species capable of flight. Macquarie has at least six, Kerguelen and South Georgia at least one each, Antarctica one, and Campbell about 30 flying species.

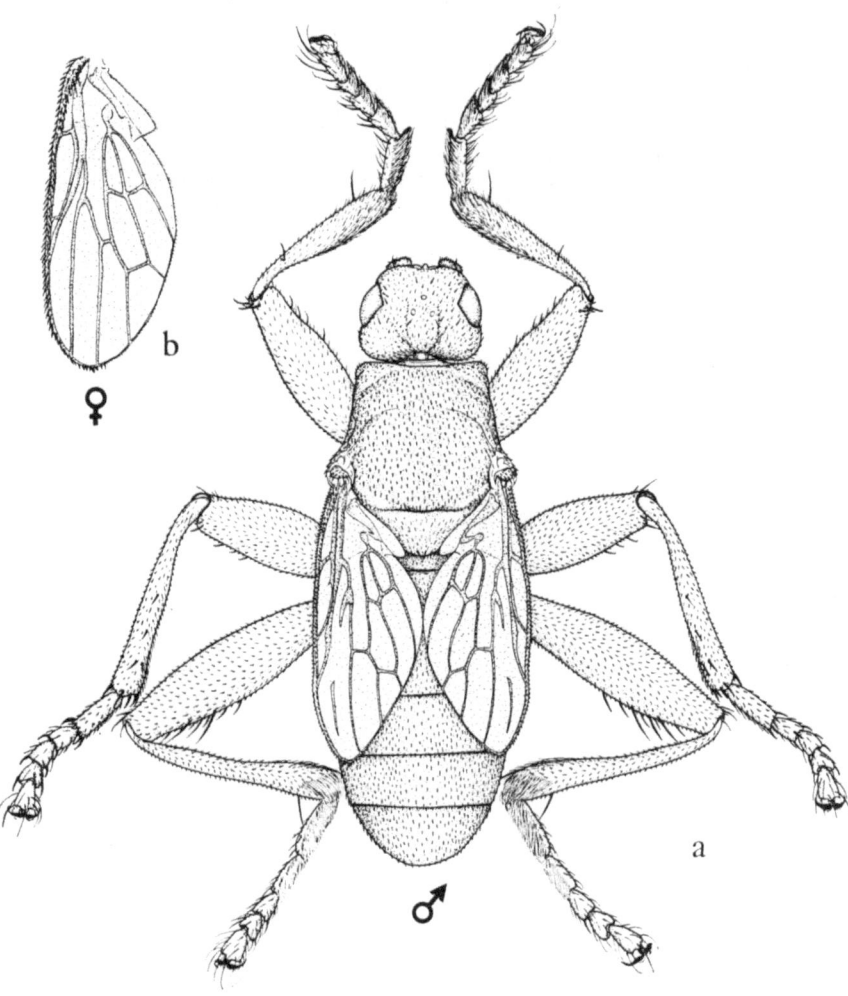

Fig. 19. Brachypterous kelp fly, *Baeopterus robustus* Lamb (Coelopidae), Campbell Island, 52°30′ S.

462

Table III.

Percentage of brachypterous or apterous insects of winged groups.

	Hemi-ptera	Lepido-ptera	Coleo-ptera	Diptera	Hymeno-ptera
South Georgia			100	0	
Marion		100	100	100	
Crozet	100	100	100	42	
Kerguelen		100	100	42	
Heard		100	100	100	
Macquarie	0	0	100	42	50
Campbell	30	33	97	18	50
Auckland	25?	6?	85	6?	75

Note: An "0" indicates order is represented by one or more winged species. Of the Hymenoptera on Campbell, 3 are wingless, 3 winged, and one has wingless females. Some of the Aucklands Hymenoptera have winged males.

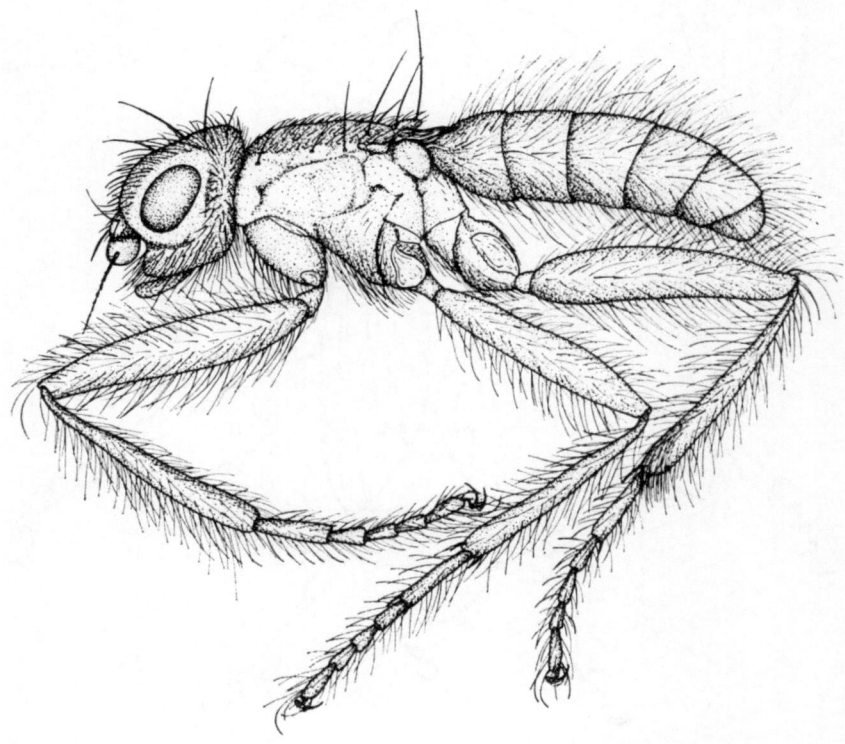

Fig. 20. Apterous kelp fly, *Icaridion nasutum* Lamb (Coelopidae), Campbell Island.

Zoogeography*

The islands surrounding Antarctica are of diverse origin geologically, and cannot all be attributable to remnants of a former common southern continent. Many of the islands on the Atlantic and Indian Ocean sides are considered oceanic, being of volcanic origin. They include Macquarie, Heard, Kerguelen, Crozet, Prince Edward, Marion and Bouvet, as well as the more northern Amsterdam, St. Paul, Tristan da Cunha and Gough. These islands may have arisen by volcanic action after present continental limits were essentially established, although JEANNEL (1940) considers those of the Marion — Crozet — Kerguelen — Heard group to be continental islands. They appear to have a primarily oceanic fauna, presumably largely the result of immigration by agency of transport in air currents or on the feathers or feet of birds. Macquarie appears to have been affected more by glaciation and other erosion than Kerguelen or Heard, and may be older, but Macquarie appears to have a younger fauna than Kerguelen. Heard appears to be younger both geologically and biologically than Kerguelen.

Another collection of islands, the South Georgia, South Sandwich and South Orkney groups, have been characterized as islands on a submarine ridge, forming the Scotia Arc, curving eastward and northward from the tip of the Antarctic Peninsula and the South Shetland Islands. Biologically, these islands may surely be considered oceanic, having received their faunae by oversea dispersal.

Nextly Campbell Island and the Auckland Islands are called islands on a subcontinental plateau. They are undoubtedly remnants of larger islands. They are very much richer than Macquarie, though not far away. The Falkland Islands are considered to be true continental islands on a continental shelf, and their fauna probably verifies this.

When the fauna, both of Antarctica and of the surrounding islands, is so imperfectly known, it is difficult to draw precise conclusions as to its origin, history and relationships. Antarctica, though now so very poor in land fauna, has possibly held a key geographical

---

* The discussion of subantarctic areas here is limited to the more southern islands: Falklands, South Georgia, Marion, Crozets, Kerguelen, Heard, Macquarie, Campbell, Aucklands. More northern islands, and South America, are largely excluded from this treatment.

Many of the statements, as well as distribution tables or maps, here presented, will undoubtedly require modification when taxonomic treatment is more complete, and when relationships of the various representatives are better understood. The fauna of South Georgia may be imperfectly known, in spite of statements to the contrary (SCHWEIGER, 1958). More groups will surely be found in the Falkland and Auckland Is., and more might be found in the southern Indian Ocean islands (Marion, Crozet, Kerguelen, Heard).

464

position in terms of evolution or spread of animals far in the past. It more or less forms a pivot point in regard to several theories of history of the earth's surface and evolution of life. Among these

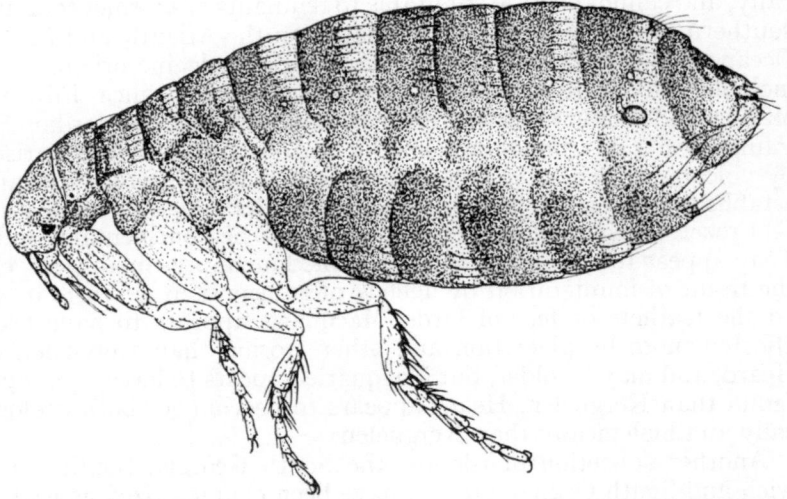

Fig. 21. Antarctic flea, *Glaciopsyllus antarcticus* SMIT & DUNNET (Ceratophyllidae), Ardery Island, near Wilkes Station, 66°22′ S. (after SMIT & DUNNET, 1962, Pacific Ins. vol. 4, no. 4).

theories is that of continents drifting apart from the south polar center; that of a larger southern continent and/or land bridges without great drift of the northern continents; and various theories of changes in the earth's axis and polar wandering. Much geological and paleontological data are rapidly accumulating, and crystallization of theories and clear approximation of chronology may be possible before long. Largely contradictory to the above theories, at least concerning their possible role in determining present animal distribution, is that of relative permanence of present continents and oceans, and the spread of most forms of life between Old and New Worlds across the Bering Arc (DARLINGTON, 1957). In the Pacific area there is much evidence for some aspects of the latter theory, as well as for dominantly Oriental relationships (partly extending on to Africa) for the insular and equatorial fauna, as well as for air dispersal both to tropical and polar areas (GRESSITT, 1961). Another theory invokes two faunal elements of different origin for groups represented by endemics in the subantarctic fauna on the South American side (often called the "Antarkto-Archiplata" region.) One element involves "paleoendemics" of ancient origin, perhaps Mesozoic, of trans-Antarctic distribution when a common southern continent is said to have existed. This involves groups represented now on both sides of the Antarctic continent. The other element involves more recent "neoendemics" which have entered the area from the north (South America) and do not have close relatives on the opposite side of Antarctica (SCHWEIGER, 1958). An example of the former is the carabid beetle tribe Migadopini, with representation in southeastern Australia, Tasmania, New Zealand, Auckland Is., South America, Tierra del Fuego and the Falkland Is. SCHWEIGER states that the South Georgia beetle fauna belongs to the paleoendemic element.

In southern Australia, Tasmania, New Zealand and southern South America, jointly or in various combinations, and to a lesser extent in South Africa, there are apparent and documented close faunal relationships. In some cases, different students have explained present distribution in the same group by opposing theories. As further correlated studies are made, particularly for southern South America and New Zealand, more evidence of relationship appears to emerge (KUSCHEL, 1960; GRESSITT, 1961; HOLDGATE, 1961). KUSCHEL feels that these relationships justify the establishment of a common southern or Austral Region (see Addendum).

So few groups are involved at present on Antarctica, that little direct light is shed on the problem from the continental fauna. With subantarctic islands, numbers of groups represented increase rapidly with distance from the continent. At this time, representation appears to a considerable extent haphazard, suggesting air dispersal in many cases. Islands on different sides of Antarctica have quite diverse faunae in some groups, but analogies or relationships in others. In some cases widespread species occur on many islands.

Some examples of apparent distribution patterns are presented on the following maps, and discussed under the respective taxonomic groups. Some of the maps show known southern limits for families, and others show ranges of genera or their southern limits, or species ranges.

Though there are considerable differences in the make-up of the faunae on different islands, there are some distinct superficial analogies in regard to types of insects represented, and ecological niches occupied. This is partly a general reflection of the rigors of the environment. Thus the superficial analogy extends to types of arthropods found in analogous environments such as coasts exposed to strong winds and waves, and high mountain peaks in temperate and tropical regions. For instance a degree of similarity in make-up was noticed between the fauna of Macquarie Island and those of the summits of Mt. Wilhelm (4600 m) and Mt. Giluwe (4250± m), highest mountains in Northeast New Guinea and Papua, respectively. General types in common included small spiders, red mites under rocks, springtails under moss or on lichens, flightless staphyli-

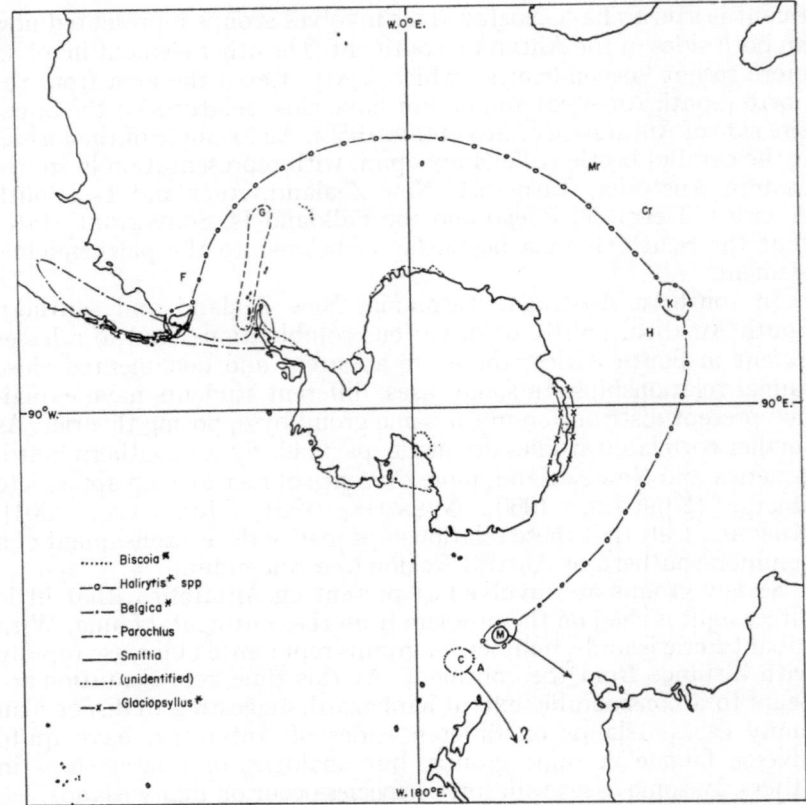

Fig. 22. Diagrammatic representation of ranges of several antarctic or subantarctic species or genera; genera asterisked are endemic to the areas shown (all flightless except *Parochlus* and unid. genus). Collembola: *Biscoia;* Diptera, Chironomidae: six genera; Siphonaptera: *Glaciopsyllus*. Code: F = Falklands, G = South Georgia, Mr = Marion, Cr = Crozets, K = Kerguelen, H = Heard, M = Macquarie, C = Campbell, A = Aucklands.

nid beetles, other flightless insects, bristly black flies, and a few moths. There are also parallels with the situation in the high Himalayas (see MANI, 1962). JEANNEL (1940) cited relationship between a beetle on Kerguelen and one at high altitude on Mt. Kilimandjaro. The fauna of the Crozet Islands was singled out by JEANNEL as the only southern Indian Ocean island with a strong African element, as well as the richest.

An interesting question is whether other groups of insects might naturally establish on subantarctic islands. Probably many additional groups, not to mention additional genera of groups already represented, could survive and find appropriate niches on some of these islands, should they be dispersed thither under favorable circumstances for reproduction. It is difficult to assess the odds against such successful establishment in a new area. Only if it could be proved that the islands are biologically young (for instance if recently ice-covered), could it be expected that additional species would frequently become established. Probably at the very least 10,000 years have elapsed since some of these islands may have been ice-covered. However, the faunal differences between some of the islands is such that it suggests different ages, in terms of ice cover. The poverty of the Heard fauna suggests both limited available habitable area (and thereby great odds against successful establishment) and perhaps a shorter period of time since the island was entirely ice-covered. Since some of the wingless species are possessed in common by Heard and Kerguelen, they might have been blown from Kerguelen to Heard since losing their wings. There have also

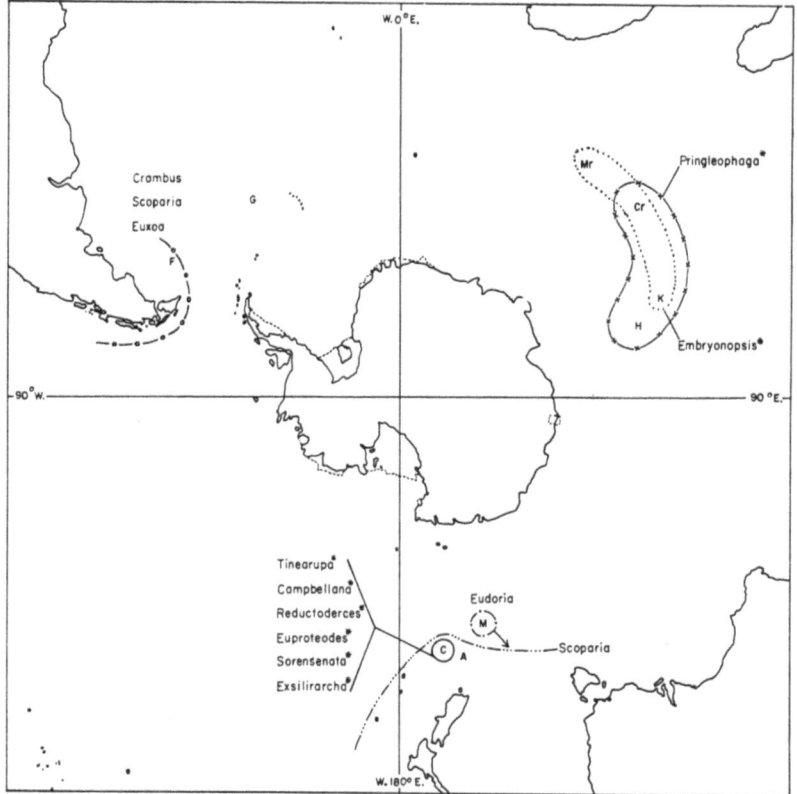

Fig. 23. Diagrammatic representation of ranges of 12 genera of Lepidoptera. Asterisked genera are flightless and endemic. Code same as in fig. 22.

468

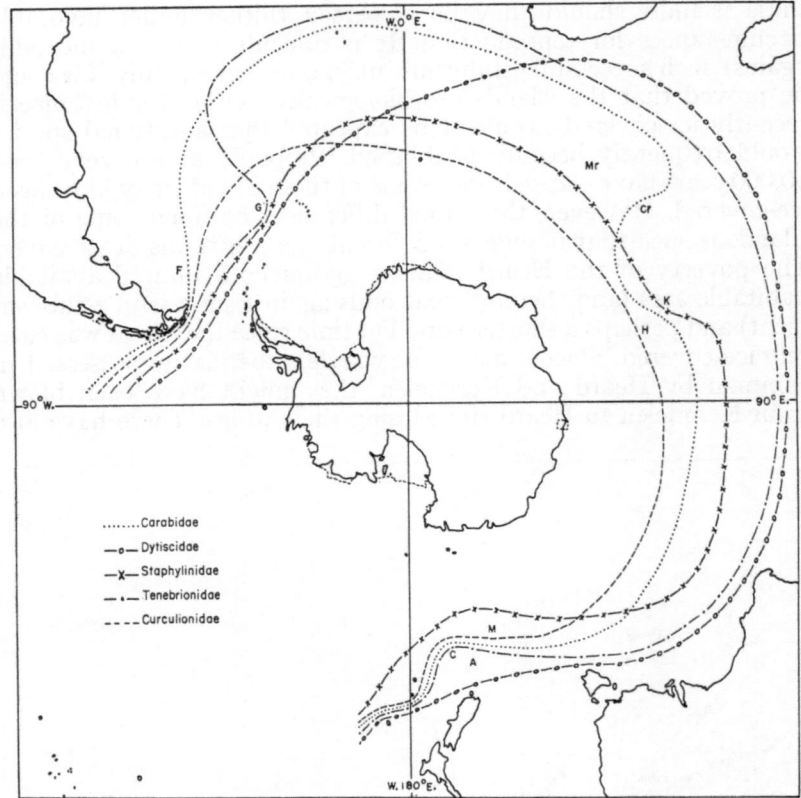

Fig. 24. Diagrammatic representation of the southern limits of five southernmost families of Coleoptera. Island code as in fig. 22.

been suggestions that flies not using their wings can lose them very rapidly.

Among the subantarctic islands, zoogeographical relationships, as far as they exist, appear to go eastward from southern South America and the south Atlantic islands, from the Falklands to South Georgia, to Marion — Crozets — Kerguelen — Heard, and then to Macquarie — Campbell — Aucklands. This appears to point to the importance of the prevailing winds and sea currents, which travel in the same direction. In the Pacific Ocean, there is a wide gap in both subantarctic and cold temperate islands. This suggests that the relationships between southern South America and New Zealand might be attributable to action of air and sea currents by the above route, or by former passage across a temperate antarctic stepping stone or general southern continent, or dispersal by birds, or a combination of these. The southernmost island of Polynesia, Rapa, is very distant, and may have played little or no part in southern dispersal, and perhaps likewise Easter Island, the easternmost. However, some very interesting groups occur on Rapa. Some so-called "antarctic" plants extend from New Zealand through parts of Polynesia to as far north as Hawaii. Perhaps more distinct insect relationships than are now apparent will come to light.

Among other interesting disparities are the lack of weevils in South Georgia and Macquarie, while a number of species occur on Kerguelen, Campbell and other islands. Also, the absence of Tenebrionidae from Kerguelen and nearby islands and the lack of Staphylinidae on Heard Island. Macquarie has nine families of flies to three

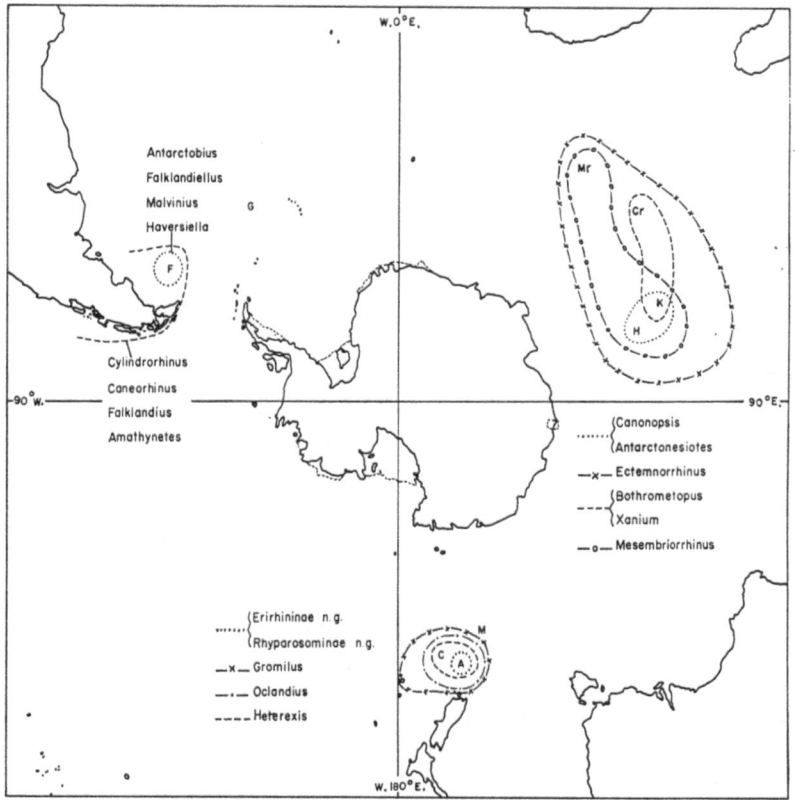

Fig. 25. Diagrammatic representation of ranges of 19 genera of weevils (Curculionidae) on subantarctic islands. Presumably all genera for Falklands and Marion — Crozet — Kerguelen — Heard area are shown; only endemic genera are shown for Campbell and Aucklands. All genera are endemic except the four indicated as common to Falklands and South America.

470

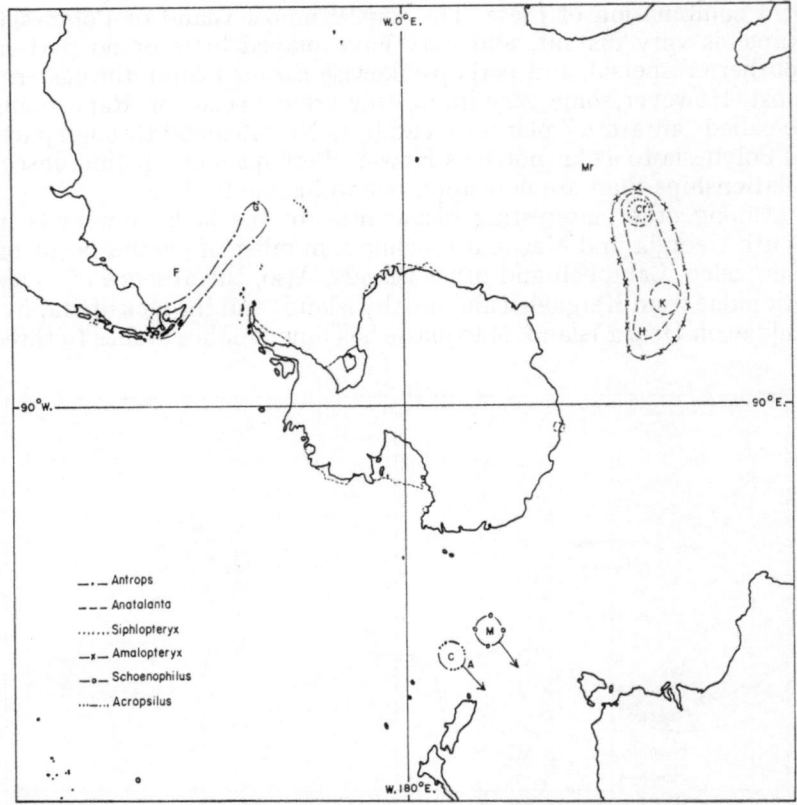

Fig. 26. Diagrammatic representation of ranges of six genera of brachypterous flies (first three are Sphaeroceridae; *Amalopteryx* is Ephydridae; last two are Dolichopodidae); all are endemic to areas shown except last two.

on Heard, whereas Macquarie has only one family of beetles to two on Heard, and 12 on Campbell. Macquarie has only one species of Lepidoptera, whereas Campbell, its nearest neighbor, has over 20 species. To be sure in some cases further field work will reduce disparities. However, the general picture seems to be of erratic distribution resulting from accidental oversea dispersal. It is unlikely that more than one native Lepidopteran occurs on Macquarie. Macquarie seems to have at least two groups (Thysanoptera, Chloropidae) not yet found on Campbell.

Some groups of islands have a number of elements or related forms in common. This is more true of Kerguelen in relation to Marion, Crozet and Heard, with some elements common to all four, and others occurring on two or three of the islands. The Ectemnorhinini weevils are a good example of speciation within the area. In other cases a species may be common to a few of the islands. In the case of Macquarie, Campbell and the Aucklands there appears to be less in common among them, or between them and other isles.

Of insect families, there appear to be six represented in the Marion — Crozet — Kerguelen — Heard group which are lacking from Macquarie, Campbell and Aucklands, one in Macquarie but not in Campbell or Aucklands, one in South Georgia but not in any of the preceding groups, and three in Campbell and Aucklands which are lacking in New Zealand. Twenty-three families occur on Macquarie, 15 on South Georgia, 13 on Heard, 50 on Campbell, 57 in the Aucklands, and 24 on Kerguelen.

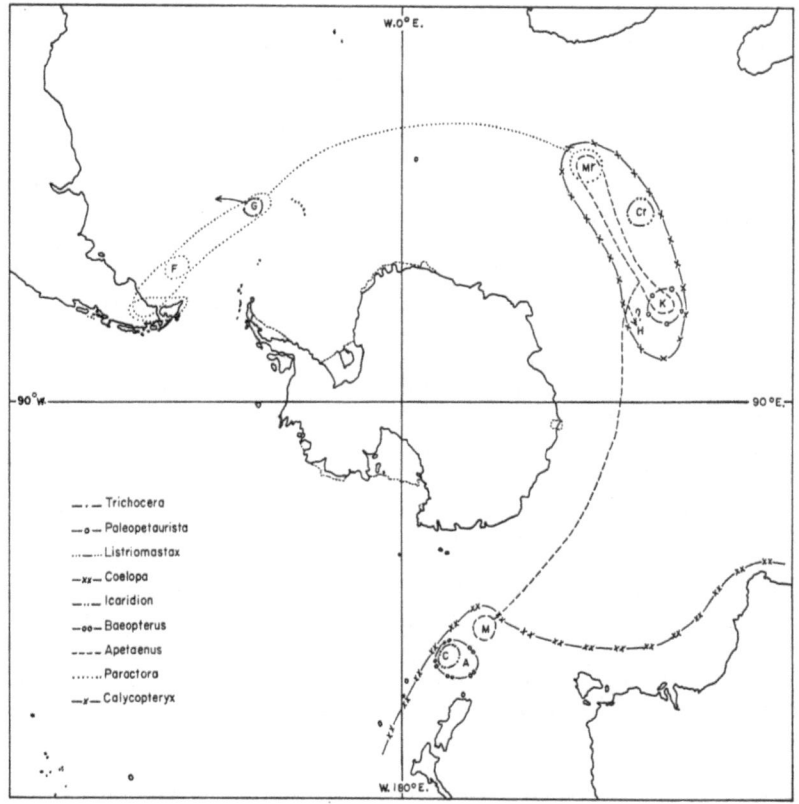

Fig. 27. Diagrammatic representation of nine genera of Diptera (first two are Trichoceridae; next five are Coelopidae; *Paractora* is Helcomyzidae; *Calypcopteryx* is Tyliidae); most are brachypterous or apterous, except *Coelopa* and *Trichocera*, and all restricted to the region except *Coelopa* and *Trichocera*.

Following is a list of certain islands with the numbers of insect families having their southern limits (not considering South America) on these islands or in Antarctica.

Antarctica	7	Falklands	10+
Macquarie	18	Aucklands	17+
South Georgia	6	Kerguelen	0?
Heard	5	Marion	1
Campbell	24	Crozet	2

At present there appear to be no families recorded from the South Shetland and South Orkney Islands which are not also known from the Antarctic continent.

In the less southern subantarctic islands, such as St. Paul, Amsterdam, Snares, Bounty, Gough and Tristan da Cunha, there are some families or genera not represented in the more southern islands. Some of them do not have close ties with elements on the more southern islands (see JEANNEL, SEGUY, HOLDGATE, KUSCHEL). Those islands are mostly smaller than the major subantarctic islands, and their faunae tend to be less endemic and specialized, although some endemic species, and even genera, do occur. Some of their elements are of considerable zoogeographical interest.

On the subantarctic islands the relative representation of various insect orders is not in the same proportion as on tropical oceanic islands. This is another indication of the selectivity of the environment. Heteroptera and Homoptera are well represented on the tropical oceanic isles, often jointly comprising 10% of the fauna, but are extremely few on subantarctic islands. Heteroptera appear to be lacking on the more southern islands, and Homoptera are represented by two species of aphids on Macquarie and a few species each of aphids and pseudococcids on Campbell. There are no records for most of the other islands. The Hymenoptera also often make up about 10% of the fauna on tropical islands, whereas they constitute a much smaller fraction on the southern islands. None are recorded from several of the islands, although Macquarie has two species and Campbell seven.

It has often been stated (HOLDGATE, 1960) that the orders Plecoptera and Ephemeroptera are not represented on any oceanic island. Interestingly, there now prove to be three species of three genera of Plecoptera on Campbell Island, and two species, if not more, from the Auckland Islands. This could of course be cited as further evidence that these islands are continental and not oceanic. Geologically they are surely continental, having various types of metamorphic rocks. However, it is possible for a geologically continental island to have an oceanic fauna (GRESSITT, 1961). This can be brought about by volcanic activity temporarily covering all of the island surface with lava or ash, temporary submersion by subsidence, or complete capping with ice. The Yap Group in the western Caroline Islands possesses a species of Ephemeroptera, and this island group also has continental rocks, while appearing to have an oceanic fauna.

One interesting matter is the presence of three families of flies (Lonchaeidae, Milichiidae, Canaceidae) in the Auckland Islands and/or Campbell Island which are apparently lacking in New Zealand. The New Zealand fauna, with its many gaps and almost oceanic disharmony, is quite a puzzle since it also harbors many isolated primitive relics. Possibly, the subantarctic families unrepresented in New Zealand are recent air-borne immigrants, or are man-imported. It would be expected that establishment would be more likely in New Zealand than on subantarctic islands from the standpoints of land size, favorable climate and appropriate niches.

Endemicity

The fauna of Antarctica proper seems to be characterized by some species of wide distribution and others of more localized occurrence. Those of wide distribution seem to be dominant on the northern fringes of the continent. Some of these are found at widely spaced localities on the continental fringes, the islands nearby, and some of the more southern subantarctic islands. Others are known only from the Antarctic Peninsula area, but often including the South Shetland Islands, the South Orkney Islands and sometimes South Georgia or even the Falkland Islands or southern South America.

More of the endemic species seem to be of local distribution and often of more southern occurrence. Some of these discoveries have been made very recently, and it seems reasonable to expect that more such local endemics may come to light in the future.

Endemicity of Antarctica has been discussed for plants and animals by CAILLEUX (1961), who cites 57% endemicity for 735 living terrestrial species of Antarctic proper. The figures given for mites and insects are 77% for 22 free-living species and 81% for 21 parasitic species (after GRESSITT & WEBER, 1960).

At present, there appear to be about five genera of mites and insects restricted to the Antarctic continent. There is the possibility that these may represent relics of the fauna of Antarctica preceding the period of maximum glaciation. Some of them do not seem to have close relatives, and some do not fit well in existing families or subfamilies; (*Biscoia* in Collembola; *Glaciopsyllus* in Siphonaptera).

Endemicity on the subantarctic islands is marked on the species level, and a number of genera are known only from one island group or from nearby islands, such as Crozet and Kerguelen, or Kerguelen and Heard. This is further discussed under the respective islands.

Geographical representation

In the following tables, numbers of genera or species which are known from antarctic areas are enumerated by families, or genera, by island groups. Brief discussion of relationships or distribution patterns is presented in the following paragraphs.

Chilopoda: Centipedes are very poorly represented on subantarctic islands. The only records from the more southern islands seem to be from Auckland and Campbell Islands, though they may occur in the Falklands. These may represent human introductions. Species have also been recorded from St. Paul Island and Gough Island.

Diplopoda: The extent of southern distribution of the millipedes is similar to that of the centipedes, though one may be endemic to Campbell (JOHNS, *in litt.*). Species probably occur only in the Aucklands, Campbell and the Falklands, of the islands being treated. There are likewise records for St. Paul Island and Gough Island.

Araneida: Spiders probably occur on all the larger subantarctic islands. Two species are found on Macquarie. There is a possibility that some small species occur on the Antarctic continent (see discussion above). The spiders of Campbell and Auckland Islands have close affinity with those of the New Zealand fauna. All of the genera are in common with New Zealand. Of the 19 species recorded from Campbell and Auckland, nine are endemic to the two island groups together. Campbell Island has nine species, none of which are endemic to it alone (after FORSTER, 1955). Several of the genera are represented on a number of the subantarctic islands, sometimes by the same species on widely separated islands. Some genera extend from Crozet to Macquarie or Campbell. This might partly be the result of air dispersal and partly by human agency. The Aucklands fauna stands out from the others with its extensive and varied nature. Table IV presents the recorded species, by families and islands (partly after FORSTER, *in litt.*).

Opiliones: The harvestmen are poorly represented on the subantarctic islands. They appear to be recorded from six of the major islands; apparently by only one species in each case, except for Campbell. They do not occur on Macquarie.

Pseudoscorpionida: There are apparently no published records of pseudoscorpions from subantarctic islands. However, a species was recently taken on Campbell Island, and the group may probably be represented on the Aucklands and Falklands.

Acarina: Ticks are represented by two species *(Ixodes auritulus* and *I. putus)* in Antarctica proper; both recorded only from the northern portions of the Antarctic Peninsula, but also from various subantarctic islands. A few additional species occur on other islands, and most of them are of wide distribution and associated with rookeries of sea birds such as cormorants, penguins and albatrosses. The

Table IV.

Numbers of species of harvestmen and spiders, by families.

	Falklands	S. Georgia	Marion	Crozet	Kerguelen	Heard	Macquarie	Campbell	Auckland
Opiliones									
Palpatores								1	
Laniatores									
Triaenonychidae				1				1	2
Gonyleptidae	1								
Araneida									
Araneomorphae									
Oonopidae									1
Amaurobiidae								1	1
Agelenidae	1	1		3	1	1	1	1	1
Hahniidae				1					1
Clubionidae	1							1	3
Amaurobioididae								1	1
Lycosidae									1
(Perissiblemmidae)								1	1
Argiopidae									
Tetragnathinae	1								
Symphytognathidae	2			1				1	2
Theridiidae									2
Linyphiidae			3	2	1		1	4	3
Attidae								1	2
Thomisidae									
Philodrominae	1								

females deposit their egg-masses among mosses near the rookeries, and apparently do not infest birds nesting where there is no moss available, such as in the middle of large penguin rookeries. Mites are the best represented group in the general area, and are discussed in the chapter by DALENIUS.

Collembola: This primitive wingless order, the springtails (figs. 8—12, 28, 29), is the dominant order of true insects on the antarctic continent, and perhaps also on some of the more southern subantarctic islands. It is well represented on all, though in the more northern islands, at least, it is apparently exceeded in numbers of species by the Diptera. Collembola have not been thoroughly collected or studied in many parts of the region, so are unequally known from different islands. The Campbell Island fauna appears to have been best covered to date, with 34 species recorded (SALMON, 1949). Ten species are recorded from the antarctic continent, but a few more are awaiting description. One genus, *Biscoia* SALMON (1962a) is known only from the continent (fig. 22) and four species are known

Table V.

Southernmost (Antarctic and insular) representation of insects, by families.

order	family	Antarctic		Subant. isles		Island groups	Islands	Southernmost record
		gen.	spp.	gen.	spp.			
Collembola	Hypogastruridae	3	4	6	20	5	G K M C A	77° S
	Onychiuridae	2	2	6	9	11	(all)	84° S
	Isotomidae	5	5	12	27	11	(all)	72° S
	Tomoceridae	*		1	1	1		Campbell
								* (1 trapped) 70° S
	Entomobryidae			4	7	3	M C A	Macquarie
	Neelidae			1	1	1		Aucklands
	Sminthuridae			4	9	5	Cr K M C A	Macquarie
Orthoptera	Tettigonioidea			3?	3?	3	F C A	Campbell
Plecoptera	Grypopterygidae			4	5	3	F C A	Campbell
Psocoptera				3+?	5?	5	Mr M C	Macquarie
Mallophaga	Menoponidae	2	2	2+?	5+	10	(all?)	69° S
	Philopteridae	6	15	8?	25?	11	(all)	78° S
Anoplura	Echinophthiriidae	3	6?	3	8?	5+	G K M C A	78° S
Odonata						1		Aucklands
Thysanoptera	Thripidae			2	1	3	Cr K M	Macquarie
Heteroptera	Enicocephalidae			1	3?	2	Cr A	Aucklands
??	??			1	1	1		Falklands
Homoptera	Aphididae			3?	4	3	M C A	Macquarie
	Pseudococcidae			1	4	2	C A	Campbell
Trichoptera				1	1	1		Falklands
Lepidoptera	Tineidae			5?	7	4	Cr K H A	Heard
	Yponomeutidae			2	2	2	C A	Campbell
	Gelechiidae			2	2	2	C A	Campbell
	Elachistidae			1	1	1		Campbell
	Tortricoidea			4	1	2	C A	Campbell
	Pyraloidea			4	10	4	F M C A	Macquarie
	Noctuoidea			4	4	1+?		Aucklands (?Falklands)
	Geometroidea			5?	9?	3	F C A	Campbell
Coleoptera	Carabidae			14	29	5	F Cr K C A	Campbell
	Dytiscidae			1	3	2	F G	South Georgia

Coleoptera (continued)

Family				Distribution	Example locality
Staphylinidae	12	24	8	F G Mr Cr K M C A	Macquarie
Catopidae	2	2	2	C A	Campbell
Silphidae	1	1	1		Falklands
Hydraenidae	1	3	5	Mr Cr K H C	Heard
Hydrophilidae	2	2	1		Aucklands
Pselaphidae	2	2	2	Cr C	Campbell
Byrrhidae	2	5	3	F C A	Campbell
Lathridiidae	2	2	2	F C	Campbell
Elateridae	1	1	1		Aucklands
Nitidulidae	1	1	1		Aucklands
Coccinellidae	1	1	2	C A	Campbell
Pythidae	1	1	1		Falklands
Tenebrionidae	7	11	4	F G C A	South Georgia
Cerambycidae	1	1	1		Aucklands
Curculionidae	25	71	7	F Mr Cr K H C A	Heard

Diptera

Family					Distribution	Example locality
Trichoceridae	3	3	3		G C A	South Georgia
Anisopodidae	1	2	1			Aucklands
Tipulidae	7	11	4		K M C A	Macquarie
Psychodidae	1	8	3		M C A	Macquarie
Cecidomyiidae	1	1	1			Campbell
Culicidae	1	1	1			Aucklands
Chironomidae	4	5	7	2	F G Mr K M C A	65° S
Mycetophilidae	2	3	2		C A	Campbell
Sciaridae	3	5	5		Cr K M C A	Macquarie
Simuliidae	2	4	4		F Cr C A	Campbell
Stratiomyiidae	4	4	1			Aucklands
Therevidae	1	1	1			Aucklands
Asilidae	2	2	1			Aucklands
Dolichopodidae	2	2	2		M C	Macquarie
Empididae	1?	2	2		C A	Campbell
Syrphidae	2	4	2		C A	Campbell
Helomyzidae	1	1	2		C A	Campbell
Coelopidae						
(Phycodromidae)						
Phoridae	7	12	9	2	F G Mr Cr K H	Macquarie
Helcomyzidae	1	1	1		M C A	Campbell
Sciomyzidae	1	4	3		F G Mr	South Georgia
Cypselidae	2	2	1		Cr K H	Aucklands
Tylidae	2	3	3		K H	Heard
Sapromyzidae	1	1	2		C A	Campbell

Table V. (continued)

order	family	Antarctic gen.	Antarctic spp.	Subant. isles gen.	Subant. isles spp.	Island groups	Islands	Southernmost record
	Lonchaeidae			1	1	1		Aucklands
	Agromyzidae			1	1	1		Aucklands
	Ephydridae			3	3	6	Cr K H M C A	Macquarie
	Milichiidae			2	2	2	M C	Macquarie
	Canaceidae			1	1	1		Campbell
	Chloropidae			2	2	1		Macquarie
	Sphaeroceridae			2	2	2	G C	South Georgia
	Calliphoridae			1	5	3	F C A	Campbell
	Muscidae			8	21	5	F Cr K C A	Campbell
	Tachinidae			2	2	1		Aucklands
	Hippoboscidae			2	2*	2	C A	*Campbell (1 introd.)
Siphonaptera	Pulicidae			1	1	1		Campbell
	Rhopalopsyllidae			2	4	6	F Mr K H M C A	Macquarie
	Pygiopsyllidae			1	2	5	G K H M A	South Georgia
	Ceratophyllidae	1	1	2	2	3	K M C	68° S
Hymenoptera				12	15	5	F Cr M C A	Macquarie
		37	52	336	574			

Notes: Southern South America is excluded. The South Shetland and South Orkney Islands are treated here as part of Antarctica. A family recorded from 11 island groups would be from these two groups in addition to the nine major and more southern subantarctic island groups listed below. Islands which are less southern (Amsterdam, St. Paul, Tristan da Cunha, Gough), less isolated (Snares, Antipodes, Bounty), and less known (South Sandwich, Bouvet), are excluded from this table. Thus it includes records from south of 50° S. less S. America, S. Sandwich and Bouvet, plus Marion, Crozet and Kerguelen.

Abbreviations:

A — Auckland Is. G — South Georgia Is.
C — Campbell I. H — Heard I.
Cr — Crozet Is. K — Kerguelen I.
F — Falkland Is. M — Macquarie I.
Mr — Marion I.

Subantarctic islands: Latitudinal order, South to North

Macquarie 54°30' Falklands 52°
South Georgia 54° Aucklands 50°30'
Heard 53° Kerguelen 49°
Campbell 52°30' Marion 47°

Crozet 46°

only from Antarctica or the South Shetland Islands. One of the species on the continent is cosmopolitan. Several species are found both on the continent and on several subantarctic islands, and others

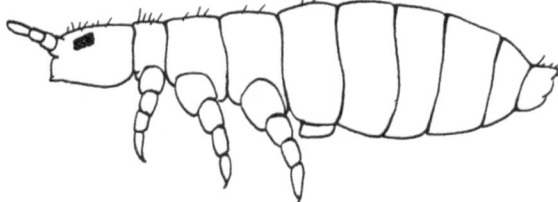

Fig. 28. Springtail *Colonavis grandis* SALMON (Hypogastruridae), Campbell Island (after SALMON, 1949, Cape Expedition 4).

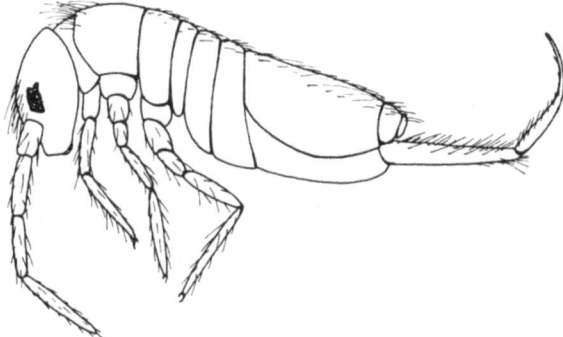

Fig. 29. Springtail *Lepidobrya violacea* SALMON (Entomobryidae), Campbell Island (after SALMON, 1949, Cape Expedition 4).

occur on several of the islands and in Tierra del Fuego or New Zealand as well. Some species are known from single islands, such as South Georgia, Kerguelen, Campbell or Auckland, but may later prove not to be endemic. In general, species of springtails appear to be widely dispersed. The zoogeographic significance of the group has been disputed, and more comprehensive studies are needed before the distribution of the group is well understood.

Some of the springtail genera occur in several zoogeographical regions, but some, like *Triacanthella* are apparently limited to the southern cold temperate zone, and others, like *Parafolsomia*, *Papillomurus*, *Pseudokatianna* and *Polykatianna* may be limited to New Zealand and its subantarctic islands, and still others, like *Subantarctica*, *Sorensia* and *Lepidiaphanus* are only recorded from Campbell Island. *Lepidobrya* is known from Macquarie, Campbell and Aucklands. The distribution of the genera is indicated in table VI for the areas discussed.

Table VI.
Collembola distribution, numbers of species.

	Antarctica	S. Shetlands	S. Orkney	S. Georgia	Falklands	Marion	Crozet	Kerguelen	Heard	Macquarie	Campbell	Auckland
Hypogastruridae												
Hypogastrurinae												
Xenella				1							1	1
Gomphiocephalus	1	1										
Triacanthella											2	1
Hypogastrura	2							2		3	4	1
Brachystomellinae												
Friesea	1	1		1							1	
Subantarctica											1	
Colonavis											1	
Anurida				1							1	
Onychiuridae												
Onychiurinae												
Onychiurus											1	
Biscoia	1											
Tullberginae												
Dinaphorura								1				
Tullbergia	1	1		2	1	1		2	2	1	2	
Isotomidae												
Isotominae												
Parafolsomia											1	2
Sorensia											2	1
Proisotomurus											1	
Archisotoma			1									1
Papillomurus											1	2
Cryptopygus	1	1	1	2			1	2	2			
Anurophorus	1											
Tomocerura					1						1	
Proisotoma				1							2	
Isotoma	1									1		
Parisotoma	1	1	1		2			2	1	1	2	2
Tomoceridae												
Lepidophorella											1	
Entomobryidae												
Entomobryinae												
Entomobrya											1	
Parasinella											1	
Lepidiaphanus											1	
Lepidobrya										1	3	1
Neelidae												
Megalothorax												1
Sminthuridae												
Sminthurinus							1				1	
Pseudokatianna											2	
Polykatianna										1		3
Stenacidia								1				

Mallophaga. Members of this order, the biting lice, are entirely ectoparasitic, and live among the feathers of birds. Apparently all species of Antarctic birds are infested with biting lice, though not necessarily every individual. Some species of birds, even one individual, may have two or three species of lice. Apparently these lice are not blood-suckers, but are principally scavengers, chewing upon the feathers and outer skin, and causing irritation to the birds. Their bodies are strongly flattened dorso-ventrally, and they generally remain attached to the bird feathers, even after death of the birds. The 18 or so species recorded from Antarctica belong to two families, of two suborders. The Menoponidae includes a species of *Austromenopon* on wandering albatross, and one of *Piagetiella* on cormorant. The Philopteridae includes species of *Austrogoniodes* on penguins; *Saemundssonia* on skua, terns, pintado petrel and snow petrel; *Docophoroides* on giant petrel; *Pseudonirmus* on snow petrel, pintado petrel and Antarctic petrel; *Perineus* on giant petrel, albatross and terns; and a species of *Naubates* on storm petrel. These lice, being on relatively large birds, vary from about 3 to 8 mm in length. Some are very slender, but some from the tops of heads of penguins are quite broad. They vary from white through pale brown and gray to largely dark.

Anoplura: Sucking lice are primarily associated with mammals. In the Antarctic, they are known only from seals. There appears to be one species of the genus *Antarctophthirus* on each species of Antarctic seal. Also there is an uncertain record of *Echinophthirus* from the leopard seal. The elephant seal (sea elephant, *Mirounga leonina*), which inhabits some of the northernmost fringes of Antarctica proper, and formerly occurred farther south, as on the Victoria Land coast, is host to *Lepidophthirus*.

An ecological study of *Lepidophthirus macrorhini* ENDERLEIN was made by MURRAY (1958) on Macquarie Island. Its habits proved to differ considerably from those of previously studied species. The louse burrows into the skin under the stratum corneum and lives in the burrow. Mating and egg-laying, as well as feeding, take place in the burrow, which may be 12 mm long. During the seal's moulting, the roof of the burrow is shed, and some lice are lost at this time. Infestation is densest on the hind flippers, and was observed to be more extensive on young seals during the long period of time they spend on the beaches. Tests showed that the skin temperature of seals on the beach was between 27 °C and 34 °C, whereas in the ocean it dropped to between 10 °C and 12.5 °C, the sea temperature being 8.5 °C. At 10 °C the lice become inactive, and when exposed to moist air at 29 °C it took them a few hours to resume activity. Oxygen consumption in air at 10 °C was about 1/15 that at 33 °C. The lice survived a 5-day immersion in sea water at 10 °C which was 7% saturated with oxygen. Thus metabolism is controlled by the skin temperature of the seal, and activity occurs when the seals are on the beaches. Tests showed that respiration of the lice at sea was primarily through the thin skin, and not through a bubble or air layer on the louse or host. The surface waters are generally 80% saturated with oxygen. The setae of the louse probably serve to keep water surrounding the louse in a burrow, to permit respiration through the skin.

The species of *Antarctophthirus*, however, are much broader-bodied and do not burrow into the skin past their head, although generally living in folds of skin, such as under flippers and around the genital or anal openings. They occur also in the fur, which apparently retains enough air to supply their needs. These lice are mostly 3 to 4 mm long when full grown, and are brown in color.

Hemiptera: This order is very poorly represented. Two aphids occur on Macquarie (EASTOP, 1962), and aphids and mealybugs on Campbell.

Thysanoptera: Thrips occur on Macquarie (STANNARD, 1962), Kerguelen and Crozet, with two endemic genera.

Psocoptera: There are few far southern records for bark-lice. There are records for Macquarie, Kerguelen and Marion, and I recently took a species on Campbell. *Austropsocus insularis* SMITHERS (1962) of Macquarie, in the Philotarsidae, has a parallel on Marion in *Antarctopsocus jeanneli* BADONNEL, of the Elipsocidae. Both have greatly reduced wings. *Rhyopsocus* occurs on Kerguelen.

Plecoptera: This freshwater group, as mentioned above, is poorly represented on islands. Three genera are found on Campbell, and two of these occur on the Aucklands. *Aucklandobius complementarius* ENDERLEIN has a fully winged adult stage, and occurs on both Campbell and the Aucklands. This is the known range of the genus, which is most closely related to *Nesoperla* of New Zealand. *Apteryoperla* has an un-named species on each of the two islands. The nymphs are aquatic and the adults are wingless. The third genus on Campbell is new, and as far as known endemic (ILLIES, *in litt.*). This latter is terrestrial both as nymph and as adult, occurring under barely damp rocks quite high on one of the higher hills. All three of the genera belong to the family Gripopterygidae, exclusively known from South America, Australia, New Zealand, Campbell and Auckland. (See Addendum).

Lepidoptera: Butterflies are completely absent from the areas considered, and moths are poorly represented on the more isolated islands. Two genera of shortwinged moths (fig. 23) are represented in parts of the Marion-Heard chain. There appear to be no records from South Georgia, and only one species *(Eudoria;* winged) occurs on Macquarie. The Falklands, Campbell and Aucklands, however, have 16 or more species each. *Scoparia* may be the only genus common to both sides. On the map, only a part of the genera are named, and for Campbell Island the interesting genera with reduced wings are listed (asterisked). *Tinearupa* (Yponomeutidae) occurs on exposed rock near the shore and at tops of cliffs. The larvae possibly feed on lichens. *Campbellana*, same family, has extremely narrow wings and is said to jump like a grasshopper. *Reductoderces* (Cosmopterygidae) has moderately narrow wings. *Euproteodes* (Elachistidae) has wings tapered and fringed, lives in tussock and jumps or shams death. *Sorensenata* (Tortricidae) has slender, tapering, fringed wings, and also lives in tussock and jumps. *Exsilirarcha* (Crambidae) also has slender tapering wings, much shorter in the female. It is stated to live in tussock (SALMON & BRADLEY, 1956), but I found it primarily on a small sedge, in which the larva is probably a leaf-miner. These six short-winged genera constitute one-third of the recorded Campbell moth fauna.

The Auckland Islands have a much richer fauna, and apparently a much lower percentage of species incapable of flight.

Coleoptera: Members of this order are in part very hardy, but are not so well adapted for oversea dispersal as are most other groups of insects. The Carabidae are moderately represented in the Falklands, the Crozets, and Kerguelen, with more species in Campbell and Aucklands. They are apparently lacking in the other southern islands (fig. 24). Dytiscidae appear to occur only in the Falklands and South Georgia.

The Staphylinidae seem to be the most widely represented family, and the southernmost, though not yet recorded from Heard. A few genera appear to be restricted to subantarctic islands. Of the five species on Macquarie, representing three genera, two appear to be introduced and one or two are found on other subantarctic islands. On Campbell, 11 species were recently taken, of which five appear to be new species. Three of these appear to represent as many new genera, of which two of the genera appear to also occur in the Auckland Islands (STEEL, *in litt.*). About eight kinds are known from the Aucklands. The relationships both for Campbell and the Aucklands are predominantly with New Zealand.

The Hydraenidae (or subfamily of Hydrophilidae) are represented by *Meropathus* on Marion, Crozets, Kerguelen, Heard and Campbell. A number of other families are represented by one or two species. Most of these families occur only in the Aucklands, Falklands, or on Campbell. Tenebrionidae are recorded from the Falklands, South Georgia, Campbell and the Aucklands, but there appear to be no genera in common among those recorded from the two sides of the antarctic continent.

The weevils are the best represented group of beetles, but apparently do not occur on Macquarie and South Georgia, the southernmost major islands. There may be said to be three faunal units of subantarctic weevils (fig. 25). First, the Falklands fauna includes about 19 species of eight genera of various subfamilies, and with none of the genera in common with the other two units, though with a few relationships on the tribal level with the second. Four of the genera appear to be restricted to the Falklands. Second comes Campbell and the Aucklands, with a much larger fauna; 14 genera and perhaps nearly 50 species (BROOKES, 1951). The five endemic genera (indicated on the map) have their closest relationships in New Zealand, and the non-endemic ones are all New Zealand genera. Thus there are no indications of very close relationships across Antarctica (partly from KUSCHEL, *in litt.*). But, the tribe Listroderini of the Cylindrorrhininae is found in both areas, as well as in Australia and southern North America (KUSCHEL, 1962). The third unit is a very interesting one, apparently constituting a tribe of shortsnouted weevils, Ectemnorrhinini, which is endemic to the Marion — Crozet — Kerguelen — Heard area. As shown on the map, most of the genera except *Ectemnorrhinus* are restricted to two or three islands. *Ectemnorrhinus* is divided into three subgenera; one largely on Kerguelen, but partly on Heard, one purely on Heard, and one on Marion and Crozet (see BROWN, in press). There are about 28 species, including those in press. This group is said to have its nearest known relatives in Europe. The group appears to have been evolving in isolation on these islands for a considerable period. On the other hand, there appears to be ample suggestion of the speed of evolution under such circumstances of isolation, division of small populations on separate islands, and further subdivision by glaciers (these weevils are flightless), in an environment of unfilled ecological niches and scarcity of natural enemies.

The presence on Kerguelen in mid-Tertiary of southern continental forests, including *Araucaria, Microcacthrys* and *Dacrydium* (HOLDGATE, 1961) does not have a clear bearing on the present land fauna. Since there appear to be no elements remaining for certain from that period, one tends to assume that the island was entirely glaciated at one or more periods in the interim. However, JEANNEL (1940) considers Kerguelen to be continental, and the Ectemnorrhininae to date from the early temperate climate period.

Diptera: The Diptera represent the only higher order among the free-living insects of the Antarctic continent, and include the southernmost winged insect. Both of the southernmost species are members of the Chironomidae: *Parochlus steineni* (GERCKE) of the Podonominae (figs. 13, 22), and *Belgica antarctica* JACOBS of the Clunioninae (figs. 14, 15, 22). The Diptera on some subantarctic islands comprise the richest order of land animals, whereas on some islands the Coleoptera, the Collembola or the Mallophaga appear to be richer. In the Falklands, Marion, Crozets, Kerguelen and Heard, Coleoptera appear to be better represented than Diptera, whereas there are almost, twice as many species of Diptera as there are of

Coleoptera on Macquarie, Campbell and the Aucklands. It may be possible that collectors on the former islands have paid better attention to beetles, whereas on the latter, the collecting has probably been more thorough and less partial. On South Georgia equivalent numbers of flies and beetles have been recorded, but probably the flies have not been adequately sampled. The Diptera are also represented in the area by the largest number of families of any order.

The Diptera represented are largely scavengers, many of them feeding in decaying vegetation, rotting kelp or in carrion. Relatively few are phytophagous, and very few are blood-sucking.

In regard to the flies of the Kerguelen area, SEGUY (1940) points out that six genera *(Sciara, Limnophyes, Paleopetaurista, Leptocera, Scaptomyza* and *Calliphora)* belong to cosmopolitan groups: three *(Halirytis,* figs. 16, 22, *Cnephia* and *Paractora)* occur also in other subantarctic areas, such as the Straits of Magellan, Falklands, or also South Georgia or Marion; while the relatives of *Calycopteryx* occur in the Orient or Australia. SEGUY points out that the endemic elements *(Apetaenus, Anatalanta, Calycopteryx, Amalopteryx* and *Listriomastax)* (figs. 26, 27) are heavily sclerotized, monophagous, and have been evolving for a long time. Their larvae have thick ectoderm, robust bodies, have the pharynx armed, long malpighian tubules, reduced gastric coecae, short hind gut, and adipose tissue well developed. The adults have the mouth well armed, mentum sclerotized, thorax short, with leg muscles only, and these poorly developed, and the abdomen often distended. The notopleural callus is gone in the apterous species and reduced in the brachypterous species.

The Cnironomidae, as mentioned above, are represented in Antarctica proper by two species of two genera. (fig. 22). The genus *Parochlus (Podonomus)* occurs also in New Zealand, South America and North America. It is related to *Podonomites* of New Zealand and the Andes, and *Podonomopsis* of the Andes. Thus this group appears to present true trans-antarctic distribution. *Belgica* occurs only on the South American side, and its subfamily, the Clunioninae, is said not to be represented in New Zealand. A winged tipulid occurs on Macquarie (ALEXANDER, 1962), and at least one wingless one on Campbell (HARRISON, 1955).

The family Culicidae is represented by one unidentified species taken in the Auckland Islands. The family Simuliidae (blackflies) is represented in Campbell and the Aucklands by *Austrosimulium*. This genus occurs primarily in Australia, Tasmania and New Zealand, but one species occurs in southern South America (WYGODZINSKY & COSCARON, 1962). One species occurs on Campbell and two in the Aucklands (HARRISON, 1955).

Table VII.

Antarctic and Subantarctic Siphonaptera (fleas)

	Antarctica	Falklands	S. Georgia	Marion	Crozet	Kerguelen	Heard	Macquarie	Campbell	Aucklands	Other
Pulicidae											
Pulex irritans 27											Tristan da Cunha
Rhopalopsyllidae											
Parapsyllus m. magellanicus 4, 14		X									
P. m. heardi 2, 6, 7, 9, 12, 13, 14, 18, 19, 26			X			X	X	X			
P. l. longicornis 2, 5											St. Paul, Amsterdam, Tristan
P. l. alginus 1, 2, 4		X									
P. l. dacunhae 16, 22											Tristan
P. l. "A" 3, 5, 18									X		
P. cardinis 7, 12, 14							X				
Listronius robertsianus 4 or 15; 2?		X									
Pygiopsyllidae											
Notiopsylla kerguelensis 7, 8, 13, 14, 16, 17, 20			X			X	X	X	X		Snares
N. enciari 4, 12, 16, 21								X		X	Snares, Antipodes
Ceratophyllidae											
Dasypsyllus stejnegeri 23, 24		X									S'n S. America
Glaciopsyllus antarcticus 10, 11	X										
Nosopsyllus fasciatus 25						X		X	X		

HOSTS

Birds
1. *Pygoscelis papua*
2. *Eudyptes* spp.
3. *Eudyptula* spp.
4. *Spheniscus magellanicus*
5. *Diomedea* spp.
6. *Phoebetria palpebrata*
7. *Macronectes giganteus*
8. *Adamastor cinereus*
9. *Daption capensis*
10. *Fulmarus glacialoides*
11. *Pagodroma nivea*
12. *Pterodroma* spp.
13. *Halobaena caerulea*
14. *Pachyptila* spp.

15. *Procellaria aequinoctialis*
16. *Puffinus griseus*
17. *Pelecanoides urinatrix*
18. *Phalacrocorax* spp.
19. *Catharacta skua lonnbergi*
20. *Larus dominicanus*
21. *Cyanoramphus* spp.
22. *Nesocichla eremita gordoni*
23. *Pezites militaris falklandicus*
24. *Passer domesticus*
Mammals
25. *Rattus* spp.
26. *Oryctolagus cuniculus*
27. *Homo sapiens*

Siphonaptera: Fleas occur on most if not all of the subantarctic islands, and one species is now known from Antarctica (SMIT & DUNNET, 1962; fig. 21). Four families are represented, and the species are listed in Table VII, cross-referenced by number with their normal hosts (after DUNNET, *in litt.*). The fact that some species or subspecies have wide ranges can be correlated with the extensive travel of their host birds.

Hymenoptera: The Hymenoptera are generally considered to be the most highly evolved order of insects. This order is among the most poorly represented in the far south. It is not known from Antarctica. It is not recorded from South Georgia, Marion, Kerguelen or Heard. Two species have been found on Macquarie Island. One of these is a small wingless diapriid wasp, described as an endemic genus, which is apparently also represented on Campbell Island. Macquarie's other species is a minute parasitic wasp. One species is recorded from the Falkland Islands, and one from the Crozet Islands.

The Hymenoptera of Campbell number at least seven species. This fauna is interesting in that three species are wingless (Ichneumonidae, Encyrtidae, Diapriidae), and a fourth (Cynipidae) has wingless females. The three fully winged species belong to the Ichneumonidae, Braconidae and Eulophidae. Some of the species parasitize fly maggots breeding in kelp on the beaches, and some probably attack the larvae of moths.

In the Auckland Islands there are also some interesting insular modifications, with a group of two or three species of ichneumonids *(Aucklandella)* with abbreviated wings in the females and apparently fully winged males.

Summary

The hostile antarctic environment excludes most types of land arthropods. Most or all of those which may have once existed there, and others which might disperse thither, could not cope with the present environmental obstacles. Furthermore, prevailing wind directions are not favorable for dispersal to the continent. Nevertheless, air dispersal is probably the most important method of colonization in the antarctic region.

Conclusions regarding the origin and relationships of the antarctic-subantarctic land fauna are difficult because so much of the past history is unknown for lack of fossil record. In many groups, the representation appears to be haphazard and the result of accidental dispersal, or to represent depauperate elements of nearby continental or subcontinental faunae. There is moderate endemism in some groups, resulting from local evolution, island-hopping, or extinction in source areas. In some portions of the area, such as the Marion — Crozet — Kerguelen — Heard area, or parts of it, there is

488

evidence of local evolution to the species, genus, or tribal level.
In many cases, there appear to be no representatives on subantarctic islands of groups known to occur in southern South
America and New Zealand and/or Australia or other southern
areas. Thus hostile environment and erratic dispersal appear to play
a great role in determining make-up of the land arthropod fauna of
subantarctic islands as well as of Antarctica itself. The possible
continental faunal nature of the Falkland, Campbell and Auckland
Islands is obscured by the ecological inhospitality of the environment. Thus many niches appear to remain unoccupied. Some species, even on the antarctic continent, do not appear to be specially
adapted to their difficult environment, and may be more recent
immigrants than those existing under the harshest conditions.

Addendum

Since the preparation of this paper, extensive new field work has been done on
Campbell I., the Auckland Isls., S. Georgia, Navarino I. and in parts of Antarctica.
Some of the new collections have not yet been studied, and other reports are in
press or in preparation. A few changes have been made in the text, where new information has been brought to light, but further changes in the general picture will be
emerging as the new data are analyzed. Just a few points are mentioned below.

Field work in several entomologically unexplored areas of the continent has proved
negative: Parts of the Ellsworth Mts., southern Queen Maud Range and Darwin
Glacier area. M. E. PRYOR collected Mallophaga, feather mites and two freeliving
mites in the Myrny and Mawson areas of East Antarctica during the period 1961—
1963. Spiders and Thysanura (p. 433) are probably lacking in the antarctic fauna.
Only 3 species of Collembola were found at C. Hallett (p. 438), but about 9 species
of freeliving mites were found near there, 1964—65. Four species of Collembola
were taken in the Terra Nova Bay area (p. 438, no. 6). Two species of prostigmatic
mites are abundant on Observation Hill, Ross I. (see page 440, paragraph 9).
They are *Stereotydeus mollis* WOMERSLEY & STRANDTMANN and *Nanorchestes antarcticus* STR. (WOMERSLEY & STRANDTMANN, 1963). What may be the latter has
now been found south of 85° S. Other new southern records were obtained in 1964—
65 (WISE & GRESSITT, in press). In late 1964 additional positive and negative
records, and ecological data, were published by GRESSITT, FEARON & RENNELL,
and WISE, FEARON & WILKES. One or two representatives of the families of flies
mentioned near the top of page 473 have recently been collected in New Zealand.

Several aspects of the subantarctic entomological picture are changed by new
information published on Campbell I. and nearby islands in GRESSITT et al. (1964).
Only a few of the changes have been incorporated into the text of this chapter. For
instance, Campbell has 10 species of Hymenoptera and not 7 (pp. 472, 478). In
regard to the stoneflies (ILLIES, 1963), the picture is somewhat different from that
described on p. 482. The "new genus" from Campbell represents another species of
Apteryoperla.

The paper on the Austral Region by KUSCHEL (p. 465) was published in GRESSITT,
Editor (1964), which includes a number of papers on Antarctic biota and biogeography. Also included is the paper by GRESSITT & YOSHIMOTO referred to above.

LITERATURE CITED

ALEXANDER, C. P. 1962. Insects of Macquarie Island. Diptera: Tipulidae. *Pacific Ins.* **4** (4): *939—944*, 8 figs.

BROOKES, A. E. 1951. The Coleoptera of the Auckland and Campbell Islands. *Cape Expedition Series, Bull.* **5**: *1—68*, 22 figs.

BROWN, K. 1964. Insects of Heard Island. ANARE Rep. B 1: *1—39*, illust.

BRUNDIN, L. 1962. Insects of Macquarie Island. Diptera: Chironomidae. *Pacific Ins.* **4** (4): *945—954*, 5 figs.

CAILLEUX, A. 1961. Endémicité actuelle et passée de l'Antarctique. *C. R. Soc. Biogeogr., Paris* **38** (334): *65—71*.

DALENIUS, PER & OVE WILSON. 1958. On the soil fauna of the Antarctic and of the Sub-Antarctic Islands. The Oribatidae (Acari). *Arkiv f. Zool.* ser. 2, **11** (23): *393—425*, 12 figs.

EASTOP, V. F. 1962. Insects of Macquarie Island. Hemiptera: Homoptera: Aphididae. *Pacific Ins.* **4** (4): *937—938*.

FALLA, R. A. 1960. Oceanic birds as dispersal agents. *Proc. Roy. Soc. Lond.*, ser. B **152**: *655—659*, 1 pl.

FORSTER, R. R. 1955. Spiders from the Subantarctic Islands of New Zealand. *Rec. Dominion Mus.* **2** (4): *167—203*, 60 figs.

GISLEN, T. 1948. Aerial plankton and its condition of life. *Biol. Rev.* **23**: *109—126*.

GRESSITT, J. L. 1961. Problems in the zoogeography of Pacific and Antarctic insects. *Pacific Ins. Monogr.* **2**: *1—94*, illust.

GRESSITT, J. L. 1962. Insects of Macquarie Island. Introduction. *Pacific Ins.* **4** (4): *905—915*, 3 figs.

GRESSITT, J. L. & R. E. LEECH, 1961. Insect habitats in Antarctica. *Polar Rec.* **10** (68): *501—504*, 1 pl.

GRESSITT, J. L., R. E. LEECH & K. A. J. WISE, 1963. Entomological investigations in Antarctica. *Pacific Ins.* **5** (1): *287—304*, 9 figs.

GRESSITT, J. L. & M. E. PRYOR, 1961. Supplement to "Bibliographic introduction to Antarctic-Subantarctic entomology". *Pacific Ins.* **3** (4): *563—567*.

GRESSITT, J. L. & N. A. WEBER, 1960. Bibliographic introduction to Antarctic-Subantarctic entomology. *Pacific Ins.* **1** (4): *441—480*.

GRESSITT, J. L. & C. M. YOSHIMOTO, 1964. Dispersal of animals in the Pacific. In GRESSITT, Pac. Basin Biogeography, *283—292*.

GRESSITT, J. L. (Editor). 1964. Pacific Basin Biogeography: A Symposium. Bishop Museum, 563 pp., many maps and illustrations.

GRESSITT, J. L. & Collaborators. 1964. Insects of Campbell Island. *Pacific Insects Monograph* **7**; 663 pp., many maps and illustrations.

HARDY, D. E. 1962. Insects of Macquarie Island. Diptera: Coelopidae. *Pacific Ins.* **4** (4): *963—971*, 4 figs.

HARRISON, R. A. 1955. The Diptera of Auckland and Campbell Islands. *Rec. Dominion Mus.* **2** (4): *205—231*, 9 figs.

HOLDGATE, M. W. 1960. The fauna of the mid-Atlantic islands. *Proc. Roy. Soc. Lond.*, ser. B, **152**: *550—567*, figs.

HOLDGATE, M. W. 1961. Biological routes between the southern continents. *New Scientist* **239**: *636—638*, 3 figs.

HOLDGATE, M. W. 1964. Ecology on the Antarctic fringe. SCAR Symposium Antarctic Biol., Paris.

HOLDGATE, M. W. & N. M. WACE. 1961. The influence of man on the floras and faunas of southern islands. *Polar Rec.* **10** (68): *475—493*.

ILLIES, J. 1963. The Plecoptera of the Auckland and Campbell Islands. *Rec. Dominion Mus.* **4** (19): *255—265*, 7 figs.

JANETSCHEK, H. 1963. On the terrestrial fauna of the Ross-Sea area, Antarctica (preliminary report). *Pacific Ins.* **5** (1): *305—311*, 2 figs.

490

JEANNEL, R. 1940a. Croisière du Bougainville aux îles australes françaises. I. Partie générale. *Mem. Mus. Hist. Nat. Paris*, ser. 2, **14**: *1—46*, 8 pls.

JEANNEL, R. 1940b. Coléoptères. Croisière du Bougainville aux îles australes françaises. l.c.: *63—201*, 280 figs.

KOHN, M. A. 1962. Insects of Macquarie Island. Diptera: Dolichopodidae. *Pacific Ins.* **4** (4): *959—962*, 1 fig.

KUBIENA, W. L. 1948. Entwicklungslehre des Bodens, Springer, Wien.

KUSCHEL, G. 1960. Terrestrial zoology in southern Chile. *Proc. Roy. Soc. Lond.*, ser. B, **152**: *540—550*, 1 fig.

KUSCHEL, G. 1962. The Curculionidae of Gough Island and the relationships of the weevil fauna of the Tristan da Cunha group. *Proc. Linn. Soc. Lond.* **173**: *69—78*, 3 figs.

MANI, M. S. 1962. Introduction to high altitude entomology. Methuen, 302 pp.

MURRAY, M. D. 1958. Ecology of the louse *Lepidophthirus macrorhini* Enderlein 1904 on the elephant seal *Mirounga leonina* (L.). *Nature, Lond.* **182** (4632): *404*.

PRYOR, M. E. 1962. Some environmental features of Hallet Station, Antarctica, with special reference to soil arthropods. *Pacific Ins.* **4** (3): *681—728*, 35 figs.

SALMON, J. T. 1949. New Sub-Antarctic Collembola. *Bull. Cape Expedition, New Zealand Sub-Antarctic Expedition*, 1941—45. **4**: *1—56*, 167 figs., 2 maps.

SALMON, J. T. 1962a. New Collembola from 83 deg. South in Antarctica. *Trans. Roy. Soc. New Zealand* **2** (18): *147—152*, 20 figs.

SALMON, J. T. 1962b. A new species and redescriptions of Collembola from Antarctica. *Pacific Ins.* **4** (4): *887—894*, 41 figs.

SALMON, J. T. & J. D. BRADLEY, 1956. Lepidoptera from the Cape Expedition and Antipodes Islands. *Rec. Dominion Mus., Wellington* **3**: *61—81*, 45 figs.

SCHWEIGER, H. 1958. Über einige von der Skottsbergexpedition im Antarkto-Archiplata-Gebiet aufgesammelte Koleopteren. *Arkiv f. Zool.* ser. 2, **12** (1): *1—43*.

SEGUY, E. 1940. Diptères. Croisière du Bougainville aux îles australes françaises. *Mem. Mus. Hist. Nat., Paris*, ser. 2, **14**: *203—267*, 139 figs.

SMIT, F. G. A. M. & G. M. DUNNET, 1962. A new genus and species of flea from Antarctica. *Pacific Ins.* **4** (4): *895—903*, 10 figs.

SMITHERS, C. N. 1962. Insects of Macquarie Island. Psocoptera: Philotarsidae. *Pacific Ins.* **4** (4): *929—932*, 6 figs.

STANNARD, L. J. 1962. Insects of Macquarie Island. Thysanoptera: Thripidae. *Pacific Ins.* **4** (4): *933—936*, 2 figs.

TAYLOR, B. W. 1954. An example of long distance dispersal. *Ecology* **35** (4): *569—572*.

TORRES, B. A. 1953. Sobre la existencia del Tendipedido "Belgica antarctica" Jacobs en el Archipiélago Melchior. *An.* (n.s.) *Zool., Mus. Ciudad Eva Peron* **1**: *1—22*, 18 figs. 3 pls.

TORRES, B. A. 1959. Primer hallazgo de Tendipedidos alados en la región Antárctica. Podonominae, una nueva subfamilia para la citada región. *An. Soc. Cient. Argent., Buenos Aires* **161** (4—6): *41—52*, 8 figs.

TYNDALE-BISCOE, H. C. 1960. On the occurrence of life near the Beardmore Glacier, Antarctica. *Pacific Ins.* **2** (2): *251—253*, map.

WOMERSLEY, H. & R. W. STRANDTMANN, 1963. On some free living prostigmatic mites of Antarctica. *Pacific Ins.* **5** (2): *451—472*, 59 figs.

WYGODZINSKY, P. & S. COSCARÓN, 1962. On the relationships and zoogeographical significance of *Austrosimulium anthracinum* (Bigot), a blackfly from southern South America. *Pacific Ins.* **4** (1): *235—244*, 4 figs.

A GENERAL REVIEW OF THE ANTARCTIC FISH FAUNA

BY

A. P. ANDRIASHEV

Zoological Institute, U.S.S.R. Academy of Sciences, Leningrad
(with 18 figs.)

Several years ago, while engaged in a study of the fish fauna of the Arctic Seas, I often wanted to make a comparison with the fauna of the far South which, together with a striking development of endemism of notothenoid fishes (outwardly resembling our northern cottoids), contains true northern fish types, such as Zoarcidae, Liparidae, Agonidae and so on. At that time the absence of collections from the Antarctic in the Russian museums made a comparative analysis difficult. That is why I was very glad to accept an invitation to take part in the Soviet Antarctic Expeditions aboard the "Ob" during the International Geophysical Year. Over several years my colleagues and myself made extensive collections of bottom and bathypelagic fishes from the Antarctic and neighbouring waters. These materials are still under initial study and results are only partially published which naturally impedes summing up. It is therefore, not without hesitation, that I accepted the amiable suggestion of the Editorial Staff of the volume "Biogeography and Ecology of Antarctica" to compile a review on Antarctic fishes; this review should, however, only be regarded as a preliminary attempt in this direction.

Historical

In a way we could say that the absence of an "ancient history" in the study of Antarctic fishes, with its usual share of doubtful data, incomplete descriptions and complicated synonymy, was fortunate. The first collections fell rightaway in "good hands", indeed they were studied by ichthyologists of the British Museum (Natural History). J. RICHARDSON, A. GÜNTHER, G. A. BOULENGER, C. T. REGAN and J. R. NORMAN; in particular the latter two, not only described more than a half of all genera and species inhabiting the Antarctic but also worked out a modern system for notothenoid fishes and made the first zoogeographical analysis of the whole of the Antarctic fish fauna.

The history of the investigations on Antarctic fishes starting after the review by DOLLO (1904) was given in details in the classical work by NORMAN (1938). Research dealing mainly with the Ross Sea was reviewed by MILLER (1961). The bibliography of Antarctic ichthyology was recently published by two American authors (ROFEN & DEWITT, 1961).

492

It is worth to note that the first ichthyological collections from the Antarctic were made by the U.S. Exploring Expedition under command of C. WILKES (1838—1842). They were passed for study to the prominent ichthyologist Louis AGASSIZ, Harvard College. However, as ROFEN & DeWITT (1961) have recently reported, neither his voluminous and well illustrated manuscript nor the collections themselves became a contribution to science because they were lost. The first specimens of fishes from the Antarctic to be reported (RICHARDSON, 1844) were obtained by the British Expedition on the vessels "Erebus" and "Terror" (1839—1843). In this case too the first fish found near the coast of the Antarctic Continent (it was a white-blooded one, supposedly a *Cryodraco*) has not been preserved as it was eaten by the ship's cat.

In the second half of the XIX century some data on fishes from peripheral parts of the Antarctic were received (Kerguelen, S. Georgia and others) but it is only at the very end of the century that the "Belgica" Expedition (1897—1899) collected the first and unique specimens preserved until now from the Antarctic shelf in the Bellingshausen Sea (DOLLO, 1904). Later on, more than thirty expeditions collected ichthyological material from different parts of the Antarctic. I shall mention only the most important ones putting the stress on recent investigations.

To begin with the early British Expeditions on the "Challenger" (1873—1876), the "Southern Cross" (1898—1900), the "Discovery" (1901—1904) and the "Terra Nova" (1910) should be mentioned. Material collected by these expeditions in the western part of the Ross Sea, near the Antarctic Peninsula (Graham Land), the Kerguelen, the Crozets, S. Georgia etc. was described and drawn in detail in the works of the ichthyologists of the British Museum (GÜNTHER, 1880, 1887; BOULENGER, 1902, 1907; REGAN, 1913, 1914, 1916). These works are the base of our present knowledge on the Antarctic fishes. A valuable addition to the list of species (and their biology) of Graham Land and the neighbouring islands was made by LÖNNBERG (1905) using the material of the Swedish Expedition "Antarctic" (1901—1902). Two French expeditions, on the "Français" (1903—1905) and on the "Pourquoi-Pas?" (1908—1910) (VAILLANT, 1906; ROULE, ANGEL & DESPAX, 1913) have worked in the same area. The first data on the East Antarctic fishes (Davis Sea) were obtained by the German Antarctic Expedition aboard the "Gauss" (1901—1903) and reported by PAPPENHEIM (1912). From the same region and eastwards interesting material was collected by the Australasian Antarctic Expedition aboard the "Aurora" (1911—1914), and the results were published by WAITE (1916). More than five hundred fishes from different parts of East Antarctica, Kerguelen and Macquarie Islands were brought back by the B.A.N.Z. Antarctic Research Expedition on the "Discovery" in 1929—1931

(NORMAN, 1937b). Still more abundant material was obtained as a result of expeditions organised over many years by the British Discovery Committee ("Discovery", "Discovery II", "William Scoresby", 1925—1936). They enabled J. R. NORMAN to write a series of valuable monographs on bottom and oceanic fishes of the southern hemisphere (NORMAN, 1930, 1935, 1937a), including his well-known general work on the Antarctic fishes (NORMAN, 1938).

In the post-war period the appearence of important works by Scandinavian authors should be mentioned. NYBELIN (1947, 1951, 1952) studied collections of fishes assembled by whalers and by several Antarctic expeditions (1904—1939, 1947—1952) including fishes from formerly unexplored areas (Peter the 1st Is., Coats Land and some other). He started the subspecific taxonomy of Antarctic fishes and gave new opinions on zoogeographical problems and on some other aspects of fish biology. The work of OLSEN (1954, 1955) was a start for the comprehensive study of the biology of notothenoid fishes. Finally, a report of great interest to general biology was made by RUUD (1954, 1958). He established that, in the Antarctic fishes of the family Chaenichthyidae examined by him, the blood is colourless since it contains neither erythrocytes nor haemoglobin.

In recent years explorations in the Antarctic are connected with the program of the International Geophysical Year. The first reports on the work conducted by scientists of Argentina, Belgium, France, Japan, New Zealand, the U.S.A. and the U.S.S.R. begin to appear. Of these we would like to mention the expedition of the N.Z. Oceanographic Institute on board the "Endeavour" (1958) which secured new data on the distribution of bottom fishes of the Ross Sea (RESECK, 1961; MILLER & RESECK, 1961). Fishes collected in 1961 by the Belgian Antarctic Epeditions near the Princess Ragnhild Coast were reported by GOSSE (1961), Shore fishes obtained by French winter teams were reported by BLANC (1961) and other in several notes.

Ichthyologists of the U.S.A. recently came forward with a wide scale program of ichthyological research (ROFEN & DEWITT, 1961; WOHLSCHLAG, 1961a and others), which is successfully realized by Californian scientists (Stanford University, Long Beach College, University of Southern California). Besides work on systematics (DEWITT & TYLER, 1960) American ichthyologists discovered giant nototheniids in the McMurdo Sound as well as the remains of similar fishes collected on the surface of the Ross Ice Shelf (SWITHINBANK, DARBY & WOHLSCHLAG, 1961). Brief data on aqualung observations in Antarctic waters were reported by NEUSHUL (1959) and the first colour photo of living Antarctic fish was published by HERALD (1961). Very important experimental investigations on the physiology and the biochemistry of live fishes were carried out for the first time in the Antarctic under the direction of D. E. WOHL-

SCHLAG (WOHLSCHLAG, 1960, 1961a, b; TYLER, 1960)[1]. At present
the U.S. explorations are continued and expanded including the
extensive and large scale oceanographical investigations carried
out on the big and well equipped vessel "Eltanin" (1961—1962).

Finally I permit myself to mention in brief the ichthyological
investigations carried out by the U.S.S.R. Although the two Russian vessels "Vostok" and "Mirny" (1819—1821) were, in their time
among the first discoverers of the Sixth Continent, our ichthyological investigations in the Antarctic practically only started in 1956
(hitherto, some fishes from S. Georgia were collected by biologists of
the whaling fleet "Slava"). The main collections were made during
three cruises of the research vessel "Ob" (1955—1958). Additional
material and observations were received from geographer E. S.
KOROTKEVITCH and ichthyologist V. M. MAKUSHOK who have wintered in "Mirny". Numerous bathypelagic fishes were caught by G. A.
SOLANIK on research whale-boats. In all nearly three thousand
bottom and oceanic fishes belonging to 85—90 species were collected
in Antarctic waters. They were caught along the coasts of the Antarctic Continent ranging from the Cape-Town meridian (20°E) to the
east up to Victoria Land; in the open waters of the S. Indian and
S. Pacific Oceans (including Amundsen and Bellingshausen Seas and
Drake Strait), and from S. Shetland, S. Georgia, Bouvet, Kerguelen,
Heard, Macquarie, Balleny and Scott Islands. All collections are
kept in the Zoological Institute, U.S.S.R. Academy of Sciences,
Leningrad, and are studied by the author. Up to now only preliminary reports by ichthyologists of the three cruises of the "Ob"
have been published (ANDRIASHEV & TOKAREV, 1958; BARSUKOV
& PERMITIN, 1960; ANDRIASHEV & PERMITIN, 1961 and others),
including preliminary station lists of fishes (ANDRIASHEV, 1958a,
1961; BARSUKOV & PERMITIN, 1959) and some scientific results
obtained from collections (BARSUKOV & PERMITIN, 1958; AN-
DRIASHEV, 1959, 1960a, b, c, d, 1962a, b).

The greatest success in ichthyological investigations of the Antarctic is achieved in the elucidation of the species composition.
However on the map of the Antarctic there are still many vast regions which are from the ichthyological point of view, "Terra incognita" such as the continental shelf from the eastern part of the Ross
Sea up to the Amundsen and Bellingshausen Seas, the Weddell Sea
with the neighbouring eastern coasts, some islands etc. The composition and distribution of the bathypelagic fauna are not studied
sufficiently and the data on the abyssal fauna are rather scanty. In

---

[1] New data on Antarctic fishes and some problems of Antarctic biology and
zoogeography have recently been reported by D.E. WOHLSCHLAG, N.B. MARSHALL,
J.C. HUREAU and other authors at the SCAR Symposium on Antarctic Biology
(*SCAR Bull.* 12, 1962).

such important fields as the life history of fishes only the first steps were made.

Composition and Distribution of the Antarctic Fish Fauna

Bottom fishes (a brief review of families)

The fish fauna of the Antarctic is very peculiar. It includes families and genera which are sometimes very similar and sometimes very different in their origin and degree of endemism, species composition, vertical and geographical distribution. A brief review on families with an indication of the most typical geographical and ecological characters, some morphological features, and the present state of knowledge may turn out to be rather useful for a better knowledge of this fauna. Accordingly a list of Antarctic coast species is given (pp. 511) as well as some data on abyssal fishes.

H a g - F i s h e s (Myxinidae). Members of this family of fish-like vertebrates are not characteristic of the Antarctic fauna — out of three species of the genus *Myxine* known from the temperate waters of the southern hemisphere, only one species (*M. australis*) was found once near the S. Shetland Islands (NORMAN, 1938). The genus *Myxine* as a whole has a bipolar distribution.

S l e e p e r s h a r k s (Dalatiidae). A shark about 2½ m long

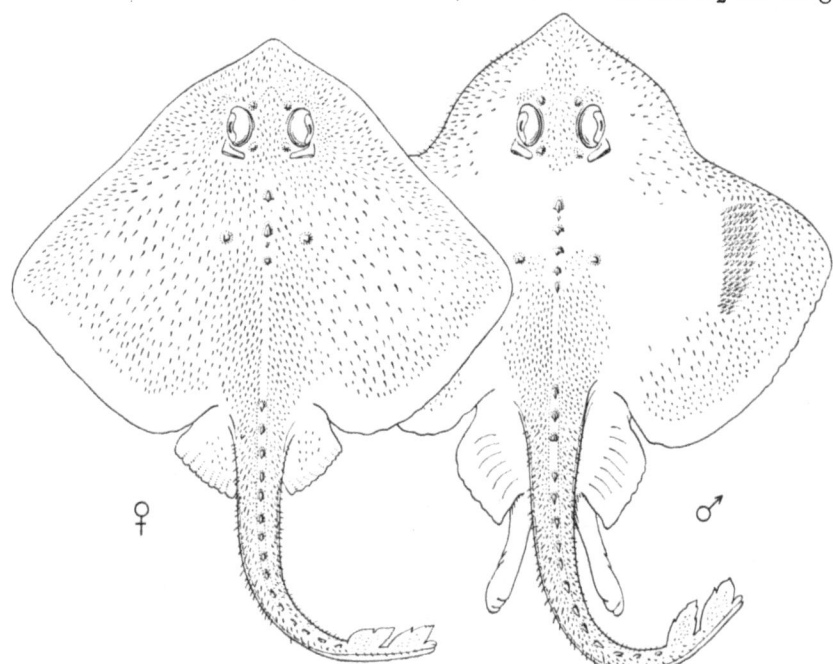

Fig. 1. Murray's skate, *Raja murrayi*, ad. ♀♂ (orig.)

was stranded in 1913 on Macquarie Island but has not been preserved. On the basis of a sketch, the preserved teeth and a piece of the skin it was referred to the genus *Somniosus* but was later separated as a new species, *S. antarcticus* WHITLEY (WAITE, 1916; GARRICK, 1960). Since no more exact data are available, I abstain from including *Somniosus* into the list of the Antarctic fauna.

Skates (Rajidae). This family of worldwide distribution is represented in the Antarctic by some inadequately studied species of the genus *Raja* s.l. (From some of them only empty egg-capsules are known). Only two species both from Kerguelen (*R. eatoni* and *R. murrayi*, fig. 1) were described on adult males and females. Taxonomic position and validity of other species are not clear. Described on a young, *R. georgiana* is very closely related to our young specimen of *R. murrayi* from Heard Island. Another species (*Raja arctowskii*) from the Bellingshausen Sea was described only on the basis of empty egg-capsules (DOLLO, 1904, pl.IX, fig. 10); they are not distinct, however, from the small egg-capsules from Kerguelen that were obtained together with an adult *R. murrayi* in the same trawl and suppose to belong to this species. Probably they all belong to the same species (*R. murrayi*). Judging by the occurrence of a very large egg-capsule similar to that of the European species, *R. batis*, one more species of skate should occur near Graham Land (NORMAN, 1938). Further study of Antarctic skates is very necessary as this may prove very advantageous of zoogeographical conclusions.

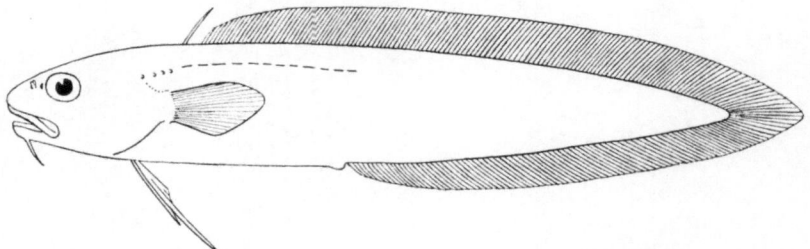

Fig. 2. Antarctic eel-cod, *Muraenolepis microps* (after REGAN).

Eel cods (Muraenolepidae). This is a rather aberrant family of the order of the Gadiformes (SVETOVIDOV, 1948, 66), in which contrary to the Gadidae, the caudal fin is completely fused with the dorsal and anal fins (fig. 2), the number of pectoral radialia is increased (10—13) and squamation is of the "parquet" type (as in many eels and ophidiids). This is the most primitive family among the Gadiformes and was separated by SVETOVIDOV (1948) as a distinct suborder. The family Muraenolepidae may be described as a primarily Antarctic one since of the four known species three live in the Antarctic Region and the last one is known only from the Patagonian-Falkland waters. Muraenolepids are near bottom inhabitants of moderate depths, one specimen, however, was caught by trawl S.E. off Heard Isl. at the depth of 1600 m. Young muraenolepids sometimes occur in the intermediate water layers over the great depths more than 100 miles off the Continent (NORMAN, 1937b). The species composition of the only genus present, *Muraenolepis*, is not well known and can unfortunately hardly be used for exact zoogeographical conclusions.

Notothenoid families (superfamily Notothenioidae = Nototheniiformes auct.). This group, consisting of 5 closely related families united into a separate superfamily of Notothenioidae (order Perciformes, suborder Percoidei, BERG, 1940) is of a great interest to us since it is well represented and distributed mainly in the Antarctic, making up nearly 3/4 of all the species of the coast fish fauna of the Antarctic Region (concerning its boundaries see p. 536). The distribution of the families is very interesting: four families (Nototheniidae, Harpagiferidae, Bathydraconidae and Chaenichthyidae; see below) are mainly Antarctic, whereas the fam. Bovichthyidae is of notal (antiboreal) nature, its species being widely distributed in temperate waters of the southern hemisphere: to the south they reach up to Tierra del Fuego, the Falkland Islands (including Burdwood Bank) and the islands to the south of New Zealand (Auckland, Campbell etc.) but never cross the line of the Antarctic Convergence (see fig. 18). All the members of the notothenoid families, although belonging to the general percoid type, differ from this type by a peculiar structure of their pectoral girdle (3 radialia); the fact that they have a single nostril on each side and usually develop two or three lateral lines, and by some other characters. Bovichthyidae are supposed to be the most generalized family (REGAN, 1914). Nearly all the species of Notothenioidae live in the near-bottom layers of the continental shelf, seldom exceeding a depth of 1000 m; few forms adapted secondarily to temporary or permanent life in the open sea but apparently all of them retain demersal eggs.

The so-called Antarctic cods (Nototheniidae) are reminiscent of the North Pacific greenlings (Hexagrammidae). The family consists of 5 genera of which 4 are well represented in the Antarctic (see below) whereas the genus *Eleginops* is limited to the temperate waters of South America. One species of this genus, *E. maclovinus*, contrary to all the southern members of the family, often occurs in well heated shallow bays and estuaries, even entering fresh waters.

The genus *Notothenia* is the most abundant in species among the notothenoid families, it contains not less than 30 species half of which are known from the Antarctic waters. Nevertheless, this genus cannot be called mainly Antarctic as the majority of its species avoid the coast of the Antarctic Continent and inhabit mainly the notal waters of the southern hemisphere (17 species) and only penetrate into the peripheral parts of the Antarctic. The greatest number of species of the genus *Notothenia* were recorded in the Patagonian-Falkland waters (16 species) while in other regions their number is considerably less: Kerguelen, Graham Land and S. Georgia have 6—7 species each; the S. Shetland, S. Orkney, S. Sandwich and Heard Islands, the Argentine coast to the north of 43° s. lat. and New Zealand with adjoining Islands have 4—5 species; the Macquarie Island, the Burdwood Bank, the Crozets Islands and Victoria Land have 2—3 species; the Bouvet, Peter 1st and Marion Islands, Adelie Land, Scott Island, the Coast of Chile to the north of 42° 30' s. lat. have 1 species each. As one can see only solitary species of the genus *Notothenia* are known in higher latitudes; they are *N. coriiceps neglecta* (Victoria Land and Adelie Land) and *N. larseni* (Peter 1st Island; apparently the same species penetrates westwards up to the Balleny Islands near Victoria Land). A rather aberrant species is *N. kempi* which previously was known only near the Palmer Archipelago but was found in 1958 in trawls taken from Scott Island (66° S and 180° E) and from Sabrine Coast (113°E.).

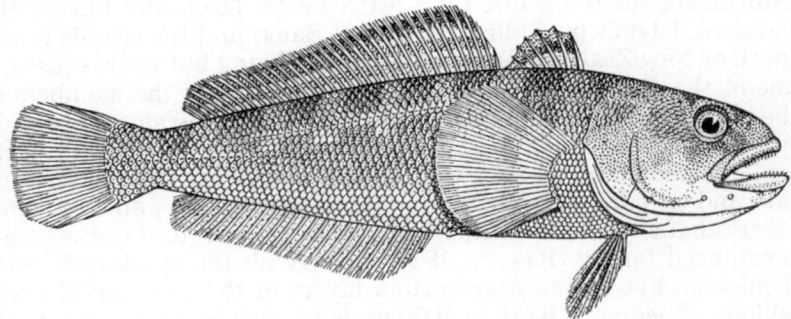

Fig. 3. Antarctic cod, *Notothenia rossi* (after Waite).

All species of *Notothenia* can be arranged into two groups according to their distribution — notal and antarctic: not one out of 13 Antarctic species comes into notal waters and only two out of 17 notal species (*N. macrocephala* and *N. colbecki*) enter peripheral parts of the Antarctic (Kerguelen, Macquarie). The distribution of Antarctic species of *Notothenia* is very representative. In general they inhabit the islands of the Scotia Arc from the S. Georgia to the S. Shetland Islands and sometimes up to the northern part of Graham Land (*N. larseni, N. nidifrons, N. gibberifrons, N. angustifrons* and, more widely distributed, *N. coriiceps* and *N. rossi*), (fig. 3), around the more easterly islands they are replaced by a few

related species: Bouvet Island: *N. larseni* (does not differ from the
typical west-antarctic species; see NORMAN, 1938, NYBELIN,
1947), Marion Island: *N. marionensis* (closely related to *N. angustifrons*), the Kerguelen: *N. squamifrons*, *N. mizops* and *N. acuta*
(closely related to *N. larseni*, *N. nudifrons* and *N. angustifrons*
respectively, see NORMAN, 1938). *N. cyaneobrancha* is typical for
the intertidal zone of the Kerguelen; this species was also reported
for Heard Island (from the stomach of *Pygoscelis papua*, see EALEY,
1954). Nearly all species of *Notothenia* are bottom dwellers of the
continental shelf, distributed from the littoral zone up to 200—500 m,
at greater depths we only met *N. larseni* and *N. kempi* (up to
400—700 m). The silvery young (0 +) of some species occur in the
open sea at a considerable distance from the coast. Certain species
make, even at the adult stage, seasonal migrations into the open
ocean when they feed on krill (*N. coriiceps*, *N. rossi*, see OLSEN,
1954), other species feed on bottom invertebrates and small fish.
Their reproduction is usually poorly known except for some species
where autumn — winter spawning was established. Eggs are demersal, with diameters of 1.2 — 2.5 mm; fecundity: from 2—3 to 10
thousand eggs (in large fishes reaching 70—90 cm length it should
be considerably higher). The best known life history is that of *N.
rossi* from S. Georgia (OLSEN, 1954).

The species of the genus *Trematomus* (fig. 4), contrary to the
numbers of the preceding genus, are found near the Antarctic
Continent, in particular along the eastern coasts. Thus all 13 species[1]
(see table I) are known to occur from Coats Land to Victoria Land
and the Ross Sea; of them only 7 species reach the West Antarctic
(Graham Land and S. Orkney Islands), and only 2 species (*T.
hansoni* and *T. bernacchii vicarius*) are found at S. Georgia Island.

No species of the genus *Trematomus* are present[2] near Bouvet
Island and all the Subantarctic Islands (including Heard Island)
revealing their high-Antarctic nature.

With the present state of our knowledge it can be supposed that
the genus *Trematomus* contains some pairs of rather closely related
species with sympatric areas. As examples *T. borchgrevinki* and *T.
brachysoma*, *T. hansoni* and *T. loennbergi*, *T. centronotus* and *T.
pennellii* can be mentioned. To a considerably less degree this can
also be said about *T. lepidorhinus* and *T. eulepidotus* which are
thought by NORMAN (1938, 40) to be also nearly related.

[1] 13 species include an undescribed new species from Victoria Land ('Ob', St, 336,
650-700 m depth), which was preliminarily identified as *T. nicolai* (ANDRIASHEV,
1961). *T. vicarius*, which was formerly considered as a separate species, should be
regarded only as a subspecies of the widely distributed *T. bernacchii*, as was suggested by NORMAN (1938).
[2] The absence of species of *Trematomus* at the S. Sandwich Islands can probably
be explained by unsufficient data from this area.

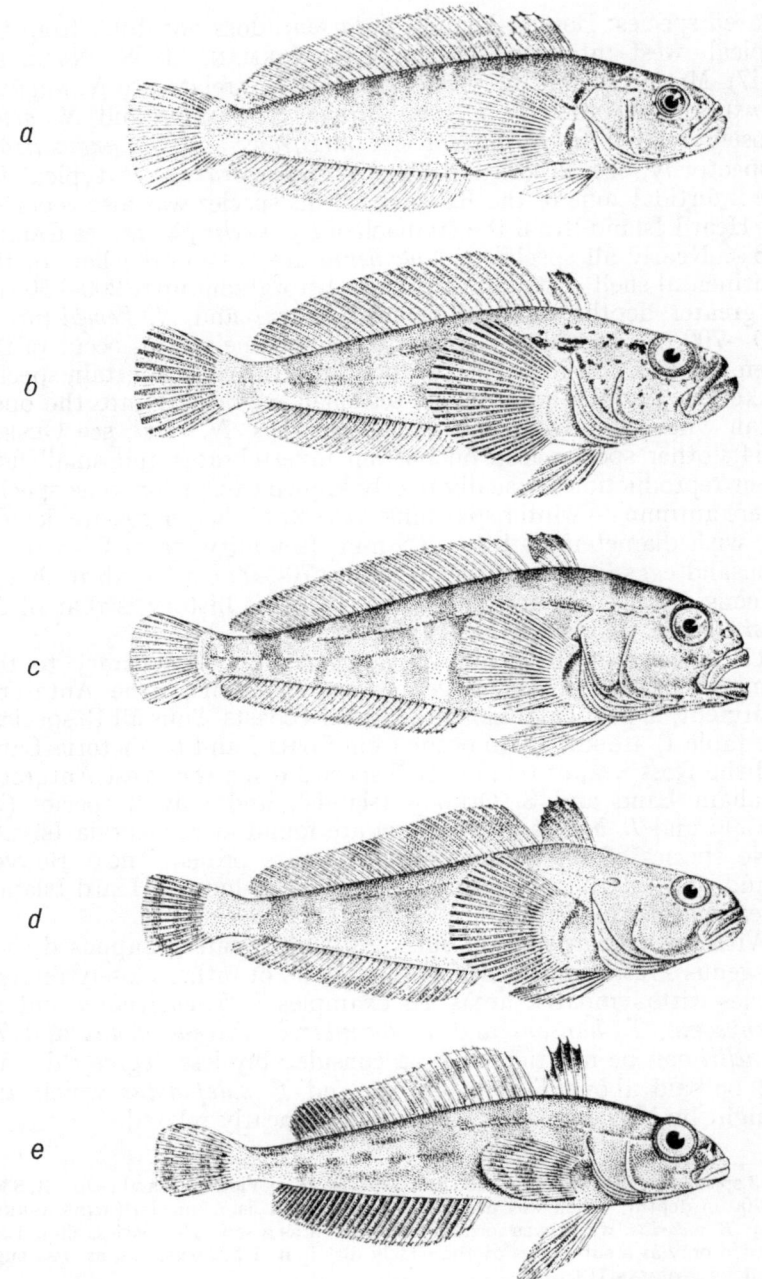

Fig. 4. Some Trematomus species: a – *T. borchgrevinki;* b – *T. nicolai;* c – *T. bernacchii;* d – *T. hansoni;* e – *T. scotti* (a, b, c, d – after BOULENGER; e – after REGAN).

It is worth to note that *T. borchgrevinki* differs considerably from other species of the genus in essential characters which indicates the necessity of its subgeneric or even generic separation (this species is related to *T. brachysoma* absent in our collections).

At the adult stage the majority of species of *Trematomus* acquire near-bottom habits and only *T. borchgrevinki*, *T. brachysoma* and *T. newnesi* occur in the midwater and surface layers above considerable depths but within the limits of the shelf. In the shallow waters near the shores *T. nicolai*, *T. hansoni* and the most common species *T. bernacchii* are also met. However, the latter two are found at considerable depths. The group of deep-water species (found at depths of up to 600—900 m) includes *T. scotti*, *T. lepidorhinus*, *T. eulepidotus*, *T. loennbergi*, the new species mentioned above and some others. Eggs from *Trematomus* are larger and less numerous than these from *Notothenia*. Thus, our new species of *Trematomus* had 1600 eggs, with a diameter of 2.6 mm, the diameter of fully ripe eggs of *T. bernacchii* from the Davis Sea was 3.7—3.8 mm and the fecundity about 1030—1060 eggs. At the age of five years this species reaches sexual maturity and has a comparatively high growth rate (WOHLSCHLAG, 1961b). Blood in *Trematomus* is red but the quantity of haemoglobin is less than in other Teleostei (TYLER, 1960).

The genus *Dissostichus* differs from *Trematomus*

a) in a clearly expressed predatory appearance (resembling *Merluccius* of the Gadidae) and a giant body size;

b) in the fact that it has very small and numerous scales;

c) in the large mouth with strong canine-like teeth.

The last feature brings *Dissostichus* close to *Notothenia canina*; from the Patagonian waters, however, the structure of the pectoral girdle shows its relationship with *Trematomus* (NORMAN, 1937b). The genus consists of only two related species: *D. eleginoides* with a length of up to 138 cm and a weight of 44 kg living in the Patagonian-Falkland area and *D. mawsoni* (fig. 5a) from the coasts of the Antarctic Continent. The latter species was found only in three well separated localities: McRobertson Land, Princess Martha Coast and the Palmer Archipelago. In our collections there are two young specimens caught near Wilkes Land and in the open part of the Amundsen Sea. Moreover I have tried to prove (ANDRIASHEV, 1962b) that the remains of the giant *"Notothenia"* found in the McMurdo Sound (BOULENGER, 1907) do not belong to *Notothenia colbecki* as BOULENGER supposed but to *Dissostichus mawsoni*. The full length of this specimen was about 150 cm, with a weight of more than 20 kg. Recently in the McMurdo Sound a large living fish of 135 cm long and 27 kg weight was taken from the mouth of a Weddell seal ("Polar Times", No. 53, 1961, photo), whereas the remains of still larger fishes were found on fast ice since the weight of each one is supposed to be 64 kg! (D. E. WOHLSCHLAG, personal communication). There are reasons to believe that these giant fishes belong to *D. mawsoni*, which is the largest fish[1] in the Antarctic as well as in the whole superfamily of the Notothenioidae.

The genus *Pleuragramma* with a single species, *P. antarcticum*, differs from other nototheniids in its typical pelagic, herring-like appearance (body shape, silvery colour of the adult, large thin scales which easily fall off, emarginate caudal fin, feebly ossified

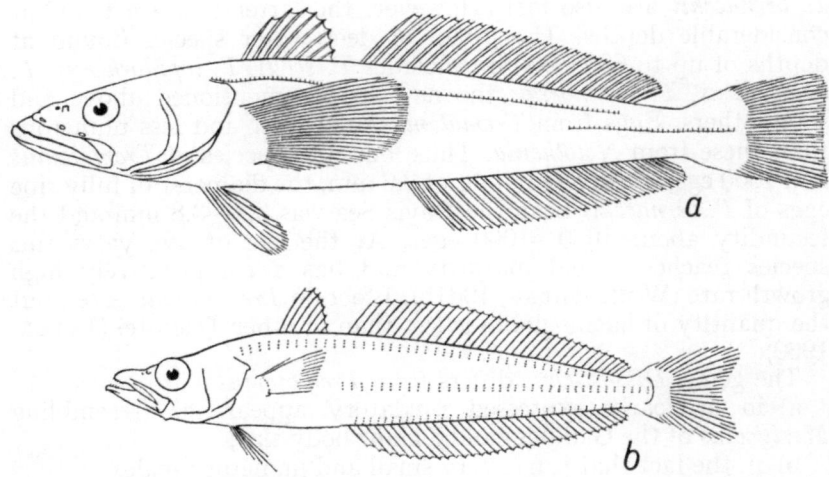

Fig. 5. Two high-antarctic nototheniids: a – Antarctic giant fish, *Dissostichus mawsoni;* b – Antarctic herring, *Pleuragramma antarcticum* (after NORMAN).

skeleton; fig. 5b). It is distributed around the Antarctic Continent and is the only true pelagic plankton-eating form among notothenoid families. *Pleuragramma*, or Antarctic herring, is very likely abundant in Antarctic waters. The Argentine ichthyologists F. GNERI and A. NANI reported (personal communication) the observation of a very great number of dead pleuragramms floating on the surface of the Bellingshausen Sea over a stretch of many miles. Spawning apparently takes place along the coast during the winter; larvae appear in the plankton around October-December reaching a length of 35—40 mm in one year (REGAN, 1961, DEWITT & TYLER, 1960). The young of *Pleuragramma* and these of *Trematomus* are the

---

[1] After this work was completed I received additional data: (1) Dr. D. E. WOHLSCHLAG kindly informed me that according to preliminary identifications the very fish from McMurdo Sound turned out to be *Dissostichus mawsoni* (2) Large fishes, very similar to *Dissostichus*, were found by biologists from the whale-boat 'Soviet Russia' in the stomachs of spermwhales harpooned near Scott Isl. The total length of the largest specimen was 175 cm!

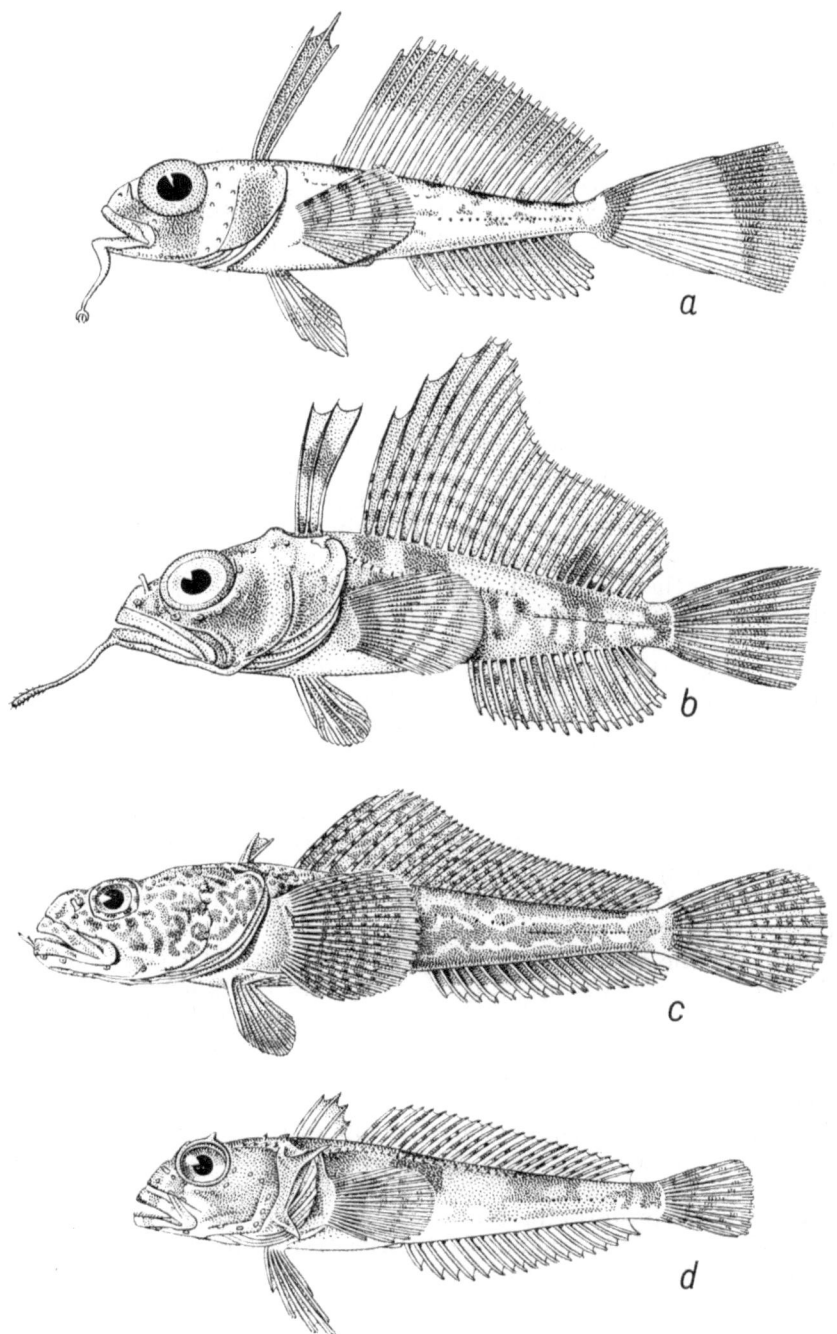

Fig. 6. Plunder fishes, Harpagiferidae: a – *Dolloidraco longedorsalis;* b – *Histiodraco velifer;* c – *Pogonophryne* sp.; d – *Harpagifer georgianus* subsp. (a – after WAITE; b – after REGAN; c, d, – orig.).

504

food of penguins and petrels, the adults are fed on by seals and probably by whales.

Plunder fishes (Harpagiferidae) consist of small bottom fishes without scales often resembling northern sculpins (Cottidae) and usually having a well developed barbel on the lower jaw (fig. 6); they are 10—15 cm long, rarely up to 25—30 cm. In the family 5 genera with 11 species are known but in the genus *Pogonophryne* containing the largest deep-water forms, some as yet undescribed species apparently exist. The plunder fishes can be arranged in two groups according to their morphology, distribution and habits. The first group contains forms with a mental barbel and a weakly armed gill cover (*Artedidraco, Dolloidraco, Histiodraco, Pogonophryne*) and is a truely Antarctic and comparatively deep-water group; its species inhabit the continental shelf of the Antarctic Continent (one species at S. Georgia) but are absent at the Kerguelen and other islands lying near the Antarctic Convergence. The second group is formed by species of the genus *Harpagifer* (fig. 6d) lacking the barbel and with a gill cover strongly armed with spines. They are very characteristic of the littoral zone (in tide pools under stones) not only on the Kerguelen, Macquarie and other peripheral islands of the Antarctic but also on the Falkland Islands and on Tierra del Fuego. In contrast to the species of the first group *Harpagifer* is absent from the coast of the Antarctic Continent (excluding the very northern part of the Graham Land); its specific and subspecific differentiation requires further study as it is very interesting from the zoogeographical point of view (NYBELIN, 1947). The eggs are demersal and large (2.6—3.0 mm), the fecundity is supposed to be very low. Recently hatched larvae, appearing in plankton in summer, are very large — for *Artedidraco scottsbergi* about 13 mm (REGAN, 1916).

Antarctic dragon fishes (Bathydraconidae). This family contains 8 genera with 15 species, and has a true Antarctic distribution; the majority of its species occur at the coast of the Antarctic Continent and only three species are known at the S. Georgia and S. Orkney Islands and at the Kerguelen-Heard submarine ridge. Bathydraconidae differ from other notothenoid families by a greatly elongated body, a usually spatulate ("pike-like") snout and by the absence of the first dorsal fin (fig. 7). However, preserved supporting elements of this fin (interneuralia), can clearly be seen in X-ray photographs (ANDRIASHEV, 1959) they show that the complete reduction of the ID is a secondary phenomenon in the evolution of this branch of the superfamily Notothenioidea.

Bathydraconidae inhabit the continental shelf up to depths of 500—700 m but some members of the genus *Bathydraco* live at depths up to 2579 m (*B. scotiae*); this is the deepest level at which Notothenoid fishes can survive.

Adults reach half a meter in length (*Cygnodraco, Parachaenichthys*) but most of the species have considerably smaller sizes. Eggs are demersal, large (2.6—3.0 mm), and scanty.

At the eastern coast of the Antarctic Continent one of the com-

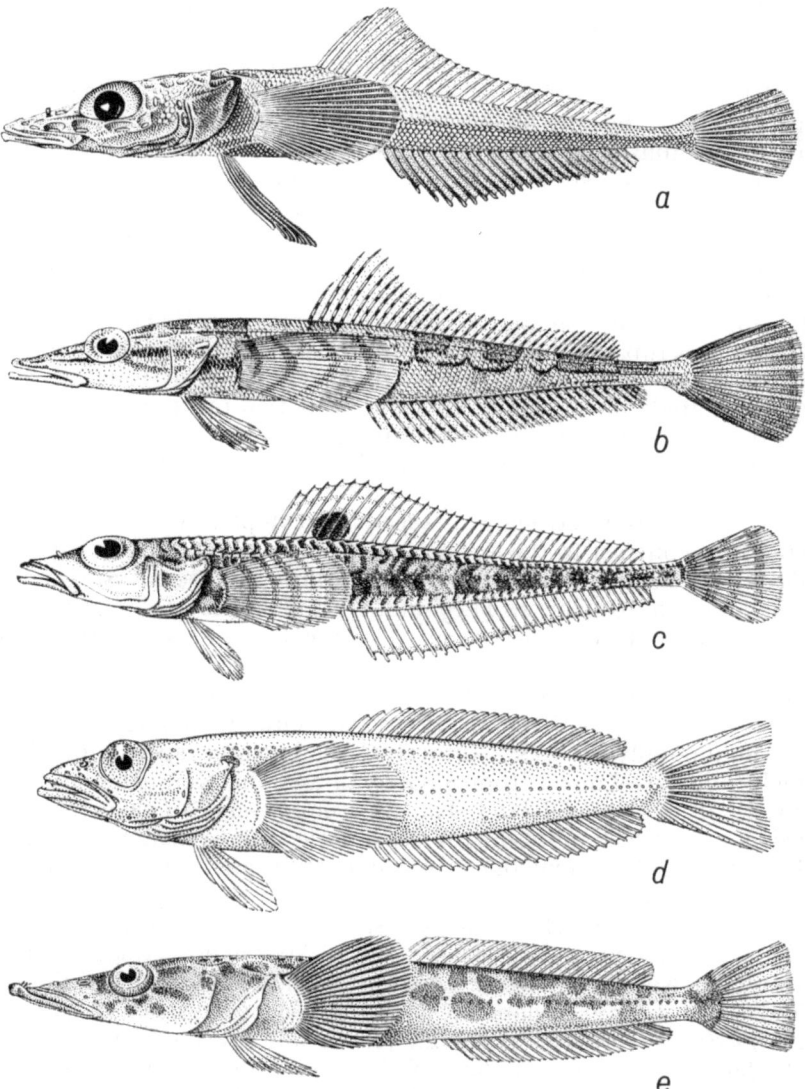

Fig. 7. Antarctic dragon fishes, Bathydraconidae: a – *Bathydraco nudiceps;* b – *Racovitzia harrissoni;* c – *Prionodraco evansi;* d – *Psilodraco brevipes;* e – *Gymnodraco acuticeps* (a – orig.; b – after WAITE, c, e – after REGAN; d – after BOULENGER).

mon forms of this family is *Gymnodraco acuticeps* (fig. 7e) which is often caught at the bottom of shallow waters but was also observed free swimming at the surface far from the coast (PAPPENHEIM, 1912).

It is of interest that in living specimens of *Prionodraco evansi*, widely distributed around the Antarctic Continent, the body is semitransparent; in its long caudal part the vertebral column and the red arteria caudalis under it are clearly visible; gill-lamellae of the same colour are seen through the gill cover and membranes.

White-blooded fishes or ice fishes (Chaenichthyidae) are the most amazing phenomenon of the Antarctic fauna. Indeed, as was established for the first time by RUND (1954), the blood of these fishes possesses a unique character among vertebrates since it is colourless and contains neither erythrocites nor haemoglobin.

In the last years direct observations on living individuals and studies on museum specimens have shown that this striking feature is characteristic of all the species of Chaenichthyidae and is therefore the distinguishing character of the whole family. Usually they are large fishes (up to 67 cm long) with a naked body on which two or three lateral lines are developed (fig. 8). The head is large with an elongated and spatulated snout and non-protractile, strongly toothed jaws; a large "pike-like" mouth usually exceeds half of the head length. The skeleton is feebly ossified. Long delicate gill arches bear gill-lamellae of a white-cream colour (when alive), gill rakers are often strongly reduced or completely absent which stands in contrast to the strongly toothed jaws. The spleen is small, flattened, nearly colourless. In this family 17 species are known belonging to 10 genera. They inhabit the whole of the Antarctic Region except for one species (*Champsocephalus esox*) from the Patagonian-Falkland waters. Host species of white-blooded fishes are found along the coast of the Antarctic Continent where 11 species are known; of these two species are circumpolar and 7 species are found only in East Antarctica. Only three species of chaenichthyids are known from S. Georgia and the Kerguelen and one species from Bouvet Island. The apparent absence of chaenichthyid fishes at the S. Sandwich, the Macquarie, the Crozets and Marion Islands might be explained by an unsufficient study of the fish fauna of these waters.

The majority of chaenichthyid species live at a depth of not less than 100—200 m (and up to 600—700 m), species occurring in shallow waters as well were observed only near the Kerguelen, the Bouvet and the S. Georgia Islands (*Chaenocephalus, Chaenichthys, Champsocephalus*). However, NYBELIN (1947) already paid attention to the fact some species of chaenichthyids not always live at the bottom but are sometimes observed in the upper layers of water often far from the coast and in the stomachs of seals and whales (*Cryodraco antarcticus, Pagetopsis macropterus, Neopagetopsis ionah*, see fig.

Fig. 8. White-blooded fishes or ice fishes, Chaenichthyidae: a – *Champsocephalus gunnari;* b– *Neopagetopsis ionah;* c – *Dacodraco hunteri;* d – *Cryodraco antarcticus;* e – *Chionodraco kathleenae* (a, b, c, e – orig.; d – after REGAN).

508

8b, d). OLSEN (1955) established summer migrations into the open
sea for feeding on krill for *Champsocephalus gunnari* and *Pseudochaenichthys georganus*. Bottom species feed on benthic crustaceans
and small fish. Spawning at the coasts of S. Georgia is carried out
in the autumn (March-April). Eggs are demersal and comparatively
large (up to 4.5 mm). The young of many species (7—14 cm long)
occur near the surface in the open sea. The growth rates of species
from S. Georgia were comparatively high (6—10 cm per year). *Ch.
gunnari* (fig. 8a) reaches sexual maturity at the age of 4 years. In
migrating species females and males are similar in length, whereas
in bottom species (*Ch. aceratus*) the females are considerably larger
than the males and can reach the age of 17 years (OLSEN, 1955).

Eel pouts (Zoarcidae) in contrast to the five above mentioned families are not aborigens of the southern hemisphere.
This family is very rich in species in the Arctic and in the northern
parts of the Atlantic and Pacific Oceans, the Okhotsk and Bering
Seas in particular. In boreal and arctic waters nearly 150 species of
zoarcid fishes live, belonging to more than 25 genera and several
subfamilies. In the Antarctic this family is represented by 5 genera[1]

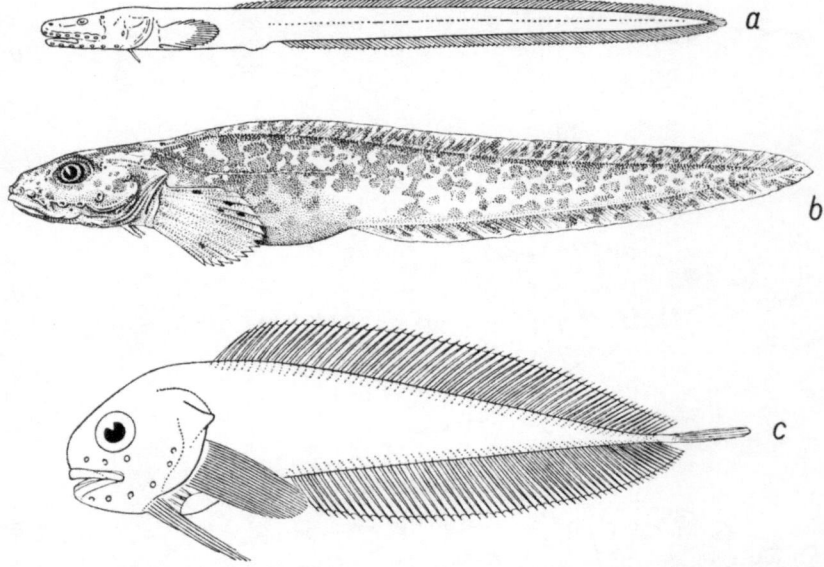

Fig. 9. Antarctic species belonging to families of northern origin, Zoarcidae and
Liparidae: a – *Lycenchelys antarcticus*; b – *Lycodichthys antarcticus*; c – *Paraliparis
terrae-novae* (a – after NORMAN; b – orig.; c – after REGAN).

---

[1] Dr. H. H. DeWITT (Stanford University) kindly informed me that zoarcid
fishes belonging to a new genus and species (*Rhigophila dearborni*) were found in the
McMurdo Sound at 585 m depth.

and 7 species (and 10 genera with 15 species live in the Patagonian-Falkland area). *Austrolycichthys* (3 species), *Lycodichthys* (1 species) and probably *Rhigophila* (1 species) are endemic Antarctic genera, while the genus *Melanostigma* is bipolar and the genus *Lycenchelys*, due to its abyssal character, is more widely distributed. It is interesting that the numerous *Lycenchelys* species of the north-western part of the Pacific are arranged into two groups according to their morphology and the depth of inhabitation: bathial (800—2500 m) and abyssal (3000—4000 m). Species of the latter group differ considerably from all bathial species by a series of characters: reduction of pores on the upper side of the head, greater number of trunk vertebrae, smaller eyes, reduction of the anterior part of the dorsal fin and ribs on the trunk vertebrae, etc.[1]. The Antarctic member of this genus (*L. antarcticus*; fig. 9a) belongs to the second group which has greater possibilities for distribution over great depths of the oceans.

Horse fishes (Congiopodidae). This small aberrant family of scorpaenoid origin contains only 4 genera with 7 species, occurring along the coasts of South America, South Africa, South Australia and New Zealand (MORELAND, 1960); in the Antarctic Region members of this peculiar notal family are known only from the Kerguelen and Macquarie Islands where an endemic genus and species (*Zanclorhynchus spinifer*) was found.

Sea snails (Liparidae). The distribution of the family of sea snails resembles that of the zoarcid fishes. The greatest diversity of species is observed in the fauna of the boreal and arctic waters. From the northern part of the Pacific only already about a hundred species are known inhabiting all the depths from the littoral up to ultraabyssal (or hadal) zones (*Careproctus amblystomopsis*, at 7579 m depth, is the deepest living bottom fish). In the Antarctic this family is represented by 5 ill-known species belonging mainly to the deep-water genera *Careproctus* and *Paraliparis* (fig. 9c).

Lefteye flounders (Bothidae) — characteristic of tropical and subtropical waters, are represented in the Far South by two peculiar genera (*Mancopsetta* and *Achiropsetta*). In contrast to other members of this family they have no pectoral fins. About their distribution many things are still unclear. The genus *Mancopsetta* with a single species *M. maculata* has been found until now only around the Falkland, the S. Georgia and the Prince Edward Islands. The genus *Achiropsetta* known only from the Magellan-Falkland area has been considered hitherto monotypic (*A. tricholepsis*). However, in 1959 in four stations to the north off the Wilkes Land G. A. SOLANIK caught, by the Isaacs-Kidd midwater trawl pelagic, nearly translucent young of flat fishes (48—63 cm

---

[1] A. P. ANDRIASHEV, *Vopr. Ikhtiol.* **11**, 1958: *171*.

long) which were described as a new species *Achiropsetta slavae* (ANDRIASHEV, 1960d; fig. 15). The origin of these young, in spite of different suppositions, at present cannot be explained satisfactorily; it is therefore better to put off discussion of this problem till new material and knowledge on the taxonomy and the distribution of "armless" flat fishes is obtained.

In conclusion of this brief review a list[1] of Antarctic coast fishes (table I) is given in which, however, we do not include the species of deep-water families (Macruridae, Moridae etc., see page 513), and occasional catches such as hag-fish and *Somniosus*. The subdivision of the Antarctic, in contrast to NORMAN's table (NORMAN, 1938, p. 95), is not given in quadrants but in aquatoria which is more natural from the zoogeographical point of view: "East Antarctica" — Antarctic continental shelf from Coats Land eastward to Victoria Land and the Ross Sea; "West Antarctica" — Antarctic peninsula (Graham Land) and neighbouring waters including Bellingshausen Sea, Palmer Archipelago, the S. Shetland and the S. Orkney Islands; "S. Georgia etc." — including the S. Sandwich Islands (and Bouvet Isl.) and "Kerguelen etc." — islands of the Indian Ocean lying near the Antarctic Convergence from Prince Edward Island eastwards to Macquarie Island including Heard Island and the Kerguelen-Heard submarine ridge. It should be kept in mind that in the table only species names are given (subspecies are not marked out but, if possible, their locations are shown with asterisk (*) while all the typical forms as well as all the species are marked with a cross).

---

[1] While this work was in press two more new Antarctic fishes were described: *Aethotaxis mitopteryx*, gen. and sp.n. (Nototheniidae) from the McMurdo Sound (H. H. DEWITT, *Copeia*, **4**, 1962, *826—833*, 4 figs.) and *Gymnodraco victori*, sp.n. (Bathydraconidae) from the Adelie Land (J. C. HUREAU, *Bull. Mus. Nat. Hist. Nat.*, 2 ser., **35**, 4, 1963, *334—342*, 4 figs., 1 pl.).

Table I.

Distribution of the Antarctic coast fishes

1	The Antarctic				Outside the Antarctic
	East Antarctic	West Antarctic	S.Georgia etc.	Kerguelen etc.	
	2	3	4	5	6
Fam. RAJIDAE					
Raja murrayi GÜNTH.	—	—	—	+	
R. georgiana NORM.	—	—	+	—	
R. arctowskii DOLLO	+?	+	—	—	
R. eatoni GÜNTH.	—	—	—	+	
Raja sp. (NORMAN, 1938)	—	+	—	—	
Fam. MURAENOLEPIDAE					
Muraenolepis marmoratus GÜNTH.	—	—	—	+	
M. microps LÖNNB.	+	—?	+	—	
M. microcephalus NORM.	+	—	—	—	Burdwood Bank
Fam. NOTOTHENIIDAE					
Notothenia kempi NORM.	+	+	—	—	
N. squamifrons GÜNTH.	—	—	—	+	
N. larseni LÖNNB.	+[1]	+	+	—	
N. gibberifrons LÖNNB.	—	+	+	—	
N. mizops GÜNTH.	—	—	—	+	
N. nudifrons LÖNNB.	—	+	+	—	
N. acuta GÜNTH.	—	—	—	+	
N. angustifrons FISCHER	—	—	+[2]	—	
N. marionensis GÜNTH.	—	—	—	+	
N. cyaneobrancha RICH.	—	—	—	+	
N. coriiceps RICH.	*[3]	*[3]	*[3]	+	
N. rossi RICH.	—	*[4]	*[4]	+	
N. normani NYBEL.	—	—	+	—	
N. macrocephala GÜNTH.	—	—	—	+	Patagonia, Falklands, S. New Zealand
N. colbecki BOUL.	—	—	—	+	Auckland, Campbell, Antipodes
Trematomus borchgrevinki BOUL.	+	+	—	—	
T. brachysoma PAPP.	+	+	—	—	
T. newnesi BOUL.	+	+	—	—	
T. nicolai (BOUL.)	+	—	—	—	
T. sp.n. (ANDRIASHEV, in litt.)	+	—	—	—	
T. bernacchii BOUL.	+	+	*[5]	—	
T. hansoni BOUL.	+	+	+	—	
T. loennbergi REGAN	+	+	—	—	
T. pennellii REGAN	+	—	—	—	
T. centronotus REGAN	+	—	—	—	
T. scotti (BOUL.)	+	+	—	—	

[1]) Found only near Balleny Ils. [2]) Subsp. *sandwichensis* NYBEL. at S.Sandwich Ils.
[3]) Subsp. *neglecta* NYBEL. [4]) Subsp. *marmorata* FISCHER. [5]) Subsp. *vicarius* LÖNNB.

512

Table I (cont.)

1	2	3	4	5	6
T. lepidorhinus (PAPP.)	+	—	—	—	
T. eulepidotus REGAN	+	+	—	—	
Dissostichus mawsoni NORM.	+	+	—	—	Related species at Patagonia and the Falkland Ils.
Pleuragramma antarcticum BOUL.	+	+	—	—	
Fam. HARPAGIFERIDAE					
Harpagifer georgianus NYBEL.	—	*6)	+	*7)	Two related species at Patagonia and the Falklands
Artedidraco mirus LÖNNB.	—	—	+	—	
A. orianae REGAN	+	—	—	—	
A. skottsbergi LÖNNB.	+	+	—	—	
A. loennbergi ROULE	+	+	—	—	
A. shackletoni WAITE	+	—	—	—	
Dolloidraco longedorsalis ROULE	+	+	—	—	
Histiodraco velifer (REGAN)	+	—	—	—	
Pogonophryne scotti REGAN	+	—	—	—	
P. marmoratus NORM.	+?	+	—	—	
Fam. BATHYDRACONIDAE					
Bathydraco antarcticus GÜNTH.	+?	—	—	+8)	
B. marri NORM.	+	—	—	—	
B. macrolepis BOUL.	+	—	—	—	
B. scotiae DOLLO	+	—	—	—	
B. nudiceps WAITE	+	—	—	—	
B. wohlschlagi DeWITT & TYLER	+	—	—	—	
Gerlachea australis DOLLO	+	+	—	·—	
Racovitzia glacialis DOLLO	—	+	—	—	
R. harrissoni (WAITE)	+	+	—	—	
Prionodraco evansi REGAN	+	+	—	—	
Cygnodraco mawsoni WAITE	+	—	—	—	
Parachaenichthys georgianus FISCHER	—	—	+	—	
P. charcoti (VAIL.)	—	+	—	—	
Psilodraco brevipes REGAN	—	—	+	—	
Gymnodraco acuticeps BOUL.	+	+	—	—	
Fam. CHAENICHTHYIDAE					
Champsocephalus gunnari LÖNNB.	—	—	+	+	Related species at Patagonia and the Falklands
Pagetopsis macropterus BOUL.	+	+	—	—	
P. maculatus BARS. & PERMIT.	+	—	—	·—	
Neopagetopsis ionah NYBEL.	+	—	·—	—	
Pseudochaenichthys georgianus NORM.	—	—	+	—	
Dacodraco hunteri WAITE.	+	—	—	—	
Chaenichthys rugosus REGAN	—	—	—	+	
Ch. rhinoceratus RICH.	—	—	—	+	
Chaenocephalus aceratus (LÖNNB.)	—	+	+	—	
Ch. bouvetensis NYBEL.	—	—	+9)	—	

6) Subsp. *antarcticus* NYBEL. 7) Subsp. *kerguelensis* NYBEL. 8) To the south-east of Heard Isl. at 1600 m depth. 9) At Bouvet Island.

Table I (cont.)

1	2	3	4	5	6
Cryodraco antarcticus DOLLO	+	+	—	—	
C. pappenheimi REGAN	+	—	—	—	
Chionodraco myersi DEWITT et TYLER	+	—	—	—	
Ch. kathleenae REGAN	+	—	—	—	
Ch. hamatus (LÖNNB.)	—	+	—	—	
Chaenodraco wilsoni REGAN	+	—	—	—	
Fam. ZOARCIDAE					
Lycenchelys antarcticus REGAN	—	+	—	—	
Austrolycichthys brachycephalus (PAPP.)	+	—	—	—	
A. concolor (ROULE et DESP.)	+	+	—	—	
A. bothriocephalus (PAPP.)	+	—	—	—	
Lycodichthys antarcticus PAPP.	+	—	—	—	
Rhigophila dearborni DEWITT	+	—	—	—	
Melanostigma gelatinosum GÜNTH.	—	—	+	—	Strait of Magellan etc.
Fam. CONGIOPODIDAE					
Zanclorhynchus spinifer GÜNTH.	—	—	—	+	
Fam. LIPARIDAE					
Careproctus georgianus LÖNNB.	—	—	+	—	
C. (?) steineni (FISCHER)	—	—	+	—	
Paraliparis antarcticus REGAN	+	—	—·	—	
P. terrae-novae REGAN	+	—	—	—	
P. gracilis NORM.	—	—	+	—	
Fam. BOTHIDAE					
Mancopsetta maculata (GÜNTH.)	—	—	+	+	Falkland Islands
Achiropsetta slavae ANDR.[10]	+	—	—	—	Related species in Magellan-Falkland area

On abyssal fishes of the Antarctic. Little is known about the bottom fish fauna at great depths in the Antarctic. To the south of the Antarctic Convergence at 2—3000 meters and deeper only few species are known. The deepest living bottom fish here is *Bassogigas brucei* (fam. Brotulidae) from the Weddell Sea (4571 m). The greater part of abyssal species belong to the fam. Macruridae. Two species of the genus *Chalinura* are to be considered as the most characteristic Antarctic elements: *Ch. ferrieri* is known only from the coasts of the Antarctic Continent (Coats Land, Wilkes Land, off Balleny Island) at 2579—3109 m depth. Another species *Ch. whitsoni* was also caught at considerable depths (1590—2579 m) near the Continent and to the south of Heard Island, but in some regions (Kemp Coast, Scott I.) it ascends to 600—800 m depth. Within the limits of the Antarctic Region also *Nematonurus lecontei* was found (2800—3246 m) reaching just as *Ch. whitsoni* the Antarctic Convergence. Other species of macrurids

---

[10]) Only pelagic young are known.

514

were found chiefly in peripheral (northern) areas of the Antarctic.
They are therefore not typical for the Antarctic abyssal fauna as
well as certain members of the families Halosauridae, Synapho-
branchidae and others. Although widely distributed in the oceans,
Antimora rostrata (fam. Moridae) had not yet been found southwards of the Antarctic Convergence (GREY, 1956) but was caught
in 1956 in the southern part of the Kerguelen-Heard submerged
ridge (1580—1620 m) together with *Bathydraco* cf. *antarcticus*.

The Bathypelagic Fish Fauna of the Antarctic

The bathypelagic fauna of the Antarctic differs very much from
that of the Arctic. In the Arctic Ocean the true bathypelagic types
of fishes are absent or occur only in its peripheral parts[1], in the
Antarctic waters, however, (to the south of the Antarctic Convergence) a considerably rich bathypelagic fauna is known containing
about 50 species and nearly twenty families (ANDRIASHEV, 1962a).
In the open sea of the Antarctic members of the following families
are regularly found: Bathylagidae, Gonostomatidae, Paralepididae,
Myctophidae, Macruridae possibly also Scopelarchidae, Melam-
phaidae and Trichiuridae (fig. 10). Moreover, in the northern parts
of the Antarctic, members of the families Stenoptychidae, Astro-
nesthidae, Idiacanthidae, Stomiatidae, Notosudidae, Anotopteri-
dae, Moridae, Oreosomatidae, Cerathiidae, and young of Bothidae
and some families of eel-like fishes are known.

In the Antarctic waters the family Myctophidae is represented
by about 14 species, belonging to the genera *Protomyctophum* (3
species), *Electrona* (3), *Lampanyctus* (1), *Ceratoscopelus* (1), *Gymnoscopelus* (4-5 species), *Notoscopelus* (1). It should be noted that endemic species of microphids are absent in the Arctic while in the Antarctic nearly half of all species are endemic or mainly Antarctic
forms. Speaking about the endemism of this fauna we should mention that two genera (*Electrona* and *Gymnoscopelus*) and one subgenus (*Protomyctophum* s.str.) are in general connected with the
Antarctic and notal waters of the southern hemisphere. In these
zones the most primitive genera of the subfamily Myctophinae
(*Protomyctophum* and *Electrona*) develop but this fact cannot be
considered sufficiently strong to confirm the Antarctic origin of the
whole family (FRASER-BRUNNER, 1949) all the diversity of which
was linked to warm waters (in the Tertiary Period as well as in present time).

The most abundant bathypelagic species of the Antarctic are the
following: *Electrona antarctica, Gymnoscopelus braueri, Bathylagus*

---

[1] Only *Benthosema glaciale* apparently penetrates farther into the Arctic (WALTERS,
V. *Bull. Amer. Mus. Nat. Hist.* **106**, 1955).

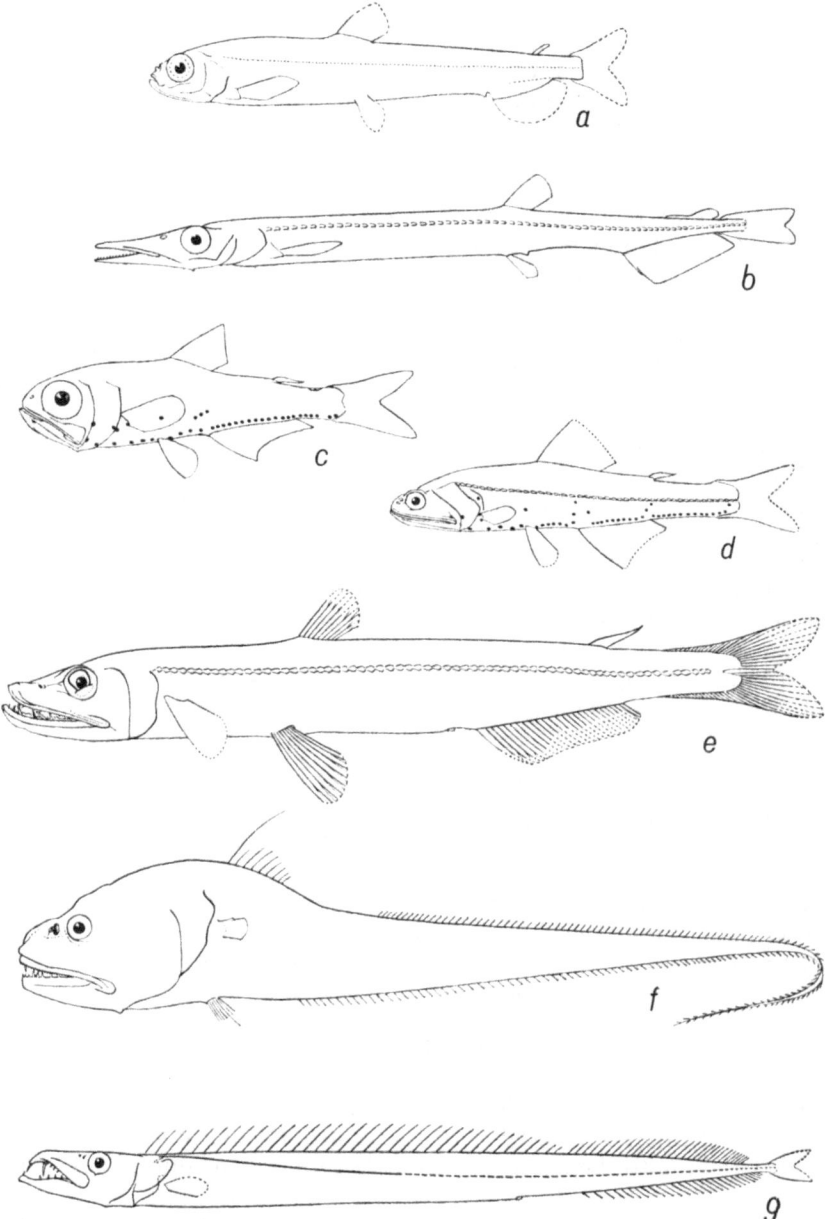

Fig. 10. Bathypelagic species caught off the Antarctic Continent: a – *Bathylagus antarcticus* (Bathylagidae); b – *Notolepis coatsi* (Paralepididae); c – *Protomyctophum bolini* (Myctophidae); d – *Gymnoscopelus braueri* (Myctophidae); e – *Neoscopelarchoides elongatus* (Scopelarchidae); f – *Cynomacrurus piriei* (Macruridae); g – *Paradiplospinus antarcticus* (Trichiuridae) (a, b – after NORMAN, c, d, e, g – after ANDRIASHEV; f – after REGAN).

516

antarcticus, Notolepis coatsi (up to 500 m depth) and *Cyclothone microdon* (at more than 500 m depth).

In the Antarctic there are no true epypelagic families of fishes and that, to a considerable degree, is also characteristic of the Arctic[1]. Fishes which at the present time live in the open sea of the Antarctic (both at the surface and in the deep waters) are descendants from three cold adapted faunal groups different in their origins. The first and most important group includes the species of the bathypelagic families mentioned above. Some of them apparently adapted long ago to the cold waters of the Antarctic and formed there a series of endemic-antarctic species. The majority of representatives of bathypelagic families only penetrate in the peripheral (northern) waters of the Antarctic for the abundancy of food but in that region they reproduce in warmer (northern) waters. The second group consists of species which originated from the bottom dwelling Antarctic families (Nototheniidae, Chaenichthyidae) and secondarily adapted temporarely or permanently to a mid-water life (see page 526). The species of this group are not so widely distributed in the open sea as the bathypelagic species because they depend on the continental shelf and the islands for their spawning. Finally the third group consists of bathypelagic species descending from deep-water *bottom* families, for instance Macruridae (*Cynomacrurus piriei*) or Moridae (*Melanonus gracilis*).

Recently many interesting papers, dealing with the distribution of phyto- and zooplankton in function of frontal zones and different waters appeared[2]. Some preliminary data were also reported about pelagic fishes (ANDRIASHEV, 1958b; 1962a).

It was recorded ("Ob", April 1958, in the southern part of the Pacific) that the line of the Antarctic Convergence nearly coincides with the distribution border of two common species *Electrona antarctica* and *E. subaspera* (fig. 11). The first of them is widely distributed around the Antarctic Continent in a zone about 900—1000 miles wide (meridionally) and practically never occurs north of the Antarctic Convergence, living mainly at temperatures lower than 3° (up to sub-zero-degrees). *E. antarctica* usually occurs in the upper 100 m layer; according to our own observations, it is usually not at night attracted to the surface by electric lights.

Just to the north of the Antarctic Convergence *E. antarctica* is replaced by another species of the same genus, *E. subaspera*. This species occupies a vast zone in the South Pacific (600—800 miles wide reaching to the north 46—47° s.lat. (in the autumn), i.e. nearly

[1] A. P. ANDRIASHEV. The fish fauna of the Arctic Seas and its origin. MS (1951). Zool. Inst. Acad. Sci. U.S.S.R., Leningrad.
[2] D. D. JOHN, *Disc. Rep.*, XIV, 1936; N. TEBBLE, *Disc. Rep.*, XXX, 1960; K. V. BEKLEMISHEV, *Rep. Soviet Antarct. Exp.*, 7, 1960; K.A. BRODSKY, *Rep. Oceanogr. Comm. Acad. Sci. USSR*, X, 4, 1960, and other.

up to the Subtropical Convergence. Surface temperatures at the fishing sites varied at from 2.8° to 12.1° (usually 6.1—8.7°) and at 200 m depth from 5.0—6.2°. The absence of a considerable vertical temperature gradient is supposed to favour the mass appearance of *E. subaspera* individuals near the surface where they gathered at night attracted by electric lights at each of the stations between 61—58° and 47—46° s.lat. Sometimes *Symbolophorus boops* occurred together with *E. subaspera*.

In the open ocean to the north of 46° s.lat. *E. subaspera* and *S.*

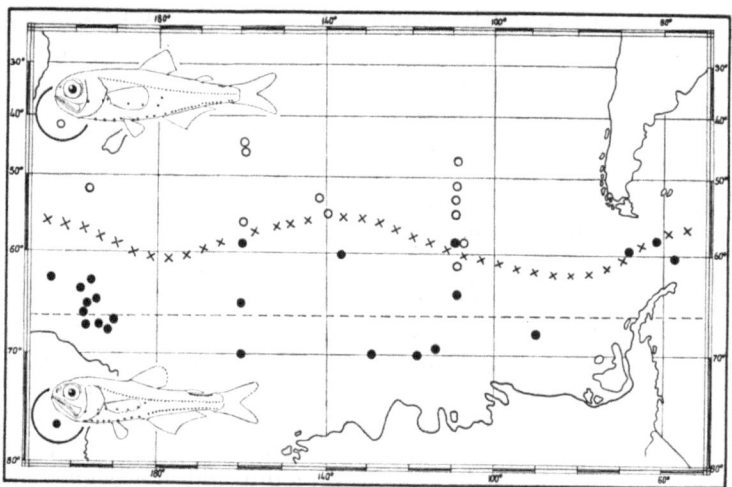

Fig. 11. Distribution of two myctophid species in the South Pacific during 1958-1960: an antarctic species, *Electrona antarctica* (black circles) and a notal species, *Electrona subaspera* (open circles). Line of Antarctic Convergence is indicated by crosses (after ANDRIASHEV, modified).

boops are replaced by *Scomberesox forsteri* (= ? *saurus*). In these waters the surface temperature varied from 9.7 to 19.5°, with a mean of 12—18°. The northern range of *Scomberesox* has not been traced by us.

Thus, inspite of the fragmentary character of our data there is reason to distinguish three main zones based on the distribution of pelagic species of fishes in the open waters of the South Pacific.

1. The Antarctic Zone ("Zone of *Electrona antarctica*") — between the Antarctic Continent and the Antarctic Convergence.

2. The Notal Zone[1] ("Zone of *Electrona subaspera*") — between the Antarctic and Subtropical Convergences.

---

[1] It corresponds approximately to the sub-antarctic or cold temperate zone of English authors; other names for this zone are: antiboreal, austral, southern temperate.

518

3. The Subtropical Zone ("Zone of *Scomberesox*") — to the north
of the Subtropical Convergence (probably up to the Southern
Tropical Convergence).

As we already mentioned (ANDRIASHEV & PERMITIN, 1961) off
Chile the southern elements penetrate further to the north under
the influence of the Humboldt current than in the open ocean and
the boundaries of the zones are shifted considerably to the north.
At the latitude of Talcahuano and Valparaiso notal and subtropical
species (*Symbolophorus boops* and *Scomberesox*) occur together.

Some biological peculiarities and puzzles of the Antarctic Fish Fauna

We know very little about the biology of Antarctic fishes. It is
the least developed part of the ichthyological research in the Antarctic. Therefore it is too early to make general conclusions or to draw
comparisons. We can only try to touch certain problems connected
with peculiarities of environments and life history in order to draw
the attention of the scientists to some interesting problems and to
show the difficulties connected with their study.

Peculiarities of the Antarctic continental shelf and the vertical distribution of fishes

When speaking of the coastal bottom fauna of the Antarctic one
always mentions an insignificant area of its continental shelf. In
fact, if we take "shelf" in its usual sense (0—200 m) then its area
free of ice turns out to be nearly negligible when seen on the chart
(BROCH[1]). Nevertheless, the shelf is developed around the whole of
the sixth continent but it is somewhat unusual. To begin with, it
has a "sunken" character with depths of more than 200 m and has a
continental edge at 400—500 m depth or even more (fig. 12). The
width of this sunken shelf varies for East Antarctica from 40 to
150 miles and occupies greater areas only in the Ross and Weddell
Seas. The second feature of the Antarctic continental shelf is the
presence within its limits of innershelf depressions and narrow
trenches often oriented along the shore of the continent and reaching
depths of 1000 m and more. The most important of them is a great
innershelf depression bordering the eastern part of the Antarctic
Continent from the Davis Sea eastward to Victoria Land thus extending for over 2000 miles. The greatest depths were recorded by
geologists from the "Ob" in the Davis Sea (1400 m) and near the
King Georg V Land (1600 m)[2]. These innershelf depressions make
the conditions of life for the bottom fauna rather peculiar and complicate the analysis of its vertical distribution.

The first attempt to throw light on the bathymetric distribution

---

[1]) H. BROCH, *Sci. Res. Norw. Antarct. Exp.* 1927-1928, 38, 1961, fig. 1.
[2]) A. V. ZHIVAGO & A. P. LISITSIN, *Izvestia Acad. Sci. U.S.S.R.* (geogr.), No. 1,
1957, *19-35*; No. 2, 1958, *9-21*.

of species in the Antarctic was made by NYBELIN (1947), who arranged them in four main groups:

A. Stenobath shallow-water species distributed from the shore and down to a lower limit of about 40—60 m depth (8 species and subspecies).

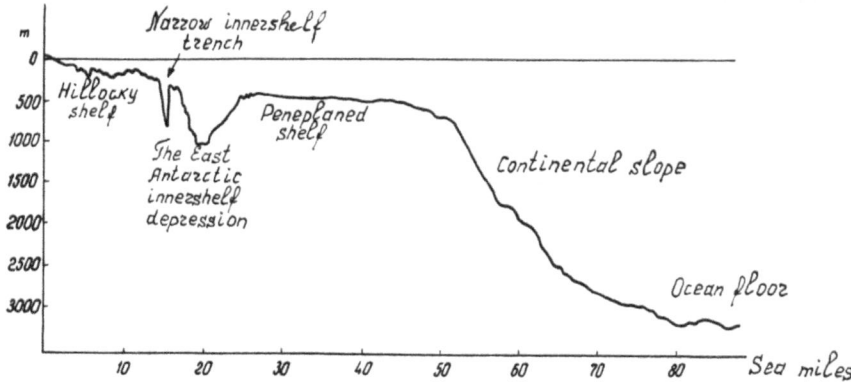

Fig. 12. Representative vertical section of the Antarctic shelf, Wilkes Land (after ZHIVAGO).

B. Eurybath shallow-water species with upper limit of 0—40 m (lower limit 250—230 m, seldom to 650 m) (19 species and subspecies).

C. Species from intermediary depth with two subgroups:

    a. Upper limit 80—230 m (23 species),

    b. Upper limit 300—610 m (12 species).

D. Deep-water species known from more than 2000 m depth (3 species).

It has been recorded that in connection with the peculiarities of the structure of the Antarctic continental shelf (average depth, continental edge at 400—500 m depth and the deep innershelf depressions) usual schemes of vertical zonation[1] can hardly be applied to the Antarctic fauna (ANDRIASHEV & TOKAREV, 1958 : 202). Thus, e.g. *Austrolycichthys brachycephalus* living at 385—1040 m depth should be defined as a bathial (or archibenthal) species. But such conclusion is erroneous since all known catches of *A. brachycephalus* were not made on the continental slope (the bathyal or archibental zone) but within the limits of the Antarctic shelf and its innershelf depressions, where temperature, oxygen and other conditions differ considerably from those at corresponding depths on the continental slope. On the other hand a deep-sea macrurid

---

[1]) P. V. USCHAKOV, *Doklady Ac. Sci. U.S.S.R.*, XVIII, No. 4, 1949; L. A. ZENKEVICH & J. A. BIRSTEIN, *Vestnik Mosc. Univ.*, No. 4-5, 1955.

520

fish *Chalinura whitsoni* can be found along the continental slope, when the waters are comparatively warm, up to 600 m depth, in the innershelf depressions, however, where temperature falls to —1.8° this species has never been caught. In order to work out a scheme of vertical zonation of Antarctic benthos it is only necessary not to accumulate data but also to investigate whether there is any differ-

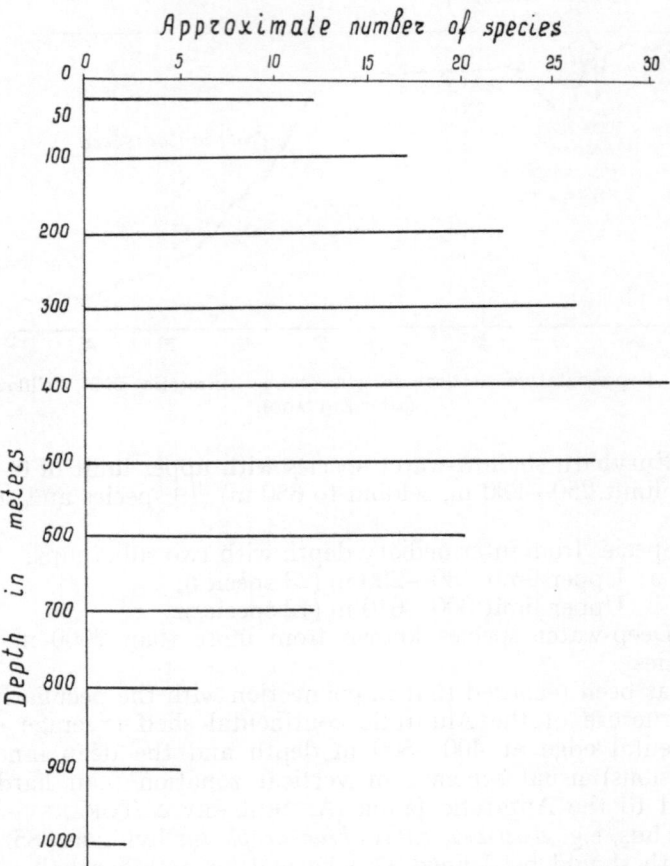

Fig. 13. Change in the number of the East Antarctic species of fishes according to depth (54 species in all) (orig.).

ence between the composition of the fauna of the innershelf depressions (the *pseudoabyssal* zone[1] in DERJUGIN'S sense) and of the continental slope (the bathial zone).

Since we only have few data concerning the depths at which Ant-

---

[1] K. M. DERJUGIN, Explor. des meis d'U.R.S.S., No 7-8, 1928.

arctic fishes live we can only trace certain features of their bathopaty. As a result of the negative effect of continental — and fast ice the Antarctic fish fauna is characterized by a nearly complete absence of true littoral forms. They are completely absent from the shores of the continent (except for one species in the very northern part of Graham Land) and only in the intertidal zone of the S. Georgia, the Kerguelen and other islands can some littoral species (*Harpagifer*, *Notothenia cyaneobrancha*, probably *N. angustifrons* and some little known liparids) be found.

The coastal line of the Antarctic Continent, especially the Eastern part, is characterized by a weak development of the sublittoral fish fauna because the coastal bottom areas (sometimes down to 100 — 200 m depth) are often covered with slipping continental ice.

It is worth noting that the greatest species diversity in bottom fishes in East Antarctica is found at depths from 200—300 m to 500—600 m where 80% of the known species from the Antarctic live. In fig. 13, in spite of the rather approximate calculations, one may see an abnormal vertical distribution of the species. Whereas in the North the number of bottom species usually decreases sharply at depths of more than 100—200 m[1], this phenomenon is shifted to depths of 500 to 600 m at the eastern part of the Antarctic Continent. This bathymethric anomaly is thought to be related to the sunken shelf and innershelf depressions. Among the Antarctic bottom fish fauna there are many species with a considerably wide vertical distribution (500—700 m).

Out of the majority of the typical shelf families deep-water species evolved secondarily and penetrated in layers of more than 500 m. For instance, Chaenichthyidae occur down to 655 m (*Pagetopsis*, *Dacodraco*), Harpagiferidae down to 850 m (*Pogonophryne*), Nototheniidae down to 920 m (*Trematomus lepidorhinus*), Bathydraconidae down to 2579 m (*Bathydraco scotiae*) and Zoarcidae down to 1040 m (*Austrolycichthys*) and even down to 3248 m (*Lycenchelys*).

Freezing temperatures and the phenomenon of cold adaptation

Many areas of the Antarctic are characterized by the lowest possible water temperatures (for a given salinity). NYBELIN (1947) said that all the catches of fish at the East Antarctic coast were made at temperatures lower than the freezing point or in the limits from —0.06[2] to —1.85° C; for west-antarctic endemics the same indices are: from —0.64 to —1.55, for fishes of S. Georgia from —0.24 to

---

[1]) A. P. ANDRIASHEV. The fishes of the Bering Sea and neighbouring waters, its origin and zoogeography. Leningr. Univ. Press, 1939, 27, fig. 12.
[2]) Such a high (close to 0°) temperature is not characteristic of the near-bottom water layers in East Antarctica. Therefore the samples must have been taken near the surface or on the continental slope.

522

2.75. Near the Kerguelen (or near other islands of the Antarctic Convergence) the temperature conditions are more favourable for coast fishes (about 2—6°).

A very severe temperature regime is typical for East Antarctica. Hydrologists of the wintering party at the "Mirny" observatory have registered temperatures below or near freezing point for over 9—10 months (from March to December) in the ice-covered coastal waters of the Davis Sea[1]. Near-bottom temperatures do not change considerably within the continental shelf, remaining during the summer at about —1.8° for depths between 200—500 m. The moderate heating of the surface layers (when the ice is broken) usually does not spread to the deeper layers of the shelf. Thus the majority of species nearly constantly live at temperatures close to the freezing point. It is remarkable that due to the fact that the shore-waters of the Davis Sea remain supercooled for many months, mid-water and bottom ice crystals can be formed very suddenly by crystallization around submerged objects. Sometimes ice-crystal layers of 4 m thick can pile up under the fast ice. At such a moment V. M. MAKUSHOK (1959) registered a mass mortality of bottom polychaetes worms and the ceasing of fish biting. It is quite possible that in such periods fishes go into greater depths to avoid the harmful effect of bottom ice formation.

The adaptation of fishes to life at such low temperatures is worthy of a very careful study. WOHLSCHLAG (1960) was the first to organize a well equipped physiological laboratory at McMurdo (77° 51'S, 166° 38'E.) and carried out an experimental work on fish metabolism. He mesured the oxigen metabolic rate of *Trematomus bernacchii* at temperatures from —2.0° to 1.5°. This fish nearly always lives under fast ice on the bottom at temperatures of about —1.9°. WOHLSCHLAG supposed that the phenomenon of cold adaptation, i.e. the higher metabolic level known for cold stenotherm forms (as compared with the standard) should be expressed very sharply in Antarctic species. The experiments proved him right. It was established that the oxygen consumption rate in *T. bernacchii* is increased with the increase of water temperature only in the limits from — 1.9 to 0° and tends to decrease if the temperature rises above this level. This fact characterizes this species as an extremely cold stenothermic one. The comparison of the metabolic level (at 0°) of the Arctic whitefish (*Coregonus sardinella*) and *T. bernacchii* shows a higher level for the Antarctic species therefore possessing a higher degree of metabolic cold adaptation. Analysing this phenomenon WOHLSCHLAG (1960, 1961a) came to some interesting conclusions: (a) that cold adapted fishes probably have a relatively more efficient mechanism for converting energy into

---

[1] Personal communication by N. P. SHESTERIKOV.

growth than fishes of warmer waters and (b) that these mechanisms, if understood, could be applied to increase production in warmer waters.

Reproduction

The data on the reproductive pattern of the species are of special value as biological characteristics of a fish fauna. Unfortunately our knowledge of the Antarctic species in this field is very scarce. Our limited data are in good agreement with MARCHALL's statements (1958) and enable us to draw the following conclusions: the vast majority of Antarctic species produce demersal, relatively large,

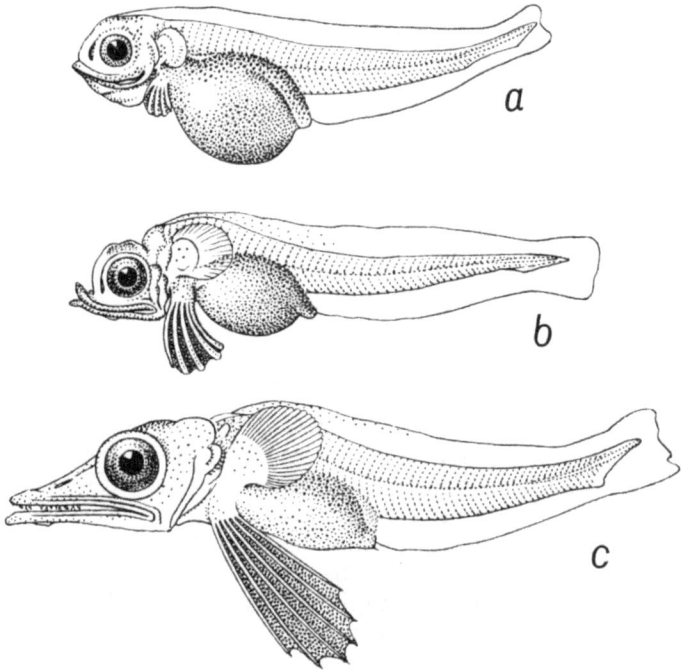

Fig. 14. Pelagic larvae of white-blooded fish, *Pagetopsis macropterus:* a – 14 mm, b – 15 mm, c – 19 mm long (after REGAN).

yolky eggs with a long embryonal period of development. Fecundity of many species is comparatively low (often a few hundreds or thousands of eggs). Spawning in some species is carried out during the autumn and the winter, but observations are solitary and often discrepant. Spawning, as judged by the size of ovarial eggs, is single (not by portions). Larvae are comparatively large and in many species have a long pelagic stage. The young of some species have very large fins (especially ventral one, see fig. 14) which ap-

524

parently helps them to stay in mid-water since they have no swim
bladder. Adaptation to pelagic life can clearly be seen in the young
of *Achiropsetta slavae*, e.g. by the weak ossification of the skeleton,
the colourless semitransparent body, the peculiar hairy scale cover
on both sides of the body which also helps the young to keep in
mid-water (ANDRIASHEV, 1961d).

If one studies the peculiarities of the reproduction in the fish
fauna from high-arctic to boreal seas (European), a series of regularities can be noted. Thus in the northern hemisphere the percentage of species with demersal eggs decreases southwards on account
of the increase of the number of pelagophilous species; the average
diameter of eggs decreases; fecundity increases; periods of incubation become shorter and sizes of newly hatched larvae — smaller.
(RASS[1]; ANDRIASHEV[2]; MARSHALL, 1953).

Table II.
Geographical changes in the number of species spawning demersal eggs (% of the
total number of species in the fauna of a given sea)

Seas	Kara Sea, the Laptev Sea, East-Siberian Sea	Barents Sea without south-western part	South-western part of the Barents Sea	Norway
Species with demersal eggs (% %)	91	85	70	43

As one can see in table II the percentage of species spawning
demersal eggs of the fauna of high-arctic seas decreases towards the
boreal seas. Analogous figures are obtained if one studies the fecundity of northern fishes (table III).

We suppose that certain similar features take place in the fish
fauna of the southern hemisphere as was first recorded by MARSHALL
(1953). Though data are incomplete it seems to hold true for the
fecundity and egg sizes. If one compares species of the related genera
Trematomus and *Notothenia*, more or less similar in body length, the
following series can be traced. East-Antarctic *Trematomus* species
(T. bernacchii and *Trematomus* n. sp.) have the largest eggs
(2.6—3.8 mm) and the lowest fecundity (1030—1660 eggs). In the
less cold-stenotherm genus *Notothenia* the two West-Antarctic
species (*N. larseni* and *N. squamifrons*) have smaller eggs (1.6—2.0
mm) and higher fecundity, varying between 2—3 and 5—6 thousand
eggs. In the warmer waters from the Kerguelen *N. mizops* has eggs

[1]) TH. S. RASS. Analogous or parallel variations in structure and development of fishes in Northern and Arctic Seas. Moscow, 1941, 1-60, 9 figs.
[2]) A. P. ANDRIASHEV. The fish fauna of the Arctic Seas and its origin. MS (1951).
Zool. Inst. Acad. Sci. D.S.S.R., Leningrad.

525

Table III.
Geographical changes of the average fecundity in the fish fauna of the northern hemisphere (% of total number of species of a given sea)

Fecundity	Siberian Seas	Barents Sea (South-western part excepted)	South-western part of the Barents Sea	Black Sea
low (tens or hundreds of eggs)	72[1]	65	37	20
medium (thousands or tens of thousands)	28	30	39	56
high (hundreds of thousands or millions)		5	24	24

with a diameter of 1.5—1.6 mm and a fecundity of about 10 000 eggs. Finally, for the two notal species from the Falkland Islands (*N. squamiceps* and *N. sima*) MARSHALL (1953) gives an egg diameter of 1.2 — 1.4 mm (the fecundity is unknown).

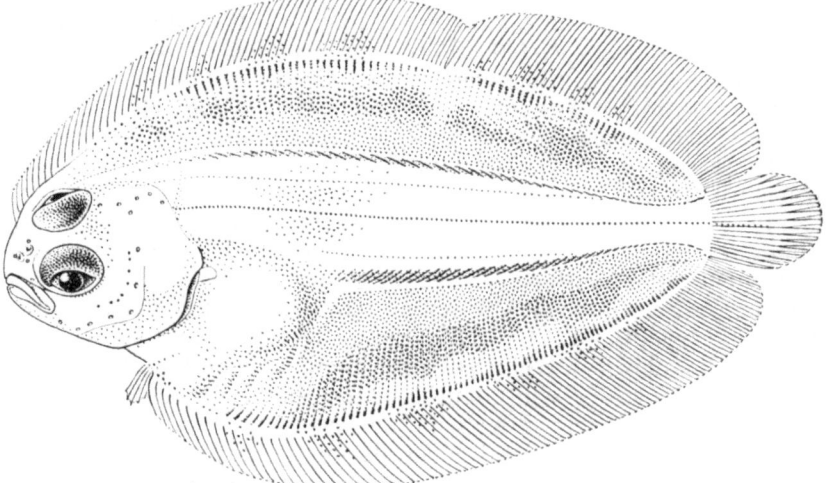

Fig. 15. Pelagic young flatfish, *Achiropsetta slavae*, 56 mm long (after ANDRIASHEV).

[1]) The figure shows that in the Siberian Seas 72% of the species have a low fecundity etc.

526

The accumulation of data on the egg sizes, the fecundity, the spawning periods and the incubation period of Antarctic and notal fish species (and first of all of the family Nototheniidae) is very desirable. These data would enable us in the future to elucidate and to refine some interesting ecological-geographical principles which are important as zoogeographical characteristics of the different faunas.

Feeding

There still are no special researches devoted to food and feeding habits of Antarctic fishes. From the few data published the following can be taken into consideration. In the fauna of the Antarctic coast fishes benthofagous forms dominate, they do not apparently differ by narrow specialization and feed chiefly on bottom crustaceans (Amphipoda, Isopoda, Decapoda etc.), polychaete worms, echinoderms, molluscs etc. In species reaching considerably large sizes the share of fish food increases. Giant *Dissostichus mawsoni* is a typical predator dangerous even for rather large fishes. The main plankton consumers belong to bathypelagic species but planktonic crustaceans serve as food for the young of many bottom species and to such pelagic fish as *Pleuragramma antarcticum*. NYBELIN (1947) paid attention to the fact that individuals of certain species from bottom families were repeatedly observed swimming near the surface over great depths and were caught by pelagic nets or found on drifting ice fields and in stomachs of whales and seals. The occurrence in the open sea is now known for many Nototheniidae (*Pleuragramma antarcticum, Notothenia rossi, Trematomus borchgrevinki, T. brachysoma,* etc.), Chaenichthydae (*Champsocephalus gunnari, Pseudochaenichthys georgianus, Neopagetopsis ionah,* etc.), Bathydraconidae (*Gymnograco acuticeps*), Muraenolepidae and so on.

The reason of the partial transition of some Antarctic bottom fishes to a temporary life in the open sea was regarded by NYBELIN (NYBELIN, 1947, 1948, 1955) as an adaptation allowing use of the abundant food potentials of the krill zone which exceeds the productivity of the benthos. As a result of quantitative work with the bottom sampler in East Antarctica, biomasses of bottom fauna: at 100—200 m depth – 1245—1347 g/m² average (maximum 2850), and at 200—500 m depth — 239—319 g/m² average, were found ,i.e. more than in the corresponding depths in the Barents and Bering Seas. However, the highest indices of biomass in the Antarctic waters (100—500 m depth) depend generally on the development of such benthos groups as sponges, ascidians and bryozoans[1] making nearly 60—90% of the total weight of all organisms which are nonedible by fishes. Nevertheless, we should be careful in considering

___

[1] G. M. BELJAEV & P. V. USCHAKOV, *Doklady Ac. Sci. U.S.S.R.*, v. 114, No. 1, 1957, *139*; PH. A. PASTERNAK & A. V. GUSEV, *Soviet Antarctic Exp.*, 7, 1960, *140*.

the insufficiency of the benthos food because this probably requires a correction with respect to a quantitative evaluation of movable organisms such as amphipods which were caught by our baited fish traps in hundreds and thousands but were very rare in bottom samplers. Further complex investigations on food and food habits of the Antarctic fishes is very promising in many respects.

Size, growth, age

Antarctic fishes are of very different sizes. The greatest length of the adult of some species does not excend 9—12 cm (*Harpagifer*, the majority of *Artedidraco, Notothenia acuta* etc.) but such large fishes as *Chaenocephalus aceratus* (67 cm), *Notothenia rossi* (90 cm), *Dissostichus mawsoni* (about 175 cm) are known. In this fauna species of small size up to 25 cm long dominate (56%), half as many species reach 25—45 cm (30%), and only about a dozen species attain the length of 1/2 m or more. It is remarkable that the greatest size is characteristic of the family of white-blooded fishes. Thus, if we calculate approximately a "mean species length" in a family (the sum of total lengths of the greatest specimen of each species in the family divided by the number of species) this would give for the fam. Harpagiferidae about 15 cm, for the Bathydraconidae — 26 cm, for the Nototheniidae — 36 cm and for the Chaenichthyidae — 43 cm.

Growth rates of fishes living along the coasts of the Antarctic Continent are studied at present only in *Trematomus bernacchii* (WOHLSCHLAG, 1961b). In spite of extremely severe conditions (ice all the year round, bottom temperatures of about — 1.9°) *T. bernacchii* has turned out to be a fast growing form reaching sex-maturity when 5 years old. Females grow considerably quicker than males do, and reaching on age of 10 years and having a 2—3 times greater weight. Determination of the age is rather difficult since year rings especially in adults are poorly visible. Fishes from S. Georgia differ because they show a considerably faster growth rate (OLSEN, 1954, 1955).

White-bloodedness of chaenichthyid fishes

It is one of the most interesting puzzles which the Antarctic has put before the biological science in the last years. Norwegian whalers have known for many years that at S. Georgia there were fishes with colourless blood; they called them "bloodless fishes", "ice fishes", "white crocodile-fishes". A biologist of the Norwegian Expedition (1927-1928) D. RUSTAD recorded that the blood of a crocodile-fish from Bouvet Isl. is colourless (NYBELIN, 1947, p. 51[1]).

---

[1] The first record on Antarctic fishes with colourless blood was published by MATHEWS (1931, p. 36).

However, this striking fact did not attract at once due attention of scientists. It was not before 1951 that the Norwegian ichthyologist S. OLSEN brought back fixed and frozen blood of chaenichthyids from S. Georgia in which Prof. P. A. OWREN could not detect erythrocytes. In 1954 a well-known paper by J. RUUD (1954) appeared in which he wrote that the blood of *Chaenocephalus aceratus* lacks red blood pigment (haemoglobin) and is nearly translucent with a white-cream tint, that erythrocytes are completely absent, whereas the leucocyte sediment amounts to less than 1% in volume, that the iron content is low (less than 1 mg% as compared with 20 and more mg% in other fishes), and that the oxygen capacity of the blood is very low (less than 1% in volume while in neighbouring nototheniids it makes about 6%). The question whether all chaenichthyids possess colourless blood was answered rather quickly. To the five known white-blooded species in 1956 two Kerguelen species of *Chaenichthys* were added (ANDRIASHEV & TOKAREV, 1958) and in the summer of 1957 BARSUKOV & PERMITIN (1960) discovered colourless blood in 6 species caught in East Antarctica including a new species, *Pagetopsis maculatus*. Further on it was ascertained that white-bloodedness can be recognized also in museum specimens (fixed long ago) by the flattened and a pale spleen (ANDRIASHEV & TOKAREV, 1958) or by the nature of the blood clot from the bulbus arteriosus (RUUD, 1958). As a result of all these observations the supposition that white-bloodedness is peculiar to all known species of Chaenichthyidae and therefore is a characteristic feature of the whole family is confirmed. Further understanding came from histological studies. WALVIG (1958) described different forms of leucocytes and thrombocytes in spleen and sections of *Ch. aceratus* and confirmed the absence of erythrocytes in the blood of this species. MARTSINKEVITCH (1958, 1961) investigated the same problem. On the material collected during the second cruise of the "Ob" she described a great variety of elements of the blood leucocytar content in 7 species of chaenichthyids thus determining the following leucocytar formula in *Chaenichthys rugosus*: lymphocytes — 74.2%, monocytes — 5.1%, granular leucocytes (granulocytes and myelocytes) — 16.5%, nongranular leucocytes with a segmented nucleus — 4.2%. Furthermore, MARTSINKEVITCH had found cells similar to erythrocytes in the blood vessels of internal organs but these cells were considerably less oxyphilous; the number of such cells is rather negligible and therefore they can hardly play any significant role in the oxygen exchange, they probably point to a rudimentary form of erythropoesis. These factual data exhaust for the present our knowledge on whitebloodedness of chaenichthyids, but to understand this phenomenon some circumstantial data can also be used. Thus, TYLER (1960) studied blood (red!) of two Antarctic species of the family Nototheniidae and recorded two important facts: in the blood of the species studied he found (1) a decreased number of erythrocytes (0.38—0.39 × 10⁶ mm³ in the bottom type *Notothenia larseni* and 0.66—0.80 × 10⁶ mm³ in the more moving type *Trematomus borchgrevinki* (the standard being 1—2 × 10⁶ mm³ in other fishes) and (2) a low haemoglobin content (3.5—4.0 g/100 cm³ as compared with the norm of nearly 7—12 g/100 cm³ [1]). These observations enable us to suppose that other Antarctic fishes have a tendency to decrease the haemoglobin and erythrocytes content of their blood and that their respiration proceeds by O_2 dissolved in blood plasma (TYLER, 1961). Possibly this is the tendency of the evolution of white-bloodedness in chaenichthyids. A comparatively large surface of gill-lamellae and special adaptations to respiration through the skin of the body and the fins (WALVIG, 1960) would favour the possibility of a saturation of the blood plasma with oxigen in waters with very low temperatures and high oxygen content. Prof. H. MONRO FOX, in his afterword to RUUND's paper (1954), recalled the experiments of M. NICLOUX (1923) whereby fishes (carp, pike and eels) did not experience any inhibition although 90% of their haemaglobin was transformed to carboxyhaemoglobin and thus incapable of oxygen transport. MONRO FOX thinks that "many fishes when swimming quietly get enough oxygen for their needs in solution in the blood plasma, and they probably only require an additional supply with the aid of haemoglobin when they are moving actively". The existence in nature of some adult individuals of different species of fish-like vertebrates and fishes lacking haemoglobin[2] speaks in favour of such a supposition. It is also known that oversaturation of water with oxygen causes delay of haemoglobin formation, slow growth of gill laminae and anomaly in spleen development in fish larvae.[3]

In further considering white-bloodedness of chaenichthyids and discussing the possible reasons for this interesting phenomenon we should keep in mind that this family can be considered on many grounds to be a biologically successfull one and well differentiated from the ecological point of view. The majority of white-blooded species live along the coasts of the Antarctic Continent at constant temperatures close to the freezing point (up to — 1.95° C) and in well aerated waters (usually 70—90% O_2 in near-bottom layers of 200—500 m depth), but endemic species of chaenichthyids live also near S. Georgia, the Kerguelen and Patagonia where the temperature of the bottom layers is considerably higher (up to 5—8°) and the oxygen content may be lower. Moreover, all species of white-blooded

---

[1]) Up to 18.5 g/100 cm³ in skipjack (*Katsuwonus pelamis*).
[2]) D. W. EWER, *Nature, Lond.* **183**, No 4656, 1959, *271;* A. STOLK, *Nature, Lond.* **185**, No 4713, 1960, *526;* B. RYBAK, *Nature, Lond.* **185**, No 4715, 1960, *777*.
[3]) I. A. SADOV, Rybn. Khoz., No.1, 1948, p. 43.

fishes are big and they feel comfortable at depths ranging from shore waters up to 600—700 m; among them are typical bottom species but also species migrating into the open sea, such as benthofagous, krill-eating and predatory forms. In order to understand the origin of white-bloodedness and its role in the evolution of the Chaenichthyidae further special investigations are necessary, especially from the ecological-physiological and biochemical point of view. At present we can only guess that whitebloodedness, as a unique biological feature of chaenichthyids, originated long ago during the early stages of evolution of this group and before its ecological-morphological differentiation.

Do fishes live under permanent ice?

The giant ice shelves of many different types, and at places the fast ice several years old, occupy large areas along the coasts of the Antarctic Continent, amounting to about one million km². These giant ice slabs, up to 200—300 m thick, completely separate the ocean from the atmosphere and sun light, and are thought to create an extremely adverse environment. A direct study of the bottom surface under the giant ice shelf is probably feasible only by means of a bathyscaph and this has not yet been used for investigations in the Antarctic. Large cracks, known to exist within some ice shelves, have also not been used to study the fauna. The problem of the possibility of life under this cover has recently been put forward by the U.S. ichthyologists (ROFEN & DEWITT, 1961) but much is still left to be solved.

We have no sufficient reasons to believe that under the ice shelf the bottom of the ocean is lifeless. Hydrological observations near the Shackleton ice shelf have shown an active water exchange under it (V. G. LEDENEV, personal communication). Moreover, southward of this massive ice shelf bays, free from ice were found within the Banger "oasis". In these fiords the same bottom fauna and the same species of *Trematomus* and *Gymnodraco* were found as near Mirny (KOROTKEVITCH, 1958). However, even if these facts speak in favour of the existence of benthos and fishes under the Shackleton Ice Shelf, it is questionable whether the same preliminary conclusions apply to considerably larger ice shelves such as the Ross, Filchner or Amery Ice Shelves.

In connection with this question it is appropriate to point to an interesting finding of the American investigators (SWITHINBANK, DARBY & WOHLSCHLAG, 1960). On the surface of an ice shelf in the McMurdo Sound they detected well preserved remains of benthic invertebrates (numerous pelecypods, gastropods, brachiopods, siliceous sponges and anthozoan corals) and nearly 50 specimens of partially decomposed fish. These fishes, as Dr. D. E. WOHLSCHLAG and his colleagues reported, belong to at least two genera of the family Nototheniidae. Notable were their very large sizes — the greatest fish was 142 cm long, and some detached heads appeared to belong to a still larger fish. The absolute age of these remains (defined by the C^{14} method) was about 1100 years.

How could this bottom fauna get on to the upper surface of the ice shelf? To explain an analogous but less rich finding by the "Terra Nova" expedition, F. DEBENHAM suggested in 1920 that fauna remains frozen long ago to the bottom surface of a glacier were gradually lifted up through the ice due to thawing of the upper part of the glacier and new ice growing from below. Other observations indeed suggest that the discovered fauna has existed under the ice shelf. The study of the remains from McMurdo still continues and we do hope that we shall very soon possess more precise informations and explanations about this mysterious finding. In any case the question of how far and which fauna penetrates to the south under the Ross ice shelf will remain unanswered for sometime.

Contribution to the Zoogeography of the Antarctic Region
On the boundaries of the Antarctic Region (on the distribution of coast fishes)

The main problems of the zoogeography of the Antarctic (on the bottom fishes) have been worked out by REGAN (1914), NORMAN (1938) and NYBELIN (1947, 1951, 1952). EKMAN (1953) used ichthyological data among others for a zoogeographical subdivision of the southern hemisphere. Giving the general characteristics of the Antarctic fauna this prominent biogeographer said that "No other large faunal region in the world can match the Antarctic in the sharpness of its boundaries" (p. 221). C. T. REGAN also paid attention to the sharp delimitation of the Antarctic Region (Zone) but accepted other boundaries than EKMAN did. So this question is still under discussion now.

The first scientific definition of the boundaries of the Antarctic Region (Zone) based on the distribution of coast fishes was given by REGAN (1914), and later defined more precisely by NORMAN (1938). These experts of the fish fauna of the extreme South thought that the Antarctic Zone includes "the coasts of the Antarctic Continent and islands lying on or to the south of the Antarctic Convergence". The fish fauna of this vast region is characterised by "the complete absence of South Temperate types, by the absence of Bovichthyidae and the great development of the other Nototheniiformes. It is more sharply marked off than any other zones, the percentage of peculiar genera being extremely high and that of species range beyond its limits very low" (REGAN, 1914, p. 33). This point of view was accepted by the prominent Russian geographer and ichthyologist L. S. BERG (1961). EKMAN (1953) held another opinion and took such three areas as the Antarctic, the Kerguelen with islands, and the antiboreal waters of S. America for zoogeographical subdivision of one and the same rank (i.e. regions). Recently NYBELIN (1947, p. 60) using ichthyological data came to the conclusion that "the Kerguelen — Macquarie District must be removed from the Antarctic Zone and be assigned to the Subantarctic Zone as a third district equivalent to the Magellan and Antipode Districts".

In the present review as well as in earlier publications (ANDRIASHEV & TOKAREV, 1958), I am inclined to accept REGAN-NORMAN'S views and shall try to give them some additional foundations[1] when dealing with the zoogeographical subdivision of this region.

Types of distribution

For the zoogeographical analysis of the Antarctic fish fauna the distinguishing of types of distribution suggested by NYBELIN (1947) has turned out rather useful. Proceeding from his general principals but taking into consideration the Antarctic boundaries in a broader sense I shall attempt to give only the main types of distribution without including any particular case.

I. The Circumpolar-Antarctic type includes nearly twenty species distributed around the Antarctic Continent. Most of these species are distributed circumcontinentally (fig. 16) and only two of them reach S. Georgia *(T. hansoni, T. bernacchii vicarius)*. *Notothenia coriiceps* (including subspecies *neglecta)* has a wider (nearly Panantarctic) distribution area.

II. The East Antarctic type (fig. 17) includes thirty-odd species which show the extremely high degree of endemism of the East Antarctic fauna. From the 27 species grouped in this type by NYBELIN (1947) we have to exclude now *Gymnograco acuticeps* but add to it some new species found only at the Victoria Land and westwards *(Pagetopsis maculatus,* our new species of *Trematomus, Chionodraco myersi, Bathydraco wohlschlagi* and others). It is worth of noting that the East Antarctic distribution type includes as much as 5 genera *(Histiodraco, Neopagetopsis, Dacodraco, Chaenodraco, Lycodichthys).*

III. The West Antarctic type (fig. 17) defined by NYBELIN is poor in species since from the 8 species mentioned for the Graham Land and neighbouring waters the following species may be acknowledged now: *Racovicia glacialis, Parachaenichthys charcoti, Chionodraco hamatus* and, possibly, the abyssal species *Lycenchelys antarcticus* and two little-known species of *Raja.*

IV. A small but peculiar group consists of species distributed

[1] I would like to stress once more that the zoogeographical conclusions stated herein are based only on the analysis of distribution of coast fishes. The most complete modern data on phytogeography and on zoogeography of invertebrates of the South Ocean are given in a paper by G. A. KNOX *(Proc. Roy. Soc.* B, **152**, 1960, *577-627)* and in other papers of the same volume.

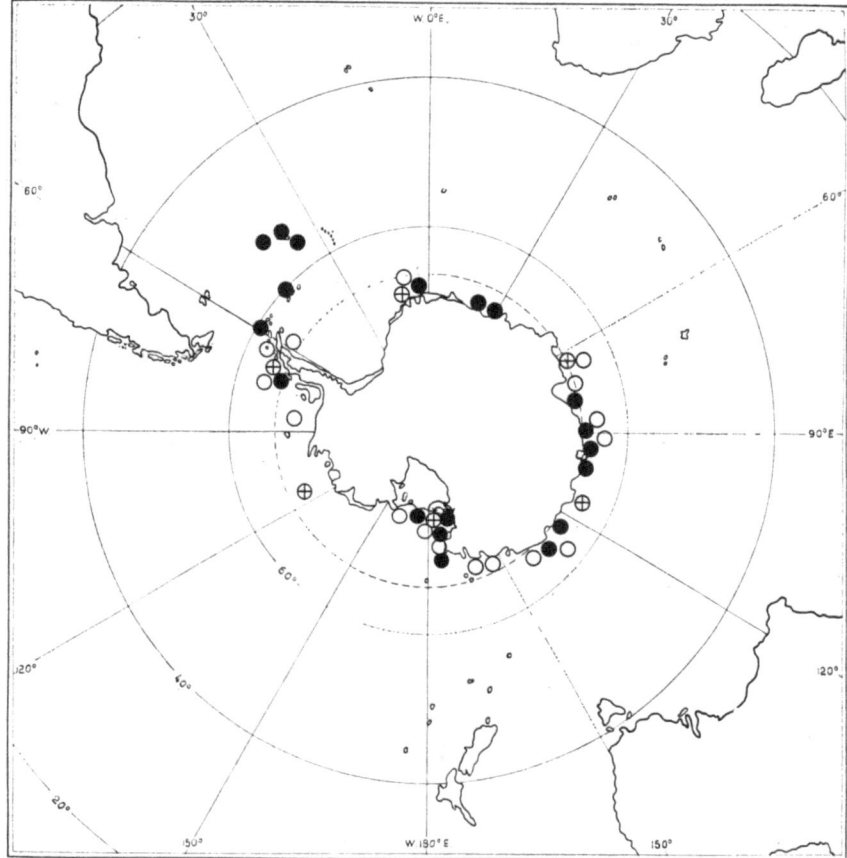

Fig. 16. Circumpolar-Antarctic distribution of three nototheniid species: a – *Pleuragramma antarcticum* (open circles); *Dissostichus mawsoni* (circles with crosses); *Trematomus hansoni* (black circles) (orig.).

exclusively in *West Antarctica* and *South Georgia (Notothenia gibbe-rifrons, N. nudifrons, Chaenocephalus aceratus* and *N. larseni).* These are West Antarctic species in a wide sense.

V. The South Georgian type includes 9—10 endemic species (partly at the S. Sandwich Isl.). To this type of area two endemic genera (*Psilodraco* and *Pseudochaenichthys*) have to be added.

VI. The Kerguelen type (not distinguished by NYBELIN) may unite nearly 11 species with local areas, from Prince Edward Isl. eastward to Macquarie Isl. For a zoogeographical analysis of these species it is very important to study the degree of their taxonomic divergence around each island and their relationships with Antarctic and notal species.

534

VII. The Antarctic-Notal or mainly notal type consists of Patagonian and Southern New-Zealand species penetrating the peripheral parts of the Antarctic (Macquarie, Kerguelen, Prince Edward, S. Georgia) where they form no separate species, e.g. *Notothenia macrocephala, N. colbecki, Melanostigma gelatinosum, Mancopsetta maculata* and other alien or often occasional elements in the Antarctic fish fauna.

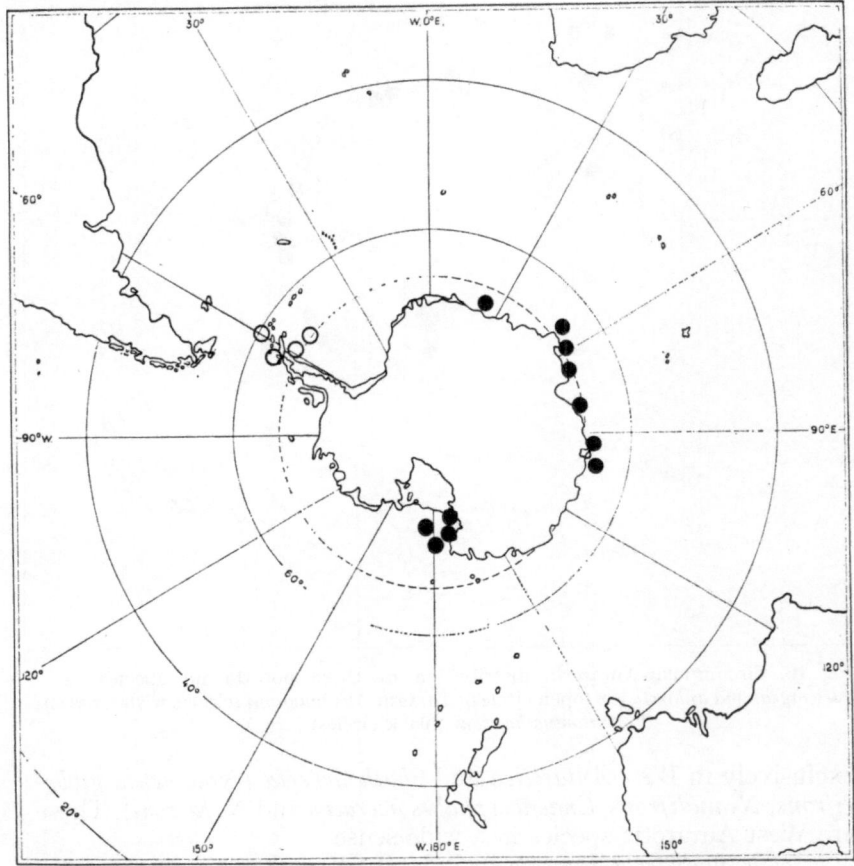

Fig. 17. Representative East-Antarctic type of distribution (*Chionodraco kathleenae*, black circles) and West-Antarctic type of distribution (*Ch. hamatus*, open circles) (orig.).

It would be interesting to show the connection of Bouvet Isl. with West Antarctica and S. Georgia and of S. Georgia with the Kerguelen and others but these are particular cases requiring more precise information on additional materials.

535

On the zoogeographical subdivision of the Antarctic Region

REGAN (1914) distinguished two faunal units within the Antarctic Zone: the Glacial District (coasts of the Antarctic Continent with neighbouring islands including S. Sandwich, S. Georgia and Bouvet) and the Kerguelen District[1] (islands of the Indian Ocean near the Antarctic Convergence). REGAN was not sure about the necessity of including the Macquarie Isl. in the last district as NORMAN (1938) did later. The latter, as well as REGAN, considered the available data insufficient for any divisions within the Glacial District. However, NYBELIN (1947) later excluded the Kerguelen Region from the Antarctic and adduced some grounds to distinguish between the East and West Antarctica Subregions by the faunal composition. As a whole the zoogeographical scheme of NYBELIN (for REGAN's Glacial District) was as follows:

I. High-Antarctic Region
   A. East Antarctic Subregion
   B. West Antarctic Subregion
II. Low-Antarctic Region, including S. Georgia; NYBELIN thought that the fauna of the S. Sandwich and the Bouvet Islands is equally similar to the fauna of S. Georgia as to the fauna of West-Antarctica.

A careful analysis of the distribution of Antarctic fish species shows that the distinctions mentioned by NYBELIN between the East (large) and West (small) parts of the Antarctic Continent are very significant in connection with the composition of species and can hardly be explained at present due to insufficient knowledge. Borders between these areas are still not defined also because of lack of data. NYBELIN's recognition of two Antarctic regions seems rather doubtful. Out of 23 species inhabiting S. Georgia 9 species at least (nearly 40%) are known also from the coasts of the Antarctic Continent (in particular the West part); if one takes into consideration other allied forms (*Raja arctowskii* — *R. georgiana*, species related to the genera *Parachaenichthys* and *Artedidraco* and the possible occurrence of muraenolepid fishes in West Antarctica, etc.), the resemblance will become even more obvious. The relationships between the fauna of S. Georgia and that of the neighbouring parts of the Antarctic Continent is thus so considerable that we see no grounds on which to define S. Georgia as a separate zoogeographical Region or even Subregion (as EKMAN does). Nevertheless, because of endemism, the presence of some notal elements and some negative features characteristic of local faunas of an insular type, S. Georgia deserves a certain degree of zoogeographical separation.

[1] NORMAN (History of Fishes, 1931, 260) suggested the name 'Periglacial Region' but in later works (NORMAN, 1938) he did not use the term.

536

Thus, taking into account the data of British authors as well as those from NYBELIN, we can conclude that there are four more or less isolated faunal units. The correlation between them can be expressed in the following scheme (fig. 18).

Antarctic Region
I. Glacial Subregion
   1. Continental Province
      A. East Antarctic District
      B. West Antarctic District
   2. S. Georgian Province.
II. Kerguelen Subregion

The grounds and boundaries of these subdivisions are included in their characteristics given below.

The *Antarctic Region* includes the shelf and continental slope of the Antarctic Continent and the islands lying to the south of and near the Antarctic Convergence. The fish fauna (39 genera with 94 species) is characterized: by the very strong development of notothenoid families (Nototheniidae, Harpagiferidae, Bathydraconidae, Chaenichthyidae) to which 3/4 of the species of the whole fauna[1] belong; Muraenolepids as a primarily Antarctic suborder of fishes ought to be mentioned too; by the complete absence of the most generalized notothenoid family: the Bovichthyidae[2]; by the negligable admixture of notal (antiboreal) species and this only in the peripheral parts of the Region; by the very high generic (70%) and specific (95%) endemism of the whole fauna and of notothenoid families in particular (97% of the species and 89% of the genera are peculiar). The Antarctic Region is naturally devided into two subregions: The Glacial and The Kerguelen one (REGAN's terms).

I. The Glacial Subregion embraces the coasts of the Antarctic Continent, the neighbouring islands of the Scotia Sea including S. Georgia and the Bouvet Islands; in other words the whole area within the extreme limits of the pack-ice. Endemic for this region are the family Bathydraconidae (15 species) and the genera *Trematomus* (13 species) and *Artedidraco* (5 species); characteristic is the dominating development of the Harpagiferidae (not less than 10 species). In the Glacial Subregion 81 species are known of which only 7 (9%) occur beyond the limits of this subregion. As mentioned above the following faunal units can be distinguished: (1) East Antarctic, (2) West Antarctic and (3) S. Georgian. It is remarkable that from the first area to the third one a natural change of indices

---

[1] In the Patagonian-Falkland fauna less than 1/5 of all the species belong to notothenoids.
[2] Moreover, just beyond the boundaries of the Antarctic region (Falkland, Auckland, Campbell ect.) members of such families as Clupeidae, Galaxiidae, Gadidae, Pleuronectidae, Ophidiidae, Tripterygiidae, Syngnathydae and many others, absolutely alien to the Antarctic families, appear.

occurs: the territory of the areas decreases (from the enormous extent of East Antarctica to the small surface of S. Georgia); severe conditions of existence grow softer (ice condition, temperature); in the intertidal zone life appears and develops; the number of spe-

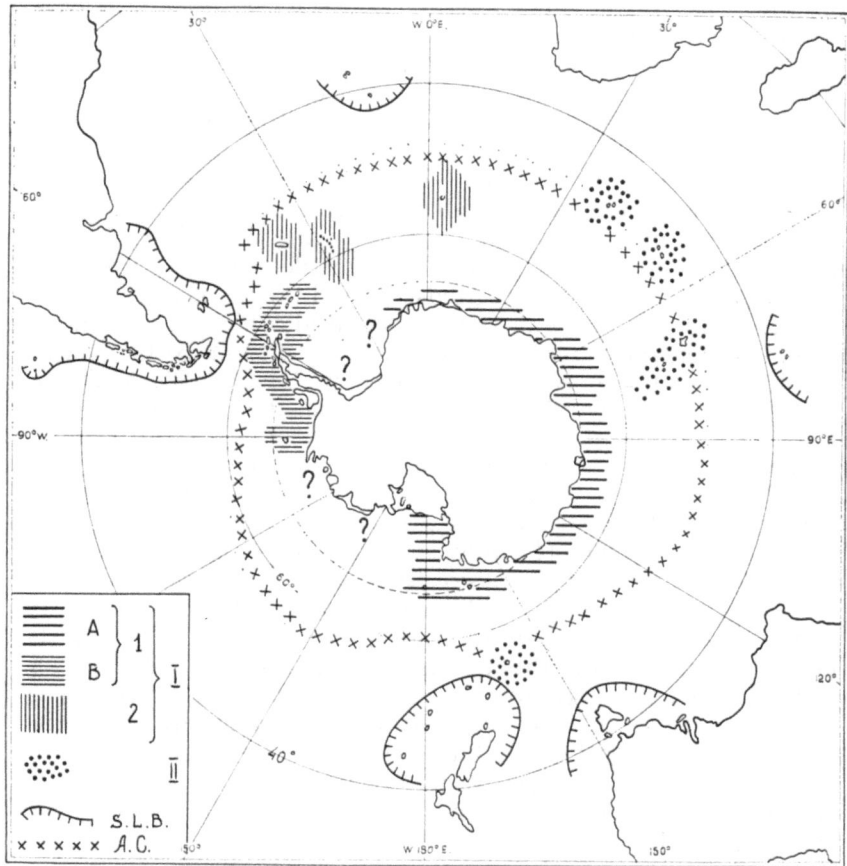

Fig. 18. Zoogeographic subdivision of the Antarctic Region based on the distribution of coast fishes: I – Glacial Subregion; 1 – Continental Province; A – East Antarctic District; B – West Antarctic District; 2 – South Georgian Province; II – Kerguelen (transitional) Subregion; S.L.B. – southern limit of distribution of Bovichthyidae A.C. – mean position of the Antarctic Convergence showing the northern boundaries of the Antarctic pelagic zone (orig.).

cies in each area *decreases* (56-34-23 species respectively); the affinity with the Kerguelen Subregion increases; the specific diversity of the genus *Trematomus* decreases on account of the increase of the number of species of *Notothenia*. It is not very easy to define the zooge-ographical rank and co-ordination of these areas. Waters connected with Graham Land are faunistically and ecologically intermediate ones. 22 species, i.e. 2/5 of the East Antarctic fish fauna, are found in West Antarctica since they have circumcontinental ranges. However, 9 species common to West Antarctica and S. Georgia make up also nearly 2/5 of the fauna of S. Georgia. This is why we have reasons to consider these three areas as zoogeographical units of a similar rank (i.e. provinces) among which West Antarctica occupies an intermediate position and enables us to unite into one subregion such different faunas as those from East Antarctica and of S. Georgia. Other ways to solve the problem exist, such as to stress the similarity of the fauna of all S. Antill Islands and that of the northern part of Graham Land both belonging to the West-Antarctic fauna in a broad sense. However, incomplete study of vast areas and insufficient clarity of relationships and ranges of many species and genera make us to abstain from considerable changes and to accept in general the schema suggested by REGAN-NORMAN with the details quoted by NYBELIN, i.e. to unite East and West Antarctica (into a Continental Province) opposing it to the S. Georgian Province.

1. CONTINENTAL PROVINCE (High-Antarctic Region of NYBELIN) includes the coastal waters of the whole Continent with the S. Shetland and the S. Orkney Islands. 11 genera are endemic for the *whole* of the Province (*Pleuragramma*; *Dolloidraco, Pogonophryne*; *Gerlachea, Racovitzia, Prionodraco, Gymnodraco*; *Pagetopsis, Cryodraco, Chionodraco*; *Austrolycichthys*) and about 19 species out of 67 species known (27%). Notothenoid species make up 81% of the fauna. The Continental Province is devided into two districts.

A. The East-Antarctic District has the most severe thermic regime (near-bottom temperatures close to the freezing point dominate). Nevertheless, its fauna shows the greatest diversity of all the Antarctic areas (56 species and 25 genera) and has a high degree of endemism — more than a half of all the species do not appear beyond the boundaries of the district (IInd type of distribution). Five to six endemic genera are known. The great diversity of species in the genus *Trematomus* and the smallest diversity in *Notothenia* is typical. Littoral fauna is completely absent. Sublittoral fauna is strongly depressed in the majority of areas (continental ice, ice shelves, formation of bottom ice). The greatest diversity of species in the fauna is found at 300—600 m depth. Forty percent of the species are common to both East and West Antarctica, whereas only 9% are common to S. Georgia.

B. The West-Antarctic District has been investigated in its northern part only (Graham Land with neighbouring islands), this is why its borders with the East-Antarctic District are not yet clearly defined[1]. Only 34 species are known within this district, 2/3 of which also live in the East Antarctic District. Nine species (27%) are common to the S. Georgia District (it is possible that this number will increase in the future). No endemic genera, and only few endemic species, are known, some of which are doubtful (IIIrd type of distribution). As compared with the previous district the diversity of species in the genus *Trematomus* decreases, becoming nearly equal to the number of species of *Notothenia*. A littoral species (*Harpagifer*) appears.

2. THE SOUTH GEORGIAN PROVINCE ("Low-Antarctic Region" of NYBELIN). Twenty three species are found in this province not taking into account the macrurids, skates and liparids which are poorly studied. The genera *Psilodraco* and *Pseudochaenichthys* and nearly 40% of the species are endemic. The diversity of the species of *Notothenia* is greater than that of *Trematomus*. Forty per cent of the species are also found in the West-Antarctic District, at least 25% in the Kerguelen Subregion[2] and nearly 20% in the East Antarctic District. A few species of notal origin penetrate from the north but faunistic differences with the Falkland Isl. (including Burdwood Bank) are very sharp. *Harpagifer georgianus* and apparently *Notothenia angustifrons* are very common in the intertidal zone. Temperatures in the shelf are highly above freezing point and the water is, as a rule, free of ice throughout the year. The S. Sandwich Islands can be included into the S. Georgian Province (owing to the presence of *N. angustifrons*) as an area with a very impoverished fauna. Near Bouvet Island only 2 species were found (and a third unidentified *Notothenia* sp.) showing nearly equal links with S. Georgia as with Graham Land (*Notothenia larseni* and *Chaenocephalus bouvetensis*, which is very close, if not identical, to *Ch. aceratus*).

II. The Kerguelen Subregion includes the Marion, Crozet, Kerguelen, Heard (with neighbouring submerged ridge) and Macquarie Islands. They have an impoverished mosaic fauna of an insular type (total of 19 species) with a considerable percentage of local endemics (more than 50% of the species). Two genera are endemic: *Chaenichthys* (2 species) belonging to the mainly Antarctic family Chaenichthyidae and *Zanclorhynchus* (= *Xanclorhynchus*) of the primarily notal family Congiopodidae. As compared with the Glacial Subregion it is characterized by negative features — the absence of

---

[1]) It should be noted that in the vicinity of Victoria Land (including the Balleny and Scott Islands) species also present in West Antarctica were found, these species were, however, never met in East Antarctica westward of Adelie Land *(Notothenia larseni, N. coriiceps neglecta* and others).

[2]) In spite of certain resemblances between S.Georgia and the Kerguelen (with respect to fauna and environment) their zoogeographical unity suggested by us in a preliminary report is hardly justified (ANDRIASHEV & TOKAREV, 1958).

540

Harpagiferidae (except for the genus *Harpagifer*), Bathydraconidae (except one deepwater species *Bathydraco*) and of the typical glacial-Antarctic genus *Trematomus*. The admixture of northern (notal) elements is greater (*Notothenia macrocephala*, *N. colbecki*, *Mancopsetta maculata*), some of which produced endemic forms (species of skates, genus *Zancolorhynchus*). Nevertheless, the faunistic border with the neighbouring southern islands of New Zealand (Auckland, Campbell — 360 miles off Macquarie) is very sharp since only two common species are known. However, several families, alien to the Antarctic appear, e.g. warm-water forms as the luminous shark *Euprotomycrus*, *Sadinops*, tuna fish (*Allothunnus fallai*[1]), *Tripterygion* (*Forsterygion*), *Rhombosolea*, pipefishes, *Bovichthys* and others. 5 species (26%) also live in S. Georgia and 6 other species (5 species of *Notothenia* and one of *Muraenolepis*) point at a relationship with the Glacial District (REGAN, 1914, NORMAN, 1938). The Kerguelen skate, *Raja murrayi*, is more closely related to the Antarctic forms — *R. georgiana* — *R. arctowskii* than to the Patagonian species *R. macloviana*. Moreover, the genus *Harpagifer* shows some local forms in the Kerguelen Subregion which, however, do not belong to notal species from Tierra del Fuego (*H. bispinis*) or from the Falkland Islands (*H. palliolatus*), but rather to *H. georgianus* (S. Georgia, subspecies on Graham Land, NYBELIN, 1947). The occurrence of a member of the high-antarctic genus *Bathydraco* at the south-eastern part of the Kerguelen-Heard submerged ridge (1600 m) shows to a certain degree the links between the Kerguelen and Antarctic faunas.

The environment in the Kerguelen Subregion is more favourable as compared with other parts of the Antarctic — above freezing point temperatures (3—6° at 50—150 m depth), absence of ice throughout the year, well developed algae belt in the sublittoral and littoral zones where *Harpagifer* and *Notothenia cyaneobrancha* are commonly found under stones in tide pools. The fauna of the Kerguelen subregion bionomically resembles that of the Commander and Kuril Islands[2] but is considerably poorer in number of species.

Some problems in connection with the origin of the Antarctic fish fauna

The problems of the origin and the history of formation of the fish fauna of the Antarctic are very interesting but also very difficult to discuss. We shall give only a brief survey of some of them.

Ancient and peculiar features of the Antarctic fish fauna were noted by REGAN (1914, 40) and his opinion on this subject was cited

[1] A. W. PARROT, *Rec. Dominion Mus., Wellington*, vol. 3, pt. 2, 1958.
[2] See P. V. USCHAKOV, *Priroda (Nature)*, No. 3, 1958.

by NORMAN (1938, 102) without any additions: "The distinctive features of the fish-fauna of the coasts of the Antarctic are that nearly all the genera and species are peculiar and that they nearly all belong to a single group, the Nototheniiformes, which is characteristic of and almost restricted to the Antarctic and Subantarctic Zones. In the Antarctic Zone this group has developed into a large number of types that differ greatly in structure, appearance and habits. These facts seem to point to the conclusions that Antarctica may have long been isolated and that its coasts have been washed by a cold sea for a long time, probably throughout the Tertiary Period." (REGAN, 1914, 40).

This statement made fifty years ago seems to hold true until now. Indeed it must have taken a long time for this uniform fauna to evolve and produce tens of different species and genera and even morphologically distinct families.

Members of the Antarctic fauna (notothenoid families in particular) got adapted to various ecological niches in the process of divergent evolution. Although they are typical coast fishes, they can be found at all depths which points to a secondary adaptation to the pseudoabyssal depths of the shelf, the continental slope and the neighbouring abyssal areas. As a near-bottom group notothenoid families turned out to be very plastic in this respect. They developed both to forms closely connected with the bottom (burying in the mud, hiding under stones), and actively swimming species temporarily or even permanently inhabiting the open sea. Northern members of two families penetrated estuaries and fresh waters (*Pseudaphritis, Eleginops*). One of the branches of the superfamily Notothenioidae acquired a character unique among fishes: white-bloodedness; subsequent evolutionary development has led to the formation of ten different genera. Significant are adaptations with respect to food differentiation since there are benthophagous species of different types (including those searching for food by means of external taste organs), sluggish bottom predators of ambushing type and actively swimming krill-feeding and predatory forms.

The availability of such a variety of morphobiological types in one small group of percoid fishes indicates indirectly its prolonged evolution in conditions of isolation from a rich notal fauna which has left few traces in the recent Antarctic fish fauna.

The ancient features of the native fauna of the Antarctic is confirmed also by the scanty paleontological data which reveal that ancestors of recent nototheniids lived along the coasts of Graham Land since the beginning of the Tertiary Period (SMITH-WOODWARD, 1908).

Sources of fauna formation. In spite of the exceptional uniformity of the Antarctic fish fauna we may try to show that different elements or faunal groups took part in its formation.

Thus the basis of the fauna consists of ancient autochthonic elements (Nototheniidae, Harpagiferidae, Bathydraconidae, Chaenichthyidae and Muraenolepidae) the specific and to a considerable degree generic separation of which was carried out in the Antarctic waters. Their distribution is mainly Antarctic and they reached a high degree of endemism. By their ancient roots they are closely connected with the cold temperate fauna of South America.

Forms non-specific for the Antarctic with a more northern origin belong to the notal family Congiopodidae or the families widely distributed in the oceans (Rajidae, Bothidae). They formed endemic species in the Antarctic (all skates and the genus *Zanelorhynchus*) related to the notal fauna.

Moreover, some notal species have probably only recently penetrated the peripheral parts of the Antarctic and have not yet formed any endemic forms (*N. macrocephala, N. colbecki, Mancopsetta maculata, Melanostigma gelatinosum*).

Endemic-Antarctic species and genera belonging to the families Zoarcidae and Liparidae form a small but very interesting faunal group. They have an evident relationship with the Patagonian fauna but their primary origin from the boreal fauna of the northern hemisphere can hardly cause any doubt (particulars on page 544).

Finally we can mention one more small faunal group with a deep-bottom origin. These are members of widely distributed true deep-water families (Macruridae, Moridae and others) which penetrated the Antarctic. They can be found in the upper horizons of the continental slope. Some of them adapted long ago to deep Antarctic waters and formed endemic species.

From the above it can be seen that deep faunistic links exist between the Antarctic and the Patagonian-Falkland faunas which have 10 genera in common (*Raja, Muraenolepis, Notothenia, Dissostichus, Harpagifer, Champsocephalus, Melanostigma, Careproctus, Mancopsetta, Achiropsetta*). It is worth noting that some genera are represented in both regions by related species (*Dissostichus mawsoni — D. eleginoides, Champsocephalus gunnari — Ch. esox*; species of the genera *Harpagifer, Raja*) or even by identical species (*Mancopsetta maculata, Melanostigma gelatinosum* and probably *Muraenopelis microps*).

Thus, by its ancient roots and to a lesser extent by modern influence the Antarctic fauna of bottom fishes is connected with the cold-temperate (subantarctic) South American fauna. Considerably weaker are the links with the New Zealand fauna the influence of which can be traced only in the Kerguelen-Macquarie endemic family Congiopodidae (*Zanclorhynchus*) and in one or two species of *Notothenia*. There is practically no relation to other regions of the southern hemisphere (Australia, S. Africa, Tristan da Cunha, Gough, Saint Paul and New Amsterdam). This is in good agreement with REGAN's conclusion that "neither marine nor fresh-water fishes support the theory that the Antarctic Continent connected America with Australia during the Tertiary Period" (REGAN, 1914, 41).

Probable reasons for the similarity in fish fauna between antarctic oceanic islands. Many examples are known of islands of the southern hemisphere, separated by vast ocean distances, with common or related species of bottom coast fishes, e.g. *Champsocephalus gunnari* (S. Georgia—Kerguelen), *Zanclorhynchus spinifer* (Kerguelen — Macquarie), *Notothenia larseni* (Bouvet — S. Georgia), *Mancopsetta maculata* (Falklands — S. Georgia — Prince Edward), *N. rossi* (S. Georgia — Kerguelen — Macquarie and others), *Chaenocephalus* (S. Georgia — Bouvet), skates (S. Georgia — Kerguelen) and others. Related species and local forms of the littoral *Harpagifer* are found in all the islands from Tierra del Fuego to the Falkland and S. Georgia islands and to the east up to Macquarie Island. One and the same species or related allopatric forms exist at remoted islands separated by "principal isolating abyssal distances" (EKMAN, 1953, 237). To explain the faunal similarity between these islands some authors drew hypothetic land bridges in the past but this is hardly necessary.

Everyone who navigated in 40° to 50° of southern latitude will have seen large "islands" of brown algae drifting to the east under the influence of the ruling winds and powerful circumglobal current, the West Wind Drift. Ed. A. SMITH[1]) is supposedly the first who attracted attention to the possibility of transfer of shore species of animals (molluscs) from island to island by means of these drifting kelps. Probably developing eggs and the young of some sluggish shore fishes (e.g. *Harpagifer, Raja)* can be transported in this way. In connection with this hypothesis it would be very important to catch drifting kelps and to examine them carefully after collecting the fauna associated with them by washing. To ascertain the actual speed of kelp drift modern technical means (radio-beacons, self recording logs, different markings etc.) could be applied.

We should keep in mind that among species of fishes common to distant islands there are good swimmers capable of independent migrations, thus the possibility of active overcoming of oceanic obstacles is not excluded in some cases. It is very significant that the pelagic silvery young of some species sometimes live far off in the ocean. Such silvery young, 6—7 cm long, were found for *N. rossi, N. coriiceps* and others. Some interesting catchings were made by us at the end of April, 1958, on the occasion of a long oceanographic section (along the meridian of Easter Isl., 109° W.lon.). In four stations ("Ob", st. 409, 411, 413, 415) several silvery young of *Notothenia*

---

[1]) Ed. A. SMITH, *Proc. Malacol. Soc. London*, III, N 1, 1898, 20-25.

sp. (38—55 mm long) were caught related to the New Zealand — Patagonian group of species *(macrocephala — microlepidota)*. The distance between the extreme stations where those young were caught made 560 miles over the meridian (from 64° 36' to 55° 18'S.). The catching points were situated 2500 miles away from New Zealand and more than a thousand miles from Patagonia. The most probable supposition is that the young have reached the 109th meridian with the West Wind Drift from the southern New Zealand Islands. If the young perish before they reach the coastal waters of S. America then this part of the population ought to be referred to "sterile expatriation area" (EKMAN, 1953, 317). However, we cannot deny the possibility that these young actually reach Patagonian waters. Most likely the presence at the coasts of New Zealand and S. America of such species as *N. macrocephala, N. microlepidota* and others common to both widely separated areas can be explained by just a similar drift.

The same type of distribution in the stage of pelagic young (only in an earlier age) can explain the area of *Mancopsetta maculata* found at Falkland, S. Georgia and Prince Edwards Islands. Well transformed pelagic young of another species *(Achiropsetta slavae)* as mentioned earlier (see page 509) were found in the high latitudes of the Antarctic, at Wilkes Land but their origin can still not be explained (ANDRIASHEV, 1960d).

In order to explain the faunistic similarity between remoted islands, EKMAN (1953 : 237) attached great importance to deep-water migrations. However, for fishes of the Antarctic Region such way of penetration is of less significance as it is accessible only to certain species of true deep-water families (Macruridae, Moridae and others) and perhaps to some of secondary deep-water forms *(Bathydraco,* Zoarcidae, Liparidae).

The faunal elements of northern (boreal) origin. The problem on the ways of penetration and distribution of the families of northern origin (Zoarcidae including Lycodapodidae, Liparidae, Agonidae and others) in the southern hemisphere is of great interest for general zoogeography. Several tens of genera and hundreds of species of these families are widely distributed in the northern parts of the Pacific and Atlantic Oceans and in the Arctic as well; they are absent in tropical waters or occur somewhere at abyssal depths by way of exception only. In the southern hemisphere, in temperate latitudes and in the Antarctic they are again met as endemic genera and species particularly at the continental shelf. Many elements of northern origin are present in the Patagonian fauna: nearly 15 species of zoarcids, several species of liparids, the only member of the family Agonidae to be present in the southern hemisphere is *Agonopsis chiloensis.* Considerable taxonomic separation of the majority of these forms enables us to suppose more ancient faunistic links between the north and the south than predicted according to the DARWIN-BERG theory for boreal-subtropical bipolar (antitropical) forms (since the occurrence of many related or even identical but discontinuously distributed species in the temperate and subtropical latitudes is explained by this theory as a result of migration through the tropics during the cool Quaternary Period; see BERG, 1933 and HUBBS, 1952).

In the case of the discontinuously distributed zoarcids and other forms of families of northern origin ORTMANN's theory of deep-water migrations is more applicable (see EKMAN, 1953 : 246, 259 etc.). REGAN (1914 : 33) said: "The Magellan District is the head-quarter of the Southern Zoarcidae, which may have reached it originally from the North along the American Coast, perhaps migrating in rather deep and cold water". To specify this question a careful comparative study of the taxonomic composition, endemism, rank, relationships, bathymetric and geographical ranges of the northern and southern forms is necessary. Nevertheless, a preliminary analysis confirms REGAN's point of view and enables us to make the following supposition (ANDRIASHEV & PERMITIN, 1961). The ancestors of recent endemic zoarcids and liparids of the southern hemisphere could have penetrate from the north most likely at comparatively great (bathyal-abyssal) depths mostly along the western outlying regions of North and South America, and Europe and West Africa as well.

Near South Africa, owing to the comparatively high near-bottom temperatures, zoarcids and liparids could not have gone into smaller depths with a rich fauna of subtropical type. Therefore northern migrants keep at depths not less than 1—1½ thousand metres where bottom temperatures do not exceed 2—4°, while passing along South Africa.

Along the coasts of S. America northern deep-water migrants apparently found more suitable thermic conditions and in the limits of the shelf they formed a secondary centre of speciation inhabiting small depths up to the sublitoral and littoral zones. Relative poverty of the specific composition of the Patagonian fauna and the absence of species similar to zoarcids in the occupied ecological niches favoured this speciation to a considerable extent.

If our considerations are confirmed in the future by further investigations the conclusion that the coastal fauna of the Patagonian Region (and partly of the Antarctic continental shelf) were formed to a certain extent by deep-water elements, will be strengthened. This would represent an exceptional example amongst fishes of the formation of a littoral fauna from a bathial-abyssal one.

Future research objectives

In this brief review we have tried to show the unique character of

546

the Antarctic fish fauna which in this respect does not yield to some other wonderful groups of Antarctic animals. The study on adaptations of this peculiar fauna to the life in extremely severe conditions is of considerable theoretical interest and perhaps, as WOHLSCHLAG (1961a) supposes, will help to solve some problems of practical importance. Therefore we would like very much to support the first steps undertaken by SCAR for conservation of the Antarctic fauna and we express our hope for future developments of the international scientific collaboration in complex exploration of the Antarctic.

In this respect I would like to say a few words about the possible trends of future investigations in the field of the ichthyology of the Antarctic.

First of all we should keep up our usual works of gathering ichthyological collections necessary for further study of the taxonomic questions (infraspecific variability in particular), the ecology and the geography of the Antarctic fishes. Special efforts should be made to study the bottom fauna of the Weddell, Bellingshausen, Amundsen and Ross Seas and of all the islands. The broader use of modern fishing equipment with due regard for the specific conditions of the Antarctic fishing is desirable.

The comprehensive study of the life history of the Antarctic fishes, with application of special techniques (hydroacoustics, underwater photography and television, aqualungs, bathyscaphs, etc.) in order to investigate the behaviour of fishes and the peculiarities of their life under the ice shelf is of great interest.

Many valuable data can be expected from the development of physiological and biochemical work, first of all a study of the properties of the fish blood, the metabolism at very low temperatures, the mechanism of cold adaptation, the threshold of the cell warm-resistance[1] and other experimental works which can throw light on the problem of the adaptive evolution of the Antarctic fauna. Very tempting in this respect is the solution of the problem of fish whitebloodedness both from physiological and from evolutionary point of view.

In order to study the pelagic ichthyofauna it is necessary to make catchings on the standard horizons by large closing mid-water trawls (of the Isaacs-Kidd type) to establish seasonal and diurnal vertical migrations and to determine the association of species with different oceanic-waters and frontal zones, particularly in the region of the Antarctic Convergence by using electric light to assist catching. Special attention should be paid to the collecting and the study of the distribution of pelagic larvae and the young of coastal fishes in the open sea and also to the study of the fauna associated with accumulations of drifting algae in the zone of the West Winds Drifts.

---

[1] B. P. USHAKOV, *J. gen.Biol., Moscow,* 17, No 2, 1956, *154-160.*

Within the years to come it would be desirable to carry out intensive trawlings in the abyssal region of the Antarctic and to make an attempt to get data on the ultraabyssal (hadal) fauna of the high and temperate latitudes of the southern hemisphere (S. Sandwich, West Australian and Atakama Trenches).

Jointly with oceanographers working in different fields a program for studying Antarctic environment of bottom and pelagic fishes should be elaborated. Problems to be solved are: the degree of isolation of deep innershelf depressions from the depths of the continental slope, the temperature and the oxygen tension of near-bottom waters of these areas; the seasonal distribution of supercooled waters; the study of the formation of deep-water ice-crystals and their influence on the fauna; the seasonal variations in the latitudinal position of frontal zones (the Antarctic Convergence in particular); latitudinal changes of seasonal maxima in plankton development and the food productivity of the plankton and benthos.

Accumulation of data along the lines above mentioned will increase our understanding of some problems in ecology, zoogeography and evolution not only of high latitudinal fauna of the southern hemisphere but also of the comparative aspects of biological symmetry of the oceans (including the bipolarity problem in broad sense).

REFERENCES[1]

ABE, T., 1957. Notes on fishes from the stomachs of whales taken in Antarctic. I. Xenocyttus nemotoi, a new genus and new species of Zeomorph fish of the subfamily Oreosominae Goode et Bean, 1895. *Sci. Rep. Whales Res. Inst.* **12**, 225—233, 2 pls.

ANDRIASHEV, A. P., 1958a. List of the ichthyological stations. Rep. Compl. Antarct. Exp. Acad. Sci. U.S.S.R., "Hydrol., hydrochem., geolog. and biol. studies R.S. "Ob" 1955—1956", *199—204*.

ANDRIASHEV, A. P., 1958b. Ichthyological investigations of the Soviet Antarctic Expedition and some problems of the zoogeography in the Antarctic. *Inform. Bull. Soviet Antarct. Exp.* **3**, 63—66.

ANDRIASHEV, A. P., 1959. On the number of vertebrae and some osteological features of the Antarctic fishes. *Probl. Ichthyol.* **12**, 3—7.

ANDRIASHEV, A. P., 1960a. On the occurrence of a rare Chaenichthyid fish near the coast of the Antarctic Continent. *Probl. Ichthyol.* **14**, *18—21*.

ANDRIASHEV, A. P., 1960 b, c, d. Families of fishes new to the Antarctic. 1. Paradiplospinus antarcticus, gen. et sp. n. (Pisces, Trichiuridae). 2. Pearleyed fishes (Scopelarchidae). 3. Pelagic young flatfishes (Pisces, Bothidae) near off the Antarctic Continent. *Zool. Zhurn.* **XXXIX**, 2, 4, 7, *244—249*, 2 figs.; *563—565*, 2 figs.; *1056—1060*, 3 figs.

ANDRIASHEV, A. P., 1961. List of the ichthyological stations. *Rep. Soviet Antarct. Exp.* **22**, 227—233.

---

[1]) This list of references mainly cites papers on the fishes of the Antarctic. References to papers on oceanography, general biology, zoogeography ect. are given in the text and in footnotes.

548

ANDRIASHEV, A. P. 1962a. Bathypelagic fishes of the Antarctic. 1. Myctophidae. Explor. Fauna Seas, I (IX), *216—294*, 36 figs.
ANDRIASHEV, A. P. 1962b. On the systematic position of the giant Nototheniid fish from the McMurdo Sound, Antarctica. *Zool. Zhurn.* XLI, 7, *1048—1050*.
ANDRIASHEV, A. P. & PERMITIN, YU, E., 1961. Ichthyological investigations. *Rep. Soviet Antarctic. Exp.* 19, *261—273*.
ANDRIASHEV, A. P. & TOKAREV, A. K., 1958. Ichthyofauna. Rep. Compl. Antarct. Exp. Acad. Sci. U.S.S.R., "Descr. Exp. R.V. "Ob" 1955—1956", *195—207*.
BARSUKOV, V. V. & PERMITIN, YU. E., 1958. A new species of the genus Pagetopsis (family Chaenichthyidae). *Zool. Zhurn.* XXXVII, 9, *1409—1411*, 2 figs.
BARSUKOV, V. V. & PERMITIN, YU. E., 1959. Ichthyological collection list. (*Rep.*) *Soviet Antarct. Exp.* 6, *379—387*.
BARSUKOV, V. V. & PERMITIN, YU. E., 1960. Ichthyological research. (*Rep.*) *Soviet Antarct. Exp.*, 7, *97—106*.
BERG, L. S., 1933. Die bipolare Verbreitung der Organismen und die Eiszeit. *Zoogeographica*, I, *449—484*.
BERG, L. S., 1940. Classification of fishes, both recent and fossil. *Trav. Inst. Zool. Acad. Sci. U.S.S.R.* V, 1, *1—517*, 190 figs.
BERG, L. S., 1961. Fishes. Selected Works, IV, **7-58**, 37 figs. (1-st. ed., 1941).
BLANC, M., 1961. Les poissons des terres australes et antarctiques Françaises. *Mém. Inst. Sci. Madagascar*, ser. F., IV, *109—157*, 30—51 figs.
BOULENGER, G. A., 1902. Pisces. Rep. Coll. Nat. Hist. Antarct. Reg. Voy. "Southern Cross", London, *174—189*, pls. XI-XVIII.
BOULENGER, G. A., 1907. Fishes. Nation. Antarct. Exp., Nat. Hist., II, Vertebrata, IV, *1—5*, 2 pls.
DEWITT, J. H. & TYLER, J. C., 1960. Fishes of the Stanford Antarctic research-program 1958—1959. *Stanf. ichthyol. Bull.* **7**, 4, *162-199*, 6 figs.
DOLLO, L., 1904. Poissons. Exp. Antarct. Belge, Rés. S.Y. Belgica 1897—1899, Zool., *1—240*, 12 pls.
EALEY, E. H. M., 1954. Analysis of stomachs contents of some Heard Islands bird. *Emu, Melbourne*, **54**, 3, *204-210*.
EKMAN, SV., 1953. Zoogeography of the sea, London, XIV+417, 121 figs.
FRASER-BRUNNER, A., 1949. A classification of the fishes of the family Myctophidae. *Proc. zool. Soc. London* **118**, IV, *1019—1106*, figs.
GARRICK, J. A. F., 1960. Studies on New Zealand Elasmobranchii, Part XII. The species of Squalus from New Zealand and Australia; and a general account and key to the New Zealand Squaloidea. *Trans. Roy. Soc. New. Zeal.* **88**, 3, *519—557*, 6 figs.
GOSSE, J.-P., 1961. Poissons antarctiques recoltés par l'expédition Belge Iris 1961. *Bull. Inst. Roy. Sci. Nat. Belgique* **XXXVII**, 29, *1—10*.
GREY, M., 1956. The distribution of fishes found below a depth of 2000 metres. *Fieldiana (Zool.)* **36**, 2, *73—337*.
GÜNTHER, A., 1880. Report on the shore fishes procured during the voyage H.M.S. "Challenger" in the years 1873—1876. Rep. Sci. Res. Challenger, I, 6, *1—82*.
GÜNTHER, A., 1887. Report on the deep-sea fishes collected by H.M.S. "Challenger" during the years 1873—1876. Rep. Sci. Res. Challenger, XXII, *1—268*, 68 pls.
HERALD, E. S., 1961. Living fishes of the world. New York, *1—304*, 300 figs. incl. 145 in full colour.
HUBBS, C. L., 1952. Antitropical distribution of fishes and other organisms. Seventh Pacific Sci. Congr., III, *1—6*.

KOROTKEVITCH, E. S., 1958. Geographical characteristic of the area covered by the Soviet Antarctic Expedition. *Izvestia geogr. Soc. U.S.S.R.* **90**, 3, *220—243*, 16 figs.

LÖNNBERG, A. J. E., 1905. The fishes of the Swedish South Polar Expedition. *Wiss. Erg. Swed. Südpol.-Exp.*, V, 6, *1—69*, 5 pls.

MAKUSHOK, V. M., 1959. About the biological collections and observations in Mirny Observatory in 1958. *Inform. Bull. Soviet Antarct. Exp.* **6**, *40—42.*

MARSHALL, N. B., 1953. Egg size in Arctic, Antarctic and deep-sea fishes. *Evolution,* **7,** 4, *328—341.*

MARTSINKEVITCH, L. D., 1958. Cell contents of blood of whiteblooded fishes in the Antarctic. *Inform. Bull. Soviet Antarct. Exp.* **3**, *67—68.*

MARTSINKEVITCH, L. D., 1961. Some characteristics of blood in whiteblood fish. *Arch. Anat., Histol., Embryol., Leningr.* **XLI**, 12, *75—78*, 4 colour figs.

MATTHEWS, L. H., 1931. South Georgia, the British Empire's Subantarctic outpost. Wright & Marshall, London (not seen).

MILLER, R. G., 1961. A sketch history of ichthyological investigations of the Ross Sea. *New Zeal. J. Sci.* **19**, 1, *9—10, 12—14.*

MILLER, N. G. & RESECK, J. JR., 1961. Chionodraco markhami, a new Antarctic fish of the family Chaenichthyidae. *Copeia,* **1**, *50—53.*

MORELAND, J., 1960. A new genus and species of Congiopodid fish from Southern New Zealand. *Rec. Dominion Mus.* **3**, 3, *241—246*, 7 figs.

NEUSHUL, M., 1959. Biological collecting in Antarctic waters. *The Veliger,* **2**, *15—17.*

NORMAN, J. R., 1930. Oceanic fishes and flatfishes collected in 1925—1927. *Disc. Rep.,* II, *261—370*, 47 figs., 1 pl.

NORMAN, J. R., 1935. Coast fishes. Part I, The South Atlantic. *Disc. Rep.,* XII, *1—58*, 15 figs.

NORMAN, J. R., 1937a. Coast fishes. Part II. The Patagonian Region. *Disc. Rep.,* XVI, *1—150*, 76 figs., 5 pls.

NORMAN, J. R., 1937b. Fishes. *Rep. B.A.N.Z. Antarct. Res. Exp.,* 1929—31, Ser. B, **I**, 2, *50—88*, 11 figs.

NORMAN, J. R., 1938. Coast fishes. Part III. The Antarctic Zone. *Disc. Rep.,* XVIII, *1—105*, 62 figs., 1 pl.

NYBELIN, O., 1947. Antarctic fishes. *Sci. Res. Norw. Antarct. Exp.* 1927— 1928, 26, *1—76*, 2 figs., 12 maps, 6 pls.

NYBELIN, O., 1951. Subantarctic and Antarctic fishes. *Sci. Res. Bratteg Exp.,* 1947—48, 2, *1—32.*

NYBELIN, O., 1952. Fishes collected during the Norwegian-British-Swedish Antarctic Expedition 1949—1952. *Göteborgs Kungl. Vet. Vitterh. Samhäll, Handl.,* Sjätte följd., Ser. B. **6**, 7, *1—13*, 4 maps.

OLSEN, S., 1954. South Georgian Cod, Notothenia rossi, marmorata Fischer. *Norsk Hvalfangst-Tidende,* **43**, 7, *373—382.*

OLSEN, S., 1955. A contribution to the systematics and biology of Chaenichthyid fishes from South Georgia. *Nytt. Magaz. for Zool.* **3**, *79—93.*

PAPPENHEIM, P., 1912. Die Fische der Deutschen Südpolar-Expedition 1901 —1903. I. Die Fische der Antarktis und Subantarktis. Deutsche Südpolar Exp., XIII, *160—182*, 10 Fig., Taf. IX—X.

REGAN, C. T., 1913. The Antarctic fishes of the Scottish National Antarctic Expedition. *Trans. Roy. Soc. Edinburgh,* **XLIX**, *229—292*, 6 figs., 11 pls.

REGAN, C. T., 1914. Fishes. British Antarct. ("Terra Nova") Exp. 1910, Nat. Hist. Rep., Zool., I, *1—54*, 8 figs., 13 pls.

REGAN, C. T., 1916. Larval and postlarval fishes. Ibid., *125—156*, 10 pls.

RESECK, J. JR., 1961. A note on fishes from the Ross Sea, Antarctica. *New Zeal. J. Sci.* **4**, 1, *107—115.*

550

RICHARDSON, J., 1844. Ichthyology of the voyage H.M.S. Erebus and Terror under the command of Captain Sir J.C. Ross, London, 1844—1848. VIII + 139, 60 pls.

ROFEN, R. R. & DeWITT, H. H., 1961. Antarctic fishes. Science in Antarctica. Part I. Rep. U. S. Comm. Polar Res., *94—112*.

ROULE, L., ANGEL, F. & DESPAX, R., 1913. Poissons, Deuxième Exp. Antarct. Française (1908—1910) comm. par le Dr. J. CHARCOT. Paris, *1—24*, 4 pls.

RUUD, J. T., 1954. Vertebrates without erythrocytes and blood pigment. *Nature, Lond.* **173**, No. 4410, *848—850*.

RUUD, J. T., 1958. Vertebrates without blood pigment; a study of the fish family Chaenichthyidae. *XVth Int. Congr. Zool.*, Sect. VI, Paper 32, *1—3*.

SMITH-WOODWARD, A., 1908. On fossil fish remains from Snow Hill and Seymour Island. Wiss. Erg. Schwed. Südpolar Exp., 1901—1903, III, Geol. u. Palaont., (1916), *1—4*, 5 figs., 1 pl.

SVETOVIDOV, A. N., 1948. Gadiformes. Fauna U.S.S.R., Pisces, IX, 4, *1—222*, 39 fig., 72 pls.

SWITHINBANK, CH. W., DARBY, D. G. & WOHLSCHLAG, D. E., 1961. Faunal remains on an Antarctic ice shelf. *Science*, **133**, No. 3455, *764—766*.

TYLER, J. C., 1960. Erythrocyte counts and haemoglobin determinations for two Antarctic Nototheniid fishes. *Stanf. ichthyol. Bull.* **7**, 4, *119—201*.

VAILLANT, L., 1906. Poissons. Exp. Antarct. Française (1903—1905) command. par le Dr. J. CHARCOT. Paris, *1—52*, 4 figs.

WAITE, E. R., 1916. Fishes. Sci. Rep. Australasian Antarct. Exp. 1911—1914, Ser. C., III, 1, *1—92*, 16 figs., 5 pls., 2 maps.

WALVIG, F., 1958. Blood and parenchymal cells in the spleen of the Icefish Chaenocephalus aceratus Lönnb. *Nytt. Magaz. Zool.* **6**, *111—120*, 2 colour pls.

WALVIG, F., 1960. The integument of the Icefish Chaenocephalus aceratus (Lönnberg). *Nytt. Magaz. Zool.* **9**, *31—36*, 3 pls.

WOHLSCHLAG, D. E., 1960. Metabolism of an Antarctic fish and the phenomenon of cold adaptation. *Ecology*, **41**, 2, *787—792*.

WOHLSCHLAG, D. E., 1961a. General ecology and physiology of Antarctic fishes. Science in Antarctica. Part I. Rep. U.S. Comm. Polar Res., *113—114*.

WOHLSCHLAG, D. E., 1961b. Growth of an Antarctic fish at freezing temperatures. *Copeia*, **1**, *11—18*, 5 figs.

ECOLOGIE DES MANCHOTS ANTARCTIQUES

J. PREVOST & J. SAPIN-JALOUSTRE

Expéditions Polaires Françaises, 47 Av. Maréchal Fayolle, Paris XVIe

LE MILIEU

Généralités

Les limites de l'Antarctique ont été fixées différemment par les astronomes (cercle polaire), les baleiniers (40me parallèle), les océanographes (ligne de convergence antarctique variant entre le 48me et le 60me parallèle) (fig. 1). Mais du point de vue écologique, il nous semble légitime de définir l'Antarctique: continent antarctique avec ses côtes, ses îles et les formations glaciaires d'origine continentale ou côtière entourant le continent et pouvant s'étendre vers le nord jusqu'au 60me parallèle.

La zone (ou région) écologique ainsi définie possède une unité géographique et une unité climatologique. Mais, du point de vue biologique, il faut insister d'emblée sur l'opposition entre le continent presque complètement azoïque et la côte où existe au contraire une vie intense d'origine essentiellement marine.

Le continent antarctique avec une surface de 14.000.000 km², une longueur de côtes de 32.000 km environ, une altitude moyenne de 2.175 m des montagnes dépassant 6.000 m (fig. 1), une calotte glaciaire dont l'épaisseur atteint parfois 4.000 m, a une forme grossièrement circulaire profondément échancrée par la Mer de Ross et la Mer de Weddell et pousse vers l'Amérique du Sud la péninsule de la Terre de Graham prolongée par les îles de l'arc de la Scotia. Il porte l'appareil glaciaire le plus formidable du globe. A la fin de l'hiver, la ceinture de glaces flottantes entourant le continent jusqu'au 60me parallèle environ double la surface continentale (ALT, 1960) et l'on a pu estimer que 91% des glaces de notre planète appartiennent à l'Antarctique (BAUER, 1961).

Parmi les 17 espèces de Manchots ne peuvent être considérées comme véritablement antarctiques que les espèces vivant et se reproduisant sur les côtes du continent ou des îles avoisinantes. Le Manchot empereur (Aptenodytes forsteri) non seulement se reproduit dans les différents secteur de la côte antarctique et jusqu'aux hautes latitudes du fond de la mer de Ross, mais encore séjourne le long de la côte en hiver de mars à janvier et ne dépasse pratiquement jamais la limite nord des glaces. Il est le plus spécifiquement antarctique des Manchots et le seul homéotherme aérien subissant l'hiver antarctique.

Le Manchot Adélie (Pygoscelis adeliae) se reproduit également sur la totalité des côtes rocheuses antarctiques avec accès possible, mais ne passe que l'été au bord du continent, émigrant sur les glaces plus au nord pendant l'hiver tout en restant constamment dans les limites de la région antarctique précédemment définie. Le cas du *Manchot antarctique (Pygoscelis antarctica)* est différent. S'il est certain qu'il possède des colonies intriqués avec celle du Manchot Adélie dans les Orcades du Sud notamment, s'il possède quelques colonies sur le continent lui-même (Péninsule de Palmer), il faut remarquer que ses lieux de nidification se trouvent surtout dans *la partie septentrionale de l'Antarctique Sud-américaine qui peut être considérée comme une région de passage entre l'antarctique et les régions sub-antarctiques.* Le Manchot antarctique n'est rencontré dans d'autres secteurs de la côte continentale qu'à titre de visiteur exceptionnel. Mais son aire de répartition, d'après des travaux récents semble s'étendre actuellement et deux couples ont été observés dans les îles Balleny (SLADEN, 1960).

Le milieu géographique des Manchots antarctiques

L'aire de distribution géographique des Manchots Antarctiques est une zone circum-polaire limitée au sud par la côte du continent. Les Manchots Adélie, les seuls ayant leurs colonies sur la terre ferme, ne s'éloignent pratiquement jamais de la mer où se trouve leur nourriture et les lieux de reproduction sont tous à quelques centaines de mètres du rivage ou à quelques kilomètres au maximum. Ce n'est que tout à fait exceptionnellement que des Manchots isolés ont été rencontrés à 22 km de l'eau libre (GAIN, 1914) et à 97 km à l'intérieur de la Barrière de Ross (WILSON, 1907).

La limite nord est beaucoup moins précise mais elle n'atteint certainement pas la ligne de convergence des eaux antarctiques située entre les 48me et 60me parallèle (fig. 1) ou entre les 50me et 58me parallèle (ALT, 1960) suivant les secteurs et la saison. Malgré le petit nombre d'observations hivernales à la limite des glaces, on peut affirmer que *les Manchots empereurs et Adélies ne s'éloignent jamais de la limite nord des glaces dérivantes et qu'ils apparaissent comme véritablement liés à la banquise antarctique.* Cette zone circumpolaire s'étend donc sur 5 à 6° de latitude dans les secteurs où la côte est au nord du cercle polaire et sur 15 à 18° au niveau des échancrures de la mer de Weddell et de la mer de Ross.

Dans cette vaste région, les Manchots sont en rapport avec trois biotopes particuliers: la côte, les eaux antarctique, les glaces.

La côte

La côte antarctique avec son développement de 32.000 km environ est évidemment très variable à tous points de vue suivant les longitudes.

Très schématiquement l'antarctide de l'est, socle pré-cambrien ayant appartenu au continent de Gondwana, possède surtout des côtes basses avec affleurements rocheux et moraines émergeant par endroits de la calotte glaciaire; la Terre de Graham est une chaîne alpine, prolongement des Andes; la grande cassure de la mer de Ross est bordée de montagnes élevées avec des volcans atteignant 4.000 m (fig. 1). *Mais, du point de vue des biotopes offerts aux Manchots antarctiques, on peut dire que toute la côte présente l'alternance de trois faciès bien individualisés.*

1. Les affleurements rocheux ne représentent qu'une faible partie de la longueur de la côte, sauf en Terre de Graham. De nature géologiques très variable (granits et gneiss du socle pré-cambrien, roches sédimentaires, roches éruptives, etc.....) de pente variable mais souvent faible, d'altitude fréquemment basse le long du rivage, ils forment des côtes rocheuses découpées, rarement des plages de galets ou de sable. Ils constituent le seul accès possible au continent et les seuls lieux de nidification des Manchots Adélies. Les possibilités d'atterrissage, la présence de cailloux pour la construction des nids paraissent les éléments le plus importants. Ces surfaces rocheuses augmentent d'étendue pendant tout le printemps du fait de la fonte et portent souvent le guano de colonisations anciennes. Elles ne sont jamais utilisées par les Manchots empereurs.

2. Les falaises de glace terminaison dans le mer de l'inlandsis continental occupent la plus grande partie de la côte. Le plateau glaciaire avance d'une centaine de mètres par an en moyenne, tombe verticalement dans l'eau en été et sur la glace de mer en hiver d'une hauteur variant de quelques mètres à plusieurs dizaines de mètres et libère les grands icebergs tabulaires caractéristiques de l'Antarctique. Cette côte de glace marque parfois la limite nord des immenses langues de glaces (ice-shelf) comme celle de la mer de Ross, ou les langues de glace Larsen, Filchner etc. ... Toute cette partie de la côte est inaccessible aux Manchots mais les falaises fournissent souvent un abri du vent pour les colonies de Manchots empereurs établies sur la glace de mer à leur pied.

3. Les déversoirs des glaciers, soit glaciers de montagne comme en Terre de Graham, soit surtout glaciers de l'inlandsis se présentent comme un enchevêtrement d'icebergs plus ou moins détachés, flottants ou échoués, emprisonnés plus ou moins longtemps dans la banquise côtière, possédant toujours entre eux des passes, des défilés, que les Manchots peuvent franchir. Ces déversoirs ne sont évidemment pas habitables pour les Manchots, mais la présence et

554

les mouvements des grands icebergs provoquent des fissures dans la banquise, des rivières et des zones d'eau libre relativement rapprochées de la côte pendant l'hiver, donc des régions de pêche possible. Et l'on observe que beaucoup de colonies de Manchots empereurs sont situées à proximité des déversoirs des glaciers antarctiques.

Les eaux antarctiques

Les premiers éléments ayant conduit à la connaissance des eaux antarctiques et à la découverte de la *"ligne de convergence"* (fig. 1) datent des mesures de DUMONT D'URVILLE et ROSS en 1840, mettant en évidence une "zone circulaire de température uniforme" des eaux de l'océan antarctique. La description si vivante de D. JOHN en 1934, "c'était comme si, en un pas, on passait de l'hiver au printemps . . .", les études d'océanographie biologiques bien résumées par MURPHY (1936) permettant de reconnaitre immédiatement la ligne de convergence par les espèces ramenées par les filets, ont bien montré ce qu'était *cette frontière entre les eaux antarctiques et les eaux sub-antarctiques de surface.* Mais la conception classique simple (les eaux antarctiques froides s'enfoncent sous les eaux sub-antarctiques plus chaudes et plus légères) parait devoir être remaniée et compliquée par l'intervention des courants, de certains mélanges des masses d'eau, de mouvements des masses d'eau le long d'un "front polaire océanique". Toutefois, cette ligne reste une frontière biologique certaine et sa position sur les différents méridiens parait remarquablement stable au long des années.

Entre la ligne de convergence et la côte, DEACON (1959) décrit, dans le secteur atlantique, la superposition de trois masses d'eau différentes par leur température, leur salinité, leur densité: une masse antarctique de surface de salinité inférieure à 34,51‰ coulant vers le nord jusqu'à la convergence; un courant chaud (plusieurs degrés plus chaud) au-dessous d'elle se dirigeant vers le sud; un courant froid antarctique au-dessous roulant sur le fond vers le nord, de salinité supérieure à 34,51‰. Ce schéma est modifié par le profil du fond, l'apport d'eau douce, le régime des vents et les mesures sont donc variables avec la position géographique, la profondeur, la saison.

Mais, de la côte à la ligne de convergence, *les eaux antarctiques de surface sont caractérisées avant tout par leur température voisine de zéro quelle que soit la saison.* La fusion des glaces immergées joue le rôle de régulateur thermique (et par ailleurs les glaces continentales constituent un apport d'eau douce). La première expédition française a enregistré les chiffres suivants devant la côte de Terre Adélie, pour les eaux de surface:

limite nord de la banquise, eau libre, au sud du 64me parallèle,

entre le 23 décembre 1949 et le 5 janvier 1950: température entre
—1,9 et —1,0 et salinité entre 33,5 et 34,1 g.

banquise dérivante entre le 6 et le 24 février 1949: température
entre —1,8 et +1,0; salinité entre 33,0 et 33,8 g.

cuvette côtière, été 1949—1950: température entre —0,95 et
+1,9.

eau le long de la côte: —1,9 en été; —2,0 en hiver dans les trous
de pêche creusés dans la glace de mer.

D'une manière générale, entre la ligne de convergence et la côte,
les eaux antarctiques sont froides variant en hiver de 0,5 à la limite
nord jusqu'à — 1,8 près de la côte, en été de 3,5 à —1° et de salinité
plutôt faible tout au long de l'année, pouvant descendre à 33,00‰ en
été, plus élevée en hiver quand l'apport d'eau douce diminue. Ces
eaux sont extrêmement riches en sels minéraux nutritifs: nitrites et
nitrates atteignant 0,055 g par m³, phosphates toujours au-dessus
de 0,050 g par m³, oxygène à 90—95% de la saturation en hiver
avec fréquente sursaturation en été (MURPHY, 1936). Les conditions
sont donc très favorables pour le plancton.

Les glaces

Ce troisième biotope prend une importance toute particulière
pour les Manchots, non seulement parce que les Sphéniscidés antarctiques y passent une partie très importante de leur vie (et pour les
Manchots empereurs leur période de reproduction toute entière)
mais aussi parce que *les glaces commandent les possibilités de passage
entre l'air où les Manchots respirent et l'eau où les Manchots trouvent
leur nourriture.*

Les glaces de terre ou d'eau douce ont une origine continentale,
soit icebergs généralement tabulaires venant de l'inlandsis des
déversoirs et des langues de glace, soit icebergs venant des glaciers
de montagne. De dimensions souvent colossales avec des dizaines
de kilomètres de longueur, emportés par la dérive générale, autour
du continent, immobilisés dans la banquise solide de l'hiver, reprenant leur voyage dans la banquise lâche de l'été, échoués pendant
des mois ou des années, puis flottants à nouveau, sculptés, déformés,
déséquilibrés et basculés par l'érosion et la fonte aériennes et sous-
marines, chargés de neige fraîche, fissurés, éclatés et morcelés,
entassés dans les baies puis emportés vers le nord, ils finissent une
vie sans doute très longue en blocs de plus en plus minuscules.

Les glaces de mer proviennent de la banquise côtière qui se forme
pendant les périodes de calme même en été sous les hautes latitudes
et qui atteint 1 m à 1,50 m d'épaisseur dès le milieu de l'hiver,
constituant d'immenses champs de glace attachés à la côte. Les
chutes de neige ou la glace déposée par les blizzards y ajoutent des
couches de glace douce, et l'ensemble est très hétérogène avec des

556

lits de sel, des couches douces, du névé, de l'air. Brisés par la débâcle, les énormes champs de glace dérivent vers le nord, se fragmentent, sont repris et remaniés dans la banquise des hivers suivants et deviennent finalement les petites tables de glace, puis les glaçons de la limite nord des glaces.

Dans la banquise dérivante, les glaces des deux origines sont mélangées en un chaos inextricable. Ces formations complexes et constamment transformées suivent la dérive générale vers l'ouest, sauf en Mer de Weddell et Mer de Bellinghausen où la dérive semble se faire vers l'est près du rivage et le long des côtes de la Terre de Graham où elle se fait vers le nord.

Mais, du point de vue de l'écologie des Manchots, ce qui est beaucoup plus important que l'origine ou l'organisation des glaces, c'est leur cycle annuel.

Pendant l'hiver, sur la presque totalité de la longueur de la côte *existe une banquise soudée* s'articulant avec la côte par une "zone de pression" crevassée et tourmentée mobilisée par les marées et formant une barrière entre l'air et l'eau. L'étendue de cette banquise fixée qui peut remonter jusqu'au 60me parallèle et doubler la surface de l'Antarctique à la fin de l'hiver (ALT, 1960) n'est interrompue que par des fissures souvent très étroites et colmatées par le regel, par les douves d'eau libre ou de glace mince autour des icebergs et les rivières ou flaques qu'entretiennent les mouvements des déversoirs des glaciers. Il semble bien qu'en certains points de la côte, pour des raisons encore mal précisées (déversoirs de glaciers avec icebergs mobiles, courants marins, etc . . .) des régions de glace mince et de zones d'eau libre existent d'une manière quasi permanente. La plupart des colonies de Manchots empereurs actuellement connues sont établies au voisinage de telles régions privilégiées (PRÉVOST, 1951).

La banquise soudée forme la plaine immense où s'effectuent la migration annuelle des Manchots Adélies, la reproduction hivernale des Manchots empereurs et leurs voyages "alimentaires" d'hiver et de printemps. Mais elle n'offre que de très rares accès à l'eau et donc à la nourriture: les Manchots Adélies ne s'aventurent jamais sous la banquise soudée et il semble que les Manchots empereurs ne voyagent qu'assez exceptionnellement sous la glace.

Pendant l'été il s'agit d'un autre monde. L'eau est libre devant la côte, portant quelques champs de glace dérivants et des surfaces variables de banquise fracassée et perméable à distance du rivage. Les Manchots Adélies peuvent pêcher au pied de leurs colonies, les Manchots empereurs et leurs poussins partent vers le nord avec la banquise morcelée.

La durée des deux périodes opposées est variable suivant les secteurs et suivant les années:

Terre Adélie, Port-Martin 1950 et 1951, prise de la glace fin mars, débâcle début janvier.

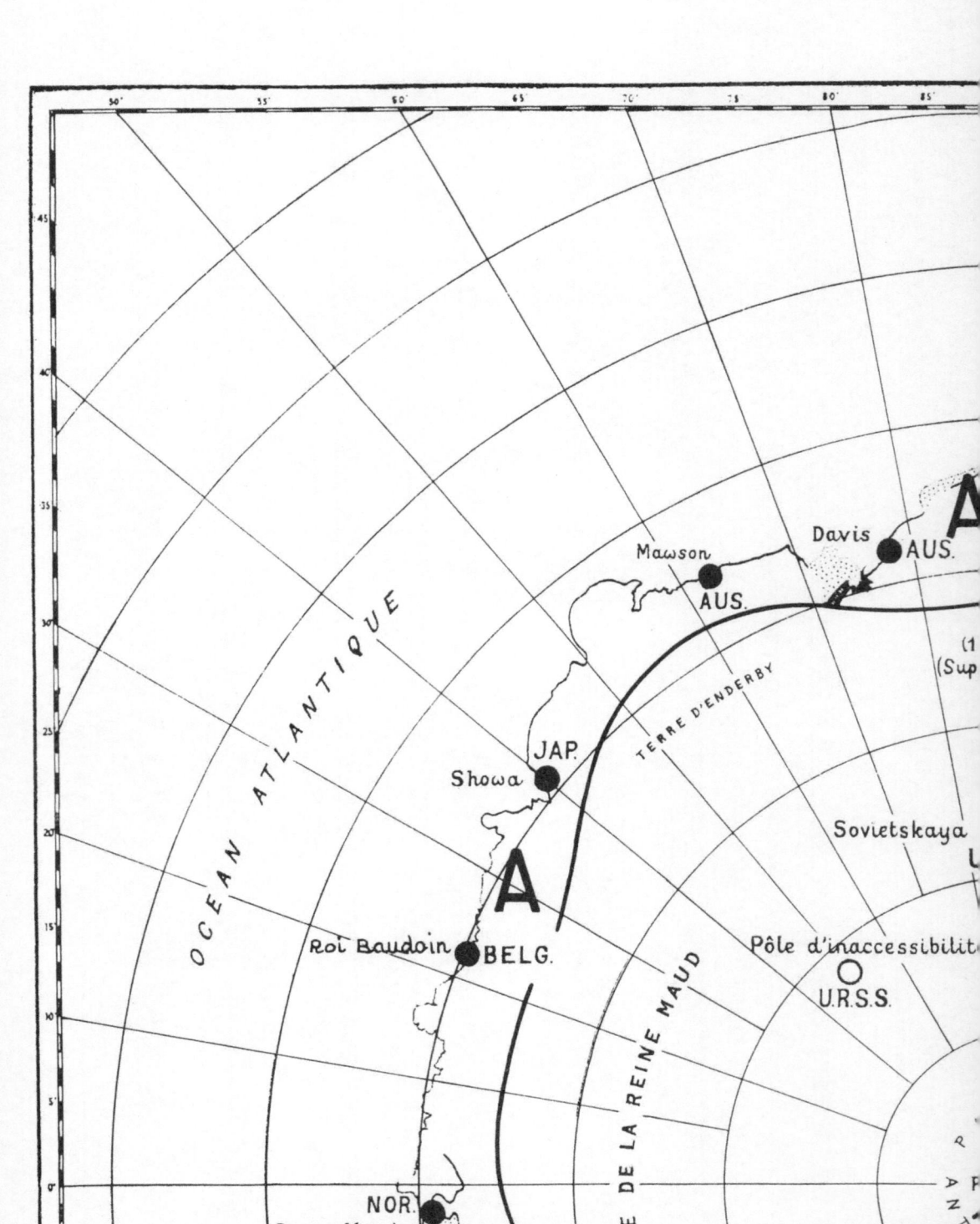

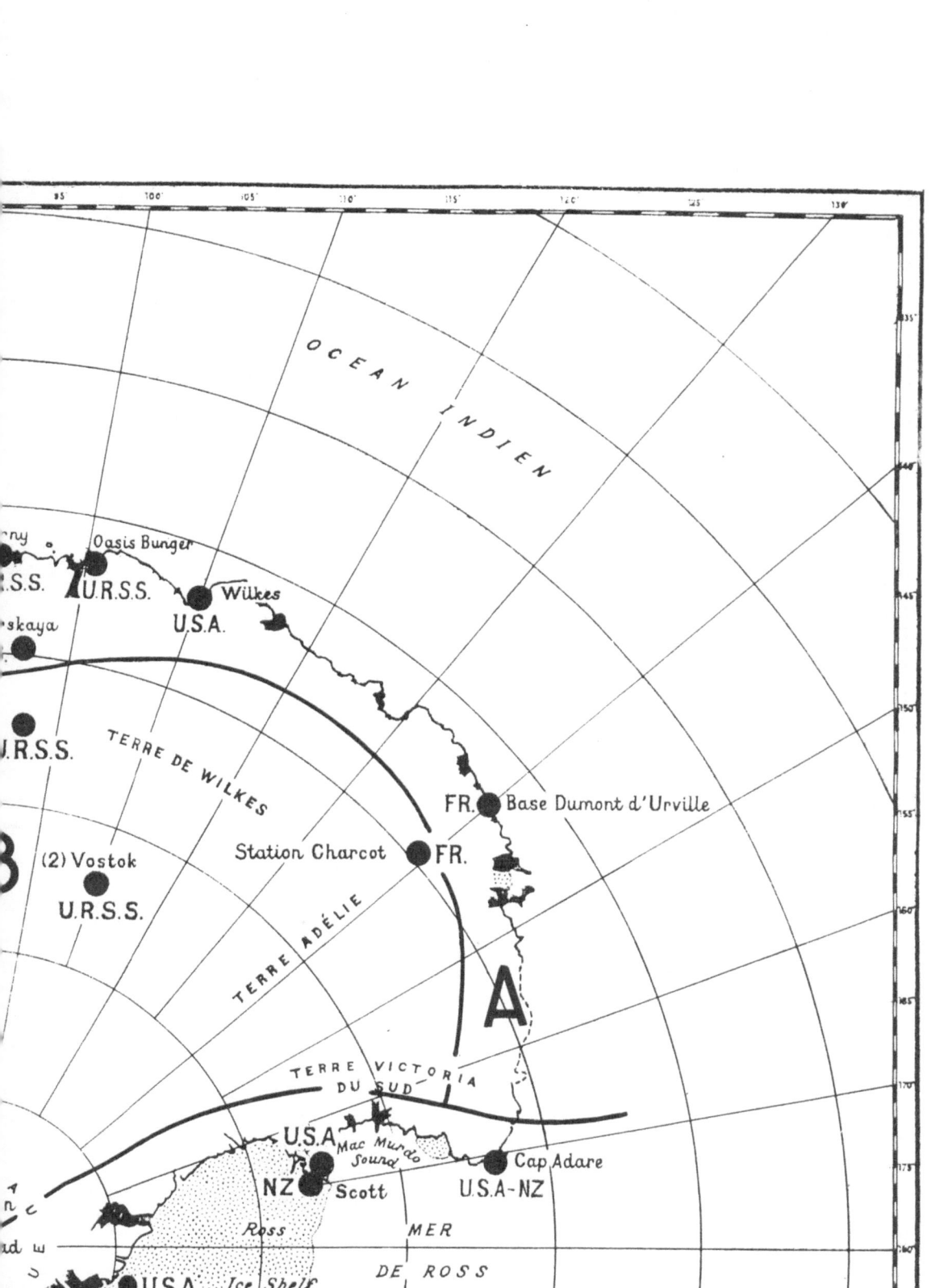

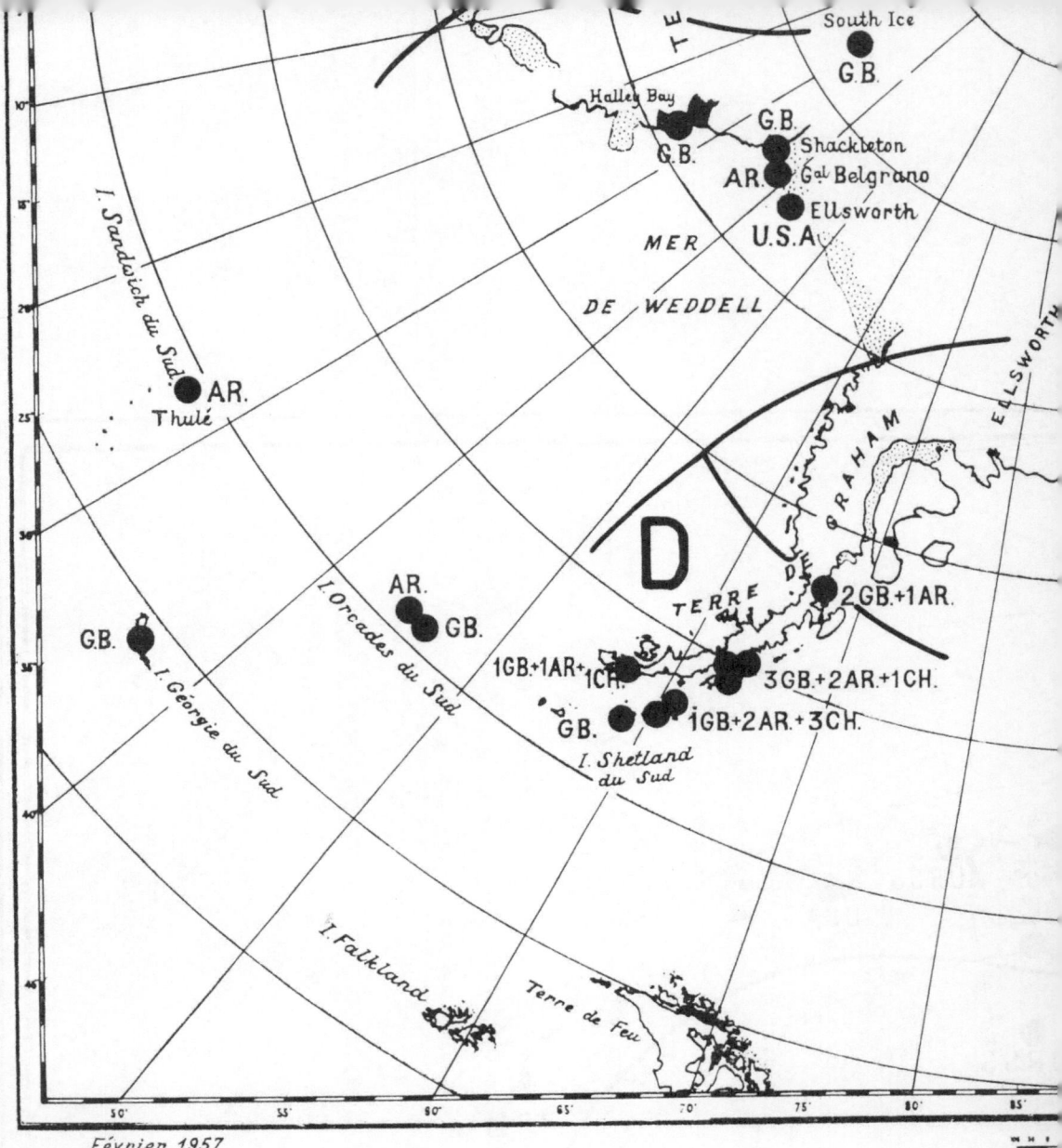

South Ice
G.B.

Halley Bay
G.B.
G.B.
Shackleton
AR. G͏al Belgrano
Ellsworth
U.S.A

MER

DE WEDDELL

I. Sandwich du Sud

AR.
Thulé

D

TERRE DE GRAHAM

2GB.+1AR.

AR.
GB.

1GB.+1AR.+1CH.

3GB.+2AR.+1CH.

G.B.

GB.

1GB.+2AR.+3CH.

I. Shetland du Sud

I. Orcades du Sud

G.B.

I. Géorgie du Sud

I. Falkland

Terre de Feu

Février 1957

Revue et corrigée en Avril 1958

Fig. 1b. Stations de l'Anné

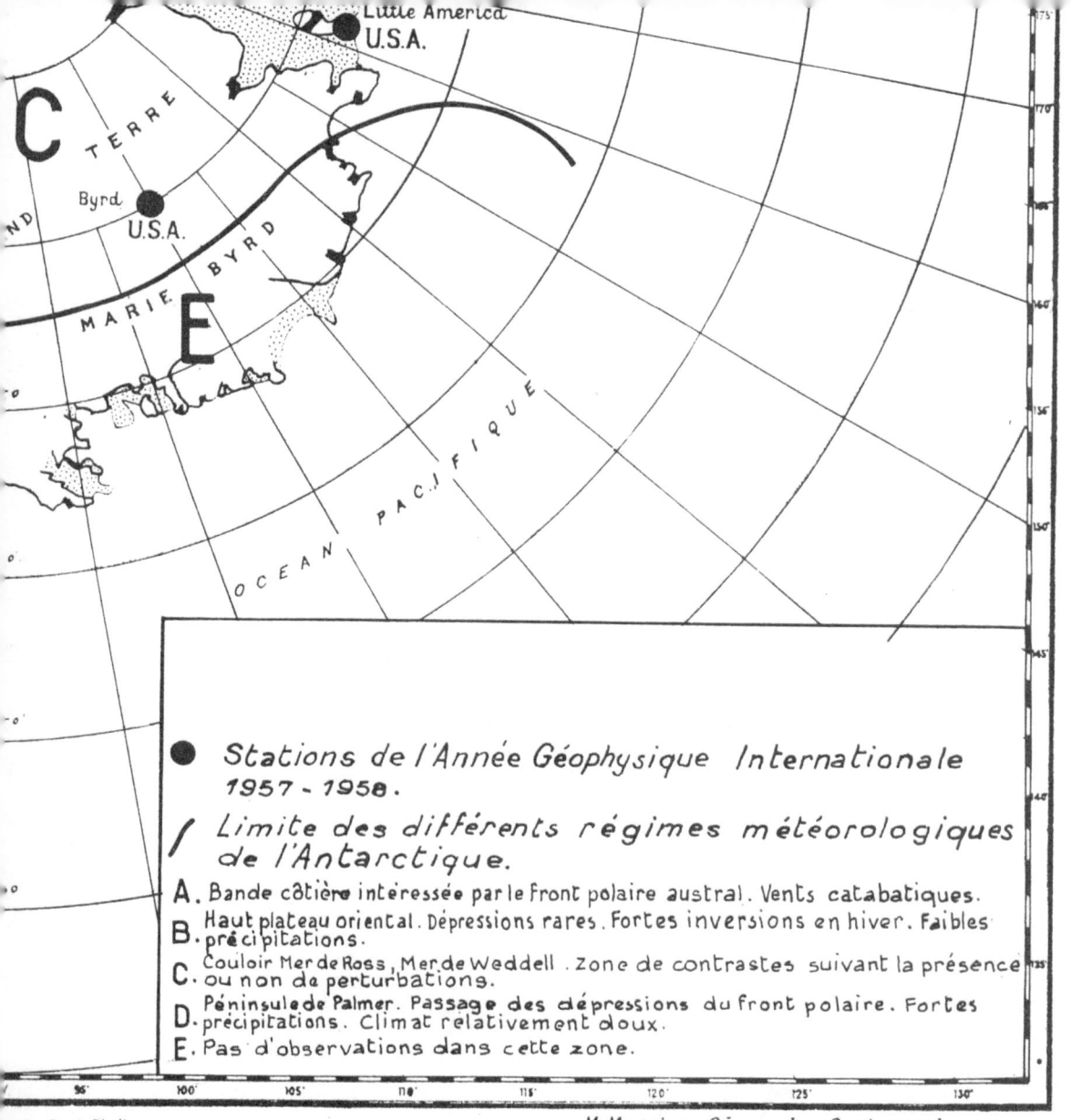

Stations de l'Année Géophysique Internationale 1957 - 1958.

Limite des différents régimes météorologiques de l'Antarctique.

A. Bande côtière intéressée par le front polaire austral. Vents catabatiques.

B. Haut plateau oriental. Dépressions rares. Fortes inversions en hiver. Faibles précipitations.

C. Couloir Mer de Ross, Mer de Weddell. Zone de contrastes suivant la présence ou non de perturbations.

D. Péninsule de Palmer. Passage des dépressions du front polaire. Fortes précipitations. Climat relativement doux.

E. Pas d'observations dans cette zone.

M. Mauvier, Géographe - Cartographe
EXPEDITIONS POLAIRES FRANÇAISES

sique Internationale 1957–1958

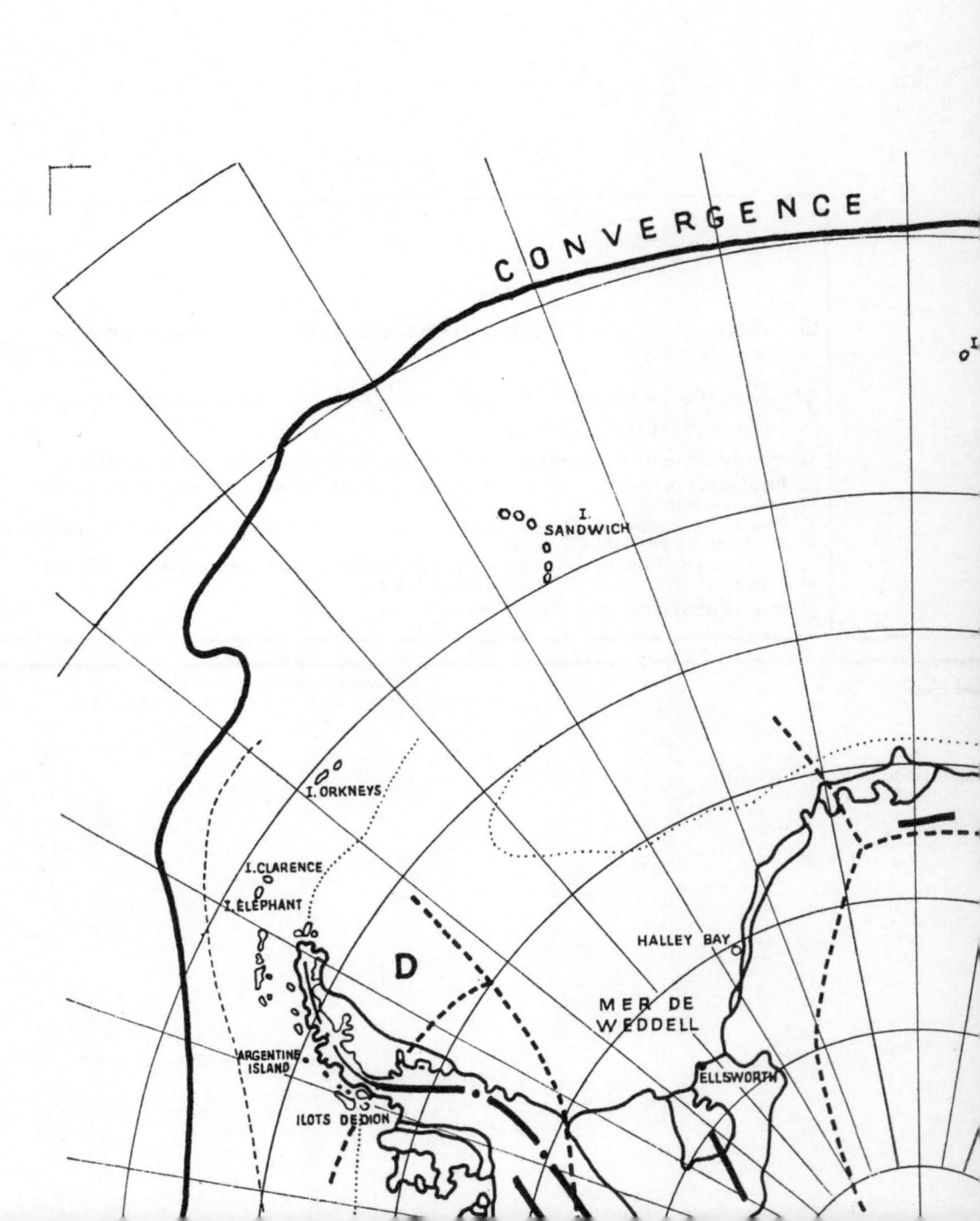

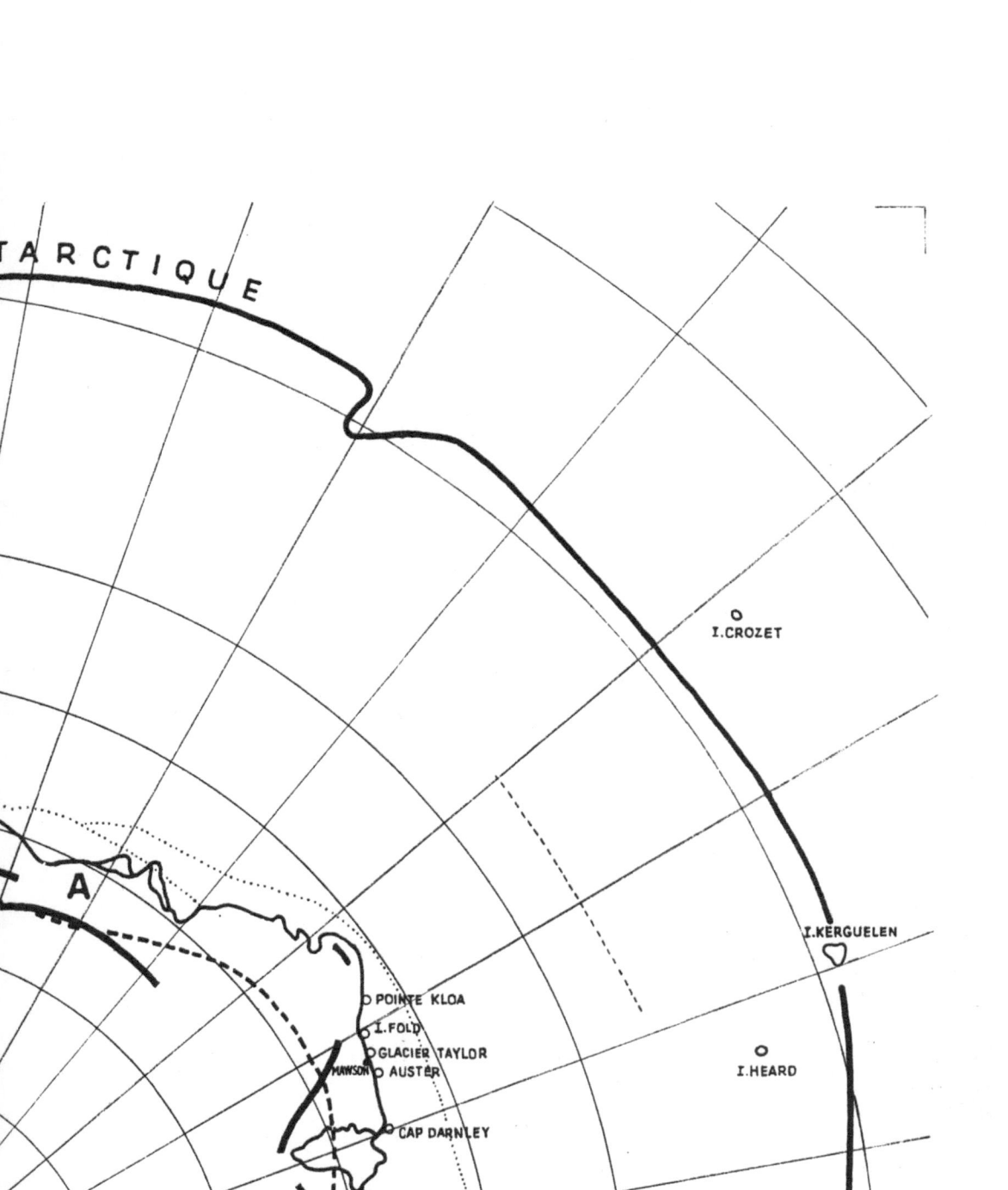

TARCTIQUE

A

I.CROZET

I.KERGUELEN

POINTE KLOA
I.FOLD
GLACIER TAYLOR
MAWSON AUSTER

I.HEARD

CAP DARNLEY

BAIE AMANDA
DAVIS

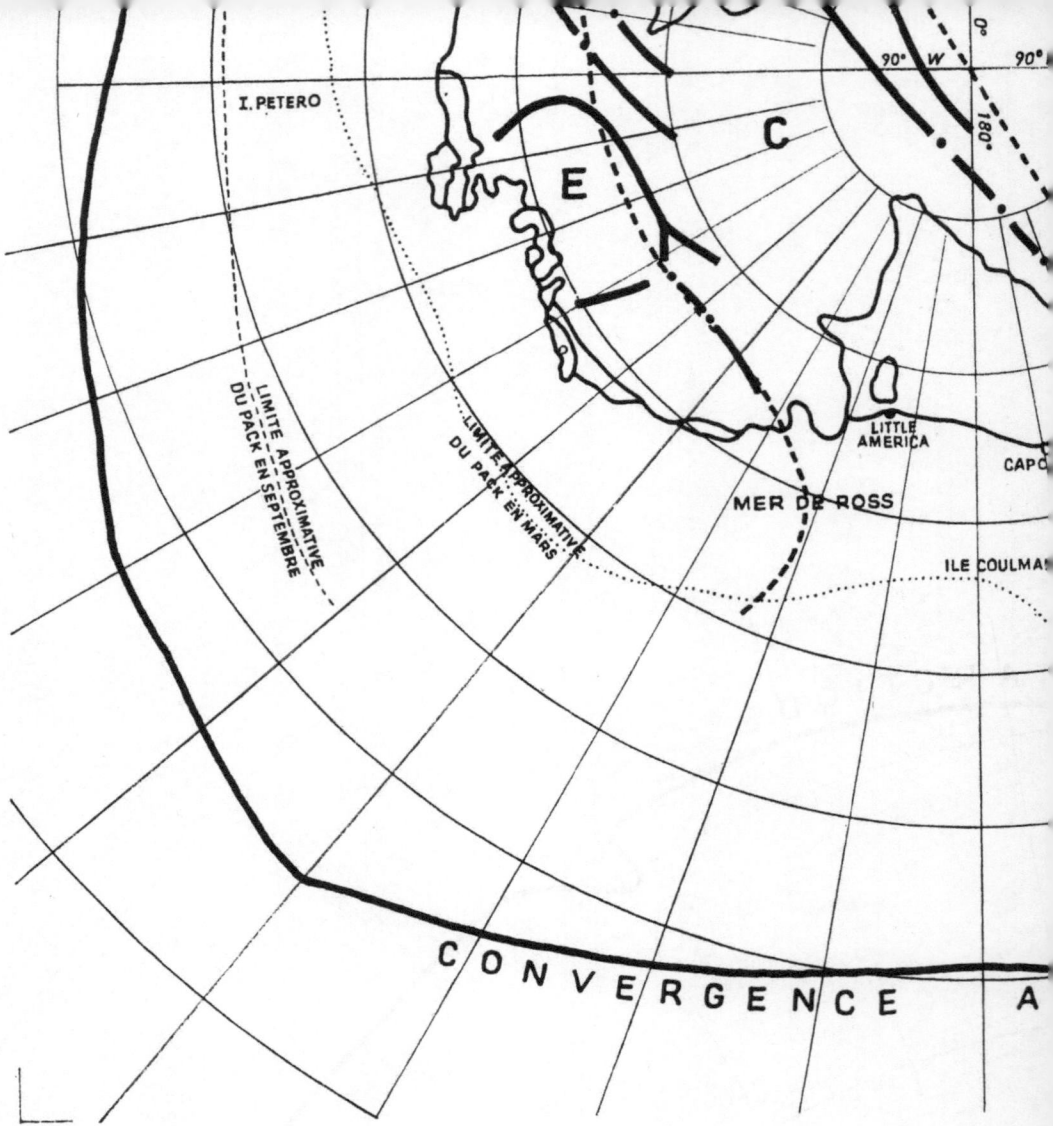

I. PETERO

E

C

90° W   90°
180°
0°

LITTLE
AMERICA

CAPO

LIMITE APPROXIMATIVE
DU PACK EN SEPTEMBRE

LIMITE APPROXIMATIVE
DU PACK EN MARS

MER DE ROSS

ILE COULMA

C O N V E R G E N C E   A

Fig. 1a. La dist

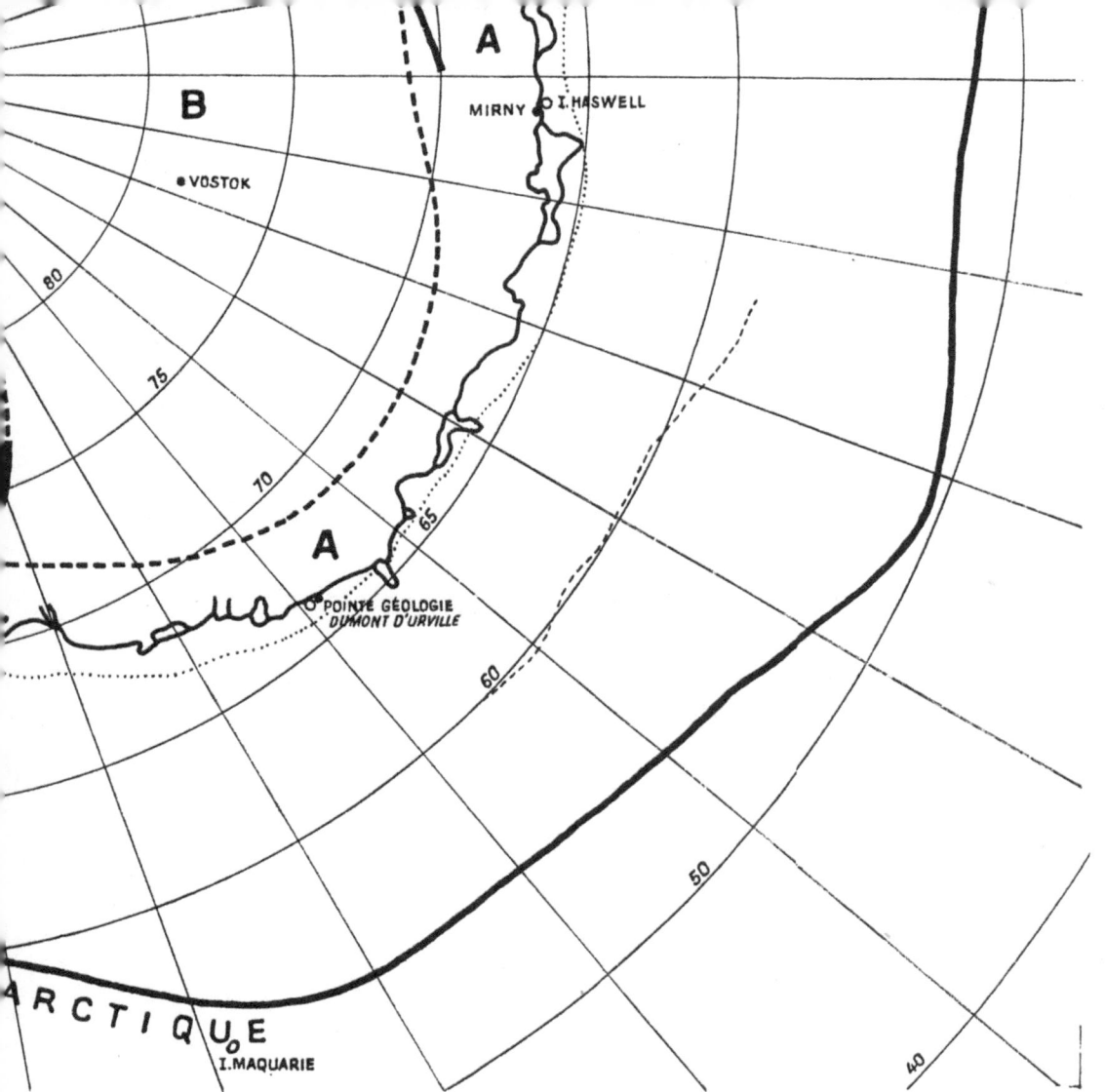

les Manchots empereur

Terre Adélie, Géologie 1952, débâcle en septembre, puis nouvelle banquise disloquée fin décembre.

Terre Adélie, Commonwealth Bay 1911—1913, jamais de glace soudée dans la baie, 1950, glace soudée jusqu'à fin décembre.

Orcades du Sud 1947: glace soudée de juillet à septembre; 1949: glace soudée de mars à fin novembre.

Nord de la Terre de Graham: années fréquentes sans longue période de glace soudée.

La débâcle des glaces peut donc survenir en un même point à des dates très variables. Si elle semble n'offrir que des avantages pour les Manchots Adélies, elle peut être fatale pour les poussins des Manchots empereurs si elle se produit avant la fin de leur développement et les colonies semblent toujours établies dans des régions de glace côtière "stable" protégées par des caps ou des glaciers (PRÉVOST, 1961).

Le climat

Confirmant, précisant et prolongeant les notions déjà acquises, les observations des dernières années permettent de définir un *climat antarctique* bien individualisé et se différenciant schématiquement de son symétrique arctique par:

des températures moyennes inférieures de plus de 10° à celles de la même latitude dans l'Arctique,

l'absence de température vraiment positive pendant l'été,

un hiver "dépourvu de centre" avec des moyennes mensuelles très voisines pendant 6 à 8 mois.

ALT (1960) décrit *cinq régions climatiques*, 3 nous intéressent ici (fig. 1 & fig. 2).

La climat côtier est celui des côtes voisines du cercle polaire entre les méridiens 0 et 170 Est. La température peut dépasser parfois 0 en été et ne descend guère au-dessous de —35° en hiver avec une moyenne annuelle de l'ordre de —10° (Année 1958: Terre Adélie Mx. +5, Min. —36,8 Moyenne annuelle — 11,3; Station Wilkes Max. +8, Min —37,8, Moyenne annuelle —9,7).

Les vents accompagnés de blizzards d'origine surtout catabatique de secteur sud-est sont particulièrement violents et fréquents (fig. 2). Dans certains points de la côte où existent des colonies de Manchots Adélies comme à Port-Martin le blizzard existe pendant 300 jours par an et les moyennes mensuelles et annuelles de vents sont impressionnantes (fig. 2).

Le climat des mers de Ross et de Weddell par des latitudes voisines de 75° est caractérisé par les grands écarts de température et une moyenne annuelle de l'ordre de —20° à —25° (1958 Little America Max. +0,6, Min. —58,3, Moyenne annuelle —23,1; Ellsworth Max. +1,1, Min. —55,6, Moyenne annuelle —23,5), la fréquence des

558

perturbations traversant le continent, des vents beaucoup moins violents et moins fréquents que sur la côte orientale (fig. 2).

Le climat des côtes de la partie nord de la Terre de Graham est relativement doux avec des températures parfois positives en été, une nébulosité forte et des précipitations abondantes, des vents variables.

Si l'on ne considère que le facteur température, tous ces climats sont bien moins sévères que celui des plateaux continentaux où le maximum estival ne dépasse pas —20 et où l'on a enregistré le record du froid sur la Terre: —88,3 à Vostok II le 28 août 1960 (fig. 2).

Mais ces données schématiques, avec la température de l'air sous abri comme élément prépondérant ne rendent pas réellement compte des relations des homéothermes avec les facteurs climatiques. D'une part, *le vent et le blizzard* jouent dans les échanges thermiques des corps chauds dans les ambiances antarctiques le rôle essentiel; d'autre part, les Manchots ne vivent pas dans les conditions étudiées par les météorologistes mais dans *des climats régionaux et des micro-climats bien particuliers.*

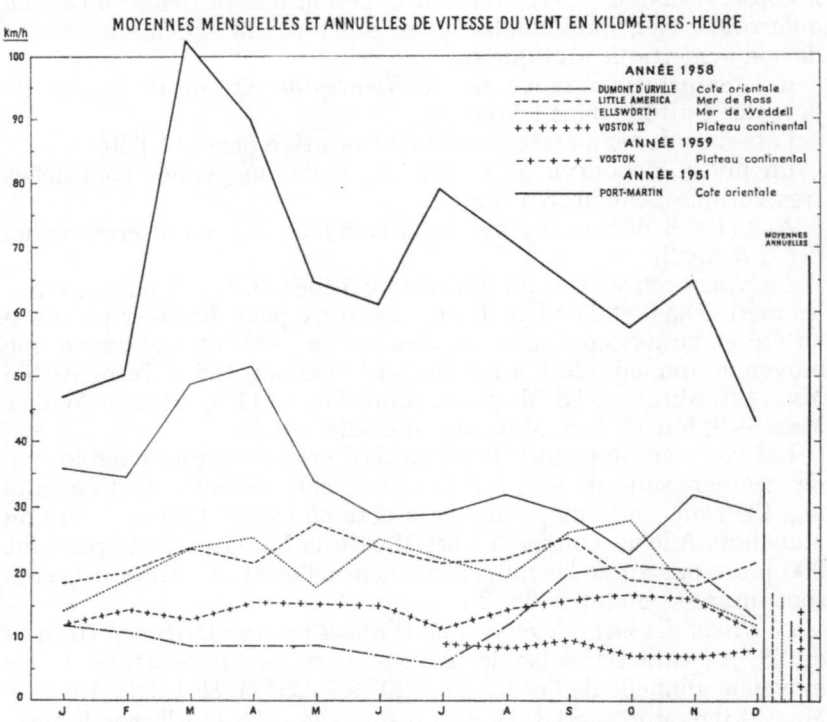

Fig. 2a. Moyennes de vent dans les climats antarctiques.

Le pouvoir de refroidissement des ambiances antarctiques

L'effet thermique de l'ambiance n'est pas défini par la température
de l'air mais par une notion plus complexe le *"pouvoir de refroidissement"* qui, pour un corps donné, à une température donnée, est
défini par "la quantité de chaleur perdue dans l'unité de temps par
unité de surface du corps considéré". Les mesures faites en Terre
Adélie (Sapin–Jaloustre, 1960) (fig. 3) montrent que *le pouvoir
de refroidissement dans l'air calme à —20° est multiplié par 20 dans
un blizzard de 100 kmh, et que dans l'air calme à —180° le pouvoir
de refroidissement est très inférieur à celui d'un grand blizzard avec
une température de —20° et un vent de 140 kmh.* Cet effet thermique
extraordinaire est dû non seulement au vent lui-même mais à la
glace transportée sous forme de balles arrondies de 200 μ de diamètre environ, très fortement chargée électriquement. Loewe
(1956) a pu montrer que, dans un blizzard de 126 kmh, 10 tonnes
de glace était transportée en 1 heure à travers un surface de 1 m²
perpendiculaire à la direction du vent.

La glace transportée double le pouvoir de refroidissement de l'air

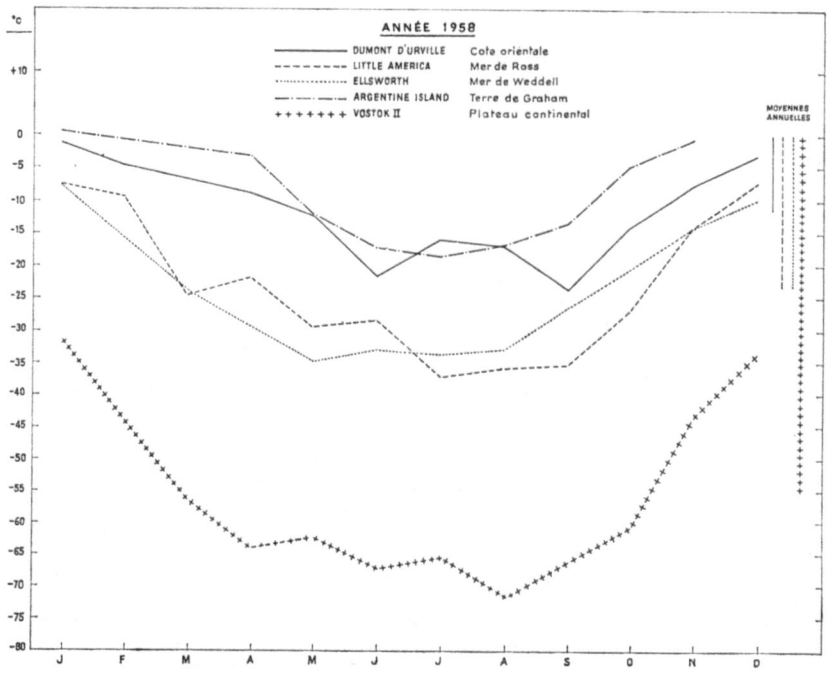

Fig. 2b. Moyennes de température dans les climats antarctiques.

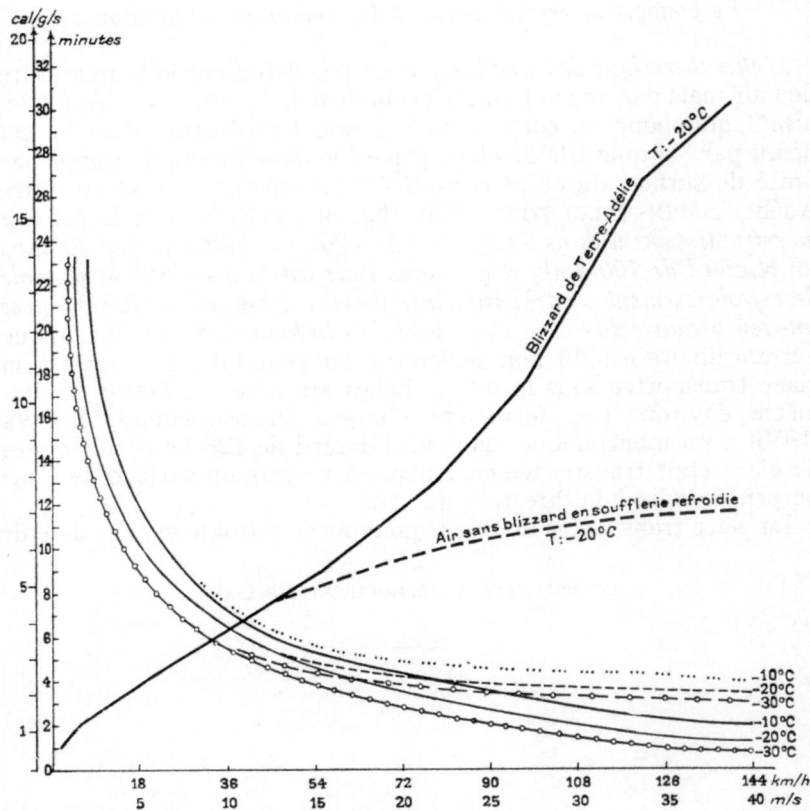

Fig. 3. Mesures du ,,pouvoir de refroidissement'' en fonction du vent, de la température et du blizzard.

Les courbes descendantes représentent le temps de refroidissement de 40 à 20° C de l'appareil utilisé (cylindre métallique rempli d'eau) en fonction de la vitesse du vent et pour les températures de—10°, —20°, —30°. Les courbes en trait plein ont été établies d'après les mesures faites en Terre Adélie dans le blizzard. Les courbes en trait pointillé ont été établies dans l'air sans blizzard de la soufflerie refroidie des établissements Chausson. L'écart entre les deux types de courbes mesure l'effet de la glace transportée par le blizzard. Les temps sont en minutes et secondes, les vitesses de vent en m/s et km/h.

Les courbes ascendantes représentent le ,,pouvoir de refroidissement'' pour l'appareil utilisé et pour la température de —20°, dans le blizzard de Terre Adélie (courbe en trait plein) et dans l'air sans blizzard (courbe en trait pointillé). Le ,,pouvoir de refroidissement'' est ici défini: nombre de cal/g perdues par seconde par l'appareil de mesure. L'appareil perdait 1730 cal/g en passant de 40 à 20°. Les points des courbes de ,,pouvoir de refroidissement'' ont été obtenus en divisant 1730 par le nombre de secondes indiqué, pour chaque vitesse de vent, par les courbes de temps de refroidissement pour la température de —20°. ,,Pouvoir de refroidissement'' en cal/g/s, vent en m/s et km/h.

On voit que le ,,pouvoir de refroidissement'' est de 2,1 cal/g/s pour un vent de 3 m/s; de 4,2 cal/g/s pour un vent de 10 m/s; de 7,7 cal/g/s pour un vent sans blizzard de 40 m/s et de 19,2 cal/g/s pour le même vent de 40 m/s très fortement chargé de glace dans un grand blizzard de Terre Adélie.

sec à même température, dans les mesures effectuées par la première expédition française en Terre Adélie (fig. 3).

Il est donc possible d'affirmer que, *pour les homéothermes soumis aux climats antarctiques, le vent et le blizzard sont les facteurs prépondérants.* De plus, les blizzards font pénétrer de la glace, souvent en quantité impressionnante dans le duvet des poussins, et, dans certaines conditions, dans le plumage des adultes. Enfin, la glace transportée ou la neige, par vent faible, ou dans les endroits abrités sont, par enneigement ou "englacement" des nids ou des poussins, d'importants facteurs de mortalité. Aussi les moyennes de vent sont-elles plus importantes que les moyennes de température et l'on voit (fig. 2) que c'est la côte orientale qui est la plus défavorisée à ce point de vue. Les Manchots antarctiques vivent donc en permanence sur des surfaces à température largement négatives (sauf pour les Manchots Adélies sur les rochers au soleil de l'été), dans les eaux antarctiques à température négative ou voisine de zéro, et dans des atmosphères à "pouvoir de refroidissement" exceptionnellement élevé.

Le climat réel subi par les Manchots

Mais les recherches récentes prouvent que les Manchots antarctiques ne subissent pas réellement les conditions climatiques extrêmes définies par les mesures des stations météorologiques. Ils sont en contact avec *des climats régionaux, des micro-climats naturels, des micro-climats artificiels beaucoup moins sévères.*

Pour les Manchots Adélies qui ne passent sur la côte que la saison d'été, il semble bien que les *climats régionaux* aient peu ou pas d'influence sur le choix d'emplacement des colonies. A Port-Martin (SAPIN-JALOUSTRE, 1960) les colonies occupaient des surfaces rocheuses recevant de plein fouet les vents dominants et les vallées dans l'axe du vent, les îlots plats, les croupes rocheuses quel que soit leur climat régional et il parait en être de même tout autour du continent.

Par contre, les Manchots empereurs qui subissent les conditions extrêmes de l'hiver, sont établis en des points de la côte présentant des conditions climatiques particulières: abri d'un glacier et d'un archipel un peu distant de la côte comme à Pointe Géologie, abri de la falaise continentale comme à Cape Crozier etc.... et PRÉVOST (1961) a fait une analyse opposant les climats régionaux de Port-Martin et de Pointe Géologie, ce dernier échappant largement au vent "catabatique" et fournissant donc aux Manchots empereurs des conditions d'échanges thermiques nettement moins défavorables (fig. 4).

Les micro-climats près du sol dans lesquels vivent réellement en permanence les Manchots antarctiques ont été étudiés en Terre Adélie par SAPIN-JALOUSTRE (1960), en ce qui concerne la température et le vent. Des mesures de température et de vent à des niveaux fixes entre le sol et 150 cm ont été pratiquées dans les "sites" caractéristiques où se déroule la vie des Manchots: glace de mer sans abri, nids de Manchot Adélie abrité par obstacle au vent ou sous le vent, croupe rocheuse sans abri, etc. . . . et ont permis de préciser les gradients verticaux de température et de vent et le "pouvoir de refroidissement" à chaque niveau (fig. 5, 6, 7, 8). Cette étude, qui devrait être reprise avec des moyens plus importants a fourni les résultats suivants:

Dès que les vents deviennent forts — comme c'est le cas dans le climat côtier oriental — *le gradient vertical de température est très faible* et l'on ne trouve que des différences négligeables entre la température près du sol, les différents niveaux et les enregistreurs de la station météorologique.

Le vent, au contraire varie beaucoup avec l'altitude, le relief, l'existence d'un déflecteur en amont ou en aval et il y a toujours une

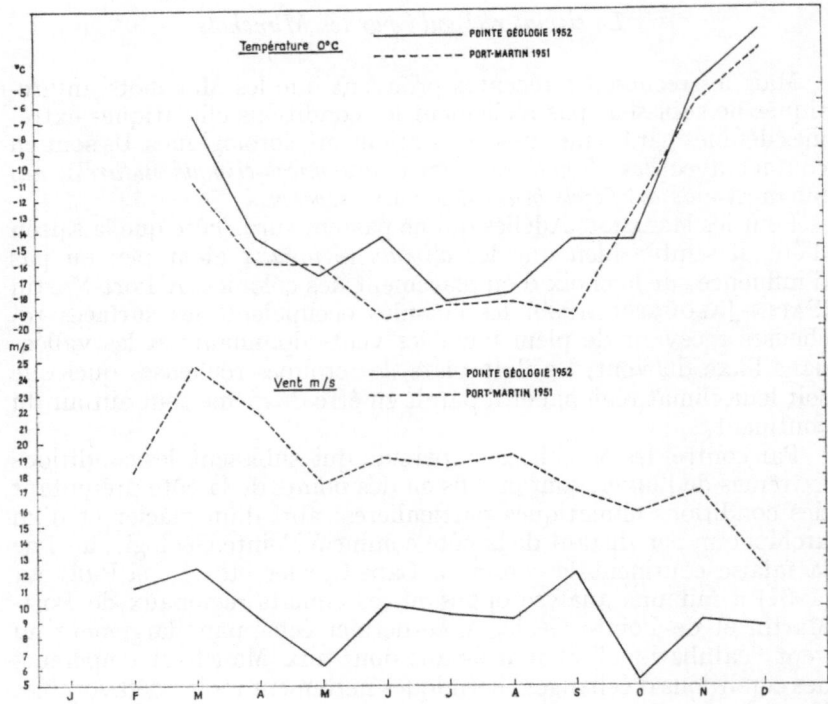

Fig. 4. Moyennes de vent et de température à Pte Géologie et à Port-Martin.

différence importante entre le vent à l'altitude où vivent les Manchots et le vent de la station météorologique. Très schématiquement, le vent entre le sol et 50 cm dans un nid bien abrité est quatre ou cinq fois plus faible que le vent mesuré par les tours météorologiques; sur un champ de glace sans abri, le vent entre le sol et 25 cm est inférieur à la moitié du vent enregistré par la station.

Surtout par le fait du gardient vertical de vent, le *"pouvoir de refroidissement" des micro-climats près du sol est toujours largement inférieur à celui défini par les mesures de la météorologie synoptique.* Pour un Manchot couché au sol sur un champ de glace sans aucun abri, le "pouvoir de refroidissement" est environ la moitié de celui existant au niveau des enregistreurs météorologiques et, dans un endroit bien abrité, il devient le tiers ou le quart du "pouvoir de refroidissement" du climat régional. Il est donc possible d'affirmer que les Manchots antarctiques vivent dans des micro-climats naturels beaucoup moins sévères que les climats régionaux étudiés par les stations météorologiques (fig. 5, 6, 7, 8).

Enfin, les Manchots antarctiques utilisent largement les micro-climats artificiels qu'ils sont capables de créer. Chez le Manchot Adélie il s'agit surtout du *micro-climat "individuel" réalisé pendant* l'incubation des oeufs et des jeunes poussins entre la face dorsale des pattes et le ventre de l'oiseau, au contact de la plaque incubatrice transmettant par la peau nue la chaleur du corps. Ce microclimat artificiel maintient les oeufs entre 20 et 30° en moyenne. Les jeunes poussins du Manchot Adélie, dans les jours qui suivent l'abandon du nid, se groupent par mauvais temps, serrés les uns contre les autres, tête contre tête, chacun bénéficiant de l'abri et de la chaleur des voisins, chacun n'ayant plus qu'une partie de sa surface exposée, chacun bénéficiant *d'un véritable micro-climat artificiel collectif.* Mais, c'est chez le Manchot empereur, dans les conditions très sévères de l'hiver que *le micro-climat artificiel collectif créé par la "formation en tortues"* prend toute son importance. PRÉVOST (1961) a étudié cette méthode collective de diminution de la thermolyse, ses conséquences sur la température centrale, sur l'amaigrissement, au sein de groupements de plusieurs milliers d'oiseaux imbriqués les uns dans les autres en une masse impénétrable au vent.

Ce sont vraisemblablement ces "corrections" successives du climat antarctique qui permettent aux Manchots de résoudre les problèmes énergétiques posés à des homéothermes par les ambiances antarctiques.

Les autres facteurs du climat sont, encore actuellement, mal connus. L'humidité physiologique est certainement constamment très basse du fait de la température. Nous ne savons rien des effets de la pression barométrique, de l'état électrique, etc. . . .

La durée de la photo-période (BOURLIÈRE & PRÉVOST, 1953; PRÉ-
VOST & BOURLIÈRE, 1955) comparée à l'état des gonades montre que
chez le Manchot Adélie comme chez les autres oiseaux antarctiques,
le maximum de développement des glandes sexuelles se place un peu
avant le solstice d'été alors que chez le Manchot empereur au cycle
inverse, ce maximum est atteint avant le solstice d'hiver. Il est
important de remarquer que certaines colonies de Manchot empereur sous hautes latitudes connaissent la nuit polaire complète
pendant une partie de l'hiver (Cape Crozier, Halley Bay, Ile Coulman), mais nous n'en connaissons pas les conséquences pour les

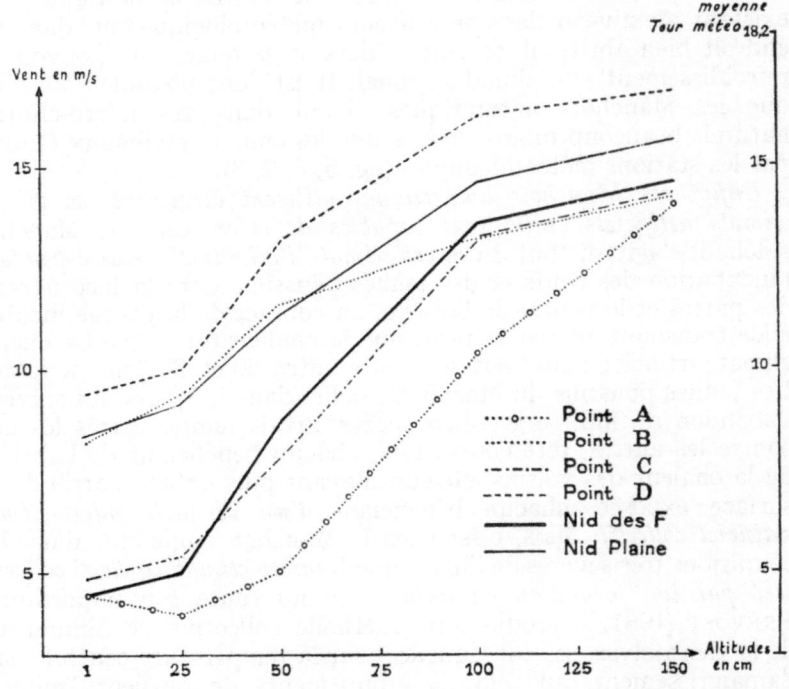

Fig. 5. Microclimats de Terre Adélie. Vitesse du vent en fonction de l'altitude pour
les 6 points caractéristiques. Moyennes de 5 séries de mesures entre le 13 novembre
et le 5 décembre 1950. Vent en m/s, altitude en cm. Différence considérable entre
la courbe du point A bien protégé et celle du point D sans abri. Gradient de vent
toujours important, spécialement dans les endroits abrités. Le vent est le facteur
essentiel des microclimats.

Point A: Creux de rocher très abrité en amont et en aval
Point B: Rocher sans aucun abri
Point C: Rocher avec déflecteur en aval
Point D: Champ de glace sans aucun abri
Nid F: Nid très abrité
Nid Plaine: Nid sans abri

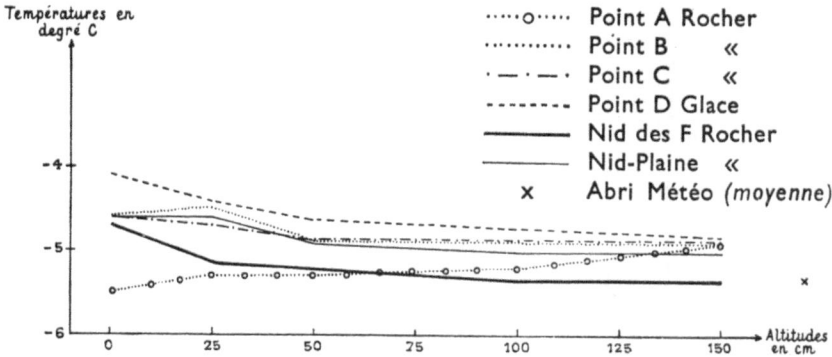

Fig. 6. Microclimats de Terre Adélie. Température de l'air en fonction de l'altitude pour les 6 points caractéristiques. Moyennes de 5 séries de mesures entre le 13 novembre et le 5 décembre 1950. Température en degrés C, altitude en cm. Gradient de température très faible. Différences de quelques dixièmes entre l'altitude de 1 cm et celle de 100 ou 150 cm; température un peu plus haute près du rocher, un peu plus basse près de la glace, par rapport à la température de l'air entre 100 cm et 150 cm. La température joue donc très peu dans les microclimats.

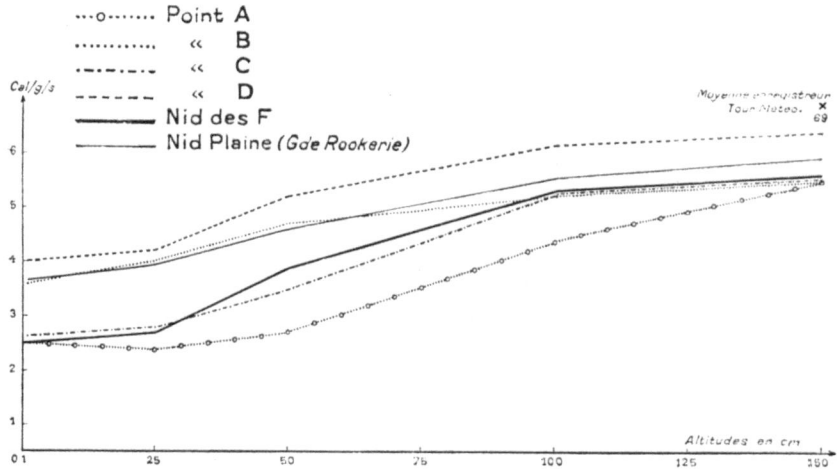

Fig. 7. Microclimats de Terre Adélie. ,,Pouvoir de refroidissement'' en fonction de l'altitude et pour la température de —20°, pour les 6 points caractéristiques. Courbes établies d'après les moyennes de la figure 5 et le courbes du ,,calorimètre déperditeur'' (fig. 3). ,,Pouvoir de refroidissement'' en cal/g/s, altitudes en cm. Le ,,pouvoir de refroidissement'' est de 2,5 cal/g/s en A ou dans le nid du couple F entre le sol et 25 cm quand il est de 7 cal/g/s au niveau des enregistreurs de la Tour météorologique.

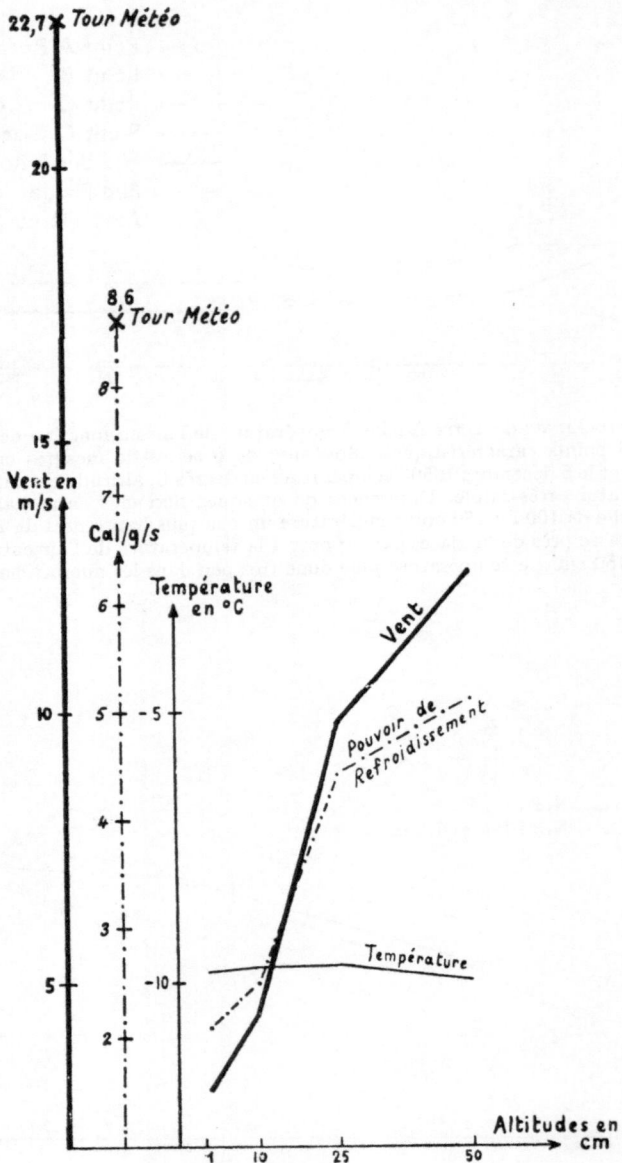

Fig. 8. Au nid du couple F, moyenne des lectures de vitesse de vent en fonction de l'altitude. Moyenne des lectures de température correspondante. Pouvoir de refroidissement pour la température de —20° et pour l'appareil de la figure 3, correspondant aux moyennes de vitesse de vent. Mêmes éléments au niveau des enregistreurs de la Tour météorologique. Altitudes en cm, vent en m/s, température en degré C, ,,pouvoir de refroidissement'' en cal/g/s. La vitesse du vent est de 4,4 m/s à 10 cm au-dessus du sol dans le nid du couple F quand elle est de 22,7 m/s à l'enregistreur de la Station météorologique. Le ,,pouvoir de refroidissement'' est de 2,5 cal/g/s à 10 cm au-dessus du sol dans le nid du couple F quand il est de 8,6 cal/g/s au niveau de l'enregistreur de la Station météorologique.

oiseaux. Par contre, aussi bien pour le Manchot Adélie que pour le Manchot empereur, *le cycle annuel dépend certainement de la latitude, donc du climat et de la photo-période, il est retardé* (parfois d'une manière importante puisqu'il semble y avoir un décalage de deux mois entre Géologie et Cape Crozier, PRÉVOST, 1961) *pour les hautes latitudes et pour les climats sévères.*

Le milieu vivant

Le continent antarctique est recouvert par une énorme calotte de glace dont l'épaisseur va s'amenuisant dans les régions côtières, laissant apparaître çà et là quelques émergences rocheuses. De plus pendant la quasi totalité de l'année une large bande de glace de mer ceinture ses côtes, interdisant leur accès pendant toute la période hivernale. Une autre ceinture, moins apparente sépare peut-être plus encore l'Antarctique des continents avoisinants. Elle est formée à la fois par les vents, la profondeur importante des océans, les courants marins les glaces et icebergs détachés de la côte qui entretiennent la température de l'eau à un niveau assez bas voisin de 0° C.

Cet isolement géographique n'est cependant pas aussi total qu'on pourrait le croire. Des voies de passage sous-marines ou aériennes le relient au régions de latitudes moins basses. L'une d'elle, formée par la Péninsule de Palmer — la seule réellement terrestre — rattache l'Antarctide au continent sud-américain par l'intermédiaire des Iles formant l'Arc de la Scotia. Plus à l'Est, le pont de jonction Kerguelen-Gaussberg dont le rôle terrestre est plus réduit du fait de la rareté des îles qui le forment, reste important de par la faible profondeur des plateaux sous-marins sous-jacents. Enfin le troisième est formé par les hauts-fonds du Sud de la région Néo-zélandaise.

Cette position particulière de l'Antarctique dans un climat apparemment défavorable, ne signifie pas pour autant que toute vie animale est proscrite de cette partie du monde.

L'on est en effet frappé dès l'abord par les immenses ressources de la mer qui contrastent vivement avec l'extrême pauvreté de la vie continentale. Cette dernière, cantonnée dans une bande côtière très étroite, est presque totalement tributaire du milieu marin pour son alimentation. Par contre, le qualificatif de désert paraît convenir tout particulièrement à l'Inlandsis. Bien d'autres régions du globe terrestre qui portent ce nom sont loin de posséder un nombre aussi faible de représentants animaux ou végétaux. Ceci ne veut pas dire pour autant que l'Antarctique soit complètement azoïque. Cette conception, souvent avancée du fait de l'extrême rigueur du climat, ne peut être admise qu'avec les plus expresses réserves. Car la vie paraît exister sous la forme d'organismes les plus simples même au sein des neiges continentales et l'étude systématique de certaines régions intérieures au climat priviligié, découvertes ces dernières années, apportera des précisions sur les organismes qui y végètent.

Enfin le Continent paraît être traversé entre la Mer de Ross et la Mer de Weddell par quelques oiseaux et notamment par des Skuas et peut-être les Pétrels de Wilson.

Le milieu marin

a. la flore océanique est tout d'abord représentée par le *microplancton végétal*, base de la pyramide alimentaire. Sa croissance est étroitement fonction des facteurs physico-chimiques des eaux dans lesquelles il évolue. Les eaux antarctiques océaniques malgré leur basse température, ont une productivité qui est parmi les plus fortes de toutes les mers. Ceci est dû à leur richesse en substances minérales, comme les nitrates et phosphates, en oligo-éléments et en facteurs de croissance telles les vitamines du groupe B. L'importance et la répartition des sédiments marins, leur brassage et leur transport par les courants, influent considérablement sur la distribution du plancton (NEAVERSON, 1934). FOXTON (1956) a montré que la répartition géographique du plancton était irrégulière. Ainsi la mer de Weddell a-t-elle une productivité faible comparée à celles d'autres régions comme le détroit de Gerlache.

De cette observation l'on peut déduire que dans les endroits fermés où l'action de courants est réduite et où une épaisse couche de glace interpose un écran opaque à la lumière solaire, la croissance et le métabolisme du plancton sont considérablement ralentis; par ailleurs, certains auteurs comme HART (1942) ont montré que l'abondance du plancton était maximum à une profondeur allant de 6 à 30 mètres environ, selon la transparence des eaux. La plupart de ces études ont été poursuivies durant les mois d'été après la débâcle des glaces et nous ne possédons que des donnés éparses sur la période hivernale. L'on sait cependant que le plancton est peu abondant entre les mois de mai et octobre (FOXTON, 1956).

Les espèces planctoniques ont pour la plupart une répartition circumpolaire et BAKER (1954) a montré que certaines d'entre elles comme *Corethron criophilum, Choetoceras atlanticum, C. dichoeta* etc ... sont très abondamment représentées, alors que d'autres cantonnées à une zone géographique restreinte sont assez pauvres en effectifs.

En dehors de leur rôle alimentaire, certains organismes végétaux ont un pouvoir anti-bactérien non négligeable. BUNT (1955) et SIEBURTH (1959) ont étudié la flore bactérienne des animaux et ce dernier a montré le pouvoir anti-bactérien de la nourriture des Euphausiidés. L'on sait aussi qu'une algue — le *Phaeocystis* — inhibe plusieurs bactéries notamment *Staphylococcus aureus* et *Escherichia coli.* Ceci souligne l'intérêt et l'ampleur des problèmes biologiques des régions antarctiques.

Enfin parmi les végétaux les algues ne comprennent pas moins de 30 genres et 200 espèces appartenant à la région antarctique auxquelles il faut joindre les espèces déjà connues dans d'autres régions.

b. Le Zooplancton constitue la base primaire du milieu animal marin. La plupart des groupes d'Invertébrés sont représentés. Le plus caractéristique est sans conteste celui des Crustacés Euphausiidés. Les différentes espèces antarctiques de cette famille occupent chacune une zone définie, circum-polaire. Ainsi *E. cristallorophias* est plutôt cantonnée aux eaux continentales froides; elle sert d'aliment aux baleines bleues dans la Mer de Ross (MARR, 1956). *E. superba* qui forme le "Krill" dont s'alimentent la plupart des Mammifères marins, vit dans des eaux voisines de 0°, à moins de 10 mètres de profondeur. Plusieurs auteurs ont montré que les 3 groupes d'âge de cette espèce, ont une richesse en lipide différente et ne sont pas tous présents au même moment pendant la période estivale (RUUD, 1932; PETERS, 1955; NEMOTO, 1959). Enfin *E. triacantha* d'habitat plus septentrional, voisin de la ligne de convergence (BAKER, 1959), est à la fois commune aux régions antarctiques et subantarctiques.

Divers autres Crustacés servent d'aliments aux mammifères marins parmi lesquels *Thysanoessa macrura* et *T. vicina, Calanus acutus* etc . . .

En dehors des Crustacés d'autres Invertébrés sont représentés en assez grand nombre dans les zones côtières littorales. Citons entre autres les Echinodermes et les Mollusques. Ces derniers comprennent plusieurs familles dont beaucoup dérivent d'un petit nombre de familles cosmopolites. Selon POWEL (1951) l'on peut supposer que la plupart des Mollusques des hautes latitudes proviennent de la région côtière Ouest du continent américain. Leur dispersion s'est effectuée à partir de la Péninsule de Palmer dont nous avons déjà relaté le rôle fondamental.

c. Parmi les Vertébrés: Les Poissons ne comptent pas moins de 60 espèces. Dans la région littorale antarctique selon EKMAN (1953) 90% des espèces et 65% des genres sont endémiques, ces pourcentages étant respectivement de 73 et 27% pour les Echinodermes. Cet auteur fait remarquer que cette proportion élevée des espèces endémiques, que l'on observe par ailleurs dans d'autres groupes, souligne l'isolation de la faune de ces régions.

Dans les eaux littorales, la famille des Nototheniidès avec les genres *Notothenia* et *Trematomus* constitue l'essentiel des populations. Les effectifs des autres familles sont beaucoup plus réduits. C'est le cas des Harpagiféridès, des Bathydraconidès et des Chaenichthyidae.

Les très importantes recherches des missions russes de ces dernières années compléteront sans doute largement nos connaissances sur ce sujet. Il est à signaler que ces biologistes ont découvert entre autres des espèces dont le sang ne contient pas de pigments d'où leur nom de "Poissons à sang incolore". (Voir chapitre précédent).

Les Vertébrés aériens dont les espèces sont peu nombreuses, mais en général richement représentées comprennent les Mammifères marins avec les Cétacés et les Pinnipèdes et les Oiseaux appartenant pour la plupart aux Sphéniscidés et aux Procellariens.

Le fait que de nombreux Cétacés à fanons viennent habiter les eaux antarctiques en été est dû pour une grande part à l'abondance et à la vaste répartition des Euphausies et notamment d'*E. Superba* (MACKINTOSH, 1942). PETERS (1955) et NEMOTO (1959) ont montré la préférence de tel ou tel Cétacé pour une dimension de "Krill" déterminé. Leur teneur en lipide, les migrations verticales, la répartition différente des groupes d'âge selon l'époque, influent sur l'alimentation des Baleines. Parmi les plus répandus de ces mammifères marins, citons surtout les balénoptères avec le Rorqual bleu, *B. musculus* et le Rorqual commun *B. physalus* et en moins grand nombre la Baleine Jubarte *Megaptera novae angliae.*

Un Cétacé Odontocète au régime alimentaire presque exclusivement carnivore *l'Orque (Orcinus orca)* espèce cosmopolite, est largement répandu dans les mers antarctiques. C'est un prédateur féroce qui se nourrit de phoques — selon ESCHRICHT (1862) il y en avait pas moins de 14 dans l'estomac d'un de ces Cétacés — de Manchots, de Poissons de grande taille et, selon certains auteurs, il dévorerait même la langue de certaines baleines (*E. gibosus* notamment). L'influence de ce Cétacé carnivore sur l'équilibre de certaines populations de Vertébrés est donc considérable.

4 espèces de *Pinnipèdes* vivent autour du continent antarctique.

Le moins répandu de tous, le *Phoque de Ross (Ommatophoca rossi)* vit probablement en solitaire dans la zone de rupture de la glace de mer. Peu d'observateurs ont eu le privilège de l'apercevoir; l'on pense que son alimentation est surtout à base de Poissons et de Céphalopodes.

Le Phoque Crabier (Lobodon carcinophagus) habite aussi le pack-ice hiver comme été. Cependant les visites de quelques individus à la côte ne sont pas exceptionnelles comme nous avons pu l'observer. Son comportement est assez grégaire et sa nourriture est formée principalement d'Euphausies.

Le Léopard de mer (Hydrurga leptonyx) est assez largement distribué tant dans les mers antarctiques que dans le régions sub-antarctiques. Il paraît toutefois fréquenter régulièrement les hautes latitudes de l'hémisphère sud pendant la période de reproduction des Manchots dont il se nourrit abondamment. Il complète son régime en y ajoutant Poissons et Céphalopodes. Enfin le plus antarctique de tous les Pinnipèdes: *le Phoque de Weddell (Leptony-*

chotes weddelli) vit à la côte sur la glace de mer où dès le mois d'octobre les femelles s'assemblent dans les endroits abrités pour mettre bas. Pendant le plein hiver (mai à août) les visites à la côte sont l'exception et il est alors probablement confiné plus au Nord dans le pack-ice, tout au moins en Terre Adélie.

Les Oiseaux, bien que se reproduisant à la côte, sont tributaires du milieu marin pour leur alimentation. Les Manchots constituent la population la plus caractéristique de cette région et des trois espèces antarctiques, deux sont circumpolaires et réparties sur la côte du continent antarctique, la troisième étant surtout abondante dans la péninsule de Palmer.

Il nous faut aussi citer pour mémoire le Manchot papou *(Pygoscelis papua)*. Bien qu'ayant colonisé le Nord de la Terre de Graham qui constitue la limite méridionale de son aire de répartition, il ne franchit pas le cercle polaire. Aussi est-il préférable de le rattacher à la région dite des îles antarctiques plutôt qu'à la zone antarctique proprement dite.

Le Manchot antarctique (Pygoscelis antarctica), malgré son nom, est le plus septentrional de tous. Cantonné dans le quadrant antarctique américain, il paraît s'étendre progressivement au point qu'il peut être considéré actuellement comme circumpolaire. SLADEN & GOLDSMITH (1960) ont confirmé sa présence aux Iles Balleny et nous avons observé nous-mêmes des visites régulières de quelques individus en Terre Adélie.

Le Manchot Adélie (Pygoscelis adeliae) est le plus répandu des Sphéniscidés antarctiques. Ses colonies souvent fort populeuses occupent la plupart des surfaces rocheuses dégagées de neige pendant la belle saison. Une adaptation imparfaite au climat antarctique hivernal le contraint à séjourner à la limite du pack pendant le plein hiver. Son alimentation peu variée est formée presque exclusivement d'Euphausies. Il paraît difficile d'évaluer l'effectif de ses populations, mais on peut l'estimer à plusieurs centaines de milliers d'oiseaux. Pour le seul Archipel de Pointe-Géologie (Terre Adélie 66°40'S 140°01'E) on en compte 38.000 environ (ISEL, 1958).

Enfin le *Manchot empereur (Aptenodytes forsteri)* le plus extrême de tous les Sphéniscidés, se reproduit en plein hiver sur la glace de mer pré-côtière et dans quelques cas sur la glace continentale. La colonie la plus septentrionale de l'espèce est située en deça du Cercle Polaire (Ile Haswell 66°33'S, 92°50'E) la plus méridionale par 77°29' de latitude Sud, au fond de la mer de Ross (Cap Crozier). La population totale de l'espèce dépasse très largement 200.000 oiseaux répartis en une vingtaine de colonies, la plus peuplée étant celle de l'île de Coulman alors que celle des îlots de Dion qui ne comprend guère plus de 300 oiseaux, est probablement la plus petite.

Outre les Manchots, la côte antarctique est peuplée en été par différentes espèces de Procellariens et par une espèce de Stercoraire: le Skua.

Le Skua (Catharacta maccormicki) se reproduit principalement dans les régions antarctiques au Sud du 65ème parallèle à l'exception de la Péninsule de Palmer où son aire de distribution s'étend quelques degrés plus au Nord. Il est communément répandu le long des côtes antarctiques tout particulièrement à proximité des colonies de Manchots Adélies dont les oeufs et les poussins forment l'essentiel de son alimentation pendant la période de reproduction. Le placenta des femelles de phoques ayant mis bas est aussi pour lui un aliment non négligeable en octobre. Son régime paraît plutôt orienté vers l'ichtyophagie (Poissons et Crustacés) durant le plein hiver au cours duquel il séjourne semble-t-il dans le pack-ice.

Les espèces de Procellariens sont relativement nombreuses:

Le Pétrel géant (Macronectes giganteus) assez bien représenté dans les îles subantarctiques nidifie sur le continent antarctique et la colonie la plus extrême semble être celle de Pointe-Géologie, Terre Adélie (66°40' Sud) (PRÉVOST, 1953). Les premiers oiseaux arrivent en août et les derniers immatures quittent la colonie fin avril. Ce prédateur se nourrit des cadavres d'animaux de toutes espèces, mais il leur préfère semble-t-il le contenu stomacal des poussins de Manchots empereurs (PRÉVOST, 1958). Ceci tend à montrer que son régime est plus ichtyophage que carnivore et ceci tout particulièrement pendant les mois d'hiver.

Le Pétrel des neiges (Pagodroma nivea) peuple de nombreux affleurements rocheux côtiers. Il vit en petites colonies groupées dans les éboulis de rochers. Bien qu'ichtyophage il lui arrive souvent de s'attaquer à des cadavres d'animaux pour se repaître de leur graisse.

Le Damier du cap (Daption capensis) est assez commun. Ses colonies, plus importantes que celles de l'espèce précédente, sont établies sur les versants rocheux exposés aux vents dominants. Il paraît cependant redouter les climats trop extrêmes des hautes latitudes.

Le Fulmar antarctique (Fulmarus glacialoides) quoique distribué tout autour du continent antarctique est moins commun que les précédents. Ses colonies sont installées dans des falaises rocheuses abruptes, bien ensoleillées. Cette exposition présente l'inconvénient d'être à l'origine de fréquents enneigements des nids. Aussi l'oiseau consacre-t-il de longues heures au début du printemps à nettoyer son territoire à l'aide du bec.

Le Pétrel antarctique (Thalassoica antarctica) a une répartition à peu près analogue à celles du précédent. Quelques colonies sont très importantes comme celles du cap Hunter alors que certaines régions comme la Terre Adélie n'en comportent aucune.

Ces 3 espèces ont un régime alimentaire très voisin formé principalement d'Euphausies et accessoirement de Céphalopodes et de Poissons de petite taille et elles se reproduisent à la côte entre les mois de novembre et mars. L'incubation de l'oeuf est longue de 38 à 46 jours environ, et l'élevage du poussin a une durée sensiblement équivalente.

Pour en terminer avec les Procellariens, il nous faut mentionner *le Pétrel de Wilson (Oceanites oceanicus)* de la famille des Hydrobathidès, le plus petit de tous. Distribué dans tout l'antarctique, il niche isolément dans des galeries assez profondes dont les fréquents enneigements sont à l'origine d'une forte mortalité.

D'autres espèces se reproduisent sur le continent antarctique dans la Péninsule de Palmer. Le fait qu'elles sont distribuées dans les îles antarctiques et subantarctiques ne nous semble de nature à les classer dans ces régions géographiques plutôt que de les qualifier réellement d'Antarctiques.

Le milieu terrestre

En opposition à cette vie marine abondante la vie proprement terrestre limitée le plus souvent aux émergences rocheuses côtières est assez réduite.

A l'exception du Nord la Péninsule de Palmer où les conditions climatiques sont, du fait de la latitude, sensiblement différentes, la vie terrestre est exclusivement représentée par des Algues, des Lichens et des Mousses.

En dehors des Diatomées collectées dans les eaux douces du Continent pendant l'été (CARLSON, 1913; FRITSCH, 1917) de nombreuses Chlorophycées et Myxophycées ont été trouvées principalement dans le secteur de la Mer de Ross. Des découvertes plus récentes ont montré le développement de nombreuses algues d'eau douce au cours de l'été. Parmi elles, citons *Nostoc commune, Prasiola crispa, Phormidium autumnale.* Le nombre des espèces de lichens actuellement connus est faible, alors que la Péninsule de Palmer en compte à elle seule 218. Cela est dû à la latitude élevée mais aussi à ce que la position géographique de ce territoire a favorisé l'installation de nombreuses missions et facilité les recherches systématiques. A titre de comparaison, l'inventaire actuel de la Terre Adélie ne fait état que de 7 espèces. Selon BURKHOLDER (1961) une espèce de lichen du genre *Ramalina* aurait une activité antibiotique et inhiberait les bactéries Gram-négatif.

Les Mousses n'ont été étudiées que dans certaines régions antarctiques. La Péninsule de Palmer en récèle le plus grand nombre, mais des espèces comme *Ceratodon purpureus, Distichium capillaceum, Sarconeurum glaciale, Bryum antarcticum* par exemple figure parmi l'inventaire de régions continentales plus méridionales que la précédente et leur répartition paraît assez large.

Le règne animal terrestre ne comprend que des Arthropodes terrestres couramment répandus dans la péninsule de Palmer. L'on ne connaît par contre pas de relations de leur présence dans les latitudes plus basses de la côte antarctique. Selon SAPIN-JALOUSTRE (1960), ils sont totalement absents à Port-Martin et nous n'en avons pas trouvé nous-mêmes à Pointe-Géologie. Ces Arthropodes vivent dans les Mousses et les Lichens des régions les plus clémentes. Ce sont des Diptères (Padurellae) et des Collemboles représentés par 9 genres. Il faudra attendre de plus amples recherches dans ce domaine pour avoir un aperçu de la distribution de ces espèces.

Exception faite des formes de la vie animale ou végétale côtière, on peut considérer que l'intérieur du Continent, l'Inlandsis, comprend que des formes de vies très élémentaires, limitées tant par leur nombre que par leur répartition.

On peut donc conclure que la calotte glaciaire antarctique balayée par la neige et le vent est à peu près azoïque alors que les mers avoisinantes abritent une vie intense dont dépendent étroitement les Vertébrés aériens qui viennent se reproduire à la côte.

LE MANCHOT ADELIE

A quelle époque le Manchot Adélie a-t-il colonisé la côte du continent antarctique? A Cape Hallet, dans la Mer de Ross, HARRINGTON & McKELLAR (1958) ont pu dater par le radio-carbone un corps de Manchot Adélie paraissant appartenir à la première colonisation en ce point, puisque reposant sur des couches sans débris organiques, et trouvé la date de 550 de notre ère. Mais ne pouvant que formuler des hypothèses sur la phylogénèse des Sphénisciformes, sans chaînons intermédiaires entre les fossiles du Miocène et les Manchots modernes, nous ne savons évidemment pas si les Manchots antarctiques ont toujours vécu sous les hautes latitudes dans les conditions écologiques actuelles, ou si, venus du nord, ils ont réussi à s'y adapter.

L'Espèce

Certainement observé par FORSTER (1777) au cours du grand voyage circum-polaire de COOK (1773—1775), le Manchot Adélie fut décrit comme espèce distincte par HOMBRON & JACQUINOT (1841) naturalistes de l'expédition DUMONT D'URVILLE (1837—1840) et malgré les remarques de FALLA (1937) et de ROBERTS (1940) qui réclament pour lui le genre *Pucheramphus*, la plupart des auteurs lui conservent avec Sladen (1958) le nom de *Pygoscelis adeliae* et sa place parmi les Pygoscélidés.

Les caractéristiques générales de l'oiseau

Les caractéristiques générales de l'oiseau sont bien connues: poids variant de 4 à 6 kg 5, longueur entre 55 et 75 cm, envergure entre 45 et 55 cm, bec entre 30 et 45 mm, plumage spécial des Sphénisciformes recouvrant un duvet épais, dos du corps et des ailerons noir, tête noire, menton noir, face ventrale du corps et des ailerons blanches, dos des pattes gris rosé clair, bec brun rouge, paupières blanches, queue à 14 plumes. L'immature est immédiatement distingué par son menton blanc et ses paupières noires.

Cette morphologie est non seulement adaptée à la vie marine qui est celle de tous les Manchots (forme profilée en "torpille", plumage serré imperméable à l'eau, forme et musculature des ailerons utilisés comme nageoires, etc . . .) mais aussi au milieu polaire. La surface est faible par rapport au poids ce qui diminue les pertes de chaleur. Les narines protégées par des plumes contre l'entrée de la glace s'opposent à celle du *Pygoscelis papua* (Murphy, 1936), vivant plus au nord. L'importance du pannicule adipeux sous-cutané (qui est un facteur d'isolement thermique) et de la graisse péritonéale permettent les longs jeunes de printemps nécessités par la présence de la glace de mer devant la côte. Les tissus plantaires des pattes sont capables de supporter le contact indéfini avec des surfaces à des températures très basses et peuvent donc descendre eux-mêmes à ces températures sans dommage et sans évidence de congélation physique, grâce à des phénomènes circulatoires mais aussi sans doute à des facteurs tissulaires.

Il n'y a pas de dimorphisme sexuel évident chez le Manchot Adélie et l'identification des sexes est plus délicate que ne l'avaient cru les premiers observateurs.

D'une manière générale on peut affirmer que les mâles sont plus grands, plus lourds, que les femelles, mais il faut tenir compte du décalage des cycles annuels du poids dans les deux sexes (à la fin de l'incubation les femelles qui viennent de s'alimenter alors que les mâles jeûnent encore ont un poids supérieur à celui de leurs conjoints) et de nombreuses exceptions. L'allure plus élancée, le bec plus fin, le bassin plus large des femelles ne constituent par des critères absolus.

Les différences classiques du comportement (timidité plus grande, agressivité plus faible, assiduité de l'incubation plus marquée chez les femelles) ne fournissent que des indications souvent démenties par la vérification anatomique.

Il faut bien admettre que l'examen extérieur rapide d'un oiseau inconnu ne permet habituellement pas de diagnostiquer le sexe avec certitude.

576

Par contre, la perception de l'oeuf dans l'oviducte avant la ponte, la dilatation du cloaque après la ponte, la copulation, le comportement d'incubation dans les premières semaines, l'observation du couple pendant quelques semaines fournissent des éléments formels. Peut-être l'étude des sons émis, comme chez le Manchot empereur, apportera-t-elle des moyens de diagnostic.

Les oeufs sont ovoïdes, blanc jaunâtre uniforme, de 61 à 78 mm de long sur 46 à 60 mm de large, pesant entre 70 et 150 g.

Les poussins à la naissance mesurent entre 13 et 16 cm de long, pèsent entre 50 et 130 g, sont couverts d'un duvet fin et relativement brillant dont la couleur varie du gris argenté très clair au gris foncé. La tête et une partie du cou sont gris très foncé, les paupières gris rosé.

La croissance est essentiellement fonction de l'alimentation et beaucoup de poussins mal nourris meurent rapidement (fig. 9).

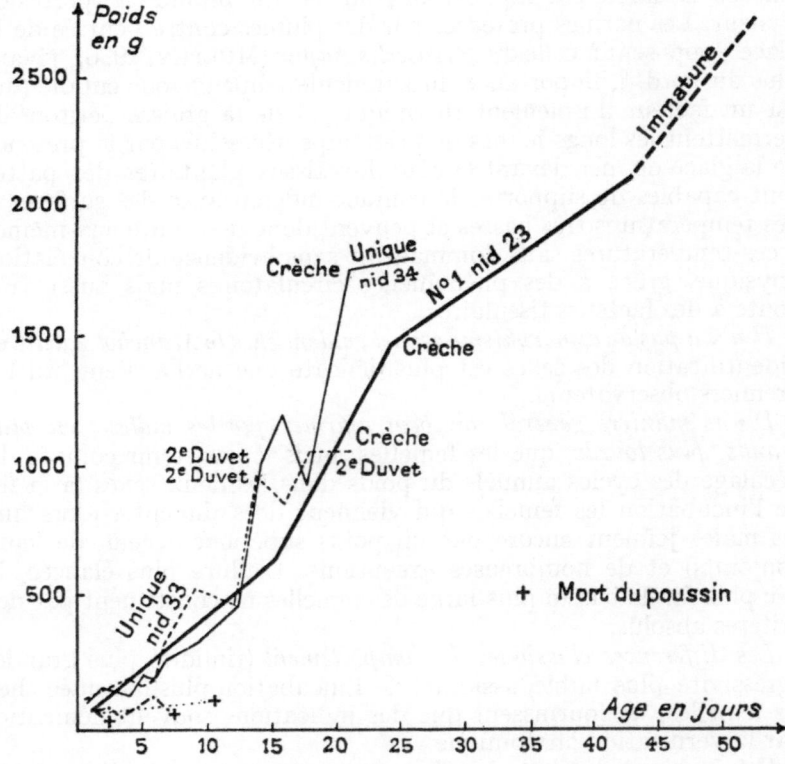

Fig. 9. Courbes de poids d'un certain nombre de poussins avec indication de la date du changement de duvet et de la date de départ en crèche. + = mort du poussin. Importantes différences entre les poids des poussins du même âge.

Les repas fournis par les parents peuvent être très abondants dès le deuxième jour (poids du poussin passant de 112 à 145 g, de 120 à 148 g). *La croissance pondérale est très irrégulière* (fig. 9): un poussin pèse 526 g à 9 jours, 478 g à 12 jours et 950 g à 14 jours; deux poussins de 6 jours pèsent, l'un 220 g, l'autre 110 g; deux poussins du même nid ayant un jour de différence pèsent l'un 290 g, l'autre 1.475 g. On peut admettre les moyennes suivantes, pour la colonie de Port-Martin:

à la naissance: 14 cm et 80 à 90 g
15 jours   26 à 28 cm et 600 à 800 g
22 jours: 34 à 36 cm et 1.400 à 1.600 g
40 jours: 40 à 45 cm et 2.000 à 2.500 g

et la croissance des poussins uniques est plus régulière et plus rapide que celles des deux poussins normaux d'un nid (fig. 9 & 10). *Le développement des poussins est donc étroitement lié à la façon dont les parents sont capables de les nourrir et donc, en définitive, aux conditions écologiques et en particulier à l'état de la glace de mer:* si la glace de mer reste soudée à la côte tard dans l'été, le développement des poussins est très sérieusement compromis.

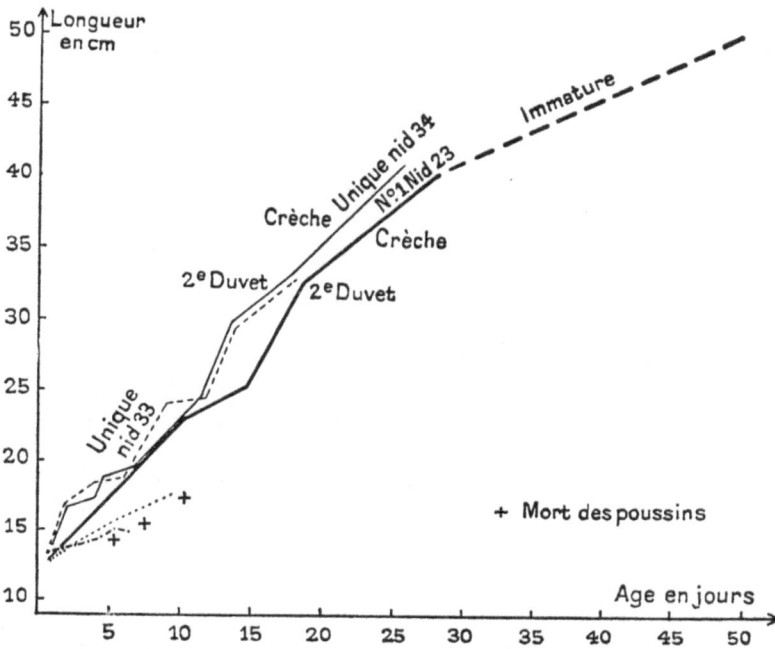

Fig. 10. Courbe de croissance staturale d'un certain nombre de poussins avec indication de la date du changement de duvet et de la date de départ en crèche. + = mort du poussin. Faibles différences entre les longueurs des poussins du même âge.

La croissance staturale, beaucoup plus régulière (fig. 10) dépend, au contraire, plus d'une facteur interne que de l'alimentation.

Entre le 15me et le 20me jour, le duvet gris clair et brillant est remplacé par un deuxième duvet terne, grossier, bourru de couleur brune uniforme mais un peu plus foncé au niveau de la tête.

Vers le 30me jour commence la mue qui, en une dizaine de jours aboutit au plumage de l'immature semblable à celui de l'adulte mais avec menton blanc et paupières foncées, permettant au jeune d'aller à l'eau et de pêcher et *lui donnant son autonomie alimentaire au moment de l'année où l'eau est libre pratiquement partout le long de la côte.*

Pendant la période d'octobre à mars passée sur les lieux de nidification, la société de Manchots Adélie est du type "mosaïque de territoires juxtaposés" et la vie sociale dominée par le comportement territorial jusqu'au moment où les poussins quittent le nid.

Les rapports du Manchot Adélie avec le milieu extérieur

Activités et Types de Comportement

Les activités du Manchot Adélie, facile à observer de très près et qui ne parait pas troublé par la présence de l'homme, ont frappé tous les explorateurs antarctiques. L'interprétation des faits observés a été complètement renouvelée par les travaux de ROBERTS (1940), de RICHDALE (1951) et de SLADEN (1953, 1954, 1955, 1958). En schématisant, il est possible de proposer le classement suivant:

activités "élémentaires"

les moyens de communications intra ou extra-spécifiques, activités manifestement en rapport avec un stimulus du milieu extérieur

les attitudes qui, actuellement, ne répondent à aucun stimulus externe apparent

I. - Les Activités élementaires

La locomotion, chez le Manchot Adélie très mobile et couvrant pendant sa migration et ses voyages "alimentaires" des centaines ou des milliers de kilomètres, est adaptée remarquablement aux divers milieux antarctiques.

Sur les surfaces solides, glace, rochers, relativement planes le mode de déplacement le plus habituel est *la marche debout, les ailerons collés au corps* qui donne une vitesse moyenne de 4 à 6 kmh. En terrain inégal ou en pente, les ailerons sont utilisés comme balanciers pour maintenir l'équilibre. *La glissade sur le ventre (tobogganing)* en poussant avec les pattes et en ramant avec les ailerons est utilisé surtout sur la glace, soit en cas d'urgence pour accélérer l'allure, soit sur des surfaces difficiles (glace accidentée, sasstruggis, neige fraiche) soit simplement pour alterner avec la marche debout pendant les longs voyages sur la banquise. *Le saut à pieds joints* est adopté pour franchir les étroites fissures de la glace ou les fentes de rocher. Les Manchots Adélie sont capables de *surprenantes escalades* sur des pentes abruptes de glace ou de rochers en se servant à la fois des pattes, des ailerons et du bec qui fait un piolet d'une efficacité remarquable et il n'est guère de lieu des faciès antarctiques qui soient hors de leur portée.

Dans l'eau la nage et les évolutions sont remarquablement aisées, les ailerons servant semble-t-il surtout à la propulsion et les pattes surtout à la direction, la vitesse de 6 noeuds en nage normale est donnée par LEVICK (1915) mais elle peut certainement être largement dépassée.

Souvent la trajectoire décrit une sinusoïde avec une partie sous-marine et une partie aérienne comme chez les marsouins (porpoising). Les plongeons, les atterrissages sur des banquettes de glace ou sur des rochers l'oiseau jaillissant de l'eau pour se retrouver debout au point visé parfois deux mètres plus haut, ont fait l'étonnement de tous les observateurs.

La vitesse et surtout l'agilité dans l'eau sont des armes essentielles pour échapper aux prédateurs, Orques ou Léopards de mer, infiniment plus puissants, certainement plus rapides, mais moins "maniables".

Les attitudes de repos observées sont au nombre de trois et paraissent dépendre des conditions climatiques. *Dans la position de repos couché*, l'oiseau est allongé sur ses pattes repliées, les ailerons collés au corps, la tête rentrée dans les épaules, prenant largement appui au sol par la face ventrale de l'abdomen et du thorax. On peut dire que la surface de convection minimum pour les échanges thermiques est ainsi réalisée, les "radiateurs" formés par la tête et les pattes étant pratiquement effacés. Aussi cette position, longueur de l'oiseau orientée dans le lit du vent de façon à offrir le plus petit maître-couple possible, corps aussi près du sol que possible pour bénéficier du micro-climat des basses couche, est-elle surtout adoptée dans les grands blizzards. *Dans la position de repos debout*, l'oiseau est vertical, les ailerons collés au corps, la tête tombant sur une épaule et le bec enfoui sous le bord postérieur de la racine de l'aileron. On note beaucoup d'oiseaux dans cette position par beau temps et au début de l'été. *Une troisième attitude* tout à fait différente se voit parfois chez les adultes et souvent chez les poussins *uniquement sous le soleil relativement chaud du plein été par temps calme.* Le Manchot est complètement à plat ventre, les pattes allongées derrière le corps, plantes en l'air, les ailerons à plat sol, écartés du corps, souvent en croix, tête et cou allongés et reposant sur le sol. C'est la position qui permet un relachement musculaire complet, mais surtout *qui place l'oiseau dans les conditons d'échanges thermiques maximum avec l'ambiance:* augmentation de la surface de convection, développement maximum des "échangeurs" de chaleur que constituent les ailerons la tête et les pattes, surface maximum et sombre pour absorber le rayonnement solaire.

II. Les moyens de communication avec le milieu extérieur

Sont classés dans cette rubrique les types de comportement manifestement en rapport avec une cause exérieure, congénère, ennemi réel ou supposé: ce sont soit des comportements amoureux ou amicaux, donc pouvant être groupés sous le terme de *comportements pacifiques*, soit des attitudes d'agression ou de défense, donc des *comportements de combat, des comportements agonistiques*.

a) *Les comportements pacifiques* dans l'état actuel des observations sur le Manchot Adélie ne semblent pas aussi nombreux et complexes que chez le *Megadyptes antipodes* si longuement et minutieusement étudié par RICHDALE (1951).

La parade mutuelle (photo 7) (mutual epigamic display de ROBERTS, mutual display de SLADEN) et ses formes mineures constituent l'essentiel des comportements pacifiques. Dans la forme typique, les deux oiseaux debout face à face, têtes dressées, becs pointés vers le ciel balancent leur tête à droite et à gauche en un mouvement de métronome, mais les deux métronomes sont décalés et les têtes se croisent, menton contre menton, à chaque mouvement, en traversant le plan sagittal médian. Souvent il y a aussi un mouvement d'avant en arrière de chaque tête et l'ensemble évoquant une succession d'accolades ou une manoeuvre d'aiguisage du fusil du boucher contre la lame du couteau. Le mouvement répété de 5 à 20 fois s'accompagne de "Ka-Ka-Ka" éclatants des deux oiseaux qui deviennent plus doux et finissent en une sorte de ronronnement.

Dans les formes mineures souvent silencieuses, les deux têtes peuvent s'incliner ensemble dans la même direction ou même simplement se dresser ensemble face à face vers le ciel. Ce sont les "quiet display" de SLADEN. Les ailerons sont toujours collés au corps.

La parade mutuelle se produit dans des circonstances multiples: avec une fréquence considérable pendant la période de formation des couples et pendant la période précédant la ponte entre des oiseaux formant des couples provisoires ou définitifs, comme "cérémonie apaisante" vis à vis d'un oiseau adoptant un comportement agonistique, après un incident quelconque venu troubler un couple (bataille, intervention de l'observateur etc.) au retour au nid d'un membre d'un couple, elle existe entre adultes et poussins, elle est tout à fait exceptionnelle en dehors des nids et disparait complètement pendant la mue.

ROBERTS (1940) *l'interprète comme un moyen de stimulation sexuelle. Pour* SLADEN *elle est un moyen de reconnaissance individuelle,* de confirmation de l'identité du partenaire, et une sorte de rituel d'accueil, ce qui cadre bien avec la plupart des circonstances de production. Mais, surtout parce qu'elle n'a lieu qu'entre oiseaux sur un nid ou, en tous cas sur un territoire, parce qu'elle disparait à la fin de la période d'organisation territoriale, elle semble avoir la valeur *d'un comportement d'apaisement de la réaction territoriale.* Enfin, elle constitue parfois une "activité de substitution" dans le sens que TINBERGEN (1951) a précisé.

La partie sonore de la parade mutuelle joue peut-être un rôle important. En particulier il est possible qu'elle soit un moyen de communication entre les oiseaux et un moyen de reconnaissance individuelle comme cela existe chez le Manchot empereur (PRÉVOST, 1961) mais actuellement nous n'avons pas d'études permettant de le prouver.

Les "courbettes" et "révérences" (photo 8) (bowing) ont été décrites par ROBERTS (1940) comme un type de comportement caractéristique des Pygoscelidés, faisant en particulier partie du cérémonial de relève du couveur. SLADEN (1958) les considère à la fois comme un mouvement, une position (posture) pour atteindre un matériau au sol, pour examiner un nid et comme une véritable parade avant et après la copulation ou encore comme réponse de la femelle à une "position extatique" du mâle. Il décrit sous le nom de "bill to axilla" une parade qu'il apparente à la "position extatique". Dans les observations de Terre Adélie, des "flexion-torsion" de la tête avec battements des ailerons et parfois avec grattage du bas-ventre par le bec sont assez souvent notées sans participation d'un congénère, des courbettes sont également enregistrées, mais elles apparaissent comme un mouvement de reconnaissance d'un caillou ou d'un oeuf et ne semblent pas provoquer de réaction spéciale de la part du conjoint. Il semble donc difficile à l'heure actuelle d'affirmer que ces comportements constituent réellement une parade individualisée et encore plus de leur assigner une signification, qui, si elle existe, ne parait pouvoir être que pacifique.

Les transports et "offrandes" de cailloux qui ont frappé tous les observateurs par leur fréquence sont classés parmi les comportements pacifiques parce qu'ils ont été d'emblée considérés comme des "offrandes nuptiales", des "propositions" ou "déclarations" jouant un rôle important dans la formation des couples. Les matériaux transportés et "offerts" sont surtout des cailloux (pouvant peser 243 g) mais parfois des os et LEVICK (1915) a montré que les oiseaux semblaient préférer les cailloux peints de couleurs vives aux cailloux naturels. Les objets transportés, parfois sur des centaines de mètres (photo 9), sont déposés doucement sur le sol, le plus souvent au pied d'un autre Manchot, conjoint réel ou potentiel. Le dépôt n'est pas suivi d'une mimique particulière de la part du transporteur, mais parfois il y a parade mutuelle entre les deux oiseaux, parfois l'oiseau "receveur" range le caillou, parfois il ne réagit pas. Pour les auteurs du début du siècle, seuls les mâles transporteraient les cailloux vers la femelle qui les rangerait dans le nid, acceptant ainsi le mâle pour conjoint, mais on peut actuellement affirmer que les deux sexes transportent des matériaux. De plus on observe des transports de caillou sur des nids vides, sur des nids contenant un cadavre, sur des emplacements quelconques, aux pieds d'un explorateur, etc. . . Le stimulus "oiseau sur nid" déclenchant le transport de matériau (ROBERTS, 1940) lequel transport déclencherait le rangement du matériau est très loin d'être constamment retrouvé. Des transports existent à la fin de la saison, en février à une époque où les gonades sont en pleine involution ce qui sépare ce comportement des activités de reproduction.

Par contre, il semble bien que toutes les observations de transport et "offrandes" de matériau s'expliquent immédiatement si l'on admet qu'il s'agit *d'un comportement territorial* ayant la signification d'une prise de possession ou de l'affirmation de la possession d'un territoire et ceci qu'un conjoint ou non se trouve sur ce territoire. On comprend alors les transports à des nids vides, à des emplacements quelconques, et aussi l'existence de ce comportement en dehors de la période de reproduction, mais à une époque où, comme nous le verrons plus loin, certaines catégories d'oiseau conservent des territoires et les défendent. Les transports de neige ou de glace d'un oiseau à son conjoint couveur qui ont été décrits et discutés notamment par LEVICK (1915) entrent probablement dans le cadre des transports "territoriaux" et ne sont sans doute pas un apport conscient de liquide gelé à un conjoint assoiffé, comme l'ont pu croire les premiers observateurs.

b. *Les comportements agonistiques* ont une origine évidente dans le milieu extérieur, un rôle évident de défense de l'individu et existent, plus ou moins marqués, pendant tout le séjour de l'oiseau à la colonie et même pendant la mue.

Les parades d'intimidation revêtent deux aspects principaux associés au même élément sonore. Dans le premier type, l'oiseau est debout, les ailerons rejetés en arrière comme pour prendre de l'élan, les plumes de la crête frontale dressées, le bec entr'ouvert dirigé vers l'ennemi. Dans le second type (photo 10), le Manchot est penché en avant sur ses pattes crispées, les ailerons collés au corps, la crête frontale dressée. Toujours les yeux sont grands ouverts, fixant l'ennemi, souvent la tête tourne d'un côté mettant en évidence la paupière blanche et l'aspect de l'oeil qui rappelle le "glare" de RICHDALE (1951). Les sons émis sont soit des "Ka-Ka-Ka" furieux, soit des grondements sourds de la gorge.

Dans les batailles, le Manchot Adélie utilise trois armes: le bec qui sert à donner des coups de pointe mais surtout à pincer, les ailerons qui frappent à cadence très rapide de leur face interne et de leur bord d'attaque, la poitrine qui donne des coups de bélier et qui pousse l'adversaire en arrière.

Dans le type habituel de bataille qui parait opposer des oiseaux de même sexe, les adversaires commencent par se lancer des coups de becs qui sont assez violents pour arracher des touffes de plumes et pour déchirer la peau humaine. Puis les ailerons entrent en action asśénant des coups très rapides qui font parfois saigner le bord d'attaque. En même temps les oiseaux se repoussent du plastron ou se lancent l'un contre l'autre pour culbuter l'adversaire en arrière. Le combat peut durer plusieurs minutes avec avances et reculs successifs de chaque oiseau et il est souvent difficile de désigner un vainqueur. Pendant toute la période territoriale où les batailles sont très fréquentes, les déplacements des combattants sont limités mais à la fin de la saison, on assiste parfois à de longues poursuites à travers la colonie (photo 11). Beaucoup plus rarement, les combattants sont les oiseaux d'un couple, l'un d'eux, la femelle d'après la taille, s'enfuit, est rattrapée, plaquée au sol et reçoit une volée de coups de bec et d'ailerons. Il ne semble pas y avoir de blessures graves. Entre oiseaux couveurs sur les nids les comportements agonistiques sont réduits à des cris et des grondements, des parades d'intimidation, et des coups de bec n'arrivant généralement pas à destination.

Dans la plupart des cas, à la colonie, les comportements batailleurs sont provoqués par un congénère et la cause déclenchante est l'envahissement d'un territoire plus ou moins précis, le vol de caillou, la défense d'un conjoint peut-être, tard dans la saison, l'intrusion d'un poussin étranger, mais *tout objet-vivant ou non-pénétrant sur le territoire* (homme, chien, bloc de rocher déplacé, morceau de bois ou de papier etc. . .) suscite immédiatement la réaction agonistique. *En dehors des territoires* nous n'avons pratiquement pas d'observations de comportements batailleurs dirigés contre des congénères, mais par contre, l'approche d'un ennemi réel ou supposé, homme, chien, déclenche des attaques souvent acharnées. *La présence des Skuas,* principaux prédateurs des colonies de Manchot Adélie, provoque immanquablement l'attaque des Manchots quand le Skua s'approche du territoire. En dehors des territoires le comportement des Manchots parait très variable. Si l'on a enregistré des attaques répétées de certain Manchot, loin de leur nid, contre des Skuas, il est également certain que les prédateurs séjournent habituellement sur de petits rochers entre les nids sans être inquiétés.

"L'agressivité" des Manchots est essentiellement individuelle, généralement plus marquée chez le mâle. Elle varie d'un jour à l'autre peut-être sous des influences météorologiques, diminuant nettement dans les grands blizzards. Elle est plus marquée pendant les premières périodes de la vie à la colonie quand le comportement territorial est à son maximum et elle disparait presque complètement pendant la mue.

En dehors du rôle évident de défense contre les prédateurs, les comportements agonistiques auraient d'après les auteurs du début du siècle une *fonction sexuelle*, les batailles entre les "prétendants" puis les batailles contre les "séducteurs" jouant un rôle de sélection. Mais beaucoup de ces combats peuvent s'expliquer simplement par la défense du territoire, le "prétendant" ou le "séducteur" étant aussi un envahisseur. *La fonction territoriale des comportements agonistiques apparait prépondérante.* Tous les auteurs ont remarqué l'ardeur au combat, l'efficacité, la victoire habituelle de l'oiseau qui se bat sur son territoire, la moindre combativité de l'envahisseur; les oiseaux circulant entre les nids pendant l'incubation ne répondent même pas aux attaques des occupants des nids et essayent seulement d'éviter les coups en zigzaguant. *La fonction familiale* de défense du conjoint, des oeufs, des poussins est difficile à séparer de la fonction de défense du territoire: le Manchot ne défend plus ni ses oeufs, ni ses poussins quand ils se trouvent en dehors du nid. La fonction de protection et de garde des crèches de poussins qui avait été invoquée semble bien, après les travaux de SLADEN (1953, 1958) n'être qu'un comportement territorial "hors de saison", de la part d'oiseaux appartenant à certains groupes d'âge. Peut-on attribuer aux comportements agonistiques une *fonction sociale* aboutissant à une certaine hiérarchie comme cela existe dans d'autres espèces? Il semble bien que toutes les observations données en faveur de cette théorie puissent s'expliquer elles aussi, simplement par le comportement de défense territoriale.

III. Les types de comportement sans stimulus extérieur apparent ne paraissent pas, dans l'état actuel de nos connaissances, en relation avec un évènement du monde extérieur; aussi leurs causes et leurs fonctions restent-elles actuellement hypothétiques.

a. *La "position extatique".* — (photo 12) — L'oiseau est debout, les pattes un peu écartées, le corps et le cou en extension complètes, le bec pointé vers le ciel. Les ailerons s'élèvent en abduction à l'horizontale puis exécutent des battements d'avant en arrière à une cadence régulière et lente. En même temps, le cou et la tête sont animés de mouvements cadencés de rythme rapide, s'élevant et s'abaissant sur un axe vertical en une sorte de vibration de faible amplitude. L'élément sonore est une succession de sons sourds et de courts silences au même rythme que la vibration du cou qui se termine par un cri éclatant rappelant le chant de la poule qui vient de pondre. Il existe des "positions extatiques" dégradées et incomplètes et SLADEN (1953, 1958) considère la parade "bill to axilla" comme une forme mineure de ce comportement.

Si ce comportement parait le plus souvent absolument spontané, son caractère "contagieux" a été noté par de nombreux observateurs car elle est adoptée en même temps par de nombreux oiseaux, mais pas par les deux oiseaux d'un même couple. Après GAIN (1914) qui pense que la position extatique n'est réalisée que par les mâles, SLADEN (1953) affirme que ce comportement appartient exclusivement aux mâles jusqu'à l'éclosion des oeufs puis qu'il est également adopté par les femelles. Il semble bien que cette parade ne disparaisse que pendant la mue.

Comme nous ne pouvons pas relier cette parade à un stimulus extérieur certain, l'interprétation en reste aléatoire. SLADEN pense qu'elle sert à attirer un oiseau de sexe opposé et à repousser un oiseau de même sexe et lui attribue donc un rôle dans la formation des couples. Mais elle est observée tard dans la saison après que les couples ont disparu et même selon LEVICK (1915) sur la banquise. GAIN et LEVICK l'ont considérée comme une expression de "bien-être" à l'égal des positions d'étirement bien connues chez les Mammifères, mais sa complexité parait bien grande pour une activité aussi élémentaire.

b. *Le baillement* se produit à toutes les périodes de la vie à la colonie et même pendant la mue et parait donc indépendant du cycle reproducteur. L'oiseau debout, la tête rentrée dans les épaules, les ailerons au corps ou repoussés en arrière, ouvre largement le bec sans émettre aucun son. Ses fonctions éventuelles sont inconnues.

c. La *"position angélique"* se voit chez des oiseaux isolés ou en couple à toutes les périodes de la vie à la colonie sauf peut-être pendant la mue. Le Manchot, debout, regarde droit devant lui, le bec horizontal et fait quelques petits mouvements avec les ailerons placés en abduction et retro-pulsion. Comme les autres parades de ce chapitre, cette attitude peut, en attendant de nouvelles informations, être considérée comme une attitude stéréotypée répondant sans doute à une motivation interne sans rapport direct avec le milieu extérieur.

Il semble devoir en être de même des "jeux" et "ébats" sur les plages, dans les baignoires d'eau de fonte ou dans la neige fondante qui apparaissent comme étant des comportements individuels plutôt que manifestation de vie sociale.

Le cycle annuel du Manchot Adélie dans le cycle annuel antarctique

Dans le cadre du cycle annuel de l'univers côtier antarctique tel qu'il a été brièvement défini au début de ce travail avec ses caractéristiques (opposition entre l'hiver et l'été, hiver "depourvu de centre", période de la glace soudée et période de l'eau libre, variation extrême de la durée du jour, etc . . .) *la vie du Manchot Adélie peut être découpée en six périodes bien individualisées.* Mais il est important de remarquer immédiatement que *cinq de ces périodes intéressant les activités de reproduction sont passées sur la côte,* sont relativement connues par les observations faites à partir des bases côtières des expéditions antarctiques et *se placent entre octobre et mars* pendant la "belle saison" antarctique, *tandis que la sixième se déroule à la limite de la banquise antarctique pendant l'hiver de mars à octobre* occupant sept mois de l'année et, du fait des difficultés de séjour à la limite nord des glaces pendant l'hiver, est encore presque complètement inconnue. *Si l'on envisage maintenant non plus la position géographique mais la vie sociale* et l'organisation de la colonie, il faut immédiatement opposer *une période de "vie familiale et territoriale"* débutant avec l'installation sur les territoires de nidification à l'arrivée sur la côte en Octobre et *une période "individuelle"* qui commence aussitôt après la séparation définitive des poussins et des parents au début de février en Terre Adélie, séparation elle-même précédée de *l'abandon des territoires et donc des comportements territoriaux.* Ce cycle sera décrit avec les dates fournies par l'étude faite en Terre Adélie, puis les comportements "anormaux" de certains oiseaux et leur signification seront envisagés.

La période précédant la ponte

Cette période s'étend depuis la migration sur la banquise amenant les oiseaux sur la côte jusqu'à la ponte soit, en Terre Adélie, du début octobre au milieu de novembre.

La migration et l'arrivée à la colonie

C'est au début d'octobre sur la plus grande partie de la côte antarctique que se produit la migration depuis la limite nord des glaces jusqu'à la côte. L'eau libre se trouve à ce moment là, suivant les méridiens, entre quelques dizaines et des centaines, parfois plus d'un millier de kilomètres de la côte. La moyenne mensuelle de température varie entre —30 et —15° avec des minimas atteignant —40° dans les secteurs de la Mer de Ross et de Weddell (fig. 2). Les vents et les blizzards sont encore très violents et fréquents, moyenne mensuelle d'octobre entre 60 et 70 kmh sur la côte de la Terre Adélie (fig. 2). *Les conditions météorologiques et d'échanges thermiques sont donc terriblement sévères pour des homéothermes effectuant un long voyage sans possibilité de s'alimenter, ni de s'abriter.* Par ailleurs nous ne pouvons pas comprendre *comment naviguent et rejoignent un point précis de la côte, des oiseaux qui ne peuvent trouver aucun repère sur la banquise, qui ne peuvent utiliser le champ magnétique terrestre dont la composante horizontale est nulle en Terre Adélie, qui ont un horizon borné à quelques kilomètres.* Il faut donc supposer que la direction du soleil et peut-être des vents dominants et donc des sasstruggis leur permet de rejoindre la côte, puisqu'une mémoire topographique des différents affleurements rocheux est suffisamment précise et étendue pour leur faire reconnaître l'emplacement de leur colonie des années précédentes.

Le voyage se fait par groupes de quelques dizaines à quelques milliers d'individus, se suivant en file indienne, marchant debout ou glissant sur le ventre à une vitesse de 4 à 6 kmh, contournant les icebergs, à peine ralentis semble-t-il par les blizzards.

A leur arrivée sur les rochers de la côte, les oiseaux qui sont manifestement isolés et indépendants les uns des autres, semblent inspecter les lieux, circulent et stationnent sur les anciens territoires, puis dès les premiers jours, la construction des nids commence et des couples se forment.

La formation des couples

GAIN, puis les observations des expéditions anglaises en Terre de Graham ont montré avec certitude que, *sauf accident, d'une année à l'autre, les mêmes couples se reforment et occupent les mêmes emplacements.* Mais les membres du futur (et ancien) couple n'arrivent pas ensemble et il y a formation de couples "provisoires" ou, en tous cas, stationnement de deux oiseaux sur un nid, les vrais couples se formant avec ou sans bataille à l'arrivée du deuxième conjoint. Les oiseaux d'un même couple se reconnaissent-ils individuellement par la vue ou des signaux sonores après la séparation supposée de l'hiver? Se retrouvent-ils ensemble seulement parce qu'ils reviennent tous deux à l'ancien territoire? Peut-on faire jouer un rôle à la notion "d'affinité" définie par RICHDALE (1951)? Comment un nouveau conjoint est-il acquis après disparition de l'ancien? Quelle est la valeur de sélection sexuelle des batailles si fréquentes à cette période? Toutes ces questions sont actuellement sans réponses formelles car nous n'avons pas de marquages suffisamment nombreux suivis pendant plusieurs années, et le problème de la formation des couples n'est pas vraiment résolu. *Mais la fidélité des conjoints l'un à l'autre et au territoire pendant toute la période de reproduction est un fait absolument acquis.*

Etablissement du territoire et comportement territorial

L'organisation de la colonie en "mosaïque de territoires juxtaposés" est évidente à la première inspection pendant la période précédant la ponte (photo 3).

Les limites de chaque territoire paraissent être fixées par la distance à laquelle l'oiseau peut lancer ses coups de bec sans quitter le nid, centre du territoire, ce qui représente un cercle de 60 à 80 cm de diamètre. *La connaissance et la reconnaissance du territoire*, soit après l'hivernage sur la banquise, soit après un voyage "alimentaire" de plusieurs semaines pendant l'incubation, *sont extrêmement*

588

précises comme les marquages l'ont montré: le Manchot est capable
de retrouver son territoire même s'il est couvert de neige à l'arrivée
à la colonie, même si le conjoint, les oeufs et les cailloux du nid ont
disparu au retour d'un voyage "alimentaire" de la période d'incubation. *La fidélité au territoire* est complète en l'absence de boule-
versements artificiels du faciès. *La défense du territoire* contre toute
intrusion est opiniâtre et pathétique chez la grande majorité des
oiseaux.

*Le territoire modifie profondément la plupart des comportements
du Manchot Adélie.* La parade mutuelle, le dépôt de cailloux, la
copulation, la position extatique, sauf exception rarissime ne sont
réalisés qu'à l'intérieur du territoire. La limite du territoire marque
le renversement complet de certains comportements: la protection
et les soins des oeufs et des poussins d'une intensité extrême à
l'intérieur du territoire cessent complètement en dehors de lui. *Avant
d'essayer d'interprèter les activités du Manchot Adélie, il est indispensable d'envisager avant toute chose le problème territorial, expression
immédiate des relations de l'oiseau et du milieu.*

La construction des nids

Il semble bien aujourd'hui que les nids soient construits non
seulement par des oiseaux en couples mais aussi par des isolés des
deux sexes et celà même très tard dans la saison. Certains nids sont
construits en quelques heures et contiennent des centaines de
cailloux, mêlés parfois d'ossements, régulièrement disposés pour
former une sorte de coupe retenant bien les oeufs et assez surélevée
pour se trouver à l'abri des ruisseaux de fonte. D'autres ne comprennent que quelques cailloux mal rassemblés en plusieurs semaines. Ces différences apparaissent comme en rapport avec l'âge et
l'expérience des constructeurs.

La copulation

Les observations récentes montrent que les oiseaux reconnaissent
mutuellement leur sexe et que les copulations sont d'emblée nor-
males. Sans préambule bien certain, le mâle debout à côté de la
femelle couchée sur le ventre, grimpe sur le dos de cette dernière,
puis, les ailerons écartés faisant balancier, le corps penché en avant,
piétine sur place comme pour conserver son équilibre. La femelle
dresse son bec vers le ciel et sa queue se relève obliquement pour
découvrir le cloaque (photo 13). Le mâle recule le plus possible sur
le dos de la femelle, se courbe en avant pour caresser de son bec le
menton de sa partenaire, agite spasmodiquement sa queue dans
un plan horizontal puis la recourbe soudain vers le bas pour amener
très brièvement les cloaques au contact. L'acte qui a duré entre

vingt secondes et une minute, est terminé et le mâle saute à terre sans autre parade.

Les copulations commencent dans les jours suivant l'arrivée à la colonie, mais leur maximum de fréquence se place fin octobre début novembre en Terre Adélie, soit dans les deux semaines précédant la ponte. A ce moment les testicules des mâles pèsent une vingtaine de grammes en moyenne et les femelles ont un ovaire d'un poids oscillant entre 10 et 60 g avec un gros follicule. La ponte du deuxième oeuf marque la fin de cette parade. Mais il existe des anomalies: copulations "hors de saison", en janvier avec des testicules du mâle pesant 2 g, tentative de copulation homosexuelle ou avec des cadavres, des poussins, des rochers.

La vie de la colonie avant la ponte

Dans l'étude faite à Port-Martin en Terre Adélie entre le 20 octobre date d'arrivée du premier Manchot et le 10 novembre date de ponte probable du premier oeuf, la vie de la colonie est caractérisée par les fait suivants:

Augmentation du nombre des oiseaux présents à la colonie et augmentation du nombre de couples et donc du pourcentage des oiseaux en couples qui atteint 95% au début de Novembre (fig. 11 & 12).

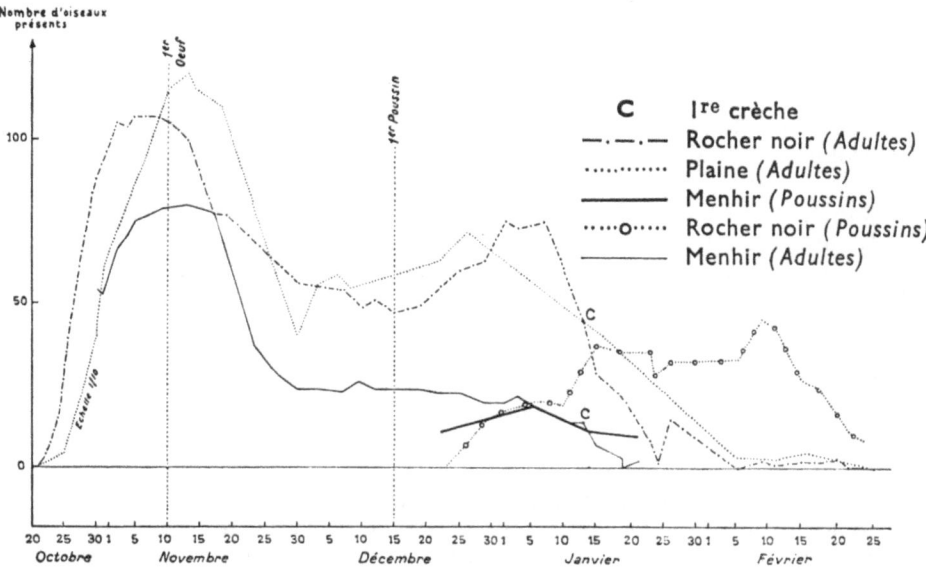

Fig. 11. Nombre d'oiseaux présents dans différents secteurs de la colonie: dans ,,La Plaine'' dénombrements directs (échelle 1/10); au ,,Menhir'', dénombrements directs; au ,,Rocher Noir'', chiffres obtenus par la photographie systématique.

590

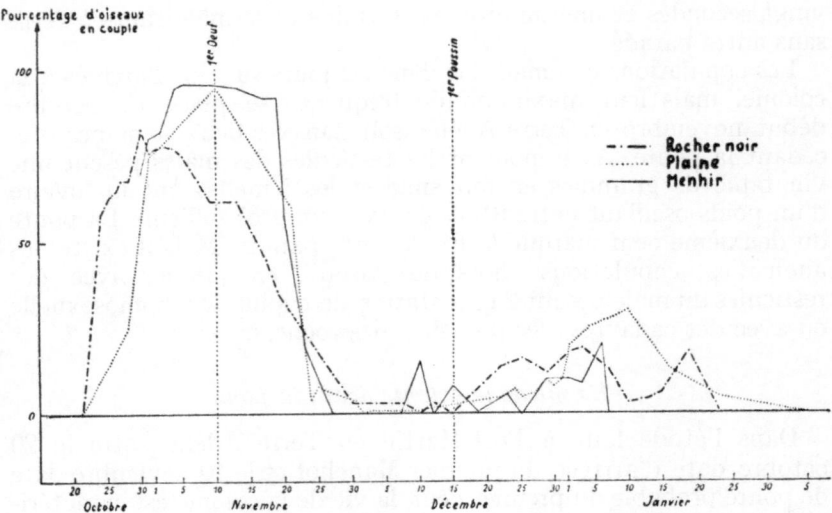

Fig. 12. Pourcentage d'oiseaux en couple dans différents secteurs de la colonie: dans „La Plaine", dénombrements directs; au „Menhir", dénombrements directs; au "Roche Noir", chiffres obtenus par la photographie systématique.

Netteté de l'organisation territoriale en mosaïque de territoires juxtaposés (photo 3) avec deux oiseaux en permanence sur chaque nid.

Activité générale très importante: fréquence des transports de cailloux pour la construction des nids ou l'acquisition de territoires, fréquence des parades mutuelles qui sont typiques, complètes et prolongées, fréquence des batailles pour des motifs territoriaux, des vols de cailloux, peut-être pour la défense ou l'acquisition de conjoints, fréquence des positions extatiques, fréquence des copulations. L'animation, le niveau sonore de la colonie sont à leur maximum, à peine diminués pendant les grands blizzards et les courtes heures de nuit.

Jeûne complet de tous les oiseaux présents attesté par la vérification des estomacs, la courbe de poids (fig. 13), les déjections vertes. Ce jeûne existe même dans les cas où les oiseaux ont de l'eau libre à proximité et ne dépend donc pas uniquement actuellement des conditions de la glace de mer, mais il *représente manifestement une adaptation écologique au milieu antarctique.*

Les conditions du milieu ne varient que faiblement pendant cette période sauf dans la partie septentrionale de la Terre de Graham. *La glace de mer* est presque partout soudée à la côte et s'étend vers le nord sur des dizaines ou des centaines de kilomètres. La température s'élève lentement, la moyenne de novembre étant de —15 à —5° suivant les secteurs. Sur la côte orientale, la moyenne des vents

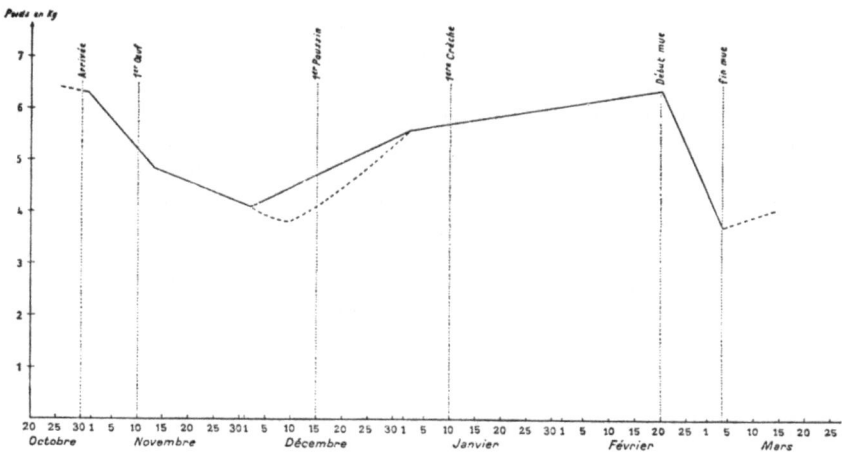

Fig. 13. Schéma de la variation de poids moyen du Manchot Adélie mâle adulte pendant le séjour à la colonie. Les chiffres pour les périodes précédant la mue ont été obtenus sur des oiseaux sacrifiés. Les chiffres pour la période de mue proviennent des pesées effectuées par J. Cendron (1953). Le pointillé représente la courbe probable à la fin du jeûne de la période d'incubation, à un moment où aucune pesée n'a pu être effectuée.

est encore aux environs de 60 kmh. Pratiquement pas de fonte ni de diminution des surfaces enneigées mais allongement rapide de la photo-période et jour perpétuel dès le début de novembre pour les régions au sud du cercle polaire.

La période d'incubation
La ponte

La ponte qui commence le 10 Novembre à Port-Martin peut être reconnue par le changement manifeste de l'aspect de la colonie: baisse du niveau sonore, calme et ordre soudainement établis, oiseaux étroitement liés à leur territoire, défense territoriale exacerbée, diminution impressionnante des comportements agonistiques réduits à des querelles de nid à nid, disparition des copulations.

La ponte s'échelonne sur une quinzaine de jours et l'intervalle entre le premier et le deuxième oeuf varie de 1 à 5 jours, 3 jours étant la durée la plus habituelle.

Les conditions du milieu

Entre le 10—20 novembre et le 15—25 décembre, limites de la période d'incubation en Terre Adélie, on assiste sur tout le pourtour du continent *à une transformation progressive et rapide des conditions écologiques*. La moyenne thermique de décembre s'élève pour se placer entre —10 et 0° suivant les secteurs, même dans le climat de la côte orientale, le vent diminue et sa moyenne mensuelle tombe à 40 à 50 kmh pour décembre. Le jour perpétuel règne partout. Le rayonnement du soleil absorbé par les rochers sombres entraine une fonte, non négligeable menaçant les oeufs d'inondation puis de regel nocturne. Les surfaces rocheuses s'agrandissent au détriment des surfaces enneigées. Surtout, *les conditions de la glace de mer sont bouleversées dans toutes les régions septentrionales:* des rivières se forment dans la banquise, souvent des flaques d'eau libre à quelques dizaines ou à quelques kilomètres de la côte rendant la pêche possible. Dans quelques secteurs, la débâcle des glaces se produit.

L'incubation

Dans les nids marqués de Port-Martin elle a duré de 30 à 37 jours mais parfois le poussin reste deux ou trois jours dans l'oeuf troué avant d'être complètement libéré. L'oeuf repose soit sur les cailloux du nid soit plus ou moins sur les pattes de l'oiseau couveur qui, en position couchée recouvre complètement les deux oeufs mais les laisse en partie découverts en position debout. La température des oeufs a varié, en Terre Adélie, entre 19 et 35°.

Le comportement d'incubation (réchauffement et protection des oeufs, défense opiniâtre des oeufs et du territoire) parait déclanché chez tous les oiseaux par le "complexe oeuf sur nid" mais pas par des oeufs isolés, donc n'existe que sur le territoire, est variable dans son intensité et son efficacité suivant les oiseaux (et probablement suivant leur âge et leur expérience), n'est pas modifié par l'enlèvement d'un des oeufs, persiste en présence d'oeufs inondés, gelés, éclatés partiellement mais cesse quand les oeufs sont vidés et écrasés. En cas de perte des deux oeufs, il n'y a pas de ponte de remplacement.

Les voyages alternés des couveurs vers l'eau libre et la nourriture

Mettant fin au jeûne physiologique de printemps avec absorption de neige, ils sont une caractéristique essentielle de cette période. *Dès la ponte du deuxième oeuf, la femelle de chaque couple quitte la colonie* après un mois de jeûne en moyenne pour aller s'alimenter en eau libre. A Port-Martin ces voyages ont duré, chez les couples marqués, de 11 à 21 jours (fig. 19). La banquise vide d'oiseaux à la fin de la période précédente connait à nouveau un trafic intense. *Les mâles restent seuls à couver et la population de la colonie baisse de moitié* (photo 4) (fig. 11 & 12). Dans les premiers jours de décembre on note les premiers retours de femelles propres, grasses, évacuant des déjections rouges. La relève des mâles couveurs se fait rapidement avec quelques parades mutuelles et les mâles de chaque couple partent à leur tour après un jeûne qui a duré par-

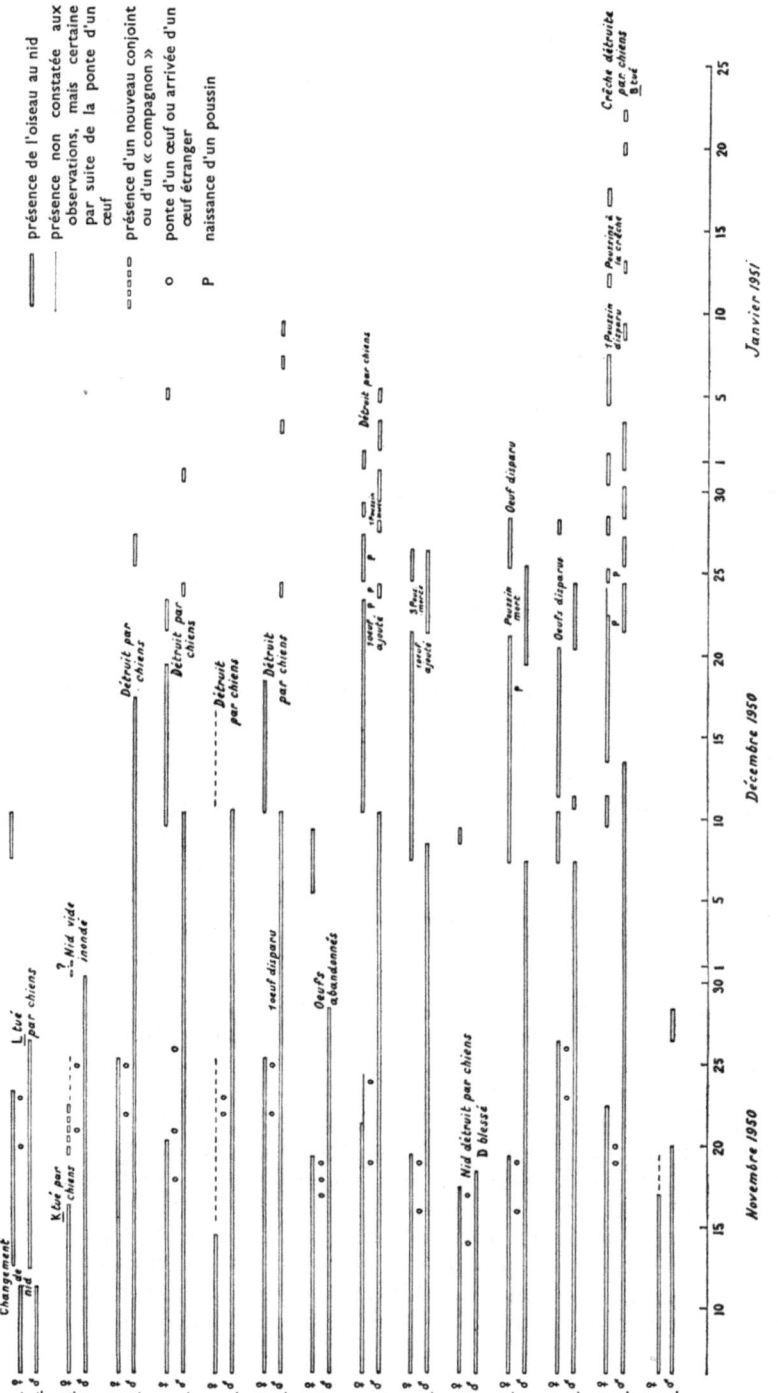

Fig. 14. Présence des oiseaux marqués à leur nid.

fois deux mois et entrainé une perte de poids de 40% (fig. 14). L'eau libre s'étant rapproché avec l'avancement de la saison l'absence des mâles est courte: 3 à 12 jours à Port-Martin. Puis les relèves se succèdent à rythme rapide (fig. 14), avec des séjours des deux oiseaux sur le nid entre les voyages alimentaires de plus en plus court avec la proximité de l'eau libre. La population de la colonie, l'activité, augmentent.

La période d'élevage du poussin au nid

L'été antarctique

Etendue du 15 décembre au 15, 25 janvier, cette période est celle du plein été. Si les moyennes mensuelles restent entre —10 et 0° on enregistre en certains secteurs, des maxima positifs. Des moyennes de vent ne dépassant guère 40 kmh pour les secteurs les plus ventés, avec de nombreux jours calmes et ensoleillés, des blizzards légers, une fonte importante au contact des rochers. *Mais surtout la débâcle de la glace de mer est générale et l'eau est libre pratiquement partout le long de la côte.*

La vie de la colonie

La vie de la colonie connait une intensité nouvelle: circulation des adultes allant à la pêche, fréquence accrue de nombreuses parades: parades mutuelles, positions extatiques, comportement agonistiques et batailles, intensité de la défense territoriale, transports de cailloux et construction de nids par des oiseaux "inemployés" sans progéniture dont le nombre s'accroit avec le temps du fait de l'importante mortalité des oeufs et des poussins et avec le retour à la colonie de certaines classes d'âge n'ayant pas eu d'activité reproductrice complète.

Les poussins, d'abord complètement recouverts par le couveur couché ou oblique, sont rapidement visibles sous le ventre de l'adulte. Ils sont réchauffés, défendus, nourris, couvés même après leur mort tant qu'ils sont sur le territoire. Mais le Manchot Adélie adulte, si habile à transporter de lourds cailloux, parait incapable d'aider son poussin égaré à regagner le nid et le laisse mourir à quelques décimètres de lui. Aussitôt après l'éclosion il accueille et protège un poussin très jeune, mais plus tard, repousse un poussin de plus d'une dizaine de jours, ce qui évoque une reconnaissance individuelle effective seulement après quelques jours.

Les nourrissages sont de plus en plus fréquents par l'un des parents qui est allé pêcher devant la colonie pendant que l'autre couvait les jeunes. Vers le 15me jour le poussin en deuxième duvet bourru, possédant déjà une bonne défense thermique, reste à côté de ses parents ne reprenant sa place sous le ventre de l'adulte que pendant les blizzards. L'organisation territoriale est toujours très nette (photo 5).

La période des crèches et la mue des poussins

Du 15 janvier au 15 février, limites schématiques de cette période à Port-Martin, *les conditions climatiques* restent celles du plein été mais la moyenne thermique baisse déjà de quelques degrés, et les vents augmentent (photo 6).

Dès le 10 janvier quelques poussins s'éloignent de leurs nids et se rassemblent par petits groupes de 3 à 4, qui deviennent les jours suivants *de véritables crèches de dix, puis vingt, puis trente à cinquante jeunes*, par fusionnement des petits groupes du début. Les poussins se serrent les uns contre les autres par mauvais temps formant des "tortues" comme celles des Manchots empereurs et se dispersent par beau temps.

Dès le 20 janvier commence *la mue des poussins* en crèche, le plumage définitif du type adulte apparaissant sous le duvet qui tombe en lambeaux découvrant d'abord la face ventrale des ailerons et du corps, pour démasquer peu à peu en une quinzaine de jours *l'habit de l'immature à menton blanc, qui va permettre la pêche et donc l'alimentation autonome.* Les immatures abandonnent les crèches et se mêlent aux adultes sur les congères et les banquettes.

Le départ des poussins du nid marque une transformation complète de la structure sociale de la colonie: l'abandon des territoires par les adultes, la fin de la période territoriale et des comportements territori-aux, la disparition des nids, la dispersion des cailloux. Le problème du "gardiennage" des crèches et de l'alimentation des poussins en crèche parait résolu grâce à SLADEN (1953, 1958): les parents viennent nourrir leurs propres poussins dans les crèches, ne nouris-sant qu'exceptionnellement des poussins étrangers; les soi-disant "gardiens" des crèches sont en fait des adultes sans poussins de certaines classes d'âge, retrouvant à cette période un comporte-ment territorial "hors de saison" et défendant leur territoire sans réellement garder les crèches comme l'avaient cru les anciens auteurs.

La mue des adultes et le départ des oiseaux

Entre le 15 février et la fin mars, *l'antarctique entre rapidement dans l'hiver.* L'obscurité nocturne allonge chaque jour, la glace de mer commence à se former les jours calmes, la température moyenne de mars est de —5 et —25° suivant la latitude, surtout la moyen-ne de vent peut atteindre des chiffres très élevés: 104 kmh en 1951 à Port-Martin, et les blizzards sont déjà très denses. *Les immatures* disparaissent rapidement vers le nord après leurs premières plongées.

Les adultes se rassemblent sur des congères ou des rochers *pour la mue* qui demande deux semaines environ. Apathiques, frileux, n'effectuant plus aucune des parades habituelles sauf de rares baillements, fébriles (CENDRON, 1953, a mesuré une élévation thermique de 1° environ) ils sont incapables d'aller pêcher, jeûnent complètement et subissent un amaigrissement extrêmement rapide atteignant environ 40% de leur poids initial (fig. 13). Dès sa mue terminée chaque oiseau part vers le nord. Certains oiseaux d'ailleurs ont quitté la colonie plus tôt pour aller muer sur la banquise.

L'hivernage

De mars à octobre, la période d'hivernage du Manchot Adélie est encore pratiquement inconnue. Elle se déroule certainement sur la banquise à la limite de l'eau libre, dans des conditions climatiques mal étudiées mais moins sévères que celles régnant sur la côte, dans un milieu mi-solide, mi-liquide permettant la pêche mais aussi le repos sur la glace. (photo 14). Les oiseaux d'une même colonie passent-ils l'hiver ensemble, dérivent-ils avec la banquise ou se maintiennent-ils sensiblement sur le méridien de leur colonie, y a-t-il une organisation sociale, les immatures sont-ils mêlés aux adultes, etc....? Il faudrait une étude de la banquise antarctique hivernale pour élucider tous ces problèmes.

Les comportements anormaux et les classes d'âge de la population

Le cycle précédemment décrit, inscrit logiquement dans le cycle annuel du milieu antarctique, est celui de la majorité de la population d'une colonie. Mais, en proportion variable suivant les années et les colonies, d'assez nombreux oiseaux présentent des comportements aberrants: oiseaux ne formant pas de couples ou se séparant rapidement de leur conjoint et devenant des "inemployés"; oiseaux ne défendant pas leurs oeufs et leur nid ou les abandonnant pendant l'incubation; oiseaux construisant des nids après la naissance des poussins; oiseaux adoptant un comportement territorial et batissant des nids à la période des crèches après l'abandon des territoires par les oiseaux ayant élevé leurs poussins, etc... Les travaux de RICHDALE (1951) et de SLADEN (1958) apportent une explication: *les différentes classes d'âge de la population ont des comportements différents.* Chez le Manchot Adélie nous pouvons admettre l'existence de 3 classes d'âge: *les reproducteurs expérimentés* (experienced or established breeders) ayant au moins 4 ou 5 ans; *les reproducteurs inexpérimentés* (inexperienced or unestablished breeders) âgés de 3 ou 4 ans; *les adultes non reproducteurs* âgés de 2 à 3 ans dont les gonades ne sont pas complètement développées qui ne construisent pas de nids, ne prennent pas de conjoint définitif et qui sont qualifiés "d'errants" (wanderers). Les oiseaux de cette dernière catégorie sont, à toutes les périodes, dans *l'état "d'inemployés"*. Les oiseaux des deux premières catégories possèdent *l'état de "parent actif"* (successful breeders) aussi longtemps qu'ils ont des oeufs ou des poussins, mais deviennent aussi des "inemployés" s'ils perdent leurs oeufs ou leurs poussins. Le nombre d'inemployés dépend donc de la mortalité des oeufs et des poussins de la saison en cours et de la proportion des non reproducteurs; la proportion entre les classes d'âge dépend de la mortalité des jeunes pendant les années précédentes. Ces notions rendent compte des comportements "anormaux", des différences d'aspect social de différentes colonies ou de la même colonie pendant des années différentes et surtout elle semble prouver que *le comportement de reproduction du Manchot Adélie est progressivement perfectionné au fur et à mesure que l'oiseau vieillit.*

Le milieu et la mortalité

Le milieu vivant par les parasites et les prédateurs, le milieu physique redoutable en dépit de l'adaptation de l'espèce représentent les causes de mortalité qui nous sont accessibles dans l'état actuel de nos connaissances sur la vie du Manchot Adélie.

Les facteurs vivants

Le facteur pathologique ou léthal représenté par *les parasites* du Manchot Adélie ne peut pas être actuellement apprécié. Mais il semble bien que *Pygoscelis adeliae* soit beaucoup moins parasité que la plupart des autres espèces: pratiquement pas d'ectoparasites, de rares Cestodes, surtout chez les immatures. *Les prédateurs sont mieux connus.* Dans l'eau, les Manchots immatures ou adultes ne craignent guère le grand prédateur des mers antarctiques *l'Orque (Orcinus orca)* trop massif et ne virant pas assez court mais ils ont à compter avec le *Léopard de Mer (Hydrurga leptonyx)* beaucoup plus redoutable. Mais, d'une part l'effectif de ce phoque est partout relativement faible, d'autre part les observations de SLADEN (1958) semblent prouver que ce sont surtout les jeunes ou les adultes en mauvaise condition physique qui ne parviennent pas à lui échapper, ce qui réalise une prédation sélective. Dans sa vie aérienne, le Manchot Adélie adulte n'a pas d'ennemi. Vivant en Terre de Graham seulement *le Chionis (Chionis alba)* est essentiellement un charognard mais s'empare aussi des oeufs mal gardés. *Le Pétrel géant (Macronectes giganteus)* est surtout un prédateur des colonies de Manchot empereur mais attaque parfois aussi les poussins Adélies abandonnés et malades. *Le vrai prédateur est le Skua (Catharacta maccormicki).* Nombreux autour de toutes les colonies, les Skuas emportent les oeufs mal surveillés, attaquent et tuent les poussins mal protégés de moins de trois semaines. SLADEN (1958) assure qu'un poussin même jeune est sauvé s'il fait face et que les Skuas choisissent les poussins affaiblis. De toute façon, les Skuas représentent une cause de mortalité certainement importante.

Les facteurs physiques

Etant donnés les conditions écologiques antarctiques précédemment étudiées, il n'est pas surprenant que les facteurs physiques jouent, dans la mortalité, un rôle très important.

Le pouvoir de refroidissement des ambiances antarctiques, malgré l'adaptation de l'espèce, est une menace permanente, sauf pour les

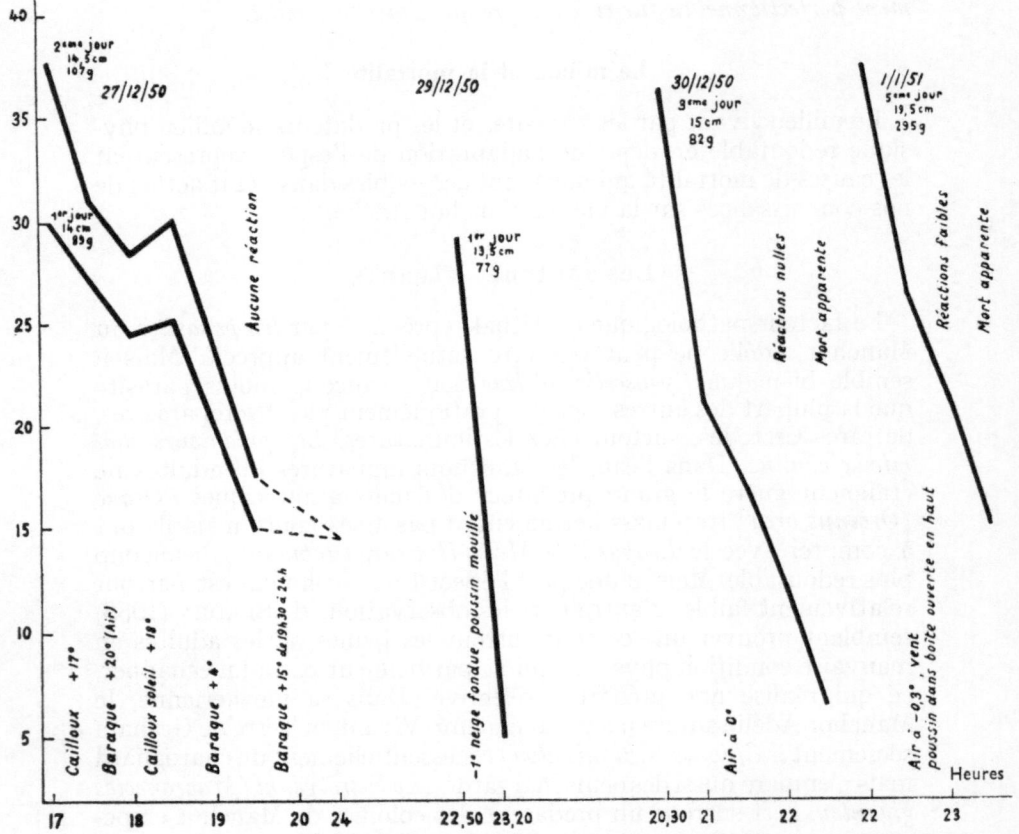

Fig. 15a. Thermo-régulation du poussin. Courbes de température cloacale de poussins âgés de 1 à 15 jours. A l'origine de chaque courbe sont indiqués la date de la mesure, l'âge en jours, la longueur en cm, le poids en g, en bas de la courbe les conditions d'exposition, à mi-hauteur les réactions observées.

adultes. Dans les conditions climatiques de la plupart des colonies pendant l'incubation, un oeuf abandonné pendant quelques minutes ou dizaines de minutes est congelé et perdu. Et l'on sait que si l'oeuf roule accidentellement hors du nid, le Manchot Adélie est incapable de l'y ramener. *Pendant l'élevage du poussin au nid*, la chaleur et la protection de l'adulte sont indispensables jusqu'à ce que le poussin possède une puissante défense thermique autonome. Une série d'expériences faites à Port-Martin (SAPIN-JALOUSTRE, 1953, 1960) ont montré que jusqu'au 9me jour le poussin se refroidit comme un corps physique et dans les conditions habituelles du climat de Port-Martin ne peut survivre plus de quelques dizaines de minutes sans la chaleur du parent couveur, ce temps tombant

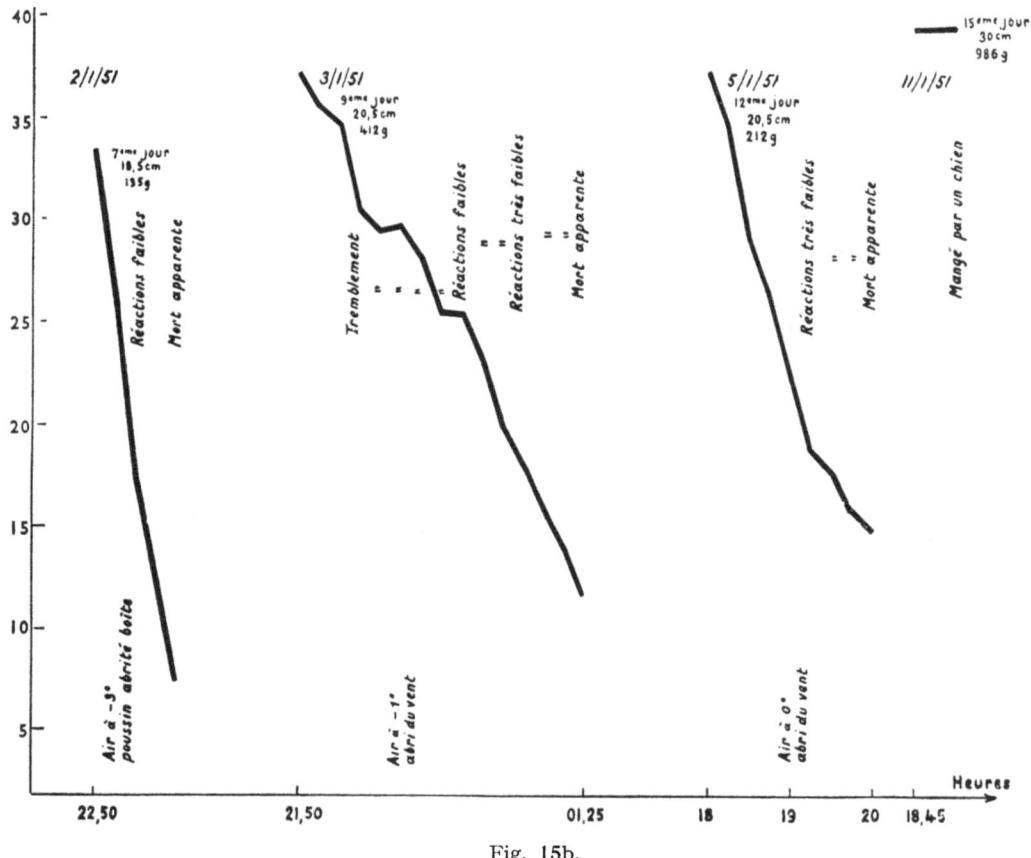

Fig. 15b.

à quelques minutes dans un grand blizzard (fig. 15). Au 9me jour, chez un poussin normalement développé, les premiers signes de thermo-régulation apparaissent: paliers dans la courbe de refroidissement, frissons thermiques. Au 12me jour, sauf dans un blizzard, le jeune peut supporter une absence du couveur d'une dizaine d'heures (fig. 16) et au 15me jour on peut considérer qu'il a acquis une thermo-régulation individuelle efficace (fig. 17) qui sera complétée par la "thermo-régulation communautaire" réalisée par la

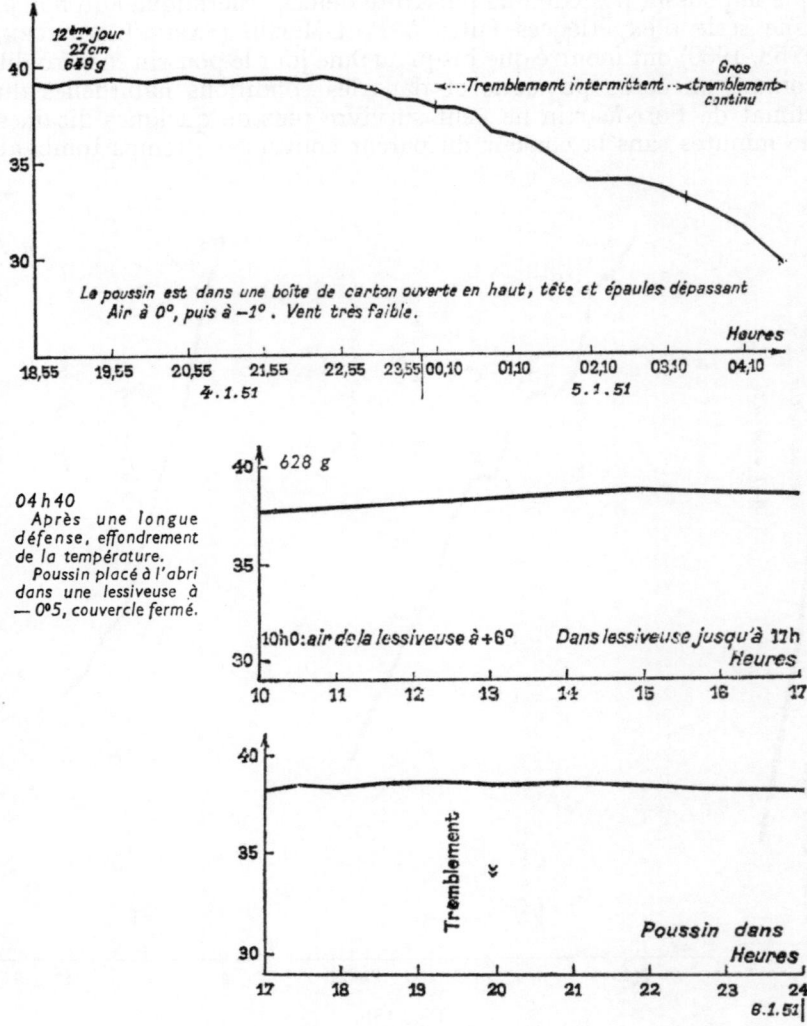

Fig. 16a. Thermo-régulation du poussin. Courbe de température cloacale du poussin no 387, âgé de 12 jours et pesant 649 g au début de l'expérience. Les conditions d'exposition et les réactions sont notées. La thermo-régulation est efficace et meilleure à une deuxième exposition.

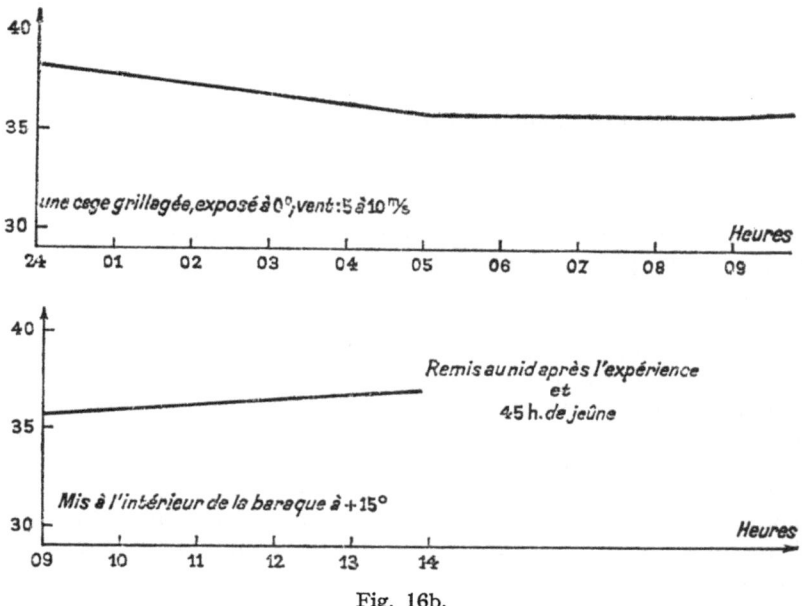

Fig. 16b.

formation des "tortues" pendant les tempêtes. Mais la thermo-régulation des poussins hypotrophiques (et leur nourriture dépend aussi, comme on le verra plus loin, des conditions physiques du milieu) est beaucoup moins bonne, le duvet des poussins laisse facilement pénétrer la glace transportée par le vent (photo 17) et *le nombre souvent impressionnant de cadavres de poussins après un blizzard atteste l'importance du pouvoir de refroidissement comme facteur de mortalité.*

Les précipitations et la fonte provoquent des pertes sérieuses d'oeufs et de poussins. *L'enneigement des nids, l'inondation des nids* par suite de la fonte diurne avec regel nocturne bien que les adultes

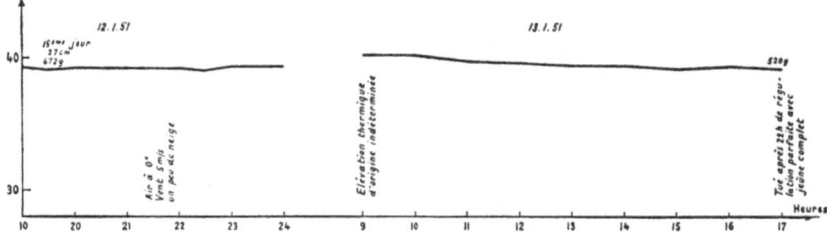

Fig. 17. Thermo-régulation du poussin. Courbe de température cloacale du poussin no. 415, âgé de 15 jours et pesant 672 g. Régulation parfaite pendant 22 h avec importante perte de poids.

continuent à couver peut amener dans certains secteurs et certaines années *une mortalité de 100% des oeufs* comme SLADEN (1958) l'a observé en Terre de Graham. Les poussins, même à la période des crèches et exceptionnellement les adultes peuvent être *enfouis dans des congères de neige dure* et incapable de se dégager (photo 15 et 16). Enfin les précipitations et les différences de température sont la cause *d'éboulements de falaises de rochers ou de glaciers qui détruisent des colonies entières* dont les vestiges ont été retrouvés en de nombreux points de l'antarctique.

Les conditions de la glace de mer jouent d'une manière indirecte mais fondamentale puisque la distance entre la colonie et l'eau libre règle la durée et la difficulté des expéditions de pêche des parents et donc *commande l'alimentation des jeunes.* Déjà pendant l'incubation, il semble que certains mâles abandonnent les oeufs si le retour des femelles tarde trop parce que l'eau libre est très loin. Plus tard, avec les besoins considérables et grandissants des poussins, une alimentation suffisante des jeunes devient impossible si les parents ont une longue distance à parcourir pour pêcher. Or, même avec l'eau libre devant la colonie, l'insuffisance alimentaire parait fréquente chez les poussins et elle aggrave d'autres facteurs de mortalité, entrainant une mauvaise défense contre le froid et contre les prédateurs.

Il n'est pas possible actuellement de chiffrer la part de tous ces facteurs écologiques de la mortalité et nous ne savons pratiquement rien de la mortalité des adultes, en particulier pendant l'hiver sur la banquise. Mais, pour les oeufs et les poussins, donc de la ponte à l'état d'immature, les différentes études montrent que la mortalité est très variable suivant les colonies et suivant les années, donc suivant les conditions du milieu, mais qu'elle est toujours élevée: SLADEN (1958) a trouvé, pour un secteur le chiffre exceptionnellement favorable de 60%, pour un autre secteur 100%, pour un autre 82%; en Terre Adélie, différentes mesures ont fourni des chiffres minimum entre 68% et 71% pour l'été 1950—1951. Ce sont ces différences de mortalité des oeufs et des poussins suivant les conditions du milieu pendant l'été qui expliquent les proportions différentes des diverses classes d'âge dans une colonie donnée pour une année donnée, et qui mettent en évidence *l'influence des facteurs géographiques et climatiques sur la structure et la destinée des colonies de Manchots Adélie.*

L'ECOLOGIE DU MANCHOT EMPEREUR

Le Manchot empereur *Aptenodytes forsteri* est le plus hautement antarctique de tous les Vertébrés aériens. La distribution de ses colonies suit assez étroitement les contours de la côte, exception faite du Nord de la Péninsule de Palmer, entre le 66ème et 78ème degré de latitude sud et si quelques représentants de l'espèce peuvent remonter beaucoup plus au Nord, ils semblent rarement franchir le 60ème parallèle. C'est dire que cet oiseau est confiné à des isothermes extrémement bas et à des biotopes plus influencés par les tempêtes continentales que par le climat océanique.

Les colonies connues

Historique

C'est probablement le navigateur russe von BELLINGSHAUSEN qui découvrit les tout premiers Manchots empereurs dans le pack-ice vers le 65ème degré de latitude Sud il y a quelques 140 années.

Par la suite, les missions se dirigeant vers l'Antarctique eurent souvent l'occasion d'apercevoir des Manchots, en plus ou moins grand nombre, au cours de leur voyage dans les glaces. Cependant, jusqu'au début de ce siècle, on s'interrogeait encore sur l'existence des colonies de reproduction. Ceci s'expliquait par le fait que les bateaux polaires ne pouvaient accéder au continent qu'au moment même où celui-ci était libéré de sa ceinture de glace de mer. Les Manchots empereurs se reproduisant sur cette dernière avaient donc déjà dérivé avec elle vers le large. Il appartenait aux missions établies à demeure en hiver, ce devait être le cas pour celle de WIL-SON et pour bien d'autres par la suite, de les découvrir, lors d'un voyage le long de la côte. C'est aussi ce qui explique que la plupart de ces colonies aient été reconnues ces toutes dernières années, au cours de l'Année Géophysique Internationale, grâce aux multiples stations et aux moyens terrestres, aériens et maritimes puissants mis à leur disposition.

La British National Antarctic Expedition de 1901 à 1904 dont le bateau, le Discovery, emprisonné par les glaces, était transformé en station d'hivernage, devait permettre au naturaliste de l'expédition, le Dr. WILSON, de trouver le plus méridional des sites de reproduction, au Cap Crozier par 77°29′ de latitude Sud. La distance qui le séparait du Discovery ne permit pas à ce naturaliste de longues observations. Il n'en ramena pas moins une ample moisson de documents originaux sur la période d'élevage des poussins.

Plus de 40 années après, un second ornithologiste, STONEHOUSE, avait le privilège d'apercevoir la colonie des Ilôts de Dion en Terre de Graham (67°52′S — 68°43′E) et ses brèves observations (5 juin au 15 août 1949) lui permettaient de réaliser une très belle étude sur le comportement des oiseaux pendant cette période du cycle.

En 1950, l'un de nous (S.J.) visitait l'emplacement d'une nouvelle colonie, celle de Pointe Géologie (66°40′S —140°01′E) qui est avec celle de la Pointe Kloa (66°37′S — 57°11′E) et de l'Ile Haswell (66°33′S — 92°50′E) une des plus septentrionales. La saison étant fort avancée, SAPIN-JALOUSTRE ne put se livrer à aucune étude suivie, mais il eût quand même le privilège d'émettre d'excellentes hypothèses sur le déroulement du cycle annuel, et surtout de s'interroger sur les multiples problèmes posés par la reproduction hivernale de l'oiseau, notamment son adaptation à l'ambiance extrême du climat antarctique.

Actuellement, on peut estimer qu'une vingtaine de colonies, représentant plus de 200.000 oiseaux, sont réparties autour du continent austral. L'effectif de l'espèce ne pourra toutefois être connu exactement que lorsque des dénombrements très précis et non des estimations, où les erreurs peuvent aller du simple au double, auront été faits dans chaque colonie.

Le micro-climat de la colonie de reproduction

Nous savons que le climat regional de Pointe Géologie défini d'après les relevés synoptiques des météorologistes est plus clément que celui d'autres régions côtières proches (Port-Martin). Cet Archipel où les Manchots empereurs et des milliers d'autres oiseaux viennent se reproduire échappe souvent, du fait de sa position insulaire, aux vents venant de l'intérieur du continent dont la vitesse est accélérée par la pente continentale pré-côtière (vents catabatiques). L'étude du micro-climat de l'emplacement de la colonie des Manchots empereurs poursuivie en 1952 et 1956 (Photo 18) nous a

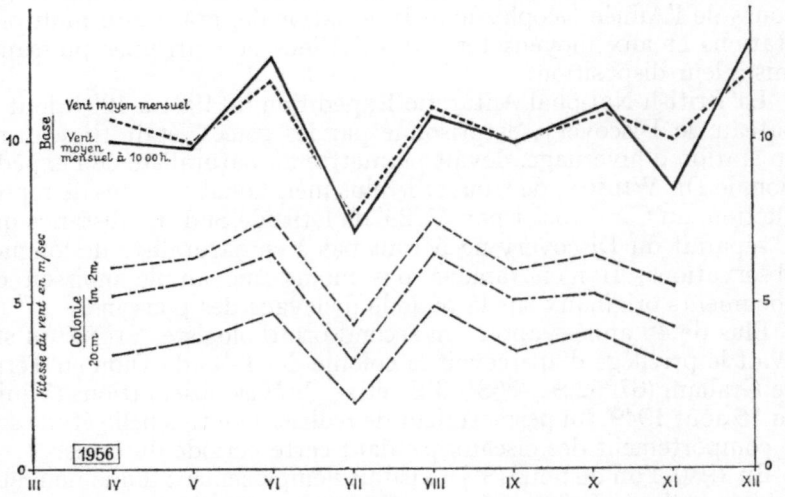

Fig. 18. Moyennes mensuelles de vent à la colonie à 0,20 m, 1 et 2 m de hauteur, et à la station, à 50 m d'altitude à 10 heures locales (00.00 heure T.U.). La courbe de moyenne mensuelle du vent obtenue d'après les six observations quotidiennes à la station météorologique est indiquée en pointillé.

montré que si la température ne différait guère de celle de la station Dumont d'Urville, le vent était par contre infiniment moins violent. C'est ainsi qu'en 1956 sa vitesse à 0,20 m de hauteur au centre de la colonie était inférieure de 60% à celle qui était mesurée à la station (anémomètre placé à 10 mètres au dessus de la station, soit 50 mètres d'altitude) (fig. 18). Ceci s'explique probablement par le fait que le territoire de la colonie est placé au niveau de la mer et abrité au Sud par la haute falaise continentale et à l'Est par celle du glacier Astrolabe. (Photo 19).

Il faut ajouter que les inégalités du sol ralentissent considérablement l'écoulement du vent à son niveau comme l'a montré SAPIN-JALOUSTRE (1955) au cours de ses études sur les micro-climats des colonies de Manchots Adélies. Cet auteur concluait que la demande thermique de ces oiseaux était inférieure de 50% à celle du climat général.

La clémence relative du climat n'est semble-t-il pas le seul facteur influencant le choix des Manchots empereurs. La présence d'une nappe d'eau libre ou de failles dans la glace de mer est indispensable aux adultes pour assurer régulièrement leur alimentation et celle de leurs poussins au cours de l'hiver. Ces "points d'eau" que recherchent les oiseaux paraissent être entretenus à proximité de la côte par l'avancement continu d'un glacier et de ses icebergs, plus au large par des courants marins et leur présence est alors indiquée par un "water-sky" plus ou moins étendu (Pointe Géologie 1956). Toutes les colonies de l'espèce sont établies près de l'un ou de l'autre voir même des deux.

Les biotopes des différentes colonies de reproduction

Il eut été particulièrement intéressant de disposer de données météorologiques et microclimatiques d'autres colonies afin de savoir si elles disposaient des mêmes privilèges que la nôtre. Nous ne possédons malheureusement à l'heure actuelle que des enregistrements effectués dans des stations d'hivernage établies à quelque distance.

Colonies du Territoire Australien

Découvertes entre les années 1954 et 1957 (WILLING, 1958) entre 66°38' et 69°15' de latitude Sud et 57°19' et 76°50' de longitude Est elles n'ont fait l'objet d'aucune mesure microclimatique. La topographie de leurs emplacements décrite par BUDD (1961) montre que toutes sont abritées par des falaises de glaces ou des collines continentales (colonie du glacier Taylor). Les données météorologiques de la station de Mawson étant à quelque chose près analogues à celles de Pointe Géologie (Température moyenne annuelle — 10°9 C, vent moyen annuel 36 km/heure) et celle de la station de Davis

606

moins sévères encore (Température — 9°5 C, vent 16,8 km/heure)
on peut penser que le micro-climat des emplacements de reproduction, est à quelque chose près voisin de celui de la colonie de Pointe-
Géologie.

La colonie de Cap Crozier

De latitude très inférieure aux précédentes (77° 29′ S) elle est
soumise à une température beaucoup plus basse de l'ordre de 20° C.
La vitesse du vent ne nous est pas connue, mais on peut supposer
qu'elle n'est guère élevée et que la température joue un rôle prépondérant. Si les conditions climatiques de cette colonie paraissent
sévères, elles n'entraînent pas, en année normale, une mortalité
importante (CAUGHLEY, 1960).

Colonie des Ilots de Dion

Etablie sur un ilot elle supporte à un climat assez voisin de celle
de Pointe-Géologie comme l'indiquent les chiffres obtenus par
STONEHOUSE au cours de son séjour (5 juin au 14 août 1949).

Colonie de l'île Haswell

Les températures enregistrées à la station de Mirny proche de
l'île Haswell sont analogue à celles de Pointe-Géologie alors que la
vitesse moyenne annuelle du vent est plus élevée (+3 m/sec). Il est
permis de penser que l'île Haswell est moins sensible à l'influence
continentale que Mirny et que le biotope de la colonie de Manchots
empereurs est plus favorable que celui de cette station.

En conclusion malgré l'absence d'informations sur les micro-
climats des sites de reproductions des *Aptenodytes forsteri*, il semble
bien que la plupart, sinon la totalité d'entre eux, soient moins
sévères que le climat environnant. Ceci nous montre tout l'intérêt
présenté par des enregistrements simultanés dans plusieurs de ces
colonies, du vent, de la température, en même temps que seraient
décrites la topographie et noté la formation, l'évolution et la dislocation de la glace de mer au cours du cycle annuel. De telles observations tout en apportant bien des éclaircissements sur l'établissement des colonies sont des préliminaires indispensables aux études
physiologiques ou écologiques.

L'espèce

Sous ce titre, nous aborderons successivement l'étude du cycle du
poids et des gonades, celle de la thyroïde et du foie ainsi que les
particularités physiologiques et éthologiques de l'espèce.

Le cycle du poids

Le poids maximum est atteint par les oiseaux des deux sexes à leur arrivée à la colonie en mars et avril. Les femelles pèsent en moyenne 28 à 32 kg, les mâles, dont le jeûne physiologique sera deux fois plus long, 35 à 40 kg.

Au mois de mai, après la ponte, les femelles quittent momentanément la colonie. Elles n'ont pris aucun aliment durant ces 45 à 50 jours et leur poids a diminué d'environ 20%. Les mâles demeurent à la colonie où ils assureront l'incubation de l'oeuf jusqu'à son terme et quelquefois au delà. La totalité de leurs réserves de graisse sera consommée et leur poids tombera à 23—24 kg, soit un amaigrissement de 35 à 45% selon les cas. Un de nos sujets marqués, le mâle no 5, maintenu alternativement en parc et dans la colonie, pesait 19.10 kg le 17 août 1956, au terme d'un jeûne observé de 93 jours, qui lui avait fait perdre plus de 50% de son poids initial. Par la suite, au cours de la période d'élevage, les variations sont beaucoup moins marquées et le poids moyen s'établit autour de 25 à 27 kg, les mâles restant toujours plus lourds que leurs partenaires (fig. 19).

A la fin du cycle annuel, la remontée de la courbe pondérale correspond à l'engraissement précédant la mue, laquelle s'accompagne d'un jeûne physiologique de l'ordre de 30 jours. Au mois de février, une nouvelle augmentation des poids individuels correspond à la préparation du jeûne physiologique du cycle annuel suivant.

Pour en terminer avec cette étude, nous mentionnerons à titre indicatif les principales caractéristiques de l'oiseau. Les mâles du plus grand des Sphéniscidés sont plus longs (117 cm) que les femelles

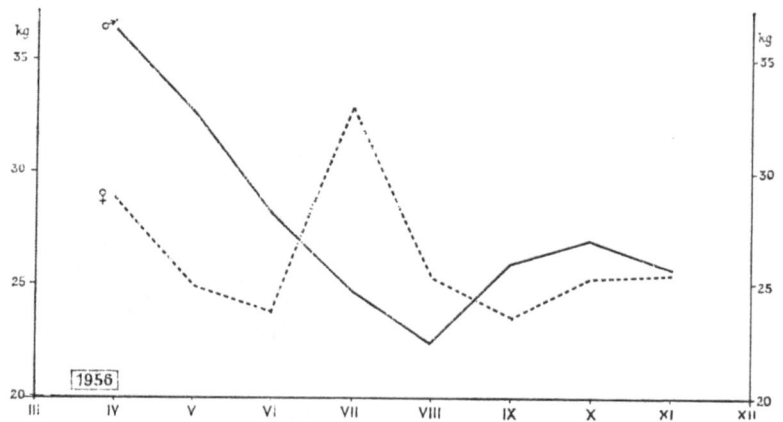

Fig. 19. Poids moyen mensuel des adultes en 1956. Dans les deux sexes, chaque point de la courbe correspond à la moyenne de cinq adultes pesés chaque mois dans la colonie.

608

(114 cm). Cette différence est trop faible pour pouvoir être appréciée chez l'oiseau vivant; elle n'est donc d'aucune utilité pour la détermination du sexe. Le périmètre thoracique comme le périmètre du bassin est fonction de l'embonpoint de l'oiseau. En règle générale, tous deux sont plus réduits chez les femelles, ce qui aide souvent à les séparer de leur partenaire. Les muscles pectoraux représentent, et de loin, la masse musculaire la plus importante du corps de l'oiseau. Leur poids moyen est supérieur à 7 kg dans les deux sexes.

Le cycle des gonades

Dans les deux sexes, les gonades atteignent leur poids maximum dans la première quinzaine de mai peu avant le solstice d'hiver. A partir de cette date, testicules et ovaires involuent très rapidement pour atteindre un minimum qui ne changera guère jusqu'au mois de décembre (fig. 20).

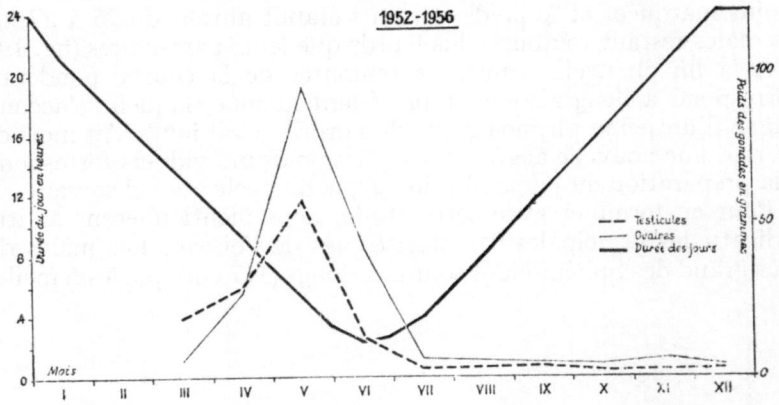

Fig. 20. Évolution pondérale des testicules et des ovaires au cours du cycle reproducteur annuel. On voit que les poids maximas des gonades sont atteints à la fin de la période de décroissance de la durée des jours.

L'étude histologique du testicule montre l'intense activité spermatogénétique des mois de mars à juin, les spermatozoïdes n'étant toutefois présents dans la lumière des tubes séminifères qu'entre la fin avril et le début de juin. Dès cette date, ces tubes régressent considérablement alors que le tissus interstitiel prend de plus en plus d'importance.

Le cycle thyroïdien

Les glandes thyroïdes sont plus légères chez les mâles (1,55 g pour 32 oiseaux) que chez les femelles (1,84 g pour 30 oiseaux). Dans les deux sexes, le poids le plus élevé est enregistré à l'arrivée à la colonie. Il régresse pendant le jeûne physiologique et se stabilise ensuite, avec de légères variations, pour augmenter à nouveau à la fin du cycle.

Les modifications histologiques de cet organe sont encore plus significatives. En mars, l'épithelium est assez haut chez les arrivants, alors que les vésicules sont plutôt de grande taille. L'épithelium prend rapidement un aspect endothéliforme alors que la réduction de taille des vésicules se fait plus lentement. La fin du jeûne physiologique s'accompagne du développement des cellules épithéliales qui deviennent cubiques, et de la diminution de volume des vésicules. (fig. 21 et 22).

Au cours de la période d'élevage les modifications seront faibles et les images d'intense activité réapparaissent à l'approche de la

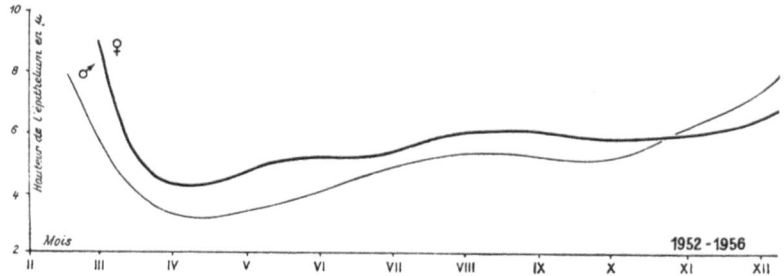

Fig. 21. Variations de la hauteur de l'épithélium vésiculaire thyroïdien au cours du cycle annuel, d'après les dimensions moyennes de 10 épithéliums par glande étudiée.

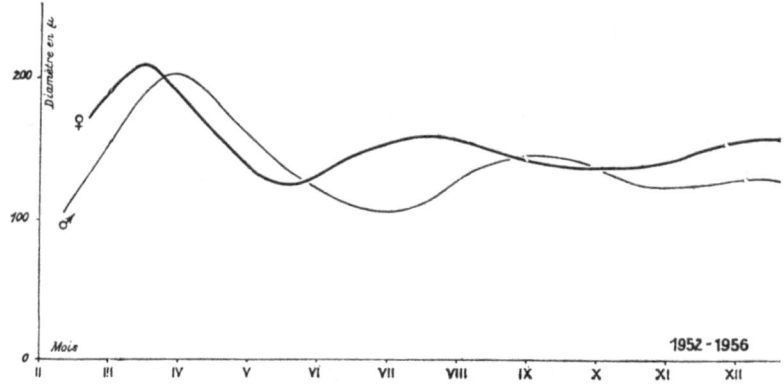

Fig. 22. Diamètre moyen mensuel des vésicules thyroïdiennes d'après l'examen des 33 glandes d'oiseaux femelles et 36 glandes d'oiseaux mâles. Pour chaque organe étudié la dimension moyenne des vésicules a été obtenue d'après les mensurations de 10 vésicules au micromètre objectif.

mue. Quelques dosages pratiqués sur la totalité de la glande thyroïde ont montré la présence d'une proportion plus ou moins élevée de sels de calcium. Le rôle de cette substance à ce niveau reste encore des plus problématiques.

Le cycle hépatique

De par sa fonction d'organe de réserve, le foie paraît jouer dans cette espèce, un rôle de tout premier plan, du fait de la durée exceptionnelle du jeûne physiologique (fig. 23).

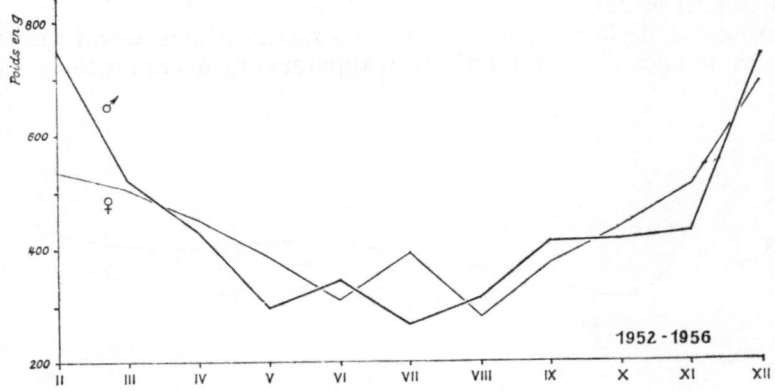

Fig. 23. Poids moyen mensuel du foie dans les deux sexes, en 1952 et 1956. (Données fournies par 41 mâles et 47 femelles).

Remarquablement lourd à l'arrivée — 750 g pour un mâle, et 536 g pour 2 femelles au mois de mars — il subit une régression spectaculaire pendant les mois suivants. Il n'atteignait en effet que 312 g chez 2 femelles au mois de juin et 267 g pour 3 mâles au mois de juillet. Au niveau de la cellule hépatique, la surcharge lipidique particulièrement importante de l'arrivée, s'atténue progressivement; elle fait place, à la fin du jeûne à une vacuolisation parfois extrême. Par la suite, le tissus hépatique reprend un aspect normal qu'il conserve jusqu'à la fin du cycle annuel.

Les particularités physiologiques

Le Manchot empereur supportant normalement des froids très vifs et des blizzards violents, il était normal de chercher à mesurer sa température, tant au niveau du corps qu'à celui des extrémités. Bien que poursuivies avec des appareils rudimentaires, ces études nous ont révélé quelques faits importants.

La température centrale

Les premièrs chiffres furent relevés sur des oiseaux capturés hors de la colonie, ou transportés à proximité de notre station d'hivernage. Ils étaient dans la plupart des cas égaux ou supérieurs à 38°, ce qui confirmait ceux obtenus par les auteurs antérieurs dont Wilson. En 1956, la plupart de nos mesures furent faites à la colonie même et les résultats furent très différents.

Température rectale au cours des principales phases du cycle de l'oiseau une fois "installé" dans la colonie (en °C)

	Mâles	Femelles	Moyenne
Pariade Avril à 20-V-1956	35,1 (9) ± 2,1	36,9 (3) ± 1,4	35,5 (12) ± 2,1
Incubation 20-V à 20-VII 1956	36,7 (98) ± 1,17	37,4 (22) ± 0,6	36,8 (120) ± 1,1
Début élevage 20-VII à 20-IX 1956	37,5 (21) ± 0,95	37,2 (13) ± 1,33	37,4 (34) ± 1,1
Fin élevage 20-IX à décembre 1956	37,5 (5) ± 0,8	37 (10) ± 0,93	37,2 (15) ± 0,9

Le niveau thermique moyen le plus élevé ne dépassait pas 37°5 alors qu'il était toujours supérieur à 38° dans nos premières mesures.

A titre de comparaison, nous avons isolé au cours de l'année 1956, 11 oiseaux à différentes périodes et les 264 mesures auxquelles ils furent soumis donnent une moyenne de 38°1. La preuve est ainsi faite que le niveau thermique du sujet isolé, expérimentalement ou non, est supérieur à celui de l'oiseau vivant au milieu de ses congénères, tout au moins pendant les mois les plus froids de l'année.

En outre, il y a, semble-t-il, une différence entre l'oiseau qui arrive à la colonie et l'installé, qui y séjourne depuis quelques temps. Chez le premier, l'activité physique, le fait qu'il vient de s'alimenter et qu'il peut de surcroit conserver une certaine quantité d'aliments dans l'estomac, jouent un rôle non négligeable. L'arrivant se comporte donc, tout au moins momentanément, comme un isolé, comme nous le montrait une expérience du mois de juin où la température de 3 d'entre eux était de 38° alors que celle des installés était de 36°5. Ces derniers, outre leur activité physique limitée, ont en général l'estomac vide et leur métabolisme paraît être plus réduit.

La température de la patte et celle de l'aileron sont très basses souvent voisines de 0° chez les installés, alors qu'elles sont notablement plus élevées chez les arrivants. Exception faite de la différence d'activité musculaire de ces deux catégories d'oiseaux, on peut penser qu'il y a probablement au niveau des membres inférieurs et supérieurs une disposition particulière des vaisseaux afférents et efférents qui permet selon les besoins, soit l'économie, soit la diffusion de la chaleur.

Quoiqu'il en soit, le fait important est la différence de température centrale entre les isolés et les installés.

Au début de l'incubation, le rôle d'isolant thermique joué par la graisse sous-cutanée s'atténue au fur et à mesure qu'elle est consommée.

En compensation, la température centrale devrait donc augmenter sensiblement. Or, si tel est le cas pour les oiseaux arrivés à l'extrême limite du jeûne physiologique, il n'en est pas ainsi pour les autres.

En fait, on constate qu'entre les mois d'avril et juillet, les tortues, ces groupes très denses où les oiseaux sont étroitement appliqués les uns contre les autres, prennent de plus en plus d'importance au cours de l'incubation (Photo 20). Ceci correspond à l'aggravation progressive des conditions météorologiques locales au cours du plein hiver, mais aussi au fait que les oiseaux peuvent grâce à elle économiser leurs réserves de graisse.

Les nombreuses mesures que nous avons effectuées à cette époque, n'avaient d'autre but que de montrer qu'il ne s'agissait pas d'une simple hypothèse.

L'observation nous a montré que les multiples tortues de la pariade s'étaient peu à peu réunies pour n'en former qu'une ou deux où les 6.000 couveurs de la colonie se trouvaient entassés pendant les tempêtes. La progression de cette masse au cours des blizzards est provoquée par le déplacement des oiseaux des endroits les plus exposés au vent vers ceux où ils bénéficient de la protection de leurs congénères.

Les oiseaux en tortue sont ceux dont la température est la plus basse, suivie par les oiseaux groupés, et enfin par les isolés la différence entre les premiers et les derniers est de l'ordre de 2° C.

A titre d'expérience, nous avons transporté un mâle capturé dans une tortue, hors de la colonie. Sa température initiale qui était de 34°5 fut en moyenne de 38° pendant les 98 heures où il fut maintenu isolé.

Restait à prouver que l'amaigrissement était effectivement moindre au sein de la colonie que chez l'isolé. Pour cela, nous avons comparé la diminution de poids d'oiseaux de la colonie, d'oiseaux pouvant former des tortues réduites dans des parcs, et enfin d'oiseaux isolés. Les résultats de ces expériences montrent les différences entre ces trois sortes de groupement. (Tableau p. 613).

On peut donc conclure que les formations en tortue de l'*Aptenodytes forsteri* constituent un véritable mécanisme de thermorégulation sociale. En ajoutant leurs effets à ceux d'un micro-climat

Amaigrissement (en g par 24 heures) d'*Aptenodytes forsteri* en fonction de leur densité de groupement.

Sexe	Oiseaux libres dans la colonie	Oiseaux en parc	Oiseux isolés hors de la colonie
Mâle	116,2 (moyenne de 5 mâles du 14 avril au 13 mai)	171,3 (moyenne de 2 mâles du 14 avril au 21 mai)	
Mâle	155,4 (moyenne de 5 mâles du 13 mai au 12 juin)	235,5 (moyenne de 2 mâles du 23 mai au 4 juin)	305 (1 mâle du 23 au 27 juin)
Femelle	133,8 (1 femelle du 14 avril du 13 mai)	205,2 (4 femelles du 14 avril au 23 mai)	297,5 (1 femelle du 11 au 14 juin)
Femelle	227 (1 femelle du 7 au 17 août)	287 (4 femelles du 3 au 31 août)	357,3 (1 femelle du 3 au 6 août)

relativement privilégié et à l'adaptation individuelle, elles permettent à l'espèce de supporter les rigueurs de l'hiver antarctique avec une moindre mortalité.

L'alimentation

L'estomac des oiseaux autopsiés à leur arrivée à la colonie était soit totalement vide, soit rempli d'une bouillie homogène inanalysable. Nous avons pu être certains cependant que le régime alimentaire comportait surtout des Crustacés (Euphausies), des Poissons et des Céphalopodes.

Ajoutons que l'estomac contient toujours une certaine quantité de graviers dont le rôle est peut-être d'aider au malaxage des proies, à moins que leur présence ne soit indispensable au maintien d'une certaine activité de cet organe au cours des jeûnes physiologiques.

Les caractéristiques éthologiques

Chez la plupart des Sphéniscidés et notamment chez le Manchot Adélie, le comportement de défense territoriale se traduit par des réactions agressives violentes à l'égard des intrus.

Chez l'*Aptenodytes forsteri*, au contraire, l'incubation ambulatoire et la formation des tortues vont de pair avec une atténuation considérable des antagonismes individuels.

Le fait de vivre dans une société nombreuse, parfois étroitement assemblée, implique aussi l'utilisation par chacun des individus qui la composent, de signaux optiques ou sonores.

Les signaux optiques

La parade mutelle est celui qui joue le rôle le plus important tant à l'échelon individuel qu'au niveau social.

Cette parade ou face à face, est ainsi nommée en raison de la position respective des deux oiseaux dont les têtes sont légèrement dressées (photo 21). Elle s'accompagne d'une immobilité quasi-totale du corps et d'un gonflement important de la région du cou. Pendant la pariade, la parade mutuelle est généralement précédée du chant de cour, alors que par la suite elle comporte quelques variantes gestuelles que ce soit avant la copulation et la ponte ou lors de l'échange de l'oeuf ou du poussin entre les deux conjoints. Nous n'avons toutefois pas jugé utile de donner un qualificatif à chacune de ces attitudes.

La fonction de cette parade est double. Elle sert tout d'abord à la reconnaissance des deux conjoints à courte distance. Outre ce rôle d'identification, elle a aussi un rôle sexuel et c'est, au sens de RICHDALE, une parade amoureuse. Le fait qu'elle soit utilisée par les couples se tenant compagnie paraît plus complexe, encore que nous sachions que ces oiseaux imitent leurs partenaires immédiats ou sont stimulés par eux.

Les comportements agonistiques

Bien qu'étant doué d'une force peu commune et disposant d'armes puissantes comme le bec et l'aileron, le Manchot empereur est un oiseau peu agressif. Les batailles sont peu fréquentes et ne concernent qu'un petit nombre d'oiseaux. Si ces rivalités individuelles peuvent être passées sous silence, il n'en est pas de même pour les luttes des "trios" et pour celles dont les poussins seront l'enjeu en juillet et août.

Les batailles à trois sont propres à la pariade. Elles débutent au moment où un oiseau, généralement de sexe femelle, s'approche d'un couple en parade et cherche à supplanter l'oiseau de son sexe (photo 22). Il s'agit peut-être de la femelle légitime, arrivée tardivement à la colonie et qui veut chasser l'intruse comme l'ont montré RICHDALE (1951) et SLADEN (1960) chez d'autres Sphéniscidés. Cela est possible, mais aucune observation précise ne nous permet de l'affirmer.

La compétition des deux femelles engendre une bataille longue de 2 jours tout au plus et à laquelle le mâle peut participer quelquefois très activement. Elle se termine par l'éviction de l'oiseau surnuméraire ou son départ brutal provoqué par l'appel d'un représentant du sexe opposé.

Les luttes pour les poussins font suite à des batailles sporadiques et limitées de quelques adultes se disputant un oeuf délaissé par un couveur au mois de juin. Elles ne débutent réellement qu'en juillet et leurs auteurs ou "inemployés" ne sont autres que d'anciens couveurs ayant perdu ou abandonné leur oeuf ou leur poussin. Ces oiseaux parcourent la colonie à la recherche des poussins échappés des pattes de leurs parents; mais ils n'hésitent pas non plus à s'attaquer à ces derniers pour s'en emparer provoquant les célèbres "mêlées de rugby" décrites par WILSON (1907).

Le chant des poussins paraît orienter leur recherche, et stimuler leur activité. Si un tel oiseau parvient à s'emparer d'un poussin, il a bien des difficultés à le conserver sur ses pattes, et l'on constate alors que son repli cutané abdominal est loin d'avoir l'ampleur de celui des reproducteurs normaux. Toutefois, ceux de ces inemployés dont l'estomac est plein, ou qui ont récemment perdu leur propre poussin, sont susceptibles d'acquérir, du moins momentanément, un comportement d'éleveur.

Ce besoin de couver, qui engendre le comportement batailleur des inemployés, est peut-être influence par un facteur hormonal (prolactine?). Un taux insuffisant de l'hormone incriminée serait peut-être de nature à justifier leur comportement incubateur incomplet, le poussin pour la capture duquel ils ont lutté avec acharnement étant abandonné quelques heures plus tard.

Mais le besoin d'imiter la majorité des parents qui les entourent et les stimulations apportées par les chants des poussins ne sont pas à négliger, et suffisent peut-être à l'expliquer.

Citons enfin pour mémoire les rares manifestations agonistiques à l'égard des prédateurs (Pétrels géants et Skuas) et des représentants d'autres espèces qui peuvent pénétrer sur l'emplacement de la colonie (Manchot Adélie).

Le baillement, les mouvements de déglutition et les mouvements contagieux

Parmi les autres attitudes observées pendant le séjour à la colonie, mentionnons le baillement, les mouvements de déglutition et les mouvements contagieux.

Les mouvements de déglutition ponctuent généralement la parade mutuelle, le chant de cour et les batailles.

Si certains mouvements contagieux répondent à un stimulus externe très perceptible, comme la chute d'un iceberg, un craquement de glace de mer, la motivation de certains autres tels que les cris et les battements collectifs d'ailerons chez les poussins échappe à l'observation.

Les signaux sonores

Parmi les cinq cris que nous avons pu individualiser dans cette espèce, nous retiendrons surtout le chant de cour dont le rôle nous a paru primer largement celui des autres.

616

Le chant de cour:

Durant toute l'émission de ce signal sonore, le bec ne s'entrouvre qu'au niveau des commissures, la tête étant largement inclinée sur la poitrine (photo 23). En 1952 l'audition répétée des chants des oiseaux de la colonie nous avait amené à conclure à une différence de voix entre les mâles et les femelles, que l'analyse spectrographique d'enregistrements sur bandes magnétiques n'a fait que confirmer. Le chant de l'oiseau adulte est long de 2 secondes environ et sa fréquence moyenne, de l'ordre de 2 KHz, est très voisine dans les deux sexes. Si les composantes fondamentales de la voix du mâle couvrent une bande de fréquence à peine plus étendue que celle de la femelle, le découpage du signal est très différent; les trains de signaux sont chez cette dernière assez brefs donc plus nombreux que ceux du mâle (Photo 24).

Le chant est utilisé dans la plupart des manifestations des couples à la colonie. Il sert à la recherche du partenaire et prélude à la majorité des parades gestuelles. Mais il est aussi souvent utilisé par des oiseaux isolés sur l'emplacement de la colonie, voire même hors de ses limites.

Outre l'existence du dimorphisme vocal, chaque signal paraît posséder une caractéristique grossière qui lui est propre. Ce serait grâce à cette faible nuance individuelle que les deux partenaires pourraient se reconnaître à distance, la confirmation étant fournie ensuite par la reconnaissance visuelle au cours du face à face. Malgré de multiples tentatives nous n'avons pas pu obtenir de preuves absolument formelles démontrant l'exactitude de cette hypothèse. Il eût fallu que nous possédions plusieurs dizaines d'enregistrements de chants isolés dont la comparaison nous aurait donné une réponse. Le seul argument dont nous disposons actuellement nous a été fourni par les oiseaux bagués. Deux membres d'un couple se retrouvent presque toujours après des appels vocaux répétés, suivis d'un face à face dès qu'ils sont en présence l'un de l'autre.

Les autres cris ont un rôle individuel et social moindre. Ce sont le "coup de trompette" sorte d'appel puissant, les cris de frayeur et de colère, et les cris émis par les oiseaux jouant dans l'eau.

Pour en terminer avec les caractéristiques de l'espèce, nous mentionnerons les activités élémentaires qui groupent la locomotion terrestre, les soins de toilette, les attitudes de repos et les jeux.

Activités élémentaires. Locomotion terrestre

La marche en station verticale est la plus utilisée. Elle est remplacée par le déplacement sur la poitrine ou toboganning, lorsque le sol est couvert d'une épaisse couche de neige, ou formé de glace vive. L'oiseau avance alors sous l'effet des impulsions données par les pattes, et à un degré moindre par les ailerons, ceux-ci ayant surtout pour mission de maintenir le corps en équilibre.

Les soins de toilette

Ils sont très fréquents pendant l'incubation, au point que par beau temps la majorité des oiseaux s'y adonnent longuement.

Le repos

L'oiseau se repose debout, le cou "rentré" entre les épaules, ou le bec glissé dans l'ailerons. (Photo 25). Mais il peut aussi s'étendre sur la glace, la position "bec dans l'aileron" étant alors l'exception.

Les jeux

Les jeux enfin ne s'observent que dans l'eau, lorsque la mer libre est proche de la colonie. Des groupes de plusieurs individus nagent en surface puis plongent et réapparaissent simultanément, émaillant leurs ébats de cris particuliers.

Le cycle annuel du comportement

Le séjour des oiseaux à la côte, long de près de dix mois, peut être divisé en six phases: la formation de la colonie et la pariade, la ponte et l'incubation, le retour des femelles et le départ des mâles couveurs, l'élevage individuel des poussins, l'émancipation des grands poussins et leur vie dans les crèches, enfin la mue et la dislocation de la colonie.

Avant d'aborder cette étude du cycle annuel, précisons tout d'abord les dates d'arrivée et de départ des oiseaux.

A la colonie de Pointe-Géologie, les premiers d'entre eux arrivent à la colonie au mois de mars, la débâcle des glaces provoquant leur départ forcé vers le milieu du mois de décembre (10—20 décembre pour ces dernières années). A la colonie de l'île Haswell, de latitude voisine, les dates sont analogues en ce qui concerne le début du cycle, mais nous ne savons rien de la date de la débâcle. A la colonie des Ilots de Dion les arrivées seraient postérieures de près d'un mois et de l'ordre de 45 jours à Cap Crozier. Ces chiffres seront probablement sujets à révision quand nous disposerons d'observations plus suivies dans ces lieux de reproduction.

La formation de la colonie et la pariade

Les premiers oiseaux regagnent la côte antarctique au moment où l'emplacement de la colonie commence à se couvrir de glace. (Photo 26).

Comment peuvent-ils retrouver ce point précis de la côte après un séjour marin estival où les recherches alimentaires et les courants les ont éloignés à de grandes distances de la côte? Ces dérives obligent-elles certains d'entre eux à se fixer dans une autre colonie, cela est possible, mais nous pouvons quand même supposer que le retour aux lieux de reproduction se fait au large d'après un repère astronomique qui pourraît bien être le soleil (KRAMER, 1948). A quelque distance du continent, cette "navigation astronomique" serait délaissée au profit de repères topographiques côtiers connus tels que glaciers, baies ou presqu'îles. Les Manchots empereurs de Pointe-Géologie accédant presque tous à la colonie par une voie parallèle au glacier Astrolabe, il se peut que ce soit à l'aide de la partie immergée des icebergs qui le constituent.

Quoiqu'il en soit de ce mécanisme d'orientation, les oiseaux arrivent d'abord isolément, puis par petits groupes et enfin par centaines à la fin du mois de mars. L'effectif ne sera complet qu'au bout d'un mois: ce décalage persistera à chaque phase du cycle de telle sorte que les poussins des plus attardés auront bien peu de chance de surmonter ce lourd handicap.

Recherche du partenaire

La plupart des oiseaux ont un comportement analogue à l'approche de la colonie naissante. Ils dressent d'abord la tête puis la frottent successivement à droite et à gauche sur la partie supérieure de l'aileron correspondant, et l'inclinent ensuite largement pour chanter. Ce frottement alternatif de la tête a pour but, pensons-nous, de dégager les conduits auditifs externes, dont l'obstruction pourrait nuire à une bonne réception des signaux sonores. Le comportement de recherche du partenaire paraît analogue dans les deux sexes, à cette exception près que l'activité vocale des femelles est plus marquée. Ainsi entre les 15 et 23 avril, 147 chants sur 160 avaient pour auteurs des oiseaux de ce sexe.

Si un oiseau est sensibilisé par l'appel sonore d'un sujet déterminé de l'autre sexe, il se déplace vers lui et tous les deux s'observent immobiles, face à face pendant quelques instants. Si dans un tel cas il y a affinité, cela ne signifie pas pour autant qu'un couple est formé. Les deux oiseaux peuvent alors se séparer et poursuivre isolément leur recherche. Nous n'avons jamais été témoins d'une union ou d'une réunion que nous ayons pu qualifier avec certitude de définitive. Aussi avons-nous supposé que les déplacements qui suivent le face à face et au cours duquel les oiseaux marchent en balançant exagérément le corps à droite et à gauche, le cou gonflé, en sont le stade ultime.

Par contre, nous avons une idée plus précise de la durée de la recherche du partenaire qui peut aller, comme nous l'ont appris nos marquages, de moins de 24 heures à 3 jours et plus.

Les couples déjà formés s'isolent légèrement des autres oiseaux, et leur activité physique et vocale réduite, contraste avec celles des trios, dont les batailles et les cris animent la colonie. Le marquage d'un certain nombre de couples nous a permis de savoir si cette union était ou non le fait du hasard et si elle était durable. Sur 73 couples marqués, 14 échappèrent à toute observation ultérieure, 5 restèrent séparés définitivement, les conjoints des 54 restants s'étant réunis dans les heures ou les jours qui suivirent le marquage. Cette opération avait eu l'énorme avantage de nous faire assister à des recherches et à la réunion de quelques uns d'entre eux.

Il nous faut toutefois insister sur les conséquences immédiates ou lointaines de la capture et du baguage d'un oiseau. La frayeur, surtout intense chez les femelles, est probablement responsable de la séparation définitive des deux partenaires. Par la suite de tels oiseaux peuvent, à la seule vue d'une forme humaine, abandonner oeuf ou poussin pour fuir rapidement.

La copulation

Les premières copulations s'observent vers le 15 avril, les dernières au début du mois de juin. Les parades mutuelles de plus en plus nombreuses annoncent la proximité du coït. Au cours de ce cérémonial, un des gestes les plus caractéristiques est celui ou l'oiseau, le mâle en général, incline la tête et place son bec au niveau de la poche incubatrice du partenaire. Cette attitude que l'on observe par la suite en d'autres circonstances paraît correspondre au "dabbling" décrit par STONEHOUSE (1960) chez le Manchot royal.

Après une ultime démonstration, la femelle s'étend sur la glace d'elle même ou à l'invite du mâle, et ce dernier grimpe sur son dos et la féconde. L'union des sexes est assez brève, mais il arrive aussi que le mâle ne puisse se maintenir en équilibre sur le dos de sa partenaire. Des chutes répétées semblent reporter la copulation de quelques heures.

La fréquence des copulations passe par un maximum dans la deuxième quinzaine d'avril. Celles que l'on observe au mois de mai sont le fait d'oiseaux retardataires alors que par la suite ce ne sont que des tentatives incomplètes, non suivies d'éjaculation.

La comparaison des poids des gonades de différents individus confirme d'ailleurs ces observations. Les testicules de la majorité des oiseaux atteignent leur développement maximum (30 à 80 g environ) dans la deuxième quinzaine d'avril; mais celui ci peut se situer dès le début du mois chez les sujets les plus précoces (32 g le 9 avril 1952) et au début du mois de juin seulement chez les derniers attardés (31 g le 6 juin 1956).

L'étude pondérale des ovaires fournit des renseignements à peu près similaires.

La ponte et l'incubation

Pendant la période qui s'étend entre la copulation et la ponte, le comportement du couple ne présente aucune particularité digne d'être relatée.

Les oiseaux stationnent côte à côte pendant la journée et se déplacent quelquefois d'un point de la colonie à un autre; le soir ils se placent dans une tortue où ils séjourneront la majeure partie de la nuit.

Disons enfin pour terminer que la formation de l'oeuf est probablement à l'origine de fréquentes absorptions de neige chez les femelles.

La ponte

Les premiers oeufs sont pondus dès le début du mois de mai, les derniers entre le 1er et le 10 juin.

L'imminence de la ponte est annoncée par les contractions qui agitent le corps de la femelle à intervalles réguliers. Les deux conjoints sont alors souvent cantonnés à un emplacement dont ils ne s'écartent pas, mais il se peut aussi que certains couples se déplacent sur l'initiative de la pondeuse (Photo 27).

Les contractions spasmodiques sont de plus en plus intenses et bientôt le pôle aigu de l'oeuf apparaît au milieu du sphincter cloacal largement ouvert (Photo 28). A ce stade, certaines femelles paraissent épuisées et restent immobiles, la tête basse et les yeux clos pendant la phase de repos séparant deux contractions.

Si ces ultimes efforts paraissent les plus douloureux, celui qui provoquera l'expulsion de l'oeuf est plus rapide. Peu avant, la femelle dont les pattes prenaient primitivement appui sur la glace par la sole plantaire antérieure et les ongles, change de position et son corps repose dès lors sur les soles plantaires postérieures. Cette attitude, qui sera celle de tous les couveurs, a aussi pour effet de rapprocher le cloaque de la surface du sol. Simultanément la pondeuse introduit sa queue entre les pattes légèrement écartées et l'oeuf tombe doucement sur les plumes caudales, puis sur la glace. Aidée du bec, la femelle le place alors sur la face supérieure des pattes (Photo 29). Exception faire des pontes douloureuses, les femelles reprennent immédiatement leurs parades avec le mâle. Ces longues démonstrations gestuelles et vocales prendront fin peu après l'échange de l'oeuf entre les deux oiseaux. Le mâle incline fréquemment la tête vers la poche incubatrice de sa partenaire, qui contracte les muscles de la paroi abdominale et "montre" l'oeuf. Les deux oiseaux chantent presque simultanément, et la femelle peut se déplacer à petits pas ou même piétiner sur place devant son conjoint, les ailerons relevés à l'horizontale. Ce dernier écarte progressivement les pattes et prend bientôt la démarche du couveur, tout en laissant apparaître souvent sa "poche incubatrice". La femelle laisse alors glisser l'oeuf sur la glace, le mâle s'en empare, son corps parfois agité de tremblements, et avec plus ou moins de difficultés le place sur ses pattes. Dès lors, sa partenaire après quelques chants et parades, va tourner autour de lui, s'éloigner pour revenir plusieurs fois de suite et le quitter définitivement pour se diriger vers la mer.

Ce départ quasi-immédiat suffit à démontrer l'improbabilité d'une ponte de remplacement. Nous avons toutefois tenu à nous assurer qu'il en était toujours ainsi. Pour cela nous avons placé dans le parc d'élevage plusieurs femelles dont l'oeuf avait été préalablement enlevé. Pendant les 21 jours que dura leur captivité aucune nouvelle ponte ne fut observée. De plus, le poids atteint par les femelles après la ponte est tel qu'elles semblent incapables d'en élaborer un second. Elles n'ont en effet pris aucun aliment depuis leur arrivée à la colonie 45 à 50 jours plus tôt. 5 d'entre elles pesaient le 14 avril, 28,85 kg, contre 22,42 kg pour 5 autres à leur départ pour la mer le 23 mai. Quatre femelles de nos parcs d'élevage perdaient de leur côté 6,5 kg à 8,5 kg au cours d'une captivité longue de 37 à 39 jours avant leur ponte. Le faible nombre des nos captifs ne permettait pas la formation d'une "tortue" les protégeant efficacement du froid, et c'est pourquoi leur amaigrissement était plus important que celui des oiseaux de la colonie. Il n'est pas besoin d'ajouter que la perte de poids des pondeuses est beaucoup plus importante que celles de femelles non reproductrices.

L'incubation

Les pontes du début de mai marquent le commencement de l'incubation qui ne se terminera qu'avec les ultimes éclosions de la première quinzaine d'août.

Le nombre des couveurs

Il croit chaque jour pour atteindre un maximum (6.000 environ) au cours du mois de juin.

La plupart des oiseaux dont les déplacements sont limités par la présence de l'oeuf se mettent alors en tortue dès la moindre aggravation du temps. Ainsi entre les 6 et 16 juin 1956 la température moyenne fut voisine de —20° C et le vent égal à 10 m/s. Pendant cette période, la quasi-totalité des oiseaux étaient étroitement groupés, et ils restent ainsi à chaque tempête pendant toute l'incubation. 4 à 8 % de la totalité des oiseaux présents — les inemployés — ne couvent pas et leur liberté de mouvement plus grande les fait se déplacer quelquefois hors des limites de la colonie.

Le sexe des oiseaux couveurs

Nos observations n'ont fait que confirmer celles de Cendron (1952) qui avait constaté que la plupart d'entre eux étaient de sexe mâle. Le 23 juin, nous ne trouvions qu'une seule femelle pour 60 mâles, et la proportion des chants des deux sexes qui était à l'avantage des premières en mai, s'inversait au profit des seconds dès le début de juin. Le 6 juin, un comptage limité à quelques minutes nous donnait 82 chants mâles pour 25 chants femelles alors qu'il n'y avait plus que 8 de ces derniers contre 100 autres pour la période du 7 au 25 du même mois 1956.

Durée de la période d'incubation.

L'extrême longueur de cette période — 100 jours environ — est dûe à la fois à la durée même de l'incubation de chaque oeuf — 62 à 66 jours — et à l'étalement des pontes sur plus d'un mois.

Le comportement des couveurs

Si lors des tempêtes les oiseaux sont groupés en tortues et restent immobiles, pendant les accalmies ils sont dispersés sur l'emplacement de la colonie et se livrent à des soins de toilette minutieux.

Certains couveurs moins sédentaires que leurs congénères quittent les lieux de reproduction pour séjourner au loin sur la glace de mer. Ces déplacements, dont la motivation nous échappe en partie, furent observés en 1951 (Cendron) en 1952 et 1956 (Prévost) à la colonie de Pointe Géologie et Stonehouse (1953) en fut aussi témoin aux Ilots de Dion.

Les mâles qui ont perdu ou abandonné leur oeuf demeurent généralement à la colonie où ils se mêlent aux inemployés; si leurs réserves de graisse sont épuisées ils se dirigent au contraire vers la mer.

L'enlèvement expérimental de l'oeuf conduit au même résultat. Il nous a permis de constater également que dans les deux heures qui suivent, le mâle le recherche, et s'il vient à le retrouver ou s'il lui est présenté il le prend sur ses pattes et continue à couver.

Passé ce délai, il y a peu de chances que le couveur retrouve un comportement incubateur normal.

La propriété de l'oeuf

Le mâle couve toujours le même oeuf et il n'y a aucun échange aucune relève de couveurs, comme l'avait supposé Wilson. Par contre, l'échange expérimental d'oeufs, même très différents d'aspect, entre deux couveurs, n'a jamais été suivi d'aucune réaction, et encore moins de refus de leur part.

Comportement des inemployés

Ces oiseaux comprennent d'anciens couveurs, des oiseaux non

appariés et peut-être de jeunes adultes non reproducteurs. Leur activité est, nous l'avons vu, plus importante que celle des couveurs. Si leur attention est souvent attirée par les oeufs abandonnés, et si même ils peuvent batailler pour s'en approcher, ils les placent rarement sur leurs pattes. Ces luttes sont la traduction d'une évolution progressive du comportement de cette catégorie d'oiseaux qui conduira aux luttes violentes pour les poussins des mois de juillet et août.

Causes d'abandon des oeufs

Nous nous bornerons à mentionner ici celles ayant trait aux variations d'intensité du comportement incubateur.

L'inaccoutumance à la présence de l'oeuf ou l'absence de comportement incubateur — fréquents chez les reproducteurs inexpérimentés — provoquent de nombreux abandons. Ces deux motifs ne peuvent guère être invoqués pour justifier l'abandon de certains oeufs en fin d'incubation. C'est alors qu'intervient l'épuisement physiologique qui contraint quelques couveurs à prendre immédiatement le chemin de la mer. Cet épuisement est semble-t-il motivé par plusieurs causes: putréfaction de l'oeuf, séjour alimentaire prolongé de la femelle, insuffisance ou consommation trop rapide des réserves lipidiques du mâle. Si la première d'entre elles conduit au maintien de l'oeuf sur les pattes au delà de la durée réelle de l'incubation, la dernière, difficilement observable à la colonie, influença directement nos couveurs du parc d'élevage. Trop peu nombreux pour pouvoir s'abriter mutuellement, les oiseaux désertèrent leurs oeufs après deux mois d'incubation et l'un d'eux le jour même de l'éclosion. Leur poids — 22 kg en moyenne — était alors inférieur à celui de la majorité des mâles quittant la colonie après la relève de leur partenaire.

Forme et dimension de l'oeuf

L'oeuf de Manchot empereur est pyriforme, plus ou moins allongé et sa surface est soit entièrement lisse, soit recouverte de petits nodules répartis pour la plupart irrégulièrement autour du pôle obtus. Sa longueur est de 117 à 132 mm pour un diamètre de 80 mm environ et un poids moyen de 447 g (313—538,5).

La température de l'oeuf

Celle ci varie en fonction de divers facteurs: exposition à l'air ambiant, température plus ou moins basse des pattes, contact avec la neige, etc. . . .; aussi diffère-t-elle sensiblement, sauf peut-être chez les oiseaux en tortue, de celle de la poche incubatrice. Nous avons trouvé pour cette dernière 35°5, CENDRON (1952) 36°6, alors que la température moyenne de l'oeuf n'était que de 31°4 (45 mesures).

Le retour des femelles et le départ des mâles couveurs

C'est à partir du 20 juin que les toutes premières femelles reviennent à la colonie de reproduction, au terme de leur séjour à la mer. Toutefois les arrivées ne débutent réellement que dans les premiers jours de juillet et atteignent leur maximum dans la dernière quinzaine du mois. 60 à 70 jours séparent donc le gros des départs après la ponte de celui des retours à la côte, ce qui ne s'accorde pas exactement avec les chiffres fournis par nos bagués dont l'absence s'échelonnait entre 73 et 84 jours.

Si la durée du séjour alimentaire varie avec chaque individu, elle paraît aussi fonction de l'embonpoint de l'oiseau à son départ de la colonie en fin de jeûne. C'est pourquoi certains de nos oiseaux bagués dont le séjour forcé en parc avait accentué et prolongé l'amaigrissement, s'absentèrent plus longtemps. Il est possible que "l'emploi" de l'oiseau influe aussi, l'absence des reproducteurs paraissant plus longue que celle des inemployés. La meilleur preuve en est que le poids de ces derniers n'augmente pas dans les mêmes proportions. Nos deux femelles marquées D 2 et 27 qui avaient déserté ou perdu leur oeuf avant le départ de la colonie, n'avaient gagné que deux kilogrammes à leur retour, alors que chez les autres oiseaux ce gain excédait 3,5 kg. Enfin l'absence de contenu stomacal, fréquente chez les inemployés, accentue également cette différence.

Quoiqu'il en soit, le poids des femelles reproductrices est presque toujours supérieur à 25 kg et dépasse même parfois 30 kg, au retour à la colonie en juillet.

Recherche du partenaire couveur

Les arrivantes recherchent leur partenaire dans la foule des oiseaux présents. La longue succession des déplacements ponctués de chants est analogue à celle du mois de mars. Son partenaire retrouvé, la femelle parade avec lui et prend bientôt sur ses pattes l'oeuf ou le poussin, selon un cérémonial identique à celui qui suivait la ponte. Cette relève est d'autant plus rapide semble-t-il que l'état de maigreur du mâle est plus avancé. La présence d'un poussin sur les pattes du mâle stimule par ailleurs davantage la femelle que la vue d'un oeuf, et dans un tel cas elle glisse fréquemment le bec au niveau du jeune empereur, esquissant une regurgitation, le bec à demi entrouvert. Un tel comportement est courant chez le Manchot royal (STONEHOUSE, 1960).

Départs des mâles et durée du jeûne physiologique

Les mâles, libérés de leur tâche, prennent le chemin de la mer où ils croisent les groupes d'oiseaux du sexe opposé qui en reviennent. Leur jeûne physiologique a duré trois mois et quelquefois plus, provoquant une perte de poids de l'ordre de 40%. Alors qu'en avril celui-ci dépassait 36 kg, il n'atteint que 23 à 24 kg en moyenne au mois de juillet. Un de nos couveurs du parc d'élevage de 1956, le mâle no. 6, avait perdu plus de 13 kg après 92 jours de captivité.

L'éclosion

Le jeune empereur naît après 62 à 66 jours d'incubation et son poids est alors de 315 g (250—383). L'éclosion a lieu sur les pattes du couveur mâle ou femelle et dure 24 à 48 heures. En 1952 et 1956 il s'écoula 61 et 63 jours respectivement entre l'apparition du premier oeuf et celle du premier poussin. A ce stade du cycle, le problème essentiel est de savoir si la femelle rejoint son partenaire de la pariade ou un oiseau quelconque. Les faits apportés par l'observation des oiseaux de la colonie et particulièrement des bagués, comme les résultats de quelques expériences, plaident tous dans le même sens, et suggèrent que la vie familiale persiste pendant toute la saison de reproduction.

Plusieurs de nos femelles baguées, entre autre C3 et C4, furent retrouvées avec leur conjoint qu'elles n'avaient probablement pas rencontré par hasard. Par ailleurs, le 25 juillet 1952, 3 femelles arrivantes furent placées dans le parc d'élevage où se trouvaient 3 couveurs mâles. Aucune parade, aucun échange d'oeuf ne se produisit, alors que les disputes entre eux étaient au contraire fréquentes.

L'élevage individuel des poussins

L'élevage du jeune Manchot empereur débute en juillet et se termine avec la débâcle des glaces au mois de décembre. Il comprend deux périodes bien distinctes: la première dite période d'élevage individuel, au cours de laquelle le poussin reste abrité dans la poche incubatrice de ses parents: l'émancipation marque le début de la seconde, dite des "crèches". Cette vie dans les crèches ne signifie pas pour autant, comme nous le verrons plus loin, l'interruption de l'élevage parental et son remplacement par un élevage collectif.

Le comportement des adultes

La population des adultes comprend grossièrement deux catégories: celle des reproducteurs assurant l'élevage de leur poussin et celle des oiseaux n'en ayant pas ou n'en ayant jamais eu, les inemployés.

Ces derniers, qui comme leur nom l'indique n'ont aucune activité précise, voient leur nombre augmenter régulièrement du fait de la mortalité des jeunes poussins. Ils cherchent comme nous l'avons vu, à s'emparer d'un jeune oiseau et sont à l'origine de toutes les batailles.

Cette période est aussi marquée par le comportement particulier des parents. Ceux d'entre eux qui sont seuls à la colonie, peuvent parader momentanément avec un adulte couveur du sexe opposé; ils se "tiennent compagnie" (keeping company) (Photo 25).

Le comportement des poussins

Dans les trois premiers jours qui suivent la naissance, les mouvements du poussin sont assez limités. Dès le 5ème jour on note déjà quelques esquisses de baillements et des mouvements d'ailerons, alors que les premières réactions agonistiques à l'égard des congénères, coups de bec ou d'ailerons, s'observent entre le 8ème et le 10ème jour.

Vers le 35ème jour, un poussin élevé par nos soins amorça une parade mutuelle analogue à celle d'un adulte, quand un sujet de son âge lui fut présenté. Les premières sorties hors des poches incubatrices s'effectuent par beau temps vers les 40ème—50ème jours. Le poussin loge difficilement sur les pattes du parent qui ne lui fournit plus qu'une protection limitée. Cette protection ne lui est d'ailleurs plus nécessaire puisqu'il peut désormais équilibrer sa température interne.

La reconnaissance vocale

Quelques expériences nous ont permis de savoir que l'adulte paraît reconnaître le signal sonore de son poussin dès le 8ème jour au moins. Inversement, ce dernier connaît le signal de l'adulte très tôt, mais nous n'avons aucune preuve de l'âge à partir duquel il en est capable.

En l'absence de tout signal sonore, le parent reconnaît son poussin par la vue, mais si plusieurs lui sont alors présentés, il est fréquent qu'il s'empare du plus proche et le place sur ses pattes. Si son propre poussin chante alors à proximité, il se dirige vers lui et l'étranger tombe alors de lui-même sur la glace. Toutefois, dans quelques cas, l'adulte semble très embarassé. Si son comportement incubateur est satisfait par une présence sur ses pattes, il paraît désorienté en entendant son propre poussin chanter à quelques pas de lui.

A l'état naturel dans la colonie, l'éventualité d'un tel choix ne se présente qu'exceptionnellement.

La nourriture du jeune poussin

L'oesophage des mâles couveurs est capable de secréter, malgré le jeûne physiologique, une substance contenant plus de 50% de protéines et une forte proportion de lipides. Cet aliment de sécurité permet au jeune poussin de se maintenir pendant quelques jours si le retour de la femelle se produit quelques jours après l'éclosion.

L'oesophage des mâles couveurs est considérablement hypertrophié à cette période, celui des femelles et des inemployés l'étant semble-t-il beaucoup moins. La sécrétion, que l'on peut rapprocher du "lait de Pigeon", apparaît sur la surface des plis de la muqueuse oesophagienne et s'accumule dans l'estomac (Photo 30).

Dans ce dernier, une autre sécrétion siégeant au niveau des glandes gastriques paraît être plus intense, tout au moins à cette période, chez les femelles. Elle est peut-être à l'origine de l'existence de cette pellicule mucilagineuse qui entoure leur contenu stomacal.

Par la suite, et jusqu'à la fin de l'élevage, le poussin est nourri avec la bouillie de Poissons et Crustacés que l'adulte rapporte de la mer. Si dans son premier âge il n'en absorbe que des quantités très réduites, il devient très vorace par la suite, mais l'irrégularité des repas ne nous a pas permis de déterminer, même approximativement, sa ration quotidienne (Fig. 24).

L'influence de ces repas irréguliers sur la croissance est par contre importante. Si le poids moyen passe rapidement de quelques 300 g de la naissance à 2 kg au moment de l'émancipation, on est frappé par les écarts entre poussins de même âge. Le 20 août 1952, le poids d'un des plus gros poussins de la colonie était de 2.910 g, alors que le plus petit n'atteignait que 560 g.

La croissance de la stature subit une évolution parallèle. De 20 à 21 cm à la naissance, la longueur totale passe à 50 cm environ au moment de l'émancipation. Celle des dénutris est toujours largement inférieure à la moyenne de la colonie.

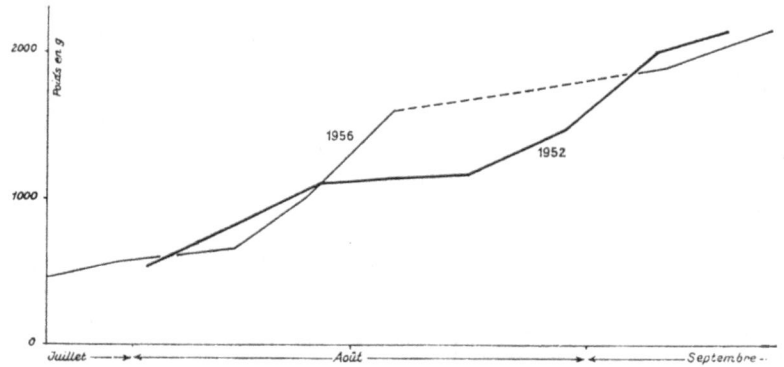

Fig. 24. Croissance pondérale moyenne des poussins pendant la période d'élevage individuel, en 1952 et 1956. (Chaque point de la courbe a été obtenu d'après la moyenne des poids de 10 sujets pris au hasard dans la colonie).

La thermorégulation des poussins

L'émancipation ne devient définitive qu'avec l'acquisition d'une bonne régulation de la température interne.

C'est vers le 20ème jour qu'apparaissent, chez le poussin momentanément privé de l'abri parental, les premiers signes de la thermorégulation.

A l'âge de 35 jours, un poussin de 1.600 g peut survivre pendant plusieurs heures dans une ambiance inférieure à —10° C avec un vent supérieur à 10 m/s.

Enfin, vers le 55ème jour, un poussin de 1.700 g supportait pendant 32 heures —20° C et un vent de 12 m/s.

Toutes les expériences ayant pour but de contrôler l'efficacité de la thermorégulation nous ont amené à conclure que le degré de développement jouait un rôle plus important que l'âge chronologique. Un poussin d'un âge donné bien développé parce que régulièrement alimenté se défend beaucoup mieux qu'un poussin plus âgé nouri sporadiquement.

La régulation thermique est autonome à partir de 45 à 50 jours; le contrôle de quelques sujets marqués nous a montré que quatre d'entre eux avaient abandonné effectivement l'abri parental entre les 46ème et 48ème jour.

L'émancipation des grands poussins et le problème des crèches

Le transfert brutal du poussin de la vie familiale à la vie en liberté dans la colonie marque le début de la deuxième période d'élevage.

Elle commence au mois de septembre et tant en 1952 qu'en 1956 une violente tempête en fut à l'origine, et si elle en accélèra le déroulement, elle en aggrava sérieusement les conséquences. Les jeunes oiseaux, délaissés par leurs parents, s'accolèrent sans ordre les uns aux autres pour se protéger du froid. Quelques jours plus tard apparaissaient des tortues régulières et circulaires, analogues à celles des adultes, dans lesquelles les poussins sous-alimentés occupaient bien souvent la périphérie (Photo 32). Chaque tempête, chaque blizzard verra se former de tels groupes çà et là au milieu des adultes. Au mois de novembre, l'élévation de la température ambiante les fera disparaître.

La température rectale des sous-alimentés est toujours plus basse que celle de la majorité des autres poussins. Elle était de 37,5° à 38° au mois de septembre, de 38,6° en octobre novembre chez la plupart des poussins, alors qu'elle était inférieure ou égale à 37 en septembre, et à 38° en novembre pour les dénutris.

Le comportement des adultes

Le père et la mère s'absentent dès lors tous les deux pour satisfaire aux besoins alimentaires grandissant du jeune empereur. La nourriture étant absorbée très rapidement, le séjour du parent à la colonie est des plus brefs, la durée du voyage alimentaire reste à peu près la même sauf peut-être en fin de cycle où le rapprochement progressif de la mer libre peut la diminuer.

Le comportement des oiseaux se tenant compagnie disparaît avec l'émancipation des poussins et du fait même de l'extrême brièveté du séjour des adultes nourrisseurs à la colonie.

Les inemployés de leur côté cessent rapidement leurs recherches et leurs luttes pour les poussins. Ils deviennent totalement passifs et se tiennent soit au milieu des poussins, soit sur les bords de la colonie. Ils entrecoupent leurs longues stations de petits déplacements, de parades mutuelles si les deux membres du couple sont présents, et de voyages alimentaires à la mer.

La reconnaissance vocale

Durant cette deuxième phase de son élevage, le poussin doit pouvoir se faire reconnaître ou être reconnu par son père ou sa mère revenant de la mer, alors qu'il n'intervenait pas ou presque dans les deux premiers mois de sa vie.

Ce sont les poussins d'adultes bagués qui nous ont montré la persistance de la vie familiale et la nature des moyens utilisés par les oiseaux pour la maintenir.

Le cas du poussin 29 nous a semblé le plus significatif. Ce jeune oiseau fut retrouvé successivement deux fois avec sa mère (Adulte 8+) et une fois avec son père (Adulte 8). Il paraît très improbable que le seul hasard ait pu permettre ces 3 rencontres dans une foule de près de 5.000 poussins.

La recherche du poussin par le parent ne paraît pas différente de celle des conjoints entre eux. Le parent se déplace dans toute la colonie et s'arrête face à chaque crèche pour chanter, jusqu'à ce qu'un poussin réponde (Photo 31). Adulte et poussin se rapprochent alors l'un de l'autre. Le poussin étranger ou les sous-alimentés que la faim pousse vers les arrivants sont alors chassés ou ignorés (Photo 33), alors que le propre poussin de l'oiseau après une esquisse de parade mutuelle et quelques chants entrecoupés de petits déplacements est reconnu définitivement et reçoit sa nourriture.

Si le signal sonore est toujours utilisé pour la première réunion, ou après une séparation à grande distance dans la colonie, les deux oiseaux paraissent au contraire à courte distance, se suivre des yeux et ils se rejoignent sans aucune émission vocale. Les quelques chants de poussins analysés au spectrographe de sons paraissent différer sensiblement les uns des autres. Leur nombre était cependant insuffisant pour nous permettre de conclure formellement à une individualité du chant de cette catégorie d'âge.

630

Le comportement des poussins

Nous avons déjà mentionné les mouvements collectifs fréquents dès la fin du mois d'octobre.

Les poussins recherchent en général la compagnie des adultes dont ils suivent les déplacements dans la colonie et plus rarement hors de ses limites. Ce besoin d'imitation est probablement à l'origine des battements d'ailerons et des cris collectifs.

Les comportements agonistiques sont rares; les plus fréquents sont ceux des poussins se tenant à côté de leur parent, à l'égard des congénères stationnant ou passant à proximité. Il s'agit peut-être là d'une manifestation de défense territoriale atténuée.

L'alimentation présente ceci de particulier que la totalité du contenu stomacal, près de 3 kg, peut être absorbée en un seul repas dès la fin novembre, et même avant, chez les plus gros poussins.

La croissance des sujets bien alimentés est très rapide et certains de ceux-ci atteignent bientôt 12 à 15 kg, et la mue commence à se manifester au niveau des ailerons et dans la région dorsale. Toutefois les différences de poids que l'on pouvait observer au cours des mois de juillet et août ne font que s'accroître et l'on peut voir encore des poussins ne pesant guère plus d'un kilogramme (fig. 25 et 26).

L'élevage est long de 5 mois environ. Le poussin le plus avancé de la colonie dépassait 15 kg le 14 novembre 1956, 133 jours après la

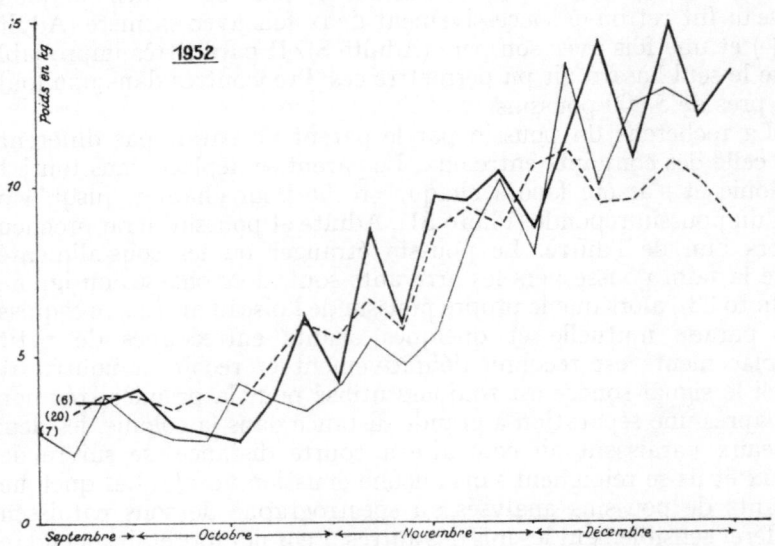

Fig. 25. Croissance pondérale de 3 poussins marqués (6, 7 et 20) au cours de l'année 1952.

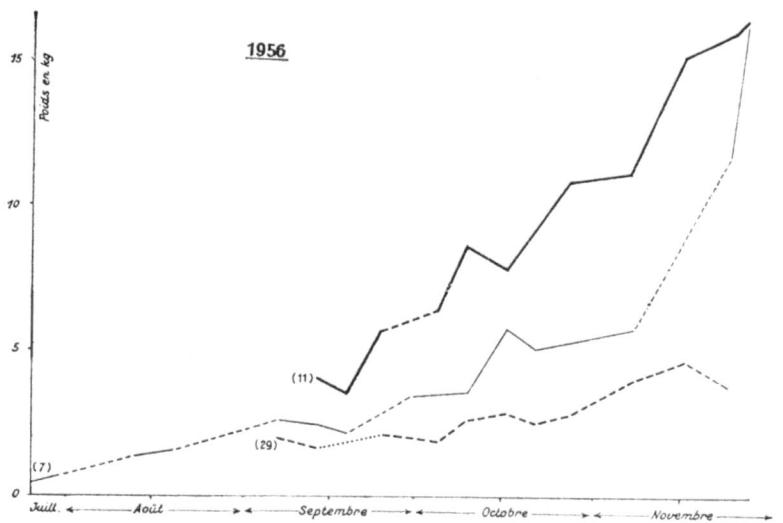

Fig. 26. Croissance pondérale de 3 poussins bagués (7, 11 et 29) en 1956 (29 représente
le cas typique du poussin sous-alimenté).

première éclosion; il se dirigeait vers la mer 16 jours plus tard, mais
son poids avait alors sensiblement diminué.

Ceci nous montre que dans la plupart des cas le poids passe par
un maximum quelques jours avant l'achèvement complet de la mue.
Le jeune oiseau n'est alors plus alimenté par ses parents et la faim
le pousse à quitter les lieux de reproduction.

La dislocation précoce de la glace de mer apporte des perturbations considérables. On peut être certain que la plupart des jeunes
oiseaux qui n'ont pas atteint un stade avancé du développement,
soit du fait de sous-alimentation, soit parce qu'ils sont nés très
tard, mourront pendant leur dérive vers le large.

Les inégalités de poids observées au départ de la colonie vont de
pair avec des différences de taille importantes. Le poussin Y âgé de
65 jours ne mesurait que 36,4 cm alors qu'un poussin de même âge,
normalement alimenté, atteignait 61 cm à la même date.

La croissance staturale est loin d'être achevée au départ de la
colonie. La longueur totale du corps est de 90 cm à 1 m, et elle
n'atteindra les chiffres de l'adulte qu'après le séjour d'un an à la mer.

Mue et dislocation de la colonie

La débâcle de la glace de mer se produisant presque toujours
avant l'achèvement complet du cycle annuel, il ne nous a pas été
possible d'observer le déroulement de la mue.

Celle des poussins se manifeste la première et elle touche d'abord les sujets les mieux développés. Il s'écoule environ un mois entre les premiers signes de mue et la chute totale du duvet. Le changement de plumage s'accompagne d'une modification de la tonalité du signal sonore et de la voix en général. Le chant de cour de l'adulte sera acquis beaucoup plus tard. Il nous a même semblé qu'il ne l'était pas encore complètement chez les immatures de deuxième année au mois de décembre.

Les poussins du cycle précédent ou immatures de deuxième année viennent muer à la côte, voire même à la colonie lors des débâcles tardives. S'ils n'ont pas encore la coloration de l'adulte, du moins en ont-ils le poids et la taille. Deux d'entre eux autopsiés en décembre 1952 donnaient respectivement 115 cm et 30,4 kg pour le mâle, et 105 cm et 24,8 kg chez la femelle. Le foie dépassait dans les deux cas 750 g. La durée de leur mue ne nous a pas parue différente de celle de l'adulte.

Les adultes changent de plumage, entre les mois de décembre et mars, soit à la colonie, soit sur des floes en dérive après la débâcle.

La mue de l'épiderme des pattes et des ornements colorés du bec débute fin octobre donc bien avant celle du plumage.

Comme chez les immatures de deuxième année, la mue est précédée d'une période de suralimentation. Les oiseaux peuvent ensuite supporter sans dommage les 30 à 35 jours pendant lesquels ils restent plus ou moins immobiles, sans pouvoir prendre aucun aliment (Photo 34).

Nous n'avons pu faire aucun contrôle de l'amaigrissement qu'ils subissent. Nous savons seulement que leur poids est de l'ordre de 24 kg quand ils regagnent la mer.

Il est probable que ce sont les inemployés qui muent les premiers suivis par les reproducteurs ayant terminé leur élevage. De ce fait, on peut penser que ces derniers seront au nombre des retardataires du cycle reproducteur suivant.

Structure et dynamique de la population

Il n'est pas possible, après deux années d'observations et des marquages réduits, de connaître même approximativement la structure d'une population de l'importance de celle de Pointe-Géologie.

Les jeunes adultes, reproducteurs ou non, séjournent à la colonie mais leur nombre et leurs catégories d'âge n'ont pas pu être déterminées. Quelques faits d'observation nous ont seulement permis de conclure à leur présence. Le mâle du couple no. 27 pesait 23,5 kg le 26 mai 1956, deux jours avant la ponte de sa partenaire plus lourde que lui (24,95 kg). Il était donc arrivé à la colonie avec des réserves de graisse très réduites et son incapacité de jeûner, ne serait-ce que quelques jours, l'avait conduit à déserter son oeuf peu après la ponte. La plupart des oeufs abandonnés au mois de mai, le furent pour des raisons identiques.

La dynamique de la population

Dans ce domaine, nos renseignements sont plus fournis, tout au moins en ce qui concerne la phase de reproduction à la colonie. Avant d'aborder cette étude, il nous paraît cependant utile de rappeler que le mode d'incubation ambulatoire de l'*Aptenodytes forsteri* n'est pas compatible avec une ponte multiple. Effectivement la ponte est unique et la fécondité réduite des femelles de l'espèce est probablement compensée par une longévité assez grande. De plus, le départ immédiat des femelles vers la mer après la ponte, le maigreur et la compression maximum du cycle annuel interdisent toute ponte de remplacement que l'on observe par ailleurs chez son proche parent le Manchot royal et chez le Manchot Adélie dont l'habitat est le même.

La mortalité
Les causes

Les causes susceptibles d'influencer la mortalité aux divers stades de la vie sont nombreuses et leur importance respective n'est pas la même d'un cycle annuel à l'autre.

Les facteurs climatiques se placent d'emblée au premier rang par le nombre de décès qu'ils provoquent chez les poussins. Comme ils ajoutent leurs effets à ceux de la dénutrition partielle ou totale de certains sujets, nous avons groupé ces deux causes en une seule.

Ce sont les mois de septembre et octobre qui fournissent le plus fort contingent de décès. L'émancipation est alors presque totale et les jeunes oiseaux directement exposés aux agressions climatiques meurent en grand nombre, si une tempête violente survient à cette période, Photo 36. Le 1er septembre 1952, au tout début de la deuxième période d'élevage, un violent blizzard — la vitesse moyenne du vent fut de 40 m/s pendant près de 36 heures — provoqua la mort de 140 poussins. Plus de 100 autres furent transportés par le vent loin de la colonie.

En 1958 la température anormalement froide du mois d'octobre fut responsable de la mort de plusieurs centaines de jeunes oiseaux. (ISEL, communication personelle).

Au stade des oeufs et des adultes, ce facteur de mortalité peut être tenu pour négligeable.

Les Prédateurs sont au nombre de deux: le *Pétrel Géant Macronectes giganteus* qui exerce ses ravages à la colonie même, et le *Léopard de mer Hydrurga leptonyx* ennemi redoutable des immatures et des adultes dans l'eau. Si le dénombrement des cadavres dépecés par le premier à la colonie est chose facile, le nombre des victimes du second, probablement important, nous échappe complètement. Les Pétrels Géants s'attaquent aux poussins dès le début de l'émancipation. Ils paraissent choisir ceux dont l'estomac est bien rebondi, les isolent de leurs congénères et, après avoir déchiré leur paroi abdominale, dévorent le contenu de l'estomac. Ils peuvent aussi dévorer les viscères mais négligent les muscles d'ailleurs fort peu développés à cet âge.

Les blessures fréquentes des oiseaux revenant de la mer témoignent de la grande activité du Léopard. Entre les 2 et 22 septembre, 6 oiseaux atteints par ses crocs furent notés parmi les arrivants.

Ajoutons que l'*Orque (Orcinus orca)* consomme probablement un certain nombre de Manchots empereurs adultes. Sa présence dans la baie de Pointe-Géologie a été confirmée tout récemment par HUREAU (Décembre 1961).

La putréfaction des oeufs en fin d'incubation n'est que la conséquence de la chute ou du gel partiel de celui-ci en cours d'incubation, mais il est aussi certain qu'un bon nombre d'entre eux sont des oeufs clairs.

Les accidents influent surtout sur la mortalité des oeufs et des poussins. Les couveurs perdent quelquefois leurs oeufs dans les petites cassures de glace de mer ou dans de simples accidents de terrain. Par la suite, ce sont les poussins qui tombent dans les crevasses de marée d'où ils ne peuvent sortir et ils meurent écrasés par la fermeture de la crevasse à la pleine mer.

Plus tard, quelques uns d'entre eux meurent noyés dans les rivières ou les lacs de la colonie.

Les maladies et la sénilité: C'est au cours du jeûne physiologique que les adultes âgés ou atteints de maladie, de tumeurs d'origine parasitaire ou autre, meurent à la colonie. Le nombre de ces décès ne dépasse pas cependant quelques unités pour chaque cycle reproducteur.

Quelques poussins sont atteints de troubles que nous avons qualifiés de "maladie de carence" dans l'impossibilité où nous étions de déterminer avec exactitude leur origine.

Les premiers symptômes de cette affection, qui paraissent s'apparenter assez étroitement à ceux de la polynévrite aviaire, commencent à se manifester à la fin du mois d'octobre chez certains sujets. Ils se traduisent par des crises passagères dont l'aggravation conduit, semble-t-il, à un affaissement total des membres inférieurs. Le poussin malade se reconnaît aisément: il ne peut se déplacer qu'en tobogannant, ce qui entraîne l'usure plus ou moins complète de son duvet ventral; de plus si on le relève, on s'aperçoit que ses deux pattes restent alors croisées l'une sur l'autre. (Photo 35).

Cette paralysie est vraisemblablement la conséquence d'une calcification de la substance fondamentale des os qui entraîne leur déformation et leur rigidité.

Les jeûnes physiologiques répétés de certains poussins entre leurs repas sont peut-être responsables d'une carence prolongée en vitamines (groupe B ou D?) à moins que les migrations verticales du plancton à cette période ne s'accompagnent d'une modification de la teneur en facteurs de croissance des aliments apportés par les parents. Il est difficile d'en décider. Ce dont nous sommes cependant certains c'est que le contenu stomacal des sujets gravement atteints était particulièrement liquide, et que la plupart d'entre eux mouraient à plus ou moins brève échéance.

Les parasites. Les parasites du tube digestif, très abondants chez les oiseaux allant régulièrement à la mer ou y séjournant longuement, disparaissent à peu près totalement pendant la durée du jeûne physiologique.

Ces endo-parasites comprennent surtout des Cestodes que l'on retrouve assez tôt chez le jeune poussin. Nous n'avons trouvé par contre aucun ecto-parasite sur les quelques sujets examinés pendant le séjour à la colonie. Il n'est pas exclu pour autant qu'il n'en existe pas et qu'ils se développent notamment pendant la vie de l'oiseau sur le pack-ice en plein été.

Les accidents consécutifs à la ponte. La plupart des femelles qui décèdent à la colonie meurent des suites immédiates ou lointaines de la ponte. Le volume important de l'oeuf, les nodules qui le parsèment, peuvent être à l'origine d'un prolapsus léger ou total du tractus génital.

Dans la première alternative, il se forme un bourgeonnement autour du cloaque et la succession d'efforts violents comme la marche ou la régurgitation, ou un frottement prolongé sur la glace entraînent une hémorragie grave ou mortelle. Cette dernière n'intervient pas obligatoirement après la ponte; elle peut survenir deux à trois mois plus tard après le séjour à la mer. L'expulsion de la totalité du tractus génital provoque par contre une hémorragie foudroyante. Deux femelles en furent les victimes les 17 mai 1952 et 15 mai 1956 respectivement. Il se peut enfin que l'oeuf trop volumineux ne puisse franchir le cloaque. La défécation ne pouvant plus se faire, l'oiseau meure par occlusion intestinale aigue, comme ce fut le cas pour une femelle le 18 mai 1956.

Les taux de mortalité

Les chiffres de nos deux années sont basés sur un recensement des poussins effectué le 26 décembre 1952. De ce dénombrement nous avons pu déduire que 6.081 oeufs avaient été pondus au mois de mai, ce qui nous permettait d'estimer la population totale de la colonie à 13.000 adultes. Bien que l'état de la glace de mer ne nous ait pas permis de répéter cette opération quatre années plus tard, quelques calculs restreints nous ont permis de savoir que l'effectif n'avait guère changé.

Au stade des oeufs

265 oeufs furent perdus ou abandonnés par les couveurs en 1952, contre 679 pour la période correspondante de 1956 (fig. 27).

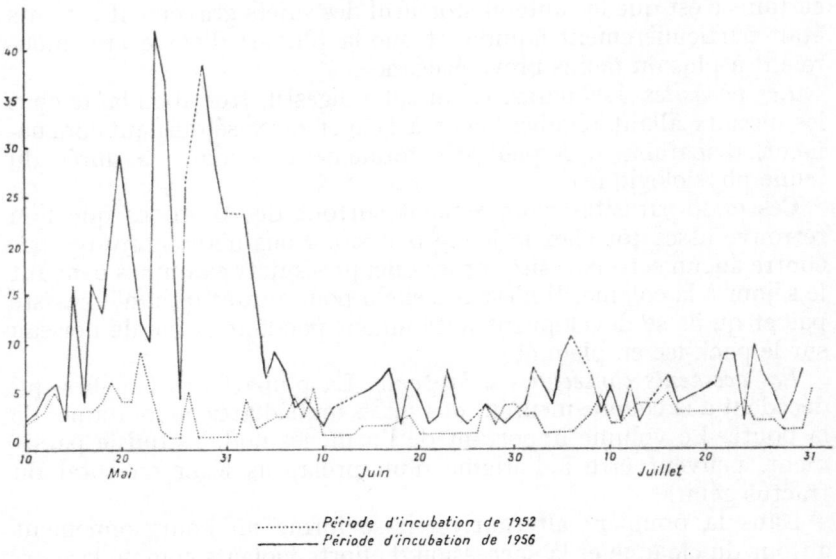

.... *Période d'incubation de 1952*
——— *Période d'incubation de 1956*

Fig. 27. Mortalité comparée des oeufs en 1952 et 1956.

La forte mortalité de cette dernière année s'explique par les abandons massifs du mois de mai —50% de la mortalité totale des oeufs — imputables, semble-t-il, à l'inexpérience d'un certain nombre de jeunes adultes.

La mortalité totale des oeufs représente respectivement 16,4% de la mortalité totale à la colonie en 1952, et 45,4% en 1956. Rapportés à la totalité des oeufs pondus, ces pourcentages ne sont plus que de 4,35% et 11,3% alors qu'ils atteignent 24 à 25% chez le Manchot Adélie (SAPIN-JALOUSTRE, 1960) et 21% chez le *Megadytes antipodes* (RICHDALE).

Au stade des poussins

En 1952, 1.352 poussins décédèrent entre les mois de juillet et décembre contre 815 pendant la période correspondante de 1956, et 1.634 entre l'éclosion et le mois de novembre en 1958.

La mortalité dûe à l'action conjuguée du froid et de la sous-alimentation représenta 80,2% (1.084) de la totalité des décès de poussins en 1952, 52% (424) en 1956, et 80% environ en 1958.

Les mois de septembre et octobre sont les plus meurtriers: 884 décès en 1952 (64,3% de la mortalité des poussins) 531 en 1956 (65,1% et 1.353 en 1958 (± 77%).

Les victimes des Pétrels géants étaient au nombre de 68 en 1952 et de 279 en 1956. Au total 23% des poussins éclos sont morts en 1952, 15,3% en 1956 et 27% environ en 1958.

Ce pourcentage est de l'ordre de 24% chez le Manchot aux yeux jaunes *(M. antipodes)* et de 45% environ chez le Manchot Adélie.

Au stade des adultes

Si la plupart des adultes de tous âges finissent leur vie en mer, victime des Prédateurs, la mortalité à la colonie est infiniment réduite. 4 décès y furent enregistrés en 1952, 9 en 1956, 2 en 1958.

Causes de Décès	1952	1956	1958
Capture ou blessure mortelle par un Léopard de mer	1 femelle (2—IV) 1 mâle (13—IX)	2 femelles (13—IV) (10—IX)	
Accidents consécutifs à la ponte	1 femelle (18—V)	4 femelles (15 et 8—V) (26—VII) (7—VIII)	1 femelle (13—VIII)
Divers (vieillesse, tumeurs et causes indéterminées)	1 mâle (5—VI)	3 mâles (10 et 30—VI) (17—VII)	1 (sexe et date indéterminées)

L'Evolution du comportement dans le genre Aptenodytes

Ce genre comprend deux espèces, *A. forsteri* étroitement cantonné à la côte antarctique dans les hautes latitudes et *A. patagonica* dont l'habitat se circonscrit aux régions et aux îles subantarctiques. La comparaison du cycle annuel du comportement de ces deux oiseaux va nous permettre de mettre en relief les adaptations qui ont permis au premier de s'établir et de se reproduire dans les climats les plus extrêmes.

Le Manchot royal

Ce manchot vit dans des colonies très populeuses sur les rivages des îles Crozet, Prince Edouard, Macquarie, etc ... Les récents

travaux de STONEHOUSE (1960) ont permis de connaître exactement le déroulement de son cycle reproducteur.

Les adultes arrivent à la colonie, se mettent en couple et choisissent un emplacement de nid voisin de celui de la précédente saison
de reproduction.

Le long étalement de la ponte (décembre à avril) découle de la
durée exceptionnelle de l'élevage du poussin et a pour effet de diviser
les reproducteurs en deux catégories bien distinctes: les reproducteurs précoces et les reproducteurs tardifs dont nous décrirons
plus loin la curieuse alternance des cycles.

L'incubation longue de 54 à 55 jours est assurée d'abord par le
mâle puis par les deux conjoints. Le poussin vit en crèches dès
l'âge de 5 à 6 semaines. Il atteint 11 à 13 kg à la fin du mois d'avril
et ses réserves de graisse sont alors importantes.

L'alimentation à base de Poissons, sporadique pendant l'hiver, est
complètée par une nourriture d'entretien fournie par l'oesophage
de l'adulte. Elle redevient régulière au mois d'octobre et entre les
mois de décembre et février le plumage juvénile remplace le duvet
et les poussins quittent alors la colonie.

Comparaison entre les cycles des deux espèces

La notion de territoire diffère fondamentalement dans chaque
espèce. Chez le Manchot royal celui-ci est topographiquement
représenté par le nid alors que chez le Manchot empereur où les
couples restent itinérants pendant tout le cycle il n'existe pour
ainsi dire pas.

La durée du cycle est de 14 à 16 mois chez le Manchot royal, et
sensiblement égale à une année chez l'*A. forsteri*.

Si dans cette dernière espèce l'adulte expérimenté paraît se reproduire chaque année, chez le premier, deux cycles occupant environ
30 mois, le troisième ne peut être accompli intégralement faute de
temps. Aussi n'y a-t-il de reproduction que deux années sur trois.

La parade mutuelle (High pointing) s'apparente à celle du Manchot
empereur et si l'attitude prise en cours du chant est voisine, il
semble qu'il y ait aussi selon MURPHY (1936) une différence entre
les voix des deux sexes.

Les quelques dix jours qui sépare les durées d'incubation dans les
deux espèces s'expliquent sans doute par leur différence de taille
et de poids. La sévérité du climat influe probablement aussi sur
la durée de celle du Manchot empereur.

Tous les deux placent l'oeuf sur leurs pattes et l'abritent en avant
par un repli cutané abdominal.

L'élevage individuel du poussin est légèrement plus court (35 à 42
jours) du fait même de la clémence relative du climat subantarctique.

Le poussin de Manchot royal atteint un poids à peu près égal à celui de l'adulte à l'âge de 3 mois et demi bien que son élevage soit loin d'être achevé, alors que le poids maximum des jeunes empereurs n'est atteint qu'à 5 mois et il est encore très inférieur aux minima de l'adulte.

L'élevage dans les crèches se poursuit pendant 3 mois à 3 mois et demi chez le Manchot empereur, soit 6 à 7/10 de sa durée totale contre 9 à 10 mois (9/10) chez le Manchot royal.

Les *A. patagonica* reproducteurs muent en dernier et s'il n'y a pas simultanéité de la mue chez les deux membres du couple, le premier oiseau revenu à la colonie pourra choisir un autre partenaire. Nous ne savons rien du comportement de l'empereur à ce stade du fait de la débâcle des glaces; mais l'existence des trios pourrait n'être que la conséquence d'un retour tardif du conjoint.

Le Manchot royal, de part sa taille et son poids, peut se déplacer assez aisément sur les rochers contrairement à son homologue antarctique.

La présence permanente d'eau libre autour de ces colonies explique la faible durée des jeûnes physiologiques alors que l'allongement de la période d'élevage est probablement motivée par la raréfaction des aliments marins en hiver.

En colonisant la côte antarctique, le Manchot empereur est devenu l'homéotherme le mieux adapté à son climat hivernal. Bien qu'il ne soit guère plus grand que le Manchot royal, le fait que son poids puisse atteindre et même dépasser 40 kg traduit une véritable adaptation morphologique à la lutte contre le froid.

Cette dernière s'est doublée à l'échelon individuel d'une adaptation physiologique marquée surtout chez les mâles par l'importance des réserves lipidiques sous cutanées et péritonéales et chez tous les oiseaux par le niveau thermique très bas des extrémités inférieures et supérieures.

L'adaptation collective enfin a pu être réalisée au détriment de la notion de territoire et des comportements agonistiques. Elle atteint sa perfection dans les tortues grâce auxquelles les oiseaux économisent leurs réserves en échappant momentanément au froid et au vent. Cette vie communautaire poussée n'a eu aucune incidence notable sur la vie familiale qui persiste grâce à la perfection des signaux sonores et visuels permettant aux oiseaux de se reconnaître dans la foule de leurs congénères. Ces tortues ne peuvent cependant se constituer que sur des surfaces planes que la glace de mer est la seule à pouvoir offrir dans bien des régions côtières.

Enfin l'inversion du cycle reproducteur fait que la ponte et l'incubation ont lieu en plein hiver. Comme l'a supposé WILSON le jeune empereur doit devenir indépendant au cours du plein été au moment de la débâcle et sa naissance au cœur de l'hiver satisfait pleinement cette condition.

Si en juillet ses besoins alimentaires sont réduits ils augmentent régulièrement jusqu'en décembre, mais en même temps la mer libre se rapproche de la colonie diminuant le trajet des parents.

Le cycle reproducteur de Manchot empereur, paradoxal à bien des égards il y a quelques années, représente en réalité la meilleure solution aux problèmes écologiques et physiologiques posés à l'espèce.

CONCLUSION

Malgré le manque de statistiques dans l'état actuel de nos connaissances, il est permis de penser que les populations des deux espèces de Manchots antarctiques ne sont pas en régression et peuvent donc être considérées comme en état d'équilibre.

Mais l'étude précédente a montré que, dans le même milieu antarctique, avec les mêmes nécessités de se nourrir dans l'eau et de vivre dans l'air la période de reproduction et d'élevage des poussins, *les deux Manchots antarctiques se trouvent en fait dans des conditions écologiques différentes.*

On peut considérer *le Manchot Adélie comme lié aux affleurements rocheux de la côte pour sa période de reproduction et libre au contraire pendant l'hiver sur la banquise dérivante.* Il n'a donc pas de problème alimentaire pendant l'hiver à la limite de l'eau libre et il est bien armé pour résister aux conditions climatiques hivernales de cette zone, beaucoup moins dures que celles des rivages continentaux. Pendant l'été, il trouve le long de la côte un climat relativement modéré et ses possibilités de jeûne de deux mois lui permettent de traverser la période pendant laquelle la glace soudée interdit la pêche. Tout au long de l'année, ses mécanismes thermo-régulateurs individuels sont habituellement suffisants. Par contre, sur des surfaces rocheuses limitées, avec des matériaux de construction de nids également limités, on observe que l'espèce possède une organisation sociale en mosaïque de territoires juxtaposés et un comportement territorial intense, amélioré par l'expérience: ceci permet la reproduction dans les meilleurs conditions des oiseaux "expérimentés" de plus de trois ans d'âge et doit donc être bénéfique pour l'espèce. Seuls les poussins, encore mal protégés individuellement contre le froid, doivent utiliser une défense "communautaire" en formant des "tortues" dans les blizzards d'automne. Pour les Manchots Adélies, *le départ précoce de la glace* de mer qui met l'eau libre et donc la pêche à proximité des colonies, *est favorable;* le danger réside dans la persistance de la banquise soudée pendant l'été.

Les deux problèmes fondamentaux de l'alimentation et de la résistance au froid sont résolus très différemment par le Manchot empereur. *Il apparait comme lié à la glace de mer* tout au long de l'année. *Pendant l'hiver*, dans les conditions terribles de la côte antarctique, *le problème de résistance au froid est essentiel.* Aussi les colonies doivent-elles rechercher les climats régionaux épargnés par les vents catabatiques, l'abri des falaises et des glaciers. Par ailleurs, sur une étendue sans limites, une technique d'incubation sans nid, une organisation sociale sans territoire et une aggressivité intra-spécifique très faible permettent la lutte communautaire contre le froid par la formation spectaculaire et généralisée des "tortues" quand les moyens individuels de thermo-régulation sont dépassés. *Le problème alimentaire* pour les adultes, (malgré l'adaptation extraordinaire représentée par le jeûne de quatre mois) et pour l'élevage des poussins est tout aussi crucial: il ne peut être résolu que par l'établissement des colonies dans des régions particulières à proximité desquelles existe de l'eau libre même pendant l'hiver. Contrairement à ce qui se passe pour le Manchot Adélie, *la débâcle précoce des glaces* qui détruit le "sol" de la colonie et entraine vers le nord des poussins encore incapables de nager, *est une catastrophe* et la prospérité annuelle d'une colonie dépend de la stabilité de la glace de mer pendant le début de l'été.

On peut donc dire que des conditions écologiques différentes accompagnent (ou expliquent) des différences d'organisation sociale et d'éthologie. On peut également affirmer que les conditions climatiques d'une année donnée, et en particulier l'état de la glace de mer et la date de la débâcle, aussi bien chez le Manchot empereur que chez le Manchot Adélie, jouent un rôle essentiel et parfois inverse sur la mortalité des jeunes de cette année-là et donc sur la structure future de la colonie et sur sa destinée même.

Il semble bien aussi que puissent s'expliquer simplement par les impératifs écologiques les cycles opposés des deux Manchots antarctiques et la reproduction hivernale apparamment paradoxale du Manchot empereur. Pour les deux espèces la période de développement du poussin correspondant à des besoins alimentaires considérables, à l'établissement d'une thermo-régulation efficace, à la possibilité de se nourrir seul en pêchant, doit se placer au moment le plus favorable de l'année, donc à la fin de l'été, période des températures maxima, des vents les plus faibles, de la dislocation générale de la banquise. La différence dans la vitesse de développement des jeunes des deux espèces explique alors les cycles opposés du Manchot Adélie et du Manchot empereur. L'écologie des Manchots antarctiques nous montre comment chaque espèce, en fonction de ses caractéristiques spécifiques fondamentales, a résolu le problème de ses relations avec le milieu antarctique.

Summary

This work is an attempt to describe and explain the relations between the two true Antarctic penguins (Adelie Penguin and Emperor Penguin) and their very particular environment.

The first chapter is a summary of what we know of Antarctic geographic and climatic conditions. The penguins live in three different biotopes: the coast, the waters south of the Antarctic Convergence line, and the pack-ice with their very important seasonal variations. Three of the five Antarctic climatic regions are inhabited by penguins, and the "total cooling power" of the Antarctic atmosphere influenced chiefly by wind and blizzard is the most important factor for homeotherms. But penguins spend their lives in micro-climates at low levels much less severe than the meteorological climate. Moreover they are able to create an artificial micro-climate by means of their social behaviour, and thus to solve a difficult energy problem.

The second chapter deals with the Adelie Penguin: general features of the species, communications and inter-action between bird and environment, life cycle of the bird within the yearly Antarctic cycle, environment and mortality, and it also shows the effect of geographical and climatic factors on the structure and destiny of Adelie Penguin colonies.

An ecological study of the Emperor Penguin is made in the third chapter, including information collected by the work of the International Geophysical Year expeditions and by their discovery of new rookeries. The individual cycle of the bird (weight, endocrine glands, temperature, etc.) and the social cycle and breeding behaviour are studied in relation to ecological conditions. This is followed by discussion of the structure and dynamic of Emperor Penguin populations and the evolution of behaviour within the genus *Aptenodytes*.

The two Antarctic species live in the same Antarctic environment with the same imperatives of breeding in the air and feeding in the water. But, in fact, they must face the two fundamental problems of food and resistance to cold under different ecological conditions. We may consider the Adelie Penguin as bound to the land along the coast during its breeding period in the relatively moderate climatic conditions of Spring and Summer. Its two-month fasting period, its social territorial organisation, the social thermo-regulation of chicks and the speed of their development are a successful answer to the coastal biotope. In winter the Adelie Penguin migrates to the North to find food and a milder climate. The Emperor Penguin seems to be bound to the pack-ice all the year round. Because of the slower development of the young, the breeding period must take place in the terrible winter conditions of the coast.

Use of very special areas with natural micro-climates and open water throughout the year, four-month fasting period, social thermo-regulation in common made possible by non-territorial organisation and very low intra-specific aggressivity, all serve to explain the survival of the species in the worst conditions on earth.

For both penguins, the ecological conditions of a given year, and above all, climate and ice conditions, have a determining effect on the mortality of the young of such a year, and therefore on the future structure of the age group and the destiny of the colony.

Legends of the text-figures

Fig. 1a. The distribution of the Emperor Penguin colonies.

Fig. 1b. Stations of the IGY 1957–1958.

Fig. 2a. Monthly and annual mean wind speed in km/hr.

Fig. 2b. Monthly and annual mean temperatures in °C.

Fig. 3. The "cooling power" in cal/g/sec as a function of wind, temperature and blizzard.

The descending curves represent the cooling time from 40 to 20° of the apparatus used (a metallic cylinder filled with water) as a function of the wind speed for the temperatures —10°, —20° and —30°. The drawn lines represent measurements in Adelie Land during a blizzard. The interrupted lines have been obtained in calm air from the cooled bellow manufactured by Messrs. Chausson. The distance between the two types of curves indicates the effect of the ice transported by the blizzard. Times in min and sec, wind speed in m/sec and km/hr. The ascending curves represent the "cooling power" for the apparatus used at a temperature of —20° C, in the blizzard in Adelie Land (drawn line) and in calm air (interrupted line). The cooling power is the number of cal/g lost per sec by the apparatus used. The apparatus lost 1730 cal/g when going from 40° to 20°. Each point on the curve has been obtained by dividing 1730 by the number of seconds indicated for each wind speed by the curves of the cooling times for the temperature —20° C. The "cooling power" is 2.1 cal/g/sec for a wind speed of 3 m/sec, 4.2 cal/g/sec for a wind speed of 10 m/sec etc.

Fig. 4. Mean wind speed (in m/sec) and temperature (in ° C) at Pte Géologie and Port-Martin.

Fig. 5. Micro-climates of Adelie Land. Wind Speed in m/sec as a function of height above ground for 6 characteristic places. Means of five series of measurements between Nov. 13 and Dec. 5, 1950. Considerable difference between the curve of the well-protected point A and that of point D without shelter. The wind gradient is always important, especially in sheltered places. The wind is the essential factor of the micro-climate.

Point A: Pit in rock very well sheltered from above and from below.

Point B: Rock without shelter.

Point C: Rock with wind-screen from below.

Point D: Ice-field without shelter.

Nid des F: Well sheltered nest.

Nid Plaine: Nest without shelter.

Fig. 6. Micro-climates of Adelie Land. Air temperature as a function of the height above ground for the 6 characteristic places. The temperature is not very important in the micro-climates.

644

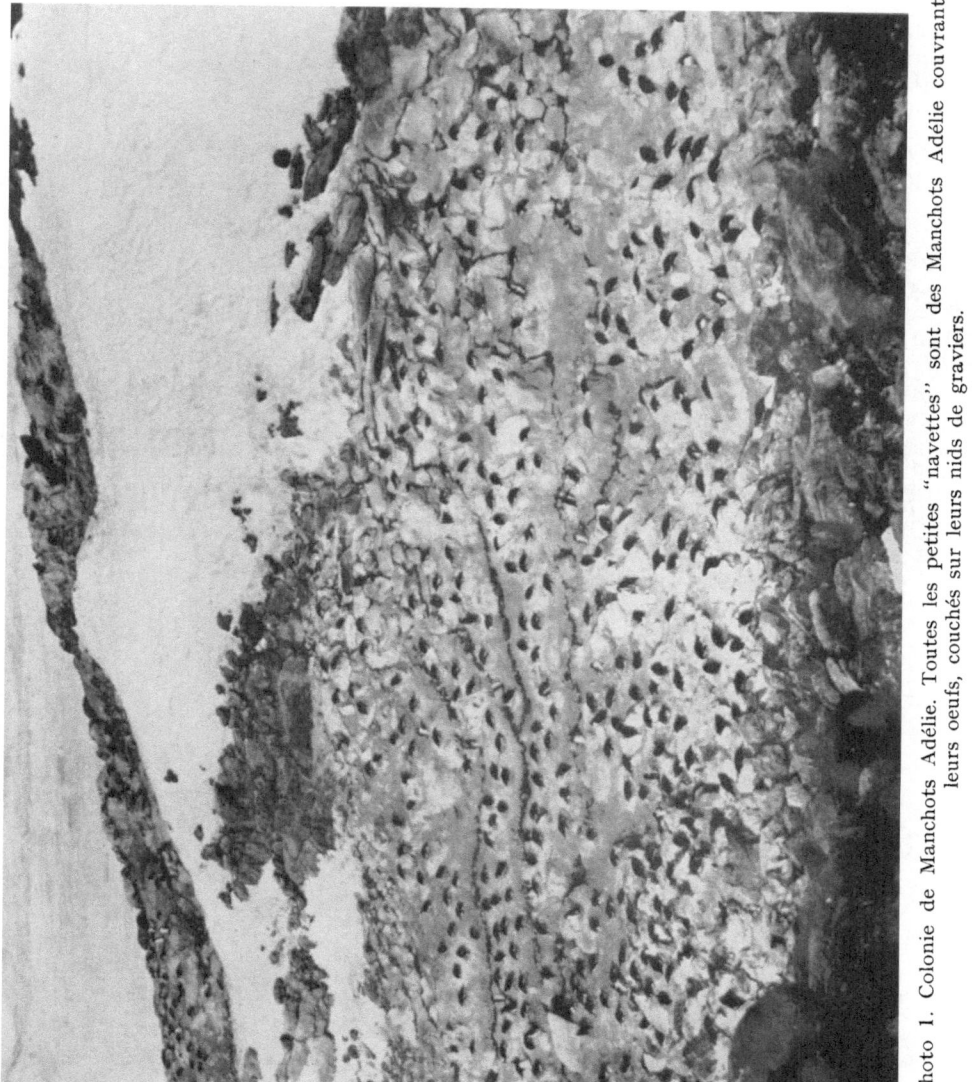

Photo 1. Colonie de Manchots Adélie. Toutes les petites "navettes" sont des Manchots Adélie couvrant leurs oeufs, couchés sur leurs nids de graviers.

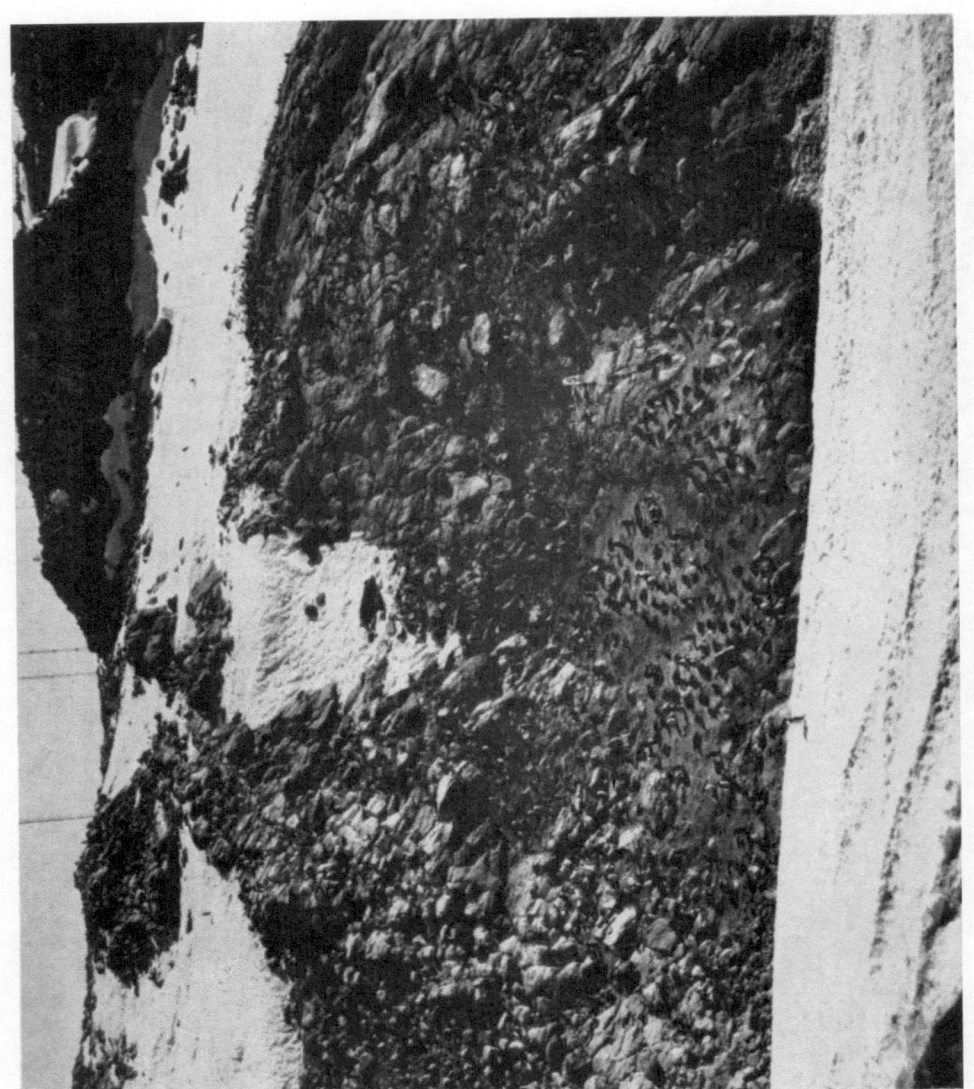

Photo 2. Colonie de Manchots Adélie pendant la période d'incubation.

Photo 3. Période de préponte par beau temps. Quatrevingt-neuf oiseaux présents, 83% d'oiseaux en couple, 10% d'oiseaux couchés. Les flèches indiquent de gauche à droite une parade mutuelle, une courbette mutuelle, et, sur les nids de la vire les nids 1, 2, 3.

IV

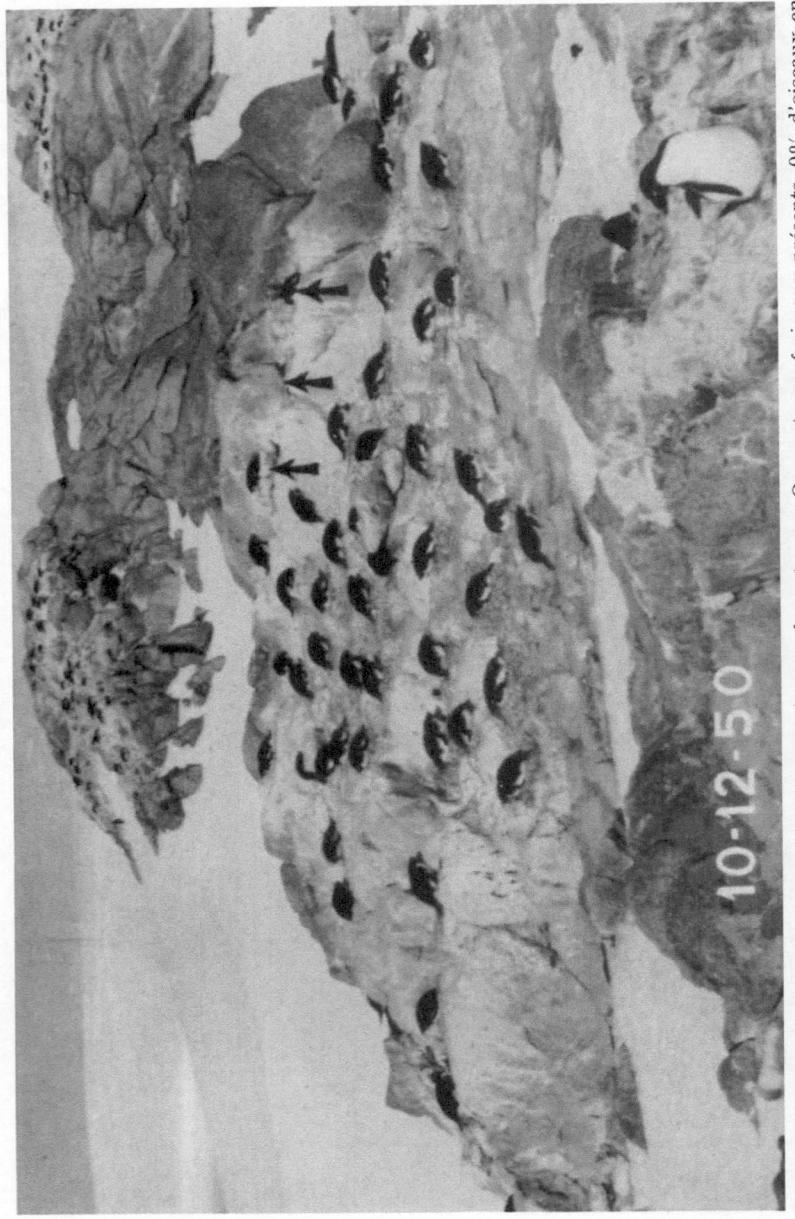

Photo 4. Le "Rocher Noir" à la période d'incubation par beau temps. Quarante-neuf oiseaux présents, 0% d'oiseaux en couple, 100% d'oiseaux en position d'incubation, 20 oiseaux sont placés tête au vent, 19 oiseaux dos au vent. Les flèches indiquent les trois nids de la vire: un couveur dans les nids 1 ct 3, nid 2 vide. Tous les oiseaux non couveurs sont à la pêche en eau libre. Diminution des surfaces enneigées par rapport à la photo précédente.

Photo 5. Fin de la période territoriale au "Rocher Noir" et début de la formation des crèches de poussins. Beau temps couvert. Quarante-cinq adultes présents, 1 couple, 30 poussins visibles dont certains en deuxième duvet. Pour la première fois, on voit des poussins isolés des adultes. Beaucoup de nids sont abandonnés. Les flèches indiquent, de gauche à droite, un bâillement à sa phase initiale, et un nourrissage au nid 1 de la vire. Le "Rocher Noir" est complètement déneigé, l'eau libre est visible dans la baie en haut et à gauche du cliché.

VI

Photo 6. Le "Rocher Noir" à la période des crèches. Huit adultes présents, 35 poussins en deuxième duvet avec, au centre, une crèche de 25 poussins. Disparition complète de toute organisation territoriale, dispersion des cailloux des nids.

Photo 7. Parade mutuelle typique avec large inclinaison latérale des oiseaux à contre-temps.

Photo 8. Courbette mutuelle.

Photo 9. Transport d'un volumineux caillou pour la construction du nid.

Photo 10. Comportement antagoniste chez deux voisins de nids. Crêtes frontales dressées chez les deux oiseaux.

Photo 11. Bataille de type "conjugal". L'oiseau poursuivi, fuyant sans se battre, ne semble pas avoir la crête dressée contrairement à son poursuivant.

Photo 12. La position extatique est souvent prise simultanément par plusieurs oiseaux voisins et a été considérée comme "contagieuse".

Photo 13. Copulation.

Photo 14. Groupe d'Adélies sur un floe dérivant au large de la côte en mars 1949.

Photo 15. Manchot couvant enfoui dans une congère après un blizzard.

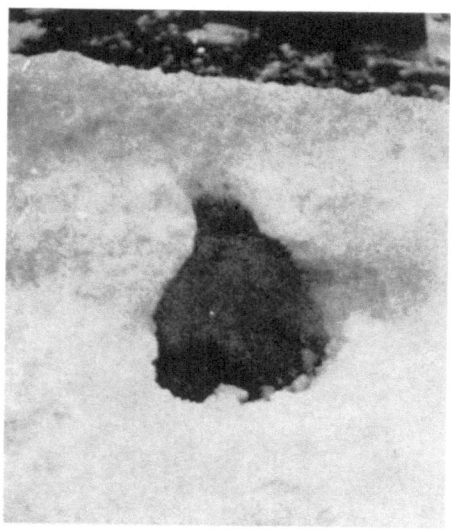

Photo 16. Grand poussin à demi enfoui dans le névé après un blizzard.

Photo 17. Une crèche pendant un blizzard. La neige colle au duvet des poussins.

Photo 18. La partie est de l'emplacement de la colonie de Manchots empereurs. Au premier plan le poste météorologique et le parc d'étude, au fond la falaise du glacier de l'Astrolabe (28-IV-1956).

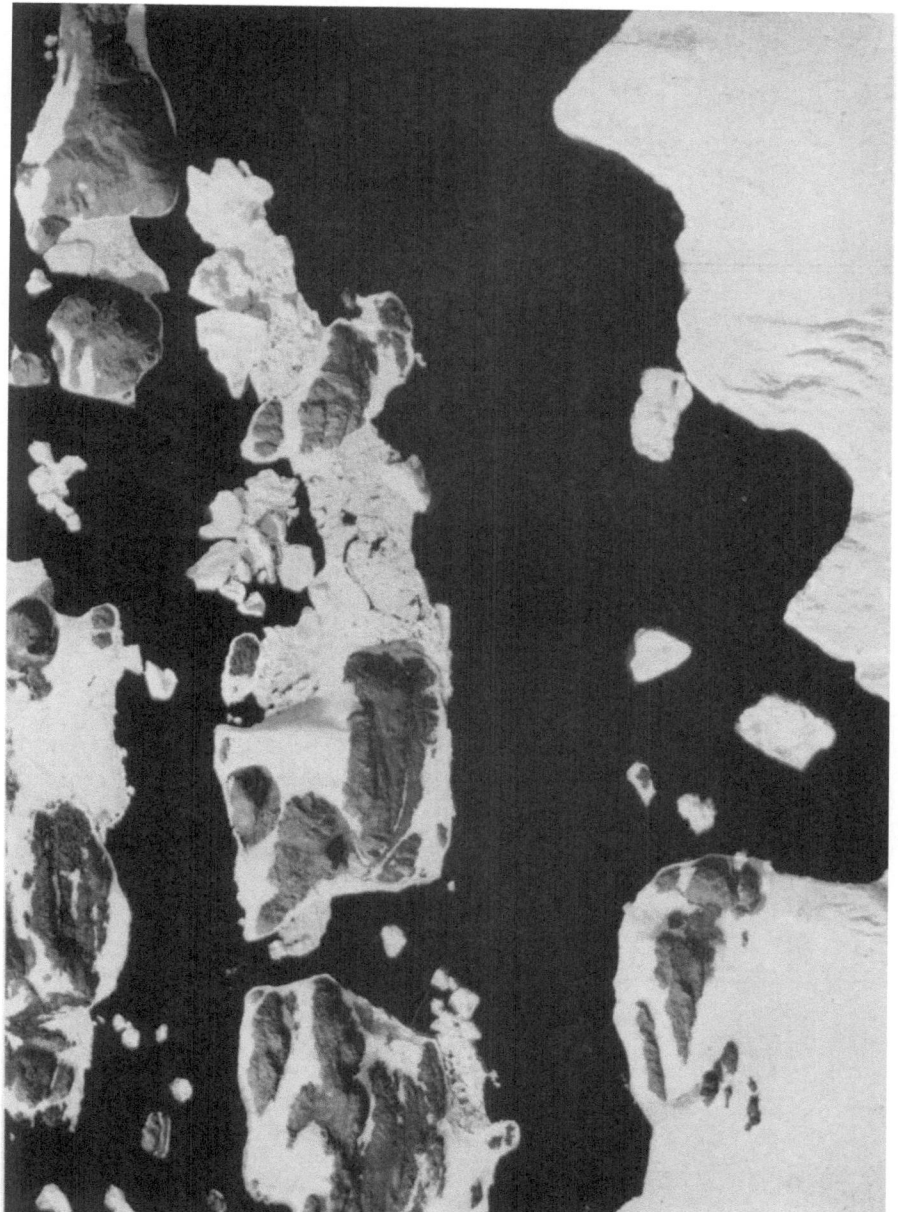

Photo 19. **Vue aérienne de l'Archipel de Pointe Géologie.** En bas le continent antarctique qui borde le territoire de la colonie limité au nord par les îles Carrel et Rostand (janvier 1957). (Photo Cdt PETITJEAN).

XIV

Photo 20. Petite "tortue" circulaire typique pendant la pariade (17-IV-1952).

Photo 21. Face-à-face mutuel; le mâle est à droite (6-IV-1952).

Photo 22. Une des femelles d'un trio amorce
un face-à-face mutuel avec le mâle (l'oiseau
de gauche) (28-IV-1956).

Photo 23. Attitude caractéristique d'un oiseau
pendant l'émission du chant (25-IV-1956).

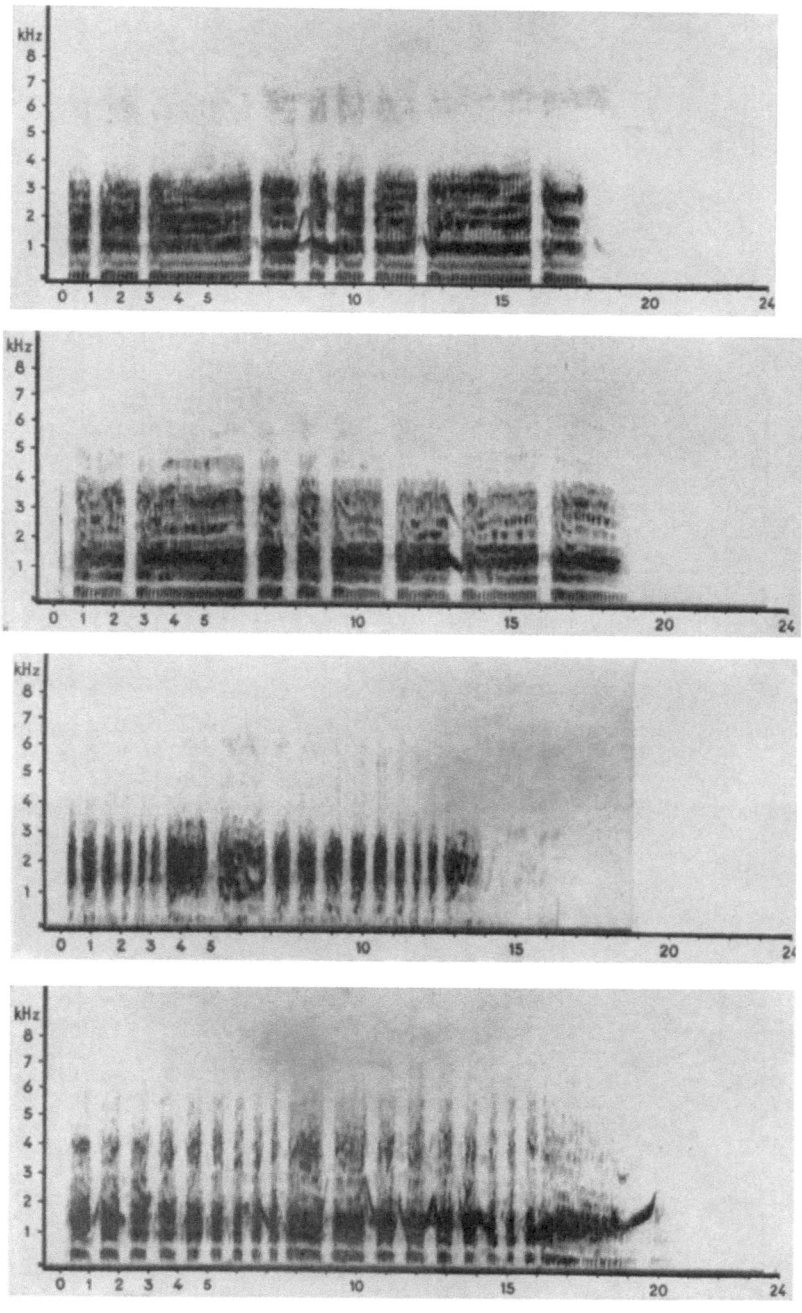

Photo 24. Deux chants d'oiseaux mâles. 1: un mâle du 15 mai 1956; 2: un mâle
du 20 novembre 1956.
Deux chants d'oiseaux femelles. 3: une femelle du 15 mai 1956; 4: une femelle
du 20 novembre 1956.

Photo 25. A gauche, attitude d'un oiseau qui sommeille le bec glissé sous l'aileron.
A droite, deux oiseaux de sexe opposé se "tenant compagnie".

Photo 26. Premiers arrivants sur la glace de mer en formation (15-III-1952).

Photo 27. Emplacement piétiné par un couple avant la ponte de la femelle (10-V-1956).

Photo 28. Une dernière contraction va expulser l'oeuf encore engagé dans le cloaque (16-V-1956).

XX

Photo 29. Les deux partenaires observent l'oeuf
après la ponte (8-V-1956).

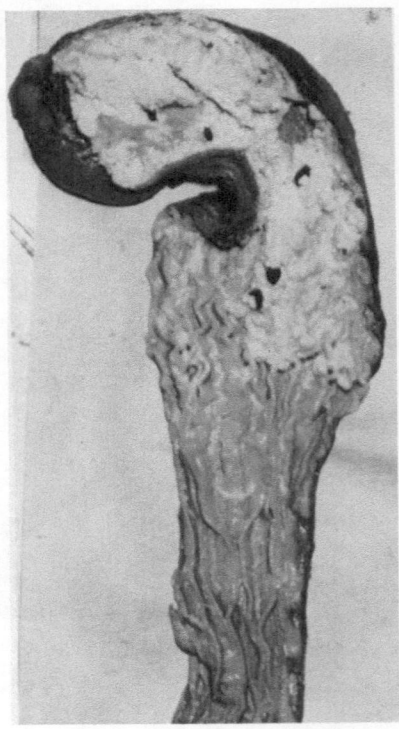

Photo 30. Estomac ouvert montrant
l'aspect de la substance sécrétée par
les couveurs mâles pour nourrir les
poussins (10-VIII-1956).

Photo 31. Un parent à la recherche de son poussin vient chanter devant un groupe (28-IX-1956).

Photo 32. "Tortue" de poussins au début de l'émancipation (8-IX-1956)

Photo 33. Adulte chassant un poussin étranger (17-IX-1952).

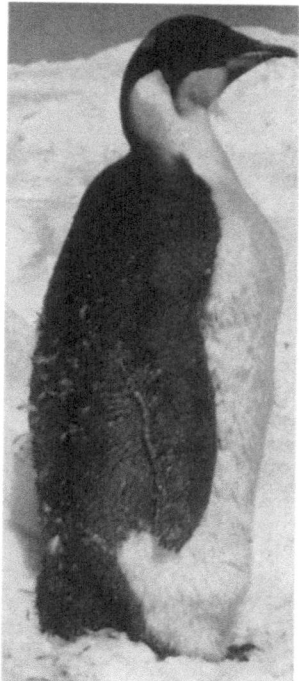

Photo 34. Immature de deuxième année perdant ses premières plumes (6-XII-1952).

Photo 35. Poussin atteint de "maladie de carence". On remarquera la position des pattes et la disparition du duvet ventral arraché par la reptation permanente sur la glace (décembre 1952).

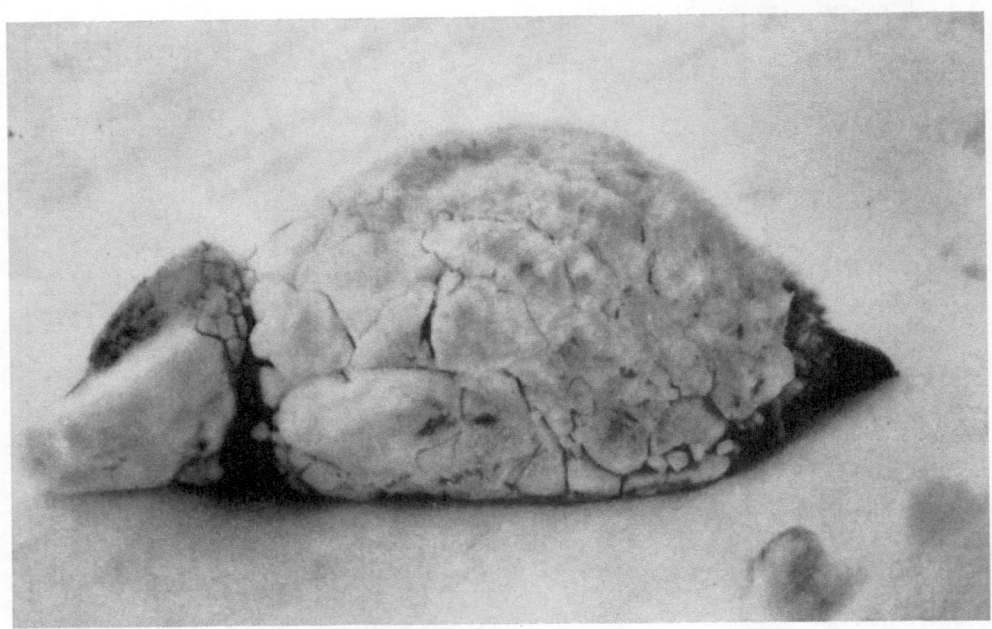

Photo 36. Poussin recouvert d'une gangue de neige dure après un blizzard (12-X-1956).

Fig. 23. Mean monthly weight of the livers of 41 male and 47 female birds in 1952 and 1956.

Fig. 24. Mean increase in weight of the chicks during the guard stage in 1952 and 1956. Each point on the curve is the mean of 10 subjects taken at random in the colony.

Fig. 25. Mean increase in weight of three marked chicks (6, 7, and 20) during 1952.

Fig. 26. Mean increase in weight of three ringed chicks (nrs. 7, 11 and 29) in 1956. Nr. 29 represents the typical case of an under-nourished chick).

Fig. 27. Comparative egg mortality in 1952 and 1956.

Legends of the photographs

The Adelie Penguin

Photo 1. Colony of Adelie Penguins. All the little black dots are birds, covering their egg while sitting on their nest of gravel.

Photo 2. Colony of Adelie Penguins during the incubation period.

Photo 3. Oct. 30, 1950. Pre-egg stage. Fine weather. 89 birds present, 83% as couples, 10% sitting. The arrows indicate from left to right: mutual display, mutual bowing, and the nests no. 1, 2 and 3.

Photo 4. Dec. 10, 1950. The "Rocher Noir" at the incubation period. Fine weather. 49 birds present, none as couple, all in incubation position. 20 birds face the wind direction, 19 do not. The arrows indicate the nests No. 1, 2 and 3: 1 and 3 are occupied, 2 is empty. All the other birds are fishing in open water. The snow cover is less than on photo 3.

Photo 5. Jan. 13, 1951. End of the territorial period on the "Rocher Noir" and beginning of the formation of the creches for the chicks. Fine, cloudy weather. 45 adults present, one couple, 30 chicks, some of which in their second down. For the first time one sees chicks isolated from their parents. Many nests have been abandoned. The arrows indicate from left to right: the initial phase of yawning, and feeding on nest 1. The snow has completely disappeared, the open water is visible in the bay (top of photo) and at left.

Photo 6. Jan, 23, 1951. The "Rocher Noir" at the creche period. Eight adults present, 35 chicks in their second down, with, in the center, a creche of 25 chicks. Any territorial organization has disappeared, the stones of the nests have been dispersed.

Photo 7. A typical mutual display with large lateral bowings at alternating times.

Photo 8. Mutual bowing.

Photo 9. Transport of a voluminous pebble-stone for the construction of the nest.

Photo 10. Agonistic behaviour of two nesting neighbours. Both birds have raised their crest.

Photo 11. "Conjugal" fight. The pursued bird, running away without fighting, seems not to have raised its crest, contrary to the pursuer.

Photo 12. The ecstatic display can often be seen simultaneously by several neighbouring birds and has been called "contagious".

Photo 13. Copulation.

Photo 14. A group of Adelie penguins on an ice-field drifting to open water in March 1949.

Photo 15. A breeding penguin dug into the snow after a blizzard.

Photo 16. Great chick half dug into last year's snow after a blizzard.

Photo 17. A creche during a blizzard. The snow sticks to the down of the chicks.

The Emperor Penguin

Photo 18. The eastern part of the rookery of Emperor Penguins. In the foreground the meteorologic station and the study area, in the background the cliffs of the Astrolabe Glacier (April 28, 1956).

646

Photo 19. Areal view of the Pointe Géologie Archipelago. Below, the Antarctic continent bordering the territory of the rookery, in the north limited by the islands Carrel and Rostand (Jan. 1957). (Photo Cdt. PETITJEAN).

Photo 20. Small circular "huddle" typical during the mating-season (Apr. 17, 1952).

Photo 21. Mutual "face-to-face". At the right the male. (Apr. 6, 1952).

Photo 22. One of the females of a trio provokes a mutual face-to-face with the male (the bird at the left). (Apr. 28, 1956).

Photo 23. Attitude of a bird during the song (Apr. 25, 1956).

Photo 24. Songs. 1: Male, May 15, 1956. 2: Male, Nov. 20, 1956. 3: Female, May 15, 1956. 4: Female, Nov. 20, 1956.

Photo 25. At the left a bird sleeping with its bill under the wing. At the right two birds of opposite sex "keeping company".

Photo 26. First arrivals at the forming sea-ice (March 15, 1952).

Photo 27. "Stamping their feet on the spot" by a couple before the laying of the egg (May 10, 1956).

Photo 28. A last contraction expels the egg, which still is half-way in the cloaca (May 16, 1956).

Photo 29. The two partners contemplate the egg after the laying (May 8, 1956).

Photo 30. Opened stomach to show the aspect of the substance which is secreted by the male incubators to feed the chickens (Aug. 10, 1956).

Photo 31. A parent, looking for his chick, sings in front of a group (Sept. 28, 1956).

Photo 32. A huddle of chicks at the beginning of the emancipation (Sept. 8, 1956).

Photo 33. An adult chasing away a strange chick (Sept. 17, 1952).

Photo 34. A second-year immature loosing its first feathers (Dec. 6, 1952).

Photo 35. A chick covered by a solid snow cover after a blizzard (Oct. 12, 1956).

Photo 36. Undernourished chick. Note the position of the paws and the disappearance of the ventral down torn out by the permanent creeping on the ice (Dec. 1952).

BIBLIOGRAPHIE SOMMAIRE

ALT, J. (1960) Quelques considérations générales sur la météorologie de l'Antarctique. *La Météorologie*, **57**: *17—41*.

BAUER, A. (1961) Nouvelle estimation du volume de la glace de l'inlandsis antarctique. Colloque sur la glaciologie antarctique. Publication n° 55 A.I.H.S.; *19—23*.

BOURLIERE, F. & PREVOST, J. (1953) Le cycle sexuel paradoxal de l'Aptenodytes forsteri. *C.R. Soc. Biol.* **147**: *1893—94.*

BUDD, G. M. (1961) The Biotopes of Emperor Penguin Rookeries. *The Emue.* **61**: *171—189.*

BUNT, J. S. (1961) Introductory studies. Hydrology and Plankton. Mawson June 1956—February 1957 — ANARE, Rep. Ser. B vol. III.

CAUGHLEY, G. (1960) The cape Crozier Emperor Penguin rookery. *Rec. Dom. Wellington* 3: *251—62.*

CENDRON, J. (1952) Médecine biologie. Expédition en Terre Adélie 1950—1952. Rapport préliminaire n° 20: *49—53*. Publications Expéditions Polaires Françaises.

CENDRON, J. (1952) Une visite hivernale à une rookerie de Manchots Empereurs. *La Terre et la Vie: 101—108.*

CENDRON, J. (1953) La mue du Manchot Adélie adulte. *Alaude* 21, 2: *77—85.*

DEACON, G. E. R. (1959) The Antarctic Ocean. *Sci. Progr.* **47**: *647—660.*

EKLUND, C. (1956) Antarctic fauna and some of its problems. *Geophys. Monog. U.S.A.*, **1**: *117—123.*

FALLA, R. A. (1937) Birds. Report of B.A.N.Z.A.R.E. 1929—1931. Series B. 2.: *1—288.*

FORSTER, G. (1777) A voyage round the world in H.M.S. Slopp "Resolution" commanded by Captain COOK 1772—1775. London.

FOXTON, P. (1956) The distribution of the standing crop of zooplancton in the Southern ocean. *Discovery Rep.*, **28**: *191—236.*

GAIN, L. (1914) Oiseaux antarctiques. Deuxième Expédition Antarctique Française 1908—1910. Masson, Paris.

GOLDSMITH, R. & SLADEN, W. J. L. (1961) Temperature regulation of some antarctic penguins. *J. Physiol.* **157**: *251—62.*

HARRINGTON, H. J. & MCKELLAR, I. C. (1958) A Radio carbon date for Penguin Colonization of Cape Hallet, Antarctica. *N. Zeal. J. Geol. Geophys.* **1, 3**: *577—6.*

HARRINGTON, H. T. (1959) Narrative of a visit to a newly discovered emperor penguin rookery at Coulman Island, Ross Sea, Antarctica. *Notornis*, **8**: *127—32.*

HART, I. J. (1934) On the phytoplankton of the South-West Atlantic and the Bellingshausen sea. *Discovery Rep.*, **8**: *1—268.*

HOMBRON, J. B. & JACQUINOT, C. H. (1841) Description de plusieurs oiseaux nouveaux ou peu connus provenant de l'expédition faite autour du monde sur les corvettes "l'Astrolabe" et "La Zélée". *Ann. Sci. Nat.* **2, 15**: *320.*

ISEL, J. (1958) Notes M.S.

LEPINEUX, G. (1955) Les observations météorologiques de Pointe-Géologie Terre Adélie 66°, 40 S 140°, 01' E. Conditions atmosphériques en surface et en altitude du 18 février 1952 au 30 décembre 1952. Expéditions Polaires Françaises. Expéditions Antarctiques. Résultats scientifiques.

LEVICK, G. MURRAY (1915) Natural History of the Adelie Penguin. British Antarctic "Terra Nova" expedition 1910. Zoology 1, 2: *55—84.*

LOEWE, F. (1956) Etudes de Glaciologie en Terre Adélie 1951—1952. Hermann, Paris.

MARR, J. W. S. (1956) *Euphausia superba* and the Antarctic surface currents. An advance note on the distribution of the Whale food. *Norsk Hvalfangst Tid.* **3**: *127—34*.

MURPHY, R. C. (1936) Oceanic Birds of South America. 2 vol. The American Museum of Natural History. New-York.

NEMOTO, T. (1959) Food of Baleen Whales with reference to Whale movements. *Sci. Rep. Whales. Res. Inst.*, **14**: *149—290*.

PREVOST, J. (1961) Ecologie du Manchot Empereur. 1 vol. Hermann, Paris.

PREVOST, J. & BOURLIERE, F. (1955) Sur le cycle reproducteur de quelques oiseaux antarctiques. *Acta XI Congr. Int. Ornithol., Basel: 248—51*.

RICHDALE, L. E. (1951) Sexual Behaviour in Penguins. 1 vol. Lawrence-University of Kansas Press.

ROBERTS, B. A. (1940) The breeding Behaviour of Penguins. British Graham Land Expedition 1934—1937. Sci. Rep. **1, 3**: *195—254*.

SAPIN-JALOUSTRE, J. (1952). Découverte et description de la rookerie de Manchots Empereurs de Pointe-Géologie (Terre Adélie). *Oiseau et R.F.O.*, **22**: *143—260*.

SAPIN-JALOUSTRE, J. (1953) L'établissement de la thermorégulation chez le Manchot Adélie. *C.R. Acad. Sci.* **237**: *1443—4*.

SAPIN-JALOUSTRE, J. (1955) Contribution à l'étude de l'acclimatation de l'homme et des vertébrés supérieurs dans l'Antarctique et plus spécialement en Terre Adélie. *Biol. méd.* **44**: *15—72 et 135—208*.

SAPIN-JALOUSTRE, J. (1960) Ecologie du Manchot Adélie. 1 vol. Hermann, Paris.

SIEBURTH, J. McN (1958) Antarctic Microbiology. *Amer. Inst. biol. Sci. Bull.* **8**: *10—12*.

SLADEN, W. J. L. (1953) The Adelie Penguin. *Nature, Lond.* **171**: *952—5*.

SLADEN, W. J. L. (1955) Some Aspects of the Behaviour of the Adelie and Chinstrap Penguins. *Acta XI Congr. int. Ornithol., Basel: 241—7*.

SLADEN, W. J. L. (1958) The Pygoscelid Penguins. *Falk. Isl. Dep. Surv., Sci. Rep.* n° **17**.

SLADEN, W. J. L. & FRIEDMAN, H. (1961) Science in Antarctica. National Academy of Science. *Nat. Res. Counc. Publ.* **839**: *62—76*.

STONEHOUSE, B. (1953) The Emperor Penguins (*Aptenodydes forsteri* Gray) I Breeding behaviour and development. *Falk. Isl. Dep. Surv. Sci. Rep.* N° **17**.

STONEHOUSE, B. (1960) The King Penguin (*Aptenodytes patagonia*) of South Georgia. I Breeding behaviour and development. *Falk. Isl. Dep. Surv. Sci. Rep.* N° **23**.

TINBERGEN, N. (1951) The Study of Instinct. University Press, Oxford.

WILSON, E. A. (1907) Natural History, Vol. 2 Zoology. Part. 2 Aves. British National Antarctic Expedition 1901—1904. Scientific Results. London, British Museum. (Natural History).

WILLING, R. L. (1958) Australian Discoveries of Emperor Penguins rookeries in Antarctica during 1954—57. *Nature, Lond.* **182**: *1393—1394*.

ANTARCTIC BIRDS

BY

K. H. VOOUS

Zoological Museum, University of Amsterdam,
Zoological Laboratory, Free University, Amsterdam

Introduction

Of the geographical distribution and the ecological conditions of birds living in the antarctic region a large amount of interesting knowledge has been collected piece-meal by a great number of antarctic explorers, all of whom deserve the deepest homage and acknowledgement for their courage and perseverance. But at least an equally large amount of information on distribution and life habits remains to be gathered and the number of problems of general zoological and ecological interest that has come to light is absorbing. Therefore in the following pages only the great lines of the distribution and the ecology of birds in the Antarctic will be described, as understood to-day. The present introductory remarks are supposed to give a summary of the kind of problems met in antarctic ornithology.

The Antarctic Convergence of surface water, as described by the marine zoologists of the "Discovery" expeditions (DEACON 1933) is here accepted as the northern boundary of the antarctic region. This circumpolar line runs approximately parallel to the 50th degree of southern latitude, but reaches considerably more northwards in the southern Atlantic and Indian Oceans than in the Pacific. The Antarctic Convergence moreover is not a fixed line, but shifts more northwards or more southwards according to seasons and years. The question can be raised whether, apart from the animal life on the antarctic continent and in the seas to the south of the convergence, the whole of the terrestrial fauna of all islands south of the convergence has to be treated as "antarctic". Diagnostic differences between "antarctic" and "sub-antarctic" terrestrial faunas apparently have not been described, lest unanimously accepted. Arguments will be given to consider the bird faunas of e.g. the islands of South Georgia, Kerguelen, and Macquarie, which are situated at approximately 54°30′, 49°, and 55°30′ southern latitude, respectively, as not yet typically antarctic. As not typically antarctic can be described a situation in which the winters are less rigid than in truly antarctic regions, permitting the wintering of terrestrial birds, like the ducks in South Georgia and Kerguelen, a small songbird (species of pipit) in South Georgia, and a big rail in Macquarie Island. The seas surrounding these islands, however, have to be considered to be inhabited by an antarctic fauna (see also: LINDSEY 1940: 456—457). In view of this discrepancy in delimiting antarctic marine and terrestrial faunas a special chapter will be devoted to land birds.

In antarctic bird life (apart from the penguins) oceanic birds are dominating. There are at least 33 species, some of these in many thousands of specimens, which wander far and wide over the antarctic seas. In size they range from birds with a wing-span of up to 345 cm and a body-weight of approximately 9000 g (Wandering Albatross, *Diomedea exulans*) to tiny creatures of 39 cm wing-span and 35 g body-weight (Wilson's Storm Petrel, *Oceanites oceanicus*). Penguins and tube-nosed sea-birds (albatrosses and petrels) form the bulk of the species; gulls and jaegers which are dominating the arctic scene are only poorly represented in the south. In view of the wide extent of the stormy seas and the inhospitality of land masses, particularly of the antarctic continent, the mapping of the geographical distribution of the birds in the antarctic region has not yet been completed; in fact, of the antarctic coast line only fragmentary records of breeding places are available. Of the most numerous and most typical of high-antarctic birds, the Antarctic Petrel *(Thalassoica antarctica)*, the first breeding place was discovered not before 1912 (NORTH 1914; MAWSON 1915 : 117) and the next following egg was found in 1940 and described in 1945 (FRIEDMANN 1945 : 308; see also, PERKINS 1945 : 275, fig. 12). Only the coasts and the nearest nunataks rising out of the land of snow and ice are inhabited by birds. Particularly Snow Petrels *(Pagodroma nivea)* and Skuas *(Stercorarius skua)* have been frequently recorded far inland, but a tiny Wilson's Storm Petrel flying in its buoyant flight over ice and snow of the Ross Shelf Ice at about 15 km from the nearest patch of open sea at the ice barrier is already a noteworthy phenomenon (SIPLE & LINDSEY 1937 : 151). No birds seem to nest at distances greater than 200 km inland of the coast of Antarctica. Still, the presence of birds has been recorded at positions in the heart of the continent, quite near the South Pole (South Polar Skua, *Stercorarius skua maccormicki;* see p. 655.

Antarctic birds spend the greater part of their lives at sea. Therefore special attention has been given to pelagic life and pelagic distribution (p. 676). There are species which have a very wide pelagic distribution during and outside the breeding season, wandering far into sub-antarctic waters to the north (Cape Pigeon, *Daption capensis)*. Others seem to keep more strongly to special types of sea-water and ice, like the Snow Petrel and the Antarctic Petrel. But all of them disperse in latitudinal zones encircling the continent, rather than sectorially, in one or more of the sections of the antarctic region. Details of the pelagic distribution, the specific differences and seasonal vicissitudes are incompletely known. These seem to be correlated with the appearance and the distribution of concentrations of superficial plankton. Interesting facts of circumpolar long-distance migration have been found as a result of the ringing scheme of the Falkland Islands Dependencies Survey of Giant Petrels *(Macronectes giganteus)*, mainly in the South Orkney Islands (p. 679). A ringing scheme of the Skua in Antarctica has been recently undertaken in connection with the activities of the International Geophysical Year (EKLUND 1959) and many interesting details about northward migration, life-span, and other facets of the intimate life of these birds can be expected from this work; preliminary results have already been published (EKLUND 1961).

The zonal arrangement of types of sea-water has induced complicate patterns of pelagic distribution in groups of sea-birds consisting of a number of species, closely resembling each other like the prions or whale-birds *(Pachyptila)*, and may possibly have been the cause of the geographical species formation in these groups (p. 670). The same situation has led to not less complicate patterns of a west-east distribution in birds like the Blue-eyed Shags *(Phalacrocorax)* and the Sheath-bills *(Chionis)* (p. 665). The extension of the southern tip of South America far southwards into sub-antarctic regions and the proximity of New Zealand and its satellite islands to the Antarctic Zone have affected the composition of the antarctic fauna (p. 665).

Not only whales are attracted to the rich feeding grounds formed by the antarctic summer plankton. Birds, both from within the antarctic region and from outside concentrate in the regions bordering the northern edge of the pack-ice where in superficial waters orange krill (mainly the adults of the crustacean *Euphausia superba*) is most abundant. Sub-antarctic birds, like the Sooty Shearwater *(Puffinus griseus)* and Peale's Petrel *(Pterodroma inexpectata)* also penetrate far southwards in summer (p. 677 and 686). There is even a non-breeding summer bird occurring in considerable numbers, which yearly arrives from the opposite side of the globe to winter in the Antarctic (Arctic Tern, *Sterna paradisaea;* p. 685).

Locally, seasonally and yearly different ice-conditions of the sea are of great influence upon the pelagic life of antarctic sea-birds. Some of these birds as a rule never seem to stray away far from sea-ice and snow, like the Snow Petrel; others migrate only somewhat northwards during the non-breeding season, whereas Wilson's Storm Petrel, which is one of the world's southernmost breeding birds, yearly makes astonishing migration flights which take these birds to the other ends of the world, mainly in the northern Atlantic and northern Indian Ocean (p. 684).

Climatic conditions during the short breeding season are hard. Hence the nesting habits of only few species manage to withstand the rigours of Antarctica. Apart from the penguins, nests of no more than 8 species have been found on the continent (p. 657); only 7 of which are regular breeding birds, viz., *Macronectes gigan-*

teus, Thalassoica antarctica, Fulmarus glacialoides, Daption capensis, Pagodroma nivea, Oceanites oceanicus, Stercorarius skua. Nesting habits will be described and discussed on p. 658—659. Comparative ecological studies of the species breeding under antarctic conditions may yield extremely interesting results. Recording the temperatures of nesting holes and burrows, of eggs during incubation and of the incubating parent birds is of great interest in view of the extremely low and changeable temperatures of the southernmost breeding places (EKLUND 1942, 1959). High mortality of eggs and young caused by freezing have been recorded by several authors.

The breeding season in the southernmost antarctic regions lasts very short. In the Ross Sea area, which penetrates farthest south into the continent, Snow Petrels usually herald the arrival of spring as early as the middle of September or the beginning of October (SIPLE & LINDSEY 1937; PERKINS 1945), but the real nesting season does not start before one month later. Even here Wilson's Storm Petrel is among the latest of the breeding birds, arriving not before the middle of December (PERKINS 1945). In the low-antarctic islands the spreading of the breeding activity is even more conspicuous. The Snow and Giant Petrels and the Cape Pigeons everywhere are the earliest, Wilson's Storm Petrel among the latest of the season (p. 667 and 684). The not truly antarctic island of Kerguelen even harbours autumn- and winter-breeders: the Grey Petrel *(Procellaria cinerea)* and the Great-winged Petrel *(Pterodroma macroptera)* (p. 668). An extremely interesting case of a pair of sibling species with different breeding times acting as isolating mechanism has been described from Kerguelen *(Sterna vittata* and *S. virgata;* p. 669).

Food is plentiful during the antarctic summer, but is not always easily available in view of adverse ice conditions. Virtually all of the antarctic birds are known to take krill, *Euphausia superba*, the basic animal food of all antarctic vertebrates. Most of the petrels and all of the albatrosses appear to be extremely partial to cephalopods of all convenient sizes. Fish and also offal from the big whaling industries are taken occasionally in large quantities. In spite of similarity in habits of all petrels and albatrosses a tendency of specialization into or preference for a particular food-item is noticeable and worth further study and discussion (p. 674). Predation by Skua and Giant Petrel causes high mortality among some of the smaller and medium-sized antarctic breeding birds and has influenced the nesting habits and daily rhythm on the breeding grounds of most of the breeding species. Predation and scavenging have been discussed in connection to each other (p. 671).

American ornithologists associated with antarctic explorations during the International Geophysical Year have developed long-term programs for the study of birds in Antarctica (EKLUND 1956, 1959, 1961). These include life history studies and colour-banding of antarctic species in order to gain more knowledge about longevity and mortality, migration pattern, local and seasonal movements, homing experiments (South Polar Skuas flown and released at South Pole Station!), non-reproductive behaviour and physiological adaptations to low temperatures (temperatures of incubating birds and incubated eggs). A similar research program of the Falkland Islands Dependencies Survey, with highly promising preliminary results, has been published by SLADEN & TICKEL (1958). Unfortunately the time has arrived already that measures should be taken to prevent the destruction of animal life in Antarctica and to preserve the birds' breeding colonies (MURPHY 1962, PANZARINI 1962).

Land Birds

In view of the great expanse of the sea and the inhospitable life conditions on the land available, land birds are extremely scarce in the antarctic region and do not occur at all in Antarctica. Obviously these birds do not belong to the true antarctic fauna, but instead are members of the terrestrial fauna of sub-antarctic island regions. Still they are of interest as they show the way of colonization from northerly regions into the Antarctic. The following species of land birds are known to occur south of the Antarctic Convergence.

Anas georgica GMELIN — South Georgia
Anas eatoni (SHARPE) — Kerguelen; Crozet Islands
Anas superciliosa GMELIN — Macquarie Island
Chloëphaga picta (GMELIN) — South Georgia
Rallus (Hypotaenidia) philippensis LINNAEUS — Macquarie Island
Gallirallus australis (SPARRMAN) — Macquarie Island
Cyanoramphus novaezelandiae (SPARRMAN) — Macquarie Island
Anthus antarcticus CABANIS — South Georgia
Carduelis flammea (LINNAEUS) — Macquarie Island
Sturnus vulgaris LINNAEUS — Macquarie Island

Both the South Georgian Teal *(Anas georgica georgica)* and the South Georgian Pipit *(Anthus antarcticus)* are only slightly deviated island forms of species with wide distributions in South America, and in fact must be considered to have reached the island accidentally from South America over sea in a not too distant past. The Teal is usually regarded as an insular relative of the Chilean, Brown, or Yellow-billed Pintail *(A. georgica spinicauda* VIEILLOT); whereas the Pipit can be considered to be a considerably diverged giant form of the South American Pipit *Anthus correndera* VIEILLOT. Presently both are resident birds which pass the winter along the ice-free and seaweed grown sea-coast and fjords; they nest throughout the island.

The Upland Goose *(Chloëphaga picta leucoptera* (GMELIN)) has been artificially introduced in South Georgia from the Falkland Islands and has succeeded to establish itself as a resident breeding

654

bird (MATTHEWS 1929 : 584). Obviously it is a sub-antarctic alien species which has found a suitable island habitat amidst antarctic waters.

The ducks from Kerguelen *(Anas eatoni eatoni)* and the Crozet Islands *(A. eatoni drygalskii* REICHENBACH) represent southern hemisphere island populations of a temperate and boreal holarctic species, the Pintail *(A. acuta)*. They are considered descendants from a direct colonization by northern birds. The present northern Pintails are migrants which are known to have reached even as remote islands as Tahiti in the south Pacific Ocean at latitude 17° south. Like South Georgia the islands of Kerguelen and the Crozets have fresh water lakes and marshes during the summer. These ducks are residents on the islands.

The land birds of Macquarie Island even more clearly show the influence of near-by land masses upon the composition of the antarctic fauna: they are New Zealand birds, which have found the antarctic region on their doorstep. Among the genuine New Zealand species the Banded Rail *(Hypotaenidia philippensis)* and the Parakeet *(Cyanoramphus novaezelandiae)* are now extinct on the island, whereas the Grey Duck *(Anas superciliosa)* is still flourishing. The Woodhen or Weka *(Gallirallus australis)* has been introduced from Stewart Island in 1872 and subsequent years and is presently a nuisance to breeding petrels. The Starlings *(Sturnus vulgaris)* and the Redpolls *(Carduelis flammea)*, though themselves European aliens in the New Zealand avifauna, are now self-introduced colonists from New Zealand in Macquarie Island.

Northern Hemisphere Colonists

Whereas arctic seas are the realm of gulls and terns of the order Laro-Limicolae, the antarctic seas are the home of petrels and albatrosses of the order of Tube-noses (Tubinares). Actually the petrel group consists of open sea birds and the gull assemblage of coastal birds. Thus, there is a correspondence between the conspicuous differences of the distribution of land and sea on the opposite polar sides of our globe. Consequently there are only few members of the gull group living in the Antarctic, numbering 4 species in total. All four species have their close relatives in the northern hemisphere and are therefore considered as colonists from the north. The species are listed below.

Family Stercorariidae: *Stercorarius* skua* (BRÜNNICH) — Southern or Great Skua
Family Laridae : *Larus dominicanus* LICHTENSTEIN — Southern Black-backed Gull
Sterna vittata GMELIN — Antarctic Tern
Sterna virgata CABANIS — Kerguelen Tern

---

* Sometimes known as *Catharacta skua.*

Southern Skua, *Stercorarius skua*. Widely and circumpolarly distributed throughout the sub-antarctic region and the islands and the continental coast of the antarctic region. Indeed, there is practically no patch of accessible coast in the Antarctic where this species is absent. It is this species of bird that has been observed nearest to the South Pole, viz. by SCOTT's sledging party at 87°20′ south, 160°40′ east, or about 200 km from the South Pole (2 January 1912) and by Sir EDMUND HILLARY at about 145 km from the South Pole (January 1958; EKLUND 1959 : 114). In most places, at all events in the antarctic region, Southern Skuas are migratory, abandoning their breeding places in winter. Throughout its range the Southern Skua exhibits some degree of geographic variation, but only the birds breeding on the coast of Antarctica are subspecifically clearly marked (by a yellowish patch on the hind neck), so that formerly they have been treated as a separate species, known as the South Polar or McCormick's Skua (now: *Stercorarius skua maccormicki* SAUNDERS). Considering the coast of Chile, the islands of the Tristan da Cunha group and the satellite islands of the South Island of New Zealand all being washed by sub-antarctic waters, the Southern Skua does not nest outside the regions of sub-antarctic and antarctic surface water. However, on their northward migration in winter, they range widely over the southern oceans towards and across the equator. For a study of the life history of the South Polar Skua, see EKLUND 1961.

The family Stercorariidae, which is very close to the Laridae, has four species, known as Jaegers or Skuas, three of which are breeding in circumpolar areas in the northern hemisphere, mainly in tundra or coastal environments. In winter they migrate far southwards, incidentally reaching sub-antarctic waters. The fourth species, the Southern Skua, has a circumpolar distribution in the southern hemisphere, but apart from that it also nests in a few restricted places in the North Atlantic (Iceland, Faeroes, Shetlands, Orkneys), in a form, which is only slightly smaller and darker than that of the sub-antarctic islands (Great Skua or Bonxie, *Stercorarius skua skua*). In view of the slight amount of outward difference and the present distribution pattern the theory has been advanced that the Great Skua is a northern colonist in the Antarctic, having found open ecological niches and consequently dispersing widely over the whole region. In this connection the present populations in the North Atlantic, instead of being relics from the time of colonization, are rather the result of successful recent re-colonization from the south (FISHER & LOCKLEY 1954 : 143; VOOUS 1960 : 103).

Southern Black-backed Gull, or Kelp Gull, *Larus dominicanus*. Mainly distributed along coasts and islands in the sub-antarctic region throughout the globe; breeding northwards to the southern

656

coast of Africa, New Zealand, and along the west coast of South America northwards to Peru. It also nests on islands in the antarctic region (South Orkneys, South Sandwich Islands, South Shetlands, Heard Island), but, if the coasts of the Antarctic Peninsula or Graham Land be excepted, it does not breed on the coast of Antarctica. It is mainly an in-shore bird, not regularly occurring outside view of land and essentially non-migratory, though individuals may occasionally wander far away. It is the only species of gull in the Antarctic, though a few other gulls occur in sub-antarctic coastal waters of South America (Dolphin Gull, *Larus scoresbii;* Belcher's Gull, *Larus belcheri).* In view of the geographic situation of South America towards Palmer Peninsula, it is likely that the Southern Black-backed Gull has penetrated into antarctic regions from South America.

The Southern Black-backed Gull belongs to the group of large gulls which in the northern hemisphere are centred around the members of the *Larus fuscus-argentatus* complex with its circumpolar distribution. Taxonomically the southern species is intermediate between the Lesser Black-backed *(L. fuscus)* and the Greater Black-backed *(L. marinus)* group, but outward appearance, breeding in colonies, and distribution pattern seem to indicate a close connection with the *fuscus*-group. The latter exhibits a great amount of geographic variation, which is absent in the southern hemisphere group. Besides, the deep and strong bill, present in the southern species, is not found in the north; it has been correlated by LÖNN-BERG (1906), MATTHEWS (1929) and others with the habit of collecting limpets and other molluscs from the rocks in the tidal zone. Northern Lesser Black-backed Gulls are highly migratory, but do not reach sub-antarctic waters. Still it is generally considered that the Southern Black-backed gulls are the descendants of a colonization of northern gulls on the southern hemisphere. In view of the resemblance to present northern Gulls and the absence of geographic variation in the southern group, it is thought that the date of colonization goes far less backwards than that of the Skuas.

Antarctic Tern, *Sterna vittata.* Mainly breeding in islands in the sub-antarctic region; the breeding range more or less resembles that of the Southern Black-backed Gull. Only in the South Orkney Islands, the South Shetland Islands, and in Graham Land does it penetrate far into the antarctic region, but it does not nest in any other part of the Antarctic Continent. Like the Southern Black-backed Gull it is virtually an in-shore and a resident bird, feeding in winter in the seas bordering ice-covered coasts and along the edges of the pack-ice. Unlike the gull it has developed a number of local geographical forms, 6 of which having been recognized by MURPHY (1938) under subspecific names.

The Antarctic Tern belongs to a group of slender, medium-sized terns to which also the Common Tern *(Sterna hirundo)* and the Arctic Tern *(S. paradisaea sive macrura)* belong. The latter two species are very widely distributed in the northern hemisphere and in winter migrate far southwards. Indeed, the Arctic Tern winters mainly in antarctic waters; wintering specimens have often been confounded with the Antarctic Tern. There does not seem to be any doubt that the Antarctic Tern originated as a colonist from the north, but, though it is still closely related to the arctic species, it must have lived in the Antarctic for a considerable time: first, because the juvenile feather plumage is quite distinct from that of both the Arctic and the Common Tern; second, because it has already developed a number of geographical forms. The reconstruction of the history of the colonization is complicated by the presence of a sibling species of tern in the sub-antarctic region, which will be dealt with below. The presence of a resembling species in southern South America, *Sterna hirundinacea*, opposite to Graham Land, may be considered to indicate the possible way of colonization.

Kerguelen Tern, *Sterna virgata*. Nesting on the Prince Edwards Islands (Marion I.), Crozet Islands, and Kerguelen. It closely resembles the Antarctic Tern and like that species it is resident. It is not truly an antarctic species and must be considered as sub-antarctic. Its origin must be sought in connection with that of the Antarctic Tern (p. 669).

Breeding Birds of the Antarctic Continent

In this review the Antarctic Peninsula or Graham Land in West-Antarctica has been excluded on geographical grounds; its fauna will be discussed in the section on the birds of the antarctic islands. The number of species of breeding birds along the coast of the continent in this restricted sense is consequently small and comprise the following 8 species (apart from 2 species of penguins):

> *Macronectes giganteus* (GMELIN) — Giant Petrel
> *Thalassoica antarctica* (GMELIN) — Antarctic Petrel
> *Fulmarus glacialoides* (SMITH) — Antarctic Fulmar
> *Daption capensis* (LINNAEUS) — Cape Pigeon
> *Pagodroma nivea* (FORSTER) — Snow Petrel
> *Pachyptila desolata* (GMELIN) — Antarctic Prion
> *Oceanites oceanicus* (KUHL) — Wilson's Storm Petrel
> *Stercorarius skua* (BRÜNNICH) — Great Skua

With the doubtful exception of *Thalassoica antarctica*, the few breeding places of which presently known are situated on the antarctic coast line, the breeding ranges of none of these species are restricted to the antarctic continent. All of them seem to have a fully circumpolar distribution as far as local opportunities permit. Geographical variation in this area is absent, or only slightly developed; of the 8 species mentioned above 6 (= 75%) do not show any geographical variation at all. Geographical variation in size (probably clinal) has been described for *Oceanites oceanicus* (see: ROBERTS 1940) and for *Pachyptila desolata* (also in shape of the bill), but the last named species nests on the antarctic islands rather than in Antarctica. The remarkable lack of geographic variation probably has to be correlated with the fact that the high-antarctic species have a very specialized life, living under similar, very severe conditions throughout their continuous range around Antarctica. Incidentally, this is in agreement with MURPHY (1936), who noticed definite faunal differences in zonal arrangement but not in sectorial arrangement in the whole of the southern region. More precisely it can be stated that during the latest geological periods high-antarctic sea-birds do not seem to have been subjected to geographical discontinuities of circumpolar ranges. The development of geographically recognizable size-forms of *Oceanites oceanicus* seems to have resulted from the segregation of winter populations: concentrations of wintering Wilson's Storm Petrels are well known from such isolated areas as the North Atlantic Ocean, the northwestern Indian Ocean, and the Humboldt Current in the eastern Pacific Ocean.

Gulls and terns do not occur in the specific list of breeding birds of the antarctic continent. Only penguins, petrels and skuas seem to have managed to withstand the hardships of the continental antarctic summer.

Unlike the penguins the above mentioned antarctic breeding birds do not show the spectacular mass breeding colonies comparable to the famous "bird rocks" of arctic regions. Still, most of them nest in colonies of some size. Each of the species has to protect itself and its nest against adverse weather conditions and against predation by Skuas and Giant Petrels. There is no need for protection against land predators as is necessary in the Arctic against Polar Foxes and Polar Bears. Hence, when conditions are favourable, antarctic birds equally well nest on the mainland as on small coastal islands.

Macronectes giganteus: nests on more open pebble or boulder beaches and on wide ledges of rocky coasts, where they hardly make any nest at all. They are least susceptible to depredations by Skuas and seem to look for places where they are as good as possible protected against wind and snow storms and at the same time have sufficient space and air current to allow to take wing.

Thalassoica antarctica, Fulmarus glacialoides, and *Daption capensis:* nest on small rock ledges, in crevices and in clefts of rocks or under projecting rocks. After the manner of Arctic Fulmars *(Ful-*

marus glacialis) breeding birds are only slightly protected against wind and snow storms. Photographic pictures of incubating birds often show these birds surrounded by snow and ice. After the Giant Petrel, the birds are second in body-size and do not have so much for themselves to fear of attacking Skuas.

Pagodroma nivea: nesting places have been described comparable to those of the previous group, but the nests are hidden in deeper rock crevices and clefts. In this way incubating birds and young are more effectively protected against Skua predation. Breeding colonies of hundreds of these birds have been found in more than one occasion on exposed mountain tops more than hundred miles inland of the icy wastes of Antarctica and at equally great distances from the nearest open water. As an example, possibly the southernmost known breeding place of the Snow Petrel, situated at about one thousand metre altitude on the summit of Mount Helen Washington, King Edward VII Land, at the southernmost shore of the Ross Sea, was discovered and visited by a party of the Second Byrd Antarctic Expedition on December 19, 1934, and was subsequently described as follows: "Winds of hurricane force sweep over the peak and prevent large accumulations of snow, while the dark rock contributes by absorbing the sun's heat and melting the snow" (SIPLE & LINDSEY 1937 : 150). In relation to all breeding places described, mention is made by all authors of the presence of bones and feathers of adults and young and of mummified and frozen young or eggs and eggshells and chalky layers of droppings of incubating birds, either in the actual nesting places or in the deep nest tunnels or thereabout. This indicates both a relatively high percentage of loss of eggs and young by predation and climate and the tenacity of these birds to their nesting sites, year after year. Similar conditions occur in the other species of antarctic petrels.

Pachyptila desolata: found nesting in King George V Land, 16 December 1913, "in a crevice, under a large rock" (FALLA 1937 : 199). This seems to be the only record of the species breeding in Antarctica.

Oceanites oceanicus: nesting in deep and narrow crevices and rock fissures (under low-antarctic and sub-antarctic conditions also in self-dug nesting burrows), where they are safe from predation by Skuas.

Stercorarius skua: nests in any place in connection with other breeding birds, usually close to penguin rookeries, but also around the breeding places of the Snow Petrel. They do not concentrate in colonies. Their nests are fairly well exposed and never hidden in rocky clefts.

Antarctic Petrel, *Thalassoica antarctica*

List of discovery of breeding places in chronological order.

date	locality	expedition and reference
28 Nov. 1912	Haswell I., Queen Mary Land	Australian Antarctic Expedition-Mawson II, 1915:117; North 1914; Falla 1937:153-5
22 Dec. 1913	Cape Hunter, King George V Land	Australian Antarctic Expedition-Falla 1937:157
13 Jan. 1930	Proclamation Rock, Enderby Land	British-Australian-New-Zealand Antarctic Research Expedition-Falla 1937:157
13 Feb. 1931	MacRobertson Land	*Idem*-Falla 1937:157
26 Feb. 1936	Scullin Monolith, MacRobertson Land	"William Scoresby"-Rayner 1940:177
18 Dec. 1940	Mt. Paterson, Rockefeller Mts., King Edward VII Peninsula	U.S. Antarctic Expedition-Perkins 1945:275

Snow Petrel, *Pagodroma nivea*

List of discovery of breeding places in chronological order.

date	locality	expedition and reference
6 Jan. 1840	Cockburn I., Palmer Land	Sir JAMES ROSS-McCORMICK, in: Ross II, 1847:34-35
25 Dec. 1882	South Georgia	German Polar Station-WILL 1884:136; PAGENSTECHER 1885:22; VON DEN STEINEN 1890:250. Not confirmed and possibility of confusion with *Fulmarus glacialoides*
20 Nov. 1899	Cape Adare, South Victoria Land	"Southern Cross"-BERNACCHI 1901:204-5; BORCHGREVINK 1901:222-3; SHARPE 1902:151-2
1902—1903	Gaussberg, Queen Mary Land	Deutsche Südpolarexpedition 1901—03-VANHÖFFEN 1905:507
1903—1904	South Orkney Islands	Scottish National Antarctic Expedition-CLARKE 1906: 170-2
Nov. 1908	Leskov I., South Sandwich Islands	"Undine", Captain C. A. LARSEN-KEMP 1931:157
7 Nov. 1912	Queen Mary Land; inland nunataks, David Island	Australian Antarctic Expedition-FALLA 1937:171
3 Dec. 1912	Cape Denison, King George V Land	idem-FALLA 1937:168-9
end 1912	Gaussberg and Haswell I., Queen Mary Land	idem-FALLA 1937:169
13 Jan. 1930	Proclamation Rock, Enderby Land	British-Australian-New-Zealand-Antarctic Research Expedition-FALLA 1937:171
Jan. 1931	Cape Denison, King George V Land	idem-FALLA 1937:171
13 Feb. 1931	Scullin Monolith, MacRobertson Land	idem-FALLA 1937:171
Jan. 1933	South Orkney Islands	"Discovery II"-ARDLEY 1936:363-4
19 Dec. 1934	Mt. Helen Washington, Rockefeller Mts., King George VII Land	Second Byrd Antarctic Expedition-SIPLE & LINDSEY 1937
26 Feb. 1936	Scullin Monolith, MacRobertson Land	"William Scoresby"-RAYNER 1940:177
27 Nov. 1940	Mt. Marujupu, Raymond Fosdick Mts., Mary Byrd Land	U.S. Antarctic Expedition-PERKINS 1945:276
18 Dec. 1940	Mt. Paterson, Rockefeller Mts., King Edward VII Land	idem-PERKINS 1945:275
1940—1941	Mt. Breckenridge, King George VII Land	idem-FRIEDMANN 1945:309
1940—1941	Alexander I Island, King George VI Sound, Palmer Land	idem-EKLUND 1945:301
Nov. 1951	Adelie Land	Expéditions Polaires Françaises-CENDRON 1953; ETCHÉCOPAR & PRÉVOST 1954:239
1958—1959	Mawson Station, Enderby Land	EKLUND 1959:116
Nov. 1958—Feb. 1959	Queen Maud Land	Norsk Polarinstitut Expedition-LÖVENSKIOLD 1960:133

In view of the inaccessibility of Antarctica's shore line, records of the breeding places of birds on the antarctic continent are accumulating very slowly and only incidentally. An attempt of the compilation of the history of the discovering of nesting sites of the Antarctic Petrel and the Snow Petrel illustrates the slow, but courageous progress in antarctic ornithological exploration (p. 660—661).

Breeding Birds of the Antarctic Islands

The number of species of breeding birds on the antarctic islands, including the northward protruding Graham Land or Palmer Peninsula of Antarctica, is considerably larger than that breeding on the continental shores proper. Apart from six or seven penguins and a few land birds mentioned earlier on p. 653, the number is 30 or 31, ten of which have also been recorded to nest in Graham Land and its off-shore islands. These birds include cormorants or shags, sheathbills, gulls, terns, albatrosses and diving petrels, all of which are absent on the continent. In the following list of islands and archipelagoes situated on or south of the Antarctic Convergence the number of breeding species recorded has been given; the full list of species is summarized in Table I.

    Graham Land — 10
    South Shetland Islands — 9
    South Orkney Islands — 12
    South Sandwich Islands — not exactly known
    South Georgia — 17
    Bouvet Island — not exactly known
    Kerguelen — 24
    Heard Island — 11
    Macquarie Island — 16
    Peter I Island — not exactly known

Probably more than in any other region in the world these birds have to protect their eggs and young against bad weather conditions and severe predation. They have tried various ways of solution of their ecological problems.

Those that are large enough to stand any predation, like the albatrosses and the Giant Petrel, simply sit on their egg or young, usually leaving these not a minute without protection. The albatrosses moreover have compact nest hummocks built of mud and peaty moss in flat, marshy country or on gently sloping coasts. These massive nests apparently must protect the egg and small young from cold winds and melting snow and ice. Some of the albatrosses nest in small communities or even in colonies, but these do not seem to serve the purpose of protection by mass, but, apart from social aspects, they merely seem to reflect the suitability of the place as a nesting site. Shags breed in colonies of sometimes hundreds of pairs

Table I.

List of the species of birds recorded to have nested on the Antarctic Islands and the Antarctic Peninsula.

	Antarctic Penins	South Shetland Islands	South Orkney Islands	South Sandwich Islands	South Georgia	Bouvet Island	Kerguelen	Heard Island	Macquarie Island	Peter I Island
Diomedeidae										
Diomedea exulans					X		X		X	
Diomedea melanophris					X		X		X	
Diomedea chrysostoma					X		?		X	
Phoebetria palpebrata					X		X	X	X	
Procellariidae										
Macronectes giganteus	X	X	X		X		X	X	X	
Daption capensis	X	X	X	X	X	X	X	X	X	
Fulmarus glacialoides	X		X	?		X				X
Pagodroma nivea	X		X	X	?					
Halobaena caerulea			?				X	?	X	
Pachyptila desolata			X	?	X	?	X	X	X	
Pachyptila belcheri					?		X			
Puffinus griseus									X	
Procellaria cinerea							X		X	
Procellaria aequinoctialis					X		X		X	
Pterodroma macroptera							X			
Pterodroma lessoni							X		X	
Pterodroma brevirostris							X			
Hydrobatidae										
Oceanites oceanicus	X	X	X		X		X	?		?
Garrodia nereis					X		X			
Fregetta tropica		X	X		X		X			
Pelecanoididae										
Pelecanoides urinatrix							X	X		
Pelecanoides georgicus					X		X	X		
Phalacrocoracidae										
Phalacrocorax albiventer						?	X			
Phalacrocorax atriceps	X	X	X	?	X			X	X	
Chionidae										
Chionis alba	X	X	X	?	X					
Chionis minor							X	X		
Stercorariidae										
Stercorarius skua	X	X	X		X	X	X	X	X	?
Laridae										
Larus dominicanus	X	X	X		X		X	X	X	
Sterna vittata	X	X	X	?	X		X	X	X	
Sterna virgata							X			
Total number of species	10	9	at least 12	?	at least 17	?	24	11	16	?

and make high nest hummocks of tussock grass, seaweed and mud, placed closely together on steep rocky coasts (e.g. on Kerguelen; ANGOT 1954, pl. II).

The larger petrels like the Antarctic Fulmar *(Fulmarus glacialoides)* and the Cape Pigeon *(Daption capensis)*, which nest on rocky ledges and in protected corners of a broken rocky coast, equally have to protect their egg and young by their mere presence from being robbed by Skuas or frozen by the cold of the air.

Other, usually smaller, petrels, like the Cape Hen *(Procellaria aequinoctialis)*, the Grey Petrel *(Procellaria cinerea)*, the prions *(Pachyptila)*, the Great-winged Petrel *(Pterodroma macroptera)*, the diving petrels *(Pelecanoides)* and the storm petrels *(Oceanites, Fregetta)*, nest well hidden in crevices and deep rock fissures, or even dig their own nest tunnels in the boggy turf layer, or among mosses and lichens, or under low herbaceous plants *(Acaena, Rosaceae)* growing over and between rock boulders. The latter situation, however, is a more regular phenomenon in sub-antarctic than in antarctic islands. Breeding sites containing hundreds or thousands of nesting burrows of Cape Hens and diving petrels have been recorded from South Georgia and Kerguelen. Nest tunnels of e.g. the Grey Petrel on Kerguelen have been stated by FALLA (1937 : 178) to measure up to 360 cm in length; whereas in South Georgia the length of the nest tunnels of the Cape Hen has been recorded to measure approximately 150 cm (GIBSON-HILL 1949) and of the Antarctic Prion *(Pachyptila desolata)* about 25—100 cm (LÖNNBERG 1906).

Where predation is very severe the smaller petrels not ony hide underground, but appear on the nesting places only by night. As these petrels generally are very clumsy on the ground and thus, when landed, are an easy prey for any predator, they avoid by their nocturnal habits a considerable degree of predation, as in these islands night predators are absent. Still, this situation works more effectively in more northerly, sub-antarctic islands than in the far south, where the period of bright polar nights is longer.

Petrels and albatrosses do not have protective egg colours; the eggs are chalky white and therefore are very conspicuous. These eggs must either be hidden in a protective place or else strictly protected and guarded by the parent birds. In contrast to this the eggs of the terns, gulls and skuas breeding in the Antarctic are of various shades of brown, largely spotted and shaded with darker tones; these eggs are highly protectively coloured against their background. Likewise, in contrast to the petrels, these birds nest in the open like their northern relatives, and apart from their aggressive behaviour largely depend for their survival on the protective coloration of their eggs and young.

Geographical Distribution and List of Antarctic Species of Sea-Birds

Geographically the present distribution of antarctic birds shows the influences of (1) the circumpolar arrangement of identical ecological conditions, (2) the proximity of South America to the antarctic region by way of Palmer Peninsula or Graham Land, (3) the infiltration of sub-antarctic elements through Kerguelen Island, and (4) the infiltration of sub-antarctic elements through the proximity of the southern New Zealand satellite islands to Macquarie Island.

(1) MURPHY (1936) in particular has stressed the importance of the zonal arrangement of species around the pole as a consequence of the unbroken zonal arrangement of ecological conditions. This is most conspicuous in respect of the antarctic zone: the sub-antarctic zone is somewhat broken by the southward extension of South America.

(2) Through the proximity of the southern tip of South America not only two species of land birds have succeeded in settling into South Georgia *(Anas georgica, Anthus antarcticus;* see, p. 653), but it is likely that also through this connection with South America the Blue-eyed Shag *(Phalacrocorax atriceps)* and the Sheath-bill *(Chionis alba)* have penetrated far southward into antarctic surroundings. By their dependence on open shores and coastal seas these species are more of a sub-antarctic than of a truly antarctic type. In spite of their relation to South America the shags at present definitely show a zonal type of geographic distribution. The southernmost breeding place of the Blue-eyed Shag on the coast of Palmer Peninsula is at about 67°S. (BEHN *c.s.* 1955); whereas elsewhere its breeding places only reach as far south as 60°30′ S. in the South Orkney Islands and 55°S. in Macquarie Island. The semi-terrestrial Sheath-bill *(Chionis alba)* has its nearest relative in the islands of the southern Indian Ocean, viz., *Chionis minor* on the Crozet Islands, Marion Island, Kerguelen and Heard Island. With the exception of Kerguelen and Heard Island all breeding places are on islands outside the zone of antarctic surface water. Actually the land fauna of Kerguelen is sub-antarctic rather than antarctic.

(3) Kerguelen and Heard Island are the only large islands situated in the great extent of the southern Indian Ocean which may serve as a breeding station for southern sea-birds. As Kerguelen is situated on the border of antarctic and sub-antarctic surface waters (the Antarctic Convergence running approximately slightly south of 49°S.) this island has attracted sea-birds from both the sub-antarctic and the antarctic zones. Apart from the land birds mentioned on p. 653, the following four species of birds nest in the antarctic region on Kerguelen Island only; they cannot therefore be considered as true antarctic breeding birds:

Pterodroma macroptera (SMITH) — Great-winged Petrel
Pterodroma brevirostris (LESSON) — Kerguelen Petrel
Phalacrocorax albiventer (LESSON) — King Shag
Sterna virgata CABANIS — Kerguelen Tern

(4) Similarly, Macquarie Island, situated almost on the Antarctic Convergence, has several species of land birds and one oceanic bird (Sooty Shearwater, *Puffinus griseus* (GMELIN)), which clearly appears to have originated from the sub-antarctic region of New Zealand and its satellite islands. The Sooty Shearwater therefore has to be discarded as a truly antarctic breeding species.

Likewise, those species that within the region of antarctic surface water do not nest otherwise than on the islands of South Georgia, Kerguelen and Macquarie Island, and in addition have breeding sites to the north of the Antarctic Convergence, cannot be regarded as genuine antarctic birds. These sub-antarctic colonists into the northern antarctic region, of which there are nine, are listed below.

List of Sub-Antarctic Colonists in the Antarctic Region
(oceanic birds)

Diomedea exulans LINNAEUS — Wandering Albatross
Diomedea melanophris TEMMINCK — Black-browed Mollymawk
Diomedea chrysostoma FORSTER — Grey-headed Mollymawk
Halobaena caerulea (GMELIN) — Blue Petrel
Pachyptila belcheri (MATHEWS) — Narrow-billed Prion
Procellaria cinerea GMELIN — Grey Petrel
Procellaria aequinoctialis LINNAEUS — Cape Hen
Pterodroma lessoni (GARNOT) — White-headed Petrel
Garrodia nereis (GOULD) — Grey-backed Storm Petrel

With the restriction of the land birds, the penguins, and the sub-antarctic colonists in South Georgia, Kerguelen, and Macquarie islands (see above), the list of present antarctic birds contains 17 species. A number of these have a wide range in the sub-antarctic region too, and therefore can be classified as pan-antarctic birds. The list includes 13 pan-antarctic and only 4 exclusively antarctic species.

List of Present Antarctic Species of Sea-Birds
(+ exclusively antarctic species)

Phoebetria palpebrata (FORSTER) — Light-mantled Sooty Albatross
Macronectes giganteus (GMELIN) — Giant Petrel
Daption capensis (LINNAEUS) — Cape Pigeon
+ *Thalassoica antarctica* (GMELIN) — Antarctic Petrel
+ *Fulmarus glacialoides* (SMITH) — Antarctic Fulmar
+ *Pagodroma nivea* (FORSTER) — Snow Petrel
Pachyptila desolata (GMELIN) — Antarctic Prion
Oceanites oceanicus (KUHL) — Wilson's Storm Petrel
Fregetta tropica (GOULD) — Black-bellied Storm Petrel
Pelecanoides urinatrix (GMELIN) — Diving Petrel
Pelecanoides georgicus MURPHY & HARPER — Antarctic Diving Petrel
Phalacrocorax atriceps KING — Blue-eyed Shag
+ *Chionis alba* (GMELIN) — Sheath-bill

Chionis minor HARTLAUB — Indian Ocean Sheath-bill
Stercorarius skua (BRÜNNICH) — Great Skua
Larus dominicanus LICHTENSTEIN — Southern Black-backed Gull
Sterna vittata GMELIN — Antarctic Tern

Breeding Seasons

In high antarctic regions the summer period is very short. Consequently there is little opportunity for great differences in the nesting period of various species. Immediately after the arrival of spring some time in October the breeding birds arrive and start nesting activities. Still, one species is earlier than the other: as an example the Giant Petrel *(Macronectes giganteus)* and the Snow Petrel *(Pagodroma nivea)* are early nesters; whereas Wilson's Storm Petrel *(Oceanites oceanicus)* is a late one. The latter, in the Argentine Islands off western Graham Land (56°15′ S.), does not start egg laying before the middle of December (ROBERTS 1940). None of the petrels lay more than one egg per clutch; the Skua *(Stercorarius skua)* generally lays two, as sometimes do Antarctic Terns *(Sterna vittata;* usually one) and the Southern Black-backed Gull *(Larus dominicanus)*. However, clutches of three have been reported of the Black-backed Gull from Kerguelen (a.o. FALLA 1937 : 238; PAULIAN 1953 : 211), from Macquarie Island (FALLA 1937 : 239) and from the South Orkney Islands (CLARKE 1906 : 179), as well as of the Antarctic Tern on Petermann Island, western Graham Land, at 65°10′S. (GAIN 1914 : 95).

In view of the shortness of the polar summer the breeding activities are in a hurry. Apart from the hazards of weather and predation the breeding success is further reduced by the fact that the incubation period is very long and that young petrels have a very prolonged nest life. Only few accurate records of incubation period have been published: Wilson's Storm Petrel 39—48, average 43 days (Argentine Islands: ROBERTS 1940), Cape Pigeon *(Daption capensis)* 42 days (South Orkney Islands: CLARKE 1906), Giant Petrel about 60 days (60: Adelie Land: ETCHÉCOPAR & PREVOST 1954; 58—61: Macquarie Island: WARHAM 1962). The period of young Wilson's Storm Petrels to stay in their nest tunnels is recorded by ROBERTS (1940) to be at the average 52 days; these birds do not fledge before the end of March.

In the more northern part of the antarctic region the mild summer period lasts longer; hence the breeding seasons in the islands of South Georgia and Kerguelen permit more specific and individual variation than in the far south. In these islands most birds start preparing their nesting burrows in October and November, but some are as a rule retarded and there are even winter-breeding species. Of the last named category none (except the Emperor Penguin) exists in high antarctic regions. The winter-breeding species occurring within the limits of the antarctic region apparently are more of a sub-antarctic faunal type. These species will be listed below.

Procellaria (sive Adamastor) cinerea GMELIN, Grey Petrel. Kerguelen: starting egg laying March-April (FALLA 1937 : 14). Incidentally, from the sub-antarctic breeding population on Tristan da Cunha fresh eggs have been recorded April-July (ELLIOTT 1957 : 561).

Pterodroma macroptera (SMITH), Great-winged Petrel. Kerguelen: preparing nesting burrows in March (MILON & JOUANIN 1953 : 16); young birds leaving the nests in November (FALLA 1937 : 180). Incidentally, the sub-antarctic breeding population on Marion Island (Prince Edward Islands) has been reported to arrive on their breeding grounds in April (RAND 1954 : 193—194; 1955 : 233); whereas from Tristan da Cunha egg laying has been reported in June and July (ELLIOTT 1957 : 564).

The advantages of a winter breeding season in contrast to summer-breeding are not apparent. But those breeding birds that manage to survive the disadvantages of a winter period either for themselves and for their young, escape inter-specific competition for nesting holes on crowded islands and find their fledged young provided with a rich food supply in early spring.

Some years in the high antarctic, spring may arrive too late (or not at all) to permit birds to start breeding. For those that still have bred, infant mortality then is extremely high (65% in Wilson's Storm Petrels on the Argentine Islands in the season 1934—35; ROBERTS 1940). The nesting success of 264 nests of the Giant Petrel on Macquarie Island in the season 1960—61 was 57% (after WARHAM 1962). Preliminary published data on the South Polar Skua *(Stercorarius skua maccormicki)* at Wilkes Land do not seem to indicate such a high mortality, as EKLUND (1961) mentions of 40 nests studied a nesting success of 70%.

Among the big albatrosses at least the Wandering Albatross *(Diomedea exulans)* does not seem to breed each year. But this species is more of a sub-antarctic than of a truly antarctic breeding type. It seems to breed regularly in alternate years. After preliminary courtship and nest building from the middle of November onwards, egg laying takes place in Kerguelen in December and January (first egg of the season recorded by HALL (1900) on 1 January 1898 and by PAULIAN (1953) on 15 December 1951), and in South Georgia towards the end of December (20—31 December, according to MATTHEWS 1929). The Albatrosses then have to take care for their only egg and growing chick throughout the whole of the summer season and the following winter. The young Wandering Albatrosses leave their nests and depart for their independant pelagic life not before the second half of November or early January.

Hence this is too late for the parents to start a new breeding cycle and it has been shown by ringed birds that after a more than 12 months' breeding season they take a leave year and do not start breeding again before the next nearest season. This remarkable situation, which, however, is quite understandable in view of the enormous size of the birds, has led to most thrilling theories on a prolonged starvation period of the young Albatross on its nest hummock during the whole winter, during which it was thought not to be visited and fed by its parents at all. RICHDALE (1954) has clearly settled accounts with this phantastic theory in relation to the Royal Albatross *(Diomedea epomophora)* and other members of the petrel family. "Facts and fiction on the breeding of the Wandering Albatross" were clearly summarized by CARRICK, KEITH & GWYNN (1960).

Breeding Seasons and Species Formation

Differences in breeding season have been reported to act as a mechanism separating the breeding populations of a pair of sibling species of terns. In Kerguelen, the Crozet Islands and the Prince Edward Islands two species of tern occur as breeding birds, viz. the Antarctic Tern *(Sterna vittata)* and the Kerguelen Tern *(Sterna virgata)*. These terns resemble each other to a considerable degree, one *(virgata)* being slightly smaller and greyer than the other *(vittata)*. They breed in exactly the same (sometimes the identical) nesting sites, but the Kerguelen Tern is a rather early (spring), the Antarctic Tern a rather late (autumn) breeder in those regions where they occur alongside each other. Where the Antarctic Tern (which is circumpolar in its breeding range, in contrast to the Kerguelen Tern which is restricted to a few islands in the southern Indian Ocean) is the only species of tern, it has a prolonged nesting period, covering those of both of the species on Kerguelen combined.

Sterna virgata, Kerguelen Tern. Kerguelen: first downy chicks, 23 November 1929 (FALLA, 1937: 257); fledged young, 4 January 1951 (MILON & JOUANIN, 1953: 40). Marion Island: nesting, October — December (RAND, 1954: 203; 1955: 234).

Sterna vittata, Antarctic Tern. Kerguelen: first downy chicks, February 1930 (FALLA, 1937: 258) and 10 January 1951 (MILON & JOUANIN, 1953: 40). Marion Island: nesting, after January (RAND, 1954: 205; 1955: 234). — South Georgia: fresh eggs, November — February (MURPHY, 1936: 111); South Orkney Islands: first eggs, 14 November 1903 and 27 november 1904 (CLARKE, 1906: 178—179); Petermann Island, western Graham Land: eggs, November — January (GAIN, 1914: 95).

One can wonder whether the reproductive isolation by difference of breeding season has been the cause or the result of the species formation. At present it is most likely that also in the case of the sibling pair of terns the island of Kerguelen has been the meeting point of two geographically separated groups of terns, an antarctic one *(vittata)* and a sub-antarctic one *(virgata)*, which have managed to keep to themselves, but under stress of inter-specific competition have ultimately divided up the time of breeding available. The Antarctic Tern being a harder (more southerly) species has got late summer and autumn as main breeding time. It fishes at sea (fish), rather than in inland pools and tarns, as does the Kerguelen Tern, which is known to feed on insects, insect larvae and spiders.

There is another pair of sibling species on Kerguelen Island, viz. the Dove Prion or Antarctic Prion *(Pachyptila desolata)* and the Slender-billed Prion *(P. belcheri)*, which resemble each other probably even more closely than the terns do. Again the sub-antarctic species *(belcheri)* has been reported to be an earlier breeder than the majority of the antarctic species *(desolata)*, but the published evidence available is not wholly conclusive (FLEMING, 1941). In the prions or whale-birds in general, differences in breeding season have been considered to act as isolating mechanism, at least among the two most confounded sibling pairs of this group. According to FLEMING (1941: 147) the time of egg-laying in these species is as follows:

P. vittata	—	Sub-tropical Convergence	—	early September
P. salvini	—	mid Sub-antarctic Zone	—	mid-November
P. belcheri	—	high Sub-antarctic Zone	—	mid-November
P. desolata	—	Antarctic Zone	—	end of December

It is not improbable that zonal distribution and the consequent shift of the appropriate period of nesting have led to a reproductive isolation in these places where the zonally separated forms met secundarily. At present Kerguelen Island, being the only island, which is situated right on the boundary of the antarctic and sub-antarctic surface waters, and being the only large island in the southern Indian Ocean, is by its situation the ideal meeting place of antarctic and sub-antarctic forms. The occurrence of two pairs of sibling species therefore is not astounding. FLEMING (1941) has called attention to the fact that at least during the pleistocene periods the vicissitudes of the boundaries of the various types of surface waters [(Antarctic Convergence, Sub-tropical Convergence) must have been considerable. They must have slowly oscillated northwards and southwards over the limited number of breeding islands with fixed geographical positions, thereby offering suitable opportunities for the origin of isolation through difference in breeding season. Recent phenomena of geographical variation and species formation are scarce in the antarctic region, but the above mentioned theory may in addition to the prions give a satisfactory explanation for the existence of two sibling species of terns (mainly sub-antarctic *Pelecanoides urinatrix* and mainly antarctic *P. georgicus*), and two vicariant species of Sooty Albatrosses (sub-antarctic *Phoebetria fusca* and antarctic *P. palpebrata*) and Cormorants or Shags (sub-antarctic *Phalacrocorax albiventer* and antarctic *P. atriceps*).

Predation, Predators, Scavengers

Whereas in the Arctic sea bird colonies are raided by Peregrine Falcons *(Falco peregrinus)*, Ptarmigan are hunted by Gyrfalcons *(Falco rusticolus)*, and lemmings and other voles are predated upon by Snowy Owls *(Nyctea scandiaca)*, in sub-antarctic and antarctic environments birds of prey and owls are conspicuous by their absence. This seems to be due to the unfavourable distribution of land and sea in the southern hemisphere, the inaccessibility of the land available and the consequent absence of small land mammals as a basic food supply. Likewise, terrestrial mammalian predators are absent, offering a situation favourable for the development of sea-birds lacking the powers of flight (penguins). Still, penguins when swimming in the water are subjected to a not inconsiderable preda-tion by Leopard Seals *(Hydrurga leptonyx)* and Orcas or Killer Whales *(Orca orca)*, but other sea birds do not seem to suffer notably, if at all from predation by these large and fierce mammals.

Instead, the ecological niche of predator is taken in the Antarctic by a member of the gull group, the Southern Skua *(Stercorarius skua)*. Egg-predation is in addition practised by the Southern Black-backed Gull *(Larus dominicanus)* and particularly by the Sheath-bills *(Chionis alba* and *Chionis minor)*. All of these species at times as well act as scavengers, to which guild must be added the Giant Petrel *(Macronectes giganteus)* and eventually other smaller petrels. We will deal with these predatory, piratic and scavenging birds in succession, but it should be reminded that the amount of predation no less than the effect of it on the population dynamics of the predated species are unknown.

Southern Skua, *Stercorarius skua*. The literature is full of remarks by antarctic explorers to the effect of the voracity of this species towards its victims. Although practically omnivorous the Skua feeds in the breeding time almost exclusively on birds and birds' eggs. It has the swift, dashing flight of the falcon and the intelligence and tenacity of the gull; still, seizing its prey with the beak and not with the talons, but holding its victim under the webbed feet in the grip of the strongly curved and sharp claws and tearing it to species with the hooked bill, or, if size permitting, swallowing it whole. There is no other limit to the potential food of the Skua than size and extreme alertness in flight. Skuas are the scourge of the penguin colonies, be it Adelie Penguins or Rockhoppers, stealing the eggs and taking the young up to fairly large size. Hardly any penguin colony is without its attendance of one or more pairs of skuas, not even excepting the King Penguin. Apart from penguins' eggs and young, Skuas predate very heavily on adults and young of whale-birds *(Pachyptila)* and diving petrels *(Pelecanoides)*, which are of relatively small size and which they catch when these birds are entering or leaving their nesting burrows, the latter usually are found in colonies by the hundreds. In addition they have been recorded feeding on other petrels and sea-birds including *Pterodroma macroptera, P. lessoni, Puffinus griseus, Halobaena caerulea,* and the eggs of cormorants, ducks and even of the Southern Black-backed Gull *(Larus dominicanus),* Giant Petrel *(Macronectes giganteus;* on South Georgia, MURPHY, 1936: 596) and the Wandering Albatross *(Diomedea exulans;* on Kerguelen, HALL, 1900). Only Southern Black-backed Gulls and Antarctic Terns seem to be able to stand their ground by their fierce and joint attack. Out at sea or among the pack-ice they practice piracy by robbing other birds off their food, whereas they are also known to feed by themselves on fish and crustaceans, seaweed and molluscs, in addition to all kinds of edible matter cast ashore antarctic sea-coasts. On carcasses of dead whales and seals they feast as if they were vultures, competing with the Giant Fulmars. They readily take advantage of all kind of offal of the whaling industry. Likewise they are attracted by the after-births of Weddell and Crab-eater Seals on the southern pack-ice. There are many witnesses in the literature of these birds committing infanticide on their own downy young, eventually also of swallowing their own eggs and of practising cannibalism on less valid, sick or wounded birds of their own kin. Out of the numerous instances mentioned in the literature the following citation of EDGAR WILSON (1907 : 73) on the predatory habits of the Southern Skua on the Adelie Penguin *(Pygoscelis adeliae)* rookery at Cape Crozier, Ross Island, has been taken as an example:

"Hanging round the rookery, with the unmistakable look of a thief, the Skua will run up to a chicken almost as big as himself, drag it by degrees away from the more crowded part of the rookery, and then gradually worry it to death; eventually tearing a ragged hole in the skin of the back over the kidneys, which are generally the first, and often the only parts that are touched. The penguin chick pipes his loudest, but the old birds standing round take very little notice . . . Literally, in a rookery such as that of Cape Crozier, one cannot walk ten yards without coming on a dead penguin chick".

And a further example, taken from FALLA (1937: 243—244) on Skuas living in Heard Island:

"Feeding was limited to the hours of dusk, when dove petrels began to fly in from the sea and alight near their burrows. This supply of food was so plentiful that only the viscera and flesh from the breast of the victims were eaten. Each nest was surrounded by a dozen or more disembowelled dove petrels which were not again touched. For the most part the victims in any given scrap heap were all of one species. Thus the remains around one nest consisted of bodies of *P. turtur* only, while in neighbouring nests *P. desolata* were the victims".

Southern Black-backed Gull, *Larus dominicanus*. Apart from a scavenger and marauder along sea-coasts, in harbours, at garbage dumps and slaughter-houses, these gulls are, like their northern relatives, egg- and nest-predators. They are, however, far less active as a predator than the Skuas. In the sub-antarctic and antarctic regions they seem in this repect to prey mostly on the eggs and young of the Antarctic Tern *(Sterna vittata)*. The animal food of these gulls moreover consists of large quantities of fish, crustaceans, limpets and other molluscs.

Giant Petrel, *Macronectes giganteus*. Nearly twice the size of the Skua and more bulky of shape and more clumsy of movements, but equipped with an enormous bill with fearful hook. It is an even greater scavenger than the Skua and represents the real vulture of antarctic and sub-antarctic seas and sea-coasts. Wherever and whenever carcasses of seals and whales are available Giant Petrels are assembling around, their beaks and feathers becoming stained with blood. They are known to feed on any animal garbage, including the placentae of seals; in addition they feed on crustaceans, molluscs, squids, and many other kinds of animals, but it seems unknown whether fish form a regular item of their diet. Because of their bulky size these birds are less expert in catching living prey, but MURPHY (1936: 595) quotes a number of observations of Giant Petrels having successfully predated upon small sea-birds, like a prion, a tern and a diving petrel. Besides they are known to catch and eat rats on South Georgia and rabbits on Kerguelen Island, whereas JEAN CENDRON (1953: 219) reports the most interesting case of Giant Petrels stealing and killing downy chicks of Emperor Penguins at Pointe Géologie, Adelie Land. Doubtless predation on life prey will prove to be more important in the Giant Petrel's diet when observations on this point have become more numerous.

Sheath-bills, *Chionis alba* and *C. minor*. Although omnivorous birds from the intertidal zones of rocky and sandy shores, these birds act as scavengers, carrion-eaters and egg-predators as well. Wherever they occur they are the constant attendants of penguin rookeries, where they have their own, simple and directs methods to get at their intended booty. They are known to steal the eggs of shags as well, whereas in addition they subsist on food remains and excrements found in any bird colony, and on dead animals, placentae, and excrements within seal nurseries. Apparently egg-predation by these birds in penguin rookeries is locally not inconsiderable, but they do not seem regularly to attack or kill valid young. The following is a quotation from a description of the predatory behaviour of *Chionis minor nasicornis* at Heard Island taken from FALLA (1937: 267):

"Egg-snatching in the penguin-rookeries was their chief regular activity during our stay, both macaroni and rockhopper penguins being the victims. Working in pairs the sheathbills divided the tasks of decoy and snatcher, the decoy baiting a penguin to such a pitch of fury that it lunged forward and uncovered the egg, which the snatcher deftly pierced and carried off on his bill. Often the upturned nasal sheath formed a useful hook on which the transfixed egg hung as the sheathbill escaped to safety. The thieves could be seen sharing their ill-gotten gains in secluded corners on the outskirts of the penguin-rookery".

Food and Feeding Habits

The majority of the antarctic birds extract their food from the pelagic animal life of the surface water of the sea. During the summer months the abundance of animal plankton and nekton not only suffices to nourish great numbers of large whales, but also serves as the basic food for hordes of sea-birds (including penguins). Thus, *Halobaena caerulea*, *Pagodroma nivea*, *Oceanites oceanicus*, and *Sterna vittata* and *Sterna paradisaea* are known largely to feed on the orange krill-shrimps *(Euphausia superba* and other euphausians), but generally any antarctic sea-bird may at times feed on krill. Other marine crustaceans (mainly *Euthemisto* and other amphipods and isopods) also appear on the food list of antarctic petrels *(Pelecanoides urinatrix* and *Oceanites oceanicus*, *Larus dominicanus* and *Chionis minor* around Kerguelen; *Fulmarus glacialoides* around Deception Island, South Shetlands). Pelagic fish (mainly nototheniids) and medium-sized squids form important food items of fairly all of the petrels, like *Thalassoica antarctica*, *Fulmarus glacialoides*, *Daption capensis*, *Pagodroma nivea*, *Procellaria aequinoctialis*, *Pterodroma macroptera*, but none of these seem to be specialized on any of these animals, although there seem to be specific differences of food preference. Albatrosses largely seem to prefer medium-sized squids. The smallest petrels seem to eat great numbers of very small cephalopods, locally also pteropods; *Pagodroma nivea* locally seems to devour large quantities of fish (BIERMAN & VOOUS 1950: 19 and 63). This has also been recorded from *Sterna vittata* and *Procellaria cinerea* (PAULIAN 1953: 171), whereas the Shags *(Phalacrocorax)* almost exclusively subsist on fish. However, Shags have also been recorded to feed on sea-urchins (spatangids; Kerguelen, PAULIAN 1953: 208) and on euphausians and decapod crustaceans (Graham Land, GAIN 1914: 77).

During the last decades whaling land stations (e.g. on South Georgia and on Deception Island, South Shetlands) and factory-ships attract thousands of sea-birds which feed on all kinds of offal from the whaling industry, including small drops of oil or fat floating on the surface. Thus, MATTHEWS (1929: 575) describes flocks of hundreds of thousands of Cape Pigeons *(Daption capensis)* around the whaling stations of South Georgia to feed on the refuse; whereas in other regions particularly Antarctic Petrels *(Thalassoica antarctica)* and Antarctic Fulmars *(Fulmarus glacialoides)* have been recorded. FALLA (1937: 211) even mentions a concentration of 5.000—6.000 Wilson's Storm Petrels *(Oceanites oceanicus)* near a factory ship in the Indian sector of the Antarctic Ocean. This seem to be quite unnatural situations, but in fact they are extensions of the habit of some petrels to follow great whales and to feed upon oily substances of whale excrements, and of the scavenging habits of the larger petrels. In fact, *Thalassoica antarctica, Halobaena caerulea, Pachyptila desolata* and even *Sterna paradisaea* have been recorded to follow herds of Little Piked or Minke Whales *(Balaenoptera acutorostrata)* through the pack-ice to feed on the excreta floating in the sea (ROUTH 1949: 600—601; BIERMAN & VOOUS 1950: 54).

Other abundant sources of food supply have been mentioned in the paragraph on predation and scavenging. Penguin's eggs and the after-births of seals form at times a rich food supply.

Littoral crustaceans, molluscs and other invertebrates from rocky shore lines may form important food items of a few species, viz. the Sheath-bill, the Southern Black-backed Gull and the Skua, but this food resource is more important in sub-antarctic than in antarctic surroundings.

Only *Sterna virgata* is known to feed inland on freshwater crustaceans, insects (diptera), insect larvae (caterpillars) and spiders living in and around the ponds of melting water and the summer marshes of Kerguelen (FALLA 1937; PAULIAN 1953).

Most of the antarctic sea-birds gather their food while soaring low over the sea and skimming the waves. Some practice stoop-diving, like the Antarctic Tern, which mainly feeds on fish, in contrast to the equally stoop-diving summering Arctic Tern which principally feeds on euphausians. Some petrels, like the Cape Pigeon, and sometimes the Antarctic Petrel have managed to practise some awkward diving technique when they are feeding while swimming and want to try to get hold of food which is just out of their reach. Only the shags and the diving petrels of the genus *Pelecanoides* exclusively get their food by diving. The last named petrels have developed a way of life comparable to that of the arctic auks: they are swimmers rather than sailing birds and are expert divers, also using their wings under water as propulsory organs. In size, general appearance, colour pattern, and way of flying and diving they so much resemble the Little Auks *(Plautus alle)* of the northern hemisphere, that Norwegian sealers and whalers know these birds as "Alkekonge" (BIERMAN & VOOUS 1950: 101—102).

In still another respect there is superficial resemblance between the feeding habits of antarctic and arctic birds: the white Snow Petrel *(Pagodroma nivea)* of the south and the equally immaculate Ivory Gull *(Pagophila eburnea)* of the high north share a prediliction for the after-births of seals; and the Southern Fulmar *(Fulmarus glacialoides)* is a similar scavenger, associating in equally large flocks with human whaling activity as the closely resembling Northern Fulmar *(Fulmarus glacialis)* with whaling and fishing activity in the north.

Pelagic Distribution

Particularly during the summer months when the antarctic waters abound with planktonic and nektonic organisms the number of bird individuals and species assembling to feed in the Antarctic Zone of the southern oceans is large. The greatest abundance of plankton is found around the edge of the pack-ice or slightly north of it, where consequently a rich bird life is concentrated in summer.

The density of pelagic bird life is not uniform throughout the whole of the antarctic region. The differences in density seem to reflect corresponding differences in the abundance of surface plankton, particularly of the larger euphausians. Pelagic birds, counting thousands of Cape Pigeons *(Daption capensis)*, Antarctic Fulmars *(Fulmarus glacialoides)* and Antarctic Petrels *(Thalassoica antarctica)*, among which huge flocks of Wilson's Storm Petrels *(Oceanites oceanicus)*, Antarctic Prions *(Pachyptila desolata)*, and Blue Petrels *(Halobaena caerulea)*, are extremely numerous e.g. in the southern Atlantic Ocean (near the entrance of the Weddell Sea), in the southern Indian Ocean and along the coast of East Antarctica. In comparison to this, birds are notably scarcer in the southern Pacific Ocean and along the coast of West Antarctica. It is most noteworthy that MARR'S map (1956: 128) of the frequency distribution of large euphausians in antarctic surface waters in January, February and March gives a similar picture.

As regards the antarctic breeding birds, one species is more closely associated with pack-ice and icebergs than the other. As the extent and the boundaries of the pack-ice fluctuate greatly throughout the season, the pelagic distribution of sea-birds is shifting northwards and southwards over hundreds of miles in the course of the year. Still, all antarctic birds show a more or less unbroken zonal distribution, particularly during the pelagic part of their year's cycle. Indeed, antarctic sea-birds pass most of their lives on the open sea, probably not breeding every year.

ROUTH (1949) has attempted to make a classification of bird species in view of the relation of their pelagic occurrence with ice conditions. Although sea-birds actually seem to be too dynamic in their distribution habits to permit a satisfactory classification in this respect, it is equally clear that specific differences exist.

The Snow Petrel *(Pagodroma nivea)* follows "the receding ice edge very closely, scarcely ever moving out of sight of it" (ROUTH 1949: 589). Indeed, the occurrence of the Snow Petrel is always associated with floating sea-ice and snow. It has often been observed resting on ice or icebergs. The Arctic Terns *(Sterna paradisaea)* wintering in the antarctic summer seem to keep strictly to the edges of the pack-ice, where they feed on euphausians in cracks and leads through the ice and find resting places on the ice during the period of their complete feather moult from January to March. There are other southern petrels which regularly occur in regions of pack-ice, but in addition widely range over open pack and drift-ice. This group includes the other species of continental birds, viz., the Antarctic Petrel *(Thalassoica antarctica)*, Southern Fulmar *(Fulmarus glacialoides)*, Cape Pigeon *(Daption capensis)*, Giant Petrel *(Macronectes giganteus)*, Wilson's Storm Petrel *(Oceanites oceanicus)*, and also the Southern Skua *(Stercorarius skua)*. These birds range slightly more northward than the Snow Petrel, but only the Antarctic Petrel generally keeps to these surroundings in the winter months too. It likes to rest on the tops of huge icebergs, often in flocks of hundreds together. The Antarctic Petrel, like the Snow Petrel, generally does not seem to occur regularly in waters with temperatures higher than 1 °C and usually less (BIERMAN & VOOUS 1950: 51 and 62, respectively). At least during the summer months the northern limit of the Snow Petrel's pelagic distribution coincides with the northern limit of large-size stages of orange krill *(Euphausia superba)*.

Apparently attracted by favourable food conditions, quite a lot of other species have been incidentally recorded from summer seas with drift ice or icebergs, including the Light-mantled Sooty Albatross *(Phoebetria palpebrata)*, the Black-browed Albatross or Mollymawk *(Diomedea melanophris)*, and the Wandering Albatross *(Diomedea exulans)*. Even breeding birds from the Sub-antarctic Zone sometimes penetrate far southward, particularly in late summer: e.g. the Sooty Albatross *(Phoebetria fusca)* and the Scaled, Mottled or Peale's Petrel *(Pterodroma inexpectata)*.

Some of the northernmost pelagic records of antarctic birds during the summer months are listed below, as well as, in contrast, some of the southernmost of birds of a mainly sub-antarctic distribution. These lists may illustrate the way in which during the summer abundance of food birds of the antarctic and sub-antarctic types of distribution mingle at sea. All of them range widely over the southern seas, irrespective of any land mass, but apparently controled by the presence of food, some of them, by the presence of drift-ice.

Selection of some of the *northernmost* pelagic summer records of *antarctic* birds. Each record gives the northernmost observation of the season recorded by the cited authority.

species	date		locality	authority
Antarctic Petrel	18 Dec.	1898	62° S 60° E	VANHÖFFEN 1901:313
Thalassoica ant-	Nov.	1901	62° S 140° E	WILSON 1907:83
arctica	17 Feb.	1903	64°18'S 23°09'W	CLARKE 1907:334
	8 Dec.	1910	63°20'S 177°22'W	LOWE & KINNEAR 1930:132
	25 Dec.	1911	62°10'S 175°37'W	*ibidem:*133
	26 Dec.	1912	63°43'S 166°36'W	*ibidem:*133
	9 Dec.	1929	61° S 77° E	FALLA 1937:154
	9 Jan.	1948	62°30'S 162°50'W	HOLGERSEN 1957:40
	2 Jan.	1960	60° S 57° W	"Protector"; Sea Swallow 14, 1961:13
Snow Petrel	18 Dec.	1840	61°03'S 146° W	Sir JAMES ROSS; cited by WILSON 1907:91
Pagodroma nivea	18 Dec.	1898	62° S 60° E	VANHÖFFEN 1901:313
	23 Dec.	1904	63° S 61°05'W	GAIN 1914:140
	8 Dec.	1910	63°20'S 177°22'W	LOWE & KINNEAR 1930:142
	29 Dec.	1912	69°28'S 166°17'W	*ibidem:*146
	9 Dec.	1929	61° S 77° E	FALLA 1937:167
	18 Dec.	1946	56°05'S 6°45'E	BIERMAN & VOOUS 1950:61
	14 Jan.	1948	63°30'S 156°50'W	HOLGERSON 1957:38
	1 Feb.	1952	63°15'S 102° E	VAN OORDT & KRUYT 1954:259
Antarctic Fulmar	9 Feb.	1903	58°57'S 33°34'W	CLARKE 1907:335
Fulmarus	7 Dec.	1910	61°22'S 179°56'E	LOWE & KINNEAR 1930:134
glacialoides	25 Dec.	1911	61°01'S 175°37'W	*ibidem:*134
	26 Dec.	1912	65°53'S 166°03'W	*ibidem:*135
	8 Dec.	1929	60°17'S 77°52'E	FALLA 1937:159
	9 Dec.	1930	60°52'S 162°13'E	FALLA 1937:159
	15 Dec.	1946	52°10'S 11°10'E	BIERMAN & VOOUS 1950:41
	15 Jan.	1948	62°40'S 153° W	HOLGERSEN 1957:30
	30 Jan.	1952	62°40'S 84°30'E	VAN OORDT & KRUYT 1954:258

Numerous observers have wondered whether in the vasts of the southern oceans individual sea-birds might encircle the globe during the pelagic period of their lives. In fact it would be not at all at variance with the flight capacities of a Wandering Albatross to travel around the pole more than once a year. Recent recoveries of ringed Albatrosses and particularly of Giant Fulmars have contributed considerably to the knowledge on this subject.

Selection of some of the *southernmost* pelagic summer records of *sub-antarctic* birds. Each record gives the southernmost observation of the season recorded by the cited authority.

species	date	locality	authority
Wandering Albatross *Diomedea exulans*	14 Feb. 1899	69°13′S	SHARPE 1902:160
	20 March 1903	61° S 43°20′W	CLARKE 1907:344
	9 March 1904	60° S 177° E	WILSON 1907:109
	14 Jan. 1910	68°30′S 89°40′W	GAIN 1914:156
	29 Jan. 1930	63° S 55° E	FALLA 1937:119
	6 Apr. 1947	62°00′S 25°30′W	BIERMAN & VOOUS 1950:34
	Apr. 1947	63°14′S	ROUTH 1949:593
	3 Jan. 1948	63°55′S 131°54′W	HOLGERSEN 1957:24
	5 March 1952	64° S 85° E	VAN OORDT & KRUYT 1954:256
Black-browed Albatross *Diomedea melanophris*	29 Feb. 1904	67°30′S 174° E	WILSON 1907:113
	16 Jan. 1910	69°20′S 102°09′W	GAIN 1914:157
	7 Dec. 1910	61°22′S 179°56′E	LOWE & KINNEAR 1930:169
	13 Feb. 1911	71°13′S 171°15′E	*ibidem*:169
	17 Dec. 1930	64°41′S 176° E	FALLA 1937:124
	11 Apr. 1947	57°20′S 3°40′E	BIERMAN & VOOUS 1950:30
	7 Feb. 1948	68°20′S 92°30′W	HOLGERSEN 1957:21
White-headed Petrel *Pterodroma lessoni*	10 March 1911	62°00′S 162°00′E	LOWE & KINNEAR 1930:136
	12 March 1912	69°23′S 177°52′E	*ibidem*:136
	30 Jan. 1931	63°40′S 95°56′E	FALLA 1937:183
	10 Apr. 1947	60°15′S 9°45′W	BIERMAN & VOOUS 1950:87
	2 Feb. 1948	63°53′S 111°52′W	HOLGERSEN 1957:45
Kerguelen Petrel *Pterodroma brevirostris*	20 March 1904	69°33′S 15°19′W	CLARKE 1907:337
	25 Feb. 1947	65°55′S 10°00′W	BIERMAN & VOOUS 1950:85

A Wandering Albatross *(Diomedea exulans)* ringed as a chick on 20 July 1952 at Kerguelen, was recovered on 1 October 1953 at Patache, Chile (ANGOT 1954). The shortest distance between these places is 13,000 km, but if the bird during the approximately 10 months of his pelagic sea life has drifted before the wind eastwards and then followed the west coast of South America northwards it may have covered a distance of at least 18,000 km.

Through combined British, Australian, and New Zealand efforts large numbers of Giant Petrels *(Macronectes giganteus)* have been ringed at the South Orkney Islands and the islands of Heard and Macquarie. At the end of the 1958—59 season the world total of ringed specimens was 15,013 (2,806 adults, 12,207 nestlings), giving a recovery rate of 1.86% (280 recoveries). The recoveries have been summarized and commented by HITCHCOCK & CARRICK (1958), SLADEN & TICKELL (1958), INGHAM (1959), CARRICK (1959) and TICKELL & SCOTLAND (1961). Nestlings ringed at Heard Island have been found during the early months of their first winter (up to and inclusive July) in South Australia, New Zealand, Chile, Argentina and South Africa. This means a dispersal throughout the whole of the Antarctic Ocean. Early first winter recoveries of birds ringed at Macquarie Island have been recorded from South Australia, New Zealand, South Africa and South Georgia. Finally, early first winter records of birds ringed in Signy Island, South Orkney Islands, have been made in South Africa, South Australia and New Zealand, but not South America. All these records substantiate the theory that these large sea-birds make extensive circumpolar journeys during their first autumn and winter with a tendency of dispersing at random, but through prevailing strong winds mainly drifting down wind in an easterly direction with an additional component towards the north.

Migration

General

In connection with the un-parallelled hazards of the weather changes of the seasons in the antarctic regions antarctic birds show movements of varying extent. During the winter months the snow- and ice-covered coasts of the continent are practically devoid of bird life, all birds having retreated to places where less severe conditions are prevailing, which for some species are the edges of the southern pack-ice a few hundred miles to the north, and for others are the temperate seas on the northern hemisphere at the opposite side of the globe. During the summer months with light polar nights antarctic bird life is enriched by non-breeding visitors from the northern hemisphere, whereas in the months of early autumn the antarctic seas are the scene of a mass migration of sub-antarctic passengers. These three aspects of bird migration in the Antarctic will be discussed below separately.

Migration of antarctic breeding birds

During the height of the southern winter the antarctic coasts do not provide attractive life conditions for any bird, how well adapted to antarctic conditions these birds may be. Hence (with the exception of the Emperor Penguin) only rarely birds are observed in winter on the continental coast of Antarctica but they are of regular occurrence around the antarctic islands. Winter observations along the coast of Antarctica seem to be known exclusively of *Pagodroma nivea* and *Macronectes giganteus.*

Arrival and departure from the breeding places not only vary with the species, but notably also with the geographic position (distance from the continent) and the weather conditions year by year.

Of the continental breeding birds the Antarctic Petrel *(Thalassoica antarctica)* and the Snow Petrel *(Pagodroma nivea)* leave their breeding sites only to retreat somewhat to the northward to pass the winter at the edge of the pack-ice and northwards as far as antarctic marine conditions prevail. They are among those species that generally do not cross the boundaries of the antarctic life-zone at any time of the year. The pelagic distribution of these species will be discussed separately; they do not show migratory movements in the strict sense. Still it is of interest to cite some dates of arrival and departure of these and some other birds on or near their breeding places on the antarctic continent. These dates more or less strictly coincide with the dates of breaking up and closing of the sea-ice (Table II and III).

Whereas some of the antarctic breeding birds appear to be innate migrants, all individual birds of which migrate wide and far like Wilson's Storm Petrel *(Oceanites oceanicus)*, other species behave as compulsory migrants, some of which staying behind and others moving over distances of greatly varying extent. Thus, the continental antarctic form of Skua *(Stercorarius skua maccormicki)* has been recorded in winter from the southern pack-ice (e.g. 49°S., 120°E.; ARDLEY, 1936: 371) on the one hand, as well as more or less regularly as a pelagic (northern) summer visitor of Japan on the other hand (Handl. Japanese Birds, 4th ed., 1958). Of in-shore birds, the Antarctic Tern *(Sterna vittata)* and the Southern Black-backed Gull *(Larus dominicanus)* have been reported as migrants in the southermost parts of their breeding range and as residents in regions where conditions in winter are less severe and at least some open water is left.

There are several antarctic species which in the non-breeding time wander widely over all sub-antarctic seas. These species incidentally follow the currents of cold surface water which flow along the west coasts of South America (Humboldt Current) and South Africa (Benguela Current) far to the northward. Without leaving waters of sub-antarctic conditions they are known to have penetrated almost to within the tropics. This is particularly noteworthy for the Southern Fulmar or Silver-grey Petrel *(Fulmarus glacialoides)*, which is one of the truly antarctic species of petrel, but in spite of this, is of regular occurrence in the Humboldt Current as far north as about 6°S. along the northern coast of Peru (MURPHY 1936: 598). The same applies to the Giant Fulmar *(Macronectes giganteus)*, the

Table II.

Dates of Arrival of Breeding Birds in Antarctica.

Species	Locality	Arrival Date	Surface Air Temperature	Reference
Pagodroma nivea (Snow Petrel)	Ross Ice Shelf	31-X -1929	—36.°4 C (1)	SIPLE & LINDSEY 1937:149
	Ross Ice Shelf	6-X -1934		*ibidem*
	Ross Ice Shelf	12-IX -1940		PERKINS 1945:275
Thalassoica antarctica (Antarctic Petrel)	Ross Ice Shelf	2-X -1929	—22.°8 C (1)	SIPLE & LINDSEY 1937:149
	Ross Ice Shelf	6-X -1934		*ibidem*
	Ross Ice Shelf	19-X -1940		PERKINS 1945:274
Oceanites oceanicus (Wilson's Storm Petrel)	Ross Ice Shelf	17-XII-1940	—11.°6 C (1)	PERKINS 1945:276
	Adelie Land	10-XI -1951		CENDRON 1953:212
Stercorarius skua (Great Skua)	Ross Ice Shelf	3-XI -1902	—11.°1 C (2)	WILSON 1907:72
	Ross Ice Shelf	25-X -1903	—21.°5 C (2)	*ibidem*
	Adelie Land	20-X -1912	—17.°5 C (1)	FALLA 1937:249
	Adelie Land	20-X -1913		*ibidem*
	Ross Ice Shelf	9-XI -1934		SIPLE & LINDSEY 1937:153
	Ross Ice Shelf	14-XI -1940		PERKINS 1945:276

(1) Mean daily surface air temperature: *Proc. Amer. Phil. Soc. Philadelphia*, **89**, 1945: *332*.

(2) Mean monthly air temperature: WILSON 1907.

683

Table III.

Dates of Departure of Breeding Birds in Antarctica.

Species	Locality	Departure Date	Surface Air Temperature	Reference
Pagodroma nivea (Snow Petrel)	Ross Ice Shelf	13-III -1935		SIPLE & LINDSEY 1937:149
Oceanites oceanicus (Wilson's Storm Petrel)	Ross Ice Shelf	2-II -1940	—11.°4 C (1)	PERKINS 1945:276
Stercorarius skua (Great Skua)	Ross Ice Shelf	30-III -1902	—13.°5 C (2)	WILSON 1907:72
	Ross Ice Shelf	7-IV -1903	—27.°2 C (2)	*ibidem*
	Adelie Land	6-IV -1912	— 6.°7 C (1)	FALLA 1937:249
	Adelie Land	9-IV -1913		*ibidem*
	Ross Ice Shelf	13-III -1935		SIPLE & LINDSEY 1937:153
	Ross Ice Shelf	17-III -1940		PERKINS 1945:276

(1) Mean daily surface air temperature: *Proc. Amer. Phil. Soc. Philadelphia*, **89**, 1945:332.

(2) Mean monthly air temperature: WILSON 1907.

684

Cape Pigeon *(Daption capensis)*, and the Antarctic or Dove Prion *(Pachyptila desolata)*, which, however, apart from having antarctic breeding places, are also widely distributed as breeding birds in sub-antarctic environments.

Not improbably the Southern Skuas *(Stercorarius skua)* have also to be arranged among the long-distance southern migrants, but so far exact or detailed records are scarce. They are awaiting the recoveries of birds ringed under the recent long-term ringing scheme (EKLUND 1959, 1961). The first results seem to indicate that at least the South Polar form of the Skua *(S.s. maccormicki)* as a rule winters within the limits of the antarctic pack-ice (EKLUND 1961: 221). Still Skuas, supposedly of one or more or the southern races, have been regularly recorded in the North Pacific along the west coast of America, as far north as the north end of the Humboldt Current and even off California and British Columbia. South Polar Skuas of the race *maccormicki* have been frequently recorded in Japanese waters (see above). In the central part of the Atlantic Ocean north of the equator, the number of records of Skuas is considerably larger, but here it is difficult to attribute these to either the southern (antarctic and sub-antarctic) or to the northern (arctic) population groups.

The most spectacular migratory movements of antarctic birds are performed by Wilson's Storm Petrel *(Oceanites oceanicus)*. The migration routes of these tiny sea-birds, the total body weight of which is about 35 grams, have been beautifully summarized and illustrated by BRIAN ROBERTS (1940). These birds nest circumpolarly apparently in any accessible place of the antarctic coastline and in antarctic and sub-antarctic islands. Seemingly without any exception the whole lot of these birds pass the winter on the northern hemisphere, thousands or millions of them having been recorded in the western parts of the North Atlantic, north to about Newfoundland and in the region of up-welling cold water along the south coast of Arabia (e.g. BRONGERSMA 1947), and also in the northern parts of the Humboldt Current, but irregularly or only accidentally farther north in the Pacific Ocean. There are striking records by ROBERTS (1940) of birds ringed on their nests in breeding holes in the Argentine Islands, western Graham Land, and caught in the same burrows the next season.

The few land-birds (South Georgian Pipit, South Georgian Teal, Kerguelen Pintail) breeding within the limits of the antarctic region are residents throughout the year. Part of the population of Sheath-bills *(Chionis alba)* from the American part of Antarctica is stationary, but the greater part is migratory. These semi-terrestrial birds then winter in Tierra del Fuego, South Patagonia, and the Falkland Islands, crossing 700 km of stormy seas of Drake Passage twice a year. In contrast, all authors which have observed the Sheath-

bill on Kerguelen and Heard Island in the southern Indian Ocean
(*Chionis minor*) have stated that this bird is an annual resident.

Summer visitors

During the summer months when the animal food supply is
notably abundant among the edges of the pack-ice antarctic bird
life is temporarily enriched with visitors from the north. Those subantarctic species which during the summer months under favourable weather conditions wander far southwards, like the White-
headed Petrel *(Pterodroma lessoni)* and the Kerguelen Petrel
(Pterodroma brevirostris) have been discussed in the paragraph on
pelagic distribution. This is, however, the place to mention the
seasonal occurrence as a non-breeding summer visitor of the Arctic
Tern *(Sterna paradisaea = S. macrura)*, which is a species not unlike the Antarctic Tern *(S. vittata)*, but breeding in arctic, subarctic and locally in temperate regions of North America, Europe,
and Asia. Its pattern of migration is the counter-part of that of
Wilson's Storm Petrel: twice a year changing arctic for antarctic
surroundings, thereby travelling as widely as is possible on this
globe. Arctic Terns enjoy the bright and long days of the polar
summers practically the whole year round, passing the northern
winter at the northern edge of the broken pack-ice of the Antarctic.
They live on the rich food supply of euphausians and have frequently been observed resting on floes at the pack edge and farther into
open pack-ice. Here they undergo a full feather moult, including
wing- and tail-feathers. It has not been without caution and reasonable doubt that the enormous extent of the migration route of
the Arctic Tern has become fully known and generally accepted
(MURPHY 1936 and 1938). At present the occurrence of Arctic Terns
in the antarctic pack-ice has been profusely proved by collected
specimens. Some of the most important of these records are summarized below. It should be stated that thousands of these terns have
been observed by members of the Scottish National Antarctic Expedition in the Weddell Sea as far south as 74°01′S., 22°0′W. on 5—13
March 1904 (CLARKE 1907: 345) and that VAN OORDT & KRUYT
(1954) observed Arctic Terns in the Ross Sea at about 71°30′S.,
174°30′W. on 16 February 1952.

Specimens of *Sterna paradisaea* collected in Antarctic Seas.

Position		Date	Authority
64°47′S	84°13′E	6 Febr. 1931	FALLA 1937:251
64°55′S	111°57′E	25 Jan. 1931	FALLA 1937:251
66°50′S	12°20′W	1 March 1947	BIERMAN & VOOUS 1950:110
70°30′S	106° W	25 Febr. 1940	FRIEDMANN 1945:312
68°50′S	90°35′W	5 Febr. 1929	HOLGERSEN 1945:69
67°22′S	11°46′E	16 March 1960	Zool. Museum Amsterdam 15435
68°32′S	12°49′W	23 March 1904	CLARKE 1907:346

The presence of a whole northern hemisphere population as summer visitors in the antarctic is illustrative of the fact that food conditions are extremely favourable and almost without limits in the antarctic during the summer months. For the study of the phenomenology, as well as for that of the origin and the meaning of bird migration, the migration of the Arctic Tern offers immense basic problems. It should be reminded here, that some authors have suggested a consanguinity of the Arctic and the Antarctic Tern. The latter, according to this hypothesis, have directly descended from those Arctic Terns that incidentally have stayed behind and in this way have given origin to an antarctic resident population-group, which have slightly altered since. Others, however, including MURPHY (1938), do not consider the morphological resemblance close enough to conform this hypothesis.

Passage migrants

By their geographic situation the antarctic regions appear among the least likely regions for passage migration. Still, there is the case of mass-migration of a mainly sub-antarctic species, the Sooty Shearwater *(Puffinus griseus)*, which nests in the New Zealand region and Macquarie Island and winters in the North Pacific as well as in considerable numbers in the North Atlantic. On its autumn migration it has been recorded to pass the antarctic seas of the southern Indian Ocean by a steady stream of migrants flying due west (FALLA 1937; VAN OORDT & KRUYT 1953). Many thousands of these birds have been observed by VAN OORDT & KRUYT (*l.c.:* 624, foot-note) in February 1952 in about latitude 64° south and between longitudes 120° and 150° east, which is off the coasts of Wilkes Land. These migration movements add to the general phenomenon of circumpolar dispersal movements observed in the majority, if not all of the birds of the Antarctic.

REFERENCES

The following list of authors is not a full bibliography; it is merely a list of references. A complete bibliography with useful cross-references has been published by BRIAN ROBERTS (1941).

ANGOT, M., 1954: Notes sur quelques oiseaux de l'archipel de Kerguelen. *Oiseau & Rev. franç. Orn.*, 24, 123—127.

ARDLEY, R. A. B., 1936: The birds of the South Orkney Islands. *Discovery Reports*, 12, 349—376.

BEHN, F., J. D. GOODALL, A. W. JOHNSON & R. A. PHILIPPI, 1955: The geographical distribution of the Blue-eyed Shags, Phalacrocorax albiventer and Phalacrocorax atriceps. *Auk*, 72, 6—13.

BERNACCHI, L., 1901: To the South Polar regions, expedition of 1898—1900. London (Hurst & Blackett).

BIERMAN, W. H., & K. H. VOOUS, 1950: Birds observed and collected during the whaling expeditions of the "Willem Barendsz" in the Antarctic, 1946—1947 and 1947—1948. *Ardea*, **37**, suppl.

BORCHGREVINK, C. E., 1901: First on the antarctic continent, being an account of the British Antarctic Expedition 1898—1900. London (George Newnes).

BRONGERSMA, L. D., 1947: Note on Oceanites oceanicus (Kuhl) in the Gulf of Aden. *Ardea*, **35**, *225—226*.

CARRICK, R., 1959: The contribution of banding to Australian ecology. In: A. KEAST c.s., Biogeography and ecology in Australia. Monographiae Biologicae, VIII, *369—382*, Den Haag (Junk).

CARRICK, R., K. KEITH & A. M. GWYNN, 1960: Fact and fiction on the breeding of the Wandering Albatross. *Nature, Lond.* **188** (no. 4745), *112—114*.

CENDRON, J., 1953: Notes sur les oiseaux de la Terre Adélie (Pétrels et Skuas). Note ornithologique 7. Expéditions polaires françaises. Missions P. E. Victor. Expédition antarctique en Terre Adélie 1950—1952. *Oiseau & Rev. franç. Orn.*, **23**, *212—220*.

CLARKE, W. EAGLE, 1906: Ornithological results of the Scottish National Antarctic Expedition. II. On the birds of the South Orkney Islands. *Ibis*, *145—187*.

CLARKE, W. EAGLE, 1907: Ornithological results of the Scottish National Antarctic Expedition. III. On the birds of the Weddell and adjacent seas, Antarctic Ocean. *Ibis*, *325—349*.

DEACON, G. E. R., 1933: A general account of the hydrology of the South Atlantic Ocean. *Discovery Reports*, **7**, *171—238*.

EKLUND, C. R., 1942: Body temperatures of Antarctic birds. *Auk*, **59**, *544—548*.

EKLUND, C. R., 1945: Condensed ornithology report, East Base, Palmer Land. *Proc. Amer. philos. Soc. Philadelphia*, **89**, *299—304*.

EKLUND, C. R., 1956: Antarctic fauna and some of its problems. Geophys. Monogr. 1, *Publ. Amer. Geophys. Union* **462**, *117—123*.

EKLUND, C. R., 1959: Antarctic ornithological studies during the IGY. *Bird Banding*, **30**, *114—118*.

EKLUND, C. R., 1961: Distribution and life history studies of the South-Polar Skua. *Bird Banding*, **32**, *187—223*.

ELLIOTT, H. F. J., 1957: A contribution to the ornithology of the Tristan da Cunha group. *Ibis*, **99**, *545—586*.

ETCHÉCOPAR, R. D., & J. PRÉVOST, 1954: Données oologiques sur l'avifaune de Terre Adélie. Note ornithologique 12. Expéditions Polaires Françaises. Missions Paul-E. Victor. Expéditions antarctiques en Terre Adélie, 1949—1953. *Oiseau & Rev. franç. Orn.*, **24**, *227—247*.

FALLA, R. A., 1937: Birds. Reports B.A.N.Z. Antarctic Research Expedition 1929—1931. Ser. B. II. Adelaide.

FISHER, J., & R. M. LOCKLEY, 1954: Sea-Birds. The New Naturalist. London (Collins).

FLEMING, C. A., 1941: The phylogeny of the Prions. *Emu*, **41**, *134—155*.

FRIEDMANN, H., 1945: Birds of the United States Service Expedition 1939—1941. *Proc. Amer. philos. Soc. Philadelphia*, **89**, *305—313*.

GAIN, L., 1914: Oiseaux antarctiques. Deuxième Expédition antarctique française (1908—1910). Paris.

GIBSON-HILL, C. A., 1949: Notes on the Cape Hen Procellaria aequinoctialis. *Ibis* **91**, *422—426*.

HALL, R., 1900: Field-notes on the birds of Kerguelen Island. *Ibis*, *1—34*.

HITCHCOCK, W. B. & R. CARRICK, 1958: First report of banded birds migrating between Australia and other parts of the world. *C.S.I.R.O. Wildlife Res.*, **3**, *54—70*.

688

HOLGERSEN, H., 1945: Antarctic and Sub-Antarctic Birds. Scient. Res. Norweg. Antarctic Exped. 1927—1928 et sqq. 23.

HOLGERSEN, H., 1957: Ornithology of the "Brategg" Expedition. Scient. Res. "Brategg" Exped. 1947—48, 4. Publ. Komm. Christensens Hvalfangstmus. 21.

INGHAM, S. E., 1959: Banding of Giant Petrels by the Australian National Antarctic Expeditions, 1955—1958. Emu, 59, 189—200.

KEMP, S., 1931: The South Sandwich Islands. Discovery Reports, 3, 133—190.

LINDSEY, A. A., 1940: Recent advances in antarctic bio-geography. Quart. Rev. Biol., 15, 456—465.

LÖNNBERG, E., 1906: Contributions to the fauna of South Georgia. I. Taxonomic and biological notes on Vertebrates. Kungl. Svenska Vet. Akad. Handl. 40 (5), 1—104.

LØVENSKIOLD, H. L., 1960: The Snow Petrel, Pagodroma nivea, nesting in Dronning Maud Land. Ibis, 102, 132—134.

LOWE, P. R., & N. B. KINNEAR, 1930: Nat. Hist. Rep. British Antarctic ("Terra Nova") Expedition, 1910. Zoology, 4 (5), Birds. 103—193.

MARR, J. W. S., 1956: Euphasia Superba and the Antarctic surface currents. Norsk Hvalfangst-Tid., 45, 127—134.

MATTHEWS, L. H., 1929: The birds of South Georgia. Discovery Reports, 1, 561—592.

MAWSON, Sir DOUGLAS, 1915: The home of the blizzard, being the story of the Australasian Antarctic Expedition, 1911—1914. II. London (W. Heinemann).

MILON, PH., & CHR. JOUANIN, 1953: Contribution à l'ornithologie de l'île Kerguelen. Oiseau & Rev. franç. Orn., 23, 4—53.

MURPHY, R. C., 1936: Oceanic Birds of South America. I—II. New York (Amer. Mus. Nat. Hist.).

MURPHY, R. C., 1938: On Pan-Antarctic Terns. Birds collected during the Whitney South Sea Expedition. 37. Amer. Mus. Nov. 977.

MURPHY, R. C., 1962: Antarctic conservation. Science, 135 (3499), 194—197.

NORTH, A. J., 1914: (Exhibit of skin and eggs of the Antarctic Petrel and the Silver-grey Petrel). Proc. Linn. Soc. New South Wales, 38, (1913), 255.

OORDT, G. J. VAN, & J. P. KRUYT, 1953: On the pelagic distribution of some Procellariiformes in the Atlantic and southern oceans. Ibis, 95, 422—426.

OORDT, G. J. VAN, & J. P. KRUYT, 1954: Birds, observed on a voyage in the South Atlantic and Southern Oceans in 1951—1952. Ardea, 42, 245—280.

PAGENSTECHER, D., 1885: Die Vögel Süd-Georgiens, nach der Ausbeute der deutschen Polarstation in 1882 und 1883. Jb. Hamburg. wiss. Anst., 2, 1—27.

PANZARINI, R. N., 1962: Antarctica, international land of science. Wild life. Unesco Courier, 15, 36—40.

PAULIAN, P., 1953: Pinnipèdes, Cétacés, oiseaux des îles Kerguelen et Amsterdam. Mém. Inst. sci. Madagascar. Sér. A, 8, 111—234.

PERKINS, J. E., 1945: Biology at Little America III, the West Base of the United States Antarctic Service Expedition 1939—1941. Proc. Amer. philos. Soc. Philadelphia, 89, 270—284.

RAND, R. W., 1954: Notes on the birds of Marion Island. Ibis, 96, 173—206.

RAND, R. W., 1955: The birds on Marion. Lantern, J. Adult Educ. (Pretoria), 4, 231—234.

RAYNER, G. W., 1940: MacRobertson Land and Kemp Land, 1936. Discovery Reports, 19, 165—179.

ROBERTS, B., 1940: The life cycle of Wilson's Petrel, Oceanites oceanicus (Kuhl). Scient. Rep. British Graham Land Exped. 1934—37, 1(2), 141—194.

RICHDALE, L. E., 1954: Duration of parental attentiveness in the Sooty Shearwater. *Ibis*, **96**, *586—600*.

ROBERTS, B., 1941: A bibliography of antarctic ornithology. Scient. Rep. British Graham Land Exped. 1934—37, 1 (9), *337—367*.

ROSS, Sir JAMES CLARK, 1847: Voyage of discovery and research in the Southern Antarctic Regions, during the years 1839—43. II. London (John Murray).

ROUTH, M., 1949: Ornithological observations in the Antarctic Seas 1946—47. *Ibis*, **91**, *577—606*.

SHARPE, R. B., 1902: Report on the collections of natural history made in the antarctic regions during the voyage of the "Southern Cross". 4. Aves. *106—173*.

SIPLE, P. A., & A. A. LINDSEY, 1937: Ornithology of the Second Byrd Antarctic Expedition. *Auk*, **54**, *147—159*.

SLADEN, W. J. L., & W. L. N. TICKELL, 1958: Antarctic bird-banding by the Falkland Islands Dependencies Survey, 1945—1957. *Bird-Banding*, **29**, *1—26*.

STEINEN, K. VON DEN, 1890: Allgemeines über die zoologische Thätigkeit und Beobachtungen über das Leben der Robben und Vögel auf Süd-Georgien. In: G. NEUMAYER, Die internationale Polarforschung 1882—1883. Die Deutschen Expeditionen und ihre Ergebnisse. 2. Hamburg. *194—279*.

TICKELL, W. L. N., & C. D. SCOTLAND, 1961: Recoveries of ringed Giant Petrels, Macronectes giganteus. *Ibis*, **103a**, *260—266*.

VANHÖFFEN, E., 1901: Bericht über die bei der deutschen Tiefseeexpedition beobachteten Vögel. *J. f. Orn.*, **49**, *304—322*.

VANHÖFFEN, E., 1905: Bericht über die bei der deutschen Südpolarexpedition beobachteten Vögel. *J. f. Orn.*, **53**, *500—515*.

VOOUS, K. H., 1960: Atlas of European Birds. London-Edinburgh (Nelson).

WARHAM, J., 1962: The biology of the Giant Petrel, Macronectes giganteus. *Auk*, **79**, *139—160*.

WILL, H., 1884: Das Exkurzionsgebiet der Deutschen Polarstation auf Süd-Georgien in geognostischer, floristischer und faunistischer Beziehung. *Deutsche Geogr. Bl. (Bremen)*, **7**, *116—144*.

WILSON, E. A., 1907: On the whales, seals and birds of Ross Sea and South Victoria Land. In: R. F. SCOTT, The voyage of the "Discovery". 2, Appendix II. *352—374* (London).

WILSON, E. A., 1907: Nat. Antarctic Expedition 1901—1904. Nat. Hist. 2, (II). Zoology, Aves. *1—121*.

HUMAN ADAPTATION TO LIFE IN ANTARCTICA

BY

OVE WILSON*

Medical Officer of the Norwegian-British-Swedish Antarctic Expedition 1949—52 to Queen Maud Land, Antarctica

(with 1 fig.)

Introduction

The Antarctic is a geographically and ecologically well defined area and is unique in many respects. It is the only land mass that has had no inhabitants up to the turn of this century, and it still has no settled population. Man is a newcomer and a short time visitor. Only a few men have spent more than three years of their life on this continent, which was first sighted in 1820 by FABIAN VON BELLINGSHAUSEN, two days before EDWARD BRANSFIELD discovered Graham Land. BRANSFIELD was the first to chart a portion of the Antarctic mainland[4]. The Antarctic Peninsula (Graham Land) was also seen by NATHANIEL PALMER at this time.

The first men to set foot on the Antarctic continent itself were two Norwegians, Captain LEONARD KRISTENSEN[14] and CARSTEN BORCH-GREVINK, who together jumped ashore from a small boat in the struggle to be the first one. This landing was made in 1895 at Cape Adare ($71°\frac{1}{2}$ S). BORCHGREVINK[9] was also the first to make a planned wintering on the continent with nine other men in 1899—1900 at Cape Adare, although an expedition under ADRIEN DE GERLACHE unintentionally had wintered on board the ship "Belgica" a year earlier[11]. The vessel had been trapped by pack ice and froze in at $71°\frac{1}{2}$ S in the Bellingshausen Sea. Thus the age of man on the Antarctic continent began less than 70 years ago, but much the same situations and problems that man met in this region had previously been encountered by explorers in the Arctic for many hundreds of years.

Although the climatic environment seems most unfavourable to human life, it does not present a serious obstacle to civilized man with his present knowledge. He is fast learning how to surmount the difficulties and is now establishing himself more permanently on the continent. The aspects of man living in the Antarctic have changed considerably since he first arrived at this distant and desolate place, which is so difficult to approach. Great advances in science and enormously increased technological resources have made possible an invasion of this area to an extent that could not have been foreseen by the most imaginative of early explorers. But the

* Department of Hygiene, University of Lund, Sweden.

hazards have not decreased, they are still there, although they have changed. The evolution that has occurred in antarctic exploration has admirably been reviewed by Sir RAYMOND PRIESTLEY[16] in a presidential address, which is very recommendable reading.

Exploration and habitation

One can discern three different eras in the history of exploration in Antarctica. The first period, which can be said to end with SHACKLETON's unsuccessful but epic expedition into the Weddell Sea in 1914—17, has been named the *"heroic era"*. The dominating health hazard of this time was that of vitamin deficiencies. The risk of scurvy and also of beriberi was constantly present on expeditions and the cause of repeated tragedies. Living on trail rations completely devoid of vitamin C while sledging for many months broke down the health and endurance of many of the field parties and seriously hampered exploration of the interior. Not all mastered the swift and efficient polar travel of AMUNDSEN.

The main feature of this early period was geographical exploration with a strong element of natural science. Only a few medical studies were conducted in connexion with health control and consisted of simple investigations such as blood counts, blood pressure, body weight, rate of nail and hair growth, and also some bacteriological studies. The results of some investigations were never published due to their shortcomings. With a few exceptions (EKELÖF[73, 74], McLEAN[116]), published medical papers from this time contain only accounts of health conditions and general aspects of life on an antarctic expedition from a doctor's point of view[67, 76, 79, 84-90, 92].

Then came BYRD's first expedition, 1928—30[10], and with it began the *"modern era"*. BYRD showed the effectiveness of mechanical transport in antarctic exploration, and British and Norwegian expeditions helped to develop the technique. Aviation entered the continent. Although unknown land was still explored and mapped, much of the effort was concentrated on investigating the scientific secrets of this vast, untouched area. The use of modern techniques to study the upper atmosphere of the continent and the ice sheet covering it was still more advanced by the Norwegian-British-Swedish Antarctic Expedition, 1949—52[13]. The scientific results achieved by this first truly international antarctic expedition were regarded as outstanding, but it also showed the need of a much more comprehensive attack on the scientific problems of Antarctica.

During this period the first attempts to conduct planned medical, physiological studies on the members of the wintering parties were made. But such efforts were still greatly hampered by primitive working conditions and were set back by the priority research programmes of the natural sciences. Mostly they were carried out as side projects by the medical officer, who also took turns in assisting in other work going on at the base. Pioneers in this field were LOCKHART and FRAZIER (Little America III, 1939—41), followed by BUTSON (Graham Land, 1946—48), SAPIN-JALOUSTRE (Adélie Land, 1949—51) and WILSON (Maudheim, 1949—52). WILSON was the first to conduct tests (blood counts, blood sugar) in the field during strenous sledging in the interior. Only fairly simple investigations could be carried out at a wintering station and these included blood counts, determination of haemoglobin, blood sugar and adrenalin, measurements of blood pressure, basal metabolic rates, body weight, and rate of nail and hair growth, collection of faeces and urine for analysis, and studies of micro-climate and atmospheric cooling. More complicated and delicate laboratory work was doomed to fail as a consequence of the limited resources and the difficulties that prevailed on polar expeditions of this time. The paucity of published results from the early studies is most certainly due to experimental deficiencies and insufficient data.

The work of the Norwegian-British-Swedish Antarctic Expedition, 1949—52, and its spirit of international co-operation marked the transition to the *"era of expansive exploration"* now under way. With the antarctic phase of the International Geophysical Year (IGY) in 1957—1958 a gigantic invasion of the Antarctic has taken place. Twelve nations dispatched expeditions to the South and some 40 stations were manned on the mainland and on the antarctic islands on its fringes. It was soon recognized that the highly successful work of the IGY was bound to be carried on, so the occupation of many of the bases has continued. Today nine nations maintain some 30 stations in the Antarctic. The number of stations vary because many are temporary or seasonal. During the dark season the antarctic population amounts to several hundred men wintering at the stations. But when the austral summer comes, there is an influx of some thousand people, counting men engaged in work at sea and in supply and support functions, as well as scientists carrying out short term investigations. The summer population is almost never subject to the problems of the wintering groups, although traversing parties and air crews may encounter cold and adverse conditions on the central continent.

The rapid advance of scientific work in the Antarctic has also encouraged an increase in medical studies, but unfortunately enough the efforts in this field have been rather sporadic and incoordinated and have not yet reached the importance of other sciences. This is in part due to the fact that physiological research still receives rather low priority. It is up to individual initiative to undertake such work, and there is no organization for international co-ordination in planning and implementing medical research, as is the case of other sciences. There is also the well-known difficulty of persuading subjects to co-operate in what might sometimes be rather unpleasant experiments, especially in the face of a stress that is expected to be considerable in this extremely cold climate.

Climatic factors

The polar climate of the South is much more severe than that of the North. The mean temperature of the year is considerably lower, and the austral summer is much colder. The mean temperature of the warmest month is usually below 0° C. The weather is characterized by low atmospheric pressures and great windiness. In some extreme places the mean wind speed of the year may amount to as much as 19 m/sec. South of the Antarctic Circle there is, of course, the same yearly alternating periods of continuous polar night in the winter and midnight sun in the summer as in the Arctic, but the intensity of the sunlight is much greater in the Antarctic. The extremely clear air and the low water vapour content (due to the low temperature) permits the solar irradiation to reach exceptionally high values. The brilliantly white snow surface, which covers more than 95% of the continent, reflects 80—90% of the total incoming solar radiation. The physical property which gives Antarctica its distinctive climate is in fact radiation. For a more detailed description of the general aspects of the antarctic climate, reference is made to RUBIN[27] and to Chapter II, also by RUBIN, in the present volume.

Wind Chill

The climate of Antarctica provides meteorological conditions which must be regarded as a severe environmental stress to man and his heat regulation. It is, however, difficult to evaluate and measure the degree of the climatic stress to which man is exposed during a sojourn in the Antarctic. It cannot simply be expressed by a temperature curve or by giving mean values of meteorological data. The atmospheric cooling power at low temperatures is greatly increased by air movement. There is also the heat gain from solar radiation. The environmental stress of a cold climate is a combination of air temperature, wind, and radiation, which determine the relative comfort sensation and may cause injury, such as frostbite, snowblindness, damage to the skin, etc., or death due to hypothermia. The main factor concerned with relative discomfort and health hazard is heat loss. The mechanisms governing heat loss are described in detail by BURTON & EDHOLM[8]. The important factor causing uncomfortable or deleterious heat loss is convective cooling. Various formulas to express the dry convective cooling power of the atmosphere have been proposed. The only one based on actual observations in an extremely cold environment and correlating atmospheric cooling to stages of human comfort sensations and the freezing of exposed skin, is the one developed by SIPLE & PASSEL[31] from experiments conducted at Little America III, Antarctica. This "wind chill" formula was calculated from observations of the cooling rate and freezing of water sealed in a plastic cylinder, and related to 33° C (neutral skin temperature). It measures the cooling power of wind and temperature on shaded, dry human skin without regard to evaporation. It does not take into account heat gain from solar radiation. The term wind chill is applied to a scale of heat loss extending from an index of 50 (hot) to 2500 (intolerably cold). The formula should not be employed to express actual amounts of heat loss, but should be considered as an empirical table, and the values used as index numbers on a relative scale. As such it provides an index corresponding quite well with the discomforts and tolerances of man in the cold, as it applies to the unprotected parts of the body, such as the face and bare hands, where the pathological effects of cooling first will appear.

In a recent paper by the author[32] the antarctic climate has been described in terms of wind chill as experienced at Maudheim (71° S, 11° W) and on sledging trips into the interior of the continent. The occurrence of frostbites was correlated to wind chill index and gave evidence in support of the wind chill scale. It confirmed the assumption of SIPLE & PASSEL[31] that exposed flesh begins to freeze at an index of 1400. In evaluating the amount of climatic stress encountered in a certain place, it is important to consider the percentage distribution of the various wind chill conditions measured during the stay. Cumulative frequencies of wind chill for each month showed that at Maudheim during the months of May to September the daytime wind chill index will be higher than 1400 during more than 85% of the time, and that a wind chill index of 1400 or more will be encountered during half of the time in the months of April and October. It was found that more than half of the days during two years at Maudheim had a wind chill index of more than 1400 measured at 09.00 hours.

A general impression of the relative atmospheric cooling effect of the climate in different parts of Antarctica can be had by comparing the mean wind chill at different stations. The relative levels of the mean wind chill indices for the year, and for the coldest and warmest months, at a number of interior and coastal stations all around the antarctic continent are given in Table I. Remarkable is the high wind chill encountered at Cape Denison (67° S) and Port Martin close by. Cape Denison is famed for its windiness, with a yearly mean wind speed of 19 m/sec. and has earned the name of "Home of the Blizzard". At this low latitude such a high wind chill is typical only for a limited area (including Port Martin) with local fierce gales. The stations in the interior of the continent, of course, show very high wind chill values, which partly is due to the decrease in temperature with high altitude. The South Pole has a yearly mean index of 2410, which is exceeded by Vostok II ($78\,^\circ\tfrac{1}{2}$ S) with an index of 2420. Although positioned much further to the north this station is situated in an intensely cold area near the "Pole of Cold".

These very high wind chill indices seem to indicate intolerable climatic conditions, but with modern technology and proper clothing they can be endured even for long periods. However, the consider-

Table I.

Mean wind chill values for some antarctic stations south of 66° S.
Wind chill index computed from monthly mean temperature and wind speed. For geographical position of the stations see map (Fig. 1).
Latitude is given to the nearest half degree.

No. on map	Station (Nation)	year	latitude	annual mean	coldest month	warmest month
1	(Coastal stations) Maudheim (Norw.-Brit.-Swed.)	1950—52	71° S	1540	1860	1100
2	Lazarev (U.S.S.R.)	1959	70° S	1570	1860	1140
3	Syowa (Japan)	1957—59	69° S	1270	1560	980
4	Mawson (Australia)	1954—57	$67\,^\circ\tfrac{1}{2}$S	1430	1780	1030
5	Mirny (U.S.S.R.)	1956—58	$66\,^\circ\tfrac{1}{2}$S	1440	1740	1050
6	Oazis (U.S.S.R.)	1957	$66\,^\circ\tfrac{1}{2}$S	1220	1530	900
7	Pointe Géologie (France) Dumont d'Urville (France)	1952 1959 }	$66\,^\circ\tfrac{1}{2}$S	1420	1660	1050
	Port Martin (France) Cape Denison (Australia)	1950—52 1912—13 }	67° S	1590	1850	1200
8	Cape Adare (U.K.)	1899—1900	$71\,^\circ\tfrac{1}{2}$S	1120	1420	830
9	McMurdo Sound (U.K.) (3 stations)	1902—03 1908—09 1911—12 }	$77\,^\circ\tfrac{1}{2}$S	1490	1790	1000
10	Framheim (Norway) Little America III & V (U.S.)	1911—12 1940,1957 }	$78\,^\circ\tfrac{1}{2}$S	1610	2030	1110
11	Horseshoe Island (U.K.)	1957—58	68° S	1120	1370	910
12	Shackleton (U.K.)	1956	78° S	1670	2130	1030
13	Halley Bay (U.K.)	1957—58	$75\,^\circ\tfrac{1}{2}$S	1520	1840	1040
14	(Interior stations) Byrd (U.S.) (altitude 1500 m)	1957—59	80° S	1890	2240	1370
15	Amundsen-Scott (U.S.) (2800 m)	1957	90° S	2410	2870	1510
16	Vostok II (U.S.S.R.) (3400 m)	1958	$78\,^\circ\tfrac{1}{2}$S	2420	2880	1740
17	Pionerskaya (U.S.S.R.) (2700 m)	1957	$69\,^\circ\tfrac{1}{2}$S	2280	2610	1740

696

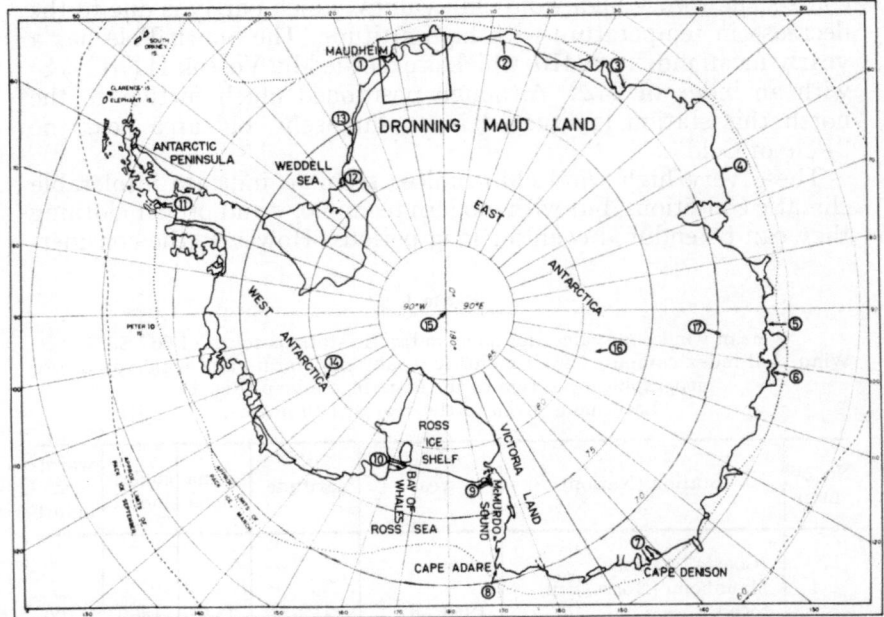

Fig. 1. Map of Antarctica. The numbers mark stations referred to in Table I. The inset shows the area traversed by the Maudheim Expedition.

ably higher wind chill values calculated for the intra-continental stations do not represent a correspondingly higher cooling power of the atmosphere. At the altitudes prevalent in the interior, the density of the air is much lower, which means a decrease in the heat-carrying capacity of the air. The term representing convective heat loss must therefore be multiplied by the square root of the ratio of density at altitude to that at sea level[8]. At the South Pole (altitude 2800 meters), as compared to Little America (sea level) where the wind chill formula was developed, this factor would be about 0.83. This agrees with the observation that men at the South Pole are able to tolerate very high wind chill values and work outdoors even in the wind at temperatures below —60 °C [20]. On one occasion two men at the South Pole[96] spent 3—4 hours out of doors in a wind chill of 3100—3200, which would be equal to an index of about 2600 at sea level. A wind chill value of this order has actually occurred at Little America[31].

The wind chill conditions at Maudheim (71° S) are typical of an antarctic coastal station with a relatively large cooling effect on man. In the Northern Hemisphere comparable wind chill conditions are only found in the winter in the coldest regions of the Arctic. The mean wind chill index of the coldest month at Maudheim (about

1875 in July) is generally found only well into the North Polar Basin at considerably higher latitudes, along part of the northern coast of Siberia (70° N), on the inland ice of Greenland, and in Central Arctic Canada north of Fort Churchill[234]. This latter area, however, extends as far south as 60° N. The wind chill index of the warmest months at Maudheim (about 1100 in December-January) is comparable to values for the coldest months in Northern Sweden and Norway, Central Iceland and on the coast of South Greenland. It must be remembered, however, that in this comparison the strong solar radiation at Maudheim in the summer has not been taken into account. It contributes greatly to lessening the effect of atmospheric cooling during this season.

Radiation

The contribution of solar radiation in Antarctica in terms of heat gain for the human body has been studied by CHRENKO & PUGH[23] using radiation values from Maudheim. Their results show that the solar heat gain in the Antarctic at the height of summer is from two to four times greater than the heat gain in the desert regions. The heat gain varies with wind speed and solar altitude and may reach values of nearly 400 kcal/m²/hr. This explains why it is possible for men in exceptionally calm weather to work stripped to the waist in subzero temperatures. When the author was sledging inland at the height of the summer in Queen Maud Land the solar heat gain was at times so great that travelling was mostly done at night to spare the dogs from the heat load and the face from severe sunburn. The rate of heat absorbtion by the dark cloth of the double-walled pyramidal tent due to solar radiation was sometimes so high that when sleeping during the daytime one could lie practically naked on top of the sleeping-bag with the tent-entrance wide open in air temperatures well below freezing. Investigations in the Antarctic by PUGH & CHRENKO[26] have shown that the effect of intense solar radiation on man inside a tent of this type may be equivalent to raising the external temperature by about 19° C.

It is important to note that the reduction of the radiation flux in dull weather with overcast skies is much smaller in the Antarctic than elsewhere, due to multiple reflexion between snow surface and cloud base. Measurements of the global radiation at Maudheim[25] have revealed that even with a dense overcast, the intensity of diffuse sky radiation is usually 50—60% of the incoming solar radiation on clear days. In temperate latitudes the corresponding value is generally less than 25%. This means that not only is there a remarkable heat gain from reflected radiation even on overcast days, but more important, there is also a great amount of ultra-violet radiation, due to multiple reflexion, when the sky is covered.

During the light season the ultraviolet radiation is so intense that it is impossible for man to stay out of doors for any longer period without being affected by snowblindness (photophthalmia), unless protected by effective sunglasses. It is essential that the protective glasses have the right qualities. An investigation by HEDBLOM[107] has shown that the relation of the energies of infrared light transmittance to visible light transmittance is of utmost importance. This is because a syndrome of heat-caused eye discomfort (calorophtalgia) can be produced by glasses that have a low visible light transmittance but allow a relatively higher proportion of infrared energy to pass and enter the dilated pupil. The risk of snowblindness is especially treacherous on overcast days when less precaution is taken in the absence of the warning glare. In the subdued light the dark glasses are apt to be removed, but must immediately be replaced by lighter glasses. With a uniform overcast there is also the hazard of the "polar whiteout"[23b, 24, 25]. The multiple reflexions between cloud base and snow surface make the light diffuse. To the eye, the light intensity is the same in all directions. All shadows disappear, everything is "white in white", and all sense of direction is completely lost. This light-condition is a great danger and has been the cause of many accidents, especially air-crashes. Vision is also impaired by heavy snow drift, which becomes a great problem in blizzards. This question has been investigated by SAPIN-JALOUSTRE[112].

Absence of light

The polar night is usually described as a period of total darkness, but this is not quite true. Most of the coastline of the antarctic continent more or less follows the 70th latitude, except for the marked indentations at the Ross and Weddell Seas, and the Antarctic Peninsula. This means that at a fairly large number of coastal stations the winter period with the sun constantly below the horizon normally will last for about two to three months. During this time it is not completely dark. At most of the places there usually will be at least four hours of twilight around noon, permitting a certain amount of outdoor activity. During unbroken spells of good weather the moon may shine continuously day and night giving working light for periods up to fourteen days. The darkness is also dispelled to some degree by starlight and southern lights. The effect of these light sources is greatly enhanced by reflexion in the snow surface. In good weather it usually is possible to distinguish dark objects against the snow at any time of the polar winter night. The period of darkness is of course still more extended at the southernmost bases and interior stations. At the South Pole there are two months of twilight and four months of winter darkness.

Humidity

The air in the Antarctic is extremely dry, below —20° C the absolute humidity is practically neglectable. Little is known of the effect on man of such low humidity, except perhaps giving rise to a dehydration of the skin, which condition might be a factor in lessening the susceptibility to frostbite[142]. In the dry air there is also an increased respiratory water loss, especially when hard work is performed. This may increase body dehydration, as supply of drinking water is frequently a problem for man living in the field.

Shelter and food

In establishing a wintering station in Antarctica every single item, housing, food, fuel and all equipment, has to be transported across the sea to the building site. The only things available on the spot are snow for water and, at the coast, seals and penguins for fresh meat and for dog food at the base. Occasional whales and many sea birds are also encountered. The wild life, however, disappears at the arrival of the winter, except in a few places where colonies of emperor penguins breed during the polar night. In contrast to the rich life of the deep waters, the coastline of Antarctica and the shores of the antarctic islands are surprisingly barren of bottom life and fish. The continental shelf lies at a depth of about 400 meters, and the waters along the ice rim of the coast are quite deep in most places. Fishing is therefore usually difficult, although practiced with success at some localities[71].

Only in great emergencies can local animal life be relied upon to supply meat for food and blubber for fuel. That it is possible to live off the land was first shown by the Swedes, under OTTO NOR-DENSKJÖLD[15]. Two parties were left stranded without provisions or gear when the expedition ship was crushed by the pack ice. They survived on seals and penguins and an occasional fish and became pioneers in living on the antarctic land. Since then several other parties have endured the same experience. The Swedish groups built stone shelters, and a marooned British party burrowed into a deep snowdrift, where they excavated a chamber[17, 213]. These are in principle the two ways to plan housing in this environment.

Housing and sanitation

The station is either built on solid snow-free land, where such can be found, or on snow, whether it be on the ice shelf, some 160—200 meters thick, or on the inland ice in the interior, where the snow cover is several kilometers thick. Erecting houses on rock or rubble permits more conventional construction techniques, but cold and lack of sand inhibits the use of concrete. The short time available for building in the summer requires prefabricated structures. Cables are essential to anchor the houses against heavy wind. Buildings on rock are kept free from snow by the winds. They need more heating than those covered by snow. Placing a station on deep snow requires different techniques. The house can be erected in a deep pit, and the snow which fills in around the building provides added insulation. The same result is attained by placing the houses on the surface and simply letting snowdrifts accumulate around the buildings until they are completely covered, which normally will take a few months' time. The houses are connected by tunnels in the snow. The New Byrd Station is housed in under-ice caverns[226a]. For their inland stations the Russians have used uniquely designed tractor quarters that are expanded at the wintering destination and re-supplied by air[21].

Housing in the Antarctic is no problem today, it is mainly a matter of logistics. Prefabricated sections are used and well designed houses are erected in a few days. Formerly *heating*, insulation and condensation were a problem. Very great temperature gradients existed indoors. Floor temperatures were often below freezing, while temperatures as high as $+25°$ C occurred at the ceiling. SAPIN-JALOUSTRE[29] made a study of the micro-climate in a French station in 1950, showing the extent of the gradient. He found it necessary to elevate the lower sleeping-bunks more than half a meter, thus increasing the temperature level at the bunk about $7°$ C. Inside the quarters at most bases the temperature is usually maintained around 15—18° C, which is a comfortable shirt-sleeve level (in a wool shirt over heavy underwear), but there are great variations according to insulation and preference. The Americans often thrive in indoor temperatures around 25—30° C, while Europeans mostly prefer temperatures around 15° C[29, 179a] or lower [71, 94, 153]. At Maudheim[186] the mean indoor temperature at the 1 meter level was 14,5° C in the first year and 13,5° C in the second year. A higher temperature could easily have been maintained, but the heating was modified to suit the average of the individual preferences. During the night a lower temperature was preferred for sleeping. The *relative humidity* is very low, normally about 20—35%, and complaints of throat irritation and coughs have been noticed[183]. At the South Pole Station[96, 109] humidity was estimated to be as low as between 5 and 15%, while at some coastal stations [23a, 29, 99] 35—65% have been measured.

Ventilation for buried camps will often become a problem, due to icing of condensation and obstruction of ventilator outlets by drifting snow. Drift-clogged chimneys or exhaust pipes frequently are the cause of carbon monoxide poisoning, which is an important hazard in Antarctica. Exhaust fumes constitute an additional risk in mechanized travel when long hours are spent in motorized vehicles with cabin windows closed. Primus stoves in tents are also a source of carbon monoxide. Symptoms of poisoning have been described from many expeditions. PUGH[110] has recently investigated the problem on the Trans-Antarctic Expedition.

Fire is perhaps the greatest hazard to men in polar camps. Fires seem to occur on the coldest or windiest days, which increases the danger as men in their haste often rush to the emergency with insufficient time to clothe themselves properly, especially in the night. The most disasterous antarctic fires occurred at the Russian station Mirny, where eight lives were lost, and at the American station McMurdo Sound, where damage for several hundred thousand dollars was caused. Two British, one French, one Argentine, two Chilean stations, and a Japanese station hut have been destroyed by fire, and several lives were lost on the British expeditions due to this hazard. The writer's tent and sleeping-bag caught fire during a storm in the Antarctic, while he was alone on a sledging trip inland. Luckily adequately dressed but without boots, he was able to dig out a spare tent in a depot close by. Special precautions have to be taken when building a base. The houses must be separated, letting thick snow firebreaks form between them, and food, fuel, and vital equipment is divided between the main buildings. An emergency hut stocked with provisions for survival is usually erected at a safe distance.

In spite of the fact that an unlimited amount of frozen water is always available, the *provision of water* offers certain difficulties[100]. Snow melters of sufficient capacity have to be supplied and must be kept continuously working. Near large bases contamination of snow-gathering areas in the summer represents a problem and facilities for purification have to be provided. Local supplies of snow are soon exhausted and the collecting trips have to be extended. At the South Pole Station[20] the problem of clean snow for drinking water was solved by digging a deep snow mine in the inland ice. Everybody at the base took turns in chipping ice for water, which is quite an exhausting job at high altitude and below —50° C. During a year it extended 90 feet (27 m) deep and 275 feet (90 m) long. At Byrd Station a sub-surface water system is installed, by which water is obtained from a pool at the bottom of a shaft melted vertically downwards into impermeable ice structure. Further melting is produced by continuous introduction of heat, and the the water is pumped up[231]. In the near future the heat will be supplied by nuclear power.

Waste disposal is not a problem in most cases. When poured out on the inland ice sewage disappears through the porous layers of the snow into the more impermeable layers and freezes. Refuse left on the surface freezes immediately and is quickly covered by drifting snow. At the coast it is dumped into the sea or on the bay ice, where it drifts out. If there is no immediate access to the sea, disposal may become a problem in the summer when freezing does not occur. There are no insects in the Antarctic, but skua gulls in large numbers visit the refuse dumps and may constitute an epidemiological hazard. Latrine refuse is usually collected in sawed off fuel drums, which are easy to dispose of. At Maudheim a large urinal chamber was dug out under the snow, where the temperature never rose above —5°. The size was calculated according to the number of men and the time of sojourn, and was never filled. In the future sewage disposal into inland ice wells will be facilitated with nuclear power.

The first American nuclear power plant in the Antarctic, on Ross Island at McMurdo Sound, went into operation in March 1962[225, 226a]. Another reactor has been authorized for that base and one each for the New Byrd and the New South Pole Stations. *Radio-active wastes* from the plants will be processed and returned to the United States for disposal.

Clothing

The most indispensable asset in man's struggle to survive and succeed in the Antarctic is efficient clothing. The general principles governing the selection of clothing for cold climates have been discussed by SIPLE[232], and research by the Quartermaster Corps[236] have contributed to further advances. There is a wide variety in clothing used at different stations, and sometimes an equal diversity from man to man in the same station, but the basic principles are about the same. No amount of clothing will keep an inactive man protected for more than an hour or two in the winter, but if his heat production is increased by activity, proper clothing may keep him warm even below —70° C. Perspiration is the greatest problem in the cold, as it will condense and freeze in the clothing and greatly lessen the insulating properties. There may be considerable sweating during activity, even at very low temperatures, but the experienced man keeps warm because he has learned how to avoid perspiration by adjusting his clothing and ventilating. The most difficult parts of the body to protect are the feet, hands, and face. Because of the law of diminishing returns for insulation over cylindrical shapes (i.e., the surface for losing heat increases more rapidly than insulation is gained in thickness), the approximate optimum thickness of insulation is as follows: fingers — less than 1 cm; head, hands, and feet — 2 to 3 cm; legs and arms — 3 to 5 cm; torso and sleeping bags — 10 to 20 cm[21]. In midwinter the men at Vostok and Sovietskaya had 40-watt batteries and electric heaters that warmed their hands, feet, and chest. The devise of a satisfactory face mask that will not become iced up by condensation has hitherto been an unsolved problem. However, it seems as if the Americans have been able to produce one that can be used[19]. A face mask that provides electrically heated air has been tried at the Russian intra-continental stations, as well as unheated types[232a]. Many prefer a beard for face protection, but it is usually messy and troublesome because of icing up. The value of a beard has been a subject for many heated discussions, and it still remains a matter of personal preference and pride, as colorimetric measurements on heat loss from skin with beards and without beards have shown no practical difference[100].

With experience of a cold climate, there is a noticeable tendency for the men to wear less clothing[142,135,29,186,23a,192a-b], suggestive of an adaptation to cold. This has led to studies of the amount of clothing worn at different times of the year. GOLDSMITH[144,145] and PALMAI[170] showed by the use of clothing records that men exposed to cold do get accustomed to wearing fewer clothes for the same cold stress and still remain comfortable. The hypothesis that cold adapted men are comfortable at lower skin temperatures than unadapted subjects was not supported by measurements made by ADAM[130] on men during field journeys in the Antarctic, although NORMAN[162] and BUDD[133b] have found that lower skin temperatures beneath the clothing are maintained during sledging than at the base.

Modern technology has also made possible the development of protective clothing for under water swimming while conducting biological studies in antarctic waters. In this equipment it is possible to make dives in water temperatures as low as $-1.5°$ C among drifting ice and spend over an hour in the water[227a,229].

Nutrition

The question of food on an expedition has always been a matter much discussed. *Vitamin* deficiencies in the early period constituted an important health hazard and many cases of scurvy occurred on wintering expeditions. There was much confusion as to the cause of this feared condition, and the attitude towards the nutritional problems of this time varied considerably as can be read in reports by expedition doctors[41,43,44]. The occurrence of scurvy on British antarctic expeditions during this period has recently been discussed in a paper by KENDALL[47]. As soon as the cause of the common vitamin deficiencies had been discovered and the knowledge had spread, this threat to health ceased to be a problem for wintering parties, although the latent risk of hypovitaminosis[94,34a-b,51,169a] must not be disregarded. A recent study by ADAMS, STANMEYER & HARDING[33] indicates that in the stresses and restrictions of the antarctic environment there is an increased and unsatisfied need for vitamin B complex and ascorbic acid, resulting in an increased occurrence of oral lesions in men on unsupplemented diet. PALEEV [169a] also noted an acclimatization hypovitaminosis B and C, and WIGG[183d] has made a study of suspected vitamin B deficiency in Antarctica. Food is now generally supplemented with vitamins. Great varieties of foodstuffs are available with modern conservation and packing techniques. Fresh meat, fruits, and vegetables are transported by air and are kept frozen for use. Greenhouses have even been tried with success[36, 71]. A cow has been kept at Little America II and at Mirny a herd of pigs solved both the garbage disposal and fresh meat problem[21]. And there is a collection of special recipes[39] for antarctic cooks, who want to try their hand at the local delicacies available along the coast!

The *food intake* at antarctic bases varies considerably, the mean values ranging from 3300 to 5000 kcal daily, depending upon the changing level of activity during different seasons[40, 48a, 57, 59, 68, 139a,d,140]. Energy output varies accordingly and is also very high. Values around 4100 kcal were found during the winter at Shackleton Base by ROGERS[140,151c] for both intake and output. The energy expenditures recorded by ROGERS appear to indicate larger expenditures than might have been predicted. A detailed study of nutrition and energy expenditure in scientists and Navy personnel was made at Little America V by MILAN & RODAHL[52-54]. For the scientists the average daily expenditure was about 3775 kcal and the consumption 3400 kcal per man in the autumn, 3370 and 4396, respectively, in the winter, and 4175 and 4285, respectively, in the spring. The Navy personnel expended 3660 kcal and consumed 4925 kcal when studied in the spring. On this regimen all of the subjects gained weight during the antarctic stay. All of the figures for intake and expenditure are high compared with surveys in a temperate climate. The percentage of calories furnished by protein, fat, and carbohydrate was not significantly different from those reported from temperate regions.

The often quoted observation[30, 142] that men seem to have a greater craving for fat in the cold climate, has led to studies[37, 56, 242] of the desire for, and the tolerance and utilization of dietary fat. ORTON[56] found no evidence of an overall shift in the fat absorbtion-excretion mechanism favouring greater or less utilization of fat by the body, nor has an increased avidity for fat over the year been proven[52].

Food requirements for sledging have been estimated from 3500 to 5500 kcal[29, 35, 42, 48a, 50, 57, 59,141a]. A nutritional study by MASTERTON, LEWIS & WIDDOWSON[242] showed an intake of about 4800 kcal daily with a mean energy expenditure of about 5100 kcal during sledging in Greenland. Similar figures for expenditure during sledging were found by NORMAN[162] in the Antarctic, although the intake was slightly less than 4500 kcal. GOLDSMITH (quoted by EDHOLM[140]) estimated the intake of two men during three separate sledging journeys under varying conditions in the Antarctic and found it to be 4200 kcal. ORR[55] found that 5000 kcal were needed to maintain body weight during sledging. Investigations by ADAM (quoted by GOLDSMITH[146]) suggest that although travel with motor vehicles is obviously less strenuous, the daily energy output actually approaches that of sledging because of the longer day.

Body weight and skinfold thickness

Generally there is a seasonal change in *body weight* in the Antarctic with a marked increase (up to 5 kg) during the sedentary dark season and a loss in the spring, observed on most expeditions[73, 116, 68, 131, 133b, 156a, 161, 183b] and studied by WILSON[186]. The rise during the polar night is mainly due to inactivity, but it has also been observed by WYATT[192] in men engaged in the heavy work of sledging during the winter. PERLITSH et al.[126] found that the weight gain for an indoor group was 3.8 kg as compared to 0.9 kg for the outdoor men. The regularity of the seasonal change suggests a possible biological rather than a sociological basis. MILAN & ROHDAHL[52, 53] found that the increase in weight over the winter was three times larger in a group of Navy personnel than in a group of scientists. It is suggested that psychological as well as physiological reasons lay behind the high caloric intakes seen in the former group and that eating served to alleviate the tedium of long isolation.

The body weight does not always increase in the winter. GOLD-SMITH[78] on the Advance Party at Shackleton Base found that the body weight fell or remained steady in the winter months, because on this expedition, owing to adverse circumstances, the men were forced to live in inadequate shelter[77] with too little fuel and had to work outdoors throughout the polar night. In the second year, however, the seasonal rise and fall was observed[140].

At the South Pole Station (altitude 2800 meters) and at Vostok (altitude 3400 meters) in the interior of the continent there was a marked decrease in body weight during the winter. In some men the loss was considerable, amounting to as much as 10 to 16% of body weight during the first few months[96, 97]. At the Russian base a mean weight loss of 5 kilograms was observed during the first 3 months[183, 183a]. The weight did not increase again until the end of the dark season, when it slowly attained initial values. This difference to the seasonal change at the coast is due to the process of acclimatization to high altitude inland.

Very rapid body weight regains after field trips, observed by WILSON[186], MASSEY[153], ORR[165] and WYATT[191, 192], are indicative of dehydration during sledging. WYATT found that there was a significant loss in mean skinfold thickness during sledging. At return to base the weight returned to pre-sledging values within the first 36 hours, but skinfold thickness did not until at least 2 weeks has passed. The early rapid regain was attributed to an increase in total body water after a reduction during sledging, the decrease in skinfold thickness indicated a negative calorie balance during the trip. Energy output usually exceeds caloric intake during sledging. ORR[55], on the other hand, found no evidence of dehydration during sledging and his results suggest that weight changes in the field mainly reflect changes in calorie balance and fitness, while DAVIES' results[139a] point to changes both in fluid and energy balance.

Medical aspects

The individuals living in the Antarctic are not representative of an average population. They are selected males, physically healthy, mainly in the age of 20 to 40 years. They are highly motivated persons, who have volunteered to come here and are ready to accept a strenuous time in a strange surrounding for a restricted period. It is therefore not surprising that health conditions on antarctic expeditions are exceptionally good, which has been attested in numerous accounts by expedition doctors[65-104]. The men have gone through careful medical examinations before leaving for the South, and under the conditions of the new environment the risk of falling ill is very small.

Infective diseases

There have been very few infective diseases diagnosed in Antarctica[80,106]. Ordinary colds are extremely rare during wintering over, but upper respiratory infections become frequent when the summer invasion of new groups arrive. The occurrence of mild colds have often been observed when the relief ships come in, but now that long range communication by air is common, more severe outbursts of epidemic character have been reported. At the South Pole Station there was a "cold siege" introduced by the first visitors and lasting for several weeks with a high degree of contagiousness suggesting an epidemic[96]. It might have been aggravated by the altitude.

There is a common theory that newcomers infect the wintering over party en masse, and that the wintering party loses its immunity to the common cold during its long isolation. Detailed statistics kept in Antarctica at McMurdo Sound[80] show that the rate of acute upper respiratory infection of any and all types among the wintering over group was less than one-fourth of the in-coming reliefs and summer visitors, who brought infection with them or developed it on the ice. When the station members who had wintered

a year at Halley Bay in the Weddell Sea came on board the British
relief ship, many caught severe colds. Several days later the ship
with the cold-infected men aboard went on to Shackleton Base.
Surprisingly enough, none of the men who had wintered on this
base caught a cold when they came in contact with the men who
were still suffering from their colds (GOLDSMITH, personal communication). These observations do not favour the myth that personnel
wintering over are more susceptible to cold after their long isolation
and will become infected as soon as they come in contact with
outside visitors. A long-term investigation of upper respiratory
infection in the Antarctic is being carried out by SLADEN[117a–120],
who has reported[119] respiratory pathogens to occur appreciably less
in isolated men after wintering over, which partly confirms the
early findings of McLEAN[116]. In an earlier study, however, SLADEN
[117b] found that respiratory pathogens were consistently carried in
the noses of men isolated in Antarctica for 9 months. ADAMS & STAN-
MEYER[115] studied the effect of prolonged cold on a bacterial strain,
both oral and intestinal, in a human host by monthly collections
at Little America V. While the bacterial counts remained unchanged
in the stools over the whole wintering period, the oral counts were
reduced during the coldest months of continued activity. They
concluded that the reduced bacterial counts were due to the lowered
oral temperatures during this cold period.

If one lists all the different ailments that have occurred on expeditions in the Antarctic and which are enumerated in the various
medical reports from this region, the list will be quite long and
varied, but the main part are only occasional stray cases not
peculiar to this environment. Several cases of appendicitis have
been reported[30, 76,100,114], some successfully treated
nonsurgically[30,100,108]. Appendectomy prior to going to the Antarctic is usually not a prerequisite, but on the Australian expeditions
prophylactic appendectomy of the medical officer is required. At the
Chilean stations all have previously undergone appendectomy as
was also the case on a French expedition[114]. The author did not
regret having had his appendix voluntarily removed before leaving.
During the first winter in the Antarctic he heard over the radio an
SOS for help coming from another expedition doctor suffering from
appendicitis on Heard Island. The doctor was eventually rescued
by an Australian cruiser[114], as the passage to this subantarctic base
was not obstructed by ice pack. Such an evacuation had been
impossible from a base on the antarctic continent. The Russian
doctor L. I. ROGOZOV[111a] was forced to remove his own appendix
in the Antarctic. A caecostomy has been carried out by the French
doctor J. CENDRON[70], the author performed an eye operation at
Maudheim[13], and recently a successful craniotomy for a subarach-
noid haemorrhage was carried out by the Australian doctor R.

708

PARDOE[67b, 109a]. Although the medical officer must be prepared
to cope with any surgical situation (including gynecological intervention in a husky bitch) or disease, even tropical ones acquired
by the men on the outward journey in some port, the medical
problems in the Antarctic are few.

Antarctic ailments

Ailments caused by the environment itself are snowblindness,
chapped lips, and sunburn from ultraviolet radiation, frostbites[113,
142,188], nose drip from condensate, coughs and irritated throats
from breathing very cold air, and also might be added carbon
monoxide poisoning[110] and sleeplessness in the dark season[203, 217,
220, 224]. Psychic disorders will be discussed later. Coughs with
expectoration of blood is a common experience when breathing
heavily at temperatures below —45° C. This is attributed to
congestion of the mucous membrane in the posterior oral pharynx[142]
or ruptured capillaries in the bronchial trachea[96]. It is generally
agreed that frosting of the lung does not occur[96, 244]. HOUK[109] has
described a syndrome of transient injury to the bronchial tree
caused by deep breathing in temperatures below —57° C, observed
at the South Pole Station. The symptoms usually consisted of a
4- or 5-hour period of transitory dyspnea and hacking productive
cough, associated with copious, thin, blood-streaked sputum. With
prolonged exposure the symptoms were more severe and might
incapacitate the individual for several days. This condition was
probably also observed by FRAZIER[142] at Little America III, and
described by SMITH[244] in the North. With precaution snowblindness
need never happen. Severe frostbites are rare, occurring almost
exclusively in connexion with accidents under adverse circumstances,
and very few amputations have been necessary[114]. Light frostbites
or "frost nips", on the other hand, are a constant feature of the cold
periods, when wind chill index stays below 1400[32]. The skin on
finger-tips, cheeks, or nose-tip of men often exposed are continuously
pealing as sequelae of small and often unnoticed frostbites, which
at times are so common as to cause no comment. Although the
frequent occurrence of frostbites often impedes activity during
high wind chill conditions, it usually never becomes a serious hazard,
due to the constant attention which is paid to its prevention.

Injury and death

The great majority of all cases handled by the doctor are injuries.
Injuries and death by accidents are by far the greatest health
hazards in the Antarctic. Since the first wintering over in 1898, up
to 1962, 88 persons have lost their lives in Antarctica (HEDBLOM[80],
and personal communication). Of these only 4 men have died of

disease (all before 1903). The other losses have been 25 in plane crashes, 16 drownings (6 from sea ice breakup, 6 from vehicles going over ice shelf or through sea ice, 4 from varied causes), 10 in fires, 9 of scurvy or starvation, 7 in crevasses, 5 frozen to death in blizzards, 3 climbing falls, 3 ship accidents, 2 run over by vehicles, and 1 each by explosion, asphyxia, metyl alcohol poisoning, and suicide.

Morbidity

In addition to the men wintering over, there are in the Antarctic during the light season also quite a number of air crews and the numerous summer visitors, including the ships' personnel. The medical aspects for these latter groups are somewhat different. For the airborne personnel there is the added risk of aircraft crashes, which have taken quite a toll as evident above. Antarctic flying is eight times more deadly than average Navy flying[80]. Health conditions in ships in antarctic waters[81], on the other hand, do not differ greatly from temperate regions, although the rate of injuries is higher. HEDBLOM[80] has analyzed complete medical statistics on the first five years of American Navy operations within the Antarctic Circle, including wintering groups, air crews, and Navy personnel. Of 10,000 sick calls the main causes reported were injuries (29%), respiratory (28%), dermatology (14%), and headache (7.5%). Neuropsychiatry was less than 1%. The main causes of admission to sick ward were injuries (35%), respiratory (18%), and genitourinary (12%). It must be noted that 88% of the genito-urinary admissions were for circumcision. As HEDBLOM remarks: "Where, better than Antarctica, for such a belated procedure?"

Dentistry

Probably the most usual single malady encountered in the Antarctic is teeth trouble, which is testified in reports from many expeditions. Both FRAZIER[142] and DALGLIESH[71] remark that the greatest amount of medical work at their stations were caused by teeth, and that a large number of temporary fillings were made. It has been believed[142, 71, 19] that the loss of fillings from teeth was caused by the difference of coefficient of contraction of tooth substance and filling material when exposed to severe cold.

Aware of this problem, the author[105] underwent thorough training in dentistry prior to leaving for the Antarctic, and had all expedition members go through a complete dental check up. During two years at Maudheim about 30 permanent restorations, including pulp treatment and root filling of two teeth, were performed on the fifteen men[104]. No marginal leakage through contraction was observed. There was no indication whatsoever that broken restora-

710

tions or secondary decay had been caused by the cold, nor were there complaints of toothache due to inhalation of very cold air. Ten extractions were also made, but none in the wintering group. Beginning in 1955 dentists have been assigned to the American bases at Little America V and McMurdo Sound. This has made possible clinical observations and dental research by trained dentists on a large number of wintering men during several subsequent expeditions. A number of reports [33,115,122-123,126-129a] from the period 1955 to 1959 have been published from these investigations. No support was found[123,129a] for the theory that cold would cause loss of restorations because of differential reaction between teeth and fillings. The restorations that did fall out, fell out because of some pre-existing condition such as secondary decay, undermined enamel, or trauma. Occasional fracturing that did occur was accounted for by a combination of two factors: rapid thermal change in the tooth caused by drinking hot liquid immediately after coming in from the cold plus trauma or masticatory pressure. A common effect of cold was toothache as a result of thermal conductivity through metallic restorations lacking or insufficient in intermediate cement bases. Alveolar osteitis ("dry socket")[123,162] was a problem during the first year and occurred once after every 4 or 5 extractions, an average of over 22%. During the second year[127] the men were not allowed to go out in the cold for 24 hours after an extraction and this reduced the incidence of alveolar osteitis to 4%. This is still considerably over the 0.9% as reported to be the national incidence in the United States.

STANMEYER & ADAMS[129] found that initial cariogenic activity continues despite the severe cold, and observed a measurable difference in the occurrence of new carious lesions between indoor and outdoor workers, the latter group having a slightly lower cariogenic rate. This is due to the decrease in the acid producing potential of oral bacteria in severe cold. This was determined by STANMEYER & ADAMS[128] by measuring acid production rates in dental plaques after carbohydrate rinses when working outdoors at temperatures as low as —50° C. The number of oral lactobacillae also decrease during these cold work periods[115]. PIDGEON[126a] found surprisingly low incidence of caries, even though in many cases the standard of oral hygiene had been low. Poor oral hygiene has repeatedly been reported on past expeditions. A good state of oral cleanliness is difficult to maintain unless a closely supervised oral hygiene program is established by the expedition dentist[122a].

PERLITSH, NIELSEN & STANMEYER[126] studied ascorbic acid and gingival health during wintering over. Determinations of the ascorbic acid plasma levels revealed that outdoor workers had a significantly lower plasma level, showing a progressive decline, while the indoor workers had a relatively high and comparatively

constant level. No significant difference existed in the gingival
tissue health between the groups. It was concluded that the ascorbic
acid intake of the outdoor group was insufficient to meet their
higher requirements during rigorous labor in extreme cold. CACCIA-
VILLANI[68] reported cases of gingivitis attributed to vitamin C
deficit, which were improved by large doses of vitamin.

Physiological adaptation

Man living in the Antarctic is faced by an unusual and stressful
environment. It is characterized by a great cooling effect of the
atmosphere, yearly alternating periods of continuous light or
darkness, intense solar radiation during the summer, prolonged
winter confinement in small station communities without women,
and a long period of isolation from the outer world. To this may be
added a strange and perhaps frightening surrounding with greater
risk for accidents, and even death. In many persons this situation
induces a vague, oppressive feeling of a hidden and unknown danger.

It is amazing how easily man adapts to this strenuous environment and how difficult it is to prove that an adaptation has occurred.
Man can live here for a considerable time, carry out hard muscular
work and travel long distances under difficult climatic conditions,
show great mental activity, and accomplish astonishing feats in all
fields of human interests. Yet there are no visible effects of acclimatization that can be clearly demonstrated and accurately measured.

For a very long time human adaptation in the Antarctic has been
synonym with adaptation to cold, and for many it still is. This is
quite natural as the dominant feature of this climate is the severe
cold. The early medical investigations were performed to find out
what bodily changes might occur while living in the Antarctic. The
changes that were demonstrated were usually attributed to the
effect of cold, or to the influence of the long polar night with absence
of light, and were regarded as a sign of acclimatization. Changes
were first sought in the elements of the blood.

Red cell count and haemoglobin

The nature of the "polar anaemia" frequently reported had long
interested the early expedition doctors. It probably originated as a
symptom of scurvy or other nutritional deficiencies, but was regarded as a separate form of anaemia peculiar to the polar regions
and caused by the long absence of light during the winter. COOK[11]
described the appearance of this condition (in Amundsen) at the
end of the wintering on the ship "Belgica" in 1899 as follows: "The
skin had a sickly, jaundiced colour, green and yellow, and muddy...
as if not washed for months".

Red cell counts and haemoglobin determinations by EKELÖF[73, 74] in 1902—03 and McLEAN[116] in 1912 showed that there was no support for the conception of such an effect of the dark period. EKELÖF reported the number of red cells to be about $6 \times 10^6/mm^3$ at the end of the second winter and repeated examinations of the haemoglobin showed no diminution during the dark season. McLEAN found that the number of red cells increased after the arrival in the Antarctic throughout the winter. At the end of the dark period the mean of the red cell counts was so high, about $7 \times 10^6/mm^3$, as to doubt the validity of the counts, especially as the colour index tended to be low with these counts. The low atmospheric pressures characteristic of the Antarctic cannot in any manner have contributed to this increase.

Monthly red cells counts by FRAZIER[142] during one year and WILSON[184] during two consecutive years demonstrated no difference from normal values at home for the same subjects, nor any change during the time in the Antarctic. In agreement were intermittent counts by MUTO[161] and by KAGEYAMA[151b, 183b]. VAN DER MERWE[51] observed a successive increase in red cells (attributed to the addition of vitamin C in the diet at the station), but his values at the arrival in the Antarctic and during the first months of the stay were indicative of an initial anaemia.

Determination of haemoglobin at Maudheim by WILSON[184] gave indication of a seasonal change with the highest values occurring during the polar night and a fall in the spring. The results of McLEAN, MUTO and VAN DER MERWE also showed an increase in haemoglobin values during the winter. There is thus no foundation for the myth of a "polar anaemia", on the contrary, higher haemoglobin values seem to occur during the polar night.

After two months in the Antarctic, PACE et al.[168] found haematocrit to be normal and not different from measurements made earlier in a temperate climate, indicating that there was no haemoconcentration. Only a moderate rise in the number of red cells was found by BARSOUM[130b] during normal chronic cold exposure in the Antarctic, but on acute exposure during midwinter there was a passing sharp rise in the red cell count, with concomitant rise in haemoglobin and haematocrit, suggesting a transient haemoconcentration.

Basal measurements of body temperature, respiration, central circulation, reflexes, sensory impairment

Investigations by McLEAN[116], LOCKHART[152], WILSON[185], MASSEY[153], GESINO & LEON[143], CACCIAVILLANI[68], PACE et al.[168], MILAN et al.[159, 160], MUTO[161], VAN DER MERWE[156], and VEGA et al.[103] have revealed no significant differences in the basal measurements of rectal[152, 168, 160] and oral[153, 183b, 103] temperature, respiratory rate

152,185,143,168,103, vital capacity[143,156], pulse rate[152,143,168,103], systolic blood pressure[116,152,68,143,168,169a,183b,103] and reflexes[103] in the Antarctic as compared to a temperate climate. Some changes related to these functions have been noted, though. PACE et al.[168] found that the diastolic blood pressure was appreciably elevated in the Antarctic. The significance of this change is not clear, especially as no such increase was observed by MUTO[161], KAGEYAMA[151b,183b] or by GESINO & LEON[143]. On the other hand, these investigators found the systolic pressure to be lower during the stay in the Antarctic. TICHOMIROV[182a,183a] reported a marked decrease in both the systolic and diastolic blood pressure at Vostok. He also found a gradual decline in the systolic pressure response to a standard work load. PALMAI[171,172a] observed that a seasonal variation in skinfold thickness, rising to a maximum in the winter and falling again in the spring, was accompanied by a closely parallel change in systolic and diastolic blood pressures. Increases in blood pressure were reported by FRAZIER[142] and BUTSON[135] in response to a cold stress. MILAN et al.[159,160] found the mean body, average skin, and foot temperatures to have increased during the winter at Little America V. Neither rectal nor finger temperatures were altered. MASSEY[155] reported that resting finger temperature tended to be lower after the prolonged exposure of the sledging season at the end of the year. There was a statistically significant difference between first and second year men in the first six months[153], the finger temperature being higher in the newcomers, but this was not evident in the latter half of the year[155].

PACE et al.[168] noted a substantial decrease in the visual flicker fusion frequency, indicating that some measure of sensory-cortical impairment or fatigue had occurred in their subjects in Antarctica. They also observed a higher threshold for painful electrical stimulus to the earlobe. The tendency was for a higher voltage to be required in Antarctica in order to elicit a painful response in the earlobe. Flicker fusion frequency tests have also been done by BROWN[131a].

Basal metabolic rate

The first attempts to measure the basal metabolic rate (BMR) were made by LOCKHART[152] and BUTSON[134,135], but their results were incomplete and equivocal[187]. Monthly determinations of BMR by WILSON[185] at Maudheim during two years with summer intermissions showed that the mean level of BMR in the Antarctic was not different from the mean of the same men measured in a temperate climate. WILSON also demonstrated a seasonal change in BMR with decreasing values during the inactive period of the polar night and a rise in the spring with resumed outdoor activity. His results have been confirmed by the recent investigations of

MUTO[161], VAN DER MERWE[156a], and GESINO & LEON[143] (personal communication from Instituto Antártico Argentino), who all reported nearly identical decreases during the winter period with a subsequent rise in the spring. The same seasonal fall and rise in BMR was observed at Vostok by TICHOMIROV[183,183a]. MILAN et al.[159,160], on the other hand, observed no change in the mean BMR of eight subjects measured in the fall, winter, and spring. All their mean values were remarkably low. Four of their men were characterized as the "outdoor group", and they were presumably engaged in more active outdoor work in the fall and spring than the "indoor group". The mean BMR values of the outdoor group were higher in the fall and spring and lower in the winter. Thus the results of MILAN et al. do not seriously contradict a seasonal change.

Body fat insulation

As previously mentioned there is a seasonal change in body weight, but over and above the seasonal variation there is also a general increase over the year[45,68,161,53,171,103]. WILSON[186] found the mean body weight to be higher during the second year in the Antarctic. The general increase was still evident after the return to a temperate climate, being 3 kg above the mean home value before the departure. LEWIS et al.[240] observed an identical weight gain of 3 kg in 2 years after a Greenland expedition. It was six times the annual increase in weight in a "standard" population in England. This might in some part be due to a gain in lean body mass as a result of the increased amount of physical work that is performed on a polar expedition.

LEWIS et al.[240] also found a close relationship in the seasonal changes of body weight and skinfold thickness in Greenland. This has also been observed by a number of investigators in the Antarctic[131,139a,148a-b,178a,192], who found that an increase usually occurs at midwinter and that the changes in body weight were significantly correlated with changes in skinfold thickness. Diverging results have been observed at other bases[55,80a,139d,153,171,172a], where no doubt different conditions have prevailed. As can be seen, measurements of body weight and skinfold thickness have now been conducted at a great number of bases, and it is extremely difficult to evaluate the varying trends unless the patterns of activity and life at the various bases are strictly comparable.

Changes in adipose fat composition have been studied by EASTY [139d], who found a clear tendency for the oleic acid to increase from the summer to the winter season during the first year.

Serum lipids

Monthly serum lipid levels were studied by EASTY[139c-d,130a] during wintering over. No seasonal trends were evident. Changes in the cholesterol levels and the α/β ratio suggested that they are more dependent upon levels of energy expenditure than on alterations of the dietary intake. Serum cholesterol is also being studied by HICKS[150] and MURRAY[160a].

The increase in body weight in the Antarctic is clearly associated with a thicker layer of subcutaneous fat, and because of the high insulating properties of fat, it may in effect mean a possible increase in body insulation. This may be an adaptation and contribute to the tendency of cold acclimatized men to wear less clothing. It may also be purely an effect of overeating. ROHRER[220] suggests that there is a compensatory factor involved in eating. The deprivation of more basic gratifications of an individual and boredom are possible factors responsible for compensatory eating.

Adrenalin and blood sugar

FRAZIER[142] surmised that there is a diminished tolerance to adrenalin in the Antarctic because of an increased amount of adrenalin in the blood stream in the cold. He concluded this from having observed repeated cases of adrenalin shock following the injection of local anaesthetics containing adrenalin. BUTSON[135] investigated the possibility of an increased sensitivity to adrenalin in the Antarctic by injecting adrenalin intravenously. In general no increased sensitivity was found in most of his subjects, although two subjects showed a marked reaction[134]. Only one subject was tested outside the Antarctic.

To investigate this further, an attempt to measure the blood adrenalin was made by WILSON[184] at Maudheim. Repeated determinations failed to demonstrate any measurable amounts of adrenalin by fluorescence according to the absorbtion method. No case of adrenalin shock was noted by WILSON in administring local anaesthetics containing adrenalin, nor has this been reported by dentists at the American stations in Antarctica[123]. Recently studies[150,183c,183d] on sensitivity to adrenalin and noradrenalin have been made, but the results are not yet available.

FRAZIER[142] has stated that there is a higher blood sugar level concomitant with increased secretion of adrenalin in the cold climate and quotes blood sugar studies by LOCKHART at Little America III showing an increase of 13% above normal. BUTSON[134,135] reported that the average fasting blood sugar determined once in nine men in the Antarctic was 109 mg%, as compared to an average of 99 mg% in a set of six controls in a temperate climate. Only one subject was the same in both climates. WILSON[184,189] measured the fasting blood sugar of twelve expedition members at home and each month during two consecutive years in Antarctica. The mean level in the Antarctic was 107 mg%, which was about 9% higher than the mean level found for the same men in a temperate climate before and after the antarctic sojourn. This difference was found to be statistically significant. Blood sugar tests made by WILSON in five men during several months of sledging in the interior showed also a statistically significant difference between the blood sugar levels at the base and in the field, being higher while sledging. During the second month in the field (February) the fasting blood sugar rose to a peak, approximately 27% above base level, then fell again. On arriving back at the base Maudheim, the level was still about 11% higher, but soon attained base values. The nature of this increase is difficult to evaluate. In the blood are present other reducing substances (such as glutathion, creatinine), which are included in the determination of blood sugar by the reduction method. Whether it is a true increase in blood sugar that has been demonstrated, or whether it has been caused by the increase of another fraction in the blood, remains to be investigated. PACE et al.[168] determined plasma glucose levels in seven men in a temperate climate and again within two months after having arrived in Antarctic. No appreciable difference was found, the values being about 75 mg %. However, WILSON[184] found that the blood sugar level on arriving in the Antarctic was low and that it did not rise until after several months' stay.

White cell counts

Regarding the *total leucocyte count* there is some conflicting evidence. McLEAN[116], FRAZIER[142] and WILSON[184,189] observed a clearly evident fall from a temperate climate to the Antarctic, with a subsequent rise during the stay to (approx.) previous values. McLEAN and WILSON found this increase to take place fairly rapidly during the first three months, while FRAZIER's counts increased irregularly during the whole stay. A rather marked increase during the first months was also noted by BARSOUM[130b]. MUTO[161], VAN DER MERWE[156], KAGEYAMA[151b,183b], and PACE et al.[168] observed no such initial fall. With the exception of FRAZIER's successive increase and the rise noted by BARSOUM, no significant trend in leucocyte levels during the stay in the Antarctic was demonstrated by any of the investigators, including an Argentine study at Ellsworth Station (personal communication from Instituto Antártico Argentino).

Generally it can be said that the mean level of the leucocyte count does not seem to be significantly different in the Antarctic, although VAN DER MERWE and MUTO at the return to a temperate climate found higher control values at home than prior to leaving for the Antarctic.

Diverging findings have been reported from Vostok (altitude 3400 meters) by TICHOMIROV[182,183,183a], who found a leucopenia, and accordingly lymphopenia and neutropenia, during the winter and the spring. TICHOMIROV regarded this as a result of the absence of bacteria in the air and in the food, and an increased strain on the haematopoiesis at high altitude, and an inadequate amount of ultraviolet light.

Conflicting evidence concerning white cell differential counts is presented by several investigators in preliminary reports. WILSON [184,189] was the first to make regular differential counts in the Antarctic. The preliminary calculations indicate no change throughout the stay in the mean monthly level of the calculated *total number of lymphocytes* (per mm³) made in the afternoon in twelve subjects at Maudheim. There was no significant difference in the mean level of five men during several months of sledging. The mean total lymphocyte level was moderately lower than at home in a temperate climate. MUTO[161], VAN DER MERWE[156], KAGEYAMA[151b,183b] and BARSOUM[130b], on the contrary, reported slight to moderate increases during the stay in the Antarctic in the total number of lymphocytes, which was higher than control level at home or on the ship.

In regard to the calculated *total number of neutrophils* (per mm³), WILSON's preliminary values indicate no clearly significant changes in the mean level of the afternoon counts throughout the stay at Maudheim, although there was a tendency for the neutrophils to vary inversely with the outdoor temperature. During sledging the total number of neutrophils was not increased above the level at the base in the beginning of the summer period, but at the end of the field season and at the return to the base the level was moderately higher. There was no difference in the mean values in the Antarctic and at home. VAN DER MERWE[156] and MUTO[161] have not reported any significant changes in the total number of neutrophils during the antarctic sojourn. While little difference was noted between control values at home prior to leaving and the mean level in the Antarctic, a marked increase in neutrophils was observed by both investigators at home after the return. It is possible that this was an effect of returning to a warm climate (S.Africa, Japan). KAGE-YAMA[151b,183b] noted a decrease during the antarctic winter.

Because of the normal diurnal variations that occur in the number of the different types of leucocytes, it is essential to know at what time of the day the counts have been made. Divergent times of the sampling might account for some of the discrepancies found between the results of the different investigators. No information concerning this important factor is available, except for the counts made by WILSON[184]. His total counts were made in the morning as well as in the afternoon. Blood smears were taken in the afternoon during the whole stay, and also in the morning during the last winter and spring. Throughout the whole period at Maudheim the total leucocyte count was always higher in the afternoon. In comparing the *morning and afternoon differential counts*, WILSON found from

718

his preliminary values that the total number of lymphocytes were always lower in the afternoon, except for the last month, January, which was the warmest to occur. The same was the case with the eosinophils. Normally the total number of lymphocytes rises during the day in a temperate climate. In January and during the return home to a temperate climate the condition was reversed to normal. The total number of neutrophils on the other hand was significantly higher in the afternoon than in the morning in the Antarctic, being solely responsible for the afternoon increase in the total leucocyte count. Investigations by SIMPSON[178] have shown that there is a depression in the total number of eosinophils during the day in the Antarctic as compared to the usual diurnal rise during the afternoon in a normal environment.

Thus there seems to be an indication of a neutrophil leucocytosis with a relative lymphopenia and a depression in the eosinophil count in the afternoon in the Antarctic. If one attempts to explain these findings it is most natural to suggest that the changes might be indicative of a stress reaction. It is well known that stress is accompanied by a similar leucocyte response. This would mean a tendency towards an increased adrenal cortex secretion in the afternoon, while in a temperate climate there is normally a declining adrenal secretion during the day. The main factor in eliciting such a response would presumably be the cold, as the effect appears to be less marked during the warmest months. Another factor could be hard physical work, but such is not performed to any great extent throughout the polar night.

During two years in the Antarctic SIMPSON[177,178] made a study of *eosinophil counts* to investigate the stressing effect of the environment. He found no significant difference in the total number of eosinophils (per mm^3) in winter and summer at the station, but during the working day in the Antarctic he observed a depression in the number of eosinophils. Living in a tent showed significantly lower levels than in the base, and the day after sledging the level was still lower. During a manhauling sledging journey with counts made every second day he observed a chronic eosinopenia for the 18 days, with a prompt recovery afterwards. These diminutions in the eosinophil counts were associated with a physical stress. Marked eosinopenia was also found by SIMPSON in individuals under periods of mental strain, as when travelling over dangerously crevassed areas, having the responsibility of a difficult leadership, and receiving distressing or exciting messages over the radio. Diminutions in the eosinophil counts thus seem to be associated with both physical and mental stress in the Antarctic, but to what extent the cold contributes to the stress is difficult to evaluate. KAGEYAMA[151b,183b] noted a slight depression in the number of eosinophils, while BARSOUM[130b] found an elevation.

Adrenocortical activity

PACE et al.[168,166,167] measured the steroid levels in plasma and the excretion rate in urine of twelve men in a study of the physiological stress in men working in the Antarctic during Operation Deepfreeze I. Tests were made prior to leaving the United States and after one to two months' stay in Antarctica. They found in the Antarctic an elevated plasma level of free 17-hydroxycorticoids, and a moderately but significantly elevated excretion rate of 17-ketosteroids and 17-hydroxycorticoids in the urine. Although neither of the increases were large, the results taken together were regarded as a clear indication of the presence of moderately heightened adrenocortical activity in the subjects, when tested in Antarctica. Cold was by no means the main stress factor as the temperature was not especially low during this period[167].

STAQUET[179a-b] investigated the variation in 17-hydroxycorticoids and 17-ketosteroids in the urine during the Belgian Antarctic Expedition. He found that the men living at the base had a normal steroid excretion. In men staying out of doors for some time at low temperatures but not engaged in physical work, the excretion of steroids was diminished. Hard physical work did not significantly lower the steroid excretion. His findings merely indicate a change in the steroid metabolism in the cold, and do not support an assumption of the occurrence of an increased adrenocortical activity in the Antarctic. KAGEYAMA[151b,183b] found no difference in the amount of urinary 17-ketosteroids in the Antarctic as compared to Japan, and the seasonal variation followed the same pattern as found in Japan.

In addition to the determination of adrenocortical steroids, PACE et al.[168,166,167] studied many other physiological parameters in the blood (sodium, chloride, urea, glucose) and in the urine (phosphate, chloride, sodium, potassium, ammonia, urea, uric acid, creatinine, glucose) at home and in the Antarctic. These measurements were designed to assess adrenocortical function in the antarctic personnel and were an attempt to establish a pattern of response related to the stresses met in the Antarctic. It is evident from the findings that only a moderate degree of physiological stress was detectable in the men examined. The results of these metabolic patterns or profiles from blood and urine constituents were more indicative of the physical work stress during the base building than of cold exposure.

Studies of saliva by DAVIES[139a] have not indicated any seasonal change in mineral corticoid production.

Thyroid activity

Little is known about the changes that might be induced by the

720

cold climate on the function of the thyroid. Seasonal variations in the basal metabolic rate indicate that there might be an effect of cold on the thyroid activity. Studies of protein-bound iodine have been conducted by STAQUET[180] and at the Argentine Ellsworth Station, but no results are as yet available.

Diuresis and urinary excretion

There is evidence of a somewhat higher level of diuresis[146,156a,183b] in the Antarctic, which may partly be accounted for by a decreased sweat secretion[156a], and partly by cold diuresis[146], although it might be expected that adaptation to cold may decrease the diuretic response to cold. In the field the cold diuresis may be counteracted by a dehydration due to an increased respiratory water loss in the dry, cold air and a decreased water intake because of difficulty in supplying drinking fluid. All water has to be obtained by melting snow and thus is not readily available. Measurements[162,179a] have shown that urine outputs below 1000 ml/24 hrs are often found when sledging and living in tents.

WYATT[192] concluded from his investigations that the occurrence of dehydration during sledging was only a temporary phenomenon and that water intakes were usually adequate to maintain normal urine output. His results suggested that cold diuresis occurred only rarely during journeys. Further studies by DAVIES[139a] showed that fluid intake and urine volume decreased during sledging. The changes in volume and electrolyte composition did not give evidence of a cold diuresis but were regarded as an effect of exercise. Measurements of specific gravity of early morning urine made by ORR[165] support these results.

VAN DER MERWE & HOLEMANS[156a] have reported a continuous decline in the excretion of urinary nitrogen over ten months in the Antarctic. The regression of the urinary nitrogen on time was found to be statistically very significant. There was no parallel decline in the creatinine values. They had no acceptable explanation for this continuous decrease in nitrogen catabolism with time.

PACE et al.[168] observed no clearly significant differences in the excretion of the urinary constituents determined in the Antarctic, except for the adrenocortical steroids mentioned earlier. The diurnal variation in urinary excretion rates was also examined in the Antarctic by PACE et al.[168]. They concluded from their results that despite continuous daylight in Antarctica, normal physiological diurnal variation was generally maintained. This is in agreement with the results of LEWIS & LOBBAN[237], who carried out a thorough study of excretory rhythms on abnormal time routines during the polar day in the Arctic.

Diurnal rhythms during polar day and night

The pattern of light and darkness undergoes marked cyclic changes and must have a more profound effect on the physiology of man in the Antarctic than has been generally recognized. The disrupted sleep rhythm and the resulting irregular pattern of rest and activity, which is common in the Antarctic, is likely to affect the diurnal rhythm of many bodily functions. Although the excretory rhythm for some of the urinary constituents is not markedly affected, dissociation of various diurnal rhythms occurs during abnormal time routines, as shown by LEWIS & LOBBAN[238] in the Arctic. The diurnal rhythms which are most easily changed are heart rate, blood pressure, and body temperature.

Gradual shifts in the diurnal rhythm of the oral temperature measured 8 times daily at Vostok were observed by TICHOMIROV[183a].

OGATA[164] found that in the antarctic summer the oral temperature curve had a tendency to become flat and to be slightly higher.

PALMAI[172] demonstrated a seasonal rhythm consisting of a rise in oral temperature in spring and summer and a fall in autumn and winter, in addition to the diurnal variation.

The changes in sleep and wakefulness patterns on a polar expedition have been studied by LEWIS & MASTERTON[239] in the Arctic. They found that there was a disruption of the normal pattern both in the periods of continuous darkness and continuous light. In the spring there was a return to the more usual pattern, but this was once again disturbed during the summer, when much of the activity went on in the midnight sun. The time of going to sleep and of getting up varied greatly, especially during midwinter and midsummer, and without special preference to a particular hour. There was no great difference in the average number of hours of sleep between months of outdoor activity and months when the men were confined to the hut. Though the men were at liberty to sleep almost as long as they liked, the mean duration of sleep was 7.9 hours.

Social factors are of great importance in determining sleep patterns. Under a strictly regulated daily routine in the Russian Arctic SEMAGIN[243] found that the diurnal sleep-waking rhythm was not determined by the light factor, but that the deciding factor for man is his social relationships.

Sleep rhythms have been studied in Antarctica by GOLDSMITH[227], GRAHAM[147, 148b] and EVANS[141a]. The mean duration was not found to differ during the various seasons of the year, and the average sleep per night was 8.2 hours. The disruption of sleep rhythm that occurs during the periods of continuous light and darkness varies considerably from base to base, and is markedly affected by the social organization. When some of this social pressure was released by the departure of the base leader at one station, the average nightly sleep went up to above 9 hours[147].

Metabolic and thermal responses to a standard cold stress

MILAN et al.[159,160] investigated the metabolic and thermal responses of men exposed nude for two hours to a standard cold stress (17° C) after varying intervals of time in the Antarctic. They found a significant decrease in the metabolic rate response to the same thermal demands of the cold stress from autumn to winter, with a return towards original levels in the spring. The same trends were seen in the temperature measurements, which showed increased mean skin and foot temperatures. It was suggested that this was an acclimatization response, confirming the statement of CARLSON et al.[233] that "adaptation in human beings involves a change in body economy whereby the individual maintains a relatively lower level of metabolism for a given heat loss".

A similar antarctic study was conducted by WYNDHAM et al.[192a-b], who investigated the same responses but at a range of cold stress temperatures (5°, 10°, 15° and 27° C). At 27°C the metabolism was the same as in unacclimatized controls, but at the lower air temperatures there was a reduction in metabolic rate as compared to controls, and toe temperatures were higher.

A different thermal response to a similar standard cold stress (10° C) was demonstrated by BUDD[132a-b]. His investigations made in Australia and during different antarctic seasons showed a fall in rectal temperature during the experimental cold exposure in the temperate climate, but an increase in the Antarctic in response to the cold stress. Metabolic rates were measured during the exposure, but the results are not yet available. This increase in temperature was not observed by MILAN et al. nor by WYNDHAM et al. BUDD regarded the enhanced ability to maintain deep body temperature in response to the experimental cold exposure observed during the antarctic stay as an effect of acclimatization, which was lost again at the return to Australia. BUDD's study is at the present being repeated and extended[133a,183c].

Finger numbness

From the various investigations described above it can be seen that no clearcut and unequivocal evidence of a physiological adaptation to the climate has been presented. No definite change in the *general response* of the body to cold has been demonstrated. Better evidence has been obtained of *local acclimatization* of the hands to cold.

In the Arctic MACKWORTH[241] demonstrated that the hands of

persons adapted to living and working at low temperatures did not become as numb on exposure to cold as the hands of unadapted persons. An investigation in the Antarctic by MASSEY[153-155] confirmed these results. A group of new arrivals in Antarctica and a group staying a consecutive year were observed by MASSEY over a period of one year. Initially there was significant difference of cold acclimatization in finger sensitivity between the two groups, which slowly diminished until there was no evident difference after 6 weeks. Thus there seems to be a clear indication of a local adaptation to the climate, but it remains to be investigated how much of this change improves the ability to work in the cold. Two-point discrimination tests have recently been conducted by SPARKE[178b]. The preliminary results do not seem to be quite in agreement with MASSEY's findings.

Diverse hand reactions

Changes in the rate of heat elimination from the hand was investigated in four subjects by HAMPTON[149a-b,151c]. Soon after the arrival in the Antarctic there was a marked decline in the heat eliminated, but later in the season, when the weather got considerably colder, the heat elimination increased and reached a peak at the coldest time of the year. It seems as if the tendency to cold vasoconstriction was released by acclimatization. But the evidence is equivocal as the increased peripheral blood flow persisted only in two of the subjects while in the other two it decreased to pre-exposure levels. Further work has been done by SPARKE[178b] on cold vasodilatation in finger, but has not yet been published.

Changes in palmar sweating were studied by DAVIES[139a-b], who concluded that the responses were more indicative of emotional stress than evidence of exposure to cold.

Nail and hair growth

One of the very first physiological investigations to be conducted in the Antarctic was MCLEAN's study[116] of the rate of nail and hair growth. He concluded from his observations that a nail appeared to grow ten times slower and hair almost forty times slower in the Antarctic than in a temperate climate. Nearly 40 years later SAPIN-JALOUSTRE[29,175,176] made a follow-up study of MCLEAN's observations. He found that the slowing down of the growth of nails and hair was very pronounced during the first weeks of the stay in Adélie Land, when the men were working and sleeping without heating while the base was being built. No exact determinations were made during this time, however. The measurements were conducted in the winter and showed the speed of growth for both nails and hair in Adélie Land to be six-tenths of the speed in France. The differences between the results of the two studies are considerably reduced if the same reference standard of growth rate in a temperate climate is used. The slower rate of growth in hair observed by McLean is explained by the considerably lower hut temperatures prevalent on his expedition. Jones[151,151a] also found a slowing down of hair growth from the chin during the first summer in the Antarctic, which was of the same order as observed by Sapin-Jaloustre, but the rate increased during the winter to values above those found in a temperate climate and the decrease next summer was not significant. Massey[153], using measurements made by Precious[173a-c] in the Antarctic, compared the nail growth of men staying a second year in Antarctica with that of the 1st-year men and found no significant difference in the rate of growth at any time of the year.

A similar investigation, carried out by Geoghegan et al.[235] under less severe conditions in the Arctic, showed the speed of nail growth to be nine-tenths of that in a temperate climate. It was found to be a statistically highly significant difference.

The slowing down of the growth of nails and hair in a cold climate is a sign of diminution of the metabolism of the body's surface. The decrease is related to a lowering of the surface temperature and may be due to the normal physiological adjustment of peripheral circulation to climate. The author noted at Maudheim frequent and intermittent changes in the appearance of the finger nails with uneven grooves and nodular irregularities, that did not occur at the same time in the different expedition members. A plausible explanation is that these changes were caused by intermittent frostbites or numbing of the fingers with reduced or altered circulation to the finger nail matrix. This observation is consistant with the effect of temperature on the rate of nail growth that has been demonstrated in cold climates.

Wound healing

It has generally been observed in polar regions that superficial wounds and cuts seem to take longer to heal. This could be a consequence of lower skin temperature and the diminution of the metabolism of the body's surface with decreased peripheral circulation. Recently Catty[136a] investigated the rate of wound healing in the Antarctic by use of experimental wounds. Control experiments are made on soldiers at home. Preliminary results indicate no significant difference in healing time.

Cold exposure time and sub-clothing temperature (micro-climate)

In attempting to obtain evidence of acclimatization to cold in men residing in the Antarctic, one is likely to forget to investigate and measure the actual degree and the duration of the exposure to the cold. In adapting to a very cold climate man makes use of modern technology and effectively protects himself with adequate clothing and well isolated base huts. With proper equipment he can live and travel and much of the time maintain a subtropical micro-climate under his clothing. Living at a wintering station in the Antarctic is not necessarily the same as cold exposure.

NORMAN[162,163] at Halley Bay, and CUMMING[138] at the Argentine Islands off Graham Land, have conducted studies of exposure time, sub-clothing temperature, and the climatic conditions in which men live at static bases. No long sledging journeys were made at these stations. The subjects were studied throughout the year for a day each month. A complete time and motion record was kept of activities and time spent indoors and outdoors. Each subject wore a vest consisting of a resistance thermometer woven into a garment so the sub-clothing temperature could be measured at frequent intervals. The micro-climate temperature beneath the clothes, which is almost identical with skin temperature[130], was found in general to remain high, averaging 32.2° C at Halley Bay and 32.5° C in the Argentine Islands. NORMAN[162] observed that the sub-clothing temperature was maintained at a lower level during sledging journeys than was typical at the base, showing a shift from 33—32° C at the base to 25—23° C in the field. BUDD[133b] has also reported lower skin temperatures beneath the clothing during field work. The results of NORMAN and CUMMING showed that the average time spent outside was 13% in the summer and 5% in the winter at Halley Bay. Only 1% of the time was spent at a temperature below —25° C. In the Argentine Islands, outside time was 15% in the summer and 8% in the winter. Though the time out of doors was related to outside temperature, it was more closely related to wind chill. As soon as the wind speed reached 8 m/sec., outside activity usually ceased. Even at bases were considerable travelling occurred, the total time spent outside hut and tent did not amount to more than 15—17%, as shown by WYATT[192] and DAVIES[139b], although the proportion of time spent outside the tent on sledging days averaged 33%.

Similar exposure times were observed by MILAN et al.[159,160] at Little America V. A group of scientists spent 13—15% of the time out of doors in the fall and spring, but only 6% in the winter. A group of Navy men acting as support personnel spent more time outside, averaging 20% of the total time in the spring.

MILAN[157,158] investigated the thermal balance in two well-clothed men at Little America V in the winter. The men were active out of doors for periods of 44 to 165 minutes' duration at low temperatures (—32° to —47° C) with wind chill indices ranging from 1220 to 2370. Despite the protective clothing worn and relatively high heat productions, the total heat debt was fairly large, and MILAN concluded that the men were moderately cold stressed in spite of high rectal temperatures. There was a fall in mean skin temperature from 32.5° C to about 26.7° C.

The results indicate that at static bases, even when some travelling is done, the time spent out of doors is short and is mostly related to wind chill. The men are well protected and maintain a fairly warm micro-climate under their clothing, although the thermal demand of the environment may be relatively high even on seemingly adequately clothed men. Under such circumstances the degree of cold exposure is probably not large enough to induce an adaptation response of the body.

However, at many antarctic stations where active field work is carried out, much longer and much more severe exposure is experienced by the men taking part in the journeys into the interior. At Maudheim, for instance, nine men were engaged in the sledging trips inland[13]. During a period of one year the author spent 41% of the days sledging and living in a tent. Most of the other field men spent even longer periods of their time sledging. Two men were in the field for more than 330 days of the 694 days in the Antarctic, which is 48% of the total time. Three men were away from the base for 163 consecutive days of dog sledging inland and arrived back at Maudheim at the end of May, when the sun had been below the horizon for almost 2 weeks. DAVIES[139b] states that he spent 40% of his time in the Antarctic away from base. Under these conditions it is clear that the men are exposed to the cold of the climate to a much greater extent than at a static base.

Much more extreme climatic conditions have been encountered by the men at the intra-continental stations, such as the South Pole Station and the Vostok Station. Especially during the penetration with land convoys deep into the interior to establish the Russian inland bases, extremely difficult and almost unbearable conditions were experienced[22].

Influence of altitude

TICHOMIROV[181-183a] has reported on physiological studies made at the Russian intra-continental station Vostok in 1959. The climatic conditions in this region are extreme with an annual mean temperature of —55° C. In the winter —80° C was often encountered. Wind speeds around 5—10 m/sec. were common, and calms were very rare. Even in special clothing it was only possible for the men to work out of doors for very short periods. In extreme cold no one was allowed to go out alone. Length of time outside was limited to 15 minutes when colder than —80° C, and to 10 minutes when colder than —85° C. The altitude at Vostok is 3400 meters, but owing to the low atmospheric pressure characteristic of Antarctica regardless of altitude, the oxygen tension prevalent at this altitude corresponds to about 4000 meters in other parts of the world.

In general the results of TICHOMIROV's investigations[182-183a] at Vostok agree with the physiological changes normally found in adaptation to high altitude of this order. There was the expected increase in haemoglobin, in number of red cells and in blood viscosity. The initial increase in heart rate and pulse pressure, and the usual respiratory changes were observed. The mean values for the partial pressures of the alveolar air was found to be 55 mm Hg for oxygen and 28 for carbon dioxide at a mean atmospheric pressure of 468 mm Hg. The resting haemoglobin oxygen saturation was 82—86%. There was a marked loss of body weight and a lymphopenia, as an effect of the high altitude stress. During the stay at Vostok the heart rate returned to normal after about 4 months, and the respiration rate and the respiratory minute volume decreased. But even at its minimum, the tidal volume remained above normal, and was about 1 liter per inspiration. The rather marked fall observed in systolic and diastolic blood pressure may be taken as a sign of increased physical fitness, as the return to normal in the rate of recovery of pulse rate and blood pressure. No change was observed in sedimentation rate of red cells. The seasonal changes in basal metabolic rate and body temperature have been mentioned earlier.

Thus the investigations by TICHOMIROV have produced evidence of an acclimatization to altitude, but have shown no marked effect of cold upon the physiological functions of the body.

The human laboratory

Taken together, all the physiological investigations conducted in the Antarctic have not yet produced any convincing evidence that a definite process of physiological acclimatization will take place in man residing in Antarctica. Man evidently adjusts adequately to all climatic conditions encountered here by bringing with him his own micro-climate. With help of artificial protection and innumerable contrivances man has successfully broken through the impressive climatic barrier of this forbidding continent and has ventured far into the interior of an extreme environment, which may only be surpassed by space.

However, there is some positive evidence, showing that at least local adaptation to cold occurs in the hands of man, and it is probable that similiar changes occur in the feet and possibly in the face. There are also positive findings indicating that minor physiological adjustments take place in the body in favour of an adaptation to the climate. The demonstration of a tendency to wear less clothing and the observation that sleep in severe cold gradually improves, indicate a changed response to sensory stimuli. This taken together with the evidence of less numbing of the adapted fingers point to an improvement of man's ability to work and live in the cold.

Other consistent findings are the seasonal changes observed in body weight and subcutaneous fat thickness and in basal metabolic rate during the winter period. It is probable that these are mainly due to social rather than climatic factors, although an adaptive mechanism cannot be excluded. The changes are only indirectly caused by the environment, in a sense that outdoor activity, exposure to cold, hard work, etc., all show a seasonal pattern owing to the specific nature of the antarctic climate and the periodical shift in polar day and night.

After discussing and summarizing all the work and effort that has gone into showing what goes on when man acclimatizes to the antarctic environment one is apt to forget that there might also be a process of reacclimatization when man returns from this region to the temperate or warm environment at home. In fact hardly any work has been carried out to investigate this and very little information of what happens is available. Observations by the author on the return trip from Maudheim after two years in the Antarctic show that some changes occur, at least in the white cell count[189], in blood sugar[184], and in body weight[186], but the changes are far from spectacular. As a rule the reacclimatization sets in quietly and seems to last 2—2½ months after return from the Antarctic[169a].

Work continues in the Antarctic in an attempt to obtain evidence of physiological adaptation to cold in man. The results of many studies have not yet been published although a number of preliminary reports are available. Work going on includes studies of microclimatology[149b,139c], skin temperature of hand (personal communication from British Antarctic Survey), cold vasodilatation of finger and two-point discrimination[178b], metabolic and thermal responses to cold stress[133a,183c], thermal sensation and clothing[133b,150], haematology, sera electrolytes, and blood chemistry[124,126], sensitivity to adrenalin and noradrenalin[150,183c,183d], protein-bound iodine[180], dietary fat and serum lipids[150,160a], rate of secretion from sebaceous glands (personal communication from British Antarctic Survey), hair growth[151a], wound healing[136a], vitamin deficiency[183d], electrocardiographical studies[72a], flicker fusion frequency and stress [131a] and upper respiratory infections[120]. It is hoped that the accumulation of data from these and future investigations carried out on antarctic expeditions will help to clarify the problem of how man adapts physiologically to this cold climate.

It is of great importance to keep in mind that the most valuable aspect of physiological investigations in this region is not just to demonstrate whether acclimatization to cold takes place or not, but to study many other processes going on over an extended period, as diurnal rhythms, sleep patterns, endocrinological shifts, energy balance, activity studies, bacteriology and virology, stress factors, effect of isolation, and related subjects. These can effectively be studied for an unbroken period in an isolated group of men in a field laboratory where environment, food, and other factors are equal and can be controlled. In fact, the investigator is able to observe and live with his subjects continuously throughout the whole investigation. An antarctic expedition offers an unequalled opportunity for controlled long time studies. GOLDSMITH & LEWIS[227] have aptly labled "polar expeditions as human laboratories".

Selection of men

The selection of the wintering groups is a problem much discussed by expedition leaders and doctors from the beginning of exploration. A chapter on the selection of their men can be read in most accounts of the early antarctic expeditions. Also many doctors in their medical reports[76] dwell on the factors governing the choice of men and how they stand up to the stress of the long polar night. SMITH[221] has discussed the different opinions as to the personality characteristics previously looked for in these men. For the most part the members of previous expeditions have been chosen by their leaders. Generally they picked men they knew or who had proven themselves, or relied on recommendations of trusted friends. In some cases, only a few applied with the right kind of training, just enough to fill the various positions in the party. This has been the procedure up to recent times.

With the IGY program in the Antarctic the conditions have changed. A great number of bases, some of them quite large with more than a hundred men, are maintained year after year, with an annual exchange of personnel. While the scientific program at the station almost exclusively is carried out by civilian scientists, a great part of the personnel responsible for logistic support at the American bases are provided by the Navy, for example technicians, doctors, cooks, etc. Two leaders are appointed, one scientific and one military. The arrangement with split command and two organizational subgroups of different composition frequently is a source of conflicts and group disruption[201, 204, 214]. The Australians advertise for new men on all posts and run the station as a classless society. The men for the Australian stations are chosen through a personal interview[200], the American personnel is selected by a selection board that base their ratings on questionnaires completed by candidates' references[221]. Now, not only new men but also leaders have to be appointed. The selection of leaders is a new and important problem. The morale and performance of a wintering group often depends on the leadership style[193b,207].

The great number of men going to the Antarctic has increased the need of neuropsychiatric assessment in the selection of personnel for the bases. Such a program has been developed by the U.S. Navy and is described by NARDINI, HERRMANN & RASMUSSEN[206]. From their report it appears that the program has been effective in identifying and eliminating individuals who would have developed serious or incapacitating emotional illness under the stress of isolation, as no documented psychiatric break-downs of psychotic proportion have occurred since the initial psychiatric screening.

It is not difficult to name the desirable and undesirable characteristics of antarctic personnel[193b]. OWENS said at a conference of the Australian Branch of the Psychological Society in Perth, August 1962, (quoted by LAW[200]):

"If it were possible to select only men who were competent, goal directed, versatile at tasks, diverse of interests, physically capable and energetic, independent but considerate of others, aware of the need for organizational structure and co-ordination, of even temperament, insensitive to criticism, reliable and responsible, yet without any inadequate or inappropriate defence mechanisms or tendency to narcissistic character disorder or rigid intolerance, our problems in selecting men would be almost completely resolved, even those of social interactions."

The difficulty lies in designing screening and rejection tests which are reliable and show a significant positive correlation between the initial psychiatric prediction and subsequent performance[195,196,211b]. A study of the predictive relationship of test data with criteria of adjustment to the antarctic conditions was made by WEYBREW, MOLISH & YOUNISS[223a-b]. According to their findings the characteristics of those men who adjust most adequately are: to have high intelligence test scores, to have control of hostile and aggressive impulses, to be single and over 25 years of age, and to have less than a college level of education.

Adjustment and psychological problems

It should be emphasized that spending a year in the Antarctic gives rise to different types of adjustment problems than those that have been reported for the Arctic. This difference is due to the fact that men in the Antarctic have no hope of escaping from the time the last plane or ship leaves for the winter, whereas in the Arctic there has always been the opportunity in case of emergencies for a plane to evacuate men from these stations. This "total commitment to isolation" for a definite period of time that exists in the Antarctic, is of critical importance in understanding the psychodynamics of the men so isolated[220].

A considerable amount of research on the psychological adaptation of men in small antarctic stations has been carried out, especially by American investigators. The research findings agree in outlining the general stress factors of station life, the emotional responses that ensue in individuals and groups, and the general criteria of psychological adaptation.

Mental stress factors

GUNDERSON & NELSON[197] describe the most stressful elements of wintering as follows: "1. *Confined isolation*: geographical, social, and emotional remoteness with limited space and absence of an opportunity to withdraw or escape from the situation. 2. *Continuous presence of same associates*: continuous proximity of others with lack of interpersonal choice; knowing that one must get along with others. 3. *Tension control*: necessity of controlling aggressive and emotional impulses; inability to relieve anxiety; lack of heterosexual object. 4. *Boredom, monotony*: sameness of physical surrounds, faces, work tasks, conversations; lack of stimulus variety. 5. *Physical hardship*: hard and heavy work, cold weather, darkness, certain food deprivations, having to work to attain minimal standards for health and safety. 6. *Status limitations*: status leveling, role overlap, lack of immediate status rewards."

The stresses of isolation appear to have a far greater impact on individuals at small stations (12—40 men) than at large stations (more than 50 men)[214].

Psychological reactions

Responses and emotional states characteristically observed among station personnel from different countries are summarized by NELSON[209] as follows: "(a) an experience of increased insomnia during the winter months, although this may be of a more qualitative than quantative nature in terms of hours of sleep actually obtained; (b) an increased incidence of headaches during the winter months, which may be as much a function of the indoor lighting and gas fumes as of psychological origin[202]; (c) an increased desire throughout the year for privacy; (d) a tendency to exaggerate experiences which in a normal environment might have gone unnoticed; (e) a tendency for aggressive feelings to be pent up; due to knowledge of the disruptive outcome of overt aggression this may be released through hard work, through form of verbal interaction and other types of abreaction; (f) a tendency for group members to shut themselves off psychologically from the outside world during the winter months; (g) a tendency for group morale to be lowest during the midwinter period; and (h) sexual deprivation does not appear to be a problem with a disruptive influence on station life in

732

so far as can be determined by manifest behaviour[211b, 224]." MULLIN & CONNERY[204] in their study did not find that isolation from women in itself was a serious problem for any of the men wintering over. Sex dreams, nocturnal emissions and masturbation showed variable frequency relationship. Increased frequency was more apparent during periods of relative inactivity and personal emotional stress, less than usual frequency occurred among the most preoccupied. There was no evidence of any overt homosexual practice, scandal, or gossip. PERRIER[212a–b] has discussed the question of homosexuality in the Antarctic and the problem of detecting latent homosexuals in the selection of men.

The question has often been raised, whether or not women should be included in the wintering parties. Irrespective of the practical disadvantages in accommodation, toilet facilities, privacy, etc., the presence of women would probably have a more disruptive influence on station life than a beneficial psychological effect, judging from experience. Two women have once wintered over on an antarctic expedition, which became noted for this incident. One has only to read the accounts written by one of the two women[12] and by the other's husband[18], to understand why this venture has not been repeated. The Russians have taken women scientists on summer expeditions to the Antarctic, but there is no record of females in their wintering groups.

ROHRER[219, 220] in a discussion of problems of human adjustment experienced by men wintering over, reports the occurrence of adjustive behavioral phenomena, apparently unique to life in Antarctica. One was that of increased sensitivity to auditory stimulation, which has been described rather vividly by the expedition leader GIAEVER[13]. The other has been labeled the "Long Eye" phenomenon; a condition (of "a twelve foot stare in a ten foot room") produced by depriving an individual of interpersonal transactions with the group. This condition should not be confused with the "Big Eye" (a term widely used in the Antarctic for the insomnia problem), which is a fairly wide-spread phenomenon confined almost exclusively to the dark winter season. The causes of this sleeplessness are not entirely clear, but seem related to such factors as the accumulation of group and personal tensions, the reduced physical activity of the dark winter period, and group suggestibility[203].

Adjustment

There is a cyclic adjustment that men make in Antarctica. According to ROHRER[220] the first phase consists of heightened anxiety on arrival in the Antarctic, lasting from a few weeks to a few months. Next is a period of varying degrees of depression affecting everyone, even the best adjusted men, although usually not of a pathological nature. This second phase covers most of the dark winter months. The last phase is characterized by the occurrence of anticipatory behavior preparatory to leaving the Antarctic.

The criteria of adaptation, as described by NELSON[208], are work proficiency, industriousness, and social compatability. There may be a level of work proficiency below which a man will not be accepted by his group members, regardless of how socially competent he might be. Above this level, however, social compatability appears to be more closely related than work proficiency to overall acceptance of an individual by group members[210].

The interest in the psychological problems of wintering personnel in the Antarctic has increased greatly and quite a few papers are now available on this subject[193-224]. When reading this accumulated psychological-psychiatric information one gets a rather morbid idea of the atmosphere in an antarctic station. In the opinion of the author and of several other wintering doctors, the impression of psychological problems being a general feature of every wintering expedition is wrong. It must be stressed that far from all men are subject to emotional disturbances. It is, however, almost impossible to find out from the reports of the various psychological investigations how large is the fraction of men who adjust well to life in the Antarctic and show no personality disorders. BEHNKE[193] reports 0.9% psychiatric cases of the total sick calls in American bases. PALMAI[211a] points out that half the visits to the surgery by members of his expedition consisted in demands for councelling with definite requests for help and advice concerning personal problems. The number of these calls were largest during the winter season. New sick calls were classified psychogenic in 3.4%, with headaches and insomnia forming another 10%. The latter may or may not have had a psychogenic origen. NARDINI, HERRMANN & RASMUSSEN[206] state that only at one of the ten investigated stations were more than half of the personnel considered to be psychiatrically inferior on the basis of initial assessment. They are of the opinion that personality disorders do constitute a rather serious problem, at least at American bases. But from personal experience and verbal communication the author asserts that there exist fortunate expedition groups that have experienced practically no disruptive problems of interpersonal adjustment, even in the case of two successive years of wintering. The proportion of the psychological problems must not be overemphasized.

After having successfully passed a year or two in the Antarctic and undergone adjustment to a polar environment with minor personality changes, man will return to civilization more mature and with a feeling that the experience has been profitable. For some there will be a problem of readjustment, which should not be neglected. Some will return to the Antarctic. Why? HEDBLOM has

734

the answer in his Polar Manual[228]: "Man fortunately remembers the
fun and forgets the hardship of an expedition. Some return because
they originally found themselves there, some because they are more
appreciated on the ice than in civilization, some for unsatisfied
ambition, some to continue a fascinating investigation, and some
because they hear Aurora's song in a polar sunset".

Considerations for the future

Antarctica is a gigantic laboratory and research field and it has
duly been recognized as such. The Antarctic Treaty, signed in 1959
by the twelve nations who had dispatched expeditions to the
Antarctic during the IGY, guarantees freedom of scientific research
in Antarctica for a period of 30 years. The members have promised
to use this vast territory only for peaceful purposes, to ban nuclear
explosions in this area, as well as the deposit of radio-active wastes,
and to exchange their programs and scientific data. A Special
Committee on Antarctic Research (SCAR) has been established
with representation provided by interested constituent scientific
unions and by the scientific academies of the twelve countries
already conducting antarctic research. SCAR has recognized the
urgent need for international scientific co-operation in antarctic
research, and in many fields it has initiated new efforts in international collaboration[230, 226]. SCAR has paid attention to the
special problems of nature conservation occasioned by continuous
human activity in the Antarctic, and has developed a series of
recommendations for international co-operation in protection and
conservation of its unique life forms.

Similar problems exist for man in this region. Not only must the
area be kept uncontaminated for scientific purposes, as not to
spoil the field for future research, but also for sanitary and health
reasons. Sewage and waste must carefully be taken care of and must
not be allowed to spread without control. Refuse can be carried over
unbelievable distances by the strong winds. Bacteria, virus, and
waste products are not destroyed in the cold but are preserved for
unlimited periods. They disappear in the snow, but may surrep-
tiously reappear when their presence is unsuspected and may then
cause great damage.

The introduction of new diseases and bacteria must be prevented.
One must not overlook the possibility of a propagation of epidemics and the appearance of endemics. Earlier there has been little
cause to consider such a probability. Expeditions have been few and
sporadic. In general all expedition members are healthy and have
undergone medical examination. Now thousands of people invade
Antarctica each summer. Although all wintering personnel are
medically selected before departure, it is probable that a large

portion of the persons taking part in the summer campaign have not passed a proper medical examination. As a rule there is no medical control on arrival and serious diseases and even epidemics may have been contracted on the journey to the South. It is becoming more and more possible for persons who have escaped all medical control to spend a short time in the Antarctic, in connexion with inspection visits, short-term operations or observations. Serious consideration should therefore be given to this problem, preferably by recommendation by SCAR, and steps be taken to ensure supervision and control through international co-operation. Systematic measures of prevention and prophylactics have already been suggested by RIVOLIER[111], and in 1962 SCAR passed a recommendation to the World Health Organization pointing out the concern felt at the risks of site contamination at antarctic bases in continuous operation.

More priority and increased resources should be given to medical research. There is evidence enough of the difficulties in carrying out physiological studies in the Antarctic. One need but to point out that much of the same fairly simple methods and investigations that were performed by FRAZIER, LOCKHART, BUTSON and WILSON, still are used and made over again by the doctors conducting physiological research at the Japanese, Russian, Chilean, Argentine and South African bases. Though in addition more advanced laboratory investigations are now also being carried out. In a number of cases, urine, blood, and other samples are collected and frozen for transport home and analysed at large laboratories. There is need not only for a medical officer but also for a trained research worker, as a physiologist, on an antarctic expedition, who could collaborate in fruitful research. Pioneers in such a teamwork were doctor MASTERTON and physiologist LEWIS on the British North Greenland Expedition 1952—54. Their studies have resulted in a number of good papers, some quoted by the author in this chapter [49, 239, 240, 242], and their work has been of great importance in stimulating further research[151c].

This teamwork was the beginning of a close association between the British Antarctic Survey (formerly F.I.D.S.) and the Medical Research Council, Division of Human Physiology, London[140, 151c, 92a]. Some of the medical officers recruited for the British Antarctic Survey are trained at the Division of Human Physiology before going to the Antarctic, where they make observations on members of their base. On return from the South, the results are worked up in the Division of Human Physiology. In this way valuable data are acquired and the continuation of promising work is assured. The Arctic Institute of North America offers grants to support individual workers with a planned scientific program for research in Antarctica, and the recently established Institute of Polar Studies, Columbus,

736

Ohio, may furnish a new avenue for initiating physiological work in this field.

However, there is no co-operation in planning medical and physiological projects between the expeditions of different nations. There is no organization for international co-ordination of medical research. The SCAR should be the right instrument for such a task, but as yet there is no working group allotted to this branch of scientific research. Medicine and physiology have been listed as "logistics".

An international collaboration in human studies would make possible a practical standardization of techniques, would allow an extension of a project over different locations during the same period and widen the scope, could help to fill gaps and avoid duplication, and assure a continuation of work showing promising results. A wish for the future is that some day a physiological research station will be established in the Antarctic, where the problems in this field may be attacked with greater resources and advanced methods, and the work conducted in international co-ordination. A step in this direction has already been made at Mc-Murdo Sound, where the Americans now have set up a complete biological and medical laboratory, which is supplied and operated by Stanford University[226, 232b]. May this be an incitement to greater activity in the study of man in Antarctica.

Acknowledgements

I offer my sincere thanks to all who have supplied me with information of work carried out on the various national expeditions in the Antarctic, and especially to my antarctic colleagues and friends who have put at my disposal preliminary accounts of unpublished work.

REFERENCES

Not all of the following references are referred to in the text, as some additional papers have been listed in order to help the reader to find out what has been published in this field and to give a more complete picture of the medical research work going on in the Antarctic. To increase the usefulness of the list, the references have been arranged provisionally under subject headings. With a few exceptions, only papers dealing with antarctic conditions have been included. The survey of literature for this chapter was concluded in Dec. 1962, but it was possible to include more recent papers by a revision of the reference list in Jan. 1964.

BIBLIOGRAPHIES

1. ANTARCTIC BIBLIOGRAPHY, NAVAER 10-35-591. Bureau of Aeronautics, Dept. of Navy, Washington, D.C., 1951.
2. ARCTIC BIBLIOGRAPHY (Annotated). Vol. 1-10 (MARIE TREMAINE, editor). Dept. of Defence, Washington, D.C., 1953—61.
3. POLAR BIBLIOGRAPHY. Air Force Manual No. 200—132. Vol. I—III. Technical Information Division, Library of Congress, Washington, D.C., 1956—59.

4. ROBERTS, B. (1958). Chronological list of antarctic expeditions. *Polar Rec.*, **9**, *97—134, 191—239*.
5. EFFECTS OF COLD ON MAN. An annotated bibliography, 1938—1951. (ELAINE CULVER, editor). *Physiol. Rev.*, **39**, Part II, Suppl. No 3, 1959.
6. CARLSON, L. D. & THURSH, H. L. (1960). Human acclimatization to cold. A selected, annotated bibliography of the concepts of adaptation and acclimatization as studied in man. *Arctic Aeromed. Lab. Techn. Rep.* 59—18. Fort Wainwright, Alaska.
7. RIVOLIER, J. (1960). Physio-pathologie du froid. Bibliographie, mise à jour au 1er janvier 1960. *Expéd. Pol. Franç., Missions P.-E. Victor,* Publ. No. 215, Paris.
8. BURTON, A. C. & EDHOLM, O. G. (1955). Man in a Cold Environment. Edward Arnold Ltd., London; Williams & Wilkins, Baltimore.

EXPEDITIONS AND EXPLORATION

9. BORCHGREVINK, C. E. (1901). First on the Antarctic Continent: Being an Account of the British Antarctic Expedition 1898—1900. Georg Newnes, London.
10. BYRD, R. E. (1930). The conquest of Antarctica by air. *Nat. geogr. Mag.,* **58**, *127—227*.
11. COOK, F. A. (1900). Through the First Antarctic Night, 1898—99. William Heinemann, London; Doubleday & McClure, New York.
12. DARLINGTON, J. (1956). My Antarctic Honeymoon: A Year at the Bottom of the World; as told to J. McIlvaine. Doubleday & Co, New York, Toronto.
13. GIAEVER, J. (1954). The White Desert. The Official Account of the Norwegian-British-Swedish Antarctic Expedition. Chatto & Windus, London; Clarke, Irwin & Co, Toronto; Dutton, New York (1955).
14. KRISTENSEN, L. (1885). Antarctic's reise til Sydishavet. Forfatterens Forlag, Tønsberg, *p. 222*.
15. NORDENSKJÖLD, N. O. G. & ANDERSSON, J. G. (1905). Antarctica: Two Years Among the Ice of the South Pole. Hurst & Blackett, London; McMillan, New York.
16. PRIESTLEY, R. (1956). Twentieth-century man against Antarctica. *Nature, Lond.,* **178**, *463—70*. Also in *Advanc. Sci.,* **13**, *3—16*, 1956.
17. PRIESTLEY, R. (1960). Antarctic exploration yesterday and today. *Polar Rec.,* **10**, *11—15*.
18. RONNE, F. (1949). Antarctic Conquest; the Story of the Ronne Expedition, 1946—1948. Putnam, New York.
19. SIPLE, P. A. (1957). We are living at the South Pole. *Nat. geogr. Mag.,* **112**, *5—35*.
20. SIPLE, P. A. (1958). Man's first winter at the South Pole. *Nat. geogr. Mag.,* **113**, *439—78*.
21. SIPLE, P. A. (1962). Living at 70° C below zero. *Courier*, **15**, (1), *21—28*.
22. SOMOV, M. M. (1962). Journey into the inaccessible. *Courier*, **15**, (1), *32—34*.

CLIMATOLOGY

23. CHRENKO, F. A. & PUGH, L. G. C. E. (1961). The contribution of solar radiation to the thermal environment of man in Antarctica. *Proc. Roy. Soc.,* **155 B**, *243—65*.
23a. GRANGE, J. J. LA (to be published). Notes on biometeorology with respect to human, animal, bird and plant life as observed on the South African National Antarctic Expedition and at Marion Island. Antarctic Office, Pretoria.

738

23b. HARKER, G. S. (1959) Whiteout — A bibliographical survey. *Bull. Amer. met. Soc.*, **40**, *225—29.*

24. KASTEN, F. (1960). Über die Sichtweite im Polar Whiteout. *Polarforschung,* Bd. V, **30**, *41—44.*

25. LILJEQUIST, G. H. (1956). Short-Wave Radiation. Energy Exchange of an Antarctic Snow-Field. *Norwegian-British-Swedish Antarctic Expedition, 1949—52. Sci. Res.,* Vol. II, Part 1 A. Norsk Polarinstitutt, Oslo.

26. PUGH, L. G. C. E. & CHRENKO, F. A. (1962). Observations of the effects of solar radiation on the thermal environment inside tents in Antarctica *Ann. occup. Hyg.,* **5,** *1—5.*

27. RUBIN, M. J. (1962). The Antarctic and the weather. *Sci. Amer.,* **207,** (3) *84—94.*

28. RIVOLIER, J. (1955). Terre Adélie 1952. Éléments de climatologie biologique dans l'Archipel de Pointe Géologie. *Expéd. Pol. Franç., Missions P.-E. Victor,* Publ. No. 47, Paris.

29. SAPIN-JALOUSTRE, J. (1955). Contribution à l'étude de l'acclimatation de l'homme et des vertébrés supérieurs dans l'Antarctique et plus spécialement en Terre Adélie. *Biol. méd., Par.,* **44,** *15—72, 113—208.* (Also Ibid., *Thèse,* Paris, 1953, 204 p.)

30. SIPLE, P. A. (1939). Adaptation of the explorer to the climate of Antarctica. Part. I—III. *Thesis,* Clark Univ., Worchester, Mass. Abstr. in *Clark Univ. Abstr. of Dissert. and Theses,* **11,** *31—35.*

31. SIPLE, P. A. & PASSEL, C. F. (1945). Measurements of dry atmospheric cooling in subfreezing temperatures. *Proc. Amer. phil. Soc.,* **89,** *177—199.*

32. WILSON, O. (1963). Cooling effect of an antarctic climate on man. With some observations on the occurrence of frostbite. *Norsk Polarinstitutt Skr.* Nr. **128,** Oslo.
    See also references No. 138, 139a, 139c, 149b, 157, 162—163, 165a, 192, 232.

NUTRITION – VITAMINS

33. ADAMS, R. J., STANMEYER, W. R. & HARDING, R. S. (1962). Antarctic stress and vitamin requirements. *J. dent. Med.,* **17,** *36—42.*

34a. ANDREW, J. D. (1947). Observations on ascorbic acid intake. Hope Bay 1946. *Falkland Islands Depend. Surv., Base Rep.* E 193/47, London (unpublished).

34b. ANDREW, J. D. (1949). Experimental ascorbic acid deficiency .*Brit. med. J.,* ii, *1273—74.*

35. BELL, M. E. (1957). Rations for the New Zealand Trans-Antarctic Expedition. *N. Z. med. J.,* **56,** *289—304.*

36. BINGHAM, E. W. (1952). A greenhouse in the Antarctic. *Polar Rec.,* **6,** *392—93.*

37. BUTSON, A. R. C. (1950). Utilisation of high-fat diet at low temperatures. *Lancet,* **258,** *993.*

38. CSORDAS, S. E. (1958). Problem of nutrition in antarctic expeditions. *J. Diet. Ass. Victoria* (Sept.), pp. *1—10.*

39. CUTLAND, G. T. (1959). Recipes for an antarctic cook. *Polar Rec.,* **9,** *562—69.*

40. EDHOLM, O. G. (1962). Food intake of men at polar bases. (Note). Document No. 24. *Conf. on Med. & Publ. Health in the Arctic & Antarctic.* WHO, Geneva (not published).

41. EKELÖF, E. (1905). Über "Präserven - Krankheiten". *Wiss. Ergebn. Schwed. Südpolar-Exped. 1901—03,* Bd. 1, Lief. 4, pp. *31—54.* Stockholm. Also in *Hygiea, Stockh.,* **4,** *1214—43,* 1904 (in Swedish).

42. FUCHS, V. E. (1952). Sledging rations of the F.I.D.S. *Polar Rec.*, **6,** *508—11.*

43. GAZERT, H. (1906). Proviant und Ernährung der Deutschen Südpolar-Expedition 1901—03. *Deutsche Südpolar-Exped. 1901—03.* (Hrsg. von E. v. Drygalski), Bd. VII, Heft 1, pp. *1—73.* Georg Reimer, Berlin.

44. GAZERT, H. (1914). Die Beriberifälle auf Kerguelen. *Deutsche Südpolar-Exped. 1901—03.* (Hrsg. von E.v. Drygalski), Bd. VII, Heft 4, pp. *353—86.* Georg Reimer, Berlin.

45. KALNENAS, K. (1951). Notes on the food rations of the Macquarie Island Antarctic Research Expedition, 1950/51. Rep. no. CM 14 (14/50/786/641—1) in library of Aust. Nat. Antarctic Res. Exped., Dept. External Affairs, Melbourne (unpublished).

46. KALNENAS, K. (1952). Ascorbic acid levels in the plasma of some antarctic birds and mammals. *Nature, Lond.,* **169,** *836.*

47. KENDALL, E. J. C. (1955). Scurvy during some British polar expeditions, 1875—1917. *Polar. Rec.,* **7,** *467—85.*

48. LAW, P. G. (1957). Nutrition in the Antarctic. *Med. J. Aust.,* **44,** i, *676—79.*

48a. LEWIS, H. E. (1963). Nutritional research in the polar regions. *Nutr. Rev.,* **21,** *353—56.*

49. LEWIS, H. E. & MASTERTON, J. P. (1958). A modern sledging ration. *Geogr. J.,* **124,** *85—88.*

50. LOCKHART, E. E. (1945). Antarctic trail diet. *Proc. Amer. phil. Soc.,* **89,** *235—48.*

51. MERWE, A. LE R. VAN DER (1962). Die Voorsiening en Verbruik van Askorbiensuur by die Suid-Afrikaanse Antarktiese Basis, 1960. *S. Afr. med. J.,* **36,** *751—54.*

52. MILAN, F. A. & RODAHL, K. (1961). Nutrition and energy expenditure at Little America V in the Antarctic. *Arctic Aeromed. Lab. Techn. Rep.* 60—11, Fort Wainwright, Alaska.

53. MILAN, F. A. & RODAHL, K. (1961). Caloric requirements of man in the Antarctic. *J. Nutr.,* **75,** *152—56.*

54. MILAN, F. A. & RODAHL, K. (1963). Caloric requirements of man in the Antarctic. *C.R. I.Symp. Biol. Antarctique,* p. *563.* Hermann, Paris. Abstr. in S.C.A.R. Bull. No. 12. *Polar Rec.,* **11,** *316,* 1962.

55. ORR, N. W. M. (1962). Food requirements of men and dogs on antarctic expeditions. *M.D. Thesis,* Cambridge Univ., Engl. (Also to be published as a British Antarctic Surv. Sci. Rep.)

56. ORTON, M. N. (to be published). Studies of dietary fat intake and absorbtion, made at Wilkes in 1961. Aust. Nat. Antarctic Res. Exped., Dept. External Affairs, Melbourne.

57. RIVOLIER, J. (1955). Froid et altitude dans leurs rapports avec l'alimentation. *Ann. Nutr. Aliment.,* **9,** *135—77.*

58. RIVOLIER, J. (1956). Lipides et froid. *Bull. Soc. sci. Hyg. aliment.,* **44,** *203—17.*

59. RIVOLIER, J. (1964). Nutrition et climat polaire. *Proc. III. Int. Biomet. Congr., Pau 1963* (S. W. TROMP & W. H. WEIHE, editors). Pergamon Press, Oxford (in press).

60. RIVOLIER, J. & LE BIDEAU, G. (1957). Le Pemmican. *Concours méd.,* **79,** *2169—74.*

SEE also references No. 29, 30, 68, 75, 94, 126, 139a, 139d, 140—141, 174, 179, 183d.

MEDICINE AND HEALTH

A profusion of general medical accounts from the Antarctic exist, many published in a popular form, others as unpublished, restricted expedition reports. Below are listed a fair number.

740

General aspects

61. ANONYM (1947). Medical report. Annex 19. In *U.S. Navy Antarctic Develop. Proj. 1947; Rep. of Operation Highjump.* Vol. III. 49 p.
62. ANONYM (1948). Medical report. Annex 10. In *U.S. Navy Antarctic Develop. Proj. 1948; Rep. of Operation Windmill.*
63. ANONYM (1957). Medical report, Deepfreeze I. Addenda B and D, Final Rep. *U.S.N. Mobile Constr. Batt. (Special), Detachment I.* Davisville, R. I.
64. ANONYM (1958). Medical report, Deepfreeze II. Addenda, Final Rep. *U.S.N. Mobile Constr. Batt. (Special), Detachment B*, Davisville, R.I.
65. BECK, E. H. (1947). Medical aspects of a wartime antarctic expedition 1944—1946, Operation Tabarin. *J. Roy. Nav. med. Serv.*, **33**, *193—97.*
66. BINGHAM, E. W. (1948). The antarctic expedition from a medical angle. *Med. Press*, **219**, *185—88.* Also in *J. Roy. Nav. med. Serv.*, **34**, *78—84.*
67. BOND, A. (1948). A medical officer in the Antarctic. *Med. Press*, **220**, *9—10.*
67a. BUDD, G. M. (1962). Select general practice — Antarctica. *Sydney Univ. med. J.*, **51**, *20—29.*
67b. [BUDD, G. M.] (1963). Doctors in the Antarctic. *Med. J. Aust.*, **50**, i, *938.*
68. CACCIAVILLANI, E. A. (1959). Contribución al conocimiento de la psicofisiología en la adaptación del organismo humano a las bajas temperaturas. Paper presented at Simposio Antártico, Buenos Aires. Abstr. in *Antarctic Symposium, Int. Union Geodesy & Geophys. Monogr. No 5*, pp. *83—84*, Paris 1960.
69. CENDRON, J. (1952). Médicine, biologie. In *Expédition en Terre Adélie 1950—1952. Expéd. Pol. Franç., Missions P.-E. Victor*, Publ. No. 20, pp. *49—53.* Paris.
70. CENDRON, J. (1953). Souvenirs médicaux de Terre Adélie. *Presse méd.*, **61**, *121—24.*
71. DALGLIESH, D. G. (1952). Two years in the Antarctic. Part. I & II. *St. Thom. Hosp. Gaz.*, **50**, *62—65, 111—117.*
72. DIGEON, F. (1961). Hygiène, santé. In *Rapport d'Activités de la IXème Expédition Antarctique Française. Terre Adélie 1958—1960. Expéd. Pol. Franç., Missions P.-E. Victor*, Publ. No. 217, pp. *116—20.* Paris.
72a. DUMAS, P. (1963). Activité médico-chirurgicale. In *Rapport d'Activités de la Xème Expédition Antarctique Française. Terre Adélie 1959—1961. Expéd. Pol. Franç., Missions P.-E. Victor*, Publ. 231, pp. *76—81* Paris.
73. EKELÖF, E. (1904). Medical aspects of the Swedish Antarctic Expedition, October 1901—January 1904. *J. Hyg., Lond.*, **4**, *511—40.* Also in *Hygiea, Stockh.*, **4**, *577—615* (in Swedish).
74. EKELÖF, E. (1905). Die Gesundheits- und Krankenpflege während der Schwedischen Südpolar-Expedition, Oktober 1901—Januar 1904. *Wiss. Ergebn. Schwed. Südpolar-Exped. 1901—03*, Bd. 1, Lief. 3, pp. *1—30.* Stockholm.
75. FUCHS, Y. (1951). Les problèmes médicaux d'une expédition antarctique. *Rev. méd. nav.*, **6**, *7—28.*
76. GAZERT, H. (1914) Ärztliche Erfahrungen und Studien auf der Deutschen Südpolar-Expedition. *Deutsche Südpolar-Exp. 1901—1903.* (Hrsg. von E. v. Drygalski), Bd. VII, Heft 4, pp. *297—352.* Georg Reimer, Berlin.
77. GOLDSMITH, R. (1957). Eight men in a crate. *St. Bart. Hosp. J.*, **61**, *288—95.*
78. GOLDSMITH, R. (1959). The Commonwealth Trans-Antarctic Expedition - medical and physiological aspects of the advance party. *Lancet*, **i**, *741—44.*

79. GOURDON, E. (1913). Un hivernage dans l'Antarctique. *Thèse*, Paris.
80. HEDBLOM, E. E. (1961). The medical problems encountered in Antarctica. *Milit. Med.*, **126**, *818—24*.
80a. HIÉLY, P. (1964). Hygiène, santé. In *Rapport d'Activités de la XIIème Expédition Antarctique Française. Terre Adélie 1961—63. Expéd. Pol. Franç., Missions P.-E. Victor*, Publ. No. 251. Paris (in press).
81. HILLENBRAND, F. K. M. (1953). In antarctic waters. *Lancet*, **265**, *246—48*.
82. ISEL, J. (1959). Hygiène, santé et biologie. I. Rapport médical. In *Rapport d'Activités des Expéditions Antarctiques Françaises 1957—1959. Année Géophys. Int.*, pp. *47—49*. Sous-comité Antarctique Français. Paris.
83. ITO, Y. (1959). Report on activities of medical subcommittee and medical team for the Japanese Antarctic Research Expedition, 1956—57. Part 1. *Antarctic Rec., Tokyo*, No. 6, pp. *54—72*.
84. JONES, S. E. (1915). Medical Report. Western Base (Queen Mary Land). Appendix V. In *The Home of the Blizzard*. (D. MAWSON, author), Vol. II, pp. *307—08*. William Heinemann, London.
85. MACKAY, A. F. (1910). Some notes on health in the Antarctic. *Edinb. med. J.* n.s. **4**, *219—22*.
86. McLEAN, A. L. (1915). Medical Report. Main Base (Adélie Land). Appendix V. In *The Home of the Blizzard*. (D. MAWSON, author), Vol. II, pp. *308—10*. William Heinemann, London.
87. MACKLIN, A. H. (1923). Medical. Appendix V. In *Shackleton's Last Voyage; the Story of the Quest*. (F. WILD, author), pp. *352—65*. Cassel, London; Stokes, New York.
88. MACKLIN, A. H. (1939). Polar exploration. Some medical aspects, with special reference to the Antarctic. *Med. Press*, **202**, *401—06*.
89. MARSHALL, E. S. (1909). Report on the health of the expedition. Appendix VI. In *The Heart of the Antarctic*. (E. H. SHACKLETON, author), Vol. II, pp. *397—99*. William Heinemann, London; Lippincott, Philadelphia.
90. MARSHALL, E. S. (1943). An antarctic episode. *Med. Press*, **210**, *359—62*.
91. MERWE, A. LE R. VAN DER (1960). Geneeskundige berigte. *Geneeskunde*, **2**, *245—48*.
92. MOUNTEVANS, E. R. G. R. E. (1937). How a sailor looks at the surgeon, and the medical aspect of polar exploration from a sailor-explorer's viewpoint. *J. Roy. Nav. med. Serv.*, **23**, *14—30*.
92a. PALMAI, G. (1963). Doctors in the Antarctic. *Med. J. Aust.*, **50**, ii, *598*.
93. RIVOLIER, J., (1954). Médecine. In *Expédition en Terre Adélie 1951—1953. Expéd. Pol. Franç., Missions P.-E. Victor*, Publ. No. 24, Part. I, pp. *32—37*. Paris.
94. ROBERTS, J. M. (1949). The Falkland Islands Dependencies Survey. *Brit. med. J.*, ii, *863—64*.
95. SAPIN-JALOUSTRE, J. (1952). Médicine-biologie. In *Expédition en Terre Adélie 1949—51. Expéd. Pol. Franç., Missions P.-E. Victor*, Publ. No. 14, pp. *76—86*. Paris.
96. SIPLE, P. A. (1960). Living on the South Polar Ice Cap. In *Cold Injury* (S.M. HORVATH, editor). Trans. Sixth Conf., **1958**, Fort Knox, pp. *89—115*. Josiah Macy, Jr. Foundation, New York.
97. SIPLE, P. A. (1960). Commentary on antarctic operations. *Fed. Proc.*, **19**, Part II, Suppl. No. 5, *10*.
97a. SLADEN, W. J. L. (1953). Medical organization of the Falkland Islands Dependencies Survey 1947—51. *M.D. Thesis*, Vol. II, London Univ.
97b. SMART, R. A. (1964). Report on health. In *Roy. Soc. IGY Antarctic Exped., Halley Bay, 1955—59*. Vol. IV (in press). The Royal Society, London.

742

98. SORIA, A. A. (1957). Ecología humana antártica. *Día méd.*, **29**, (26), *725—32*.
99. STAQUET, M. (1961). Observations physiologiques et médicales d'une expédition au Pôle Sud. *Brux.-méd.*, **41**, *741—51*.
100. TAYLOR, I. M. (1960). Medical experiences at McMurdo Sound. In *Cold Injury* (S. M. HORVATH, editor). Trans. Sixth Conf., 1958, Fort Knox, pp. *117—140*. Josiah Macy, Jr. Foundation, New York.
101. TICHOMIROV, I. I. (1961). O charaktere zabolevaemosti na stancii Vostok v zimovku 1959 g. (Illness statistics at station Vostok during wintering in 1959). *Inform. Bjull. Sovet. Antarktičeskoj Eksped.*, No. **27**, *36—39*. (Information Bull. of the Soviet Antarctic Expedition).
102. TIMOFEEV, V. V. (1960). Medicinskoe obespečenie antarktičeskoj ekspedicii 1819—1821 gg. (Medical protection of the Antarctic Expedition in 1819—1821). *Sovet. Zdravoochr.*, **19**, (12), *50—54*.
103. VEGA, J., FLÜHMANN, G., DE LA FUENTE, L., MORALES, L. & OYANGUREN, H. (1962). Medicine and public health in the Chilean Antarctic Bases. Document No. 22. *Conf. on Med. & Publ. Health in the Arctic & Antarctic*. WHO, Geneva (not published).
104. WILSON, O. (1952). Sjukvård i Antarktis. *Vårt Röda Kors.*, **7**, *80—83*.
105. WILSON, O. & HARALDSON, S. (1949). Norsk-Svensk-Brittiska Antarktisexpeditionen 1949—52. *Svenska Läkartidn.*, **46**, *1989—97*.

SEE also references No. 11, 29, 30, 116, 142.

Special problems

106. BUDD, G. M. (1962). A polio-like illness in Antarctica. *Med. J. Aust.*, **49**, i, *482—86*.
107. HEDBLOM, E. E. (1961). Snowscape eye protection. *Arch. environ. Health*, **2**, *685—704*.
108. HOUK, V. N. (1959). Appendicitis treated nonsurgically at the South Pole. *U.S. Armed Forces med. J.*, **10**, *1352—54*.
109. HOUK, V. N. (1959). Transient pulmonary insufficiency caused by cold. *U.S. Armed Forces med. J.*, **10**, *1354—57*.
109a. PARDOE, R. (to be published). A ruptured intracranial aneurysm in Antarctica. Aust. Nat. Antarctic Res. Exped., Dept. External Affairs, Melbourne. (cf. ref. 67b)
110. PUGH, L. G. C. E. (1959). Carbon monoxide hazard in Antarctica. *Brit. med. J.*, i, *192—96*.
111. RIVOLIER, J. (1962). Measures for the prevention of the propagation of new diseases in the Arctic and Antarctic. Document No. 17. *Conf. on Med. & Publ. Health in the Arctic & Antarctic*. WHO, Geneva (not published).
111a. ROGOZOV, L. I. (1962). Operacija na sebe. (Operation on oneself). *Inform Bjull. Sovet. Antarktičeskoj Eksped.*, No. **27**, *42—44*. (Information Bull. of the Soviet Antarctic Expedition).
112. SAPIN-JALOUSTRE, J. (1954). Le problème de la vision dans le blizzard. *Ann. Opt. ocul., Par.*, **3**, *36—53*. (Also in *L'Opticien belg., Brux.*, No. **19** (mars), *99—108*, 1955).
113. SAPIN-JALOUSTRE, J. (1956). Enquête sur les gelures. A propos des observations de la 1re Expédition Française en Terre Adélie 1948— 1951. *Actualités Sci. et Industr.*, No. 1248. Hermann & Cie, Paris. (Also Ibid. Expéd. Pol. Franç., Missions P.-E. Victor, Publ. No. 64. Paris.)
114. SAPIN-JALOUSTRE, J. & SAPIN-JALOUSTRE, H. (1956). La pathologie en expédition polaire antarctique. A propos des observations de la 1re Expédition en Terre Adélie 1948—51. *Presse méd.*, **64**, *579—82*, *637—40, 791—94, 821—23*.

SEE also references No. 29, 32.

HUMAN BACTERIOLOGY

115. ADAMS, R. J. & STANMEYER, W. R. (1960). Effects of prolonged antarctic isolation on oral and intestinal bacteria. *Oral Surg.*, **13**, *117—20.*
116. MCLEAN, A. L. (1919). Bacteriological and other researches. *Australasian Antarctic Exped. 1911—14, Sci. Rep.* Ser. C, Vol. VII, Part 4, Sydney.
117a. SLADEN, W. J. L. (1952). Staphylococci in nose and throat of healthy individuals in the Antarctic. Paper read at meeting of Path. Soc. Gt. Brit. & Ireland. *J. Path. Bact.*, **64**, *671*, (title only).
117b. SLADEN, W. J. L. (1953). Bacteriological work in the Antarctic. *M.D. Thesis*, Vol. I, London Univ.
118. SLADEN, W. J. L. (1961). Medical microbiology. In *Science in Antarctica*, Part. I, The Life Sciences in Antarctica, pp. 151—55. A Rep. by the Committee on Polar Res. Publ. 839. Nat. Acad. Sci. - Nat. Res. Council, Washington, D.C.
119. SLADEN, W. J. L. (1963) Upper respiratory staphylococci and streptococci in antarctic communities. *C.R. I.Symp. Biol. Antarctique*, pp. *101—04*. Hermann, Paris. Abstr. in S.C.A.R. Bull. No. 12. *Polar Rec.*, **11**, *318—319*, 1962.
120. SLADEN, W. J. L. & GOLDSMITH, R. (1960). Biological and medical research based on USS. *Staten Island*, Antarctic, 1958—59. *Polar. Rec.*, **10**, *146—48*. (cf. Operation "Snuffles". *IGY Bull.*, No. **26**, *14—15*. Nat. Acad. Sci., Aug. 1959).

SEE also reference No. 96.

DENTISTRY

121. ANONYM, (1947). Dental report. Annex 20. In *U.S. Navy Antarctic Develop. Proj. 1947; Rep. of Operation Highjump.* Vol. III, 17 p.
122. ADAMS, R. J. & STANMEYER, W. R. (1959). Antarctic 'day' of a naval dentist. *J. Amer. dent. Ass.*, **59**, *322—26.*
122a. ADAMS, R. J. & STANMEYER, W. R. (1960). The effects of a closely supervised oral hygiene program upon oral cleanliness. *J. Periodont.*, **31**, *242—45.*
123. KNOEDLER, D. & STANMEYER, W. R. (1958). Dental observations made while wintering in Antarctica, 1956—1957. *J. dent. Res.*, **37**, *614—22.*
124. LINDSAY, J. S. (1961). Report of the dental research program, Deep Freeze 60. *Bull. U.S. Antarctic Proj. Office*, Vol. II, No. 7. (March).
125. NELSON, C. (1931). Two years after. *Dent. Mag. & Oral Topics*, **48**, *44.*
126. PERLITSH, M. J., NIELSEN, A. G. & STANMEYER, W. R. (1961). Ascorbic acid plasma levels and gingival health in personnel wintering over in Antarctica. *J. dent. Res.*, **40**, *789—99.*
126a. PIDGEON, J. W. G. (1960). Dental research report, 1959—60. *Falkland Islands Depend. Surv.*, Base Rep. 165/60. London (unpublished).
127. STANMEYER, W. R. (1959). The dentist and the space age. *Milit. Med.*, **124**, *417—21.*
128. STANMEYER, W. R. & ADAMS, R. J. (1959). Reduced oral temperatures and acid production rates in dental plaques. *J. dent. Res.*, **38**, *905—09.*
129. STANMEYER, W. R. & ADAMS, R. J. (1961). Antarctic stress and the teeth. *J. Amer. dent. Ass.*, **63**, *665—70.*
129a. STANMEYER, W. R. & ADAMS, R. J. (1961). Tooth sensitivity during Operation Deepfreeze. *Dent. Progr.*, **2**, *52—54.*

SEE also references No. 33, 66, 68, 71, 104, 115, 131, 142, 162.

744

PHYSIOLOGY

130. ADAM, J. M. (1959). Subjective sensations and sub-clothing temperatures in Antarctica. *J. Physiol.*, **145**, *26—27* P.

130a. ANTONIS, A., BERSOHN, I. & EASTY, D. L. (1963). Serum lipid changes in young men in Antarctica. *J. Physiol.*, **167**, *26—27* P.

130b. BARSOUM, A. H. (1962). Some observations on blood in relation to cold acclimatization in the Antarctic. *Milit. Med.*, **127**, *719—22*.

131. BROOKER, B. K. (1959). Medical report for Halley Bay 1958. Report to the Royal Society, London (unpublished).

131a. BROWN, C. T. (1962). Physiology report. Halley Bay 1962. *British Antarctic Surv.*, Base Rep. M2/1962/Z, London (unpublished).

132a. BUDD, G. M. (1962). Acclimatization to cold in Antarctica as shown by rectal temperature response to a standard cold stress. *Nature, Lond.*, **193**, *886*.

132b. BUDD, G. M. (1964). General acclimatization to cold in men studied before, during and after a year in Antarctica. *A.N.A.R.E.Rep.* No. 70. Aust. Nat. Antarctic Res. Exped., Melbourne (in press).

133a. BUDD, G. M. (1964). The 1963 ANARE Heard Island Expedition. *Polar Rec.* (in press).

133b. BUDD, G. M. (to be published). Studies of thermal sensation and skin temperature in relation to clothing, activity, and environmental conditions, made at Mawson in 1959—60. Aust. Nat. Antarctic Res. Exped., Dept. External Affairs, Melbourne.

134. BUTSON, A. R. C. (1949). Report on cold weather acclimatization. Submitted to Falkland Islands Depend. Surv., London (unpublished).

135. BUTSON, A. R. C. (1949). Acclimatization to cold in the Antarctic. *Nature, Lond.*, **163**, *132—33*.

136. CABEZA QUIROGA, M. A. (1959). Observaciones fisiológicas sobre la conducta humana en dotaciones polares. Paper presented at Simposio Antártico, Buenos Aires. Abstr. in *Antarctic Symposium, Int. Union Geodesy & Geophys. Monogr. No. 5, p. 83.* Paris, 1960.

136a. CATTY, R. H. C. (1961). Report on medical research and human physiology. Hope Bay 1961. *Falkland Islands Depend. Surv.*, Base rep. M2/1961/D. London (unpublished).

137. CHANNON, J. E. G. (1961). Physiological investigations at Mawson in 1958. Paper presented at Symp. Physiol. & Ecolog. Probl. in Antarctica. Abstr. in Aust. & N. Zealand Ass. Advance. Sci. Brisbane Congr. Pap., Sect. N. Report with Aust. Nat. Antarctic Res. Exped., Dept. External Affairs, Melbourne (unpublished).

138. CUMMING, A. (1960) Notes on medical research. Argentine Islands 1959. *Falkland Islands Depend. Surv.*, Base Rep. M2/1959/F. London (unpublished). Cited by EDHOLM (ref. 141).

139a. DAVIES, A. G. (1962). Observations on urine, saliva, and sweat of men living in the Antarctic. *M.D. Thesis*, Univ. of St. Andrews, Engl. (Also to be published as a British Antarctic Surv. Sci. Rep.)

139b. DAVIES, A. G. (1963). Changes in palmar sweating of men in the Antarctic. *J. Physiol.*, **165**, *50*P.

139c. EASTY, D. L. (1961). Physiology report. Halley Bay 1961. *Falkland Islands Depend. Surv.*, Base Rep. M2/1961/Z. London (unpublished).

139d. EASTY, D. L. (1963). Seasonal variations in body weight, skinfold thickness, food intake, serum lipids and adipose fat composition in young men in Antarctica. *M.D. Thesis*, Manchester Univ., Engl. (Also to be published as a British Antarctic Surv. Sci. Rep.).

140. EDHOLM, O. G. (1960). Polar physiology. *Fed. Proc.*, **19**, Part II, Suppl. No. 5, *3—8*.

141. EDHOLM, O. G. (1961). Physiological problems in polar regions. In *Man Living in the Arctic* (F. R. FISHER, editor). Proc. Conf. Quartermaster Res. & Eng. Cent., 1960, Natick, Mass., pp. *91—100*. Nat. Acad. Sci.— Nat. Res. Council, Washington, D.C.

141a. EVANS, M. (1957). Medical research report. Horseshoe Island, Marguerite Bay. *Falkland Islands Depend. Surv.*, Base Rep. 16/57. London (unpublished).

142. FRAZIER, R. G. (1945). Acclimatization and the effects of cold on the human body as observed at Little America III, on the United States Antarctic Service Expedition 1939—1941. *Proc. Amer. phil. Soc.*, **89**, *249—55*.

143. GESINO, A. & LEON, R. C. (1959). Resultados estadísticos de fisiología humana obtenidos en la estación científica Ellsworth durante el año 1959. Paper presented at Simposio Antártico, Buenos Aires. Abstr. in *Antarctic Symposium, Int. Union Geodesy & Geophys. Monogr. No. 5*, 1960, Paris, *p. 86*.

144. GOLDSMITH, R. (1959). Evidence of acclimatization to cold obtained from clothing records. *J. Physiol.*, **148**, *79—80* P.

145. GOLDSMITH, R. (1960). Use of clothing records to demonstrate acclimatization to cold in man. *J. appl. Physiol.*, **15**, *776—80*.

146. GOLDSMITH, R. (1960). British symposium on polar medicine, 1959. *Polar Rec.*, **10**, *75—76*. (Cf. Polar medicine, *Lancet*, ii, *786—87*, 1959).

147. GRAHAM, J. G. (1959). Observations on sleep rhythm (Graham Land) presented at a symposium on Polar Medicine. In *Lancet*, ii, *786—87*, 1959.

148a. GRAHAM, J. G. (1959). Physiological report. Loubet Coast. *Falkland Islands Depend. Surv.*, Base Rep. 193/59. London (unpublished).

148b. GRAHAM, J. G. (unpublished). Studies on fold thickness, body weight, and sleep patterns in the Antarctic 1958. Report with Med. Res. Council Lab., Hampstead, Engl.

149a. HAMPTON, I. F. G. (1960). Changes in the rate of heat elimination from the hands of four subjects in response to cold stress. Physiology report. Hope Bay 1959. *Falkland Islands Depend. Surv.*, Base Rep. 65/60. London (unpublished).

149b. HAMPTON, I. F. G. (1960). Physiology report. Hope Bay 1960. *Falkland Islands Depend. Surv.*, Base Rep. M2/1960/D. London (unpublished).

150. HICKS, K. (in preparation). Physiological research at Wilkes Station 1963, including dietary fat, serum cholesterol, adrenal sensitivity, blood coagulability, body weight, skinfold thickness, blood pressure, reaction time, body temperature, thermal sensation and clothing. Aust. Nat. Antarctic Res. Exped., Dept. External Affairs, Melbourne.

151. JONES, D. P. M. (1958). Record of work carried out for the Medical Research Council. *Falkland Islands Depend. Surv.*, Rep. M 85/1958/F. London (unpublished).

151a. JONES, D. P. M. (personal communication). Unpublished data on hair growth in the Antarctic. Med. Res. Council Lab., Hampstead, Engl.

151b. KAGEYAMA, T. (1962). [Paper in Japanese]. *J. Tokyo Women's Med. Coll.*, **32** (3), *103—16*. (cf. ref. 183b).

151c. LEWIS, H. E. & MASTERTON, J. P. (1963). Polar physiology. Its development in Britain. *Lancet*, i, *1009—14*.

152. LOCKHART, E. E. (1941). Acclimatization in the Antarctic. *Science*, **94**, *550*.

153. MASSEY, P. M. O. (1956). Acclimatisation to cold in Antarctica. *Appl. Psychol. Res. Unit*, Rep. A.P.U. 262/56. Cambridge, Engl. Also *M.D. Thesis*, Cambridge Univ., Engl.

154. MASSEY, P. M. O. (1957). Finger acclimatisation in Grahamland. *Advanc. Sci., Lond.*, **13**, *418—19*.

746

155. MASSEY, P. M. O. (1959). Finger numbness and temperature in Antarctica. *J. appl. Physiol.*, **14**, *616—20*.
156. MERWE, A. LE R. VAN DER (unpublished) Concise report on medical research. First South African Nat. Antarctic Exped., 1960. (personal communication).
156a. MERWE, A. LE R. VAN DER & HOLEMANS, K. (1962). Observations on body weight, basal metabolic rate, urinary nitrogen excretion and diuresis of members of the First South African National Antarctic Expedition (SANAE I) February—December 1960. *S. Afr. med. J.*, **36**, *767—69*.
157. MILAN, F. A. (1961). Thermal stress in the Antarctic. *Arctic Aeromed. Lab. Techn. Rep.* 60—10. Fort Wainwright, Alaska.
158. MILAN, F. A. (1963). Maintenance of thermal balance in Arctic Eskimos and Antarctic sojourners. *C.R. I.Symp. Biol. Antarctique*, pp. *529—34*. Hermann, Paris. Abstr. in S.C.A.R. Bull. No. 12. *Polar Rec.*, **11**, *315—16*, 1962.
159. MILAN, F. A., ELSNER, R. W., & RODAHL, K. (1961). The effect of a year in the Antarctic on human thermal and metabolic responses to an acute standardized cold stress. *Arctic Aeromed. Lab. Techn. Rep.* 60—9. Fort Wainwright, Alaska.
160. MILAN, F. A., EISNER, R. W. & RODAHL, K. (1961). Thermal and metabolic responses of men in the Antarctic to a standard cold stress. *J. appl. Physiol.*, **16**, *401—04*.
160a. MURRAY, L. (in preparation). Physiological research at Macquarie Island 1963, including serum lipids in relation to diet, blood pressure, weight, cold exposure. Aust. Nat. Antarctic Res. Exped., Dept. External Affairs, Melbourne.
161. MUTO, A. (1960). Medical research. In *National Report of Japanese Antarctic Research Expeditions 1958—60* (compiled by T. NAGATA), pp. *55—58*. Japan Antarctic Office, Ministry of Education, Tokyo.
162. NORMAN, J. N. (1960). Man in the Antarctic. *M.D. Thesis*, Glasgow Univ. (Also to be published as a British Antarctic Surv. Sci. Rep.)
163. NORMAN, J. N. (1962). Micro-climate of man in Antarctica. *J. Physiol.*, **160**, *27—28P*.
164. OGATA, M. (1959). Report on physiological results of the Japanese Antarctic Research Expedition I, 1956—57. *Antarctic Rec., Tokyo*, No. 6, *346—53*.
165. ORR, N. W. M. (1960). Human physiology. Weight changes while sledging. Hope Bay. *Falkland Islands Depend. Surv.*, Base Rep. 67/60 and M2/1960/D. London (unpublished).
166. PACE, N. (1959). Metabolic patterns in stressful environments. *Trans. Amer. Soc. Heat. Refrig. Air-Cond. Engrs.*, **65**, *101—13*.
167. PACE, N. (1960). Physiological studies in the Antarctic. In *Cold Injury* (S.M. Horvath, editor). Trans. Sixth Conf., 1958, Fort Knox, pp. *141—174*. Josiah Macy, Jr. Foundation, New York.
168. PACE, N., VAUGHAN, B. E., PARKER, H. G., TIMIRAS, P. S., GRISWOLD, R. L., GULDENZOPF, J. A. & HWANG, C. A. (1959). The physiological stress produced in men during an antarctic expedition (Operation Deepfreeze I). Final report, filed with *Arctic Inst. North Amer.*, Washington, D.C.
169. PALEEV, N. R. (1959). Medicinskie issledovanija v vostočnoj Antarktide. (Medical investigations in eastern Antarctica). In *Pervaja kontinental'naja ėkspedicija, 1955—1957 gg.* (M. M. SOMOV, editor). Naučnye rezul'taty, **2**, pp. *157—62*. Leningrad. (First continental expedition, 1955—57. Scientific results).

169a. PALEEV, N. R. (1964). The influence of the central arctic and antarctic climate on the human body. *Proc. III. Int. Biomet. Congr., Pau 1963* (S. W. TROMP & W. H. WEIHE, editors), Pergamon Press, Oxford (in press).

170. PALMAI, G. (1962). Thermal comfort and acclimatization to cold in a sub-antarctic environment. *Med. J. Aust.*, **49,** i, *9—12.*

171. PALMAI, G. (1962). Skin-fold thickness in relation to body weight and arterial blood pressure. *Med. J. Aust.*, **49,** ii, *13—15.*

172. PALMAI, G. (1962). Diurnal and seasonal variations in deep body temperature. *Med. J. Aust.* **49,** ii, *989—91.*

172a. PALMAI, G. (1963). Polar physiology. *Lancet,* ii, *640.*

173a. PRECIOUS, A. (1955). Finger nail growth. Hope Bay 1954. *Falkland Islands Depend. Surv.*, Base Rep. 30/55. London (unpublished).

173b. PRECIOUS, A. (1956). Finger nail growth. Hope Bay 1955. *Falkland Islands Depend. Surv.*, Base Rep. 56/56. London (unpublished).

173c. PRECIOUS, A. (1958). Finger nail growth. Admiral Bay, King George Island, 1957. *Falkland Islands Depend. Surv.*, Base Rep. 92/58. London (unpublished).

174. ROGERS, A. F. (1957). The physiological programme for the Trans-Antarctic Expedition, 1955—58. *Advanc. Sci., Lond.*, **13,** *419—20.* (cf. *Lancet,* ii, *786—87,* 1959, and *Polar Rec.*, **10,** *75—76,* 1960).

175. SAPIN-JALOUSTRE, J. & GODDARD, T. H. (1956). A French follow-up of research carried out in 1912 by Dr. A. L. MacLean, of the Australasian Antarctic Expedition, 1911—1914, on the slowing down of the growth of hair and nails in Antarctica. *Med. J. Aust.*, **43,** ii, *639—41.*

176. SAPIN-JALOUSTRE, J. & SAPIN-JALOUSTRE, H. (1956). Le ralentissement de la vitesse de croissance des phanères dans l'Antarctique. *Presse méd.*, **64,** *901—03.*

177. SIMPSON, H. W. (1959). Eosinophils and stress. Field studies in Antarctica. *M.D. Thesis*, Univ. of Edinburgh.

178. SIMPSON, H. W. (1959). Stress: studies in Antarctica. *New Scientist*, **6,** *927—29.*

178a. SMART, R. A., BRADBURY, P. & ADAM, J. M. (1964). Aspects of human physiology. In *Roy. Soc. IGY Antarctic Exped., Halley Bay, 1955—59.* Vol. IV (in press). The Royal Society, London.

178b. SPARKE, B. R. (in preparation). Physiological research in the Antarctic, including cold vasodilatation in finger and two-point discrimination tests. Med. Res. Council Lab., Hampstead, Engl.

179a. STAQUET, M. (1961). Besoins énergétiques et activité cortico-surrénale chez les membres d'une expédition polaire. *Arch. belg. Méd. soc.*, **19,** *661—712.*

179b. STAQUET, M. (1963). Variation de l'élimination urinaire des stéroïdes en climat polaire. *C.R. I.Symp. Biol. Antarctique, p. 515.* Hermann, Paris, Abstr. in S.C.A.R. Bull. No. 12. *Polar Rec.*, **11,** *317,* 1962.

180. STAQUET, M. (to be published). Variations du PBI[127] par climat polaire. Centre Nat. Recherch. Polaires, Bruxelles.

181. TICHOMIROV, I. I. (1959). Osobennosti akklimatizacii poljarnikov na stancii Vostok-I. (Features of the acclimatization of polar workers at station Vostok-I). *Inform. Bjull. Sovet. Antarktičeskoj Èksped.*, No. 6, *43—46.* (Information Bull. of the Soviet Antarctic Expedition).

182. TICHOMIROV, I. I. (1961) Charakter sdvigov v krovi u zimovščikov na stancii Vostok. (Character of blood changes of the personnel wintering at station Vostok). *Inform. Bjull. Sovet. Antarktičeskoj Èksped.*, No. 31, *44—47.* (Information Bull. of the Soviet Antarctic Expedition).

182a. TICHOMIROV, I. I. (1963). Nabljudenija za dejatel "nost" ju serdeč-nososudistoj sistemy u zimovščikov stancii Vostok v 1959 g. (Observations on the activity of the cardiovascular systems of the winter-

748

ing-over personnel of Vostok Station in 1959). *Inform. Bjull. Sovet. Antarktičeskoj Exped.*, No. **41**, *57—60*. (Information Bull. of the Soviet Antarctic Expedition).

183. TICHOMIROV, I. I. (1962). O nekotorych fiziologičeskich sdvigach v organizme čeloveka v processe akklimatizacii vo vnutrimaterikovych rajonach Antarktidy. (On some physiological shifts in the body of man during the process of acclimatization in intracontinental regions of the Antarctic.) *Vestn. Akad. Med. Nauk.*, **17**, (3), *74—82*.

183a. TIKHOMIROV, I. I. (1963). Some physiological changes in man in the process of acclimatization in inland regions of Antarctica. *Fed. Proc.*, **22**, Transl. Suppl. No. 1, *T3—7*. (Translation of ref. 183).

183b. WATANABE, K., KAGEYAMA, T. & AZUMA, T. (1964). Physiological acclimatization in the Fourth Japanese Antarctic Research Expedition, 1959—61. *Proc. III. Int. Biomet. Congr., Pau 1963* (S. W. TROMP & W. H. WEIHE, editors). Pergamon Press, Oxford (in press).

183c. WARHAFT, N. (in preparation). Physiological research at Mawson Station 1964, including catecholamine excretion and sensitivity to intravenous noradrenalin. Aust. Nat. Antarctic. Res. Exped., Dept. External Affairs, Melbourne.

183d. WIGG, D. (in preparation). Physiological research at Mawson Station 1963, including clinical and photographic study of suspected vitamin B deficiency and sensitivity to adrenalin and noradrenalin. Aust. Nat. Antarctic Res. Exped., Dept. External Affairs, Melbourne.

184. WILSON, O. (1953). Physiological changes in blood in the Antarctic. A preliminary report. *Brit. med. J.*, **ii**, *1425—28*.

185. WILSON, O. (1956). Basal metabolic rate in the Antarctic. *Metabolism*, **5**, *543—554*.

186. WILSON, O. (1960). Changes in body-weight of men in the Antarctic. *Brit. J. Nutr.*, **14**, *391—401*.

187. WILSON, O. (1962). Basal metabolic rate of "tropical" man in a polar climate. In *Biometeorology* (S. W. TROMP, editor), pp. *411—26*. Pergamon Press, Oxford.

188. WILSON, O. (1964). Atmospheric cooling and the occurrence of frostbite in exposed skin. In *The Treatment of Frostbite*. Symposia on Arctic Biology and Medicine. Proceedings. IV. (E. G. VIERECK, editor). Arctic Aeromed. Lab., Fort Wainwright, Alaska. (in press).

189. WILSON, O. (in preparation). Changes in blood in the Antarctic.

190. WYATT, H. T. (1957). Human physiology report, Loubet Coast 1957. *Falkland Islands Depend. Surv.*, Base Rep. M74/1957/W. London (unpublished).

191. WYATT, H. T. (1960). Changes in body weight and skinfold thickness after sledging journeys. Stonington Island and Loubet Coast. *Falkland Islands Depend. Surv.*, London (to be published as a British Antarctic Surv. Sci. Rep.).

192. WYATT, H. T. (1963). Observations on the physiology of men during sledging expeditions. *M.D. Thesis*, London Univ.

192a. WYNDHAM, C. H., PLOTKIN, R. & MUNRO, A. (1963). Responses to cold. Part II. Adaptation to cold of men in the Antarctic. *Appl. Physiol. Lab.*, A.P.L. Rep. No. 4/63. Johannesburg (unpublished). (To be published in *J. appl. Physiol.*)

192b. WYNDHAM, C. H., PLOTKIN, R. & MUNRO, A. (1963). Ethnic differences in physiological reactions to cold. *C.R. I.Symp. Biol. Antarctique*, pp. *535—60*. Hermann, Paris. Abstr. in S.C.A.R. Bull. No. 13. *Polar Rec.*, **11**, *500—01*, 1963.

SEE also references No. 23, 23a, 26, 29, 32, 33, 37, 40, 45, 51, 52—53, 55, 56, 68, 72a, 73—74, 78, 80a, 96—97, 99, 103, 107, 110, 116, 117b, 119, 128, 227, 227a, 229.

PSYCHOLOGY

193. BEHNKE, A. R. (1962). Human factors in antarctic operations. Paper prepared for I. Symp. Biol. Antarctique, Paris. Abstr. in S.C.A.R. Bull. No. 12, *Polar Rec.*, **11**, *316*.

193a. CUMMING, A. (unpublished). Psychological factors involved in life on an antarctic base. Argentine Islands 1959—60. Report with Med. Res. Council Lab., Hampstead, Engl.

193b. FUCHS, V. E. (1963). The human element in exploration. *Brit. Antarctic Surv. Bull.*, No. **1**, *1—8*.

194. GOY, G. (1959). Climat psychologique de l'hivernage. *Thèse de Méd.*, Paris.

195. GUNDERSON, E. K. E. & NELSON, P. D. (1962). Adjustment criteria in Antarctica. *U.S. Navy Med. Neuropsychiat. Res. Unit. Rep.* No. 62—1, San Diego, Calif.

196. GUNDERSON, E. K. E. & NELSON, P. D. (1962). Attitude changes in small groups under prolonged isolation. *U.S. Navy Med. Neuropsychiat. Res. Unit Rep.* No. 62—2, San Diego, Calif.

197. GUNDERSON, E. K. E. & NELSON, P. D. (1962). Clinician agreement in assessing for an unknown environment. *U.S. Navy Med. Neuropsychiat. Res. Unit. Rep.* No. 62—4, San Diego, Calif.

198. LAW, P. G. (1960). Personality problems in Antarctica. *Med. J. Aust.*, **47**, i, *273—82*.

199. LAW, P. G. (1960). Some psychological aspects of life at an antarctic station. *Discovery*, **21**, *431—37*.

200. LAW, P. G. (1962). The selection of men for antarctic expeditions. Document No. 36. *Conf. on Med. & Publ. Health in the Arctic & Antarctic.* WHO, Geneva (not published).

200a. LUSH, G. R. & NORMAN, .J N. (1960). Psychology and administration of an antarctic base. Halley Bay 1959—60. *Falkland Islands Depend. Surv.*, Base Rep. 155/60. London (unpublished).

201. MCGUIRE, F. & TOLCHIN, S. (1961). Group adjustment at the South Pole. *J. ment. Sci.*, **107**, *954—60*.

202. MULLIN, C. S. (1959). Headaches in the Antarctic *J.A.M.A.*, **170**, *163—64*.

203. MULLIN, C. S. (1960). Some psychological aspects of isolated antarctic living. *Amer. J. Psychiat.*, **117**, *323—25*.

204. MULLIN, C. S. & CONNERY, H. J. M. (1959). Psychological study at an antarctic IGY station. *U.S. Armed Forces med. J.*, **10**, *290—96*.

205. MULLIN, C. S., CONNERY, H. J. M. & WOUTERS, F. W. (1958). A psychological - psychiatric study of an IGY station in the Antarctic. *Special rep. to Bureau of Med. & Surg.*, U.S. Navy.

206. NARDINI, J. E., HERRMANN, R. S. & RASMUSSEN, J. E. (1962). Navy psychiatric assessment program in the Antarctic. *Amer. J. Psychiat.*, **119**, *97—105*.

207. NELSON, P. D. (1962). Leadership in small isolated groups. *U.S. Navy Med. Neuropsychiat. Res. Unit Rep.* 62—13, San Diego, Calif.

208. NELSON, P. D. (1963). Human adaptation to antarctic station life. In *Medicine and Public Health in the Arctic and Antarctic*, pp. *138—45*. Publ. Health Pap. No. 18. WHO, Geneva.

209. NELSON, P. D. (1963). Personal communication, and in *Conference on Medicine and Public Health in the Arctic and Antarctic. WHO techn. Rep. Ser.*, No. 253, pp. *19—21*. WHO, Geneva.

210. NELSON, P. D. & GUNDERSEN, E. K. E. (1962). Analysis of adjustment dimensions in small confined groups. *U.S. Navy Med. Neuropsychiat. Res. Unit Rep.* No. 62-3. San Diego, Calif.

211a. PALMAI, G. (1963). Psychological observations on an isolated group in Antarctica. *Brit. J. Psychiat.*, **109**, *364—70*.

211b. PALMAI, G. (1963). Psychological aspects of transient populations in Antarctica. In *Medicine and Public Health in the Arctic and Antarctic*, pp. *146—58*. *Publ. Health Pap.* No. 18. WHO, Geneva.

212a. PERRIER, F. (1963). Psychopathologie en expédition polaire. *C.R. I. Symp. Biol. Antarctique*, p. 567. Hermann, Paris. Abstr. in S.C.A.R. Bull., No. 12, *Polar Rec.*, **11**, *317*, 1962.

212b. PERRIER, F. (1964). Psychopathologie et climat polaire. *Proc. III. Int. Biomet. Congr., Pau 1963* (S. W. TROMP & W. H. WEIHE, editors). Pergamon Press, Oxford (in press).

213. PRIESTLEY, R. (1921). The psychology of exploration. *Psyche*, **2**, *18—28*.

214. RASMUSSEN, J. E. (1961). Group behavior in isolation — Antarctica. Invited paper, *XIV Int. Congr. Appl. Psychol.*, Copenhagen.

215. RIVERA CARRASCO, M. (1960). El hombre en la Antártida. *Revista de Marina*, **76**, *87—93*.

216. RIVOLIER, J. (1954). De quelques problèmes posés au médecin d'une expédition polaire. *Concours méd.*, **76**, *2045—50, 2155—59*.

217. RIVOLIER, J. (1962). Some psychological aspects of the problems of adaptation to wintering in stations. Document No. 15. *Conf. on Med. & Publ. Health in the Arctic & Antarctic*. WHO, Geneva (not published).

218. ROHRER, J. H. (1958). Some impressions of psychic adjustment to polar isolation. Techn. rep. on Office of Naval Res. contract NONR 1530(06), *Neuropsychiat. Branch, Bureau of Med. & Surg., U.S. Navy*.

219. ROHRER, J. H. (1959). Human adjustment to antarctic isolation. *U.S. Nav. Res. Rev.* (June), pp. *1—5*.

220. ROHRER, J. H. (1960). Human adjustment to antarctic isolation. Techn. rep. on Office of Naval Res. contract. NONR 1530 (07), *Neuropsychiat. Branch, Bureau of Med. & Surg., U.S. Navy*.

221. SMITH, W. M. (1961). Scientific personnel in Antarctica: Their recruitment, selection and performance. *Psychol. Rep.*, **9**, *163—82*, Monogr. Suppl. 1—V9.

222. SMITH, W. M. & JONES, M. B. (1962). Astronauts, antarctic scientists, and personal autonomy. *Aerospace Med.*, **33**, *162—66*.

223a. WEYBREW, B. B. (1961). Factor analysis of ratings from two populations: Wintering over antarctic personnel and the crew of Triton during the world circumnavigation. In Tri-Service Conference on Research Relevant to Behavior Problems of Small Military Groups under Isolation and Stress (S. B. Sells, editor), pp. *117—204*. Arctic Aeromed. Lab., Fort Wainwright, Alska.

223b. WEYBREW, B. B., MOLISH, H. B. & YOUNISS, R. P. (1961). Prediction of adjustment to the Antarctic. *U.S. Nav. Med. Res. Lab. Rep.* No. 350: Vol. 20, No. 1, New London, Conn.

224. ZÚÑIGA, A. & GILL, G. (1962). Psychological and physiological aspects of life at an antarctic base. Prelim. rep. presented at *Conf. on Med. & Publ. Health in the Arctic & Antarctic*. WHO, Geneva (not published).

SEE also references No. 68, 76, 80a, 228.

MISCELLANY

Antarctic

225. ANONYM (1962). The first nuclear power plant in Antarctica. *Polar Rec.*, **11**, *300—01*.

226. CRARY, A. P. (1960). A report on Antarctica. *Bull. Atomic Sci.*, **16**, *376—81*.

226a. DUFEK, G. J. (1962). Nuclear power for the polar regions. *Nat. geogr. Mag.*, **121**, *712—30*.

227. GOLDSMITH, R. & LEWIS, H. E. (1960). Polar expeditions as human laboratories. *J. occup. Med.*, **2**, *118—22*.

227a. GRUA, P. (1964). Lutte contre le froid et adaptation rapide aux conditions de plongées dans des eaux froides — Kerguelen, 1962—1963. *Proc. III. Int. Biomet. Congr., Pau 1963* (S. W. TROMP & W. H. WEIHE, editors). Pergamon Press, Oxford (in press).

228. HEDBLOM, E. E. (1961). Polar Manual. U.S. Nav. Med. School. Nat. Nav. Med. Center, Bethesda, Maryland.

229. NEUSHUL, M. (1961). Diving in antarctic waters. *Polar Rec.*, **10**, *353—58*.

230. PEAVEY, R. C. & GOULD, L. M. (1962). Antarctica, International Land of Science. *Courier*, **15**, (1), *9—74*.

231. SCHMITT, R. P. & RODRIGUEZ, R. (1962). Glacier water supply and sewage disposal systems. Document No 32. Conf. on *Med. & Publ. Health in the Arctic & Antarctic*. WHO, Geneva (not published). (Also presented at Antarctic Logistic Sympos., Boulder, Colorado, 1962.)

232. SIPLE, P. A. (1945). General principles governing selection of clothing for cold climates. *Proc. Amer. phil. Soc.*, **89**, *200—34*.

232a. TICHOMIROV, I. I. & NIZJAEV, D. A. (1963). Podogrev vdychaemogo vozducha bez vnešnich istočnikov tepla. (Heating of the inspiration air without exterior heat sources). *Inform. Bjull. Sovet. Antarktičeskoj Exped.*, No. **41**, *51—55*. (Information Bull. of the Soviet Antarctic Expedition).

232b. WOHLSCHLAG, D. E. (1963). The biological laboratory and field research facilities at the United States "McMurdo" Station, Antarctica. *Polar Rec.*, **11**, *713—18*.

Arctic

233. CARLSON, L. D., BURNS, H. L., HOLMES, T. H. & WEBB, P. P. (1953) Adaptive changes during exposure to cold. *J. appl. Physiol.*, **5**, *672—76*.

234. FALKOWSKI, S. J. & HASTINGS, Jr., A. D. (1958). Windchill in the Northern Hemisphere. *U.S. Army Quartermaster Res. & Eng. Cent. Environ. Protect. Res. Div. Tech. Rep.* EP-82. Natick, Mass.

235. GEOGHEGAN, B., ROBERTS, D. F. & SAMPFORD, M.R. (1958). A possible climatic effect on nail growth. *J. appl. Physiol.*, **13**, *135—38*.

236. KENNEDY, S. J. (1961). Clothing and personal protection. In *Man Living in the Arctic*. (F. R. FISHER, editor), pp. *56—67*. Proc. Conf. Quartermaster Res. & Eng. Cent., Natick, Mass., 1960. Nat. Acad. Sci. — Nat. Res. Council, Washington, D.C.

237. LEWIS, P. R. & LOBBAN, M. C. (1957). The effects of prolonged periods of life on abnormal time routines upon excretory rhythms in human subjects. *Quart. J. exp. Physiol.*, **42**, *356—71*.

238. LEWIS, P. R. & LOBBAN, M. C. (1957). Dissociation of diurnal rhythms in human subjects living on abnormal time routines. *Quart. J. exp. Physiol.*, **42**, *371—86*.

239. LEWIS, H. E. & MASTERTON, J. P. (1957). Sleep and wakefulness in the Arctic. *Lancet*, **i**, *1262—66*.

240. LEWIS, H. E., MASTERTON, J. P. & ROSENBAUM, S. (1960). Body weight and skinfold thickness of men on a polar expedition. *Clin. Sci.*, **19**, *551—61*.

241. MACKWORTH, N. H. (1955). Cold acclimatization and finger numbness. *Proc. Roy. Soc.*, **143B**, *392—407*.

752

242. MASTERTON, J. P., LEWIS, H. E. & WIDDOWSON, E. M. (1957). Food intakes, energy expenditures and faecal excretions of men on a polar expedition. *Brit. J. Nutr.*, **11**, *346—58*.
243. SEMAGIN, V. N. (1961). The sleep of man in arctic regions. *Sechenov Physiol. J. USSR*, **47**, *1037—46*.
244. SMITH, S. (1944) Frostbite of lungs. *Air Surgeon's Bull.*, **1** (6), *17*.

758

762